Applied Multivariate Statistical Analysis

Richard A. Johnson
Dean W. Wichern
University of Wisconsin–Madison

Prentice-Hall, Inc., Englewood Cliffs, New Jersey 07632

Library of Congress Cataloging in Publication Data
JOHNSON, RICHARD ARNOLD.
Applied multivariate statistical analysis.

Includes bibliographical references and index.

1. Multivariate analysis. I. Wichern, Dean W. II. Title.
QA278.J63 519.5'35 81-19894
ISBN 0-13-041400-X AACR2

Editorial/production supervision
and interior design: *Kathleen M. Lafferty*
Manufacturing buyer: *John B. Hall*

Printed in the United States of America

10 9 8 7 6 5

ISBN 0-13-041400-X

Prentice-Hall International, Inc., *London*
Prentice-Hall of Australia Pty. Limited, *Sydney*
Prentice-Hall of Canada, Ltd., *Toronto*
Prentice-Hall of India Private Limited, *New Delhi*
Prentice-Hall of Japan, Inc., *Tokyo*
Prentice-Hall of Southeast Asia Pte. Ltd., *Singapore*
Whitehall Books Limited, *Wellington, New Zealand*

To my mother and to the memory of my father.

R.A.J.

To Dorothy, Michael, and Andrew.

D.W.W.

Prentice-Hall Series in Statistics

Richard A. Johnson and Dean W. Wichern, Series Editors
University of Wisconsin

Johnson and Wichern, *Applied Multivariate Statistical Analysis*

Larsen and Marx, *An Introduction to Mathematical Statistics and Its Applications*

Morrison, *Applied Linear Statistical Methods*

Remington and Schork, *Statistics with Applications to the Biological and Health Sciences*, 2nd edition

Contents

Part III Analysis of Covariance Structure 359

Part IV Classification and Grouping Techniques 459

Preface

Researchers in the biological, physical, and social sciences frequently collect measurements on several variables. *Applied Multivariate Statistical Analysis* is concerned with statistical methods for describing and analyzing these multivariate data. Data analysis, while interesting with one variable, becomes truly fascinating and challenging when several variables are involved. Modern computer packages readily provide the numerical results to rather complex statistical analyses. We have endeavored to provide readers with the supporting knowledge necessary for making proper interpretations, selecting appropriate techniques, and understanding their strengths and weaknesses. We hope our discussions will meet the needs of experimental scientists, in a wide variety of subject matter areas, as a readable introduction to the statistical analysis of multivariate observations.

Our aim is to present the concepts and methods of multivariate analysis at a level that is readily understandable by readers who have taken two or more statistics courses. We emphasize the applications of multivariate methods and, consequently, have attempted to make the mathematics as palatable as possible. We avoid the use of calculus. On the other hand, the concepts of a matrix and of matrix manipulations are important. We do not assume the reader is familiar with matrix algebra. Rather, we introduce matrices as they appear naturally in our discussions, and we then show how they simplify the presentation of multivariate models and techniques.

An introductory account of matrix algebra, in Chapter 2, highlights the more important matrix algebra results *as they apply to multivariate analysis*. The Chapter 2 supplement provides a summary of matrix algebra results for those with little or no previous exposure to the subject. This supplementary material helps make the book self-contained and is used to complete proofs. The proofs may be ignored on a first reading. In this way we hope to make the book accessible to a wide audience.

The methodological "tools" of multivariate analysis are contained in Chapters 5 through 11. These chapters represent the heart of the book but they cannot be easily assimilated without much of the material in the introductory Chapters 1 through 4. Even those readers with a good knowledge of matrix algebra or those willing to accept the mathematical results on faith should, at the very least, peruse Chapter 3 (Sample Geometry) and Chapter 4 (Multivariate Normal Distribution).

Our approach in the methodological chapters is to keep the discussion direct and uncluttered. Typically, we start with a formulation of the population models, delineate the corresponding sample results, and liberally illustrate everything with examples. The examples are of two types: those that are simple and whose calculations can be easily done by hand and those that rely on real world data and computer software. (For the most part we have used the UCLA BMDP package of multivariate analysis programs.) The reader will note the extensive collections of real data sets included in the book. These will provide an opportunity to: (1) duplicate our analyses, (2) carry out the analyses dictated by exercises, or (3) analyze the data using methods other than the ones we have used or suggested.

This book grew out of our lecture notes for an "Applied Multivariate Analysis" course offered jointly by the Statistics Department and the School of Business at the University of Wisconsin–Madison. This course is offered once a year and enrolls approximately 50 graduate students from the various colleges and departments on campus. We cannot cover all of the material in one semester (15 weeks).

The division of the methodological chapters (5 through 11) into three units allows instructors some flexibility in tailoring a course to their needs. Possible sequences for a one-semester (two-quarter) course are indicated schematically below.

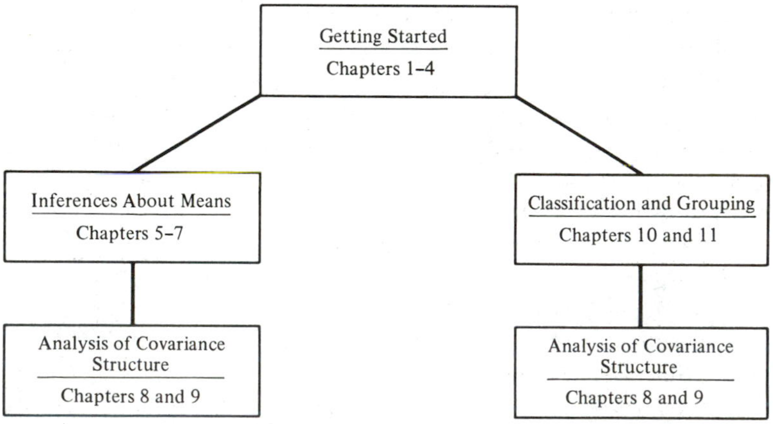

Each instructor will undoubtedly omit certain sections from some chapters in order to cover a broader collection of topics than is indicated by these two choices.

For most students, we would suggest a quick pass through the first four chapters (concentrating primarily on the material in Chapter 1, Sections 2.1, 2.2, 2.3, 2.5, 2.6, and 3.6, and the "assessing normality" material in Chapter 4) followed by a selection of methodological topics. For example, one might discuss the comparison of mean vectors, principal components, factor analysis, discriminant analysis, and clustering. The discussions could feature the many "worked out" examples included in these sections of the text. Instructors could rely on diagrams and verbal descriptions to "teach" the corresponding theoretical developments. If the students have uniformly *strong* mathematical backgrounds, much of the book can successfully be covered in one semester (two quarters).

In our attempt to make the study of multivariate analysis appealing to a large audience (both practitioners and theoreticians), we have had to sacrifice an evenness of level. Some sections are harder than others. In particular, we have summarized a voluminous amount of material on regression and path analysis in Chapter 7. The resulting presentation is rather succinct and difficult the first time through. We hope instructors will be able to compensate for the unevenness in level by judiciously choosing those sections (and subsections) appropriate for their students and by toning them down if necessary.

We have found individual data-analysis projects useful for integrating material from several of the methods chapters. Here our rather complete treatments of MANOVA, regression analysis, factor analysis, discriminant analysis, and so forth are helpful, even though they may not be specifically covered in lectures.

We thank our many colleagues who helped improve the applied aspect of the book by contributing their own data sets for examples and exercises. (These names appear next to appropriate data sets in the body of the book.) We also acknowledge the feedback of the students we have taught these past 10 years in our applied multivariate analysis course. Their comments and suggestions are largely responsible for the present iteration of the manuscript. We would like to thank the Prentice-Hall reviewers for their valuable suggestions. We must thank Steve Verrill for valuable computing assistance throughout and Alison Pollack for implementing a Chernoff faces program. We are indebted to Dr. Cliff Gilman for his assistance with the multidimensional scaling examples discussed in Chapter 11. Jacquelyn Forer did most of the typing of the draft manuscript and we appreciate her expertise and willingness to endure the cajolling of authors faced with publication deadlines. Finally, we would like to thank Robert Sickles, Kathleen Lafferty, and the rest of the Prentice–Hall staff for their help with this project.

R. A. JOHNSON
D. W. WICHERN

Madison, Wisconsin

Part I

Getting Started

1

Aspects of Multivariate Analysis

1.1 INTRODUCTION

Scientific inquiry is an iterative learning process. Objectives pertaining to the explanation of a social or physical phenomenon must be specified and then tested by gathering and analyzing data. In turn, an analysis of the data gathered by experimentation or observation will usually suggest a modified explanation of the phenomenon. Throughout this iterative learning process, variables are often added or deleted from the study. Thus the complexities of most phenomena require an investigator to collect observations on many different variables. This book is concerned with statistical methods designed to elicit information from these kinds of data sets. Because the data include simultaneous measurements on many variables, this body of methodology is called *multivariate analysis*.

The need to understand the relationships between many variables makes multivariate analysis an inherently difficult subject. Often, the human mind is overwhelmed by the sheer bulk of the data. Additionally, more mathematics is required to derive multivariate statistical techniques for making inferences than in a univariate setting. We have chosen to provide explanations based upon algebraic concepts and to avoid the derivations of statistical results that *require* the calculus of many variables. Our objective is to introduce several useful multivariate techniques in a clear manner, making heavy use of illustrative examples and a minimum of mathematics. Nonetheless, some mathematical sophistication and a desire to think quantitatively will be required.

Most of our emphasis will be on the *analysis* of measurements obtained, without actively controlling or manipulating any of the variables on which the measurements are made. Only in Chapters 6 and 7 shall we treat a few experimental plans (designs) for generating data that prescribe the active manipulation of important variables. Although the experimental design is ordinarily the most important part of a scientific investigation, it is frequently impossible to control the generation of appropriate data in certain disciplines. (This is true, for example, in business, economics, ecology, geology, and sociology.) You should consult [5] and [6] for detailed accounts of design principles which, fortunately, also apply to multivariate situations.

It will become increasingly clear that many multivariate methods are based upon an underlying probability model known as the multivariate normal distribution. Other methods are ad hoc in nature and are justified by logical or common sense arguments. Regardless of their origin, multivariate techniques must, invariably, be implemented on a computer. Recent advances in computer technology have been accompanied by the development of rather sophisticated statistical software packages, making the implementation step easier.

Multivariate analysis is a "mixed bag." It is difficult to establish a classification scheme for multivariate techniques that is both widely accepted and also indicates the appropriateness of the techniques. One classification distinguishes techniques designed to study interdependent relationships from those designed to study dependent relationships. Another classifies techniques according to the number of populations and the number of sets of variables being studied. Chapters in this text are divided into sections according to inference about treatment means, inference about covariance structure, and techniques for sorting or grouping. This should not, however, be considered as an attempt to place each method into a slot. Rather, the choice of methods and the types of analyses employed are largely determined by the objectives of the investigation. Below, we list a small number of practical problems designed to illustrate the connection between the choice of a statistical method and the objectives of the study. These problems, plus the examples in the text, should provide you with an appreciation for the applicability of multivariate techniques across different fields.

The objectives of scientific investigations, for which multivariate methods most naturally lend themselves, include the following:

1. *Data reduction or structural simplification*. (The phenomenon being studied is represented as simply as possible without sacrificing valuable information. It is hoped that this will make interpretation easier.)
2. *Sorting and grouping*. (Groups of "similar" objects or variables are created, based upon measured characteristics. Alternatively, rules for classifying objects to well-defined groups may be required.)
3. *Investigation of the dependence among variables*. (The nature of the relationships among variables is of interest. Are all the variables mutually independent or is one or more variable dependent on the others? If so, how?)

4. *Prediction.* (Relationships between variables must be determined for the purpose of predicting the values of one or more variables on the basis of observations on the other variables.)
5. *Hypothesis construction and testing.* (Specific statistical hypotheses, formulated in terms of the parameters of multivariate populations, are tested. This may be done to validate assumptions or reinforce prior convictions.)

We conclude this brief overview of multivariate analysis with a quotation from F. H. C. Marriott [14], page 89. The statement was made in a discussion of cluster analysis, but we feel it is appropriate for a broader range of methods. You should keep it in mind whenever data analyses are attempted. It allows one to maintain a proper perspective and not to be overwhelmed by the elegance of some of the theory.

> If the results disagree with informed opinion, do not admit a simple logical interpretation, and do not show up clearly in a graphical presentation, they are probably wrong. There is no magic about numerical methods, and many ways in which they can break down. They are a valuable aid to the interpretation of data, not sausage machines automatically transforming bodies of numbers into packets of scientific fact.

1.2 APPLICATIONS OF MULTIVARIATE TECHNIQUES

Statistical techniques are an integral part of any scientific inquiry and, consequently, their use is widespread. In particular, multivariate methods have been regularly applied to problems arising in the physical, social, and medical sciences.

It seems appropriate at this point to briefly describe real-world situations where multivariate techniques have proven valuable. This will allow us to indicate the wide variety of problems that can be addressed with multivariate methods.

Medicine

1E.1. A study (see Exercise 1.15) was conducted to investigate the reactions of cancer patients to radiotherapy. Measurements were made on 6 reaction variables for 98 patients. Because it was difficult to interpret observations on all 6 reaction variables at the same time, a simpler measure of patients' response was required. Multivariate analysis was used to construct a simple measure of patient response to radiotherapy, yet one that still contained much of the available sample information. (The objective here was data reduction.)

1E.2. Evoked responses to visual stimuli, such as flashes and patterns, can be recorded from the scalp of a human subject by computer averaging. These responses are referred to as the *Visual Electroencephalographic Computer Analysis* (VECA) profile of the subject. In a medical study (see Exercise 1.14) of the effects of multiple sclerosis on the visual system, multivariate analysis was used to examine whether the use of VECA is a practical and reliable means for diagnosing visual pathology. (The objective here was sorting or classification; that is, the development of a numerical rule for separating people suffering from a multiple sclerosis caused visual pathology from those not suffering from the disease.)

Sociology

1E.3. Competing current sociological theories suggest that the structure of American occupations is determined by one strong socioeconomic dimension and a few minor unexplored dimensions, or three well-defined dimensions called (1) requirements, (2) routines, and (3) rewards. Measurements on 25 variables for 583 occupations were analyzed (see [11] and [15]) using multivariate methods in order to provide support for one of the two positions mentioned above. (Here an initial objective was hypothesis verification; that is, can occupations be regarded as one-dimensional or three-dimensional?)

1E.4. In a study of mobility, counts of the number of foreign-born and second-generation U.S. residents in 1970 were tabulated by country of origin and state of residence. Multivariate methods were used (see [4]) to group states on the basis of comparable distributions of nationalities. (Here the objective was to find natural homogeneous groupings.)

Business and Economics

1E.5. Measurements on six accounting and financial variables were used in developing a multivariate model to help insurance regulators identify potentially insolvent property-liability insurers (see [17]). Using the model, an insurance company could be classified as *solvent* or *distressed* and remedial steps could then be taken to prevent bankruptcy of the distressed firm. (Here the objective was to obtain a classification rule for distinguishing solvent firms from distressed firms.)

1E.6. Knowledge of the relationships among policy instruments and goals for underdeveloped countries can aid the process of national development and modernization. Data from 74 non-Communist underdeveloped countries allowed an investigator (see [1]) to find the subsets of goals and instruments most closely associated with each other and to estimate the nature of the simultaneous relationships between the two subsets. The results indicated which national goals could be most reliably achieved with the set of policy instruments at the disposal of developing nations. (Here the objective was to determine the dependence between two sets of variables corresponding to goals and instruments.)

Education

1E.7. Scholastic Aptitude Test (SAT) scores and high school academic performance are often used as indicators of academic success in college. Measurements on five precollege *predictor* variables, (SAT verbal and quantitative scores, high school grade-point averages for the junior and senior years, and number of high-school extracurricular activities) and four college performance *criterion* variables (grades in courses in four different subject matter areas) were used to determine the association between the predictor and criterion scores. The study (see [8]) was concerned with substantiating the usefulness of test scores and high-school achievement as predictors of college performance. (Here the objective was prediction of the college performance variables based on the set of predictor variables. This may also be extended to a rule for classifying students as apt to succeed or not to succeed in college.)

1E.8. Motor performance, motor activities, and motor skills are popular targets for multivariate analysis. Specifically, track-and-field athletic events have frequently been analyzed with the hope of identifying primary skills within the various events. Data from 8 different Olympic decathlon championships were subjected to a multivariate analysis in order to identify the underlying physical factors responsible for the outcomes in the 10 decathlon events (see [12]). The outcomes could apparently be explained in terms of 4 physical factors: running speed, arm strength, running endurance, and leg strength. (Here the objective was to determine the dependence of observed variables [track and field outcomes] on fewer latent variables [physical factors].)

Biology

1E.9. In plant breeding it is necessary, after the end of one generation, to select those plants that will be the parents of the next generation. The selection is to be done in such a way that the succeeding generation will be improved in a number of characteristics over that of the previous generation. Many characteristics are often measured and evaluated. The plant breeder's goal is to maximize the genetic gain in the minimum amount of time. Multivariate techniques were used in a bean-breeding program (see [10]) to convert measurements on several variables relating to yield and protein content into a "selection index." Scores on this index were then used to determine parents of the subsequent families of beans. (Here the objectives were data reduction—that is, construction of an index to replace measurements on many variables—and the development of a sorting rule.)

1E.10. Two species of chickweed have proved difficult to identify. Measurements on four variables for chickweed plants, known to belong to the two species, were used to construct a function whose values allowed one to separate the two groups (see [3]). Consequently, the function could be used to classify a new candidate plant as belonging to one species or the other. (Here the objective was sorting or classification.)

Environmental Studies

1E.11. The atmospheric concentrations of air pollutants in the Los Angeles area have been extensively studied. In one study (see Exercise 1.6), daily measurements on seven pollution-related variables were recorded over an extended period of time. Of immediate interest was whether the levels of air pollutants were roughly constant throughout the week or whether there was a noticeable difference between weekdays and weekends. A secondary objective was to see if the mass of data available could be summarized in a readily interpretable manner. (Here the objectives were hypothesis testing and data reduction.)

Meteorology

1E.12. A study (see [9]) was initiated to quantify the relationships between tree-ring chronologies and various climatic parameters. It was of interest to determine the type of climatic information each tree ring contained and to then reconstruct climatic anomalies dating back to A.D. 1700. Multivariate techniques were used to reduce the tremendous amount of data available to a manageable size. The few new variables created in the process made subsequent analysis and interpretation much easier. (Here the initial objective was data reduction.)

Geology

1E.13. Multivariate analysis was used in a study (see [7] and [13]) of the size-class distribution of sediments in order to construct two linear functions of 10 size-class variables that would allow geologists to distinguish among 5 depositional environments. The results allow one to considerably reduce the laboratory work necessary to differentiate among the different sediment types. (Once again, the objectives were data reduction and classification.)

Psychology

1E.14. A study (see [16]) was undertaken to investigate risk-taking behavior. As part of the study, students were randomly assigned to receive one of three different sets of directions, or "treatments." They were then administered two parallel forms of a test given under high and low penalties for incorrect responses. The high- and low-penalty test scores were then related to the experimental treatments in order to investigate the students' reactions to the risk implied by the directions. (Here the objective was hypothesis testing; that is, to test whether the nature of the directions makes any difference in the perceived risk as quantified by the test scores.)

The scenarios above offer glimpses into the use of multivariate methods in widely diverse fields. Although the settings are different, it is clear that many of the data-analysis problems we have discussed are the same or very similar. Multivariate analysis, like most collections of statistical techniques, is not restricted to particular subject-matter areas.

1.3 THE ORGANIZATION OF DATA

Throughout this text, we are going to be concerned with analyzing measurements made on several variables or characteristics. These measurements (commonly called *data*) must frequently be arranged and displayed in various ways. For example, graphs and tabular arrangements are important aids in data analysis. Summary numbers, which quantitatively portray certain features of the data, are also necessary to any description.

We now introduce the preliminary concepts underlying these first steps of data organization.

Arrays

Multivariate data arise whenever an investigator, seeking to understand a social or physical phenomenon, selects a number $p \geq 1$ of *variables* or *characters* to record. The values of these variables are all recorded for each distinct *item*, *individual*, or *experimental trial*.

We will use the notation x_{ij} to indicate the particular value of the ith variable that is observed on the jth item, or trial. That is,

$$x_{ij} = \text{measurement of the } i\text{th variable on the } j\text{th item}$$

Consequently, n measurements on p variables can be displayed as follows:

	Item 1	Item 2	$\cdots$	Item j	$\cdots$	Item n
Variable 1:	x_{11}	x_{12}	$\cdots$	x_{1j}	$\cdots$	x_{1n}
Variable 2:	x_{21}	x_{22}	$\cdots$	x_{2j}	$\cdots$	x_{2n}
$\vdots$	$\vdots$	$\vdots$		$\vdots$		$\vdots$
Variable i:	x_{i1}	x_{i2}	$\cdots$	x_{ij}	$\cdots$	x_{in}
$\vdots$	$\vdots$	$\vdots$		$\vdots$		$\vdots$
Variable p:	x_{p1}	x_{p2}	$\cdots$	x_{pj}	$\cdots$	x_{pn}

or as a rectangular array, called $\mathbf{X}$, of p rows and n columns, where

$$\mathbf{X} = \begin{bmatrix} x_{11} & x_{12} & \cdots & x_{1j} & \cdots & x_{1n} \\ x_{21} & x_{22} & \cdots & x_{2j} & \cdots & x_{2n} \\ \vdots & \vdots & & \vdots & & \vdots \\ x_{i1} & x_{i2} & \cdots & x_{ij} & \cdots & x_{in} \\ \vdots & \vdots & & \vdots & & \vdots \\ x_{p1} & x_{p2} & \cdots & x_{pj} & \cdots & x_{pn} \end{bmatrix}$$

The array $\mathbf{X}$ then contains the data consisting of all of the observations on all of the variables.

Example 1.1

A selection of four receipts from a university book store was obtained in order to investigate the nature of book sales. Each receipt provided, among other things, the number of books sold and the total amount of each sale. Let the first variable be total dollar sales and the second variable be number of books sold. Then we can regard the corresponding numbers on the receipts as four measurements on two variables. Suppose the data in tabular form are:

Variable 1 (dollar sales):	42	52	48	58
Variable 2 (number of books):	4	5	4	3

Using the notation introduced above, we have

$$x_{11} = 42 \qquad x_{12} = 52 \qquad x_{13} = 48 \qquad x_{14} = 58$$

$$x_{21} = 4 \qquad x_{22} = 5 \qquad x_{23} = 4 \qquad x_{24} = 3$$

and the data array $\mathbf{X}$ is

$$\mathbf{X} = \begin{bmatrix} 42 & 52 & 48 & 58 \\ 4 & 5 & 4 & 3 \end{bmatrix}$$

with two rows and four columns. ■

Considering data in the form of arrays facilitates exposition and allows numerical calculations to be performed in an orderly and efficient manner. The efficiency is twofold as gains are attained both in (1) *describing* numerical calculations as operations on arrays, and (2) their *implementation* on computers, where many languages and statistical packages now perform array operations. We consider the manipulation of arrays of numbers in Chapter 2. At this point, we are concerned only with their value as devices for displaying data.

Descriptive Statistics

A large data set is bulky and its very mass poses a serious obstacle to any attempt at the visual extraction of pertinent information. Much of the information contained in the data can be assessed by calculating certain summary numbers, known as *descriptive statistics.* For example, the arithmetic average, or sample mean, is a descriptive statistic that provides a measure of location; that is, a "central value" for a set of numbers. As another example, the average squared distance of each number from the mean provides a measure of spread, or variation, in the numbers.

We shall rely most heavily on descriptive statistics that measure location, variation, and linear association. The formal definitions of these quantities follow.

Let $x_{11}, x_{12}, \ldots, x_{1n}$ be n measurements on the first variable. The arithmetic average of these measurements, denoted by $\bar{x}_1$, is given by

$$\bar{x}_1 = \frac{1}{n} \sum_{j=1}^{n} x_{1j}$$

If the n measurements represent a subset of the full set of measurements that might have been observed, then $\bar{x}_1$ is also called the *sample mean* for the first variable. We adopt this terminology because the bulk of this book is devoted to procedures designed for analyzing samples of measurements from larger collections.

The sample mean can be computed from the n measurements on each of the p variables so that, in general, there will be p sample means:

$$\bar{x}_i = \frac{1}{n} \sum_{j=1}^{n} x_{ij} \qquad i = 1, 2, \ldots, p \tag{1-1}$$

A measure of spread is provided by the *sample variance*, defined for n measurements on the first variable as

$$s_1^2 = \frac{1}{n} \sum_{j=1}^{n} (x_{1j} - \bar{x}_1)^2$$

where $\bar{x}_1$ is its sample mean. In general, for p variables, we have

$$s_i^2 = \frac{1}{n} \sum_{j=1}^{n} (x_{ij} - \bar{x}_i)^2 \qquad i = 1, 2, \ldots, p \tag{1-2}$$

Two comments are in order. First, many authors define the sample variance with a divisor of $n - 1$ rather than n. Later we shall see that there are theoretical reasons for doing this and it is particularly appropriate if the number of measurements, n, is small. The two versions of the sample variance will always be differentiated by displaying the appropriate expression.

Second, although the s^2 notation is traditionally used to indicate the sample variance, we shall eventually consider an array of quantities in which the sample variances lie along the main diagonal. In this situation, it is convenient to use double subscripts on the variances in order to indicate their positions in the array. Therefore we introduce the notation s_{ii} to denote the same variance computed from measurements on the ith variable and we have the notational identities

$$s_i^2 = s_{ii} = \frac{1}{n} \sum_{j=1}^{n} (x_{ij} - \bar{x}_i)^2 \qquad i = 1, 2, \ldots, p \tag{1-3}$$

The square root of the sample variance, $\sqrt{s_{ii}}$, is known as the *sample standard deviation*. This measure of variation is in the same units as the observations.

Consider n pairs of measurements on each of variables 1 and 2,

$$\begin{bmatrix} x_{11} \\ x_{21} \end{bmatrix}, \begin{bmatrix} x_{12} \\ x_{22} \end{bmatrix}, \ldots, \begin{bmatrix} x_{1n} \\ x_{2n} \end{bmatrix}$$

That is, x_{1j} and x_{2j} are observed on the jth experimental item ($j = 1, 2, \ldots, n$). A measure of linear association between the measurements of variables 1 and 2 is provided by the *sample covariance*

$$s_{12} = \frac{1}{n} \sum_{j=1}^{n} (x_{1j} - \bar{x}_1)(x_{2j} - \bar{x}_2)$$

or the average product of deviations from their respective means. If large values for one variable are observed in conjunction with large values for the other variable and the small values also occur together, s_{12} will be positive. If large values from one variable occur with small values for the other variable, s_{12} will be negative. If there is no particular association between the values for the two variables, s_{12} will be approximately zero.

The *sample covariance*

$$s_{ik} = \frac{1}{n} \sum_{j=1}^{n} (x_{ij} - \bar{x}_i)(x_{kj} - \bar{x}_k) \qquad i = 1, 2, \ldots, p, \quad k = 1, 2, \ldots, p \tag{1-4}$$

measures the association between the ith and kth variables. We note that the covariance reduces to the sample variance when $i = k$. Moreover, $s_{ik} = s_{ki}$ for all i and k.

The final descriptive statistic considered here is the *sample correlation coefficient* (or *Pearson's product moment correlation coefficient*; see [2]). This measure of the linear association between two variables does not depend on the units of measurement. The sample correlation coefficient, for the ith and kth variables, is defined as

$$r_{ik} = \frac{s_{ik}}{\sqrt{s_{ii}}\sqrt{s_{kk}}} = \frac{\sum_{j=1}^{n} (x_{ij} - \bar{x}_i)(x_{kj} - \bar{x}_k)}{\sqrt{\sum_{j=1}^{n} (x_{ij} - \bar{x}_i)^2}\sqrt{\sum_{j=1}^{n} (x_{kj} - \bar{x}_k)^2}} \tag{1-5}$$

for $i = 1, 2, \ldots, p$ and $k = 1, 2, \ldots, p$. Note $r_{ik} = r_{ki}$ for all i and k.

The sample correlation coefficient is a standardized version of the sample covariance, where the product of the square roots of the sample variances provides the standardization. Notice that r_{ik} has the same value whether n or $n - 1$ is chosen as the divisor for s_{ii}, s_{kk}, and s_{ik}.

The sample correlation coefficient, r_{ik}, can also be viewed as a sample *covariance*. Suppose the original values x_{ij} and x_{kj} are replaced by *standardized* values $(x_{ij} - \bar{x}_i)/\sqrt{s_{ii}}$ and $(x_{kj} - \bar{x}_k)/\sqrt{s_{kk}}$. The standardized values are commensurable because both sets are centered at zero and expressed in standard deviation units. The sample correlation coefficient, r_{ik}, is just the sample covariance of the standardized observations.

Although the signs of the sample correlation and the sample covariance are the same, the correlation is ordinarily easier to interpret because its magnitude is bounded. To summarize, the sample correlation, r, has the following properties:

1. The value of r must be between -1 and $+1$.
2. Here r measures the strength of the linear association. If $r = 0$, this implies a lack of linear association between the components. Otherwise, the sign of r indicates the direction of the association: $r < 0$ implies a tendency for one value in the pair to be larger than its average when the other is smaller than its average, and $r > 0$ implies a tendency for one value of the pair to be large when the other value is large and also for both values to be small together.
3. The value of r_{ik} remains unchanged if the measurements of the ith variable are changed to $y_{ij} = ax_{ij} + b$, $j = 1, 2, \ldots, n$, and the values of the kth variable

are changed to $y_{kj} = cx_{kj} + d$, $j = 1, 2, \ldots, n$, provided the constants a and c have the same sign.

The quantities s_{ik} and r_{ik} do not, in general, convey all there is to know about the association between two variables. Nonlinear associations can exist that are not revealed by these descriptive statistics. Covariance and correlation provide measures of linear association, or association along a line. Their values are less informative for other kinds of association. On the other hand, these quantities can be very sensitive to "wild" observations ("outliers") and may indicate association when, in fact, little exists. In spite of these shortcomings, covariance and correlation coefficients are routinely calculated and analyzed. They provide cogent numerical summaries of association when the data do not exhibit obvious nonlinear patterns of association or wild observations are not present.

Suspect observations must be accounted for by correcting obvious recording mistakes and by taking actions consistent with the identified causes. The values of s_{ik} and r_{ik} should be quoted both with and without these observations.

The sum of squared deviations from the mean and the sum of cross product deviations are often of interest themselves. These quantities are

$$W_{ii} = \sum_{j=1}^{n} (x_{ij} - \bar{x}_i)^2 \qquad i = 1, 2, \ldots, p \tag{1-6}$$

and

$$W_{ik} = \sum_{j=1}^{n} (x_{ij} - \bar{x}_i)(x_{kj} - \bar{x}_k) \qquad i = 1, 2, \ldots, p, \quad k = 1, 2, \ldots, p \tag{1-7}$$

The descriptive statistics computed from n measurements on p variables can also be organized into arrays.

Arrays of Basic Descriptive Statistics

Sample means
$$\bar{\mathbf{x}} = \begin{bmatrix} \bar{x}_1 \\ \bar{x}_2 \\ \vdots \\ \bar{x}_p \end{bmatrix}$$

Sample variances and covariances
$$\mathbf{S}_n = \begin{bmatrix} s_{11} & s_{12} & \cdots & s_{1p} \\ s_{21} & s_{22} & \cdots & s_{2p} \\ \vdots & \vdots & & \vdots \\ s_{p1} & s_{p2} & \cdots & s_{pp} \end{bmatrix} \tag{1-8}$$

Sample correlations
$$\mathbf{R} = \begin{bmatrix} 1 & r_{12} & \cdots & r_{1p} \\ r_{21} & 1 & \cdots & r_{2p} \\ \vdots & \vdots & & \vdots \\ r_{p1} & r_{p2} & \cdots & 1 \end{bmatrix}$$

The sample mean array is denoted by $\bar{\mathbf{x}}$, the sample variance and covariance array is denoted by the capital letter $\mathbf{S}_n$, and the sample correlation array by $\mathbf{R}$. The subscript n on the array $\mathbf{S}_n$ is a mnemonic device used to remind you that n is employed as a divisor for the elements s_{ik}. The size of all of the arrays is determined by the number of variables, p.

The arrays $\mathbf{S}_n$ and $\mathbf{R}$ consist of p rows and p columns. The array $\bar{\mathbf{x}}$ is a single column with p rows. The first subscript on an entry in arrays $\mathbf{S}_n$ and $\mathbf{R}$ indicates the row; the second subscript indicates the column. Since $s_{ik} = s_{ki}$ and $r_{ik} = r_{ki}$ for all i and k, the entries in symmetric positions about the main northwest–southeast diagonals in arrays $\mathbf{S}_n$ and $\mathbf{R}$ are the same and the arrays are said to be *symmetric*.

Example 1.2

Consider the data introduced in Example 1.1. Each receipt yields a pair of measurements, total dollar sales and number of books sold. Find the arrays $\bar{\mathbf{x}}$, $\mathbf{S}_n$, and $\mathbf{R}$.

Since there are four receipts, we have a total of four measurements (observations) on each variable.

The sample means are:

$$\bar{x}_1 = \tfrac{1}{4} \sum_{j=1}^{4} x_{1j} = \tfrac{1}{4}(42 + 52 + 48 + 58) = 50$$

$$\bar{x}_2 = \tfrac{1}{4} \sum_{j=1}^{4} x_{2j} = \tfrac{1}{4}(4 + 5 + 4 + 3) = 4$$

$$\bar{\mathbf{x}} = \begin{bmatrix} \bar{x}_1 \\ \bar{x}_2 \end{bmatrix} = \begin{bmatrix} 50 \\ 4 \end{bmatrix}$$

The sample variances and covariances are:

$$s_{11} = \tfrac{1}{4} \sum_{j=1}^{4} (x_{1j} - \bar{x}_1)^2 = \tfrac{1}{4}\big((42 - 50)^2 + (52 - 50)^2 + (48 - 50)^2 + (58 - 50)^2\big) = 34$$

$$s_{22} = \tfrac{1}{4} \sum_{j=1}^{4} (x_{2j} - \bar{x}_2)^2 = \tfrac{1}{4}\big((4 - 4)^2 + (5 - 4)^2 + (4 - 4)^2 + (3 - 4)^2\big) = .5$$

$$s_{12} = \tfrac{1}{4} \sum_{j=1}^{4} (x_{1j} - \bar{x}_1)(x_{2j} - \bar{x}_2)$$

$$= \tfrac{1}{4}\big((42 - 50)(4 - 4) + (52 - 50)(5 - 4) + (48 - 50)(4 - 4) + (58 - 50)(3 - 4)\big) = -1.5$$

$$s_{21} = s_{12}$$

and

$$\mathbf{S}_n = \begin{bmatrix} 34 & -1.5 \\ -1.5 & .5 \end{bmatrix}$$

The sample correlation is:

$$r_{12} = \frac{s_{12}}{\sqrt{s_{11}}\sqrt{s_{22}}} = \frac{-1.5}{\sqrt{34}\sqrt{.5}} = -.36$$

$$r_{21} = r_{12}$$

so

$$\mathbf{R} = \begin{bmatrix} 1 & -.36 \\ -.36 & 1 \end{bmatrix}$$ ■

Graphical Techniques

Plots are important, but frequently neglected, aids in data analysis. Although it is impossible to simultaneously plot *all* the measurements made on several variables and study the configurations, plots of individual variables and plots of pairs of variables can still be very informative. Sophisticated computer programs and display equipment allow one the luxury of visually examining data in one, two, or three dimensions with relative ease. On the other hand, many valuable insights can be obtained from the data by constructing plots with paper and pencil. Simple, yet elegant and effective, methods for displaying data are available in [18]. It is good statistical practice to plot pairs of variables and to visually inspect the pattern of association. Consider, then, the seven pairs of measurements on two variables given below:

Variable 1 (x_1):	3	4	2	6	8	2	5
Variable 2 (x_2):	5	5.5	4	7	10	5	7.5

These data are plotted as seven points in two dimensions (each axis representing a variable) in Figure 1.1 (see page 16). The coordinates of the points are determined by the paired measurements: $(3, 5), (4, 5.5), \ldots, (5, 7.5)$. The resulting two dimensional plot is known as a *scatter diagram*.

Also shown in Figure 1.1 are separate plots of the observed values of variable 1 and the observed values of variable 2, respectively. These plots are called (*marginal*) *dot diagrams*. They can be obtained from the original observations or by projecting the points in the scatter diagram onto each coordinate axis.

The information contained in the single-variable dot diagrams can be used to calculate the sample means $\bar{x}_1$ and $\bar{x}_2$ and the sample variances s_{11} and s_{22} (see Exercise 1.1). The scatter diagram indicates the orientation of the points, and their coordinates can be used to calculate the sample covariance s_{12}. In the scatter diagram of Figure 1.1, large values of x_1 occur with large values of x_2 and small values of x_1 with small values of x_2. Hence s_{12} will be positive.

Dot diagrams and scatterplots contain different kinds of information. The information in the marginal dot diagrams is not sufficient for constructing the scatterplot. As an illustration, suppose the data preceding Figure 1.1 had been

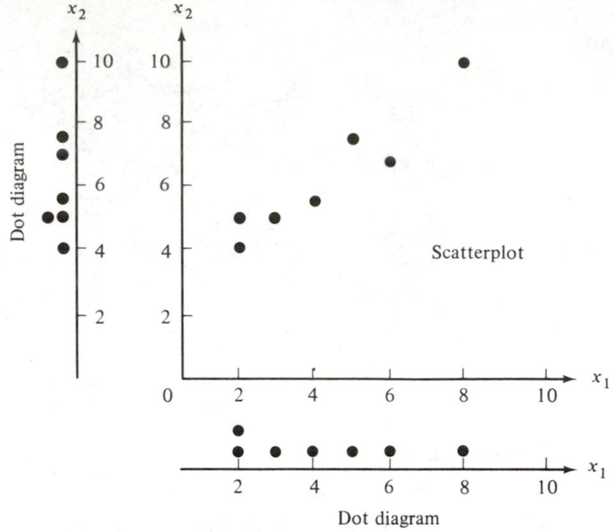

Figure 1.1 A scatterplot and marginal dot diagrams.

paired differently, so that the measurements on the variables x_1 and x_2 are as follows.

Variable 1 (x_1):	5	4	6	2	2	8	3
Variable 2 (x_2):	5	5.5	4	7	10	5	7.5

(We have simply rearranged the values of variable 1.) The scatter and dot diagrams for the "new" data are shown in Figure 1.2. Comparing Figures 1.1 and 1.2, it is evident that the marginal dot diagrams are the same but that the scatter diagrams are decidedly different. In Figure 1.2 large values of x_1 are paired with small values of x_2 and small values of x_1 with large values of x_2. Consequently the descriptive statistics for the individual variables $\bar{x}_1$, $\bar{x}_2$, s_{11}, and s_{22} remain unchanged, but the sample covariance s_{12}, which measures the association between pairs of variables, will now be negative.

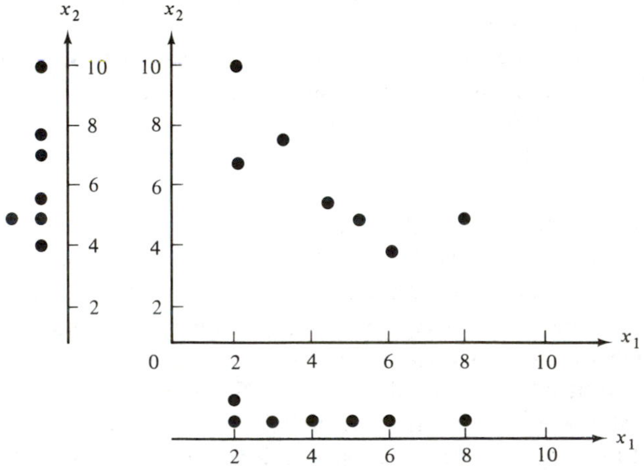

Figure 1.2 Scatterplot and dot diagrams for rearranged data.

The different orientations of the data in Figures 1.1 and 1.2 are not discernable from the marginal dot diagrams alone. At the same time, the fact that the marginal dot diagrams are the same in the two cases is not immediately apparent from the scatterplots. The two types of graphical procedures complement one another. They are not competitors.

The next two examples further illustrate the information that can be conveyed by a graphic display.

Example 1.3

Some financial data for the 25 largest securities firms appeared in an article in *Fortune* magazine on May 22, 1978. The data for the pair of variables x_1 = total capital and x_2 = securities revenues is graphed in Figure 1.3. We have labeled two "unusual" observations. Merrill Lynch, the largest firm in terms of numbers of offices and employees, is considerably larger than the other firms in *both* the x_1 and x_2 values. Salomon Brothers has "typical" securities revenues but comparatively large total capital.

The sample correlation coefficient computed from the values of x_1 and x_2 is

$$r_{12} = \begin{cases} .94 & \text{for all 25 firms} \\ .78 & \text{for all firms but Merrill Lynch} \\ .96 & \text{for all firms but Salomon Brothers} \\ .92 & \text{for all firms but Merrill Lynch and} \\ & \text{Salomon Brothers} \end{cases}$$

It is clear that atypical observations can have a considerable effect on the sample correlation coefficient. ■

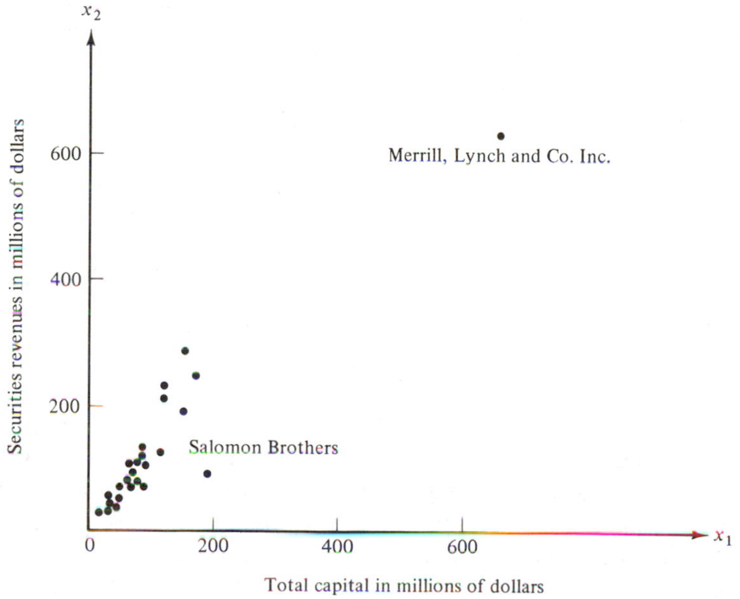

Figure 1.3 Securities revenues and total capital for 25 securities firms.

Example 1.4

In a July 17, 1978, article on money in sports, *Sports Illustrated* magazine provided data on x_1 = player payroll for National League East baseball teams. We have added data on x_2 = won/lost percentage for 1977. The results are given in Table 1.1.

TABLE 1.1 1977 SALARY AND FINAL RECORD FOR THE NATIONAL LEAGUE EAST

Team	x_1 = player payroll	x_2 = won/lost percentage
Philadelphia Phillies	3,497, 900	.623
Pittsburgh Pirates	2,485,475	.593
St. Louis Cardinals	1,782,875	.512
Chicago Cubs	1,725,450	.500
Montreal Expos	1,645,575	.463
New York Mets	1,469,800	.395

The scatterplot in Figure 1.4 supports the claim that a championship team can be bought. Of course, this cause-effect relationship cannot be substantiated because the experiment did not include a random assignment of payrolls. Statistics cannot answer the question: Could the Mets have won with $4 million to spend on player salaries? ■

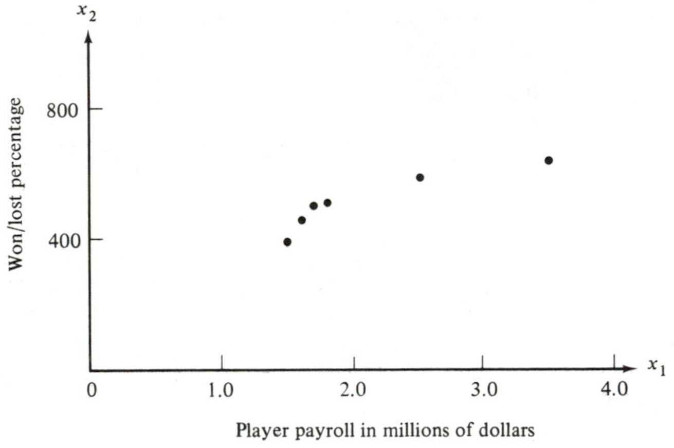

Figure 1.4 Salaries and records from Table 1.1.

To construct the scatterplot in Figure 1.4, for example, we have regarded the six paired observations in Table 1.1 as the coordinates of six points in two-dimensional space. The figure allows us to visually examine the grouping of teams with respect to the variables total payroll and won/lost percentage.

In the general multiresponse situation, p variables are simultaneously recorded on n items. Scatterplots should be made for pairs of important variables and, if the task is not too great to warrant the effort, for all pairs.

Limited as we are to a three-dimensional world, we cannot always picture the entire set of data. However, two further geometric representations of the data provide an important conceptual framework for viewing multivariable statistical methods. In cases where it is possible to capture the essence of the data in three dimensions, these representations can actually be graphed.

n points in p dimensions (p-dimensional scatterplot). Consider the natural extension of the scatterplot where the p measurements

$$\begin{bmatrix} x_{1j} \\ x_{2j} \\ \vdots \\ x_{pj} \end{bmatrix}$$

on the jth item represent the coordinates of a point in p-dimensional space. The coordinate axes are taken to correspond to the variables, so that the jth point is x_{1j} units along the first axis, x_{2j} units along the second, ..., x_{pj} units along the pth axis. The resulting plot with n points will not only exhibit the overall pattern of variability, but will show similarities (and differences) among the n items. Groupings of *items* will manifest themselves in this representation.

p-points in n-dimensions. The n observations of the p variables can also be regarded as p points in n-dimensional space. Each row of $\mathbf{X}$ determines one of the points. The ith row,

$$[x_{i1}, x_{i2}, \ldots, x_{in}]$$

consisting of all n measurements on the ith variable determines the ith point.

In Chapter 3 we show how the closeness of points in n dimensions can be related to measures of association between the corresponding *variables*.

1.4 DISTANCE

Although they may at first appear formidable, most multivariate techniques are based upon the simple concept of distance. Straight-line, or Euclidean, distance should be familiar. If we consider the point $P = (x_1, x_2)$ in the plane, the straight line distance, $d(O, P)$, from P to the origin $O = (0, 0)$ is, according to the Pythagorean theorem,

$$d(O, P) = \sqrt{x_1^2 + x_2^2} \tag{1-9}$$

The situation is illustrated in Figure 1.5. In general, if the point P has p coordinates

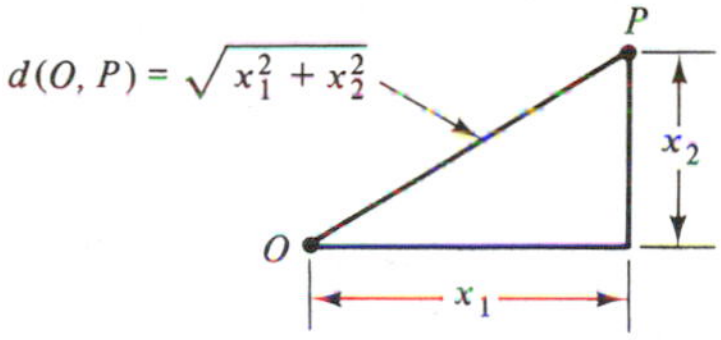

Figure 1.5 Distance given by the Pythagorean theorem.

so that $P = (x_1, x_2, \ldots, x_p)$, the straight-line distance from P to the origin $O = (0, 0, \ldots, 0)$ is

$$d(O, P) = \sqrt{x_1^2 + x_2^2 + \cdots + x_p^2} \qquad (1\text{-}10)$$

(see Chapter 2). All points $(x_1, x_2, \ldots, x_p)$ that lie a constant squared distance, such as c^2, from the origin satisfy the equation

$$d^2(O, P) = x_1^2 + x_2^2 + \cdots + x_p^2 = c^2 \qquad (1\text{-}11)$$

Because this is the equation of a hypersphere (a circle if $p = 2$), points equidistant from the origin lie on a hypersphere.

The straight-line distance between two arbitrary points P and Q with coordinates $P = (x_1, x_2, \ldots, x_p)$ and $Q = (y_1, y_2, \ldots, y_p)$ is given by

$$d(P, Q) = \sqrt{(x_1 - y_1)^2 + (x_2 - y_2)^2 + \cdots + (x_p - y_p)^2} \qquad (1\text{-}12)$$

Straight-line, or Euclidean, distance is unsatisfactory for most statistical purposes. This is because each coordinate contributes equally to the calculation of Euclidean distance. When the coordinates represent measurements that are subject to random fluctuations of differing magnitudes, it is often desirable to weight coordinates subject to a great deal of variability less heavily than those that are not highly variable. This suggests a different measure of distance.

Our purpose now is to develop a "statistical" distance that accounts for differences in variation and, in due course, the presence of correlation. Because our choice will depend upon the sample variances and covariances, at this point, we use the term *statistical distance* to distinguish it from ordinary Euclidean distance. It is this kind of distance (see below) that is fundamental to multivariate analysis.

To begin, we take as *fixed* the set of observations graphed as the p-dimensional scatterplot. From these we shall construct a measure of distance from the origin to a point $P = (x_1, x_2, \ldots, x_p)$. In our arguments, the coordinates $(x_1, x_2, \ldots, x_p)$ of P can vary to produce different locations for the point. The data that determine distance will, however, remain fixed.

To illustrate, suppose we have n pairs of measurements on two variables. Call the variables x_1 and x_2 and assume that the x_1 measurements vary independently of the x_2 measurements.[1] In addition, assume the variability in the x_1 measurements is larger than the variability in the x_2 measurements. A scatterplot of the data would look something like the one pictured in Figure 1.6.

Glancing at Figure 1.6, we see values that are a given deviation from the origin in the x_1 direction are not as "surprising" or "unusual" as are values equidistant from the origin in the x_2 direction. This is true because the inherent variability in the x_1 direction is greater than the variability in the x_2 direction. Consequently, large x_1 coordinates (in absolute value) are not as unexpected as large (in absolute value) x_2 coordinates. It seems reasonable, then, to weight an x_2 coordinate more heavily than an x_1 coordinate of the same value when computing the "distance" to the origin.

[1] At this point, "independently" means the x_2 measurements cannot be predicted with any accuracy from the x_1 measurements, and vice versa.

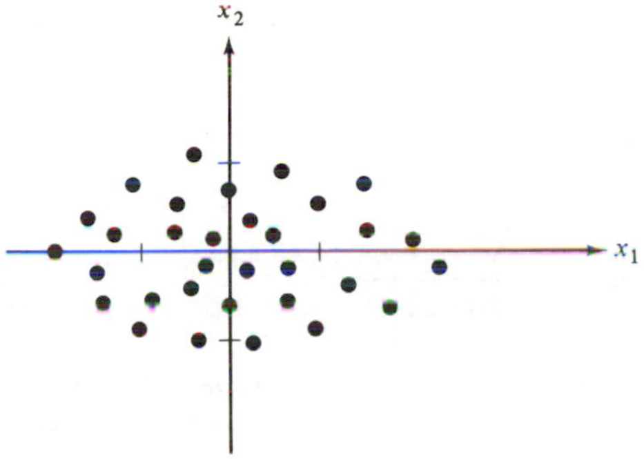

Figure 1.6 A scatterplot with greater variability in the x_1 direction than in the x_2 direction.

One way to proceed is to divide each coordinate by the sample standard deviation. Therefore, upon division by the standard deviations, we have the "standardized" coordinates $x_1^* = x_1/\sqrt{s_{11}}$ and $x_2^* = x_2/\sqrt{s_{22}}$. The standardized coordinates are now on an equal footing with one another. After taking the differences in variability into account, we determine distance using the standard Euclidean formula.

Thus a statistical distance of the point $P = (x_1, x_2)$ from the origin $O = (0, 0)$ can be computed from its standardized coordinates, $x_1^* = x_1/\sqrt{s_{11}}$ and $x_2^* = x_2/\sqrt{s_{22}}$ as

$$d(O, P) = \sqrt{(x_1^*)^2 + (x_2^*)^2}$$

$$= \sqrt{\left(\frac{x_1}{\sqrt{s_{11}}}\right)^2 + \left(\frac{x_2}{\sqrt{s_{22}}}\right)^2} = \sqrt{\frac{x_1^2}{s_{11}} + \frac{x_2^2}{s_{22}}} \tag{1-13}$$

Comparing (1-13) with (1-9), we see that the difference between the two expressions is due to the weights $k_1 = 1/s_{11}$ and $k_2 = 1/s_{22}$ attached to x_1^2 and x_2^2 in (1-13). Note that if the sample variances are the same, $k_1 = k_2$, and x_1^2 and x_2^2 will receive the same weight. In cases where the weights are the same, it is convenient to ignore the common divisor and use the usual Euclidean distance formula. In other words, if the variability in the x_1 direction is the same as the variability in the x_2 direction and the x_1 values vary independently of the x_2 values, Euclidean distance is appropriate.

Using (1-13), all points that have coordinates (x_1, x_2) and are a constant squared distance, c^2, from the origin must satisfy

$$\frac{x_1^2}{s_{11}} + \frac{x_2^2}{s_{22}} = c^2 \tag{1-14}$$

Equation (1-14) is the equation of an ellipse centered at the origin, whose major and minor axes coincide with the coordinate axes. That is, the statistical distance in (1-13) has an ellipse as the locus of all points a constant distance from the origin. This general case is shown in Figure 1.7 on the following page.

Example 1.5

A set of paired measurements (x_1, x_2) on two variables yields $\bar{x}_1 = \bar{x}_2 = 0$, $s_{11} = 4$, and $s_{22} = 1$. Suppose the x_1 measurements are unrelated to the x_2

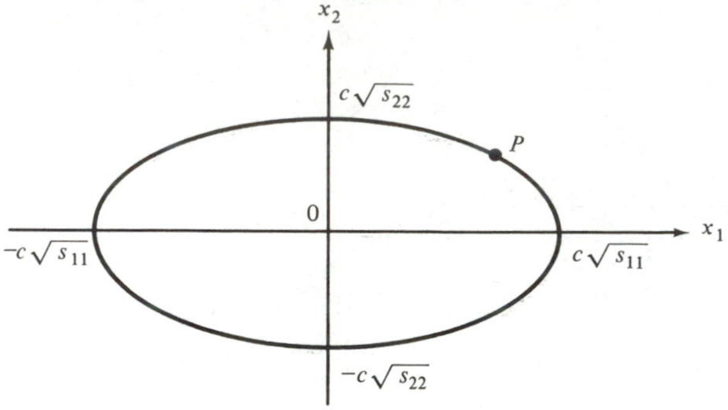

Figure 1.7 The ellipse of constant statistical distance $d^2(O, P) = x_1^2/s_{11} + x_2^2/s_{22} = c^2$.

measurements; that is, measurements within a pair vary independently of one another. Since the sample variances are unequal, we measure the squared distance of an arbitrary point $P = (x_1, x_2)$ to the origin $O = (0,0)$ by

$$d^2(O, P) = \frac{x_1^2}{4} + \frac{x_2^2}{1}$$

All points (x_1, x_2) that are a constant distance 1 from the origin satisfy the equation

$$\frac{x_1^2}{4} + \frac{x_2^2}{1} = 1$$

The coordinates of some points a unit distance from the origin are presented in the table below.

Coordinates: (x_1, x_2)	Distance: $\frac{x_1^2}{4} + \frac{x_2^2}{1} = 1$
$(0, 1)$	$\frac{0^2}{4} + \frac{1^2}{1} = 1$
$(0, -1)$	$\frac{0^2}{4} + \frac{(-1)^2}{1} = 1$
$(2, 0)$	$\frac{2^2}{4} + \frac{0^2}{1} = 1$
$(1, \sqrt{3}/2)$	$\frac{1^2}{4} + \frac{(\sqrt{3}/2)^2}{1} = 1$

A plot of the equation $x_1^2/4 + x_2^2/1 = 1$ is an ellipse centered at $(0,0)$, whose major axis lies along the x_1 coordinate axis and whose minor axis lies along the x_2 coordinate axis. The half lengths of these major and minor axes are $\sqrt{4} = 2$ and $\sqrt{1} = 1$, respectively. The ellipse of unit distance is plotted in Figure 1.8. All points on the ellipse are regarded as being the same distance from the origin; in this case, a distance of 1 from the origin. ■

The expression in (1-13) can be generalized to accommodate the calculation of statistical distance from an arbitrary point $P = (x_1, x_2)$ to any *fixed* point $Q = (y_1, y_2)$. If we assume the coordinate variables vary independently of one another,

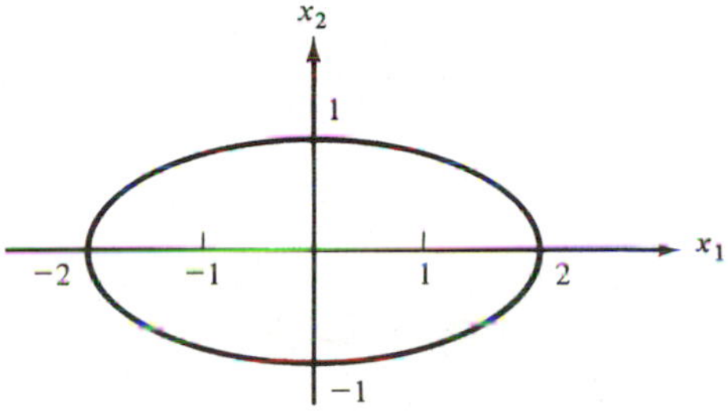

Figure 1.8 Ellipse of unit distance, $\frac{x_1^2}{4} + \frac{x_2^2}{1} = 1$.

the distance from P to Q is given by

$$d(P, Q) = \sqrt{\frac{(x_1 - y_1)^2}{s_{11}} + \frac{(x_2 - y_2)^2}{s_{22}}} \tag{1-15}$$

The extension of this statistical distance to more than two dimensions is straightforward. Let the points P and Q have p coordinates so that $P = (x_1, x_2, \ldots, x_p)$ and $Q = (y_1, y_2, \ldots, y_p)$. Suppose Q is a fixed point [it may be the origin $O = (0, 0, \ldots, 0)$] and the coordinate variables vary independently of one another. Let $s_{11}, s_{22}, \ldots, s_{pp}$ be sample variances constructed from n measurements on $x_1, x_2, \ldots, x_p$, respectively. The statistical distance from P to Q is

$$d(P, Q) = \sqrt{\frac{(x_1 - y_1)^2}{s_{11}} + \frac{(x_2 - y_2)^2}{s_{22}} + \cdots + \frac{(x_p - y_p)^2}{s_{pp}}} \tag{1-16}$$

All points P that are a constant squared distance from Q lie on a hyperellipsoid centered at Q whose major and minor axes are parallel to the coordinate axes. We note the following:

1. The distance of P to the origin O is obtained by setting $y_1 = y_2 = \cdots = y_p = 0$ in (1-16).
2. If $s_{11} = s_{22} = \cdots = s_{pp}$, the Euclidean distance formula in (1-12) is appropriate.

The distance in (1-16) still does not include most of the important cases we shall encounter because of the assumption of independent coordinates. The scatterplot in Figure 1.9 depicts a two-dimensional situation in which the x_1 measurements do not vary independently of the x_2 measurements. In fact, the coordinates of the pairs (x_1, x_2) exhibit a tendency to be large or small together and the sample correlation coefficient is positive. Moreover, the variability in the x_2 direction is larger than the variability in the x_1 direction.

What is a meaningful measure of distance when the variability in the x_1 direction is different from the variability in the x_2 direction and the variables x_1 and x_2 are correlated? Actually, we can use what we have already introduced, provided we look at things in the right way. From Figure 1.9 we see that if we rotate the original coordinate system through the angle θ while keeping the scatter fixed and label the rotated axes $\tilde{x}_1$ and $\tilde{x}_2$, the scatter in terms of the new axes looks very much

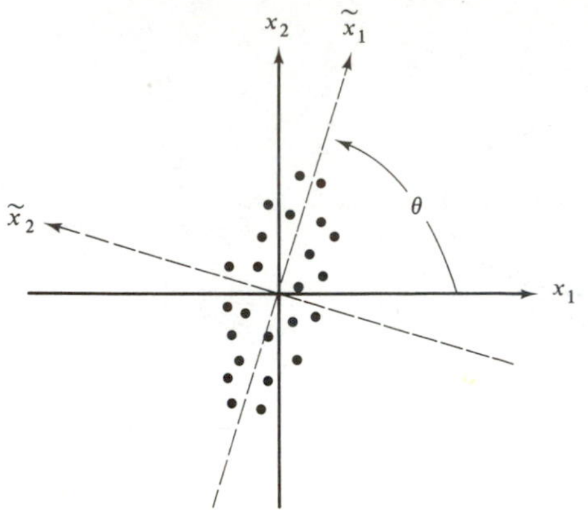

Figure 1.9 A scatterplot for positively correlated measurements and a rotated coordinate system.

like that in Figure 1.6 (you may wish to turn the book to place the $\tilde{x}_1$ and $\tilde{x}_2$ axes in their customary positions). This suggests that we calculate the sample variances using the $\tilde{x}_1$ and $\tilde{x}_2$ coordinates and measure distance as in Equation (1-13). That is, with reference to the $\tilde{x}_1$ and $\tilde{x}_2$ axes, we define the distance from the point $P = (\tilde{x}_1, \tilde{x}_2)$ to the origin $O = (0, 0)$ as

$$d(O, P) = \sqrt{\frac{\tilde{x}_1^2}{\tilde{s}_{11}} + \frac{\tilde{x}_2^2}{\tilde{s}_{22}}} \tag{1-17}$$

where $\tilde{s}_{11}$ and $\tilde{s}_{22}$ denote the sample variances computed with the $\tilde{x}_1$ and $\tilde{x}_2$ measurements.

The relation between the original coordinates (x_1, x_2) and the rotated coordinates $(\tilde{x}_1, \tilde{x}_2)$ is provided by

$$\begin{aligned} \tilde{x}_1 &= x_1 \cos(\theta) + x_2 \sin(\theta) \\ \tilde{x}_2 &= -x_1 \sin(\theta) + x_2 \cos(\theta) \end{aligned} \tag{1-18}$$

Given the relations in (1-18), we can formally substitute for $\tilde{x}_1$ and $\tilde{x}_2$ in (1-17) and express the distance in terms of the original coordinates.

After some straightforward algebraic manipulations, the distance from $P = (\tilde{x}_1, \tilde{x}_2)$ to the origin $O = (0, 0)$ can be written in terms of the original coordinates x_1 and x_2 of P as

$$d(O, P) = \sqrt{a_{11}x_1^2 + 2a_{12}x_1x_2 + a_{22}x_2^2} \tag{1-19}$$

where the a's are numbers such that the distance is nonnegative for all possible values x_1 and x_2. Here a_{11}, a_{12}, and a_{22} are determined by the angle θ and s_{11}, s_{12}, and s_{22} all calculated from the original data.[2] The particular forms for a_{11}, a_{22}, and a_{12} are not important at this point. What is important is the appearance of the cross-product term $2a_{12}x_1x_2$ necessitated by the nonzero correlation r_{12}.

---

[2]Specifically,

$$a_{11} = \frac{\cos^2(\theta)}{\cos^2(\theta)s_{11} + 2\sin(\theta)\cos(\theta)s_{12} + \sin^2(\theta)s_{22}} + \frac{\sin^2(\theta)}{\cos^2(\theta)s_{22} - 2\sin(\theta)\cos(\theta)s_{12} + \sin^2(\theta)s_{11}}$$

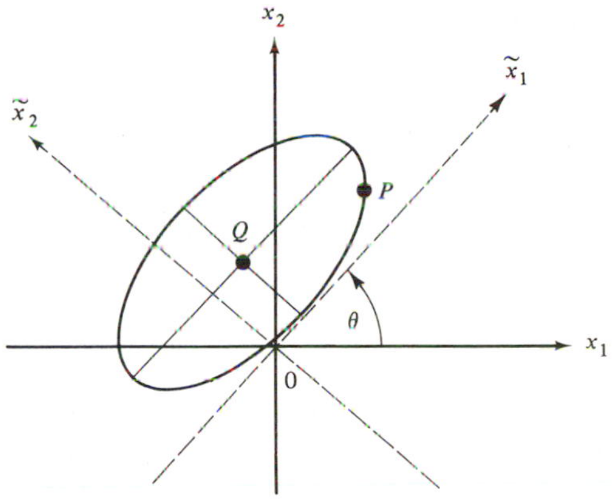

Figure 1.10 Ellipse of points a constant distance from the point Q.

Equation (1-19) can be compared with Equation (1-13). The expression in (1-13) can be regarded as a special case of (1-19) with $a_{11} = 1/s_{11}$, $a_{22} = 1/s_{22}$, and $a_{12} = 0$.

In general, the statistical distance of the point $P = (x_1, x_2)$ from the *fixed* point $Q = (y_1, y_2)$ for situations where the variables are correlated has the general form

$$d(P, Q) = \sqrt{a_{11}(x_1 - y_1)^2 + 2a_{12}(x_1 - y_1)(x_2 - y_2) + a_{22}(x_2 - y_2)^2} \tag{1-20}$$

and can always be computed once a_{11}, a_{22}, and a_{12} are known. In addition, the coordinates of all points $P = (x_1, x_2)$ that are a constant squared distance, c^2, from Q satisfy

$$a_{11}(x_1 - y_1)^2 + 2a_{12}(x_1 - y_1)(x_2 - y_2) + a_{22}(x_2 - y_2)^2 = c^2 \tag{1-21}$$

By definition, this is the equation of an ellipse centered at Q. The graph of such an equation is displayed in Figure 1.10. The major (long) and minor (short) axes are indicated. They are parallel to the $\tilde{x}_1$ and $\tilde{x}_2$ axes. For the choice of a_{11}, a_{12}, and a_{22} in Footnote 2, the $\tilde{x}_1$ and $\tilde{x}_2$ axes are at an angle θ with respect to the x_1 and x_2 axes.

The generalization of the distance formulas of (1-19) and (1-20) to p dimensions is straightforward. Let $P = (x_1, x_2, \ldots, x_p)$ be a point whose coordinates represent variables that are correlated and subject to inherent variability. Let $O = (0, 0, \ldots, 0)$ denote the origin and let $Q = (y_1, y_2, \ldots, y_p)$ be a specified *fixed*

$$a_{22} = \frac{\sin^2(\theta)}{\cos^2(\theta)s_{11} + 2\sin(\theta)\cos(\theta)s_{12} + \sin^2(\theta)s_{22}} + \frac{\cos^2(\theta)}{\cos^2(\theta)s_{22} - 2\sin(\theta)\cos(\theta)s_{12} + \sin^2(\theta)s_{11}}$$

and

$$a_{12} = \frac{\cos(\theta)\sin(\theta)}{\cos^2(\theta)s_{11} + 2\sin(\theta)\cos(\theta)s_{12} + \sin^2(\theta)s_{22}} - \frac{\sin(\theta)\cos(\theta)}{\cos^2(\theta)s_{22} - 2\sin(\theta)\cos(\theta)s_{12} + \sin^2(\theta)s_{11}}$$

point. Then the distances from P to O and from P to Q have the general forms

$$d(O, P) = \sqrt{a_{11}x_1^2 + a_{22}x_2^2 + \cdots + a_{pp}x_p^2 + 2a_{12}x_1x_2 + 2a_{13}x_1x_3 + \cdots + 2a_{p-1,p}x_{p-1}x_p} \tag{1-22}$$

and

$$d(P, Q) = \sqrt{\begin{aligned}&[a_{11}(x_1 - y_1)^2 + a_{22}(x_2 - y_2)^2 + \cdots + a_{pp}(x_p - y_p)^2 + 2a_{12}(x_1 - y_1)(x_2 - y_2)\\ &\quad + 2a_{13}(x_1 - y_1)(x_3 - y_3) + \cdots + 2a_{p-1,p}(x_{p-1} - y_{p-1})(x_p - y_p)]\end{aligned}} \tag{1-23}$$

where the a's are numbers such that the distances are always nonnegative.[3]

We note that the distances in (1-22) and (1-23) are completely determined by the coefficients (weights) a_{ik}; $i = 1, 2, \ldots, k$, $k = 1, 2, \ldots, p$. These coefficients can be set out in the rectangular array

$$\begin{bmatrix} a_{11} & a_{12} & \cdots & a_{1p} \\ a_{12} & a_{22} & \cdots & a_{2p} \\ \vdots & \vdots & & \vdots \\ a_{1p} & a_{2p} & \cdots & a_{pp} \end{bmatrix} \tag{1-24}$$

where the a_{ik}'s with $i \neq k$ are displayed twice since they are multiplied by 2 in the distance formulas. Consequently, the entries in this array specify the distance functions. The a_{ik}'s cannot be arbitrary numbers. They must be such that the computed distance is nonnegative for every pair of points (see Exercise 1.10).

Contours of constant distances computed from (1-22) and (1-23) are hyperellipsoids. A hyperellipsoid resembles a football when $p = 3$. They are impossible to visualize in more than three dimensions.

The need to consider statistical rather than Euclidean distance is illustrated heuristically in Figure 1.11. Figure 1.11 depicts a cluster of points whose center of gravity (sample mean) is indicated by the point Q. Consider the Euclidean distances from the point Q to the points P and the origin O. The Euclidean distance from Q to P is larger than the Euclidean distance from Q to O. However, P appears to be more like the points in the cluster than does the origin. If we take into account the variability of the points in the cluster and measure distance by the statistical distance in (1-20), then Q will be closer to P than to O. This result seems reasonable given the nature of the scatter.

Other measures of distance can be advanced (see Exercise 1.12). At times it is useful to consider distances that are not related to circles or ellipses. Any distance measure, $d(P, Q)$, between two points P and Q is valid provided it satisfies the

[3]The algebraic expressions for the *squares* of the distances in (1-22) and (1-23) are known as *quadratic forms* and, in particular, *positive definite quadratic* forms. It is possible to display these quadratic forms in a simpler manner using matrix algebra; we shall do so in Section 2.3 of Chapter 2.

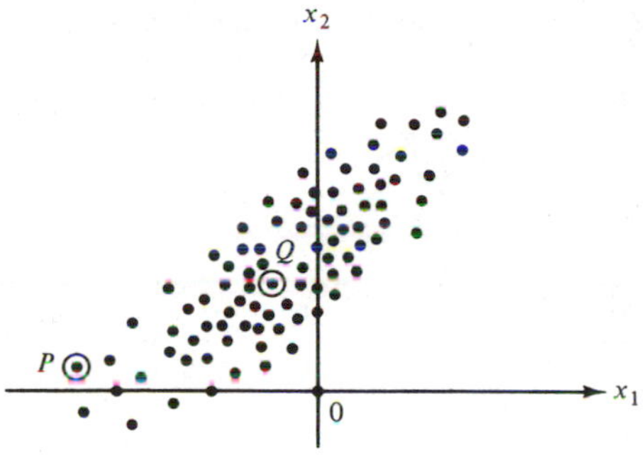

Figure 1.11 A cluster of points relative to a point P and the origin.

following properties, where R is any other intermediate point.

$$
\begin{aligned}
&d(P,Q) = d(Q,P) \\
&d(P,Q) > 0 \text{ if } P \neq Q \\
&d(P,Q) = 0 \text{ if } P = Q \\
&d(P,Q) \leq d(P,R) + d(R,Q) \qquad \text{triangle inequality}
\end{aligned}
\tag{1-25}
$$

1.5 FINAL COMMENTS

We have attempted to motivate the study of multivariate analysis and to provide you with some rudimentary but important methods for organizing and summarizing data. In addition, a general concept of distance has been introduced that will be used repeatedly in later chapters.

EXERCISES

1.1. Consider the seven pairs of measurements (x_1, x_2) plotted in Figure 1.1.

x_1	3	4	2	6	8	2	5
x_2	5	5.5	4	7	10	5	7.5

Calculate the sample means $\bar{x}_1$ and $\bar{x}_2$, the sample variances s_{11} and s_{22}, and the sample covariance s_{12}.

1.2. A morning newspaper lists the following used car prices for a foreign compact with age x_1 measured in years and selling price x_2 measured in thousands of dollars.

x_1	3	5	5	7	7	7	8	9	10	11
x_2	2.30	1.90	1.00	.70	.30	1.00	1.05	.45	.70	.30

(a) Construct a scatterplot of the data and marginal dot diagrams.
(b) Infer the sign of the sample covariance s_{12} from the scatterplot.
(c) Compute the sample means $\bar{x}_1$ and $\bar{x}_2$ and sample variances s_{11} and s_{22}. Compute

the sample covariance s_{12} and the sample correlation coefficient r_{12}. Interpret these quantities.

(d) Display the sample mean array, $\bar{\mathbf{x}}$, the sample variance-covariance array $\mathbf{S}_n$ and the sample correlation array $\mathbf{R}$ using (1-8).

1.3. Given the five measurements on variables x_1, x_2, and x_3,

x_1	9	2	6	5	8
x_2	12	8	6	4	10
x_3	3	4	0	2	1

find the arrays $\bar{\mathbf{x}}$, $\mathbf{S}_n$, and $\mathbf{R}$.

1.4. The 10 largest U.S. industrial corporations yield the following data. (Source: "Fortune 500," *Fortune*, **97**, pp. 238–272, May 8, 1978.)

Company	x_1 = assets (millions of dollars)	x_2 = net income (millions of dollars)	x_3 = stockholder equity (millions of dollars)
G.M.	26.7	3.3	15.8
Exxon	38.4	2.4	19.5
Ford	19.2	1.7	8.4
Mobil	20.6	1.0	8.2
Texaco	18.9	.9	9.4
Std. Oil	14.8	1.0	7.6
IBM	19.0	2.7	12.6
Gulf	14.2	.8	7.3
G.E.	13.7	1.1	5.9
Chrysler	7.7	.2	2.9

(a) Plot the scatter diagram and marginal dot diagrams for variables x_1 and x_2. Comment on their appearances.

(b) Compute $\bar{x}_1$, $\bar{x}_2$, s_{11}, s_{22}, s_{12}, and r_{12}. Interpret r_{12}.

1.5. Use the data in Exercise 1.4.

(a) Plot the scatter diagrams and dot diagrams for (x_2, x_3) and (x_1, x_3). Comment on the patterns.

(b) Compute the $\bar{\mathbf{x}}$, $\mathbf{S}_n$, and $\mathbf{R}$ arrays for (x_1, x_2, x_3).

1.6. The data in Table 1.2 are 42 measurements on air-pollution variables recorded at 12:00 noon in the Los Angeles area on different days.

(a) Plot the marginal dot diagrams for all the variables.

(b) Construct the $\bar{\mathbf{x}}$, $\mathbf{S}_n$, and $\mathbf{R}$ arrays and interpret the entries in $\mathbf{R}$.

1.7. You are given the following $n = 3$ observations on $p = 2$ variables.

Variable 1: $x_{11} = 2$	$x_{12} = 3$	$x_{13} = 4$
Variable 2: $x_{21} = 1$	$x_{22} = 2$	$x_{23} = 4$

(a) Plot the pairs of observations in the two-dimensional "variable space." That is, construct a two-dimensional scatterplot of the data.

(b) Plot the data as two points in the three-dimensional "item space."

1.8. Evaluate the distance of the point $P = (-1, -1)$ to the point $Q = (1, 0)$ using the Euclidean distance formula in (1-12) with $p = 2$ and using the statistical distance in

TABLE 1.2 AIR-POLLUTION DATA

Wind (x_1)	Solar rad. (x_2)	CO (x_3)	NO (x_4)	NO_2 (x_5)	O_3 (x_6)	HC (x_7)
8	98	7	2	12	8	2
7	107	4	3	9	5	3
7	103	4	3	5	6	3
10	88	5	2	8	15	4
6	91	4	2	8	10	3
8	90	5	2	12	12	4
9	84	7	4	12	15	5
5	72	6	4	21	14	4
7	82	5	1	11	11	3
8	64	5	2	13	9	4
6	71	5	4	10	3	3
6	91	4	2	12	7	3
7	72	7	4	18	10	3
10	70	4	2	11	7	3
10	72	4	1	8	10	3
9	77	4	1	9	10	3
8	76	4	1	7	7	3
8	71	5	3	16	4	4
9	67	4	2	13	2	3
9	69	3	3	9	5	3
10	62	5	3	14	4	4
9	88	4	2	7	6	3
8	80	4	2	13	11	4
5	30	3	3	5	2	3
6	83	5	1	10	23	4
8	84	3	2	7	6	3
6	78	4	2	11	11	3
8	79	2	1	7	10	3
6	62	4	3	9	8	3
10	37	3	1	7	2	3
8	71	4	1	10	7	3
7	52	4	1	12	8	4
5	48	6	5	8	4	3
6	75	4	1	10	24	3
10	35	4	1	6	9	2
8	85	4	1	9	10	2
5	86	3	1	6	12	2
5	86	7	2	13	18	2
7	79	7	4	9	25	3
7	79	5	2	8	6	2
6	68	6	2	11	14	3
8	40	4	3	6	5	2

SOURCE: Data courtesy of Professor G. C. Tiao.

(1-20) with $a_{11} = 1/3$, $a_{22} = 4/27$, and $a_{12} = 1/9$. Sketch the locus of points that are a constant squared statistical distance 1 from the point Q.

1.9. Consider the eight pairs of measurements on two variables x_1 and x_2 given below.

x_1	-6	-3	-2	1	2	5	6	8
x_2	-2	-3	1	-1	2	1	5	3

(a) Plot the data as a scatter diagram and compute s_{11}, s_{22}, and s_{12}.

(b) Using (1-18), calculate the corresponding measurements on variables $\tilde{x}_1$ and $\tilde{x}_2$, assuming the original coordinate axes are rotated through an angle of $\theta = 26°$ [given $\cos(26°) = .899$ and $\sin(26°) = .438$].

(c) Using the $\tilde{x}_1$ and $\tilde{x}_2$ measurements from (b), compute the sample variances $\tilde{s}_{11}$ and $\tilde{s}_{22}$.

(d) Consider the *new* pair of measurements $(x_1, x_2) = (4, -2)$. Transform these to measurements on $\tilde{x}_1$ and $\tilde{x}_2$ using (1-18) and calculate the distance $d(O, P)$ of the new point $P = (\tilde{x}_1, \tilde{x}_2)$ from the origin $O = (0, 0)$ using (1-17). [*Note:* You will need $\tilde{s}_{11}$ and $\tilde{s}_{22}$ from (c).]

(e) Calculate the distance from $P = (4, -2)$ to the origin $O = (0, 0)$ using (1-19) and the expressions for a_{11}, a_{22}, and a_{12} in Footnote 2. [*Note:* You will need s_{11}, s_{22}, and s_{12} from (a).] Compare the distance calculated here with the distance calculated using the $\tilde{x}_1$ and $\tilde{x}_2$ values in (d). (The numbers should, within rounding error, be the same.)

1.10. Are the following distance functions valid for distance from the origin? Explain.

(a) $x_1^2 + 4x_2^2 + x_1x_2 = (\text{distance})^2$

(b) $x_1^2 - 2x_2^2 = (\text{distance})^2$

1.11. Verify that distance defined by (1-20) with $a_{11} = 4$, $a_{22} = 1$, and $a_{12} = -1$ satisfies the first three conditions in (1-25). (The triangle inequality is more difficult to verify.)

1.12. Define distance from the point $P = (x_1, x_2)$ to the origin $O = (0, 0)$ as

$$d(O, P) = \max(|x_1|, |x_2|)$$

(a) Compute the distance from $P = (-3, 4)$ to the origin.

(b) Plot the locus of points whose squared distance from the origin is 1.

(c) Generalize the distance expression above to points in p dimensions.

1.13. A large city has major roads laid out in a grid pattern, as indicated in the diagram. First through fifth streets run north–south (NS) and streets A through E run east–west (EW). Suppose there are retail stores located at intersections $(A, 2)$, $(E, 3)$ and $(C, 5)$. Assume the distance along a street between two intersections in either the NS or EW directions is 1 unit. Define the distance between any two intersections (points) on the grid to be the "city block" distance. (For example, the distance between intersections

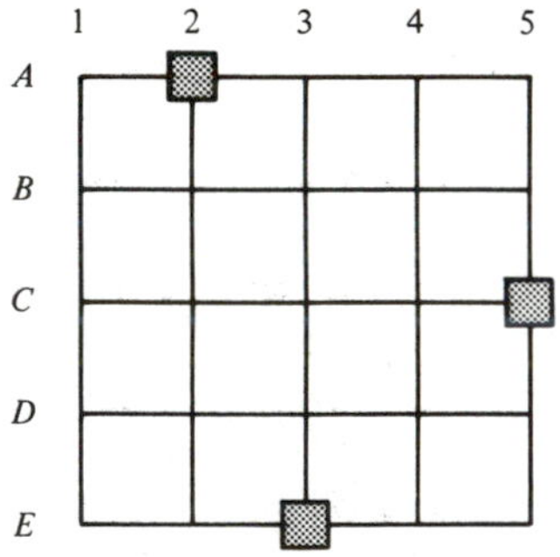

$(D, 1)$ and $(C, 2)$, which we might call $d((D, 1), (C, 2))$, is given by: $d((D, 1), (C, 2)) = d((D, 1), (D, 2)) + d((D, 2), (C, 2)) = 1 + 1 = 2$. Alternatively, $d((D, 1), (C, 2)) = d((D, 1), (C, 1)) + d((C, 1), (C, 2)) = 1 + 1 = 2$.)

Locate a supply facility (warehouse) at an intersection such that the sum of the distances from the warehouse to the three retail stores is minimized.

The following exercises contain fairly extensive data sets. *A computer may be necessary for the required calculations.*

1.14. Table 1.3 contains the raw data discussed in 1E.2. Two different visual stimuli ($S1$ and $S2$) produced responses in both the left eye (L) and the right eye (R) of subjects in the study groups. The values recorded in the table include: x_1 (subject age); x_2 (total response of both eyes to stimulus $S1$, that is, $S1L + S1R$); x_3 (difference between responses of eyes to stimulus $S1$, $|S1L - S1R|$); and so forth.

(a) Plot the two-dimensional scatterplot for the variables x_2 and x_4 for the multiple-sclerosis group. Comment on its appearance.

(b) Compute the $\bar{\mathbf{x}}$, $\mathbf{S}_n$, and $\mathbf{R}$ arrays for the non–multiple-sclerosis and multiple-sclerosis groups separately.

TABLE 1.3 MULTIPLE-SCLEROSIS DATA

Non–Multiple-Sclerosis Group Data

Subject number	x_1 (Age)	x_2 $(S1L + S1R)$	x_3 $\|S1L - S1R\|$	x_4 $(S2L + S2R)$	x_5 $\|S2L - S2R\|$
1.	18.	152.0	1.6	198.4	.0
2.	19.	138.0	.4	180.8	1.6
3.	20.	144.0	.0	186.4	.8
4.	20.	143.6	3.2	194.8	.0
5.	20.	148.8	.0	217.6	.0
6.	21.	141.6	.8	181.6	.8
7.	21.	136.0	1.6	180.0	.8
8.	21.	137.6	1.6	185.6	3.2
9.	22.	140.4	3.2	182.0	3.2
10.	22.	137.2	.0	181.8	.2
11.	22.	125.4	1.0	169.2	.0
12.	22.	142.4	4.8	185.6	.0
13.	22.	150.4	.0	214.4	3.2
14.	22.	145.6	1.6	203.6	5.2
15.	23.	147.2	3.2	196.8	1.6
16.	23.	139.2	1.6	179.2	.0
17.	24.	169.6	.0	204.8	.0
18.	24.	139.2	1.6	176.0	3.2
19.	24.	153.6	.0	212.0	.8
20.	25.	146.8	.0	194.8	3.2
21.	25.	139.2	1.6	198.4	3.2
22.	25.	136.0	1.6	181.6	2.4
23.	26.	138.8	1.6	191.6	.0
24.	26.	150.4	.0	205.2	.4
25.	26.	139.0	1.4	178.6	.2
26.	27.	133.8	.2	180.8	.0
27.	27.	139.0	1.8	190.4	1.6
28.	28.	136.0	1.6	193.2	3.6
29.	28.	146.4	.8	195.6	2.8

TABLE 1.3 (*continued*)

Non–Multiple-Sclerosis Group Data

Subject number	x_1 (Age)	x_2 $(S1L + S1R)$	x_3 $\|S1L - S1R\|$	x_4 $(S2L + S2R)$	x_5 $\|S2L - S2R\|$
30.	29.	145.2	4.8	194.2	3.8
31.	29.	146.4	.8	208.2	.2
32.	29.	138.0	2.8	181.2	.4
33.	30.	148.8	1.6	196.4	1.6
34.	31.	137.2	.0	184.0	.0
35.	31.	147.2	.0	197.6	.8
36.	32.	144.0	.0	185.8	.2
37.	32.	156.0	.0	192.8	2.4
38.	34.	137.0	.2	182.4	.0
39.	35.	143.2	2.4	184.0	1.6
40.	36.	141.6	.8	187.2	1.6
41.	37.	152.0	1.6	189.2	2.8
42.	39.	157.4	3.4	227.0	2.6
43.	40.	141.4	.6	209.2	1.6
44.	42.	156.0	2.4	195.2	3.2
45.	43.	150.4	1.6	180.0	.8
46.	43.	142.4	1.6	188.8	.0
47.	46.	158.0	2.0	192.0	3.2
48.	48.	130.0	3.6	190.0	.4
49.	49.	152.2	1.4	200.0	4.8
50.	49.	150.0	3.2	206.6	2.2
51.	50.	146.4	2.4	191.6	2.8
52.	54.	146.0	1.2	203.2	1.6
53.	55.	140.8	.0	184.0	1.6
54.	56.	140.4	.4	203.2	1.6
55.	56.	155.8	3.0	187.8	2.6
56.	56.	141.6	.8	196.8	1.6
57.	57.	144.8	.8	188.0	.8
58.	57.	146.8	3.2	191.6	.0
59.	59.	176.8	2.4	232.8	.8
60.	60.	171.0	1.8	202.0	3.6
61.	60.	163.2	.0	224.0	.0
62.	60.	171.6	1.2	213.8	3.4
63.	60.	146.4	4.0	203.2	4.8
64.	62.	146.8	3.6	201.6	3.2
65.	67.	154.4	2.4	205.2	6.0
66.	69.	171.2	1.6	210.4	.8
67.	73.	157.2	.4	204.8	.0
68.	74.	175.2	5.6	235.6	.4
69.	79.	155.0	1.4	204.4	.0

TABLE 1.3 *(continued)*

Multiple-Sclerosis Group Data

Subject number	x_1	x_2	x_3	x_4	x_5
1.	23.	148.0	.8	205.4	.6
2.	25.	195.2	3.2	262.8	.4
3.	25.	158.0	8.0	209.8	12.2
4.	28.	134.4	.0	198.4	3.2
5.	29.	190.2	14.2	243.8	10.6
6.	29.	160.4	18.4	222.8	31.2
7.	31.	227.8	90.2	270.2	83.0
8.	34.	211.0	3.0	250.8	5.2
9.	35.	204.8	12.8	254.4	11.2
10.	36.	141.2	6.8	194.4	21.6
11.	39.	157.4	3.4	227.0	2.6
12.	42.	166.4	.0	226.0	.0
13.	43.	191.8	35.4	243.6	40.8
14.	44.	156.8	.0	203.2	.0
15.	44.	202.8	29.2	246.4	24.8
16.	44.	165.2	18.4	254.0	46.4
17.	45.	162.0	5.6	224.4	8.8
18.	45.	138.4	.8	176.8	4.0
19.	45.	158.4	1.6	214.4	.0
20.	46.	155.4	1.8	201.2	6.0
21.	46.	214.8	9.2	290.6	.6
22.	47.	185.0	19.0	274.4	7.6
23.	48.	236.0	20.0	328.0	.0
24.	57.	170.8	24.0	228.4	33.6
25.	57.	165.6	16.8	229.2	15.6
26.	58.	238.4	8.0	304.4	6.0
27.	58.	164.0	.8	216.8	.8
28.	58.	169.8	.0	219.2	1.6
29.	59.	199.8	4.6	250.2	1.0

SOURCE: Data courtesy of Dr. G. G. Celesia.

1.15. The data described in 1E.1. are listed in Table 1.4. The data consist of average ratings over the course of treatment for patients undergoing radiotherapy. Variables measured included: x_1 (number of symptoms, such as sore throat or nausea); x_2 (amount of activity on a 1–5 scale); x_3 (amount of sleep on a 1–5 scale); x_4 (amount of food consumed on a 1–3 scale); x_5 (appetite on a 1–5 scale); and x_6 (skin reaction on a 0–3 scale).

(a) Construct the two-dimensional scatterplot for variables x_2 and x_3 and the marginal dot diagrams (or histograms). Do there appear to be any errors in the x_3 data?

(b) Compute the $\bar{\mathbf{x}}$, $\mathbf{S}_n$, and $\mathbf{R}$ arrays. Interpret the pairwise correlations.

TABLE 1.4 RADIOTHERAPY DATA

x_1 Symptoms	x_2 Activity	x_3 Sleep	x_4 Eat	x_5 Appetite	x_6 Skin reaction
.889	1.389	1.555	2.222	1.945	1.000
2.813	1.437	.999	2.312	2.312	2.000
1.454	1.091	2.364	2.455	2.909	3.000
.294	.941	1.059	2.000	1.000	1.000
2.727	2.545	2.819	2.727	4.091	.000
3.937	1.250	1.937	2.937	3.749	1.000
2.786	1.714	2.357	2.071	2.000	2.000
5.231	2.692	1.077	1.846	2.539	1.000
1.150	1.100	.950	2.000	1.000	1.000
6.500	2.562	1.749	2.562	2.499	1.000
.800	1.000	2.200	2.267	2.466	2.000
4.600	2.000	3.000	2.500	3.400	1.000
3.500	1.286	2.714	1.286	1.285	3.000
3.444	2.556	2.388	2.389	3.000	1.000
4.071	1.000	1.000	2.357	1.572	1.000
3.692	1.000	2.538	2.154	2.615	1.000
5.167	3.000	1.000	2.667	3.666	.000
.500	1.000	1.000	2.000	1.000	.000
2.385	1.923	2.539	2.154	2.461	1.000
2.100	1.300	1.300	1.800	2.600	1.000
5.000	3.250	3.125	2.375	3.375	.000
4.571	1.214	3.286	2.571	3.572	1.000
2.733	1.133	2.600	1.933	1.667	1.000
4.235	2.294	2.706	2.176	1.883	1.000
.000	1.000	1.941	2.000	2.000	.000
.750	1.125	3.000	1.875	2.000	3.000
3.077	1.462	2.384	2.000	1.846	2.000
1.600	1.200	2.950	2.000	2.750	1.000
6.273	3.636	1.182	2.545	3.364	.000
2.625	1.000	2.438	1.937	2.062	2.000
1.250	1.000	2.000	2.000	3.000	1.000
2.437	2.062	1.687	1.875	1.375	1.000
4.454	1.727	2.637	2.636	3.546	1.000
.133	1.000	1.000	2.000	1.000	.000
.222	1.222	1.445	2.000	1.000	1.000
2.467	2.667	2.200	1.933	1.800	3.000
4.000	1.000	4.000	2.167	2.500	.000
5.385	3.154	2.384	2.846	2.539	1.000
.773	1.000	2.273	1.909	2.091	.000
3.786	2.000	1.571	1.786	1.285	3.000
1.923	1.615	1.693	2.000	1.846	1.000
1.000	1.333	1.834	2.000	1.917	1.000
5.800	2.600	3.000	2.800	4.200	1.000
6.062	1.000	1.562	2.375	1.750	.000
3.706	1.235	1.530	2.118	2.294	1.000
2.444	2.333	1.223	2.444	1.776	3.000
6.111	2.222	2.889	2.889	3.555	2.000
2.533	1.067	1.600	2.000	1.333	1.000
2.167	1.000	2.167	2.000	2.500	1.000
2.375	1.062	2.375	2.000	2.125	3.000

TABLE 1.4 *(continued)*

x_1 Symptoms	x_2 Activity	x_3 Sleep	x_4 Eat	x_5 Appetite	x_6 Skin reaction
1.875	1.312	2.188	2.125	2.062	2.000
1.750	1.333	1.167	1.750	1.000	1.000
7.333	1.333	1.459	1.958	1.542	3.000
5.250	1.375	2.812	2.125	2.563	3.000
5.182	2.000	2.727	2.818	4.000	2.000
1.875	2.000	2.250	2.813	2.437	2.000
5.400	2.000	1.200	1.800	1.400	2.000
1.154	1.000	1.923	1.846	2.462	1.000
6.375	2.250	2.500	2.125	3.000	.875
9.454	2.727	3.818	2.455	3.272	3.000
1.000	1.000	1.917	1.833	2.167	1.000
1.444	1.111	2.000	2.111	2.000	1.000
1.800	1.100	3.100	2.200	2.600	1.000
2.818	2.000	1.955	2.045	2.546	2.000
10.461	2.154	2.769	2.000	2.923	.000
4.143	1.929	2.642	2.429	3.142	3.000
1.227	1.182	1.091	2.227	3.182	1.000
5.667	3.000	1.667	2.667	5.000	1.000
4.111	2.556	2.222	2.778	3.778	1.000
4.444	1.667	2.222	2.000	2.444	.000
3.714	3.857	2.643	2.286	3.285	.000
7.400	3.700	3.100	2.500	4.200	1.000
3.182	2.455	1.636	2.273	3.000	1.000
5.200	2.600	.800	1.800	2.000	.000
2.333	1.667	.666	1.667	2.166	.000
3.333	1.917	2.083	1.917	3.000	1.000
5.250	2.750	2.500	2.000	4.000	.000
7.714	4.000	3.071	2.929	4.428	3.000
3.846	2.615	3.000	2.692	3.693	2.000
2.444	1.111	1.000	2.111	1.667	2.000
5.333	1.917	3.000	2.250	1.917	1.000
1.556	1.778	3.444	2.667	3.333	1.000
3.182	1.545	1.910	2.273	3.000	1.000
6.222	2.444	3.889	2.444	3.445	1.000
7.231	1.000	3.154	2.308	4.384	2.000
3.857	1.071	3.000	2.071	2.286	1.000
3.778	1.944	1.612	1.611	1.945	1.000
6.000	1.400	2.067	2.267	2.866	2.000
2.333	3.583	2.334	2.333	2.667	2.000
7.571	2.143	3.143	2.571	3.929	1.000
3.667	2.000	2.111	2.778	4.000	3.000
3.600	2.933	2.067	2.200	2.867	.000
3.364	1.273	1.810	2.000	2.273	.000
4.100	1.900	2.800	2.000	2.600	2.000
.125	1.062	1.437	1.875	1.563	.000
6.231	2.769	1.462	2.385	4.000	2.000
3.000	1.455	2.090	2.273	3.272	2.000
.889	1.000	1.000	2.000	1.000	2.000

SOURCE: Data courtesy of Mrs. Annette Tealey, R. N. Values of x_2 and x_3 less than 1.0 are due to errors in the data collection process. Rows containing values of x_2 and x_3 less than 1.0 may be omitted.

REFERENCES

[1] Adelman, I., M. Greer, and C. T. Morris, "Instruments and Goals in Economic Development," *American Economic Review*, **59**, no. 2 (1969), 409–426.

[2] Bhattacharyya, G. K., and R. A. Johnson, *Statistical Concepts and Methods*, New York: John Wiley, 1977.

[3] Bliss, C. I., "Statistics in Biology," *Statistical Methods for Research in the Natural Sciences*, vol. 2, New York: McGraw-Hill, 1967.

[4] Cleveland, W. S., and D. A. Relles, "Clustering by Identification with Special Application to Two-Way Tables of Counts," *Journal of the American Statistical Association*, **70**, no. 351 (1975), 626–630.

[5] Cochran, W. G., *Sampling Techniques* (3rd ed.), New York: John Wiley, 1977.

[6] Cochran, W. G., and G. M. Cox, *Experimental Designs* (2nd ed.), New York: John Wiley, 1957.

[7] Davis, J. C., "Information Contained in Sediment Size Analysis," *Mathematical Geology*, **2**, no. 2 (1970), 105–112.

[8] Dunham, R. B., and D. J. Kravetz, "Canonical Correlation Analysis in a Predictive System," *Journal of Experimental Education*, **43**, no. 4 (1975), 35–42.

[9] Fritts, H. C., T. J. Blasing, B. P. Hayden, and J. E. Kutzbach, "Multivariate Techniques for Specifying Tree-Growth and Climate Relationships and Reconstructing Anomalies in Paleoclimate," *Journal of Applied Meteorology*, **10**, no. 5 (1971), 845–864.

[10] Halinar, J. C., "Principal Component Analysis in Plant Breeding," unpublished report based on data collected by Dr. F. A. Bliss, University of Wisconsin, 1979.

[11] Klatzky, S. R., and R. W. Hodge, "A Canonical Correlation Analysis of Occupational Mobility," *Journal of the American Statistical Association*, **66**, no. 333 (1971), 16–22.

[12] Linden, M., "Factor Analytic Study of Olympic Decathlon Data," *Research Quarterly*, **48**, no. 3 (Oct. 1977), 562–568.

[13] Mather, P. M., "Study of Factors Influencing Variation in Size Characteristics in Fluvioglacial Sediments," *Mathematical Geology*, **4**, no. 3 (1972), 219–234.

[14] Marriott, F. H. C., *The Interpretation of Multiple Observations*, London: Academic Press, 1974.

[15] Spenner, K. I., "From Generation to Generation: The Transmission of Occupation," Ph.D. dissertation, University of Wisconsin, 1977.

[16] Timm, N. H., *Multivariate Analysis with Applications in Education and Psychology*, Monterey, Calif.: Brooks/Cole, 1975.

[17] Trieschmann, J. S., and G. E. Pinches, "A Multivariate Model for Predicting Financially Distressed P-L Insurers," *Journal of Risk and Insurance*, **40**, no. 3 (1973), 327–338.

[18] Tukey, J. W., *Exploratory Data Analysis*, Reading, Mass.: Addison-Wesley, 1977.

2

Matrix Algebra and Random Vectors

2.1 INTRODUCTION

We saw in Chapter 1 that multivariate data can be conveniently displayed as an array of numbers. In general, a rectangular array of numbers with, for instance, p rows and n columns is called a *matrix* of dimension $p \times n$. The study of multivariate methods is greatly facilitated by the use of matrix algebra.

The matrix algebra results presented in this chapter will enable us to concisely state statistical models. Moreover, the formal relations expressed in matrix terms are easily programmed on computers to allow the routine calculation of important statistical quantities.

We begin by introducing some very basic concepts that are essential to both our geometrical interpretations and algebraic explanations of subsequent statistical techniques. If you have not been previously exposed to the rudiments of matrix algebra, you may prefer to follow the brief refresher in the next section by the more detailed review provided in Supplement 2A.

2.2 SOME BASICS OF MATRIX AND VECTOR ALGEBRA

Vectors

An array $\mathbf{x}$ of n real numbers $x_1, x_2, \ldots, x_n$ is called a *vector* and it is written as

$$\mathbf{x} = \begin{bmatrix} x_1 \\ x_2 \\ \vdots \\ x_n \end{bmatrix} \quad \text{or} \quad \mathbf{x}' = [x_1, x_2, \ldots, x_n]$$

where the prime denotes the operation of *transposing* a column to a row.

A vector $\mathbf{x}$ can be represented geometrically as a directed line in n-dimensions with component x_1 along the first axis, x_2 along the second axis, ..., x_n along the nth axis. This is illustrated in Figure 2.1 for $n = 3$.

A vector can be *expanded* or *contracted* by multiplying by a constant c. In particular, we define the vector $c\mathbf{x}$ as

$$c\mathbf{x} = \begin{bmatrix} cx_1 \\ cx_2 \\ \vdots \\ cx_n \end{bmatrix}$$

That is, $c\mathbf{x}$ is the vector obtained by multiplying each element of $\mathbf{x}$ by c. [See Figure 2.2(a)].

Two vectors may be added. *Addition* of $\mathbf{x}$ and $\mathbf{y}$ is defined as

$$\mathbf{x} + \mathbf{y} = \begin{bmatrix} x_1 \\ x_2 \\ \vdots \\ x_n \end{bmatrix} + \begin{bmatrix} y_1 \\ y_2 \\ \vdots \\ y_n \end{bmatrix} = \begin{bmatrix} x_1 + y_1 \\ x_2 + y_2 \\ \vdots \\ x_n + y_n \end{bmatrix}$$

so that $\mathbf{x} + \mathbf{y}$ is the vector with ith element $x_i + y_i$.

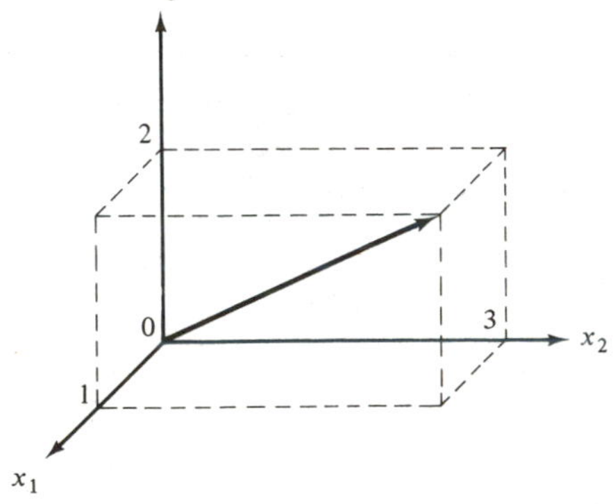

Figure 2.1 The vector $\mathbf{x}' = [1, 3, 2]$.

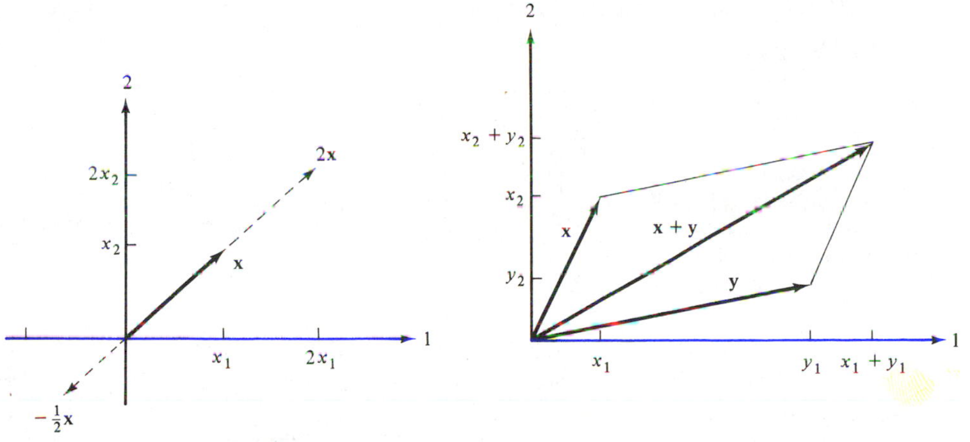

(a) Multipication of a vector by a constant

(b) Addition of two vectors

Figure 2.2 Scalar multiplication and vector addition.

The sum of two vectors emanating from the origin is the diagonal of the parallelogram formed with the two original vectors as adjacent sides. This geometrical interpretation is illustrated in Figure 2.2(b).

A vector has both direction and length. In $n = 2$ dimensions, we consider the vector

$$\mathbf{x} = \begin{bmatrix} x_1 \\ x_2 \end{bmatrix}$$

The length of $\mathbf{x}$, written $L_{\mathbf{x}}$, is defined to be

$$L_{\mathbf{x}} = \sqrt{x_1^2 + x_2^2}$$

Geometrically, the length of a vector in two dimensions can be viewed as the hypothenuse of a right triangle. This is demonstrated schematically in Figure 2.3.

The *length* of a vector $\mathbf{x}' = [x_1, x_2, \ldots, x_n]$, with n components, is defined by

$$L_{\mathbf{x}} = \sqrt{x_1^2 + x_2^2 + \cdots + x_n^2} \tag{2-1}$$

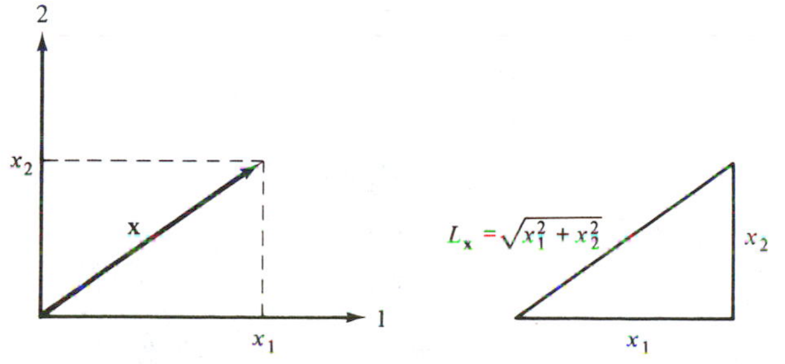

Figure 2.3 Length of $\mathbf{x} = \sqrt{x_1^2 + x_2^2}$.

Multiplication of a vector $\mathbf{x}$ by a scalar c changes the length. From Equation (2-1),

$$L_{c\mathbf{x}} = \sqrt{c^2x_1^2 + c^2x_2^2 + \cdots + c^2x_n^2}$$
$$= |c|\sqrt{x_1^2 + x_2^2 + \cdots + x_n^2} = |c|\,L_{\mathbf{x}}$$

Multiplication by c does not change the direction of the vector $\mathbf{x}$ if $c > 0$. However, a negative value of c creates a vector with a direction opposite that of $\mathbf{x}$. From

$$L_{c\mathbf{x}} = |c|\,L_{\mathbf{x}} \tag{2-2}$$

it is clear that $\mathbf{x}$ is expanded if $|c| > 1$ and contracted if $0 < |c| < 1$. [Recall Figure 2.2(a).] Choosing $c = L_{\mathbf{x}}^{-1}$ we obtain the *unit vector* $L_{\mathbf{x}}^{-1}\mathbf{x}$, which has length 1 and lies in the direction of $\mathbf{x}$.

A second geometrical concept is *angle*. Consider two vectors in a plane and the angle, θ, between them, as in Figure 2.4. From Figure 2.4, θ can be represented as the difference between the angles θ_1 and θ_2 formed by the two vectors and the first coordinate axis. Since, by definition,

$$\cos(\theta_1) = \frac{x_1}{L_{\mathbf{x}}} \qquad \cos(\theta_2) = \frac{y_1}{L_{\mathbf{y}}}$$

$$\sin(\theta_1) = \frac{x_2}{L_{\mathbf{x}}} \qquad \sin(\theta_2) = \frac{y_2}{L_{\mathbf{y}}}$$

and

$$\cos(\theta) = \cos(\theta_2 - \theta_1) = \cos(\theta_2)\cos(\theta_1) + \sin(\theta_2)\sin(\theta_1),$$

the angle θ between the two vectors $\mathbf{x}' = [x_1, x_2]$ and $\mathbf{y}' = [y_1, y_2]$ is specified by

$$\cos(\theta) = \cos(\theta_2 - \theta_1) = \left(\frac{y_1}{L_{\mathbf{y}}}\right)\left(\frac{x_1}{L_{\mathbf{x}}}\right) + \left(\frac{y_2}{L_{\mathbf{y}}}\right)\left(\frac{x_2}{L_{\mathbf{x}}}\right) = \frac{x_1y_1 + x_2y_2}{L_{\mathbf{x}}L_{\mathbf{y}}} \tag{2-3}$$

We find it convenient to introduce the *inner product* of two vectors. For $n = 2$ dimensions, the inner product of $\mathbf{x}$ and $\mathbf{y}$ is

$$\mathbf{x}'\mathbf{y} = x_1y_1 + x_2y_2$$

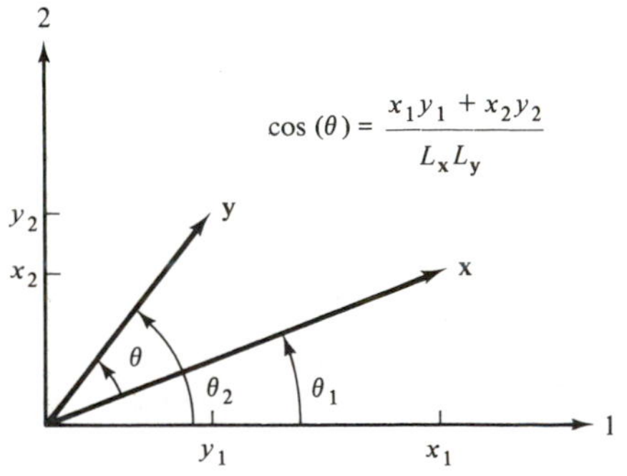

Figure 2.4 The angle θ between $\mathbf{x}' = [x_1, x_2]$ and $\mathbf{y}' = [y_1, y_2]$.

With this definition and Equation (2-3),

$$L_{\mathbf{x}} = \sqrt{\mathbf{x}'\mathbf{x}} \qquad \cos(\theta) = \frac{\mathbf{x}'\mathbf{y}}{L_{\mathbf{x}} L_{\mathbf{y}}} = \frac{\mathbf{x}'\mathbf{y}}{\sqrt{\mathbf{x}'\mathbf{x}}\sqrt{\mathbf{y}'\mathbf{y}}}$$

Since $\cos(90°) = \cos(270°) = 0$ and $\cos(\theta) = 0$ only if $\mathbf{x}'\mathbf{y} = 0$, $\mathbf{x}$ and $\mathbf{y}$ are perpendicular when $\mathbf{x}'\mathbf{y} = 0$.

For an arbitrary number of dimensions, n, we define the inner product of $\mathbf{x}$ and $\mathbf{y}$ as

$$\mathbf{x}'\mathbf{y} = x_1 y_1 + x_2 y_2 + \cdots + x_n y_n \tag{2-4}$$

The inner product is denoted by either $\mathbf{x}'\mathbf{y}$ or $\mathbf{y}'\mathbf{x}$.

Using the inner product, we have the natural extension of length and angle to vectors of n components:

$$L_{\mathbf{x}} = \text{length of } \mathbf{x} = \sqrt{\mathbf{x}'\mathbf{x}} \tag{2-5}$$

$$\cos(\theta) = \frac{\mathbf{x}'\mathbf{y}}{L_{\mathbf{x}} L_{\mathbf{y}}} = \frac{\mathbf{x}'\mathbf{y}}{\sqrt{\mathbf{x}'\mathbf{x}}\sqrt{\mathbf{y}'\mathbf{y}}} \tag{2-6}$$

Since, again, $\cos(\theta) = 0$ only if $\mathbf{x}'\mathbf{y} = 0$, we say $\mathbf{x}$ and $\mathbf{y}$ are *perpendicular* when $\mathbf{x}'\mathbf{y} = 0$.

Example 2.1

Given the vectors $\mathbf{x}' = [1, 3, 2]$ and $\mathbf{y}' = [-2, 1, -1]$, find $3\mathbf{x}$ and $\mathbf{x} + \mathbf{y}$. Next, determine the length of $\mathbf{x}$, the length of $\mathbf{y}$, and the angle between $\mathbf{x}$ and $\mathbf{y}$. Also, check that the length of $3\mathbf{x}$ is three times the length of $\mathbf{x}$.

First,

$$3\mathbf{x} = 3\begin{bmatrix} 1 \\ 3 \\ 2 \end{bmatrix} = \begin{bmatrix} 3 \\ 9 \\ 6 \end{bmatrix}$$

$$\mathbf{x} + \mathbf{y} = \begin{bmatrix} 1 \\ 3 \\ 2 \end{bmatrix} + \begin{bmatrix} -2 \\ 1 \\ -1 \end{bmatrix} = \begin{bmatrix} 1-2 \\ 3+1 \\ 2-1 \end{bmatrix} = \begin{bmatrix} -1 \\ 4 \\ 1 \end{bmatrix}$$

Next $\mathbf{x}'\mathbf{x} = 1^2 + 3^2 + 2^2 = 14$, $\mathbf{y}'\mathbf{y} = (-2)^2 + 1^2 + (-1)^2 = 6$, and $\mathbf{x}'\mathbf{y} = 1(-2) + 3(1) + 2(-1) = -1$. Therefore

$$L_{\mathbf{x}} = \sqrt{\mathbf{x}'\mathbf{x}} = \sqrt{14} = 3.742 \qquad L_{\mathbf{y}} = \sqrt{\mathbf{y}'\mathbf{y}} = \sqrt{6} = 2.449$$

and

$$\cos(\theta) = \frac{\mathbf{x}'\mathbf{y}}{L_{\mathbf{x}} L_{\mathbf{y}}} = \frac{-1}{3.742 \times 2.449} = -.109$$

so $\theta = 96.3$ degrees. Finally,

$$L_{3\mathbf{x}} = \sqrt{3^2 + 9^2 + 6^2} = \sqrt{126} \quad \text{and} \quad 3L_{\mathbf{x}} = 3\sqrt{14} = \sqrt{126}$$

showing $L_{3\mathbf{x}} = 3L_{\mathbf{x}}$. ■

A pair of vectors $\mathbf{x}$ and $\mathbf{y}$ of the same dimension is said to be *linearly dependent* if there exist constants c_1 and c_2, both not zero, such that

$$c_1\mathbf{x} + c_2\mathbf{y} = \mathbf{0}$$

A set of vectors $\mathbf{x}_1, \mathbf{x}_2, \ldots, \mathbf{x}_k$ is said to be *linearly dependent* if there exist constants $c_1, c_2, \ldots, c_k$, not all zero, such that

$$c_1\mathbf{x}_1 + c_2\mathbf{x}_2 + \cdots + c_k\mathbf{x}_k = \mathbf{0} \tag{2-7}$$

Linear dependence implies at least one vector in the set can be written as a linear combination of the other vectors. Vectors of the same dimension that are not linearly dependent are said to be *linearly independent*.

Example 2.2

Consider the set of vectors

$$\mathbf{x}_1 = \begin{bmatrix} 1 \\ 2 \\ 1 \end{bmatrix} \quad \mathbf{x}_2 = \begin{bmatrix} 1 \\ 0 \\ -1 \end{bmatrix} \quad \mathbf{x}_3 = \begin{bmatrix} 1 \\ -2 \\ 1 \end{bmatrix}$$

Setting

$$c_1\mathbf{x}_1 + c_2\mathbf{x}_2 + c_3\mathbf{x}_3 = \mathbf{0}$$

implies

$$\begin{aligned} c_1 + c_2 + c_3 &= 0 \\ 2c_1 \qquad - 2c_3 &= 0 \\ c_1 - c_2 + c_3 &= 0 \end{aligned}$$

with the unique solution $c_1 = c_2 = c_3 = 0$. As we cannot find three constants c_1, c_2, and c_3, *not all zero*, such that $c_1\mathbf{x}_1 + c_2\mathbf{x}_2 + c_3\mathbf{x}_3 = \mathbf{0}$, the vectors $\mathbf{x}_1$, $\mathbf{x}_2$, and $\mathbf{x}_3$ are *linearly independent*. ■

The *projection* (or shadow) of a vector $\mathbf{x}$ on a vector $\mathbf{y}$ is

$$\text{Projection of } \mathbf{x} \text{ on } \mathbf{y} = \frac{(\mathbf{x}'\mathbf{y})}{\mathbf{y}'\mathbf{y}}\mathbf{y} = \frac{(\mathbf{x}'\mathbf{y})}{L_{\mathbf{y}}}\frac{1}{L_{\mathbf{y}}}\mathbf{y} \tag{2-8}$$

where the vector $L_{\mathbf{y}}^{-1}\mathbf{y}$ has unit length. The *length* of the projection is

$$\text{Length of projection} = \frac{|\mathbf{x}'\mathbf{y}|}{L_{\mathbf{y}}} = L_{\mathbf{x}}\left|\frac{\mathbf{x}'\mathbf{y}}{L_{\mathbf{x}}L_{\mathbf{y}}}\right| = L_{\mathbf{x}}|\cos(\theta)| \tag{2-9}$$

where θ is the angle between $\mathbf{x}$ and $\mathbf{y}$ (see Figure 2.5).

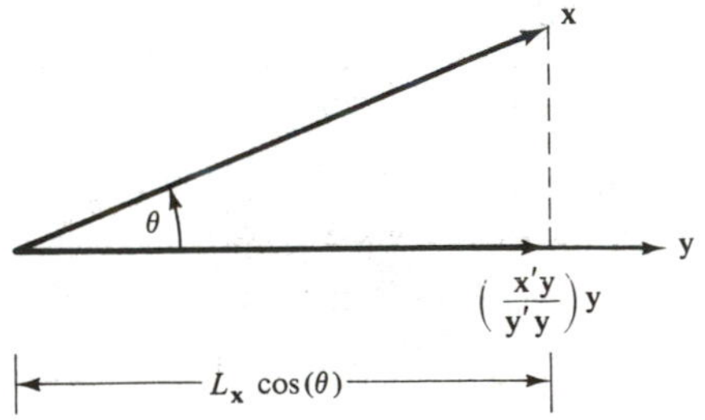

Figure 2.5 The projection of $\mathbf{x}$ on $\mathbf{y}$.

Matrices

A *matrix* is any rectangular array of real numbers. We denote an arbitrary array of p rows and n columns by

$$\underset{(p\times n)}{\mathbf{A}} = \begin{bmatrix} a_{11} & a_{12} & \cdots & a_{1n} \\ a_{21} & a_{22} & \cdots & a_{2n} \\ \vdots & \vdots & & \vdots \\ a_{p1} & a_{p2} & \cdots & a_{pn} \end{bmatrix}$$

Many of the vector concepts introduced above have direct generalizations to matrices.

The *transpose* operation, $\mathbf{A}'$, of a matrix changes the columns into rows so that the first column of $\mathbf{A}$ becomes the first row of $\mathbf{A}'$, the second column becomes the second row, and so forth.

Example 2.3

If

$$\underset{(2\times 3)}{\mathbf{A}} = \begin{bmatrix} 3 & -1 & 2 \\ 1 & 5 & 4 \end{bmatrix}$$

then

$$\underset{(3\times 2)}{\mathbf{A}'} = \begin{bmatrix} 3 & 1 \\ -1 & 5 \\ 2 & 4 \end{bmatrix} \qquad \blacksquare$$

A matrix may also be multiplied by a constant c. The product $c\mathbf{A}$ is the matrix that results from multiplying each element of $\mathbf{A}$ by c. Thus

$$\underset{(p\times n)}{c\mathbf{A}} = \begin{bmatrix} ca_{11} & ca_{12} & \cdots & ca_{1n} \\ ca_{21} & ca_{22} & \cdots & ca_{2n} \\ \vdots & \vdots & & \vdots \\ ca_{p1} & ca_{p2} & \cdots & ca_{pn} \end{bmatrix}$$

Two matrices, $\mathbf{A}$ and $\mathbf{B}$, of the same dimensions can be added. The sum $\mathbf{A} + \mathbf{B}$ has (i, j)th entry $a_{ij} + b_{ij}$.

Example 2.4

If $\underset{(2\times 3)}{\mathbf{A}} = \begin{bmatrix} 0 & 3 & 1 \\ 1 & -1 & 1 \end{bmatrix}$ and $\underset{(2\times 3)}{\mathbf{B}} = \begin{bmatrix} 1 & -2 & -3 \\ 2 & 5 & 1 \end{bmatrix}$,

$$\underset{(2\times 3)}{4\mathbf{A}} = \begin{bmatrix} 0 & 12 & 4 \\ 4 & -4 & 4 \end{bmatrix} \quad \text{and}$$

$$\underset{(2\times 3)}{\mathbf{A}} + \underset{(2\times 3)}{\mathbf{B}} = \begin{bmatrix} 0+1 & 3-2 & 1-3 \\ 1+2 & -1+5 & 1+1 \end{bmatrix} = \begin{bmatrix} 1 & 1 & -2 \\ 3 & 4 & 2 \end{bmatrix} \qquad \blacksquare$$

It is also possible to define *matrix* multiplication if the dimensions of the matrices conform in the following manner. When $\mathbf{A}$ is $(p \times k)$ and $\mathbf{B}$ is $(k \times n)$, so that the number of elements in a row of **A** is the same as the number of elements in a column of **B**, we can form the matrix product **AB**. An element of the new matrix **AB** is formed by taking the inner product of each row of **A** with each column of **B**.

The *matrix product* **AB** is

$$\underset{(p\times k)}{\mathbf{A}}\ \underset{(k\times n)}{\mathbf{B}} = \text{the } (p \times n) \text{ matrix whose entry in the } i\text{th row and } j\text{th column is the inner product of the } i\text{th row of } \mathbf{A} \text{ and } j\text{th column of } \mathbf{B}$$

or

$$(i, j) \text{ entry of } \mathbf{AB} = a_{i1}b_{1j} + a_{i2}b_{2j} + \cdots + a_{ik}b_{kj} = \sum_{\ell=1}^{k} a_{i\ell}b_{\ell j} \tag{2-10}$$

When $k = 4$, we have four products to add for each entry in the matrix **AB**. Thus

$$\underset{(p\times 4)}{\mathbf{A}}\ \underset{(4\times n)}{\mathbf{B}} = \begin{bmatrix} a_{11} & a_{12} & a_{13} & a_{14} \\ \vdots & \vdots & \vdots & \vdots \\ a_{i1} & a_{i2} & a_{i3} & a_{i4} \\ \vdots & \vdots & \vdots & \vdots \\ a_{p1} & a_{p2} & a_{p3} & a_{p4} \end{bmatrix} \begin{bmatrix} b_{11} & \cdots & b_{1j} & \cdots & b_{1n} \\ b_{21} & \cdots & b_{2j} & \cdots & b_{2n} \\ b_{31} & \cdots & b_{3j} & \cdots & b_{3n} \\ b_{41} & \cdots & b_{4j} & \cdots & b_{4n} \end{bmatrix}$$

$$\begin{array}{c} \text{Column} \\ j \end{array}$$

$$= \text{Row } i \begin{bmatrix} & \vdots & \\ \cdots & (a_{i1}b_{1j} + a_{i2}b_{2j} + a_{i3}b_{3j} + a_{i4}b_{4j}) & \cdots \\ & \vdots & \end{bmatrix}$$

Example 2.5

If

$$\mathbf{A} = \begin{bmatrix} 3 & -1 & 2 \\ 1 & 5 & 4 \end{bmatrix}, \mathbf{B} = \begin{bmatrix} -2 \\ 7 \\ 9 \end{bmatrix}, \text{ and } \mathbf{C} = \begin{bmatrix} 2 & 0 \\ 1 & -1 \end{bmatrix},$$

then

$$\underset{(2\times 3)}{\mathbf{A}}\ \underset{(3\times 1)}{\mathbf{B}} = \begin{bmatrix} 3 & -1 & 2 \\ 1 & 5 & 4 \end{bmatrix} \begin{bmatrix} -2 \\ 7 \\ 9 \end{bmatrix} = \begin{bmatrix} 3(-2) + (-1)(7) + 2(9) \\ 1(-2) + 5(7) \quad + 4(9) \end{bmatrix}$$

$$= \underset{(2\times 1)}{\begin{bmatrix} 5 \\ 69 \end{bmatrix}}$$

$$\underset{(2\times 2)}{\mathbf{C}}\ \underset{(2\times 3)}{\mathbf{A}} = \begin{bmatrix} 2 & 0 \\ 1 & -1 \end{bmatrix} \begin{bmatrix} 3 & -1 & 2 \\ 1 & 5 & 4 \end{bmatrix}$$

$$= \begin{bmatrix} 2(3) + 0(1) & 2(-1) + 0(5) & 2(2) + 0(4) \\ 1(3) - 1(1) & 1(-1) - 1(5) & 1(2) - 1(4) \end{bmatrix} = \underset{(2\times 3)}{\begin{bmatrix} 6 & -2 & 4 \\ 2 & -6 & -2 \end{bmatrix}}$$

■

Square matrices will be of special importance in our development of statistical methods. A square matrix is said to be *symmetric* if $\mathbf{A} = \mathbf{A}'$ or $a_{ij} = a_{ji}$ for all i and j.

Example 2.6

The matrix

$$\begin{bmatrix} 3 & 5 \\ 5 & -2 \end{bmatrix}$$

is symmetric; the matrix

$$\begin{bmatrix} 3 & 6 \\ 4 & -2 \end{bmatrix}$$

is not symmetric. ■

When two square matrices **A** and **B** are of the same dimension, both products **AB** and **BA** are defined, although they need not be equal (see Supplement 2A). If we let **I** denote the square matrix with ones on the diagonal and zeros elsewhere, it follows from the definition of matrix multiplication that the (i, j) entry of **AI** is $a_{i1} \times 0 + \cdots + a_{i,j-1} \times 0 + a_{ij} \times 1 + a_{i,j+1} \times 0 + \cdots + a_{ik} \times 0 = a_{ij}$, so $\mathbf{AI} = \mathbf{A}$. Similarly, $\mathbf{IA} = \mathbf{A}$ and

$$\underset{(k\times k)}{\mathbf{I}}\ \underset{(k\times k)}{\mathbf{A}} = \underset{(k\times k)}{\mathbf{A}}\ \underset{(k\times k)}{\mathbf{I}} = \underset{(k\times k)}{\mathbf{A}} \quad \text{for any } \underset{(k\times k)}{\mathbf{A}} \tag{2-11}$$

The matrix **I** acts like 1 in ordinary multiplication ($1 \cdot a = a \cdot 1 = a$), so it is called the *identity* matrix.

The fundamental scalar relation about the existence of an inverse number a^{-1} such that $a^{-1}a = aa^{-1} = 1$, if $a \neq 0$, has the following matrix algebra extension. If there exists a matrix **B** such that

$$\underset{(k\times k)}{\mathbf{B}}\ \underset{(k\times k)}{\mathbf{A}} = \underset{(k\times k)}{\mathbf{A}}\ \underset{(k\times k)}{\mathbf{B}} = \underset{(k\times k)}{\mathbf{I}}$$

then **B** is called the *inverse* of **A** and is denoted by $\mathbf{A}^{-1}$.

The technical condition that an inverse exists is that the k columns $\mathbf{a}_1, \mathbf{a}_2, \ldots, \mathbf{a}_k$ of **A** are linearly independent. That is, the existence of $\mathbf{A}^{-1}$ is equivalent to

$$c_1\mathbf{a}_1 + c_2\mathbf{a}_2 + \cdots + c_k\mathbf{a}_k = \mathbf{0} \quad \text{only if} \quad c_1 = \cdots = c_k = 0 \tag{2-12}$$

(See Result 2A.9 in Supplement 2A.)

Example 2.7

For

$$\mathbf{A} = \begin{bmatrix} 3 & 2 \\ 4 & 1 \end{bmatrix}$$

you may verify that

$$\begin{bmatrix} -.2 & .4 \\ .8 & -.6 \end{bmatrix}\begin{bmatrix} 3 & 2 \\ 4 & 1 \end{bmatrix} = \begin{bmatrix} (-.2)3 + (.4)4 & (-.2)2 + (.4)1 \\ (.8)3 + (-.6)4 & (.8)2 + (-.6)1 \end{bmatrix} = \begin{bmatrix} 1 & 0 \\ 0 & 1 \end{bmatrix}$$

so

$$\begin{bmatrix} -.2 & .4 \\ .8 & -.6 \end{bmatrix}$$

is $\mathbf{A}^{-1}$. We note that

$$c_1\begin{bmatrix} 3 \\ 4 \end{bmatrix} + c_2\begin{bmatrix} 2 \\ 1 \end{bmatrix} = \begin{bmatrix} 0 \\ 0 \end{bmatrix}$$

implies $c_1 = c_2 = 0$, so the columns of $\mathbf{A}$ are linearly independent. This confirms the condition stated in (2-12). ■

A method for computing an inverse, when it exists, is given in Supplement 2A. The routine, but lengthy, calculations are usually relegated to a computer, especially when the dimension is greater than three. Even so, you must be forewarned that if the column sum in (2-12) is *nearly* $\mathbf{0}$, for some constants $c_1, \ldots, c_k$ the computer may produce incorrect inverses due to extreme errors in rounding. It is always good to check the products $\mathbf{AA}^{-1}$ and $\mathbf{A}^{-1}\mathbf{A}$ for equality with $\mathbf{I}$ when $\mathbf{A}^{-1}$ is produced by a computer package. (See Exercise 2.10.)

Diagonal matrices have inverses that are easy to compute. For example,

$$\begin{bmatrix} a_{11} & 0 & 0 & 0 & 0 \\ 0 & a_{22} & 0 & 0 & 0 \\ 0 & 0 & a_{33} & 0 & 0 \\ 0 & 0 & 0 & a_{44} & 0 \\ 0 & 0 & 0 & 0 & a_{55} \end{bmatrix} \quad \text{has inverse} \quad \begin{bmatrix} \frac{1}{a_{11}} & 0 & 0 & 0 & 0 \\ 0 & \frac{1}{a_{22}} & 0 & 0 & 0 \\ 0 & 0 & \frac{1}{a_{33}} & 0 & 0 \\ 0 & 0 & 0 & \frac{1}{a_{44}} & 0 \\ 0 & 0 & 0 & 0 & \frac{1}{a_{55}} \end{bmatrix}$$

if all the $a_{ii} \neq 0$.

Another special class of square matrices with which we shall become familiar are the *orthogonal* matrices characterized by

$$\mathbf{QQ}' = \mathbf{Q}'\mathbf{Q} = \mathbf{I} \quad \text{or} \quad \mathbf{Q}' = \mathbf{Q}^{-1} \tag{2-13}$$

The name derives from the property that if $\mathbf{Q}$ has ith row $\mathbf{q}_i'$, then $\mathbf{QQ}' = \mathbf{I}$ implies that $\mathbf{q}_i'\mathbf{q}_i = 1$ and $\mathbf{q}_i'\mathbf{q}_j = 0$ for $i \neq j$ so the rows have unit length and are mutually perpendicular (orthogonal). According to the condition $\mathbf{Q}'\mathbf{Q} = \mathbf{I}$, the columns have the same property.

We conclude our brief introduction to the elements of matrix algebra by introducing a concept fundamental to multivariate statistical analysis. A square matrix $\mathbf{A}$ is said to have an *eigenvalue* λ, with corresponding *eigenvector* $\mathbf{x} \neq \mathbf{0}$, if

$$\mathbf{Ax} = \lambda\mathbf{x} \tag{2-14}$$

Ordinarily, we normalize $\mathbf{x}$ so that it has length one so $1 = \mathbf{x}'\mathbf{x}$. It is convenient to denote normalized eigenvectors by $\mathbf{e}$, and we do so in the sequel. Sparing you the details of the derivation (see [1]), we state the following basic result.

Let $\mathbf{A}$ be a $(k \times k)$ square symmetric matrix. Then $\mathbf{A}$ has k pairs of eigenvalues and eigenvectors:

$$\lambda_1, \mathbf{e}_1 \qquad \lambda_2, \mathbf{e}_2 \; \ldots \; \lambda_k, \mathbf{e}_k \tag{2-15}$$

The eigenvectors can be chosen to satisfy $1 = \mathbf{e}_1'\mathbf{e}_1 = \cdots = \mathbf{e}_k'\mathbf{e}_k$ and be mutually perpendicular. The eigenvectors are unique unless two or more eigenvalues are equal.

Example 2.8

Let

$$\mathbf{A} = \begin{bmatrix} 1 & -5 \\ -5 & 1 \end{bmatrix}$$

Then, since

$$\begin{bmatrix} 1 & -5 \\ -5 & 1 \end{bmatrix} \begin{bmatrix} \frac{1}{\sqrt{2}} \\ -\frac{1}{\sqrt{2}} \end{bmatrix} = 6 \begin{bmatrix} \frac{1}{\sqrt{2}} \\ -\frac{1}{\sqrt{2}} \end{bmatrix}$$

$\lambda_1 = 6$ is an eigenvalue and

$$\mathbf{e}_1 = \begin{bmatrix} \frac{1}{\sqrt{2}} \\ -\frac{1}{\sqrt{2}} \end{bmatrix}$$

is its corresponding normalized eigenvector. You may wish to show that a second eigenvalue, eigenvector pair is: $\lambda_2 = -4$, $\mathbf{e}_2' = [1/\sqrt{2}, \;\; 1/\sqrt{2}]$. ■

A method for calculating the λ's and $\mathbf{e}$'s is described in Supplement 2A. It is instructive to do a few sample calculations to understand the technique. We usually rely on a computer when the dimension of the square matrix is greater than two or three.

2.3 POSITIVE DEFINITE MATRICES

The study of the variation and interrelationships in multivariate data is often based upon distances and the assumption that the data are multivariate normally distributed. Squared distances (see Chapter 1) and the multivariate normal density can be expressed in terms of matrix products called *quadratic forms* (see Chapter 4).

Consequently, it should not be surprising that quadratic forms play a central role in multivariate analysis. In this section we consider quadratic forms that are always nonnegative and the associated *positive definite* matrices.

Results involving quadratic forms and symmetric matrices are, in many cases, a direct consequence of an expansion for symmetric matrices known as the *spectral decomposition*. The spectral decomposition of a $k \times k$ symmetric matrix $\mathbf{A}$ is given by[1]

$$\underset{(k\times k)}{\mathbf{A}} = \lambda_1 \underset{(k\times 1)}{\mathbf{e}_1} \underset{(1\times k)}{\mathbf{e}_1'} + \lambda_2 \underset{(k\times 1)}{\mathbf{e}_2} \underset{(1\times k)}{\mathbf{e}_2'} + \cdots + \lambda_k \underset{(k\times 1)}{\mathbf{e}_k} \underset{(1\times k)}{\mathbf{e}_k'} \tag{2-16}$$

where $\lambda_1, \lambda_2, \ldots, \lambda_k$ are the eigenvalues of $\mathbf{A}$ and $\mathbf{e}_1, \mathbf{e}_2, \ldots, \mathbf{e}_k$ are the associated normalized eigenvectors. Thus $\mathbf{e}_i'\mathbf{e}_i = 1$, $i = 1, 2, \ldots, k$ and $\mathbf{e}_i'\mathbf{e}_j = 0$, $i \neq j$.

Example 2.9

Consider the symmetric matrix

$$\mathbf{A} = \begin{bmatrix} 13 & -4 & 2 \\ -4 & 13 & -2 \\ 2 & -2 & 10 \end{bmatrix}$$

The eigenvalues obtained from the characteristic equation $|\mathbf{A} - \lambda\mathbf{I}| = 0$ are $\lambda_1 = 9$, $\lambda_2 = 9$, and $\lambda_3 = 18$ (Definition 2A.30). The corresponding eigenvectors $\mathbf{e}_1$, $\mathbf{e}_2$ and $\mathbf{e}_3$ are the (normalized) solutions of the equations $\mathbf{A}\mathbf{e}_i = \lambda_i\mathbf{e}_i$ for $i = 1, 2, 3$. Thus $\mathbf{A}\mathbf{e}_1 = \lambda\mathbf{e}_1$ gives

$$\begin{bmatrix} 13 & -4 & 2 \\ -4 & 13 & -2 \\ 2 & -2 & 10 \end{bmatrix} \begin{bmatrix} e_{11} \\ e_{21} \\ e_{31} \end{bmatrix} = 9 \begin{bmatrix} e_{11} \\ e_{21} \\ e_{31} \end{bmatrix}$$

or

$$\begin{aligned} 13e_{11} - 4e_{21} + 2e_{31} &= 9e_{11} \\ -4e_{11} + 13e_{21} - 2e_{31} &= 9e_{21} \\ 2e_{11} - 2e_{21} + 10e_{31} &= 9e_{31} \end{aligned}$$

Moving the terms on the right of the equals sign to the left yields three homogeneous equations in three unknowns, but two of the equations are redundant. Selecting one of the equations and arbitrarily setting $e_{11} = 1$ and $e_{21} = 1$, we find $e_{31} = 0$. Consequently, the normalized eigenvector is $\mathbf{e}_1' = [1/\sqrt{1^2 + 1^2 + 0^2},\ 1/\sqrt{1^2 + 1^2 + 0^2},\ 0/\sqrt{1^2 + 1^2 + 0^2}] = [1/\sqrt{2}, 1/\sqrt{2}, 0]$ since the sum of the squares of its elements is unity. You may verify that $\mathbf{e}_2' = [1/\sqrt{18}, -1/\sqrt{18}, -4/\sqrt{18}]$ is also an eigenvector for $9 = \lambda_2$, and $\mathbf{e}_3' = [2/3, -2/3, 1/3]$ is the normalized eigenvector corresponding to the eigenvalue $\lambda_3 = 18$. Moreover, $\mathbf{e}_i'\mathbf{e}_j = 0$, $i \neq j$.

[1]A proof of Equation (2-16) is beyond the scope of this book. The interested reader will find a proof in [5], Chapter 9.

The spectral decomposition of $\mathbf{A}$ is then

$$\mathbf{A} = \lambda_1 \mathbf{e}_1 \mathbf{e}_1' + \lambda_2 \mathbf{e}_2 \mathbf{e}_2' + \lambda_3 \mathbf{e}_3 \mathbf{e}_3'$$

or

$$\begin{bmatrix} 13 & -4 & 2 \\ -4 & 13 & -2 \\ 2 & -2 & 10 \end{bmatrix} = 9 \begin{bmatrix} \frac{1}{\sqrt{2}} \\ \frac{1}{\sqrt{2}} \\ 0 \end{bmatrix} \begin{bmatrix} \frac{1}{\sqrt{2}} & \frac{1}{\sqrt{2}} & 0 \end{bmatrix}$$

$$+ 9 \begin{bmatrix} \frac{1}{\sqrt{18}} \\ \frac{-1}{\sqrt{18}} \\ \frac{-4}{\sqrt{18}} \end{bmatrix} \begin{bmatrix} \frac{1}{\sqrt{18}} & \frac{-1}{\sqrt{18}} & \frac{-4}{\sqrt{18}} \end{bmatrix} + 18 \begin{bmatrix} 2/3 \\ -2/3 \\ 1/3 \end{bmatrix} \begin{bmatrix} 2/3 & -2/3 & 1/3 \end{bmatrix}$$

$$= 9 \begin{bmatrix} 1/2 & 1/2 & 0 \\ 1/2 & 1/2 & 0 \\ 0 & 0 & 0 \end{bmatrix} + 9 \begin{bmatrix} 1/18 & -1/18 & -4/18 \\ -1/18 & 1/18 & 4/18 \\ -4/18 & 4/18 & 16/18 \end{bmatrix}$$

$$+ 18 \begin{bmatrix} 4/9 & -4/9 & 2/9 \\ -4/9 & 4/9 & -2/9 \\ 2/9 & -2/9 & 1/9 \end{bmatrix}$$

as you may readily verify. ■

The spectral decomposition is an important analytical tool. With it we are very easily able to demonstrate certain statistical results. The first of these is a matrix explanation of distance, which we now develop.

When a $(k \times k)$ symmetric matrix $\mathbf{A}$ is such that

$$0 \leq \mathbf{x}'\mathbf{A}\mathbf{x} \tag{2-17}$$

for all $\mathbf{x}' = [x_1, x_2, \ldots, x_k]$, $\mathbf{A}$ is said to be *nonnegative definite*. If equality holds in (2-17) only for the vector $\mathbf{x}' = [0, 0, \ldots, 0]$, then $\mathbf{A}$ is said to be *positive definite*. In other words, $\mathbf{A}$ is positive definite if

$$0 < \mathbf{x}'\mathbf{A}\mathbf{x} \tag{2-18}$$

for all vectors $\mathbf{x} \neq \mathbf{0}$. Because $\mathbf{x}'\mathbf{A}\mathbf{x}$ has only squared terms x_i^2, and product terms, $x_i x_k$, it is called a *quadratic form*.

Example 2.10

Show that the quadratic form

$$3x_1^2 + 2x_2^2 - 2\sqrt{2}\, x_1 x_2$$

can be written in matrix notation as

$$[x_1 \quad x_2]\begin{bmatrix} 3 & -\sqrt{2} \\ -\sqrt{2} & 2 \end{bmatrix}\begin{bmatrix} x_1 \\ x_2 \end{bmatrix} = \mathbf{x}'\mathbf{A}\mathbf{x}$$

and that **A** is positive definite.

By Definition 2A.30, the eigenvalues of **A** are the solutions of the equation $|\mathbf{A} - \lambda\mathbf{I}| = 0$, or $(3 - \lambda)(2 - \lambda) - 2 = 0$. The solutions are $\lambda_1 = 4$ and $\lambda_2 = 1$. Using the spectral decomposition in (2-16), we can write

$$\underset{(2\times 2)}{\mathbf{A}} = \lambda_1 \underset{(2\times 1)}{\mathbf{e}_1} \underset{(1\times 2)}{\mathbf{e}_1'} + \lambda_2 \underset{(2\times 1)}{\mathbf{e}_2} \underset{(1\times 2)}{\mathbf{e}_2'}$$

$$= 4\underset{(2\times 1)}{\mathbf{e}_1} \underset{(1\times 2)}{\mathbf{e}_1'} + \underset{(2\times 1)}{\mathbf{e}_2} \underset{(1\times 2)}{\mathbf{e}_2'}$$

where $\mathbf{e}_1$ and $\mathbf{e}_2$ are the normalized and orthogonal eigenvectors associated with the eigenvalues $\lambda_1 = 4$ and $\lambda_2 = 1$, respectively. Because 4 and 1 are scalars, premultiplication and postmultiplication of **A** by $\mathbf{x}'$ and **x** respectively, where $\mathbf{x}' = [x_1, x_2]$ is any *nonzero* vector, gives

$$\underset{(1\times 2)}{\mathbf{x}'} \underset{(2\times 2)}{\mathbf{A}} \underset{(2\times 1)}{\mathbf{x}} = 4\underset{(1\times 2)}{\mathbf{x}'} \underset{(2\times 1)}{\mathbf{e}_1} \underset{(1\times 2)}{\mathbf{e}_1'} \underset{(2\times 1)}{\mathbf{x}}$$

$$+ \underset{(1\times 2)}{\mathbf{x}'} \underset{(2\times 1)}{\mathbf{e}_2} \underset{(1\times 2)}{\mathbf{e}_2'} \underset{(2\times 1)}{\mathbf{x}}$$

$$= 4y_1^2 + y_2^2 \geq 0$$

with

$$y_1 = \mathbf{x}'\mathbf{e}_1 = \mathbf{e}_1'\mathbf{x} \quad \text{and} \quad y_2 = \mathbf{x}'\mathbf{e}_2 = \mathbf{e}_2'\mathbf{x}$$

We now show that y_1 and y_2 are not both zero and, consequently, that $\mathbf{x}'\mathbf{A}\mathbf{x} = 4y_1^2 + y_2^2 > 0$, or **A** is *positive definite.*

From the definitions of y_1 and y_2 we have

$$\begin{bmatrix} y_1 \\ y_2 \end{bmatrix} = \begin{bmatrix} \mathbf{e}_1' \\ \mathbf{e}_2' \end{bmatrix}\begin{bmatrix} x_1 \\ x_2 \end{bmatrix}$$

or

$$\underset{(2\times 1)}{\mathbf{y}} = \underset{(2\times 2)}{\mathbf{E}} \underset{(2\times 1)}{\mathbf{x}}$$

Now **E** is an orthogonal matrix and hence has inverse $\mathbf{E}'$. Thus $\mathbf{x} = \mathbf{E}'\mathbf{y}$. But **x** is a nonzero vector and $\mathbf{0} \neq \mathbf{x} = \mathbf{E}'\mathbf{y}$ implies $\mathbf{y} \neq \mathbf{0}$. ■

Using the spectral decomposition it is rather easy to show that **A** is a $(k \times k)$ positive definite matrix if and only if every eigenvalue of **A** is positive (see Exercise 2.17). It is a nonnegative definite matrix if and only if all of its eigenvalues are greater than or equal to zero.

Assume for the moment that the p elements $x_1, x_2, \ldots, x_p$ of a vector **x** are realizations of p random variables $X_1, X_2, \ldots, X_p$. As we pointed out in Chapter 1, we can regard these elements as the coordinates of a point in p-dimensional space and the "distance" of the point $[x_1, x_2, \ldots, x_p]$ to the origin can, and in this case should, be interpreted in terms of standard deviation units. In this way we can account for the inherent uncertainty (variability) in the observations. Points with the same associated "uncertainty" are regarded as the same distance from the origin.

If we use the distance formula introduced in Chapter 1, [see Equation (1-22)] the distance from the origin satisfies the general formula

$$(\text{distance})^2 = a_{11}x_1^2 + a_{22}x_2^2 + \cdots + a_{pp}x_p^2 + 2(a_{12}x_1x_2 + a_{13}x_1x_3 + \cdots + a_{p-1,p}x_{p-1}x_p)$$

provided $(\text{distance})^2 > 0$ for all $[x_1, x_2, \ldots, x_p] \neq [0, 0, \ldots, 0]$. Setting $a_{ij} = a_{ji}$, $i \neq j$, $i = 1, 2, \ldots, p$, $j = 1, 2, \ldots, p$, we have

$$0 < (\text{distance})^2 = [x_1, \ldots, x_p]\begin{bmatrix} a_{11} & a_{12} & \cdots & a_{1p} \\ a_{21} & a_{22} & \cdots & a_{2p} \\ \vdots & \vdots & & \vdots \\ a_{p1} & a_{p2} & \cdots & a_{pp} \end{bmatrix}\begin{bmatrix} x_1 \\ x_2 \\ \vdots \\ x_p \end{bmatrix}$$

or

$$0 < (\text{distance})^2 = \mathbf{x}'\mathbf{A}\mathbf{x} \quad \text{for} \quad \mathbf{x} \neq \mathbf{0} \tag{2-19}$$

From (2-19) we see that the $(p \times p)$ symmetric matrix $\mathbf{A}$ is positive definite. In summary, distance is determined from a positive definite quadratic form $\mathbf{x}'\mathbf{A}\mathbf{x}$. Conversely, a positive definite quadratic form can be interpreted as a squared distance.

Comment. Let the squared distance from the point $\mathbf{x}' = [x_1, x_2, \ldots, x_p]$ to the origin be given by $\mathbf{x}'\mathbf{A}\mathbf{x}$ where $\mathbf{A}$ is a $(p \times p)$ symmetric and positive definite matrix. Then the squared distance from $\mathbf{x}$ to an arbitrary fixed point $\boldsymbol{\mu}' = [\mu_1, \mu_2, \ldots, \mu_p]$ is given by the general expression $(\mathbf{x} - \boldsymbol{\mu})'\mathbf{A}(\mathbf{x} - \boldsymbol{\mu})$.

Expressing distance in terms of a positive definite matrix $\mathbf{A}$ as the square root of a quadratic form allows us to give a geometrical interpretation based on the eigenvalues and eigenvectors of $\mathbf{A}$. For example, suppose $p = 2$. The points $\mathbf{x}' = [x_1, x_2]$ of constant distance c from the origin satisfy

$$\mathbf{x}'\mathbf{A}\mathbf{x} = a_{11}x_1^2 + a_{22}x_2^2 + 2a_{12}x_1x_2 = c^2$$

By the spectral decomposition, as in Example 2.10,

$$\mathbf{A} = \lambda_1\mathbf{e}_1\mathbf{e}_1' + \lambda_2\mathbf{e}_2\mathbf{e}_2' \quad \text{so} \quad \mathbf{x}'\mathbf{A}\mathbf{x} = \lambda_1(\mathbf{x}'\mathbf{e}_1)^2 + \lambda_2(\mathbf{x}'\mathbf{e}_2)^2$$

Now $c^2 = \lambda_1 y_1^2 + \lambda_2 y_2^2$ is an ellipse in $y_1 = \mathbf{x}'\mathbf{e}_1$ and $y_2 = \mathbf{x}'\mathbf{e}_2$ because $\lambda_1, \lambda_2 > 0$ when $\mathbf{A}$ is positive definite (see Exercise 2.17). We easily verify that $\mathbf{x} = c\lambda_1^{-1/2}\mathbf{e}_1$ satisfies $\mathbf{x}'\mathbf{A}\mathbf{x} = \lambda_1(c\lambda_1^{-1/2}\mathbf{e}_1'\mathbf{e}_1)^2 = c^2$ and $\mathbf{x} = c\lambda_2^{-1/2}\mathbf{e}_2$ gives the appropriate distance in the $\mathbf{e}_2$ direction. Thus the points at distance c lie on an ellipse whose axes are given by the eigenvectors of $\mathbf{A}$ with lengths proportional to the reciprocals of the square roots of the eigenvalues. The constant of proportionality is c. The situation is illustrated in Figure 2.6.

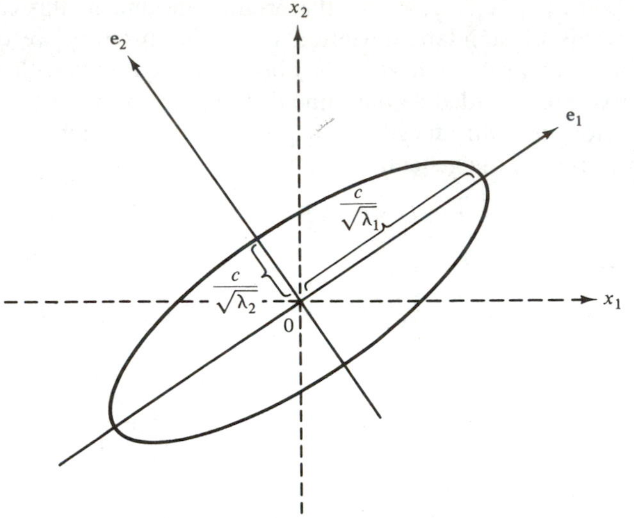

Figure 2.6 Points at constant distance c from the origin ($p = 2, 1 \le \lambda_1 < \lambda_2$).

If $p > 2$, the points $\mathbf{x}' = [x_1, x_2, \ldots, x_p]$ a constant distance $c = \sqrt{\mathbf{x}'\mathbf{A}\mathbf{x}}$ from the origin lie on hyperellipsoids $c^2 = \lambda_1(\mathbf{x}'\mathbf{e}_1)^2 + \cdots + \lambda_p(\mathbf{x}'\mathbf{e}_p)^2$, whose axes are given by the eigenvectors of $\mathbf{A}$. The half-length in the direction $\mathbf{e}_i$ is equal to $c/\sqrt{\lambda_i}$, $i = 1, 2, \ldots, p$, where $\lambda_1, \lambda_2, \ldots, \lambda_p$ are the eigenvalues of $\mathbf{A}$.

2.4 A SQUARE-ROOT MATRIX

The spectral decomposition allows us to express the inverse of a square matrix in terms of its eigenvalues and eigenvectors, and this leads to a useful *square-root matrix*.

Let $\mathbf{A}$ be a $k \times k$ positive definite matrix with the spectral decomposition $\mathbf{A} = \sum_{i=1}^{k} \lambda_i \mathbf{e}_i \mathbf{e}_i'$. Let the normalized eigenvectors be the columns of another matrix $\mathbf{P} = [\mathbf{e}_1, \mathbf{e}_2, \ldots, \mathbf{e}_k]$. Then

$$\underset{(k\times k)}{\mathbf{A}} = \sum_{i=1}^{k} \lambda_i \underset{(k\times 1)}{\mathbf{e}_i} \underset{(1\times k)}{\mathbf{e}_i'} = \underset{(k\times k)}{\mathbf{P}} \underset{(k\times k)}{\mathbf{\Lambda}} \underset{(k\times k)}{\mathbf{P}'} \tag{2-20}$$

where $\mathbf{P}\mathbf{P}' = \mathbf{P}'\mathbf{P} = \mathbf{I}$ and $\mathbf{\Lambda}$ is the diagonal matrix

$$\underset{(k\times k)}{\mathbf{\Lambda}} = \begin{bmatrix} \lambda_1 & 0 & \cdots & 0 \\ 0 & \lambda_2 & \cdots & 0 \\ \vdots & \vdots & \ddots & \vdots \\ 0 & 0 & \cdots & \lambda_k \end{bmatrix} \quad \text{with } \lambda_i > 0$$

Thus

$$\mathbf{A}^{-1} = \mathbf{P}\boldsymbol{\Lambda}^{-1}\mathbf{P}' = \sum_{i=1}^{k} \frac{1}{\lambda_i}\mathbf{e}_i\mathbf{e}_i' \tag{2-21}$$

since $(\mathbf{P}\boldsymbol{\Lambda}^{-1}\mathbf{P}')\mathbf{P}\boldsymbol{\Lambda}\mathbf{P}' = \mathbf{P}\boldsymbol{\Lambda}\mathbf{P}'(\mathbf{P}\boldsymbol{\Lambda}^{-1}\mathbf{P}') = \mathbf{P}\mathbf{P}' = \mathbf{I}$.

Next let $\boldsymbol{\Lambda}^{1/2}$ denote the diagonal matrix with $\sqrt{\lambda_i}$ as the ith diagonal element. The matrix $\sum_{i=1}^{k} \sqrt{\lambda_i}\,\mathbf{e}_i\mathbf{e}_i' = \mathbf{P}\boldsymbol{\Lambda}^{1/2}\mathbf{P}'$ is called the *square root* of $\mathbf{A}$ and is denoted by $\mathbf{A}^{1/2}$.

The square root matrix

$$\mathbf{A}^{1/2} = \sum_{i=1}^{k} \sqrt{\lambda_i}\,\mathbf{e}_i\mathbf{e}_i' = \mathbf{P}\boldsymbol{\Lambda}^{1/2}\mathbf{P}' \tag{2-22}$$

has the following properties:

1. $(\mathbf{A}^{1/2})' = \mathbf{A}^{1/2}$ (that is, $\mathbf{A}^{1/2}$ is symmetric).
2. $\mathbf{A}^{1/2}\mathbf{A}^{1/2} = \mathbf{A}$
3. $(\mathbf{A}^{1/2})^{-1} = \sum_{i=1}^{k} \frac{1}{\sqrt{\lambda_i}}\mathbf{e}_i\mathbf{e}_i' = \mathbf{P}\boldsymbol{\Lambda}^{-1/2}\mathbf{P}'$, where $\boldsymbol{\Lambda}^{-1/2}$ is a diagonal matrix with $1/\sqrt{\lambda_i}$ as the ith diagonal element.
4. $\mathbf{A}^{1/2}\mathbf{A}^{-1/2} = \mathbf{A}^{-1/2}\mathbf{A}^{1/2} = \mathbf{I}$ and $\mathbf{A}^{-1/2}\mathbf{A}^{-1/2} = \mathbf{A}^{-1}$ where $\mathbf{A}^{-1/2} = (\mathbf{A}^{1/2})^{-1}$.

2.5 RANDOM VECTORS AND MATRICES

A *random vector* is a vector whose elements are random variables. Similarly, a *random matrix* is a matrix whose elements are random variables. The expected value of a random matrix (or vector) is the matrix consisting of the expected values of each of its elements. Specifically, let $\mathbf{X} = \{X_{ij}\}$ be a $(p \times n)$ random matrix. The expected value of $\mathbf{X}$, denoted by $E(\mathbf{X})$, is the $(p \times n)$ matrix of numbers (if they exist)

$$E(\mathbf{X}) = \begin{bmatrix} E(X_{11}) & E(X_{12}) & \cdots & E(X_{1n}) \\ E(X_{21}) & E(X_{22}) & \cdots & E(X_{2n}) \\ \vdots & \vdots & & \vdots \\ E(X_{p1}) & E(X_{p2}) & \cdots & E(X_{pn}) \end{bmatrix} \tag{2-23}$$

where, for each element of the matrix[2]

$$E(X_{ij}) = \begin{cases} \displaystyle\int_{-\infty}^{\infty} x_{ij} f_{ij}(x_{ij})\, dx_{ij} & \text{if } X_{ij} \text{ is a continuous random variable with probability density function } f_{ij}(x) \\ \displaystyle\sum_{\text{all } x_{ij}} x_{ij} p_{ij}(x_{ij}) & \text{if } X_{ij} \text{ is a discrete random variable with probability function } p_{ij}(x) \end{cases}$$

Example 2.11

Suppose $p = 2$ and $n = 1$ and consider the random vector $\mathbf{X}' = [X_1, X_2]$. Let the discrete random variable X_1 have the following probability function

x_1	-1	0	1
$p_1(x_1)$	.3	.3	.4

Then $E(X_1) = \sum_{\text{all } x_1} x_1 p_1(x_1) = (-1)(.3) + (0)(.3) + (1)(.4) = .1$.

Similarly, let the discrete random variable X_2 have the probability function

x_2	0	1
$p_2(x_2)$	.8	.2

Then $E(X_2) = \sum_{\text{all } x_2} x_2 p_2(x_2) = (0)(.8) + (1)(.2) = .2$

Thus

$$E(\mathbf{X}) = \begin{bmatrix} E(X_1) \\ E(X_2) \end{bmatrix} = \begin{bmatrix} .1 \\ .2 \end{bmatrix}$$ ■

Two results involving the expectation of sums and products of matrices follow directly from the definition of the expected value of a random matrix and the univariate properties of expectation, $E(X_1 + Y_1) = E(X_1) + E(Y_1)$ and $E(cX_1) = cE(X_1)$.

Let **X** and **Y** be random matrices of the same dimension and let **A** and **B** be conformable matrices of constants. Then (see Exercise 2.35)

$$E(\mathbf{X} + \mathbf{Y}) = E(\mathbf{X}) + E(\mathbf{Y})$$

$$E(\mathbf{AXB}) = \mathbf{A}E(\mathbf{X})\mathbf{B} \qquad (2\text{-}24)$$

[2] If you are unfamiliar with calculus, you should concentrate on the interpretation of the expected value and, eventually, variance. Our development is based primarily on the properties of expectation rather than its particular evaluation for continuous or discrete random variables.

2.6 MEAN VECTORS AND COVARIANCE MATRICES

Suppose $\mathbf{X} = \{X_i\}$ is a $(p \times 1)$ random matrix, that is, a random vector. Each element of $\mathbf{X}$ is a random variable with its own marginal probability distribution (see Example 2.11). The marginal means, μ_i, and variances, σ_i^2, are defined as $\mu_i = E(X_i)$ and $\sigma_i^2 = E(X_i - \mu_i)^2$, $i = 1, 2, \ldots, p$, respectively. Specifically,

$$\mu_i = \begin{cases} \displaystyle\int_{-\infty}^{\infty} x_i f_i(x_i)\, dx_i & \text{if } X_i \text{ is a continuous random variable with probability density function } f_i(x_i) \\ \displaystyle\sum_{\text{all } x_i} x_i p_i(x_i) & \text{if } X_i \text{ is a discrete random variable with probability function } p_i(x_i) \end{cases}$$

$$\sigma_i^2 = \begin{cases} \displaystyle\int_{-\infty}^{\infty} (x_i - \mu_i)^2 f_i(x_i)\, dx_i & \text{if } X_i \text{ is a continuous random variable with probability density function } f_i(x_i) \\ \displaystyle\sum_{\text{all } x_i} (x_i - \mu_i)^2 p_i(x_i) & \text{if } X_i \text{ is a discrete random variable with probability function } p_i(x_i) \end{cases} \tag{2-25}$$

It will be convenient in later sections to denote the marginal variances by σ_{ii} rather than the more traditional σ_i^2 and, consequently, we shall adopt this notation.

The behavior of any pair of random variables, such as X_i and X_k, is described by their joint probability function and a measure of the linear association between them is provided by the covariance σ_{ik}, where

$$\sigma_{ik} = E(X_i - \mu_i)(X_k - \mu_k)$$

$$= \begin{cases} \displaystyle\int_{-\infty}^{\infty}\int_{-\infty}^{\infty} (x_i - \mu_i)(x_k - \mu_k) f_{ik}(x_i, x_k)\, dx_i dx_k & \text{if } X_i, X_k \text{ are continuous random variables with the joint density function } f_{ik}(x_i, x_k) \\ \displaystyle\sum_{\text{all } x_i}\sum_{\text{all } x_k} (x_i - \mu_i)(x_k - \mu_k) p_{ik}(x_i, x_k) & \text{if } X_i, X_k \text{ are discrete random variables with the joint probability function } p_{ik}(x_i, x_k) \end{cases} \tag{2-26}$$

and μ_i and μ_k, $i, k = 1, 2, \ldots, p$, are the marginal means. When $i = k$, the covariance becomes the marginal variance.

More generally, the collective behavior of the p random variables $X_1, X_2, \ldots, X_p$, or, equivalently, the random vector $\mathbf{X}' = [X_1, X_2, \ldots, X_p]$ is described by a joint probability density function $f(x_1, x_2, \ldots, x_p) = f(\mathbf{x})$. As we have already noted in this book, $f(\mathbf{x})$ will often be the multivariate normal density function (see Chapter 4).

If the joint probability $P[X_i \leq x_i \text{ and } X_k \leq x_k]$ can be written as the product of the corresponding marginal probabilities so that

$$P[X_i \leq x_i \text{ and } X_k \leq x_k] = P[X_i \leq x_i]P[X_k \leq x_k] \tag{2-27}$$

for all pairs of values x_i, x_k, then X_i and X_k are said to be *statistically independent*. When X_i and X_k are continuous random variables with joint density $f_{ik}(x_i, x_k)$ and marginal densities $f_i(x_i)$ and $f_k(x_k)$, the independence condition becomes

$$f_{ik}(x_i, x_k) = f_i(x_i)f_k(x_k)$$

for all pairs (x_i, x_k).

The p continuous random variables $X_1, X_2, \ldots, X_p$ are *mutually statistically independent* if their joint density factors as

$$f_{12\cdots p}(x_1, x_2, \ldots, x_p) = f_1(x_1)f_2(x_2)\cdots f_p(x_p) \tag{2-28}$$

for all p-tuples $(x_1, x_2, \ldots, x_p)$.

Statistical independence has an important implication for covariance. The factorization in (2-28) implies $\text{Cov}(X_i, X_k) = 0$. Thus

$$\text{Cov}(X_i, X_k) = 0 \quad \text{if } X_i \text{ and } X_k \text{ are independent.} \tag{2-29}$$

The converse of (2-29) is not true in general. There are situations where $\text{Cov}(X_i, X_k) = 0$ and X_i and X_k are not independent (see [2]).

The means and covariances of the $(p \times 1)$ random vector $\mathbf{X}$ can be set out as matrices. The expected value of each element is contained in the vector of means $\boldsymbol{\mu} = E(\mathbf{X})$ and the p variances σ_{ii} and the $p(p-1)/2$ distinct covariances σ_{ik} $(i < k)$ are contained in the symmetric variance-covariance matrix $\boldsymbol{\Sigma} = E(\mathbf{X} - \boldsymbol{\mu})(\mathbf{X} - \boldsymbol{\mu})'$. Specifically,

$$E(\mathbf{X}) = \begin{bmatrix} E(X_1) \\ E(X_2) \\ \vdots \\ E(X_p) \end{bmatrix} = \begin{bmatrix} \mu_1 \\ \mu_2 \\ \vdots \\ \mu_p \end{bmatrix} = \boldsymbol{\mu} \tag{2-30}$$

and

$$\boldsymbol{\Sigma} = E(\mathbf{X} - \boldsymbol{\mu})(\mathbf{X} - \boldsymbol{\mu})' = E\left(\begin{bmatrix} X_1 - \mu_1 \\ X_2 - \mu_2 \\ \vdots \\ X_p - \mu_p \end{bmatrix}[X_1 - \mu_1, X_2 - \mu_2, \ldots, X_p - \mu_p]\right)$$

$$= E\begin{bmatrix} (X_1-\mu_1)^2 & (X_1-\mu_1)(X_2-\mu_2) & \cdots & (X_1-\mu_1)(X_p-\mu_p) \\ (X_2-\mu_2)(X_1-\mu_1) & (X_2-\mu_2)^2 & \cdots & (X_2-\mu_2)(X_p-\mu_p) \\ \vdots & \vdots & & \vdots \\ (X_p-\mu_p)(X_1-\mu_1) & (X_p-\mu_p)(X_2-\mu_2) & \cdots & (X_p-\mu_p)^2 \end{bmatrix}$$

$$= \begin{bmatrix} E(X_1-\mu_1)^2 & E(X_1-\mu_1)(X_2-\mu_2) & \cdots & E(X_1-\mu_1)(X_p-\mu_p) \\ E(X_2-\mu_2)(X_1-\mu_1) & E(X_2-\mu_2)^2 & \cdots & E(X_2-\mu_2)(X_p-\mu_p) \\ \vdots & \vdots & & \vdots \\ E(X_p-\mu_p)(X_1-\mu_1) & E(X_p-\mu_p)(X_2-\mu_2) & \cdots & E(X_p-\mu_p)^2 \end{bmatrix}$$

or

$$\mathbf{\Sigma} = \text{Cov}(\mathbf{X}) = \begin{bmatrix} \sigma_{11} & \sigma_{12} & \cdots & \sigma_{1p} \\ \sigma_{21} & \sigma_{22} & \cdots & \sigma_{2p} \\ \vdots & \vdots & & \vdots \\ \sigma_{p1} & \sigma_{p2} & \cdots & \sigma_{pp} \end{bmatrix} \tag{2-31}$$

Example 2.12

Find the covariance matrix for the two random variables X_1 and X_2 introduced in Example 2.11 when their joint probability function, $p_{12}(x_1, x_2)$ is represented by the entries in the body of the following table.

x_1 \ x_2	0	1	$p_1(x_1)$
−1	.24	.06	.3
0	.16	.14	.3
1	.40	.00	.4
$p_2(x_2)$	.8	.2	1

We have already shown that $\mu_1 = E(X_1) = .1$ and $\mu_2 = E(X_2) = .2$ (see Example 2.11). In addition

$$\sigma_{11} = E(X_1-\mu_1)^2 = \sum_{\text{all } x_1} (x_1 - .1)^2 p_1(x_1)$$

$$= (-1-.1)^2(.3) + (0-.1)^2(.3) + (1-.1)^2(.4) = .69$$

$$\sigma_{22} = E(X_2-\mu_2)^2 = \sum_{\text{all } x_2} (x_2 - .2)^2 p_2(x_2)$$

$$= (0-.2)^2(.8) + (1-.2)^2(.2)$$

$$= .16$$

$$\sigma_{12} = E(X_1 - \mu_1)(X_2 - \mu_2) = \sum_{\text{all pairs } (x_1, x_2)} (x_1 - .1)(x_2 - .2)p_{12}(x_1, x_2)$$

$$= (-1 - .1)(0 - .2)(.24) + (-1 - .1)(1 - .2)(.06)$$

$$+ \cdots + (1 - .1)(1 - .2)(.00) = -.08$$

$$\sigma_{21} = E(X_2 - \mu_2)(X_1 - \mu_1) = E(X_1 - \mu_1)(X_2 - \mu_2) = \sigma_{12} = -.08$$

Consequently, with $\mathbf{X}' = [X_1, X_2]$,

$$\boldsymbol{\mu} = E(\mathbf{X}) = \begin{bmatrix} E(X_1) \\ E(X_2) \end{bmatrix} = \begin{bmatrix} \mu_1 \\ \mu_2 \end{bmatrix} = \begin{bmatrix} .1 \\ .2 \end{bmatrix}$$

and

$$\boldsymbol{\Sigma} = E(\mathbf{X} - \boldsymbol{\mu})(\mathbf{X} - \boldsymbol{\mu})' = E\begin{bmatrix} (X_1 - \mu_1)^2 & (X_1 - \mu_1)(X_2 - \mu_2) \\ (X_2 - \mu_2)(X_1 - \mu_1) & (X_2 - \mu_2)^2 \end{bmatrix}$$

$$= \begin{bmatrix} E(X_1 - \mu_1)^2 & E(X_1 - \mu_1)(X_2 - \mu_2) \\ E(X_2 - \mu_2)(X_1 - \mu_1) & E(X_2 - \mu_2)^2 \end{bmatrix}$$

$$= \begin{bmatrix} \sigma_{11} & \sigma_{12} \\ \sigma_{21} & \sigma_{22} \end{bmatrix} = \begin{bmatrix} .69 & -.08 \\ -.08 & .16 \end{bmatrix}$$

■

We note that the computation of means, variances, and covariances for *discrete* random variables involves summation (as in Examples 2.11 and 2.12), while analogous computations for *continuous* random variables involves integration.

Because $\sigma_{ik} = E(X_i - \mu_i)(X_k - \mu_k) = \sigma_{ki}$, it is convenient to write the matrix appearing in (2-31) as

$$\boldsymbol{\Sigma} = E(\mathbf{X} - \boldsymbol{\mu})(\mathbf{X} - \boldsymbol{\mu})' = \begin{bmatrix} \sigma_{11} & \sigma_{12} & \cdots & \sigma_{1p} \\ \sigma_{12} & \sigma_{22} & \cdots & \sigma_{2p} \\ \vdots & \vdots & & \vdots \\ \sigma_{1p} & \sigma_{2p} & \cdots & \sigma_{pp} \end{bmatrix} \tag{2-32}$$

We shall refer to $\boldsymbol{\mu}$ and $\boldsymbol{\Sigma}$ as the *population mean* (vector) and *population variance-covariance* (matrix), respectively.

The multivariate normal distribution is completely specified once the mean vector $\boldsymbol{\mu}$ and variance-covariance matrix $\boldsymbol{\Sigma}$ are given (see Chapter 4), so it is not surprising that these quantities play an important role in many multivariate procedures.

It is frequently informative to separate the information contained in variances, σ_{ii}, from that contained in measures of association and, in particular, the measure of association known as the *population correlation coefficient*, ρ_{ik}. The correlation coefficient ρ_{ik} is defined in terms of the covariance σ_{ik} and variances σ_{ii} and σ_{kk} as

$$\rho_{ik} = \frac{\sigma_{ik}}{\sqrt{\sigma_{ii}}\sqrt{\sigma_{kk}}} \tag{2-33}$$

The correlation coefficient measures the amount of *linear* association between the random variables X_i and X_k. (See, for example, [2].)

Let the population correlation matrix be the $(p \times p)$ symmetric matrix $\boldsymbol{\rho}$ where

$$\boldsymbol{\rho} = \begin{bmatrix} \frac{\sigma_{11}}{\sqrt{\sigma_{11}}\sqrt{\sigma_{11}}} & \frac{\sigma_{12}}{\sqrt{\sigma_{11}}\sqrt{\sigma_{22}}} & \cdots & \frac{\sigma_{1p}}{\sqrt{\sigma_{11}}\sqrt{\sigma_{pp}}} \\ \frac{\sigma_{12}}{\sqrt{\sigma_{11}}\sqrt{\sigma_{22}}} & \frac{\sigma_{22}}{\sqrt{\sigma_{22}}\sqrt{\sigma_{22}}} & \cdots & \frac{\sigma_{2p}}{\sqrt{\sigma_{22}}\sqrt{\sigma_{pp}}} \\ \vdots & \vdots & & \vdots \\ \frac{\sigma_{1p}}{\sqrt{\sigma_{11}}\sqrt{\sigma_{pp}}} & \frac{\sigma_{2p}}{\sqrt{\sigma_{22}}\sqrt{\sigma_{pp}}} & \cdots & \frac{\sigma_{pp}}{\sqrt{\sigma_{pp}}\sqrt{\sigma_{pp}}} \end{bmatrix}$$

$$= \begin{bmatrix} 1 & \rho_{12} & \cdots & \rho_{1p} \\ \rho_{12} & 1 & \cdots & \rho_{2p} \\ \vdots & \vdots & & \vdots \\ \rho_{1p} & \rho_{2p} & \cdots & 1 \end{bmatrix} \tag{2-34}$$

and let the $(p \times p)$ *standard deviation* matrix, $\mathbf{V}^{1/2}$, be

$$\mathbf{V}^{1/2} = \begin{bmatrix} \sqrt{\sigma_{11}} & 0 & \cdots & 0 \\ 0 & \sqrt{\sigma_{22}} & \cdots & 0 \\ \vdots & \vdots & & \vdots \\ 0 & 0 & \cdots & \sqrt{\sigma_{pp}} \end{bmatrix} \tag{2-35}$$

Then it is easily verified (see Exercise 2.21) that

$$\mathbf{V}^{1/2}\boldsymbol{\rho}\mathbf{V}^{1/2} = \boldsymbol{\Sigma} \tag{2-36}$$

and

$$\boldsymbol{\rho} = (\mathbf{V}^{1/2})^{-1}\boldsymbol{\Sigma}(\mathbf{V}^{1/2})^{-1} \tag{2-37}$$

That is, $\boldsymbol{\Sigma}$ can be obtained from $\mathbf{V}^{1/2}$ and $\boldsymbol{\rho}$, while $\boldsymbol{\rho}$ can be obtained from $\boldsymbol{\Sigma}$. Moreover, the expression of these relationships in terms of matrix operations allows the calculations to be conveniently implemented on a computer.

Example 2.13

Suppose

$$\boldsymbol{\Sigma} = \begin{bmatrix} 4 & 1 & 2 \\ 1 & 9 & -3 \\ 2 & -3 & 25 \end{bmatrix} = \begin{bmatrix} \sigma_{11} & \sigma_{12} & \sigma_{13} \\ \sigma_{12} & \sigma_{22} & \sigma_{23} \\ \sigma_{13} & \sigma_{23} & \sigma_{33} \end{bmatrix}$$

Obtain $\mathbf{V}^{1/2}$ and $\boldsymbol{\rho}$.

Here

$$\mathbf{V}^{1/2} = \begin{bmatrix} \sqrt{\sigma_{11}} & 0 & 0 \\ 0 & \sqrt{\sigma_{22}} & 0 \\ 0 & 0 & \sqrt{\sigma_{33}} \end{bmatrix} = \begin{bmatrix} 2 & 0 & 0 \\ 0 & 3 & 0 \\ 0 & 0 & 5 \end{bmatrix}$$

and

$$(\mathbf{V}^{1/2})^{-1} = \begin{bmatrix} \frac{1}{2} & 0 & 0 \\ 0 & \frac{1}{3} & 0 \\ 0 & 0 & \frac{1}{5} \end{bmatrix}$$

Consequently, using (2-37), the correlation matrix $\boldsymbol{\rho}$ is given by

$$(\mathbf{V}^{1/2})^{-1}\boldsymbol{\Sigma}(\mathbf{V}^{1/2})^{-1} = \begin{bmatrix} \frac{1}{2} & 0 & 0 \\ 0 & \frac{1}{3} & 0 \\ 0 & 0 & \frac{1}{5} \end{bmatrix} \begin{bmatrix} 4 & 1 & 2 \\ 1 & 9 & -3 \\ 2 & -3 & 25 \end{bmatrix} \begin{bmatrix} \frac{1}{2} & 0 & 0 \\ 0 & \frac{1}{3} & 0 \\ 0 & 0 & \frac{1}{5} \end{bmatrix}$$

$$= \begin{bmatrix} 1 & \frac{1}{6} & \frac{1}{5} \\ \frac{1}{6} & 1 & -\frac{1}{5} \\ \frac{1}{5} & -\frac{1}{5} & 1 \end{bmatrix} \qquad \blacksquare$$

Partitioning the Covariance Matrix

Often, the characteristics measured on individual trials will fall naturally into two or more groups. As examples, consider measurements of variables representing consumption and income or variables representing personality traits and physical characteristics. One approach to handling these situations is to let the characteristics defining the distinct groups be subsets of the *total* collection of characteristics. If the total collection is represented by a $(p \times 1)$-dimensional random vector $\mathbf{X}$, the subsets can be regarded as components of $\mathbf{X}$ and can be sorted by partitioning $\mathbf{X}$.

In general, we can partition the p characteristics contained in the $p \times 1$ random vector $\mathbf{X}$ into, for instance, two groups of size q and $p - q$, respectively. For example, we can write

$$\mathbf{X} = \begin{bmatrix} X_1 \\ \vdots \\ X_q \\ \hline X_{q+1} \\ \vdots \\ X_p \end{bmatrix} \begin{matrix} \left.\vphantom{\begin{matrix}X_1\\ \vdots \\ X_q\end{matrix}}\right\} q \\ \left.\vphantom{\begin{matrix}X_1\\ \vdots \\ X_q\end{matrix}}\right\} p-q \end{matrix} = \begin{bmatrix} \mathbf{X}^{(1)} \\ \hline \mathbf{X}^{(2)} \end{bmatrix} \quad \text{and} \quad \boldsymbol{\mu} = E(\mathbf{X}) = \begin{bmatrix} \mu_1 \\ \vdots \\ \mu_q \\ \hline \mu_{q+1} \\ \vdots \\ \mu_p \end{bmatrix} = \begin{bmatrix} \boldsymbol{\mu}^{(1)} \\ \hline \boldsymbol{\mu}^{(2)} \end{bmatrix} \tag{2-38}$$

From the definitions of transpose and matrix multiplication,

$$(\mathbf{X}^{(1)} - \boldsymbol{\mu}^{(1)})(\mathbf{X}^{(2)} - \boldsymbol{\mu}^{(2)})' = \begin{bmatrix} X_1 - \mu_1 \\ X_2 - \mu_2 \\ \vdots \\ X_q - \mu_q \end{bmatrix} [X_{q+1} - \mu_{q+1}, X_{q+2} - \mu_{q+2}, \ldots, X_p - \mu_p] =$$

$$\begin{bmatrix} (X_1 - \mu_1)(X_{q+1} - \mu_{q+1}) & (X_1 - \mu_1)(X_{q+2} - \mu_{q+2}) & \cdots & (X_1 - \mu_1)(X_p - \mu_p) \\ (X_2 - \mu_2)(X_{q+1} - \mu_{q+1}) & (X_2 - \mu_2)(X_{q+2} - \mu_{q+2}) & \cdots & (X_2 - \mu_2)(X_p - \mu_p) \\ \vdots & \vdots & & \vdots \\ (X_q - \mu_q)(X_{q+1} - \mu_{q+1}) & (X_q - \mu_q)(X_{q+2} - \mu_{q+2}) & \cdots & (X_q - \mu_q)(X_p - \mu_p) \end{bmatrix}$$

Upon taking the expectation of the matrix $(\mathbf{X}^{(1)} - \boldsymbol{\mu}^{(1)})(\mathbf{X}^{(2)} - \boldsymbol{\mu}^{(2)})'$, we get

$$E(\mathbf{X}^{(1)} - \boldsymbol{\mu}^{(1)})(\mathbf{X}^{(2)} - \boldsymbol{\mu}^{(2)})' = \begin{bmatrix} \sigma_{1,q+1} & \sigma_{1,q+2} & \cdots & \sigma_{1p} \\ \sigma_{2,q+1} & \sigma_{2,q+2} & \cdots & \sigma_{2p} \\ \vdots & \vdots & & \vdots \\ \sigma_{q,q+1} & \sigma_{q,q+2} & \cdots & \sigma_{qp} \end{bmatrix} = \boldsymbol{\Sigma}_{12} \qquad \text{(2-39)}$$

which gives all the covariances, σ_{ij}, $i = 1, 2, \ldots, q$, $j = q + 1, q + 2, \ldots, p$, between a component of $\mathbf{X}^{(1)}$ and a component of $\mathbf{X}^{(2)}$. You will note that the matrix $\boldsymbol{\Sigma}_{12}$ is not necessarily symmetric or even square.

Making use of the partitioning in Equation (2-38), it is easily demonstrated that

$$(\mathbf{X} - \boldsymbol{\mu})(\mathbf{X} - \boldsymbol{\mu})' = \begin{bmatrix} \underset{(q \times 1)}{(\mathbf{X}^{(1)} - \boldsymbol{\mu}^{(1)})} \underset{(1 \times q)}{(\mathbf{X}^{(1)} - \boldsymbol{\mu}^{(1)})'} & \underset{(q \times 1)}{(\mathbf{X}^{(1)} - \boldsymbol{\mu}^{(1)})} \underset{(1 \times (p-q))}{(\mathbf{X}^{(2)} - \boldsymbol{\mu}^{(2)})'} \\ \underset{((p-q) \times 1)}{(\mathbf{X}^{(2)} - \boldsymbol{\mu}^{(2)})} \underset{(1 \times q)}{(\mathbf{X}^{(1)} - \boldsymbol{\mu}^{(1)})'} & \underset{((p-q) \times 1)}{(\mathbf{X}^{(2)} - \boldsymbol{\mu}^{(2)})} \underset{(1 \times (p-q))}{(\mathbf{X}^{(2)} - \boldsymbol{\mu}^{(2)})'} \end{bmatrix}$$

and consequently

$$\underset{(p \times p)}{\boldsymbol{\Sigma}} = E(\mathbf{X} - \boldsymbol{\mu})(\mathbf{X} - \boldsymbol{\mu})' = \begin{matrix} q \\ p-q \end{matrix} \overset{\begin{matrix} q & p-q \end{matrix}}{\underset{(p \times p)}{\left[\begin{array}{c|c} \boldsymbol{\Sigma}_{11} & \boldsymbol{\Sigma}_{12} \\ \hline \boldsymbol{\Sigma}_{21} & \boldsymbol{\Sigma}_{22} \end{array}\right]}} \qquad \text{(2-40)}$$

$$= \left[\begin{array}{ccc|ccc} \sigma_{11} & \cdots & \sigma_{1q} & \sigma_{1,q+1} & \cdots & \sigma_{1p} \\ \vdots & & \vdots & \vdots & & \vdots \\ \sigma_{q1} & \cdots & \sigma_{qq} & \sigma_{q,q+1} & \cdots & \sigma_{qp} \\ \hline \sigma_{q+1,1} & \cdots & \sigma_{q+1,q} & \sigma_{q+1,q+1} & \cdots & \sigma_{q+1,p} \\ \vdots & & \vdots & \vdots & & \vdots \\ \sigma_{p1} & \cdots & \sigma_{pq} & \sigma_{p,q+1} & \cdots & \sigma_{pp} \end{array}\right]$$

Note that $\boldsymbol{\Sigma}_{12} = \boldsymbol{\Sigma}'_{21}$. The covariance matrix of $\mathbf{X}^{(1)}$ is $\boldsymbol{\Sigma}_{11}$, that of $\mathbf{X}^{(2)}$ is $\boldsymbol{\Sigma}_{22}$, and that of elements from $\mathbf{X}^{(1)}$ and $\mathbf{X}^{(2)}$ is $\boldsymbol{\Sigma}_{12}$ (or $\boldsymbol{\Sigma}_{21}$).

The Mean Vector and Covariance Matrix for Linear Combinations of Random Variables

Recall that if a single random variable, such as X_1, is multiplied by a constant c, then

$$E(cX_1) = cE(X_1) = c\mu_1$$

and

$$\text{Var}(cX_1) = E(cX_1 - c\mu_1)^2 = c^2\text{Var}(X_1) = c^2\sigma_{11}$$

If X_2 is a second random variable and a and b are constants, then using additional properties of expectation

$$\begin{aligned}\text{Cov}(aX_1, bX_2) &= E(aX_1 - a\mu_1)(bX_2 - b\mu_2)\\ &= abE(X_1 - \mu_1)(X_2 - \mu_2)\\ &= ab\,\text{Cov}(X_1, X_2) = ab\sigma_{12}\end{aligned}$$

Finally, for the linear combination $aX_1 + bX_2$, we have

$$E(aX_1 + bX_2) = aE(X_1) + bE(X_2) = a\mu_1 + b\mu_2$$

$$\begin{aligned}\text{Var}(aX_1 + bX_2) &= E[(aX_1 + bX_2) - (a\mu_1 + b\mu_2)]^2\\ &= E[a(X_1 - \mu_1) + b(X_2 - \mu_2)]^2\\ &= E[a^2(X_1 - \mu_1)^2 + b^2(X_2 - \mu_2)^2 + 2ab(X_1 - \mu_1)(X_2 - \mu_2)]\\ &= a^2\text{Var}(X_1) + b^2\text{Var}(X_2) + 2ab\,\text{Cov}(X_1, X_2)\\ &= a^2\sigma_{11} + b^2\sigma_{22} + 2ab\sigma_{12}.\end{aligned} \tag{2-41}$$

With $\mathbf{c}' = [a, b]$, $aX_1 + bX_2$ can be written as

$$\begin{bmatrix} a & b \end{bmatrix}\begin{bmatrix} X_1 \\ X_2 \end{bmatrix} = \mathbf{c}'\mathbf{X}$$

Similarly, $E(aX_1 + bX_2) = a\mu_1 + b\mu_2$ can be expressed as

$$\begin{bmatrix} a & b \end{bmatrix}\begin{bmatrix} \mu_1 \\ \mu_2 \end{bmatrix} = \mathbf{c}'\boldsymbol{\mu}$$

If we let

$$\boldsymbol{\Sigma} = \begin{bmatrix} \sigma_{11} & \sigma_{12} \\ \sigma_{12} & \sigma_{22} \end{bmatrix}$$

be the variance-covariance matrix of $\mathbf{X}$, Equation (2-41) becomes

$$\text{Var}(aX_1 + bX_2) = \text{Var}(\mathbf{c}'\mathbf{X}) = \mathbf{c}'\boldsymbol{\Sigma}\mathbf{c} \tag{2-42}$$

since

$$\mathbf{c}'\boldsymbol{\Sigma}\mathbf{c} = \begin{bmatrix} a & b \end{bmatrix}\begin{bmatrix} \sigma_{11} & \sigma_{12} \\ \sigma_{12} & \sigma_{22} \end{bmatrix}\begin{bmatrix} a \\ b \end{bmatrix} = a^2\sigma_{11} + 2ab\sigma_{12} + b^2\sigma_{22}$$

The results above can be extended to a linear combination of p random variables.

The linear combination $\mathbf{c}'\mathbf{X} = c_1X_1 + \cdots + c_pX_p$ has

$$\text{mean} = E(\mathbf{c}'\mathbf{X}) = \mathbf{c}'\boldsymbol{\mu}$$

$$\text{variance} = \text{Var}(\mathbf{c}'\mathbf{X}) = \mathbf{c}'\boldsymbol{\Sigma}\mathbf{c} \tag{2-43}$$

where $\boldsymbol{\mu} = E(\mathbf{X})$ and $\boldsymbol{\Sigma} = \text{Cov}(\mathbf{X})$.

In general consider the q linear combinations of the p random variables $X_1, \ldots, X_p$,

$$\begin{aligned} Z_1 &= c_{11}X_1 + c_{12}X_2 + \cdots + c_{1p}X_p \\ Z_2 &= c_{21}X_1 + c_{22}X_2 + \cdots + c_{2p}X_p \\ &\vdots \qquad\qquad\qquad\qquad \vdots \\ Z_q &= c_{q1}X_1 + c_{q2}X_2 + \cdots + c_{qp}X_p \end{aligned}$$

or

$$\underset{(q\times 1)}{\mathbf{Z} = \begin{bmatrix} Z_1 \\ Z_2 \\ \vdots \\ Z_q \end{bmatrix}} = \underset{(q\times p)}{\begin{bmatrix} c_{11} & c_{12} & \cdots & c_{1p} \\ c_{21} & c_{22} & \cdots & c_{2p} \\ \vdots & \vdots & & \vdots \\ c_{q1} & c_{q2} & \cdots & c_{qp} \end{bmatrix}} \underset{(p\times 1)}{\begin{bmatrix} X_1 \\ X_2 \\ \vdots \\ X_p \end{bmatrix}} = \mathbf{CX} \tag{2-44}$$

The linear combination $\mathbf{Z} = \mathbf{CX}$ has

$$\boldsymbol{\mu}_{\mathbf{Z}} = E(\mathbf{Z}) = E(\mathbf{CX}) = \mathbf{C}\boldsymbol{\mu}_{\mathbf{X}} \tag{2-45}$$

$$\boldsymbol{\Sigma}_{\mathbf{Z}} = \text{Cov}(\mathbf{Z}) = \text{Cov}(\mathbf{CX}) = \mathbf{C}\boldsymbol{\Sigma}_{\mathbf{X}}\mathbf{C}'$$

where $\boldsymbol{\mu}_{\mathbf{X}}$ and $\boldsymbol{\Sigma}_{\mathbf{X}}$ are the mean vector and variance-covariance matrix of $\mathbf{X}$, respectively. (See Exercise 2.26 for the computation of the off-diagonal terms in $\mathbf{C}\boldsymbol{\Sigma}_{\mathbf{X}}\mathbf{C}'$.)

We shall rely heavily on the result in (2-45) in our discussions of principal components and factor analysis in Chapters 8 and 9.

Example 2.14

Let $\mathbf{X}' = [X_1, X_2]$ be a random vector with mean vector $\boldsymbol{\mu}'_{\mathbf{X}} = [\mu_1, \mu_2]$ and variance-covariance matrix

$$\boldsymbol{\Sigma}_{\mathbf{X}} = \begin{bmatrix} \sigma_{11} & \sigma_{12} \\ \sigma_{12} & \sigma_{22} \end{bmatrix}$$

Find the means and covariance matrix for the linear combinations

$$Z_1 = X_1 - X_2$$
$$Z_2 = X_1 + X_2$$

or

$$\mathbf{Z} = \begin{bmatrix} Z_1 \\ Z_2 \end{bmatrix} = \begin{bmatrix} 1 & -1 \\ 1 & 1 \end{bmatrix} \begin{bmatrix} X_1 \\ X_2 \end{bmatrix} = \mathbf{CX}$$

in terms of $\boldsymbol{\mu}_\mathbf{X}$ and $\boldsymbol{\Sigma}_\mathbf{X}$.

Here

$$\boldsymbol{\mu}_\mathbf{Z} = E(\mathbf{Z}) = \mathbf{C}\boldsymbol{\mu}_\mathbf{X} = \begin{bmatrix} 1 & -1 \\ 1 & 1 \end{bmatrix} \begin{bmatrix} \mu_1 \\ \mu_2 \end{bmatrix} = \begin{bmatrix} \mu_1 - \mu_2 \\ \mu_1 + \mu_2 \end{bmatrix}$$

and

$$\boldsymbol{\Sigma}_\mathbf{Z} = \text{Cov}(\mathbf{Z}) = \mathbf{C}\boldsymbol{\Sigma}_\mathbf{X}\mathbf{C}' = \begin{bmatrix} 1 & -1 \\ 1 & 1 \end{bmatrix} \begin{bmatrix} \sigma_{11} & \sigma_{12} \\ \sigma_{12} & \sigma_{22} \end{bmatrix} \begin{bmatrix} 1 & 1 \\ -1 & 1 \end{bmatrix}$$

$$= \begin{bmatrix} \sigma_{11} - 2\sigma_{12} + \sigma_{22} & \sigma_{11} - \sigma_{22} \\ \sigma_{11} - \sigma_{22} & \sigma_{11} + 2\sigma_{12} + \sigma_{22} \end{bmatrix}$$

Note that if $\sigma_{11} = \sigma_{22}$, that is, if X_1 and X_2 have equal variances, the off-diagonal terms in $\boldsymbol{\Sigma}_\mathbf{Z}$ vanish. This demonstrates the well-known result that the sum and difference of two random variables with identical variances are uncorrelated. ■

Partitioning the Sample Mean Vector and Covariance Matrix

Many of the matrix results in this section have been expressed in terms of population means and variances (covariances). Results in Equations (2-36), (2-37), (2-38), and (2-40) also hold if the population quantities are replaced by their appropriately defined sample counterparts.

Let $\bar{\mathbf{x}}' = [\bar{x}_1, \bar{x}_2, \ldots, \bar{x}_p]$ be the vector of sample averages constructed from n observations on p variables $X_1, X_2, \ldots, X_p$, and let

$$\mathbf{S}_n = \begin{bmatrix} s_{11} & \cdots & s_{1p} \\ \vdots & & \vdots \\ s_{1p} & \cdots & s_{pp} \end{bmatrix}$$

$$= \begin{bmatrix} \frac{1}{n}\sum_{j=1}^{n}(x_{1j} - \bar{x}_1)^2 & \cdots & \frac{1}{n}\sum_{j=1}^{n}(x_{1j} - \bar{x}_1)(x_{pj} - \bar{x}_p) \\ \vdots & & \vdots \\ \frac{1}{n}\sum_{j=1}^{n}(x_{1j} - \bar{x}_1)(x_{pj} - \bar{x}_p) & \cdots & \frac{1}{n}\sum_{j=1}^{n}(x_{pj} - \bar{x}_p)^2 \end{bmatrix}$$

be the corresponding sample variance-covariance matrix.

The sample mean vector and covariance matrix can be partitioned in order to distinguish quantities corresponding to groups of variables. Thus

$$\underset{(p\times 1)}{\bar{\mathbf{x}}} = \begin{bmatrix} \bar{x}_1 \\ \vdots \\ \bar{x}_q \\ \hline \bar{x}_{q+1} \\ \vdots \\ \bar{x}_p \end{bmatrix} = \begin{bmatrix} \bar{\mathbf{x}}^{(1)} \\ \hline \bar{\mathbf{x}}^{(2)} \end{bmatrix} \tag{2-46}$$

and

$$\underset{(p\times p)}{\mathbf{S}_n} = \left[\begin{array}{ccc|ccc} s_{11} & \cdots & s_{1q} & s_{1,q+1} & \cdots & s_{1p} \\ \vdots & & \vdots & \vdots & & \vdots \\ s_{q1} & \cdots & s_{qq} & s_{q,q+1} & \cdots & s_{qp} \\ \hline s_{q+1,1} & \cdots & s_{q+1,q} & s_{q+1,q+1} & \cdots & s_{q+1,p} \\ \vdots & & \vdots & \vdots & & \vdots \\ s_{p1} & \cdots & s_{pq} & s_{p,q+1} & \cdots & s_{pp} \end{array}\right]$$

$$= \begin{matrix} q \\ p-q \end{matrix} \overset{\begin{matrix} q & p-q \end{matrix}}{\left[\begin{array}{c|c} \mathbf{S}_{11} & \mathbf{S}_{12} \\ \hline \mathbf{S}_{21} & \mathbf{S}_{22} \end{array}\right]} \tag{2-47}$$

where $\bar{\mathbf{x}}^{(1)}$ and $\bar{\mathbf{x}}^{(2)}$ are the sample mean vectors constructed from observations on $\mathbf{x}^{(1)} = [x_1, \ldots, x_q]'$ and $\mathbf{x}^{(2)} = [x_{q+1}, \ldots, x_p]'$, respectively; $\mathbf{S}_{11}$ is the sample covariance matrix computed from observations on $\mathbf{x}^{(1)}$; $\mathbf{S}_{22}$ is the sample covariance matrix computed from observations on $\mathbf{x}^{(2)}$; and $\mathbf{S}_{12} = \mathbf{S}'_{21}$ is the sample covariance matrix for elements of $\mathbf{x}^{(1)}$ and elements of $\mathbf{x}^{(2)}$.

2.7 MATRIX INEQUALITIES AND MAXIMIZATION

Maximization principles play an important role in several multivariate techniques. Linear discriminant analysis, for example, is concerned with allocating observations to predetermined groups. The allocation rule is often a linear function of measurements that *maximizes* the separation between groups relative to their within-group variability. As another example, principal components are linear combinations of measurements with *maximum* variability.

The matrix inequalities presented in this section will easily allow us to derive certain maximization results, which will be referenced in later chapters.

Cauchy–Schwarz inequality. Let $\mathbf{b}$ and $\mathbf{d}$ be *any* two $(p \times 1)$ vectors. Then

$$(\mathbf{b}'\mathbf{d})^2 \leq (\mathbf{b}'\mathbf{b})(\mathbf{d}'\mathbf{d}) \tag{2-48}$$

with equality if and only if $\mathbf{b} = c\mathbf{d}$ (or $\mathbf{d} = c\mathbf{b}$) from some constant c.

Proof. The inequality is obvious if either $\mathbf{b} = \mathbf{0}$ or $\mathbf{d} = \mathbf{0}$. Excluding this possibility, consider the vector $\mathbf{b} - x\mathbf{d}$ where x is an arbitrary scalar. Since the length of $\mathbf{b} - x\mathbf{d}$ is positive for $\mathbf{b} - x\mathbf{d} \neq \mathbf{0}$, in this case

$$\begin{aligned} 0 < (\mathbf{b} - x\mathbf{d})'(\mathbf{b} - x\mathbf{d}) &= \mathbf{b}'\mathbf{b} - x\mathbf{d}'\mathbf{b} - \mathbf{b}'(x\mathbf{d}) + x^2\mathbf{d}'\mathbf{d} \\ &= \mathbf{b}'\mathbf{b} - 2x(\mathbf{b}'\mathbf{d}) + x^2(\mathbf{d}'\mathbf{d}) \end{aligned}$$

The expression immediately above is quadratic in x. If we complete the square by adding and subtracting the scalar $(\mathbf{b}'\mathbf{d})^2/\mathbf{d}'\mathbf{d}$, we get

$$\begin{aligned} 0 &< \mathbf{b}'\mathbf{b} - \frac{(\mathbf{b}'\mathbf{d})^2}{\mathbf{d}'\mathbf{d}} + \frac{(\mathbf{b}'\mathbf{d})^2}{\mathbf{d}'\mathbf{d}} - 2x(\mathbf{b}'\mathbf{d}) + x^2(\mathbf{d}'\mathbf{d}) \\ &= \mathbf{b}'\mathbf{b} - \frac{(\mathbf{b}'\mathbf{d})^2}{\mathbf{d}'\mathbf{d}} + (\mathbf{d}'\mathbf{d})\left[x - \frac{\mathbf{b}'\mathbf{d}}{\mathbf{d}'\mathbf{d}}\right]^2 \end{aligned}$$

The term in brackets is zero if we choose $x = \mathbf{b}'\mathbf{d}/\mathbf{d}'\mathbf{d}$, so we conclude

$$0 < \mathbf{b}'\mathbf{b} - \frac{(\mathbf{b}'\mathbf{d})^2}{\mathbf{d}'\mathbf{d}}$$

or $(\mathbf{b}'\mathbf{d})^2 < (\mathbf{b}'\mathbf{b})(\mathbf{d}'\mathbf{d})$ if $\mathbf{b} \neq x\mathbf{d}$ for some x.

Note that if $\mathbf{b} = c\mathbf{d}$, $0 = (\mathbf{b} - c\mathbf{d})'(\mathbf{b} - c\mathbf{d})$ and the same argument produces $(\mathbf{b}'\mathbf{d})^2 = (\mathbf{b}'\mathbf{b})(\mathbf{d}'\mathbf{d})$. ■

A simple but important extension of the Cauchy–Schwarz inequality follows directly.

Extended Cauchy–Schwarz inequality. Let $\underset{(p\times 1)}{\mathbf{b}}$ and $\underset{(p\times 1)}{\mathbf{d}}$ be any two vectors and let $\underset{(p\times p)}{\mathbf{B}}$ be a positive definite matrix. Then

$$(\mathbf{b}'\mathbf{d})^2 \leq (\mathbf{b}'\mathbf{B}\mathbf{b})(\mathbf{d}'\mathbf{B}^{-1}\mathbf{d}) \qquad (2\text{-}49)$$

with equality if and only if $\mathbf{b} = c\mathbf{B}^{-1}\mathbf{d}$ (or $\mathbf{d} = c\mathbf{B}\mathbf{b}$) for some constant c.

Proof. The inequality is obvious when $\mathbf{b} = \mathbf{0}$ or $\mathbf{d} = \mathbf{0}$. For cases other than these, consider the square-root matrix $\mathbf{B}^{1/2}$ defined in terms of its eigenvalues, λ_i, and the normalized eigenvectors, $\mathbf{e}_i$, as: $\mathbf{B}^{1/2} = \sum_{i=1}^{p} \sqrt{\lambda_i}\,\mathbf{e}_i\mathbf{e}_i'$. Setting [see also (2-22)]

$$\mathbf{B}^{-1/2} = \sum_{i=1}^{p} \frac{1}{\sqrt{\lambda_i}}\mathbf{e}_i\mathbf{e}_i'$$

it follows that

$$\mathbf{b}'\mathbf{d} = \mathbf{b}'\mathbf{I}\mathbf{d} = \mathbf{b}'\mathbf{B}^{1/2}\mathbf{B}^{-1/2}\mathbf{d} = (\mathbf{B}^{1/2}\mathbf{b})'(\mathbf{B}^{-1/2}\mathbf{d})$$

and the proof is completed by applying the Cauchy–Schwarz inequality to the vectors $(\mathbf{B}^{1/2}\mathbf{b})$ and $(\mathbf{B}^{-1/2}\mathbf{d})$. ■

The extended Cauchy–Schwarz inequality gives rise to the following maximization result.

Maximization lemma. Let $\underset{(p\times p)}{\mathbf{B}}$ be positive definite and $\underset{(p\times 1)}{\mathbf{d}}$ be a given vector. Then for an arbitrary nonzero vector $\underset{(p\times 1)}{\mathbf{x}}$,

$$\max_{\mathbf{x}\neq\mathbf{0}} \frac{(\mathbf{x}'\mathbf{d})^2}{\mathbf{x}'\mathbf{B}\mathbf{x}} = \mathbf{d}'\mathbf{B}^{-1}\mathbf{d} \tag{2-50}$$

with the maximum attained when $\underset{(p\times 1)}{\mathbf{x}} = c\underset{(p\times p)}{\mathbf{B}^{-1}}\underset{(p\times 1)}{\mathbf{d}}$ for any constant $c \neq 0$.

Proof. By the extended Cauchy–Schwarz inequality, $(\mathbf{x}'\mathbf{d})^2 \le (\mathbf{x}'\mathbf{B}\mathbf{x})(\mathbf{d}'\mathbf{B}^{-1}\mathbf{d})$. Because $\mathbf{x} \neq \mathbf{0}$ and $\mathbf{B}$ is positive definite, $\mathbf{x}'\mathbf{B}\mathbf{x} > 0$. Dividing both sides of the inequality by the positive scalar $\mathbf{x}'\mathbf{B}\mathbf{x}$ yields the upper bound

$$\frac{(\mathbf{x}'\mathbf{d})^2}{\mathbf{x}'\mathbf{B}\mathbf{x}} \le \mathbf{d}'\mathbf{B}^{-1}\mathbf{d}$$

Taking the maximum over $\mathbf{x}$ gives Equation (2-50) because the bound is attained for $\mathbf{x} = c\mathbf{B}^{-1}\mathbf{d}$. ■

A final maximization result will provide us with an interpretation of eigenvalues.

Maximization of quadratic forms for points on the unit sphere. Let $\underset{(p\times p)}{\mathbf{B}}$ be a positive definite matrix with eigenvalues $\lambda_1 \ge \lambda_2 \ge \cdots \ge \lambda_p > 0$ and associated normalized eigenvectors $\mathbf{e}_1, \mathbf{e}_2, \ldots, \mathbf{e}_p$. Then

$$\begin{aligned} &\max_{\mathbf{x}\neq\mathbf{0}} \frac{\mathbf{x}'\mathbf{B}\mathbf{x}}{\mathbf{x}'\mathbf{x}} = \lambda_1 \qquad \text{attained when } \mathbf{x} = \mathbf{e}_1 \\ &\min_{\mathbf{x}\neq\mathbf{0}} \frac{\mathbf{x}'\mathbf{B}\mathbf{x}}{\mathbf{x}'\mathbf{x}} = \lambda_p \qquad \text{attained when } \mathbf{x} = \mathbf{e}_p \end{aligned} \tag{2-51}$$

Moreover,

$$\max_{\mathbf{x}\perp\mathbf{e}_1,\ldots,\mathbf{e}_k} \frac{\mathbf{x}'\mathbf{B}\mathbf{x}}{\mathbf{x}'\mathbf{x}} = \lambda_{k+1} \qquad \text{attained when } \mathbf{x} = \mathbf{e}_{k+1},\ k = 1, 2, \ldots, p-1 \tag{2-52}$$

where the symbol $\perp$ is read "perpendicular to."

Proof. Let $\underset{(p\times p)}{\mathbf{P}}$ be the orthogonal matrix whose columns are the eigenvectors $\mathbf{e}_1, \mathbf{e}_2, \ldots, \mathbf{e}_p$ and $\mathbf{\Lambda}$ be the diagonal matrix with the eigenvalues $\lambda_1, \lambda_2, \ldots, \lambda_p$ along the main diagonal. Let $\mathbf{B}^{1/2} = \mathbf{P}\mathbf{\Lambda}^{1/2}\mathbf{P}'$ [see (2-22)] and $\underset{(p\times 1)}{\mathbf{y}} = \underset{(p\times p)}{\mathbf{P}'}\underset{(p\times 1)}{\mathbf{x}}$. Consequently, $\mathbf{x} \neq \mathbf{0}$ implies $\mathbf{y} \neq \mathbf{0}$. Thus

$$\frac{\mathbf{x}'\mathbf{B}\mathbf{x}}{\mathbf{x}'\mathbf{x}} = \frac{\mathbf{x}'\mathbf{B}^{1/2}\mathbf{B}^{1/2}\mathbf{x}}{\mathbf{x}'\underbrace{\mathbf{P}\mathbf{P}'}_{\underset{(p\times p)}{\mathbf{I}}}\mathbf{x}} = \frac{\mathbf{x}'\mathbf{P}\mathbf{\Lambda}^{1/2}\mathbf{P}'\mathbf{P}\mathbf{\Lambda}^{1/2}\mathbf{P}'\mathbf{x}}{\mathbf{y}'\mathbf{y}} = \frac{\mathbf{y}'\mathbf{\Lambda}\mathbf{y}}{\mathbf{y}'\mathbf{y}} = \frac{\sum_{i=1}^{p} \lambda_i y_i^2}{\sum_{i=1}^{p} y_i^2} \le \lambda_1 \frac{\sum_{i=1}^{p} y_i^2}{\sum_{i=1}^{p} y_i^2} = \lambda_1 \tag{2-53}$$

Setting $\mathbf{x} = \mathbf{e}_1$ gives

$$\mathbf{y} = \mathbf{P}'\mathbf{e}_1 = \begin{bmatrix} 1 \\ 0 \\ \vdots \\ 0 \end{bmatrix}$$

since

$$\mathbf{e}_k'\mathbf{e}_1 = \begin{cases} 1 & k = 1 \\ 0 & k \neq 1 \end{cases}$$

For this choice of $\mathbf{x}$, $\mathbf{y}'\boldsymbol{\Lambda}\mathbf{y}/\mathbf{y}'\mathbf{y} = \lambda_1/1 = \lambda_1$ or

$$\frac{\mathbf{e}_1'\mathbf{B}\mathbf{e}_1}{\mathbf{e}_1'\mathbf{e}_1} = \mathbf{e}_1'\mathbf{B}\mathbf{e}_1 = \lambda_1 \tag{2-54}$$

A similar argument produces the second part of (2-51).

Now $\mathbf{x} = \mathbf{P}\mathbf{y} = y_1\mathbf{e}_1 + y_2\mathbf{e}_2 + \cdots + y_p\mathbf{e}_p$, so $\mathbf{x} \perp \mathbf{e}_1, \ldots, \mathbf{e}_k$ implies

$$0 = \mathbf{e}_i'\mathbf{x} = y_1\mathbf{e}_i'\mathbf{e}_1 + y_2\mathbf{e}_i'\mathbf{e}_2 + \cdots + y_p\mathbf{e}_i'\mathbf{e}_p = y_i, \qquad i \leq k$$

Therefore, for $\mathbf{x}$ perpendicular to the first k eigenvectors $\mathbf{e}_i$, the left-hand side of the inequality in (2-53) becomes

$$\frac{\mathbf{x}'\mathbf{B}\mathbf{x}}{\mathbf{x}'\mathbf{x}} = \frac{\sum_{i=k+1}^{p} \lambda_i y_i^2}{\sum_{i=k+1}^{p} y_i^2}$$

Taking $y_{k+1} = 1$, $y_{k+2} = \cdots = y_p = 0$ gives the asserted maximum. ■

For a fixed $\mathbf{x}_0 \neq \mathbf{0}$, $\mathbf{x}_0'\mathbf{B}\mathbf{x}_0/\mathbf{x}_0'\mathbf{x}_0$ has the same value as $\mathbf{x}'\mathbf{B}\mathbf{x}$, where $\mathbf{x}' = \mathbf{x}_0'/\sqrt{\mathbf{x}_0'\mathbf{x}_0}$ is of unit length. Consequently, Equation (2-51) says that the largest eigenvalue, λ_1, is the maximum value of the quadratic form $\mathbf{x}'\mathbf{B}\mathbf{x}$ for all points $\mathbf{x}$ whose distance from the origin is unity. Similarly, λ_p is the smallest value of the quadratic form for all points $\mathbf{x}$ one unit from the origin. The largest and smallest eigenvalues thus represent extremal values of $\mathbf{x}'\mathbf{B}\mathbf{x}$ for points on the unit sphere. The "intermediate" eigenvalues of the $(p \times p)$ positive definite matrix $\mathbf{B}$ also have an interpretation as extremal values when $\mathbf{x}$ is further restricted to be perpendicular to the earlier choices.

Supplement 2A: Vectors and Matrices: Basic Concepts

Vectors

Many concepts, such as a person's health, intellectual abilities, or personality cannot be adequately quantified as a single number. Rather, several different measurements $x_1, x_2, \ldots, x_m$ are required.

Definition 2A.1. An m-tuple of real numbers $(x_1, x_2, \ldots, x_i, \ldots, x_m)$ arranged in a column is called a *vector* and is denoted by a boldfaced, lowercase letter.

Examples of vectors are

$$\mathbf{x} = \begin{bmatrix} x_1 \\ x_2 \\ \vdots \\ x_m \end{bmatrix}, \quad \mathbf{a} = \begin{bmatrix} 1 \\ 0 \\ 0 \end{bmatrix}, \quad \mathbf{b} = \begin{bmatrix} 1 \\ -1 \\ 1 \\ -1 \end{bmatrix}, \quad \mathbf{y} = \begin{bmatrix} 1 \\ 2 \\ -2 \end{bmatrix}$$

Vectors are said to be equal if their corresponding entries are the same.

Definition 2A.2 (Scalar multiplication). Let c be an arbitrary scalar. Then the *product* $c\mathbf{x}$ is a vector with ith entry cx_i.

To illustrate scalar multiplication, take $c_1 = 5$ and $c_2 = -1.2$. Then

$$c_1\mathbf{y} = 5\begin{bmatrix} 1 \\ 2 \\ -2 \end{bmatrix} = \begin{bmatrix} 5 \\ 10 \\ -10 \end{bmatrix} \quad \text{and} \quad c_2\mathbf{y} = (-1.2)\begin{bmatrix} 1 \\ 2 \\ -2 \end{bmatrix} = \begin{bmatrix} -1.2 \\ -2.4 \\ 2.4 \end{bmatrix}.$$

Definition 2A.3 (Vector addition). The sum of two vectors $\mathbf{x}$ and $\mathbf{y}$, each having the same number of entries, is that vector

$$\mathbf{z} = \mathbf{x} + \mathbf{y} \quad \text{with } i\text{th entry} \quad z_i = x_i + y_i$$

Thus

$$\underset{\mathbf{x}}{\begin{bmatrix} 3 \\ -1 \\ 4 \end{bmatrix}} + \underset{\mathbf{y}}{\begin{bmatrix} 1 \\ 2 \\ -2 \end{bmatrix}} = \underset{\mathbf{z}}{\begin{bmatrix} 4 \\ 1 \\ 2 \end{bmatrix}}$$

Taking the zero vector, $\mathbf{0}$, to be the m-tuple $(0, 0, \ldots, 0)$ and the vector $-\mathbf{x}$ to be the m-tuple $(-x_1, -x_2, \ldots, -x_m)$, the two operations of scalar multiplication and vector addition can be combined in a useful manner.

Definition 2A.4. The space of all real m-tuples, with scalar multiplication and vector addition as defined above, is called a *vector space*.

Definition 2A.5. The vector $\mathbf{y} = a_1\mathbf{x}_1 + a_2\mathbf{x}_2 + \cdots + a_k\mathbf{x}_k$ is a *linear combination* of the vectors $\mathbf{x}_1, \mathbf{x}_2, \ldots, \mathbf{x}_k$. The set of all linear combinations of $\mathbf{x}_1, \mathbf{x}_2, \ldots, \mathbf{x}_k$ is called their *linear span*.

Definition 2A.6. A set of vectors $\mathbf{x}_1, \mathbf{x}_2, \ldots, \mathbf{x}_k$ is said to be *linearly dependent* if there exist k numbers $(a_1, a_2, \ldots, a_k)$, not all zero, such that

$$a_1\mathbf{x}_1 + a_2\mathbf{x}_2 + \cdots + a_k\mathbf{x}_k = \mathbf{0}$$

Otherwise the set of vectors is said to be *linearly independent*.

If one of the vectors, for example, $\mathbf{x}_i$, is $\mathbf{0}$, the set is linearly dependent (let a_i be the only nonzero coefficient in Definition 2A.6).

The familiar vectors with a one as an entry and zeros elsewhere are linearly independent. For $m = 4$

$$\mathbf{x}_1 = \begin{bmatrix} 1 \\ 0 \\ 0 \\ 0 \end{bmatrix}, \quad \mathbf{x}_2 = \begin{bmatrix} 0 \\ 1 \\ 0 \\ 0 \end{bmatrix}, \quad \mathbf{x}_3 = \begin{bmatrix} 0 \\ 0 \\ 1 \\ 0 \end{bmatrix}, \quad \mathbf{x}_4 = \begin{bmatrix} 0 \\ 0 \\ 0 \\ 1 \end{bmatrix}$$

so

$$\mathbf{0} = a_1\mathbf{x}_1 + a_2\mathbf{x}_2 + a_3\mathbf{x}_3 + a_4\mathbf{x}_4 = \begin{bmatrix} a_1 \cdot 1 + a_2 \cdot 0 + a_3 \cdot 0 + a_4 \cdot 0 \\ a_1 \cdot 0 + a_2 \cdot 1 + a_3 \cdot 0 + a_4 \cdot 0 \\ a_1 \cdot 0 + a_2 \cdot 0 + a_3 \cdot 1 + a_4 \cdot 0 \\ a_1 \cdot 0 + a_2 \cdot 0 + a_3 \cdot 0 + a_4 \cdot 1 \end{bmatrix} = \begin{bmatrix} a_1 \\ a_2 \\ a_3 \\ a_4 \end{bmatrix}$$

implies $a_1 = a_2 = a_3 = a_4 = 0$.

As another example, let $k = 3$ and $m = 3$.

$$\mathbf{x}_1 = \begin{bmatrix} 1 \\ 1 \\ 1 \end{bmatrix}, \quad \mathbf{x}_2 = \begin{bmatrix} 2 \\ 5 \\ -1 \end{bmatrix}, \quad \mathbf{x}_3 = \begin{bmatrix} 0 \\ 1 \\ -1 \end{bmatrix}$$

Then

$$2\mathbf{x}_1 - \mathbf{x}_2 + 3\mathbf{x}_3 = \mathbf{0}$$

Thus $\mathbf{x}_1, \mathbf{x}_2, \mathbf{x}_3$ are a linearly dependent set of vectors since any one of them can be written as a linear combination of the others (for example, $\mathbf{x}_2 = 2\mathbf{x}_1 + 3\mathbf{x}_3$).

Definition 2A.7. Any set of m linearly independent vectors is called a *basis* for the vector space of all m-tuples of real numbers.

Result 2A.1. Every vector can be expressed as a unique linear combination of a fixed basis. ■

With $m = 4$, the usual choice of a basis is

$$\begin{bmatrix} 1 \\ 0 \\ 0 \\ 0 \end{bmatrix}, \begin{bmatrix} 0 \\ 1 \\ 0 \\ 0 \end{bmatrix}, \begin{bmatrix} 0 \\ 0 \\ 1 \\ 0 \end{bmatrix}, \begin{bmatrix} 0 \\ 0 \\ 0 \\ 1 \end{bmatrix}$$

where these four vectors were shown to be linearly independent above. Any vector **x**, can be uniquely expressed as

$$x_1 \begin{bmatrix} 1 \\ 0 \\ 0 \\ 0 \end{bmatrix} + x_2 \begin{bmatrix} 0 \\ 1 \\ 0 \\ 0 \end{bmatrix} + x_3 \begin{bmatrix} 0 \\ 0 \\ 1 \\ 0 \end{bmatrix} + x_4 \begin{bmatrix} 0 \\ 0 \\ 0 \\ 1 \end{bmatrix} = \begin{bmatrix} x_1 \\ x_2 \\ x_3 \\ x_4 \end{bmatrix} = \mathbf{x}$$

A vector consisting of m elements may be regarded geometrically as a point in m-dimensional space. For example, with $m = 2$, the vector **x** may be regarded as representing the point in the plane with coordinates x_1 and x_2.

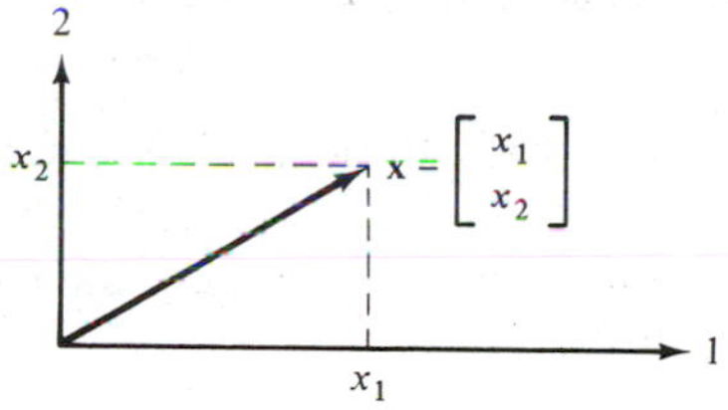

Vectors have the geometrical properties of length and direction.

Definition 2A.8. The *length* of a vector of m elements emanating from the origin is given by the Pythagorean formula:

$$\text{length of } \mathbf{x} = L_{\mathbf{x}} = \sqrt{x_1^2 + x_2^2 + \cdots + x_m^2}$$

Definition 2A.9. The *angle* θ between two vectors **x** and **y**, both having m entries, is defined as follows.

$$\cos(\theta) = \frac{(x_1 y_1 + x_2 y_2 + \cdots + x_m y_m)}{L_{\mathbf{x}} L_{\mathbf{y}}}$$

where $L_{\mathbf{x}}$ = length of **x** and $L_{\mathbf{y}}$ = length of **y**, $x_1, x_2, \ldots, x_m$ are the elements of **x**, and $y_1, y_2, \ldots, y_m$ are the elements of **y**.

Let

$$\mathbf{x} = \begin{bmatrix} -1 \\ 5 \\ 2 \\ -2 \end{bmatrix} \quad \text{and} \quad \mathbf{y} = \begin{bmatrix} 4 \\ -3 \\ 0 \\ 1 \end{bmatrix}$$

Then the length of **x**, length of **y**, and the angle between the two vectors are

$$\text{length of } \mathbf{x} = \sqrt{(-1)^2 + 5^2 + 2^2 + (-2)^2} = \sqrt{34} = 5.83$$

$$\text{length of } \mathbf{y} = \sqrt{4^2 + (-3)^2 + 0^2 + 1^2} = \sqrt{26} = 5.10$$

and

$$\begin{aligned}\cos(\theta) &= \frac{1}{L_{\mathbf{x}}}\frac{1}{L_{\mathbf{y}}}[x_1y_1 + x_2y_2 + x_3y_3 + x_4y_4] \\ &= \frac{1}{\sqrt{34}}\frac{1}{\sqrt{26}}[(-1)4 + 5(-3) + 2(0) + (-2)1] \\ &= \frac{1}{5.83 \times 5.10}[-21] = -.706\end{aligned}$$

Consequently, $\theta = -45°$.

Definition 2A.10. The *inner* (or *dot*) *product* of two vectors **x** and **y** with the same number of entries is defined as the sum of component products

$$x_1y_1 + x_2y_2 + \cdots + x_my_m$$

We use the notation $\mathbf{x}'\mathbf{y}$ or $\mathbf{y}'\mathbf{x}$ to denote this inner product.

With the $\mathbf{x}'\mathbf{y}$ notation, we may express the length of a vector and the cosine of the angle between two vectors as

$$L_{\mathbf{x}} = \text{length of } \mathbf{x} = \sqrt{x_1^2 + x_2^2 + \cdots + x_m^2} = \sqrt{\mathbf{x}'\mathbf{x}}$$

$$\cos(\theta) = \frac{\mathbf{x}'\mathbf{y}}{\sqrt{\mathbf{x}'\mathbf{x}}\sqrt{\mathbf{y}'\mathbf{y}}}$$

Definition 2A.11. When the angle between two vectors **x**, **y** is $\theta = 90°$ or 270°, we say that **x** and **y** are perpendicular. Since $\cos(\theta) = 0$ only if $\theta = 90°$ or 270°, the condition becomes

$$\mathbf{x} \text{ and } \mathbf{y} \text{ are } \textit{perpendicular} \text{ if } \mathbf{x}'\mathbf{y} = 0$$

We write $\mathbf{x} \perp \mathbf{y}$.

The basis vectors

$$\begin{bmatrix}1\\0\\0\\0\end{bmatrix}, \quad \begin{bmatrix}0\\1\\0\\0\end{bmatrix}, \quad \begin{bmatrix}0\\0\\1\\0\end{bmatrix}, \quad \begin{bmatrix}0\\0\\0\\1\end{bmatrix}$$

are mutually perpendicular. Also, each has length one. The same construction holds for any number of entries *m*.

Result 2A.2

(a) **z** is perpendicular to every vector if and only if $\mathbf{z} = \mathbf{0}$.

(b) If $\mathbf{z}$ is perpendicular to each vector $\mathbf{x}_1, \mathbf{x}_2, \ldots, \mathbf{x}_k$, then $\mathbf{z}$ is perpendicular to their linear span.

(c) Mutually perpendicular vectors are linearly independent. ■

Definition 2A.12. The *projection* (or *shadow*) of a vector $\mathbf{x}$ on a vector $\mathbf{y}$ is

$$\textit{projection of } \mathbf{x} \textit{ on } \mathbf{y} = \frac{(\mathbf{x}'\mathbf{y})}{L_{\mathbf{y}}^2}\mathbf{y}$$

If $\mathbf{y}$ has unit length so that $L_{\mathbf{y}} = 1$,

$$\textit{projection of } \mathbf{x} \textit{ on } \mathbf{y} = (\mathbf{x}'\mathbf{y})\mathbf{y}$$

Result 2A.3 (Gram–Schmidt process). Given linearly independent vectors $\mathbf{x}_1, \mathbf{x}_2, \ldots, \mathbf{x}_k$, there exist mutually perpendicular vectors $\mathbf{u}_1, \mathbf{u}_2, \ldots, \mathbf{u}_k$ with the same linear span. These may be constructed sequentially by setting

$$\begin{aligned}
\mathbf{u}_1 &= \mathbf{x}_1 \\
\mathbf{u}_2 &= \mathbf{x}_2 - \frac{(\mathbf{x}_2'\mathbf{u}_1)}{\mathbf{u}_1'\mathbf{u}_1}\mathbf{u}_1 \\
&\vdots \\
\mathbf{u}_k &= \mathbf{x}_k - \frac{(\mathbf{x}_k'\mathbf{u}_1)}{\mathbf{u}_1'\mathbf{u}_1}\mathbf{u}_1 - \cdots - \frac{(\mathbf{x}_k'\mathbf{u}_{k-1})}{\mathbf{u}_{k-1}'\mathbf{u}_{k-1}}\mathbf{u}_{k-1}
\end{aligned}$$

We can also convert the $\mathbf{u}$'s to unit length by setting $\mathbf{z}_j = \mathbf{u}_j / \sqrt{\mathbf{u}_j'\mathbf{u}_j}$. In this construction, $(\mathbf{x}_k'\mathbf{z}_j)\mathbf{z}_j$ is the projection of $\mathbf{x}_k$ on $\mathbf{z}_j$ and $\sum_{j=1}^{k-1} (\mathbf{x}_k'\mathbf{z}_j)\mathbf{z}_j$ is the *projection of* $\mathbf{x}_k$ *on the linear span of* $\mathbf{x}_1, \mathbf{x}_2, \ldots, \mathbf{x}_{k-1}$. ■

For example, to construct perpendicular vectors from

$$\mathbf{x}_1 = \begin{bmatrix} 4 \\ 0 \\ 0 \\ 2 \end{bmatrix} \quad \text{and} \quad \mathbf{x}_2 = \begin{bmatrix} 3 \\ 1 \\ 0 \\ -1 \end{bmatrix}$$

we take

$$\mathbf{u}_1 = \mathbf{x}_1 = \begin{bmatrix} 4 \\ 0 \\ 0 \\ 2 \end{bmatrix}$$

so

$$\mathbf{u}_1'\mathbf{u}_1 = 4^2 + 0^2 + 0^2 + 2^2 = 20$$

and

$$\mathbf{x}_2'\mathbf{u}_1 = 3(4) + 1(0) + 0(0) - 1(2) = 10$$

Thus

$$\mathbf{u}_2 = \begin{bmatrix} 3 \\ 1 \\ 0 \\ -1 \end{bmatrix} - \frac{10}{20}\begin{bmatrix} 4 \\ 0 \\ 0 \\ 2 \end{bmatrix} = \begin{bmatrix} 1 \\ 1 \\ 0 \\ -2 \end{bmatrix} \quad \text{and} \quad \mathbf{z}_1 = \frac{1}{\sqrt{20}}\begin{bmatrix} 4 \\ 0 \\ 0 \\ 2 \end{bmatrix}, \quad \mathbf{z}_2 = \frac{1}{\sqrt{6}}\begin{bmatrix} 1 \\ 1 \\ 0 \\ -2 \end{bmatrix}$$

Matrices

Definition 2A.13. An $m \times k$ *matrix*, generally denoted by boldface uppercase letters like **A**, **R**, $\mathbf{\Sigma}$, and so forth, is a rectangular array of elements having m rows and k columns.

Examples of matrices are

$$\mathbf{A} = \begin{bmatrix} -7 & 2 \\ 0 & 1 \\ 3 & 4 \end{bmatrix}, \quad \mathbf{B} = \begin{bmatrix} x & 3 & 0 \\ 4 & -2 & 1/x \end{bmatrix}, \quad \mathbf{I} = \begin{bmatrix} 1 & 0 & 0 \\ 0 & 1 & 0 \\ 0 & 0 & 1 \end{bmatrix}$$

$$\mathbf{\Sigma} = \begin{bmatrix} 1 & .7 & -.3 \\ .7 & 2 & 1 \\ -.3 & 1 & 8 \end{bmatrix}, \quad \mathbf{E} = [e_1]$$

In our work, the matrix elements will be real numbers or functions taking on values in the real numbers.

Definition 2A.14. The *dimension* (abbreviated *dim*) of an $m \times k$ matrix is the ordered pair (m, k); m is the row dimension and k is the column dimension. The dimension of a matrix is frequently indicated in parentheses below the letter representing the matrix. Thus the $m \times k$ matrix **A** is denoted by $\underset{(m\times k)}{\mathbf{A}}$. In the examples above the dimension of the matrix $\mathbf{\Sigma}$ is 3×3 and this information can be conveyed by writing, $\underset{(3\times 3)}{\mathbf{\Sigma}}$.

An $m \times k$ matrix, call it **A**, of arbitrary constants can be written

$$\underset{(m\times k)}{\mathbf{A}} = \begin{bmatrix} a_{11} & a_{12} & \cdots & a_{1k} \\ a_{21} & a_{22} & \cdots & a_{2k} \\ \vdots & \vdots & & \vdots \\ a_{m1} & a_{m2} & \cdots & a_{mk} \end{bmatrix}$$

or more compactly as $\underset{(m\times k)}{\mathbf{A}} = \{a_{ij}\}$, where the index i refers to the row and the index j refers to the column.

An $m \times 1$ matrix is referred to as a column *vector*. A $1 \times k$ matrix is referred to as a row *vector*. Since matrices can be considered as vectors side by side, it is natural to define multiplication by a scalar and the addition of two matrices with the same dimensions.

Definition 2A.15. Two matrices $\underset{(m\times k)}{\mathbf{A}} = \{a_{ij}\}$ and $\underset{(m\times k)}{\mathbf{B}} = \{b_{ij}\}$ are said to be *equal*, written $\mathbf{A} = \mathbf{B}$, if $a_{ij} = b_{ij}$, $i = 1, 2, \ldots, m$, $j = 1, 2, \ldots, k$. That is, two matrices are equal if:

(a) Their dimensionality is the same.
(b) Every corresponding element is the same.

Definition 2A.16 (Matrix addition). Let the matrices $\mathbf{A}$ and $\mathbf{B}$ both be of dim $m \times k$ with arbitrary elements a_{ij} and b_{ij}, $i = 1, 2, \ldots, m$, $j = 1, 2, \ldots, k$, respectively. The sum of the matrices $\mathbf{A}$ and $\mathbf{B}$ is an $m \times k$ matrix $\mathbf{C}$, written $\mathbf{C} = \mathbf{A} + \mathbf{B}$, such that the arbitrary element of $\mathbf{C}$, c_{ij}, is given by

$$c_{ij} = a_{ij} + b_{ij} \qquad i = 1, 2, \ldots, m,\ j = 1, 2, \ldots, k$$

Note that the addition of matrices is defined only for matrices of the same dimension.

For example

$$\underset{\mathbf{A}}{\begin{bmatrix} 3 & 2 & 3 \\ 4 & 1 & 1 \end{bmatrix}} + \underset{\mathbf{B}}{\begin{bmatrix} 3 & 6 & 7 \\ 2 & -1 & 0 \end{bmatrix}} = \underset{\mathbf{C}}{\begin{bmatrix} 6 & 8 & 10 \\ 6 & 0 & 1 \end{bmatrix}}$$

Definition 2A.17 (Scalar multiplication). Let c be an arbitrary scalar and $\underset{(m\times k)}{\mathbf{A}} = \{a_{ij}\}$. Then $\underset{(m\times k)}{c\mathbf{A}} = \underset{(m\times k)}{\mathbf{A}c} = \underset{(m\times k)}{\mathbf{B}} = \{b_{ij}\}$, where $b_{ij} = ca_{ij} = a_{ij}c$, $i = 1, 2, \ldots, m$, $j = 1, 2, \ldots, k$.

Multiplication of a matrix by a scalar produces a new matrix whose elements are the elements of the original matrix, *each* multiplied by the scalar.

For example, if $c = 2$,

$$\underset{c\mathbf{A}}{2\begin{bmatrix} 3 & -4 \\ 2 & 6 \\ 0 & 5 \end{bmatrix}} = \underset{\mathbf{A}c}{\begin{bmatrix} 3 & -4 \\ 2 & 6 \\ 0 & 5 \end{bmatrix}2} = \underset{\mathbf{B}}{\begin{bmatrix} 6 & -8 \\ 4 & 12 \\ 0 & 10 \end{bmatrix}}$$

Definition 2A.18 (Matrix subtraction). Let $\underset{(m\times k)}{\mathbf{A}} = \{a_{ij}\}$ and $\underset{(m\times k)}{\mathbf{B}} = \{b_{ij}\}$ be two matrices of equal dimension. Then the difference between $\mathbf{A}$ and $\mathbf{B}$, written $\mathbf{A} - \mathbf{B}$, is an $m \times k$ matrix $\mathbf{C} = \{c_{ij}\}$ given by

$$\mathbf{C} = \mathbf{A} - \mathbf{B} = \mathbf{A} + (-1)\mathbf{B}$$

That is, $c_{ij} = a_{ij} + (-1)b_{ij} = a_{ij} - b_{ij}$, $i = 1, 2, \ldots, m$, $j = 1, 2, \ldots, k$.

Definition 2A.19. Consider the $m \times k$ matrix $\mathbf{A}$ with arbitrary elements a_{ij}, $i = 1, 2, \ldots, m$, $j = 1, 2, \ldots, k$. The *transpose* of the matrix $\mathbf{A}$, denoted by $\mathbf{A}'$, is the $k \times m$ matrix with elements a_{ji}, $j = 1, 2, \ldots, k$; $i = 1, 2, \ldots, m$. That is, the transpose of the matrix $\mathbf{A}$ is obtained from $\mathbf{A}$ by interchanging the rows and columns.

As an example, if

$$\underset{(2\times 3)}{\mathbf{A}} = \begin{bmatrix} 2 & 1 & 3 \\ 7 & -4 & 6 \end{bmatrix}, \quad \text{then} \quad \underset{(3\times 2)}{\mathbf{A}'} = \begin{bmatrix} 2 & 7 \\ 1 & -4 \\ 3 & 6 \end{bmatrix}$$

Result 2A.4. For all matrices **A**, **B**, and **C** (*of equal dimension*) and scalars c and d, the following hold:

(a) $(\mathbf{A} + \mathbf{B}) + \mathbf{C} = \mathbf{A} + (\mathbf{B} + \mathbf{C})$
(b) $\mathbf{A} + \mathbf{B} = \mathbf{B} + \mathbf{A}$
(c) $c(\mathbf{A} + \mathbf{B}) = c\mathbf{A} + c\mathbf{B}$
(d) $(c + d)\mathbf{A} = c\mathbf{A} + d\mathbf{A}$
(e) $(\mathbf{A} + \mathbf{B})' = \mathbf{A}' + \mathbf{B}'$ (That is, the transpose of the sum is equal to the sum of the transposes.)
(f) $(cd)\mathbf{A} = c(d\mathbf{A})$
(g) $(c\mathbf{A})' = c\mathbf{A}'$ ■

Definition 2A.20. If an arbitrary matrix **A** has the *same* number of rows and columns, then **A** is called a *square* matrix. In the matrices given after Definition 2A.13, the matrices **Σ**, **I**, and **E** are square matrices.

Definition 2A.21. Let **A** be a $k \times k$ (square) matrix. The matrix **A** is said to be *symmetric* if $\mathbf{A} = \mathbf{A}'$. That is, **A** is symmetric if $a_{ij} = a_{ji}$, $i = 1, 2, \ldots, k$, $j = 1, 2, \ldots, k$.

Examples of symmetric matrices are

$$\underset{(3\times 3)}{\mathbf{I}} = \begin{bmatrix} 1 & 0 & 0 \\ 0 & 1 & 0 \\ 0 & 0 & 1 \end{bmatrix}, \quad \underset{(2\times 2)}{\mathbf{A}} = \begin{bmatrix} 2 & 4 \\ 4 & 1 \end{bmatrix}, \quad \underset{(4\times 4)}{\mathbf{B}} = \begin{bmatrix} a & c & e & f \\ c & b & g & d \\ e & g & c & a \\ f & d & a & d \end{bmatrix}$$

Definition 2A.22. The $k \times k$ *identity* matrix, denoted by $\underset{(k\times k)}{\mathbf{I}}$, is the square matrix with ones on the main (NW–SE) diagonal and zeros elsewhere. The 3×3 identity matrix is included in the collection above.

Definition 2A.23 (Matrix multiplication). The product **AB** of an $m \times n$ matrix $\mathbf{A} = \{a_{ij}\}$ and an $n \times k$ matrix $\mathbf{B} = \{b_{ij}\}$ is the $m \times k$ matrix **C** whose element c_{ij} is given by

$$c_{ij} = \sum_{\ell=1}^{n} a_{i\ell} b_{\ell j} \qquad i = 1, 2, \ldots, m, \; j = 1, 2, \ldots, k$$

Note that for the product **AB** to be defined, the column dimension of **A** must equal the row dimension of **B**. Then the row dimension of **AB** equals the row dimension of **A** and the column dimension of **AB** equals the column dimension of **B**.

For example let

$$\underset{(2\times3)}{\mathbf{A}} = \begin{bmatrix} 3 & -1 & 2 \\ 4 & 0 & 5 \end{bmatrix} \quad \text{and} \quad \underset{(3\times2)}{\mathbf{B}} = \begin{bmatrix} 3 & 4 \\ 6 & -2 \\ 4 & 3 \end{bmatrix}$$

Then

$$\underset{(2\times3)}{\begin{bmatrix} 3 & -1 & 2 \\ 4 & 0 & 5 \end{bmatrix}} \underset{(3\times2)}{\begin{bmatrix} 3 & 4 \\ 6 & -2 \\ 4 & 3 \end{bmatrix}} = \underset{(2\times2)}{\begin{bmatrix} 11 & 20 \\ 32 & 31 \end{bmatrix}} = \begin{bmatrix} c_{11} & c_{12} \\ c_{21} & c_{22} \end{bmatrix}$$

where

$$\begin{aligned} c_{11} &= (3)(3) + (-1)(6) + (2)(4) = 11 \\ c_{12} &= (3)(4) + (-1)(-2) + (2)(3) = 20 \\ c_{21} &= (4)(3) + (0)(6) + (5)(4) = 32 \\ c_{22} &= (4)(4) + (0)(-2) + (5)(3) = 31 \end{aligned}$$

As an additional example, consider the product of two vectors. Let

$$\mathbf{x} = \begin{bmatrix} 1 \\ 0 \\ -2 \\ 3 \end{bmatrix} \quad \text{and} \quad \mathbf{y} = \begin{bmatrix} 2 \\ -3 \\ -1 \\ -8 \end{bmatrix}$$

Then $\mathbf{x}' = [1 \quad 0 \quad -2 \quad 3]$ and

$$\mathbf{x}'\mathbf{y} = [1 \quad 0 \quad -2 \quad 3] \begin{bmatrix} 2 \\ -3 \\ -1 \\ -8 \end{bmatrix} = [-20] = [2 \quad -3 \quad -1 \quad -8] \begin{bmatrix} 1 \\ 0 \\ -2 \\ 3 \end{bmatrix} = \mathbf{y}'\mathbf{x}$$

Note that the product $\mathbf{xy}$ is undefined since $\mathbf{x}$ is a 4×1 matrix and $\mathbf{y}$ is a 4×1 matrix so the column dim of $\mathbf{x}$, 1, is unequal to the row dim of $\mathbf{y}$, 4. If $\mathbf{x}$ and $\mathbf{y}$ are vectors of the same dimension, such as $n \times 1$, both of the products $\mathbf{x}'\mathbf{y}$ and $\mathbf{xy}'$ are defined. In particular, $\mathbf{y}'\mathbf{x} = \mathbf{x}'\mathbf{y} = x_1y_1 + x_2y_2 + \cdots + x_ny_n$ and $\mathbf{xy}'$ is an $n \times n$ matrix with i, jth element x_iy_j.

Result 2A.5. For all matrices $\mathbf{A}$, $\mathbf{B}$, and $\mathbf{C}$ (of dimensions such that the indicated products are defined) and a scalar c,

(a) $c(\mathbf{AB}) = (c\mathbf{A})\mathbf{B}$

(b) $\mathbf{A}(\mathbf{BC}) = (\mathbf{AB})\mathbf{C}$

(c) $\mathbf{A}(\mathbf{B} + \mathbf{C}) = \mathbf{AB} + \mathbf{AC}$

(d) $(\mathbf{B} + \mathbf{C})\mathbf{A} = \mathbf{BA} + \mathbf{CA}$

(e) $(\mathbf{AB})' = \mathbf{B}'\mathbf{A}'$ ■

There are several important differences between the algebra of matrices and the algebra of real numbers. Two of these differences follow.

1. Matrix multiplication is, in general, not commutative. That is, in general, $\mathbf{AB} \neq \mathbf{BA}$.
Several examples will illustrate the failure of the commutative law (for matrices).

$$\begin{bmatrix} 3 & -1 \\ 4 & 7 \end{bmatrix}\begin{bmatrix} 0 \\ 2 \end{bmatrix} = \begin{bmatrix} -2 \\ 14 \end{bmatrix}$$

but

$$\begin{bmatrix} 0 \\ 2 \end{bmatrix}\begin{bmatrix} 3 & -1 \\ 4 & 7 \end{bmatrix}$$

is not defined.

$$\begin{bmatrix} 1 & 0 & 1 \\ 2 & -3 & 6 \end{bmatrix}\begin{bmatrix} 7 & 6 \\ -3 & 1 \\ 2 & 4 \end{bmatrix} = \begin{bmatrix} 9 & 10 \\ 35 & 33 \end{bmatrix}$$

but

$$\begin{bmatrix} 7 & 6 \\ -3 & 1 \\ 2 & 4 \end{bmatrix}\begin{bmatrix} 1 & 0 & 1 \\ 2 & -3 & 6 \end{bmatrix} = \begin{bmatrix} 19 & -18 & 43 \\ -1 & -3 & 3 \\ 10 & -12 & 26 \end{bmatrix}$$

Also

$$\begin{bmatrix} 4 & -1 \\ 0 & 1 \end{bmatrix}\begin{bmatrix} 2 & 1 \\ -3 & 4 \end{bmatrix} = \begin{bmatrix} 11 & 0 \\ -3 & 4 \end{bmatrix}$$

but

$$\begin{bmatrix} 2 & 1 \\ -3 & 4 \end{bmatrix}\begin{bmatrix} 4 & -1 \\ 0 & 1 \end{bmatrix} = \begin{bmatrix} 8 & -1 \\ -12 & 7 \end{bmatrix}$$

2. Let $\mathbf{0}$ denote the zero matrix, that is, the matrix with zero for every element. In the algebra of real numbers, if the product of two numbers, ab, is zero, then $a = 0$ or $b = 0$. In matrix algebra, however, the product of two *nonzero* matrices may be the zero matrix. Hence,

$$\underset{(m\times n)(n\times k)}{\mathbf{AB}} = \underset{(m\times k)}{\mathbf{0}}$$

does not imply that $\mathbf{A} = \mathbf{0}$ or $\mathbf{B} = \mathbf{0}$. For example,

$$\begin{bmatrix} 3 & 1 & 3 \\ 1 & 2 & 2 \end{bmatrix}\begin{bmatrix} 4 \\ 3 \\ -5 \end{bmatrix} = \begin{bmatrix} 0 \\ 0 \end{bmatrix}$$

It is true, however, that if either $\underset{(m\times n)}{\mathbf{A}} = \underset{(m\times n)}{\mathbf{0}}$ or $\underset{(n\times k)}{\mathbf{B}} = \underset{(n\times k)}{\mathbf{0}}$, then $\underset{(m\times n)}{\mathbf{A}}\ \underset{(n\times k)}{\mathbf{B}} = \underset{(m\times k)}{\mathbf{0}}$.

Definition 2A.24. The *determinant* of the square $k \times k$ matrix $\mathbf{A} = \{a_{ij}\}$, denoted by $|\mathbf{A}|$, is the scalar

$$|\mathbf{A}| = a_{11} \qquad \text{if } k = 1$$

$$|\mathbf{A}| = \sum_{j=1}^{k} a_{1j}|\mathbf{A}_{1j}|(-1)^{1+j} \qquad \text{if } k > 1$$

where $\mathbf{A}_{1j}$ is the $(k-1) \times (k-1)$ matrix obtained by deleting the first row and jth column of $\mathbf{A}$. Also, $|\mathbf{A}| = \sum_{j=1}^{k} a_{ij} |\mathbf{A}_{ij}| (-1)^{i+j}$ using the ith row in place of the first row.

Examples of determinants (evaluated using Definition 2A.24) are

$$\begin{vmatrix} 1 & 3 \\ 6 & 4 \end{vmatrix} = 1|4|(-1)^2 + 3|6|(-1)^3 = 1(4) + 3(6)(-1) = -14$$

In general,

$$\begin{vmatrix} a_{11} & a_{12} \\ a_{21} & a_{22} \end{vmatrix} = a_{11}a_{22}(-1)^2 + a_{12}a_{21}(-1)^3 = a_{11}a_{22} - a_{12}a_{21}$$

$$\begin{vmatrix} 3 & 1 & 6 \\ 7 & 4 & 5 \\ 2 & -7 & 1 \end{vmatrix} = 3\begin{vmatrix} 4 & 5 \\ -7 & 1 \end{vmatrix}(-1)^2 + 1\begin{vmatrix} 7 & 5 \\ 2 & 1 \end{vmatrix}(-1)^3 + 6\begin{vmatrix} 7 & 4 \\ 2 & -7 \end{vmatrix}(-1)^4$$

$$= 3(39) - 1(-3) + 6(-57) = -222$$

$$\begin{vmatrix} 1 & 0 & 0 \\ 0 & 1 & 0 \\ 0 & 0 & 1 \end{vmatrix} = 1\begin{vmatrix} 1 & 0 \\ 0 & 1 \end{vmatrix}(-1)^2 + 0\begin{vmatrix} 0 & 0 \\ 0 & 1 \end{vmatrix}(-1)^3 + 0\begin{vmatrix} 0 & 1 \\ 0 & 0 \end{vmatrix}(-1)^4$$

$$= 1(1) = 1$$

If $\mathbf{I}$ is the $k \times k$ identity matrix, $|\mathbf{I}| = 1$.

$$\begin{vmatrix} a_{11} & a_{12} & a_{13} \\ a_{21} & a_{22} & a_{23} \\ a_{31} & a_{32} & a_{33} \end{vmatrix} = a_{11}\begin{vmatrix} a_{22} & a_{23} \\ a_{32} & a_{33} \end{vmatrix}(-1)^2 + a_{12}\begin{vmatrix} a_{21} & a_{23} \\ a_{31} & a_{33} \end{vmatrix}(-1)^3$$

$$+ a_{13}\begin{vmatrix} a_{21} & a_{22} \\ a_{31} & a_{32} \end{vmatrix}(-1)^4 = a_{11}a_{22}a_{33} + a_{12}a_{23}a_{31} + a_{21}a_{32}a_{13}$$

$$-a_{31}a_{22}a_{13} - a_{21}a_{12}a_{33} - a_{32}a_{23}a_{11}$$

The determinant of any 3×3 matrix can be computed by summing the products of elements along the solid lines and subtracting the products along the dashed lines in the diagram below. This procedure is *not* valid for matrices of higher dimension but, in general, Definition 2A.24 can be employed to evaluate these determinants.

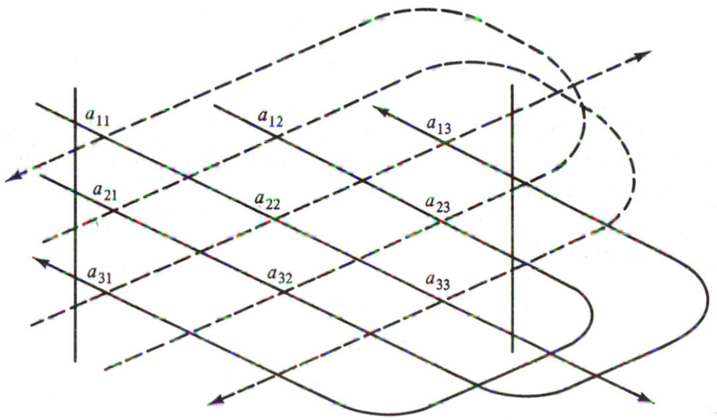

We want to state a result that describes some properties of the determinant. However, we must first introduce some notions related to matrix inverses.

Definition 2A.25. The *row rank* of a matrix is the maximum number of linearly independent rows considered as vectors (that is, row vectors). The *column rank* of a matrix is the rank of its set of columns considered as vectors.

For example, let the matrix **A** be

$$\mathbf{A} = \begin{bmatrix} 1 & 1 & 1 \\ 2 & 5 & -1 \\ 0 & 1 & -1 \end{bmatrix}$$

The rows of **A** written as vectors were shown to be linearly dependent after Definition 2A.6. Note that the column rank of **A** is also 2, since

$$-2\begin{bmatrix} 1 \\ 2 \\ 0 \end{bmatrix} + \begin{bmatrix} 1 \\ 5 \\ 1 \end{bmatrix} + \begin{bmatrix} 1 \\ -1 \\ -1 \end{bmatrix} = \begin{bmatrix} 0 \\ 0 \\ 0 \end{bmatrix}$$

but columns 1 and 2 are linearly independent. This is no coincidence, as the following result indicates.

Result 2A.6. The row rank and the column rank of a matrix are equal. ■

Thus the *rank of a matrix* is either the row rank or the column rank.

Definition 2A.26. A square matrix $\underset{(k\times k)}{\mathbf{A}}$ is *nonsingular* if $\underset{(k\times k)}{\mathbf{A}}\;\underset{(k\times 1)}{\mathbf{x}} = \underset{(k\times 1)}{\mathbf{0}}$ implies $\underset{(k\times 1)}{\mathbf{x}} = \underset{(k\times 1)}{\mathbf{0}}$. If a matrix fails to be nonsingular, it is called *singular*. Equivalently, a *square* matrix is nonsingular if its rank is equal to the number of rows (or columns).

Note that $\mathbf{Ax} = x_1\mathbf{a}_1 + x_2\mathbf{a}_2 + \cdots + x_k\mathbf{a}_k$, where $\mathbf{a}_i$ is the ith column of **A**, so that the condition of nonsingularity is just the statement that the columns of **A** are linearly independent.

Result 2A.7. Let **A** be a nonsingular square matrix of dim $k \times k$. Then there is a unique $k \times k$ matrix **B** such that

$$\mathbf{AB} = \mathbf{BA} = \mathbf{I}$$

where **I** is the $k \times k$ identity matrix. ■

Definition 2A.27. The **B** such that $\mathbf{AB} = \mathbf{BA} = \mathbf{I}$ is called the *inverse* of **A** and is denoted by $\mathbf{A}^{-1}$. In fact, if $\mathbf{BA} = \mathbf{I}$ *or* $\mathbf{AB} = \mathbf{I}$, then $\mathbf{B} = \mathbf{A}^{-1}$ and both products must equal **I**.

For example:

$$\mathbf{A} = \begin{bmatrix} 2 & 3 \\ 1 & 5 \end{bmatrix} \quad \text{has} \quad \mathbf{A}^{-1} = \begin{bmatrix} \frac{5}{7} & -\frac{3}{7} \\ -\frac{1}{7} & \frac{2}{7} \end{bmatrix}$$

since

$$\begin{bmatrix} 2 & 3 \\ 1 & 5 \end{bmatrix} \begin{bmatrix} \frac{5}{7} & -\frac{3}{7} \\ -\frac{1}{7} & \frac{2}{7} \end{bmatrix} = \begin{bmatrix} \frac{5}{7} & -\frac{3}{7} \\ -\frac{1}{7} & \frac{2}{7} \end{bmatrix} \begin{bmatrix} 2 & 3 \\ 1 & 5 \end{bmatrix} = \begin{bmatrix} 1 & 0 \\ 0 & 1 \end{bmatrix}$$

Result 2A.8.

(a) The inverse of the 2×2 matrix

$$\mathbf{A} = \begin{bmatrix} a_{11} & a_{12} \\ a_{21} & a_{22} \end{bmatrix}$$

is given by

$$\mathbf{A}^{-1} = \frac{1}{|\mathbf{A}|} \begin{bmatrix} a_{22} & -a_{12} \\ -a_{21} & a_{11} \end{bmatrix}$$

(b) The inverse of a 3×3 matrix

$$\mathbf{A} = \begin{bmatrix} a_{11} & a_{12} & a_{13} \\ a_{21} & a_{22} & a_{23} \\ a_{31} & a_{32} & a_{33} \end{bmatrix}$$

is given by

$$\mathbf{A}^{-1} = \frac{1}{|\mathbf{A}|} \begin{bmatrix} \begin{vmatrix} a_{22} & a_{23} \\ a_{32} & a_{33} \end{vmatrix} & -\begin{vmatrix} a_{12} & a_{13} \\ a_{32} & a_{33} \end{vmatrix} & \begin{vmatrix} a_{12} & a_{13} \\ a_{22} & a_{23} \end{vmatrix} \\ -\begin{vmatrix} a_{21} & a_{23} \\ a_{31} & a_{33} \end{vmatrix} & \begin{vmatrix} a_{11} & a_{13} \\ a_{31} & a_{33} \end{vmatrix} & -\begin{vmatrix} a_{11} & a_{13} \\ a_{21} & a_{23} \end{vmatrix} \\ \begin{vmatrix} a_{21} & a_{22} \\ a_{31} & a_{32} \end{vmatrix} & -\begin{vmatrix} a_{11} & a_{12} \\ a_{31} & a_{32} \end{vmatrix} & \begin{vmatrix} a_{11} & a_{12} \\ a_{21} & a_{22} \end{vmatrix} \end{bmatrix}$$

In both cases above, it is clear that $|\mathbf{A}| \neq 0$ if the inverse is to exist.

(c) In general, $\mathbf{A}^{-1}$ has j, ith entry $[|\mathbf{A}_{ij}|/|\mathbf{A}|](-1)^{i+j}$, where $\mathbf{A}_{ij}$ is the matrix obtained from $\mathbf{A}$ by deleting the ith row and jth column. ■

Result 2A.9. For a square matrix $\mathbf{A}$ of dim $k \times k$, the following are equivalent:

(a) $\underset{(k\times k)}{\mathbf{A}} \underset{(k\times 1)}{\mathbf{x}} = \underset{(k\times 1)}{\mathbf{0}}$ implies $\underset{(k\times 1)}{\mathbf{x}} = \underset{(k\times 1)}{\mathbf{0}}$ ($\mathbf{A}$ is nonsingular).

(b) $|\mathbf{A}| \neq 0$.

(c) There exists a matrix $\mathbf{A}^{-1}$ such that $\mathbf{A}\mathbf{A}^{-1} = \mathbf{A}^{-1}\mathbf{A} = \underset{(k\times k)}{\mathbf{I}}$. ■

Result 2A.10. Let $\mathbf{A}$ and $\mathbf{B}$ be square matrices of the same dimension and let the indicated inverses exist. Then the following hold:

(a) $(\mathbf{A}^{-1})' = (\mathbf{A}')^{-1}$

(b) $(\mathbf{AB})^{-1} = \mathbf{B}^{-1}\mathbf{A}^{-1}$ ■

The determinant has the following properties.

Result 2A.11. Let $\mathbf{A}$ and $\mathbf{B}$ be $k \times k$ square matrices.

(a) $|\mathbf{A}| = |\mathbf{A}'|$
(b) If each element of a row (column) of $\mathbf{A}$ is zero, then $|\mathbf{A}| = 0$.
(c) If any two rows (columns) of $\mathbf{A}$ are identical, then $|\mathbf{A}| = 0$.
(d) If $\mathbf{A}$ is nonsingular, then $|\mathbf{A}| = 1/|\mathbf{A}^{-1}|$; that is $|\mathbf{A}||\mathbf{A}^{-1}| = 1$.
(e) $|\mathbf{AB}| = |\mathbf{A}||\mathbf{B}|$
(f) $|c\mathbf{A}| = c^k |\mathbf{A}|$, where c is a scalar. ■

You are referred to [5] for proofs of parts of Results 2A.9 and 2A.11. Some of these proofs are rather complex and beyond the scope of this book.

Definition 2A.28. Let $\mathbf{A} = \{a_{ij}\}$ be a $k \times k$ square matrix. The *trace* of the matrix $\mathbf{A}$, written tr($\mathbf{A}$), is the sum of the diagonal elements; that is, $\text{tr}(\mathbf{A}) = \sum_{i=1}^{k} a_{ii}$.

Result 2A.12. Let $\mathbf{A}$ and $\mathbf{B}$ be $k \times k$ matrices and c be a scalar.

(a) $\text{tr}(c\mathbf{A}) = c\,\text{tr}(\mathbf{A})$
(b) $\text{tr}(\mathbf{A} \pm \mathbf{B}) = \text{tr}(\mathbf{A}) \pm \text{tr}(\mathbf{B})$
(c) $\text{tr}(\mathbf{AB}) = \text{tr}(\mathbf{BA})$
(d) $\text{tr}(\mathbf{B}^{-1}\mathbf{AB}) = \text{tr}(\mathbf{A})$
(e) $\text{tr}(\mathbf{AA}') = \sum_{i=1}^{k} \sum_{j=1}^{k} a_{ij}^2$ ■

Definition 2A.29. A square matrix $\mathbf{A}$ is said to be *orthogonal* if its rows, considered as vectors, are mutually perpendicular and have unit lengths; that is, $\mathbf{AA}' = \mathbf{I}$.

Result 2A.13. A matrix $\mathbf{A}$ is orthogonal if and only if $\mathbf{A}^{-1} = \mathbf{A}'$. For an orthogonal matrix, $\mathbf{AA}' = \mathbf{A}'\mathbf{A} = \mathbf{I}$, so the columns are also mutually perpendicular and have unit lengths. ■

An example of an orthogonal matrix is

$$\mathbf{A} = \begin{bmatrix} -\frac{1}{2} & \frac{1}{2} & \frac{1}{2} & \frac{1}{2} \\ \frac{1}{2} & -\frac{1}{2} & \frac{1}{2} & \frac{1}{2} \\ \frac{1}{2} & \frac{1}{2} & -\frac{1}{2} & \frac{1}{2} \\ \frac{1}{2} & \frac{1}{2} & \frac{1}{2} & -\frac{1}{2} \end{bmatrix}$$

Clearly $\mathbf{A} = \mathbf{A}'$, so $\mathbf{AA}' = \mathbf{A}'\mathbf{A} = \mathbf{AA}$. We verify that $\mathbf{AA} = \mathbf{I} = \mathbf{AA}' = \mathbf{A}'\mathbf{A}$, or

$$\underset{\mathbf{A}}{\begin{bmatrix} -\frac{1}{2} & \frac{1}{2} & \frac{1}{2} & \frac{1}{2} \\ \frac{1}{2} & -\frac{1}{2} & \frac{1}{2} & \frac{1}{2} \\ \frac{1}{2} & \frac{1}{2} & -\frac{1}{2} & \frac{1}{2} \\ \frac{1}{2} & \frac{1}{2} & \frac{1}{2} & -\frac{1}{2} \end{bmatrix}} \underset{\mathbf{A}}{\begin{bmatrix} -\frac{1}{2} & \frac{1}{2} & \frac{1}{2} & \frac{1}{2} \\ \frac{1}{2} & -\frac{1}{2} & \frac{1}{2} & \frac{1}{2} \\ \frac{1}{2} & \frac{1}{2} & -\frac{1}{2} & \frac{1}{2} \\ \frac{1}{2} & \frac{1}{2} & \frac{1}{2} & -\frac{1}{2} \end{bmatrix}} = \underset{\mathbf{I}}{\begin{bmatrix} 1 & 0 & 0 & 0 \\ 0 & 1 & 0 & 0 \\ 0 & 0 & 1 & 0 \\ 0 & 0 & 0 & 1 \end{bmatrix}}$$

so $\mathbf{A}' = \mathbf{A}^{-1}$ and $\mathbf{A}$ must be an orthogonal matrix.

Square matrices are best understood in terms of quantities called eigenvalues and eigenvectors.

Definition 2A.30. Let $\mathbf{A}$ be a $k \times k$ square matrix and $\mathbf{I}$ be the $k \times k$ identity matrix. Then the scalars $\lambda_1, \lambda_2, \ldots, \lambda_k$ satisfying the polynomial equation $|\mathbf{A} - \lambda\mathbf{I}| = 0$ are called the *eigenvalues* (or *characteristic roots*) of a matrix $\mathbf{A}$. The equation $|\mathbf{A} - \lambda\mathbf{I}| = 0$ (as a function of λ) is called the *characteristic equation*.

For example, let

$$\mathbf{A} = \begin{bmatrix} 1 & 0 \\ 1 & 3 \end{bmatrix}$$

Then

$$|\mathbf{A} - \lambda\mathbf{I}| = \left| \begin{bmatrix} 1 & 0 \\ 1 & 3 \end{bmatrix} - \lambda \begin{bmatrix} 1 & 0 \\ 0 & 1 \end{bmatrix} \right| = \begin{vmatrix} 1-\lambda & 0 \\ 1 & 3-\lambda \end{vmatrix}$$
$$= (1-\lambda)(3-\lambda) = 0$$

implies there are two roots, $\lambda_1 = 1$ and $\lambda_2 = 3$. The eigenvalues of $\mathbf{A}$ are 3 and 1. Let

$$\mathbf{A} = \begin{bmatrix} 13 & -4 & 2 \\ -4 & 13 & -2 \\ 2 & -2 & 10 \end{bmatrix}$$

Then the equation

$$|\mathbf{A} - \lambda\mathbf{I}| = \begin{vmatrix} 13-\lambda & -4 & 2 \\ -4 & 13-\lambda & -2 \\ 2 & -2 & 10-\lambda \end{vmatrix} = -\lambda^3 + 36\lambda^2 - 405\lambda + 1458 = 0$$

has three roots: $\lambda_1 = 9$, $\lambda_2 = 9$, and $\lambda_3 = 18$; that is, 9, 9, and 18 are the eigenvalues of $\mathbf{A}$.

Definition 2A.31. Let $\mathbf{A}$ be a matrix of dim $k \times k$ and let λ be an eigenvalue of $\mathbf{A}$. If $\underset{(k\times 1)}{\mathbf{x}}$ is a *nonzero* vector $\left(\underset{(k\times 1)}{\mathbf{x}} \neq \underset{(k\times 1)}{\mathbf{0}}\right)$ such that

$$\mathbf{Ax} = \lambda\mathbf{x}$$

then $\mathbf{x}$ is said to be an *eigenvector* (*characteristic vector*) of the matrix $\mathbf{A}$ associated with the *eigenvalue* λ.

An equivalent condition for λ to be a solution of the eigenvalue-eigenvector equation is $|\mathbf{A} - \lambda\mathbf{I}| = 0$. This follows since the statement that $\mathbf{Ax} = \lambda\mathbf{x}$ for some λ and $\mathbf{x} \neq 0$ implies

$$\mathbf{0} = (\mathbf{A} - \lambda\mathbf{I})\mathbf{x} = x_1\text{col}_1(\mathbf{A} - \lambda\mathbf{I}) + \cdots + x_k\text{col}_k(\mathbf{A} - \lambda\mathbf{I})$$

That is, the columns of $\mathbf{A} - \lambda\mathbf{I}$ are linearly dependent so, by Result 2A.9(b), $|\mathbf{A} - \lambda\mathbf{I}| = 0$ as asserted. Following Definition 2A.30, we have shown that the eigenvalues of

$$\mathbf{A} = \begin{bmatrix} 1 & 0 \\ 1 & 3 \end{bmatrix}$$

are $\lambda_1 = 1$ and $\lambda_2 = 3$. The eigenvectors associated with these eigenvalues can be determined by solving the following equations:

$$\begin{bmatrix} 1 & 0 \\ 1 & 3 \end{bmatrix}\begin{bmatrix} x_1 \\ x_2 \end{bmatrix} = 1\begin{bmatrix} x_1 \\ x_2 \end{bmatrix}$$
$$\mathbf{Ax} = \lambda_1\mathbf{x}$$

$$\begin{bmatrix} 1 & 0 \\ 1 & 3 \end{bmatrix}\begin{bmatrix} x_1 \\ x_2 \end{bmatrix} = 3\begin{bmatrix} x_1 \\ x_2 \end{bmatrix}$$
$$\mathbf{Ax} = \lambda_2\mathbf{x}$$

From the first expression,

$$x_1 = x_1$$
$$x_1 + 3x_2 = x_2$$

or

$$x_1 = -2x_2$$

There are many solutions for x_1 and x_2.

Setting $x_2 = 1$ (arbitrarily) gives $x_1 = -2$ and hence

$$\mathbf{x} = \begin{bmatrix} -2 \\ 1 \end{bmatrix}$$

is an eigenvector corresponding to the eigenvalue 1. From the second expression,

$$x_1 = 3x_1$$
$$x_1 + 3x_2 = 3x_2$$

implies $x_1 = 0$ and $x_2 = 1$ (arbitrarily), and hence

$$\mathbf{x} = \begin{bmatrix} 0 \\ 1 \end{bmatrix}$$

is an eigenvector corresponding to the eigenvalue 3. It is usual practice to determine an eigenvector so that it has length one. That is, if $\mathbf{Ax} = \lambda\mathbf{x}$, we take $\mathbf{e} = \mathbf{x}/\sqrt{\mathbf{x}'\mathbf{x}}$ as the eigenvector corresponding to λ. For example, $[-2/\sqrt{5}, 1/\sqrt{5}]'$ is the eigenvector for $\lambda = 1$.

Definition 2A.32. A *quadratic form*, $Q(\mathbf{x})$, in the k variables $x_1, x_2, \ldots, x_k$ is $Q(\mathbf{x}) = \mathbf{x}'\mathbf{Ax}$ where $\mathbf{x}' = [x_1, x_2, \ldots, x_k]$ and $\mathbf{A}$ is a $k \times k$ symmetric matrix.

Note that a quadratic form can be written as $Q(\mathbf{x}) = \sum_{i=1}^{k}\sum_{j=1}^{k} a_{ij}x_ix_j$. For example:

$$Q(\mathbf{x}) = [x_1 \quad x_2]\begin{bmatrix} 1 & 1 \\ 1 & 1 \end{bmatrix}\begin{bmatrix} x_1 \\ x_2 \end{bmatrix} = x_1^2 + 2x_1x_2 + x_2^2$$

$$Q(\mathbf{x}) = [x_1 \quad x_2 \quad x_3]\begin{bmatrix} 1 & 3 & 0 \\ 3 & -1 & -2 \\ 0 & -2 & 2 \end{bmatrix}\begin{bmatrix} x_1 \\ x_2 \\ x_3 \end{bmatrix} = x_1^2 + 6x_1x_2 - x_2^2 - 4x_2x_3 + 2x_3^2$$

EXERCISES

2.1. Let $\mathbf{x}' = [5, 1, 3]$ and $\mathbf{y}' = [-1, 3, 1]$.

(a) Graph the two vectors.

(b) Find (i) the length of $\mathbf{x}$, (ii) the angle between $\mathbf{x}$ and $\mathbf{y}$, and (iii) the projection of $\mathbf{y}$ on $\mathbf{x}$.

(c) Since $\bar{x} = 3$ and $\bar{y} = 1$, graph $[5 - 3, 1 - 3, 3 - 3] = [2, -2, 0]$ and $[-1 - 1, 3 - 1, 1 - 1] = [-2, 2, 0]$.

2.2. Given the matrices

$$\mathbf{A} = \begin{bmatrix} -1 & 3 \\ 4 & 2 \end{bmatrix}, \quad \mathbf{B} = \begin{bmatrix} 4 & -3 \\ 1 & -2 \\ -2 & 0 \end{bmatrix} \quad \text{and} \quad \mathbf{C} = \begin{bmatrix} 5 \\ -4 \\ 2 \end{bmatrix}$$

perform the indicated multiplications.

(a) $5\mathbf{A}$

(b) $\mathbf{BA}$

(c) $\mathbf{A}'\mathbf{B}'$

(d) $\mathbf{C}'\mathbf{B}$

(e) Is $\mathbf{AB}$ defined?

2.3. Verify the following properties of transpose when

$$\mathbf{A} = \begin{bmatrix} 2 & 1 \\ 1 & 3 \end{bmatrix}, \quad \mathbf{B} = \begin{bmatrix} 1 & 4 & 2 \\ 5 & 0 & 3 \end{bmatrix} \text{ and } \mathbf{C} = \begin{bmatrix} 1 & 4 \\ 3 & 2 \end{bmatrix}$$

(a) $(\mathbf{A}')' = \mathbf{A}$

(b) $(\mathbf{C}')^{-1} = (\mathbf{C}^{-1})'$

(c) $(\mathbf{AB})' = \mathbf{B}'\mathbf{A}'$

(d) Prove (c) for general $\underset{(m\times k)}{\mathbf{A}}$ and $\underset{(k\times \ell)}{\mathbf{B}}$.

2.4. When $\mathbf{A}^{-1}$ and $\mathbf{B}^{-1}$ exist, prove each of the following.

(a) $(\mathbf{A}')^{-1} = (\mathbf{A}^{-1})'$

(b) $(\mathbf{AB})^{-1} = \mathbf{B}^{-1}\mathbf{A}^{-1}$

(*Hint:* Part a can be proved by noting that $\mathbf{AA}^{-1} = \mathbf{I}$, $\mathbf{I} = \mathbf{I}'$, and $(\mathbf{AA}^{-1})' = (\mathbf{A}^{-1})'\mathbf{A}'$. Part b follows from $(\mathbf{B}^{-1}\mathbf{A}^{-1})\mathbf{AB} = \mathbf{B}^{-1}(\mathbf{A}^{-1}\mathbf{A})\mathbf{B} = \mathbf{B}^{-1}\mathbf{B} = \mathbf{I}$.)

2.5. Check that

$$\mathbf{Q} = \begin{bmatrix} \frac{5}{13} & \frac{12}{13} \\ -\frac{12}{13} & \frac{5}{13} \end{bmatrix}$$

is an orthogonal matrix.

2.6. Let

$$\mathbf{A} = \begin{bmatrix} 9 & -2 \\ -2 & 6 \end{bmatrix}$$

(a) Is $\mathbf{A}$ symmetric?
(b) Show that $\mathbf{A}$ is positive definite.

2.7. Let $\mathbf{A}$ be as given in Exercise 2.6.
(a) Determine the eigenvalues and eigenvectors of $\mathbf{A}$.
(b) Write the spectral decomposition of $\mathbf{A}$.
(c) Find $\mathbf{A}^{-1}$.
(d) Find the eigenvalues and eigenvectors of $\mathbf{A}^{-1}$.

2.8. Given the matrix

$$\mathbf{A} = \begin{bmatrix} 1 & 2 \\ 2 & -2 \end{bmatrix}$$

find the eigenvalues λ_1 and λ_2 and the associated normalized eigenvectors $\mathbf{e}_1$ and $\mathbf{e}_2$. Determine the spectral decomposition (2-16) of $\mathbf{A}$.

2.9. Let $\mathbf{A}$ be as in Exercise 2.8.
(a) Find $\mathbf{A}^{-1}$.
(b) Compute the eigenvalues and eigenvectors of $\mathbf{A}^{-1}$.
(c) Write the spectral decomposition of $\mathbf{A}^{-1}$ and compare with that of $\mathbf{A}$ from Exercise 2.8.

2.10. Consider the matrices

$$\mathbf{A} = \begin{bmatrix} 4 & 4.001 \\ 4.001 & 4.002 \end{bmatrix} \quad \text{and} \quad \mathbf{B} = \begin{bmatrix} 4 & 4.001 \\ 4.001 & 4.002001 \end{bmatrix}$$

The matrices are identical except for a small difference in the $(2, 2)$ position. Moreover, the columns of $\mathbf{A}$ (and $\mathbf{B}$) are nearly linearly dependent. Show that $\mathbf{A}^{-1} \doteq (-3)\mathbf{B}^{-1}$. Consequently, small changes—perhaps caused by rounding—can give substantially different inverses.

2.11. Show that the determinant of the $p \times p$ diagonal matrix $\mathbf{A} = \{a_{ij}\}$ with $a_{ij} = 0$, $i \neq j$, is given by the product of the diagonal elements; that is, $|\mathbf{A}| = a_{11}a_{22}\cdots a_{pp}$.
(*Hint:* By Definition 2A.24, $|\mathbf{A}| = a_{11}\mathbf{A}_{11} + 0 + \cdots + 0$. Repeat for the submatrix $\mathbf{A}_{11}$ obtained by deleting the first row and first column of $\mathbf{A}$.)

2.12. Show that the determinant of a square $p \times p$ matrix $\mathbf{A}$ can be expressed as the product of its eigenvalues $\lambda_1, \lambda_2, \ldots, \lambda_p$; that is, $|\mathbf{A}| = \prod_{i=1}^{p} \lambda_i$.
(*Hint:* Using (2-20), $\mathbf{A} = \mathbf{P\Lambda P'}$ with $\mathbf{P'P} = \mathbf{I}$. From Result 2A.11(e), $|\mathbf{A}| = |\mathbf{P\Lambda P'}| = |\mathbf{P}||\mathbf{\Lambda P'}| = |\mathbf{P}||\mathbf{\Lambda}||\mathbf{P'}| = |\mathbf{\Lambda}||\mathbf{I}|$ since $|\mathbf{I}| = |\mathbf{P'P}| = |\mathbf{P'}||\mathbf{P}|$. Apply Exercise 2.11.)

2.13. Show that $|\mathbf{Q}| = +1$ or -1 if $\mathbf{Q}$ is a $p \times p$ orthogonal matrix.
(*Hint:* $|\mathbf{QQ'}| = |\mathbf{I}|$. Also, from Result 2A.11, $|\mathbf{QQ'}| = |\mathbf{Q}||\mathbf{Q'}| = |\mathbf{Q}|^2$. Thus $|\mathbf{Q}|^2 = |\mathbf{I}|$. Now use Exercise 2.11.)

2.14. Show that $\underset{(p\times p)}{\mathbf{Q}'}\ \underset{(p\times p)}{\mathbf{A}}\ \underset{(p\times p)}{\mathbf{Q}}$ and $\underset{(p\times p)}{\mathbf{A}}$ have the same eigenvalues if $\mathbf{Q}$ is orthogonal.
(*Hint:* Let λ be an eigenvalue of $\mathbf{A}$. Then $0 = |\mathbf{A} - \lambda\mathbf{I}|$. By Exercise 2.13 and Result 2A.11(e), we can write $0 = |\mathbf{Q'}||\mathbf{A} - \lambda\mathbf{I}||\mathbf{Q}| = |\mathbf{Q'AQ} - \lambda\mathbf{I}|$, since $\mathbf{Q'Q} = \mathbf{I}$.)

2.15. A quadratic form $\mathbf{x'Ax}$ is said to be positive definite if the matrix $\mathbf{A}$ is positive definite. Is the quadratic form $3x_1^2 + 3x_2^2 - 2x_1x_2$ positive definite?

2.16. Consider an arbitrary $p \times n$ matrix $\mathbf{A}$. Then $\mathbf{A'A}$ is a symmetric $n \times n$ matrix. Show that $\mathbf{A'A}$ is necessarily nonnegative definite.
(*Hint:* Set $\mathbf{y} = \mathbf{Ax}$ so $\mathbf{y'y} = \mathbf{x'A'Ax}$.)

2.17. Prove that every eigenvalue of a $k \times k$ positive definite matrix $\mathbf{A}$ is positive. (*Hint:* Consider the definition of an eigenvalue where $\mathbf{Ae} = \lambda\mathbf{e}$. Multiply on the left by $\mathbf{e}'$ so $\mathbf{e}'\mathbf{Ae} = \lambda\mathbf{e}'\mathbf{e}$.)

2.18. Consider the sets of points (x_1, x_2) whose "distances" from the origin are given by

$$c^2 = 4x_1^2 + 3x_2^2 - 2\sqrt{2}\,x_1x_2$$

for $c^2 = 1$ and for $c^2 = 4$. Determine the major and minor axes of the ellipses of constant distances and their associated lengths. Sketch the ellipses of constant distances and comment on their positions. What will happen as c^2 increases?

2.19. Let $\underset{(m\times m)}{\mathbf{A}^{1/2}} = \sum_{i=1}^{m} \sqrt{\lambda_i}\,\mathbf{e}_i\mathbf{e}_i' = \mathbf{P}\mathbf{\Lambda}^{1/2}\mathbf{P}'$, where $\mathbf{PP}' = \mathbf{P}'\mathbf{P} = \mathbf{I}$. (The λ_i's and the $\mathbf{e}_i$'s are the eigenvalues and associated normalized eigenvectors of the matrix $\mathbf{A}$.) Show Properties (1)–(4) of the square root matrix in (2-22).

2.20. Determine the square root matrix, $\mathbf{A}^{1/2}$, using the matrix $\mathbf{A}$ in Exercise 2.3. Also determine $\mathbf{A}^{-1/2}$ and show $\mathbf{A}^{1/2}\mathbf{A}^{-1/2} = \mathbf{A}^{-1/2}\mathbf{A}^{1/2} = \mathbf{I}$.

2.21. Verify the relationships $\mathbf{V}^{1/2}\boldsymbol{\rho}\mathbf{V}^{1/2} = \mathbf{\Sigma}$ and $\boldsymbol{\rho} = (\mathbf{V}^{1/2})^{-1}\mathbf{\Sigma}(\mathbf{V}^{1/2})^{-1}$, where $\mathbf{\Sigma}$ is the $p \times p$ population covariance matrix [Equation (2-32)], $\boldsymbol{\rho}$ is the $p \times p$ population correlation matrix [Equation (2-34)], and $\mathbf{V}^{1/2}$ is the population standard deviation matrix [Equation (2-35)].

2.22. Let $\mathbf{X}$ have covariance matrix

$$\mathbf{\Sigma} = \begin{bmatrix} 4 & 0 & 0 \\ 0 & 9 & 0 \\ 0 & 0 & 1 \end{bmatrix}$$

Find

(a) $\mathbf{\Sigma}^{-1}$

(b) The eigenvalues and eigenvectors of $\mathbf{\Sigma}$.

(c) The eigenvalues and eigenvectors of $\mathbf{\Sigma}^{-1}$.

2.23. Let $\mathbf{X}$ have covariance matrix

$$\mathbf{\Sigma} = \begin{bmatrix} 25 & -2 & 4 \\ -2 & 4 & 1 \\ 4 & 1 & 9 \end{bmatrix}$$

(a) Determine $\boldsymbol{\rho}$ and $\mathbf{V}^{1/2}$.

(b) Multiply your matrices to check the relation $\mathbf{V}^{1/2}\boldsymbol{\rho}\mathbf{V}^{1/2} = \mathbf{\Sigma}$.

2.24. Use $\mathbf{\Sigma}$ as given in Exercise 2.23.

(a) Find ρ_{13}.

(b) Find the correlation between X_1 and $\frac{1}{2}X_2 + \frac{1}{2}X_3$.

2.25. Derive expressions for the mean and variances of the following linear combinations in terms of the means and covariances of the random variables X_1, X_2, and X_3.

(a) $X_1 - 2X_2$

(b) $-X_1 + 3X_2$

(c) $X_1 + X_2 + X_3$

(d) $X_1 + 2X_2 - X_3$

(e) $3X_1 - 4X_2$, if X_1 and X_2 are independent random variables.

2.26. Show that

$$\text{Cov}(c_{11}X_1 + c_{12}X_2 + \cdots + c_{1p}X_p, c_{21}X_1 + c_{22}X_2 + \cdots + c_{2p}X_p) = \mathbf{c}_1'\mathbf{\Sigma}_{\mathbf{X}}\mathbf{c}_2$$

where $\mathbf{c}_1' = [c_{11}, c_{12}, \ldots, c_{1p}]$ and $\mathbf{c}_2' = [c_{21}, c_{22}, \ldots, c_{2p}]$. This verifies the off-diagonal elements of $\mathbf{C\Sigma_X C}'$ in (2-45) or diagonal elements if $\mathbf{c}_1 = \mathbf{c}_2$. (See next page.)

($Hint$: By (2-43), $Z_1 - E(Z_1) = c_{11}(X_1 - \mu_1) + \cdots + c_{1p}(X_p - \mu_p)$ and $Z_2 - E(Z_2) = c_{21}(X_1 - \mu_1) + \cdots + c_{2p}(X_p - \mu_p)$, so $\text{Cov}(Z_1, Z_2) = E[(Z_1 - E(Z_1))(Z_2 - E(Z_2))] = E[(c_{11}(X_1 - \mu_1) + \cdots + c_{1p}(X_p - \mu_p))(c_{21}(X_1 - \mu_1) + c_{22}(X_2 - \mu_2) + \cdots + c_{2p}(X_p - \mu_p))]$. The product $(c_{11}(X_1 - \mu_1) + c_{12}(X_2 - \mu_2) + \cdots + c_{1p}(X_p - \mu_p))(c_{21}(X_1 - \mu_1) + c_{22}(X_2 - \mu_2) + \cdots + c_{2p}(X_p - \mu_p)) = \left(\sum_{\ell=1}^{p} c_{1\ell}(X_\ell - \mu_\ell)\right)\left(\sum_{m=1}^{p} c_{2m}(X_m - \mu_m)\right) = \sum_{\ell=1}^{p}\sum_{m=1}^{p} c_{1\ell}c_{2m}(X_\ell - \mu_\ell)(X_m - \mu_m)$ has expected value $\sum_{\ell=1}^{p}\sum_{m=1}^{p} c_{1\ell}c_{2m}\sigma_{\ell m} = [c_{11}, \ldots, c_{1p}]\boldsymbol{\Sigma}[c_{21}, \ldots, c_{2p}]'$. Verify the last step by the definition of matrix multiplication. The same steps hold for all elements.)

2.27. Consider the arbitrary random vector $\mathbf{X} = [x_1, x_2, x_3, x_4, x_5]'$ with mean vector $\boldsymbol{\mu} = [\mu_1, \mu_2, \mu_3, \mu_4, \mu_5]'$. Partition $\mathbf{X}$ into

$$\mathbf{X} = \begin{bmatrix} \mathbf{X}^{(1)} \\ \hline \mathbf{X}^{(2)} \end{bmatrix}$$

where

$$\mathbf{X}^{(1)} = \begin{bmatrix} X_1 \\ X_2 \end{bmatrix} \quad \text{and} \quad \mathbf{X}^{(2)} = \begin{bmatrix} X_3 \\ X_4 \\ X_5 \end{bmatrix}$$

Let $\boldsymbol{\Sigma}$ be the covariance matrix of $\mathbf{X}$ with general element σ_{ik}. Partition $\boldsymbol{\Sigma}$ into the covariance matrices of $\mathbf{X}^{(1)}$ and $\mathbf{X}^{(2)}$ and the covariance matrix of an element of $\mathbf{X}^{(1)}$ and an element of $\mathbf{X}^{(2)}$.

2.28. You are given the random vector $\mathbf{X}' = [X_1, X_2, \ldots, X_5]$ with mean vector $\boldsymbol{\mu}'_{\mathbf{X}} = [2, 4, -1, 3, 0]$ and variance-covariance matrix

$$\boldsymbol{\Sigma}_{\mathbf{X}} = \begin{bmatrix} 4 & -1 & \frac{1}{2} & -\frac{1}{2} & 0 \\ -1 & 3 & 1 & -1 & 0 \\ \frac{1}{2} & 1 & 6 & 1 & -1 \\ -\frac{1}{2} & -1 & 1 & 4 & 0 \\ 0 & 0 & -1 & 0 & 2 \end{bmatrix}$$

Partition $\mathbf{X}$ as

$$\mathbf{X} = \begin{bmatrix} X_1 \\ X_2 \\ \hline X_3 \\ X_4 \\ X_5 \end{bmatrix} = \begin{bmatrix} \mathbf{X}^{(1)} \\ \hline \mathbf{X}^{(2)} \end{bmatrix}$$

Let

$$\mathbf{A} = \begin{bmatrix} 1 & -1 \\ 1 & 1 \end{bmatrix} \quad \text{and} \quad \mathbf{B} = \begin{bmatrix} 1 & 1 & 1 \\ 1 & 1 & -2 \end{bmatrix}$$

and consider the linear combinations $\mathbf{AX}^{(1)}$ and $\mathbf{BX}^{(2)}$. Find

(a) $E(\mathbf{X}^{(1)})$

(b) $E(\mathbf{AX}^{(1)})$

(c) $\text{Cov}(\mathbf{X}^{(1)})$
(d) $\text{Cov}(\mathbf{A}\mathbf{X}^{(1)})$
(e) $E(\mathbf{X}^{(2)})$
(f) $E(\mathbf{B}\mathbf{X}^{(2)})$
(g) $\text{Cov}(\mathbf{X}^{(2)})$
(h) $\text{Cov}(\mathbf{B}\mathbf{X}^{(2)})$
(i) $\text{Cov}(\mathbf{X}^{(1)}, \mathbf{X}^{(2)})$
(j) $\text{Cov}(\mathbf{A}\mathbf{X}^{(1)}, \mathbf{B}\mathbf{X}^{(2)})$

2.29. Consider the vectors $\mathbf{b}' = [2, -1, 4, 0]$ and $\mathbf{d}' = [-1, 3, -2, 1]$. Verify the Cauchy–Schwarz inequality $(\mathbf{b}'\mathbf{d})^2 \leq (\mathbf{b}'\mathbf{b})(\mathbf{d}'\mathbf{d})$.

2.30. Using the vectors $\mathbf{b}' = [-4, 3]$ and $\mathbf{d}' = [1, 1]$, verify the extended Cauchy–Schwarz inequality $(\mathbf{b}'\mathbf{d})^2 \leq (\mathbf{b}'\mathbf{B}\mathbf{b})(\mathbf{d}'\mathbf{B}^{-1}\mathbf{d})$ if

$$\mathbf{B} = \begin{bmatrix} 2 & -2 \\ -2 & 5 \end{bmatrix}$$

2.31. Find the maximum and minimum value of the quadratic form $4x_1^2 + 4x_2^2 + 6x_1x_2$ for all points $\mathbf{x}' = [x_1, x_2]$ such that $\mathbf{x}'\mathbf{x} = 1$.

2.32. With $\mathbf{A}$ as given in Exercise 2.6, find the maximum value of $\mathbf{x}'\mathbf{A}\mathbf{x}$ for $\mathbf{x}'\mathbf{x} = 1$.

2.33. Find the maximum and minimum values of the ratio $\mathbf{x}'\mathbf{A}\mathbf{x}/\mathbf{x}'\mathbf{x}$ for any nonzero vectors $\mathbf{x}' = [x_1, x_2, x_3]$ if

$$\mathbf{A} = \begin{bmatrix} 13 & -4 & 2 \\ -4 & 13 & -2 \\ 2 & -2 & 10 \end{bmatrix}$$

2.34. Show that

$$\underset{(r\times s)}{\mathbf{A}} \; \underset{(s\times t)}{\mathbf{B}} \; \underset{(t\times v)}{\mathbf{C}} \text{ has } (i, j) \text{ entry } \sum_{\ell=1}^{s} \sum_{k=1}^{t} a_{i\ell} b_{\ell k} c_{kj}$$

(*Hint:* $\mathbf{BC}$ has (ℓ, j) entry $\sum_{k=1}^{t} b_{\ell k} c_{kj} = d_{\ell j}$. Then $\mathbf{A}(\mathbf{BC})$ has (i, j) element $a_{i1}d_{1j} + a_{i2}d_{2j} + \cdots + a_{is}d_{sj} = \sum_{\ell=1}^{s} a_{i\ell}\left(\sum_{k=1}^{t} b_{\ell k}c_{kj}\right) = \sum_{\ell=1}^{s}\sum_{k=1}^{t} a_{i\ell}b_{\ell k}c_{kj}$.)

2.35. Verify (2-24) that $E(\mathbf{X} + \mathbf{Y}) = E(\mathbf{X}) + E(\mathbf{Y})$ and $E(\mathbf{AXB}) = \mathbf{A}E(\mathbf{X})\mathbf{B}$.
(*Hint:* $\mathbf{X} + \mathbf{Y}$ has $X_{ij} + Y_{ij}$ as its (i, j) element. Now $E(X_{ij} + Y_{ij}) = E(X_{ij}) + E(Y_{ij})$, by a univariate property of expectation, and this last quantity is the (i, j) element of $E(\mathbf{X}) + E(\mathbf{Y})$. Next (see Exercise 2.34) $\mathbf{AXB}$ has (i, j) entry $\sum_{\ell}\sum_{k} a_{i\ell}X_{\ell k}b_{kj}$ and, by the additive property of expectation,

$$E\left(\sum_{\ell}\sum_{k} a_{i\ell}X_{\ell k}b_{kj}\right) = \sum_{\ell}\sum_{k} a_{i\ell}E(X_{\ell k})b_{kj}$$

which is the (i, j) element of $\mathbf{A}E(\mathbf{X})\mathbf{B}$.)

REFERENCES

[1] Bellman, R., *Introduction to Matrix Analysis* (2nd ed.), New York: McGraw-Hill, 1970.

[2] Bhattacharyya, G. K., and R. A. Johnson, *Statistical Concepts and Methods*, New York: John Wiley, 1977.

[3] Graybill, F. A., *Introduction to Matrices with Applications in Statistics*, Belmont, Calif.: Wadsworth, 1969.

[4] Halmos, P. R., *Finite Dimensional Vector Spaces* (2nd ed.), Princeton, N.J.: D. Van Nostrand, 1958.

[5] Noble, B., and J. W. Daniel, *Applied Linear Algebra* (2nd ed.), Englewood Cliffs, N.J.: Prentice-Hall, Inc., 1977.

3

Sample Geometry and Random Sampling

3.1 INTRODUCTION

With the vector concepts introduced in the previous chapter, we can now delve deeper into the geometrical interpretations of the descriptive statistics $\bar{\mathbf{x}}$, $\mathbf{S}_n$ and $\mathbf{R}$; we do so in Section 3.2. Many of our explanations use the representation of the rows of $\mathbf{X}$ as p points in n dimensions. In Section 3.3 we introduce the assumption that the observations constitute a random sample. Simply stated, random sampling implies (1) measurements taken on different items (or trials) are unrelated to one another and (2) the joint distribution of all p variables remains the same for all items. Ultimately, it is this structure of the random sample that justifies a particular choice of distance and dictates the geometry for the n-dimensional representation of the data. Furthermore, when data can be treated as a random sample, statistical inferences are based on a solid foundation.

Returning to geometric interpretations in Section 3.4, we introduce a single number, called *generalized variance*, to describe variability. This generalization of variance is an integral part of the comparison of multivariate means. In later sections we use matrix algebra to provide concise expressions for the matrix products and sums that allow us to calculate $\bar{\mathbf{x}}$ and $\mathbf{S}_n$ directly from the data matrix $\mathbf{X}$. The connection between $\bar{\mathbf{x}}$, $\mathbf{S}_n$, and the means and covariances for linear combinations of variables is also clearly delineated, using the notion of matrix products.

3.2 THE GEOMETRY OF THE SAMPLE

A single multivariate observation is the collection of measurements on p different variables taken on the same item or trial. As in Chapter 1, if n observations have been obtained, the entire data set can be placed in a $p \times n$ array (matrix) $\mathbf{X}$ as

$$\underset{(p \times n)}{\mathbf{X}} = \begin{bmatrix} x_{11} & x_{12} & \cdots & x_{1n} \\ x_{21} & x_{22} & \cdots & x_{2n} \\ \vdots & \vdots & & \vdots \\ x_{p1} & x_{p2} & \cdots & x_{pn} \end{bmatrix}$$

Each column of $\mathbf{X}$ represents a multivariate observation. Since the entire set of measurements is often one particular realization of what might have been observed, we say the data are a *sample* of size n from a p-variate "population." The sample then consists of n measurements, each of which has p components.

As we have seen, the data can be plotted in two different ways. For the p-dimensional scatterplot, the *columns* of $\mathbf{X}$ represent n points in p-dimensional space. We can write

$$\underset{(p \times n)}{\mathbf{X}} = \begin{bmatrix} x_{11} & x_{12} & \cdots & x_{1n} \\ x_{21} & x_{22} & \cdots & x_{2n} \\ \vdots & \vdots & & \vdots \\ x_{p1} & x_{p2} & \cdots & x_{pn} \end{bmatrix} = [\underset{\uparrow}{\mathbf{x}_1}, \mathbf{x}_2, \ldots, \underset{\uparrow}{\mathbf{x}_n}] \tag{3-1}$$

1st (multivariate) observation; nth (multivariate) observation

The column vector $\mathbf{x}_j$, representing the jth observation, contains the coordinates of a point.

The scatter of n points in p-dimensional space provides information on their locations and variability. If the points are regarded as solid spheres, the sample mean vector, $\bar{\mathbf{x}}$, given by (1-8) is the center of balance. Variability occurs in more than one direction and it is quantified by the sample variance-covariance matrix, $\mathbf{S}_n$. A *single* numerical measure of variability is provided by the determinant of the sample variance-covariance matrix. When p is greater than 3, this scatterplot representation cannot actually be graphed. Yet the consideration of the data as n points in p dimensions provides insights that are not readily available from algebraic expressions. Moreover, the concepts illustrated for $p = 2$ or $p = 3$ remain valid for the other cases.

Example 3.1

Compute the mean vector $\bar{\mathbf{x}}$ from the given data matrix $\mathbf{X}$. Plot the $n = 3$ data points in $p = 2$ space and locate $\bar{\mathbf{x}}$ on the resulting diagram.

$$\mathbf{X} = \begin{bmatrix} 4 & -1 & 3 \\ 1 & 3 & 5 \end{bmatrix}$$

The first point, $\mathbf{x}_1$, has coordinates $\mathbf{x}_1' = [4, 1]$. Similarly, the remaining two

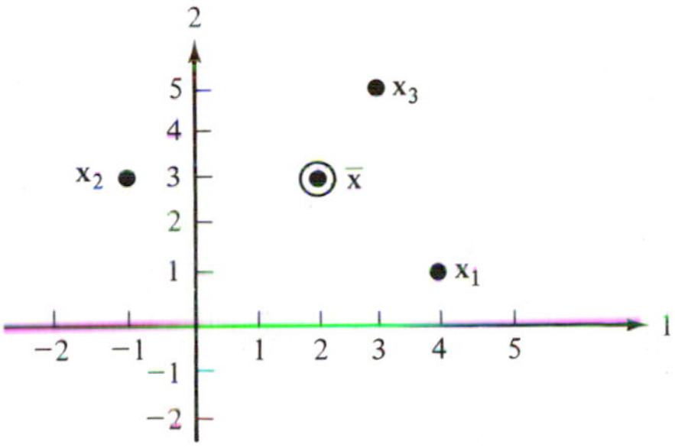

Figure 3.1 A plot of the data matrix $\mathbf{X}$ as $n = 3$ points in $p = 2$ space.

points are $\mathbf{x}_2' = [-1, 3]$ and $\mathbf{x}_3' = [3, 5]$. Finally,

$$\bar{\mathbf{x}} = \begin{bmatrix} \dfrac{4 - 1 + 3}{3} \\ \dfrac{1 + 3 + 5}{3} \end{bmatrix} = \begin{bmatrix} 2 \\ 3 \end{bmatrix}$$

Figure 3.1 shows that $\bar{\mathbf{x}}$ is the balance point (center of gravity) of the scatterplot. ■

The alternative geometrical representation is constructed by considering the data as p points in n-dimensional space. Here we take the elements of the *rows* of the data matrix to be the coordinates of points. Let

$$\underset{(p \times n)}{\mathbf{X}} = \begin{bmatrix} x_{11} & x_{12} & \cdots & x_{1n} \\ x_{21} & x_{22} & \cdots & x_{2n} \\ \vdots & \vdots & & \vdots \\ x_{p1} & x_{p2} & \cdots & x_{pn} \end{bmatrix} = \begin{bmatrix} \mathbf{y}_1' \\ \hline \mathbf{y}_2' \\ \hline \vdots \\ \hline \mathbf{y}_p' \end{bmatrix} \tag{3-2}$$

Then the coordinates of the first point $\mathbf{y}_1' = [x_{11}, x_{12}, \ldots, x_{1n}]$ are the n measurements on the first variable. In general, the ith point $\mathbf{y}_i' = [x_{i1}, x_{i2}, \ldots, x_{in}]$ is determined by the n-tuple of all measurements on the ith variable. It is convenient, in this geometrical representation, to depict $\mathbf{y}_1, \ldots, \mathbf{y}_n$ as vectors rather than points. We shall be manipulating these quantities shortly using the algebra of vectors discussed in Chapter 2.

Example 3.2

Plot the following data, $\mathbf{X}$, as $p = 2$ vectors in $n = 3$ space.

$$\mathbf{X} = \begin{bmatrix} 4 & -1 & 3 \\ 1 & 3 & 5 \end{bmatrix}$$

Here $\mathbf{y}_1' = [4, -1, 3]$ and $\mathbf{y}_2' = [1, 3, 5]$. These vectors are shown in Figure 3.2 on the next page. ■

Many of the algebraic expressions we shall encounter in multivariate analysis can be related to the geometrical notions of length, angle, and volume. This is

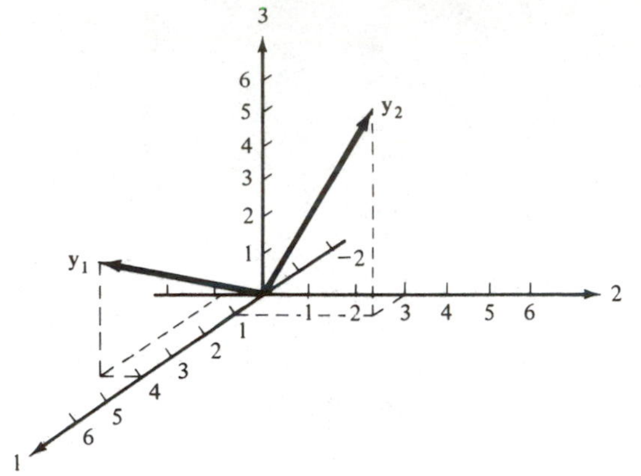

Figure 3.2 A plot of the data matrix $\mathbf{X}$ as $p = 2$ vectors in $n = 3$ space.

important because geometrical representations ordinarily facilitate understanding and lead to further insights.

Unfortunately, we are limited to visualizing objects in three dimensions and, consequently, the n-dimensional representation of the data matrix $\mathbf{X}$ may not seem like a particularly useful device for $n > 3$. It turns out, however, that geometrical relationships and the associated statistical concepts depicted for any three vectors remain valid regardless of their dimension. This follows because three vectors, even if n-dimensional, can span no more than a 3-dimensional space, just as two vectors with any number of components must lie in a plane. By selecting an appropriate 3-dimensional perspective—that is, a portion of the n-dimensional space containing the three vectors of interest—a view is obtained that preserves both lengths and angles. Thus it is possible, with the right choice of axes, to illustrate certain algebraic statistical concepts in terms of only two or three vectors of any dimension n. Since the specific choice of axes is not relevant to the geometry, we shall always label the coordinate axes 1, 2, and 3.

It is possible to give a geometrical interpretation of the process of finding a sample mean. We start by defining the $n \times 1$ vector $\mathbf{1}_n' = [1, 1, \ldots, 1]$. (To simplify the notation, the subscript n will be dropped when the dimension of the vector $\mathbf{1}_n$ is clear from the context.) The vector $\mathbf{1}$ forms equal angles with each of the n coordinate axes so the vector $(1/\sqrt{n})\mathbf{1}$ has unit length in the equal-angle direction. Consider the vector $\mathbf{y}_i' = [x_{i1}, x_{i2}, \ldots, x_{in}]$. The projection of $\mathbf{y}_i$ on the unit vector $(1/\sqrt{n})\mathbf{1}$ is, by (2-8),

$$\mathbf{y}_i'\left(\frac{1}{\sqrt{n}}\mathbf{1}\right)\frac{1}{\sqrt{n}}\mathbf{1} = \frac{x_{i1} + x_{i2} + \cdots + x_{in}}{n}\mathbf{1} = \bar{x}_i\mathbf{1} \tag{3-3}$$

That is, the sample mean $\bar{x}_i = (x_{i1} + x_{i2} + \cdots + x_{in})/n = \mathbf{y}_i'\mathbf{1}/n$ corresponds to the multiple of $\mathbf{1}$ required to give the projection of $\mathbf{y}_i$ onto the line determined by $\mathbf{1}$.

Further, for each $\mathbf{y}_i$, we have the decomposition

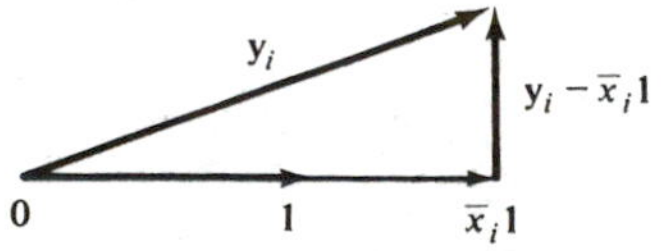

where $\bar{x}_i\mathbf{1}$ is perpendicular to $\mathbf{y}_i - \bar{x}_i\mathbf{1}$. We note that the difference vector $\mathbf{e}_i = \mathbf{y}_i - \bar{x}_i\mathbf{1}$ consists of the elements

$$\mathbf{e}_i = \mathbf{y}_i - \bar{x}_i\mathbf{1} = \begin{bmatrix} x_{i1} - \bar{x}_i \\ x_{i2} - \bar{x}_i \\ \vdots \\ x_{in} - \bar{x}_i \end{bmatrix} \tag{3-4}$$

which are the deviations of the measurements on the ith variable from their sample mean. Decomposition of the $\mathbf{y}_i$ vectors into mean components and deviation from the mean components is shown in Figure 3.3 for $p = 3$ and $n = 3$.

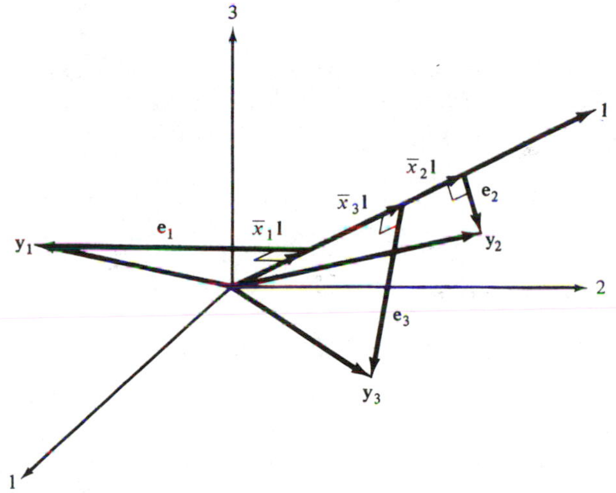

Figure 3.3 The decomposition of $\mathbf{y}_i$ into a mean component $\bar{x}_i\mathbf{1}$ and a deviation component $\mathbf{e}_i = \mathbf{y}_i - \bar{x}_i\mathbf{1}$, $i = 1, 2, 3$.

Example 3.3

Let us carry out the decomposition of $\mathbf{y}_i$ into $\bar{x}_i\mathbf{1}$ and $\mathbf{e}_i = \mathbf{y}_i - \bar{x}_i\mathbf{1}$, $i = 1, 2$, for the data given in Example 3.2.

$$\mathbf{X} = \begin{bmatrix} 4 & -1 & 3 \\ 1 & 3 & 5 \end{bmatrix}$$

Here $\bar{x}_1 = (4 - 1 + 3)/3 = 2$ and $\bar{x}_2 = (1 + 3 + 5)/3 = 3$ so

$$\bar{x}_1\mathbf{1} = 2\begin{bmatrix} 1 \\ 1 \\ 1 \end{bmatrix} = \begin{bmatrix} 2 \\ 2 \\ 2 \end{bmatrix} \qquad \bar{x}_2\mathbf{1} = 3\begin{bmatrix} 1 \\ 1 \\ 1 \end{bmatrix} = \begin{bmatrix} 3 \\ 3 \\ 3 \end{bmatrix}$$

Consequently,

$$\mathbf{e}_1 = \mathbf{y}_1 - \bar{x}_1\mathbf{1} = \begin{bmatrix} 4 \\ -1 \\ 3 \end{bmatrix} - \begin{bmatrix} 2 \\ 2 \\ 2 \end{bmatrix} = \begin{bmatrix} 2 \\ -3 \\ 1 \end{bmatrix}$$

and

$$\mathbf{e}_2 = \mathbf{y}_2 - \bar{x}_2\mathbf{1} = \begin{bmatrix} 1 \\ 3 \\ 5 \end{bmatrix} - \begin{bmatrix} 3 \\ 3 \\ 3 \end{bmatrix} = \begin{bmatrix} -2 \\ 0 \\ 2 \end{bmatrix}$$

We note that $\bar{x}_1\mathbf{1}$ and $\mathbf{e}_1 = \mathbf{y}_1 - \bar{x}_1\mathbf{1}$ are perpendicular, because

$$(\bar{x}_1\mathbf{1})'(\mathbf{y}_1 - \bar{x}_1\mathbf{1}) = [2 \quad 2 \quad 2]\begin{bmatrix} 2 \\ -3 \\ 1 \end{bmatrix} = 4 - 6 + 2 = 0$$

A similar result holds for $\bar{x}_2\mathbf{1}$ and $\mathbf{e}_2 = \mathbf{y}_2 - \bar{x}_2\mathbf{1}$. The decomposition is

$$\mathbf{y}_1 = \begin{bmatrix} 4 \\ -1 \\ 3 \end{bmatrix} = \begin{bmatrix} 2 \\ 2 \\ 2 \end{bmatrix} + \begin{bmatrix} 2 \\ -3 \\ 1 \end{bmatrix}$$

$$\mathbf{y}_2 = \begin{bmatrix} 1 \\ 3 \\ 5 \end{bmatrix} = \begin{bmatrix} 3 \\ 3 \\ 3 \end{bmatrix} + \begin{bmatrix} -2 \\ 0 \\ 2 \end{bmatrix}$$

■

For the time being, we are interested in the deviation (or residual) vectors $\mathbf{e}_i = \mathbf{y}_i - \bar{x}_i\mathbf{1}$. A plot of the deviation vectors of Figure 3.3 is given in Figure 3.4. We have translated the deviation vectors to the origin without changing their lengths or orientations.

Consider the squared lengths of the deviation vectors. Using (2-5) and (3-4),

$$L_{\mathbf{e}_i}^2 = \mathbf{e}_i'\mathbf{e}_i = \sum_{j=1}^{n} (x_{ij} - \bar{x}_i)^2 \tag{3-5}$$

$$\text{(Length of deviation vector)}^2 = \text{sum of squared deviations}$$

From (1-3) we see that the squared length is proportional to the variance of the measurements on the ith variable. Equivalently, the *length* is proportional to the *standard deviation*. Longer vectors represent more variability than shorter vectors.

For any two deviation vectors $\mathbf{e}_i$ and $\mathbf{e}_k$,

$$\mathbf{e}_i'\mathbf{e}_k = \sum_{j=1}^{n} (x_{ij} - \bar{x}_i)(x_{kj} - \bar{x}_k) \tag{3-6}$$

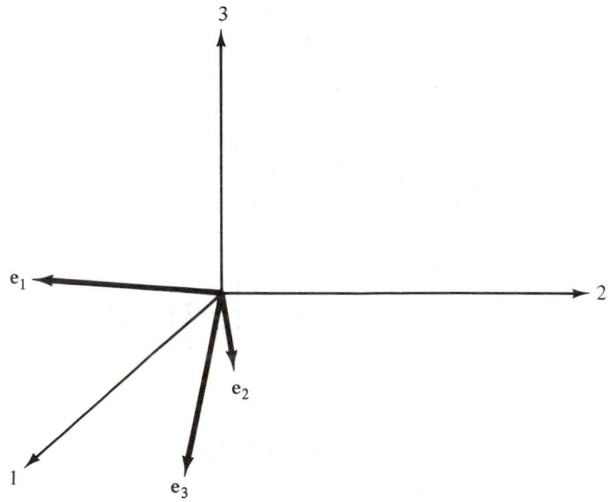

Figure 3.4 The deviation vectors $\mathbf{e}_i$ from Figure 3.3.

Let θ_{ik} denote the angle formed by the vectors $\mathbf{e}_i$ and $\mathbf{e}_k$. From (2-6)

$$\mathbf{e}_i'\mathbf{e}_k = L_{\mathbf{e}_i} L_{\mathbf{e}_k} \cos(\theta_{ik})$$

or, using (3-5) and (3-6),

$$\sum_{j=1}^{n} (x_{ij} - \bar{x}_i)(x_{kj} - \bar{x}_k) = \sqrt{\sum_{j=1}^{n} (x_{ij} - \bar{x}_i)^2} \sqrt{\sum_{j=1}^{n} (x_{kj} - \bar{x}_k)^2} \cos(\theta_{ik})$$

so that (see (1-5)),

$$r_{ik} = \frac{s_{ik}}{\sqrt{s_{ii}}\sqrt{s_{kk}}} = \cos(\theta_{ik}) \qquad (3\text{-}7)$$

The *cosine* of the angle is the sample *correlation coefficient*. Thus, if the two deviation vectors have nearly the same orientation, the sample correlation will be close to 1. If the two vectors are nearly perpendicular, the sample correlation will be approximately zero. If the two vectors are oriented in nearly opposite directions, the sample correlation will be close to -1.

Example 3.4

Given the deviation vectors in Example 3.3, let us compute the sample variance-covariance matrix $\mathbf{S}_n$ and sample correlation matrix $\mathbf{R}$ using the geometrical concepts just introduced.

From Example 3.3,

$$\mathbf{e}_1 = \begin{bmatrix} 2 \\ -3 \\ 1 \end{bmatrix} \quad \text{and} \quad \mathbf{e}_2 = \begin{bmatrix} -2 \\ 0 \\ 2 \end{bmatrix}$$

These vectors, translated to the origin, are shown in Figure 3.5.

Now

$$\mathbf{e}_1'\mathbf{e}_1 = [2 \quad -3 \quad 1] \begin{bmatrix} 2 \\ -3 \\ 1 \end{bmatrix} = 14 = 3s_{11}$$

or $s_{11} = \frac{14}{3}$.

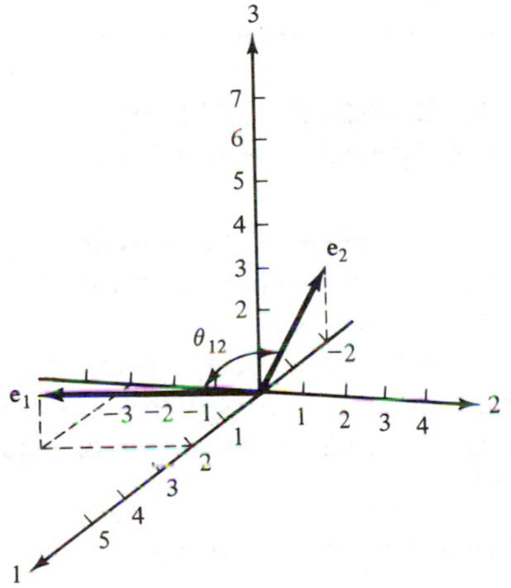

Figure 3.5 The vectors $\mathbf{e}_1$ and $\mathbf{e}_2$.

$$\mathbf{e}_2'\mathbf{e}_2 = [-2 \quad 0 \quad 2]\begin{bmatrix}-2\\0\\2\end{bmatrix} = 8 = 3s_{22}$$

or $s_{22} = \frac{8}{3}$.

$$\mathbf{e}_1'\mathbf{e}_2 = [2 \quad -3 \quad 1]\begin{bmatrix}-2\\0\\2\end{bmatrix} = -2 = 3s_{12}$$

or $s_{12} = -\frac{2}{3}$.
Consequently,

$$r_{12} = \frac{s_{12}}{\sqrt{s_{11}}\sqrt{s_{22}}} = \frac{-\frac{2}{3}}{\sqrt{\frac{14}{3}}\sqrt{\frac{8}{3}}} = -.189$$

and

$$\mathbf{S}_n = \begin{bmatrix}\frac{14}{3} & -\frac{2}{3}\\ -\frac{2}{3} & \frac{8}{3}\end{bmatrix}, \qquad \mathbf{R} = \begin{bmatrix}1 & -.189\\ -.189 & 1\end{bmatrix}$$ ■

The concepts of length, angle, and projection have provided us with a geometrical interpretation of the sample. We summarize as follows.

Geometrical Interpretation of the Sample

1. The projection of a row $\mathbf{y}_i'$ of the data matrix $\mathbf{X}$ onto the equal angular vector $\mathbf{1}$ is the vector $\bar{x}_i\mathbf{1}$. The vector $\bar{x}_i\mathbf{1}$ has length $\sqrt{n}\,|\bar{x}_i|$. Therefore, the ith sample mean, $\bar{x}_i$, is related to the length of the projection of $\mathbf{y}_i$ on $\mathbf{1}$.
2. The information comprising $\mathbf{S}_n$ is obtained from the deviation vectors $\mathbf{e}_i = \mathbf{y}_i - \bar{x}_i\mathbf{1} = [x_{i1} - \bar{x}_i, x_{i2} - \bar{x}_i, \ldots, x_{in} - \bar{x}_i]'$ after they are translated to the origin. Their squared length is ns_{ii} and the (inner) product between $\mathbf{e}_i$ and $\mathbf{e}_k$ is ns_{ik}.[1]
3. The sample correlation r_{ik} is the cosine of the angle between $\mathbf{e}_i$ and $\mathbf{e}_k$.

3.3 RANDOM SAMPLES AND THE EXPECTED VALUES OF THE SAMPLE MEAN AND COVARIANCE MATRIX

In order to study the sampling variability of statistics like $\bar{\mathbf{x}}$ and $\mathbf{S}_n$ with the ultimate aim of making inferences, we need to make assumptions about the variables whose observed values constitute the data set $\mathbf{X}$.

Suppose, then, that the data have not yet been observed, but we *intend* to collect n sets of measurements on p variables. Before the measurements are made, their values cannot, in general, be predicted exactly. Consequently, we treat them as

[1]The squared length and inner product are $(n-1)s_{ii}$ and $(n-1)s_{ik}$, respectively, when the divisor $n-1$ is used in the definitions of the sample variance and covariance.

random variables. In this context, let the (i, j) entry in the data matrix be the random variable X_{ij}. Each set of measurements $\mathbf{X}_i$ on p variables, is a random vector and we have the random matrix

$$\underset{(p \times n)}{\mathbf{X}} = \begin{bmatrix} X_{11} & X_{12} & \cdots & X_{1n} \\ X_{21} & X_{22} & \cdots & X_{2n} \\ \vdots & \vdots & & \vdots \\ X_{p1} & X_{p2} & \cdots & X_{pn} \end{bmatrix} = [\mathbf{X}_1, \mathbf{X}_2, \ldots, \mathbf{X}_n] \tag{3-8}$$

A *random sample* can now be defined.

If the column vectors $\mathbf{X}_1, \mathbf{X}_2, \ldots, \mathbf{X}_n$ in (3-8) represent *independent* observations from a *common* joint distribution with density function $f(\mathbf{x}) = f(x_1, x_2, \ldots, x_p)$, then $\mathbf{X}_1, \mathbf{X}_2, \ldots, \mathbf{X}_n$ are said to form *a random sample* from $f(\mathbf{x})$. Mathematically, $\mathbf{X}_1, \mathbf{X}_2, \ldots, \mathbf{X}_n$ form a random sample if their joint density function is given by the product $f(\mathbf{x}_1)f(\mathbf{x}_2) \cdots f(\mathbf{x}_n)$, where $f(\mathbf{x}_j) = f(x_{1j}, x_{2j}, \ldots, x_{pj})$ is the density function for the jth column vector.

Two points connected with the definition of random sample merit special attention.

1. The measurements of the p variables in a *single* trial, such as $\mathbf{X}'_j = [X_{1j}, X_{2j}, \ldots, X_{pj}]$, will usually be correlated. Indeed, we expect this to be the case. The measurements from *different* trials must, however, be independent.
2. The independence of measurements from trial to trial may not hold when the variables are likely to drift over time, as with sets of p stock prices or p economic indicators. Violation of the tentative assumption of independence can have a serious impact on the quality of statistical inferences.

The following examples illustrate these remarks.

Example 3.5

As a preliminary step in designing a permit system for utilizing a wilderness canoe area without overcrowding, a natural-resource manager took a survey of users. The total wilderness area was divided into subregions and respondents were asked to give information on the regions visited, lengths of stay, and other variables.

The method followed was to randomly select persons (perhaps using a random number table) from all those who entered the wilderness area during a particular week. All persons were equally likely to be in the sample, so the more popular entrances were represented by larger proportions of canoeists.

Here one would expect the sample observations to conform closely to the criterion for a random sample from the population of users or potential users. On the other hand, if one of the samplers had waited at a campsite far in the interior of the area and interviewed only canoeists who reached that spot, successive measurements of, for instance, length of stay would tend to be related and the independence assumption would not hold. ■

Example 3.6

The following values of X_1, the cost of crude oil in current dollars (cents per million British thermal units), and X_2, the imports of crude oil (trillion British thermal units), were obtained from the *Statistical Abstract of the United States 1977.*

TABLE 3.1 CRUDE OIL

Year	1970	1971	1972	1973	1974	1975
x_1 (cost)	54.8	58.4	58.4	67.0	116.2	137.9
x_2 (imports)	2716	3431	4541	6876	7360	8688

Should these measurements on $\mathbf{X}' = [X_1, X_2]$ be treated as a random sample of size $n = 6$? No! In fact, *both* variables are increasing over time. A drift like this would be very rare if the year-to-year values were independent observations from the same distribution. ■

As we have argued heuristically in Chapter 1, the notion of statistical independence has important implications for measuring distance. Euclidean distance appeared appropriate if the components of a vector are independent and have the same variances. Suppose we consider the location of the ith row $\mathbf{Y}_i' = [X_{i1}, X_{i2}, \ldots, X_{in}]$ of $\mathbf{X}$ regarded as a point in n dimensions. The location of this point is determined by the joint probability distribution $f(\mathbf{y}_i) = f(x_{i1}, x_{i2}, \ldots, x_{in})$. When the measurements $X_{i1}, X_{i2}, \ldots, X_{in}$ are a random sample, $f(\mathbf{y}_i) = f(x_{i1}, x_{i2}, \ldots, x_{in}) = f_i(x_{i1})f_i(x_{i2}) \cdots f_i(x_{in})$ and, consequently, each coordinate x_{ij} contributes equally to the location through the identical marginal distributions $f_i(x_{ij})$.

If the n components are not independent or the marginal distributions are not identical, the influence of individual measurements (coordinates) on location is asymmetrical. We would then be led to consider a distance function in which the coordinates are weighted unequally, as in the "statistical" distances or quadratic forms introduced in Chapters 1 and 2.

Certain conclusions can be reached concerning the sampling distributions of $\overline{\mathbf{X}}$ and $\mathbf{S}_n$ without making further assumptions regarding the form of the underlying joint distribution of the variables. In particular, we can see how $\overline{\mathbf{X}}$ and $\mathbf{S}_n$ fare as point estimators of the corresponding population mean vector $\boldsymbol{\mu}$ and covariance matrix $\boldsymbol{\Sigma}$.

Result 3.1. Let $\mathbf{X}_1, \mathbf{X}_2, \ldots, \mathbf{X}_n$ be a random sample from a joint distribution which has mean vector $\boldsymbol{\mu}$ and covariance matrix $\boldsymbol{\Sigma}$. Then $\overline{\mathbf{X}}$ is an *unbiased* estimator of $\boldsymbol{\mu}$ and its covariance matrix is

$$\frac{1}{n}\boldsymbol{\Sigma}$$

That is,

$$E(\bar{\mathbf{X}}) = \boldsymbol{\mu} \qquad \text{(population mean vector)}$$

$$\text{Cov}(\bar{\mathbf{X}}) = \frac{1}{n}\boldsymbol{\Sigma} \qquad \begin{pmatrix} \text{population variance-covariance matrix} \\ \text{divided by sample size} \end{pmatrix} \tag{3-9}$$

For the covariance matrix $\mathbf{S}_n$,

$$E(\mathbf{S}_n) = \frac{n-1}{n}\boldsymbol{\Sigma} = \boldsymbol{\Sigma} - \frac{1}{n}\boldsymbol{\Sigma}$$

Thus

$$E\left(\frac{n}{n-1}\mathbf{S}_n\right) = \boldsymbol{\Sigma} \tag{3-10}$$

so $[n/(n-1)]\mathbf{S}_n$ is an *unbiased* estimator of $\boldsymbol{\Sigma}$, while $\mathbf{S}_n$ is a *biased* estimator with (bias) $= E(\mathbf{S}_n) - \boldsymbol{\Sigma} = -(1/n)\boldsymbol{\Sigma}$.

Proof. Now $\bar{\mathbf{X}} = (\mathbf{X}_1 + \mathbf{X}_2 + \cdots + \mathbf{X}_n)/n$. The repeated use of the properties of expectation in (2-24) for two vectors gives

$$\begin{aligned}
E(\bar{\mathbf{X}}) &= E\left(\frac{1}{n}\mathbf{X}_1 + \frac{1}{n}\mathbf{X}_2 + \cdots + \frac{1}{n}\mathbf{X}_n\right) \\
&= E\left(\frac{1}{n}\mathbf{X}_1\right) + E\left(\frac{1}{n}\mathbf{X}_2\right) + \cdots + E\left(\frac{1}{n}\mathbf{X}_n\right) \\
&= \frac{1}{n}E(\mathbf{X}_1) + \frac{1}{n}E(\mathbf{X}_2) + \cdots + \frac{1}{n}E(\mathbf{X}_n) = \frac{1}{n}\boldsymbol{\mu} + \frac{1}{n}\boldsymbol{\mu} + \cdots + \frac{1}{n}\boldsymbol{\mu} \\
&= \boldsymbol{\mu}
\end{aligned}$$

Next

$$\begin{aligned}
(\bar{\mathbf{X}} - \boldsymbol{\mu})(\bar{\mathbf{X}} - \boldsymbol{\mu})' &= \left(\frac{1}{n}\sum_{j=1}^{n}(\mathbf{X}_j - \boldsymbol{\mu})\right)\left(\frac{1}{n}\sum_{\ell=1}^{n}(\mathbf{X}_\ell - \boldsymbol{\mu})\right)' \\
&= \frac{1}{n^2}\sum_{j=1}^{n}\sum_{\ell=1}^{n}(\mathbf{X}_j - \boldsymbol{\mu})(\mathbf{X}_\ell - \boldsymbol{\mu})'
\end{aligned}$$

so

$$\text{Cov}(\bar{\mathbf{X}}) = E(\bar{\mathbf{X}} - \boldsymbol{\mu})(\bar{\mathbf{X}} - \boldsymbol{\mu})' = \frac{1}{n^2}\left(\sum_{j=1}^{n}\sum_{\ell=1}^{n}E(\mathbf{X}_j - \boldsymbol{\mu})(\mathbf{X}_\ell - \boldsymbol{\mu})'\right)$$

For $j \neq \ell$, each entry in $E(\mathbf{X}_j - \boldsymbol{\mu})(\mathbf{X}_\ell - \boldsymbol{\mu})'$ is zero because the entry is the covariance between a component of $\mathbf{X}_j$ and a component of $\mathbf{X}_\ell$, and these are independent [see Exercise 3.15 and (2-29)].

Therefore

$$\text{Cov}(\bar{\mathbf{X}}) = \frac{1}{n^2}\left(\sum_{j=1}^{n}E(\mathbf{X}_j - \boldsymbol{\mu})(\mathbf{X}_j - \boldsymbol{\mu})'\right)$$

Since $\boldsymbol{\Sigma} = E(\mathbf{X}_j - \boldsymbol{\mu})(\mathbf{X}_j - \boldsymbol{\mu})'$ is the common population covariance matrix for each $\mathbf{X}_j$, we have

$$\text{Cov}(\overline{\mathbf{X}}) = \frac{1}{n^2}\left(\sum_{j=1}^{n} E(\mathbf{X}_j - \boldsymbol{\mu})(\mathbf{X}_j - \boldsymbol{\mu})'\right) = \frac{1}{n^2}\underbrace{(\boldsymbol{\Sigma} + \boldsymbol{\Sigma} + \cdots + \boldsymbol{\Sigma})}_{n \text{ terms}}$$

$$= \frac{1}{n^2}(n\boldsymbol{\Sigma}) = \left(\frac{1}{n}\right)\boldsymbol{\Sigma}$$

To obtain the expected value for $\mathbf{S}_n$, we first note that $(X_{ij} - \overline{X}_i)(X_{kj} - \overline{X}_k)$ is the (i, k) element of $(\mathbf{X}_j - \overline{\mathbf{X}})(\mathbf{X}_j - \overline{\mathbf{X}})'$. By (1-7), the matrix representing sums of squares and cross-products can then be written as

$$\sum_{j=1}^{n} (\mathbf{X}_j - \overline{\mathbf{X}})(\mathbf{X}_j - \overline{\mathbf{X}})' = \sum_{j=1}^{n} (\mathbf{X}_j - \overline{\mathbf{X}})\mathbf{X}_j' + \left(\sum_{j=1}^{n} (\mathbf{X}_j - \overline{\mathbf{X}})\right)(-\overline{\mathbf{X}})'$$

$$= \sum_{j=1}^{n} \mathbf{X}_j\mathbf{X}_j' - n\overline{\mathbf{X}}\overline{\mathbf{X}}'$$

since $\sum_{j=1}^{n} (\mathbf{X}_j - \overline{\mathbf{X}}) = \mathbf{0}$ and $n\overline{\mathbf{X}}' = \sum_{j=1}^{n} \mathbf{X}_j'$. Therefore its expected value is

$$E\left(\sum_{j=1}^{n} \mathbf{X}_j\mathbf{X}_j' - n\overline{\mathbf{X}}\overline{\mathbf{X}}'\right) = \sum_{j=1}^{n} E(\mathbf{X}_j\mathbf{X}_j') - nE(\overline{\mathbf{X}}\overline{\mathbf{X}}')$$

For any random vector $\mathbf{V}$ with $E(\mathbf{V}) = \boldsymbol{\mu}_V$ and $\text{Cov}(\mathbf{V}) = \boldsymbol{\Sigma}_V$, we have $E(\mathbf{V}\mathbf{V}') = \boldsymbol{\Sigma}_V + \boldsymbol{\mu}_V\boldsymbol{\mu}_V'$ (see Exercise 3.14). Consequently,

$$E(\mathbf{X}_j\mathbf{X}_j') = \boldsymbol{\Sigma} + \boldsymbol{\mu}\boldsymbol{\mu}' \quad \text{and} \quad E(\overline{\mathbf{X}}\overline{\mathbf{X}}') = \frac{1}{n}\boldsymbol{\Sigma} + \boldsymbol{\mu}\boldsymbol{\mu}'$$

Using these results,

$$\sum_{j=1}^{n} E(\mathbf{X}_j\mathbf{X}_j') - nE(\overline{\mathbf{X}}\overline{\mathbf{X}}') = n\boldsymbol{\Sigma} + n\boldsymbol{\mu}\boldsymbol{\mu}' - n\left(\frac{1}{n}\boldsymbol{\Sigma} + \boldsymbol{\mu}\boldsymbol{\mu}'\right) = (n-1)\boldsymbol{\Sigma}$$

and thus, since $\mathbf{S}_n = (1/n)\left(\sum_{j=1}^{n} \mathbf{X}_j\mathbf{X}_j' - n\overline{\mathbf{X}}\overline{\mathbf{X}}'\right)$, it follows immediately that

$$E(\mathbf{S}_n) = \frac{(n-1)}{n}\boldsymbol{\Sigma} \qquad \blacksquare$$

Result 3.1 shows that the (i, k) entry, $(n-1)^{-1}\sum_{j=1}^{n} (X_{ij} - \overline{X}_i)(X_{kj} - \overline{X}_k)$ of $[n/(n-1)]\mathbf{S}_n$ is an unbiased estimator of σ_{ik}. However, the individual sample standard deviations $\sqrt{s_{ii}}$, calculated with either n or $n-1$ as a divisor, are not unbiased estimators of the corresponding population quantities $\sqrt{\sigma_{ii}}$. Moreover, the correlation coefficients r_{ik} are *not* unbiased estimators of the population quantities ρ_{ik}. However, the bias $E(\sqrt{s_{ii}}) - \sqrt{\sigma_{ii}}$, or $E(r_{ik}) - \rho_{ik}$, can usually be ignored if the sample size n is moderately large.

Consideration of bias motivates a slightly modified definition of the sample variance-covariance matrix. Result 3.1 provides us with an unbiased estimator $\mathbf{S}$ of $\boldsymbol{\Sigma}$.

(Unbiased) sample variance-covariance matrix

$$\mathbf{S} = \left(\frac{n}{n-1}\right)\mathbf{S}_n = \frac{1}{n-1}\sum_{j=1}^{n}(\mathbf{x}_j - \bar{\mathbf{x}})(\mathbf{x}_j - \bar{\mathbf{x}})' \qquad (3\text{-}11)$$

Here $\mathbf{S}$, without a subscript, has (i, k) entry $(n-1)^{-1}\sum_{j=1}^{n}(X_{ij} - \bar{X}_i)(X_{kj} - \bar{X}_k)$. This definition of sample covariance is commonly used in many multivariate test statistics. Therefore it will replace $\mathbf{S}_n$ as the sample covariance matrix in most of the following material.

3.4 GENERALIZED VARIANCE

With a single variable, the sample variance is often used to describe the amount of variation in the measurements on that variable. When p variables are observed on each unit, the variation is described by the sample variance-covariance matrix

$$\mathbf{S} = \begin{bmatrix} s_{11} & s_{12} & \cdots & s_{1p} \\ s_{12} & s_{22} & \cdots & s_{2p} \\ \vdots & \vdots & & \vdots \\ s_{1p} & s_{2p} & \cdots & s_{pp} \end{bmatrix} = \left\{ s_{ik} = \frac{1}{n-1}\sum_{j=1}^{n}(x_{ij} - \bar{x}_i)(x_{kj} - \bar{x}_k) \right\}$$

The sample covariance matrix contains p variances and $\frac{1}{2}p(p-1)$ potentially different covariances. Sometimes it is desirable to assign a *single* numerical value for the variation expressed by $\mathbf{S}$. One choice for a value is the determinant of $\mathbf{S}$, which reduces to the usual sample variance of a single characteristic when $p = 1$. This determinant[2] is called the *generalized sample variance*.

$$\text{Generalized sample variance} = |\mathbf{S}| \qquad (3\text{-}12)$$

Example 3.7

Total capital (x_1) and securities revenues (x_2) for the 25 largest securities firms in the United States are shown in Figure 1.3. The sample covariance matrix $\mathbf{S}$, obtained from the data in the May 22, 1978, *Fortune* magazine article, is

$$\mathbf{S} = \begin{bmatrix} 14{,}808 & 14{,}213 \\ 14{,}213 & 15{,}538 \end{bmatrix}$$

Evaluate the generalized variance.

In this case, we compute

$$|\mathbf{S}| = (14{,}808)(15{,}538) - (14{,}213)(14{,}213) = 28.08 \times 10^6 \qquad \blacksquare$$

[2] Definition 2A.24 contains a definition of determinant and indicates one method for calculating the value of a determinant.

The generalized sample variance provides one way of writing the information on all variances and covariances as a single number. Of course, when $p > 1$, some sample information is lost in the process. A geometrical interpretation of $|\mathbf{S}|$ will help us appreciate its strengths and weaknesses as a descriptive summary.

Consider the volume (area) generated within the plane by two deviation vectors $\mathbf{e}_1 = \mathbf{y}_1 - \bar{x}_1\mathbf{1}$ and $\mathbf{e}_2 = \mathbf{y}_2 - \bar{x}_2\mathbf{1}$. Let L_1 be the length of $\mathbf{e}_1$ and L_2 the length of $\mathbf{e}_2$. By elementary geometry,

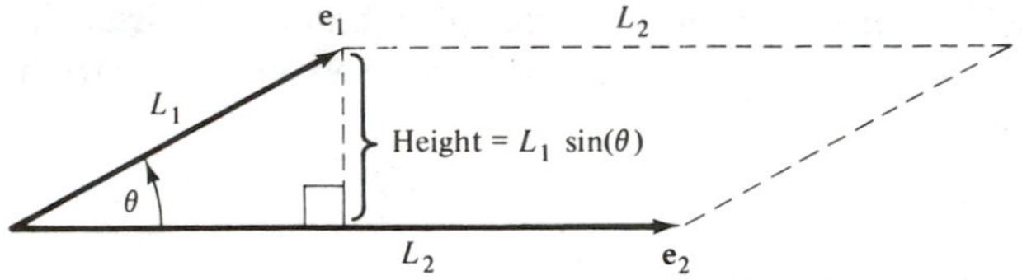

and the area of the trapezoid in the figure is $(L_1 \sin(\theta))L_2$. Since $\cos^2(\theta) + \sin^2(\theta) = 1$, we can express this area as

$$\text{Area} = L_1 L_2 \sqrt{1 - \cos^2(\theta)}$$

From (3-6) and (3-7)

$$L_1 = \sqrt{\sum_{j=1}^{n} (x_{1j} - \bar{x}_1)^2} = \sqrt{(n-1)s_{11}}$$

$$L_2 = \sqrt{\sum_{j=1}^{n} (x_{2j} - \bar{x}_2)^2} = \sqrt{(n-1)s_{22}}$$

and

$$\cos(\theta) = r_{12}$$

Therefore

$$\text{Area} = (n-1)\sqrt{s_{11}}\sqrt{s_{22}}\sqrt{1 - r_{12}^2} = (n-1)\sqrt{s_{11}s_{22}(1 - r_{12}^2)} \tag{3-13}$$

Also,

$$|\mathbf{S}| = \left| \begin{bmatrix} s_{11} & s_{12} \\ s_{12} & s_{22} \end{bmatrix} \right| = \left| \begin{bmatrix} s_{11} & \sqrt{s_{11}}\sqrt{s_{22}}\, r_{12} \\ \sqrt{s_{11}}\sqrt{s_{22}}\, r_{12} & s_{22} \end{bmatrix} \right|$$

$$= s_{11}s_{22} - s_{11}s_{22}r_{12}^2 = s_{11}s_{22}(1 - r_{12}^2) \tag{3-14}$$

If we compare (3-14) with (3-13), we see that

$$|\mathbf{S}| = (\text{area})^2 / (n-1)^2$$

Assuming now that $|\mathbf{S}| = (n-1)^{-(p-1)}(\text{volume})^2$ holds for the volume generated in n space by the $p - 1$ deviation vectors $\mathbf{e}_1, \mathbf{e}_2, \ldots, \mathbf{e}_{p-1}$, we can establish the following general result for p deviation vectors by induction (see [1], p. 167).

$$\text{Generalized sample variance} = |\mathbf{S}| = (n-1)^{-p}(\text{volume})^2 \tag{3-15}$$

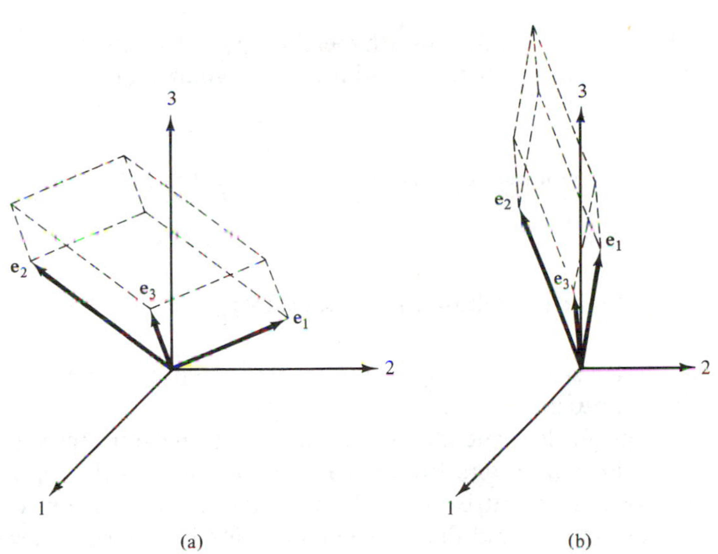

Figure 3.6 (a) "Large" generalized sample variance for $p = 3$. (b) "Small" generalized sample variance for $p = 3$.

Equation (3-15) says that the generalized sample variance, for a fixed set of data, is proportional to the square of the volume generated by the p deviation vectors[3] $\mathbf{e}_1 = \mathbf{y}_1 - \bar{x}_1\mathbf{1}$, $\mathbf{e}_2 = \mathbf{y}_2 - \bar{x}_2\mathbf{1}, \ldots, \mathbf{e}_p = \mathbf{y}_p - \bar{x}_p\mathbf{1}$. Figure 3.6(a) and (b) show trapezoidal regions, generated by $p = 3$ residual vectors, corresponding to "large" and "small" generalized variances.

For a fixed sample size, it is clear from the geometry that volume, or $|\mathbf{S}|$, will increase when the length of any $\mathbf{e}_i = \mathbf{y}_i - \bar{x}_i\mathbf{1}$ (or $\sqrt{s_{ii}}$) is increased. In addition, volume will increase if the residual vectors of fixed length are moved until they are at right angles to one another, as in Figure 3.6(a). On the other hand, the volume, or $|\mathbf{S}|$, will be small if just one of the s_{ii} is small or one of the deviation vectors lies nearly in the (hyper) plane formed by the others, or both. In the second case, the trapezoid has very little height above the plane. This is the situation in Figure 3.6(b) where $\mathbf{e}_3$ lies nearly in the plane formed by $\mathbf{e}_1$ and $\mathbf{e}_2$.

Generalized variance also has interpretations in the p-space scatterplot representation of the data. The most intuitive interpretation concerns the spread of the scatter about the sample mean point $\bar{\mathbf{x}}' = [\bar{x}_1, \bar{x}_2, \ldots, \bar{x}_p]$. Consider the measure of distance given in the comment below (2-19) with $\bar{\mathbf{x}}$ playing the role of the fixed point $\boldsymbol{\mu}$ and $\mathbf{S}^{-1}$ playing the role of $\mathbf{A}$. With these choices, the coordinates $\mathbf{x}' = [x_1, x_2, \ldots, x_p]$ of the points a constant distance c from $\bar{\mathbf{x}}$ satisfy

$$(\mathbf{x} - \bar{\mathbf{x}})'\mathbf{S}^{-1}(\mathbf{x} - \bar{\mathbf{x}}) = c^2 \tag{3-16}$$

(When $p = 1$, $(\mathbf{x} - \bar{\mathbf{x}})'\mathbf{S}^{-1}(\mathbf{x} - \bar{\mathbf{x}}) = (x_1 - \bar{x}_1)^2/s_{11}$ is the squared distance from x_1 to $\bar{x}_1$ in standard deviation units.)

[3] If generalized variance is defined in terms of the sample covariance matrix $\mathbf{S}_n = [(n-1)/n]\mathbf{S}$, then Result 2A.11 implies $|\mathbf{S}_n| = |[(n-1)/n]\mathbf{I}_p\mathbf{S}| = |[(n-1)/n]\mathbf{I}_p||\mathbf{S}| = [(n-1)/n]^p|\mathbf{S}|$. Consequently, using (3-15) we can also write the following: generalized sample variance $= |\mathbf{S}_n| = n^{-p}(\text{volume})^2$.

Equation (3-16) defines a hyperellipsoid (an ellipse if $p = 2$) centered at $\bar{\mathbf{x}}$. It can be shown using integral calculus that the volume of this hyperellipsoid is related to $|\mathbf{S}|$. In particular,

$$\text{Volume of}\{\mathbf{x}: (\mathbf{x}-\bar{\mathbf{x}})'\mathbf{S}^{-1}(\mathbf{x}-\bar{\mathbf{x}}) \le c^2\} = k_p |\mathbf{S}|^{1/2} c^p \tag{3-17}$$

or

$$(\text{Volume of ellipsoid})^2 = (\text{constant})(\text{generalized sample variance})$$

where the constant k_p is rather formidable.[4] A large volume corresponds to a large generalized variance.

Although the generalized variance has some intuitively pleasing geometrical interpretations, it suffers from a basic weakness as a descriptive summary of the sample covariance matrix $\mathbf{S}$. To illustrate the deficiency, consider the three sample covariance matrices and the derived correlation coefficients shown below.

$$\mathbf{S} = \begin{bmatrix} 5 & 4 \\ 4 & 5 \end{bmatrix} \qquad \mathbf{S} = \begin{bmatrix} 5 & -4 \\ -4 & 5 \end{bmatrix} \qquad \mathbf{S} = \begin{bmatrix} 3 & 0 \\ 0 & 3 \end{bmatrix}$$

$$r_{12} = \frac{s_{12}}{\sqrt{s_{11}}\sqrt{s_{22}}} = \frac{4}{\sqrt{5}\sqrt{5}} = .8 \qquad r_{12} = \frac{-4}{\sqrt{5}\sqrt{5}} = -.8 \qquad r_{12} = \frac{0}{\sqrt{3}\sqrt{3}} = 0$$

Each of these covariance matrices has the *same* generalized variance—namely, $|\mathbf{S}| = 9$—and yet they possess distinctly different correlation (covariance) structures. Different correlation structures are not detected by $|\mathbf{S}|$. The situation for $p > 2$ can be even more obscure.

Consequently, it is often desirable to provide more than the single number $|\mathbf{S}|$ as a summary of $\mathbf{S}$. From Exercise 2.12, $|\mathbf{S}|$ can be expressed as the product $\lambda_1\lambda_2 \cdots \lambda_p$ of the eigenvalues of $\mathbf{S}$. Moreover, the mean centered ellipsoid based on $\mathbf{S}^{-1}$ [see (3-16)] has axes whose lengths are proportional to the square roots of the λ_i's (see Section 2.3). These eigenvalues then provide information on the variability in all directions in the p-space representation of the data. It is useful, therefore, to report their individual values, as well as their product. We shall pursue this topic later when we discuss principal components.

Situations where the Generalized Sample Variance Is Zero

The generalized sample variance will be zero in certain situations. A generalized variance of zero is indicative of extreme degeneracy in the sense that at least one row

[4]For those who are curious, $k_p = 2\pi^{p/2}/p\Gamma(p/2)$, where $\Gamma(z)$ denotes the gamma function evaluated at z.

of the matrix of deviations,

$$\begin{bmatrix} \mathbf{y}_1' - \bar{x}_1\mathbf{1}' \\ \mathbf{y}_2' - \bar{x}_2\mathbf{1}' \\ \vdots \\ \mathbf{y}_p' - \bar{x}_p\mathbf{1}' \end{bmatrix} = \begin{bmatrix} x_{11} - \bar{x}_1 & x_{12} - \bar{x}_1 & \cdots & x_{1n} - \bar{x}_1 \\ x_{21} - \bar{x}_2 & x_{22} - \bar{x}_2 & \cdots & x_{2n} - \bar{x}_2 \\ \vdots & \vdots & & \vdots \\ x_{p1} - \bar{x}_p & x_{p2} - \bar{x}_p & \cdots & x_{pn} - \bar{x}_p \end{bmatrix}$$

$$= \underset{(p\times n)}{\mathbf{X}} - \underset{(p\times 1)}{\bar{\mathbf{x}}}\ \underset{(1\times n)}{\mathbf{1}'} \tag{3-18}$$

can be expressed as a linear combination of the other rows. As we have shown geometrically, this is a case where one of the deviation vectors—for instance, $\mathbf{e}_i' = (x_{i1} - \bar{x}_i, \ldots, x_{in} - \bar{x}_i)$—lies in the (hyper) plane generated by $\mathbf{e}_1, \ldots, \mathbf{e}_{i-1}, \mathbf{e}_{i+1}, \ldots, \mathbf{e}_p$.

Result 3.2. The generalized variance is zero when and only when at least one deviation vector lies in the (hyper) plane formed by all linear combinations of the others; that is, when the rows of matrix of deviations in (3-18) are linearly dependent.

Proof. If the rows of the deviation matrix $(\mathbf{X} - \bar{\mathbf{x}}\mathbf{1}')$ are linearly dependent, there is a linear combination of its rows or, equivalently, of the columns of $(\mathbf{X} - \bar{\mathbf{x}}\mathbf{1}')'$ such that

$$\mathbf{0} = \ell_1 \text{col}_1(\mathbf{X} - \bar{\mathbf{x}}\mathbf{1}')' + \cdots + \ell_p \text{col}_p(\mathbf{X} - \bar{\mathbf{x}}\mathbf{1}')'$$

$$= (\mathbf{X} - \bar{\mathbf{x}}\mathbf{1}')'\boldsymbol{\ell} \qquad \text{for some } \boldsymbol{\ell} \neq \mathbf{0}$$

But then as you may verify, $(n-1)\mathbf{S} = (\mathbf{X} - \bar{\mathbf{x}}\mathbf{1}')(\mathbf{X} - \bar{\mathbf{x}}\mathbf{1}')'$ and

$$(n-1)\mathbf{S}\boldsymbol{\ell} = (\mathbf{X} - \bar{\mathbf{x}}\mathbf{1}')(\mathbf{X} - \bar{\mathbf{x}}\mathbf{1}')'\boldsymbol{\ell} = \mathbf{0}$$

so the same $\boldsymbol{\ell}$ corresponds to a linear dependency, $\ell_1\text{col}_1(\mathbf{S}) + \cdots + \ell_p\text{col}_p(\mathbf{S}) = \mathbf{S}\boldsymbol{\ell} = \mathbf{0}$, in the columns of $\mathbf{S}$. By Result 2A.9, $|\mathbf{S}| = 0$.

In the other direction, if $|\mathbf{S}| = 0$ then there is some linear combination $\mathbf{S}\boldsymbol{\ell}$ of the columns of $\mathbf{S}$ such that $\mathbf{S}\boldsymbol{\ell} = \mathbf{0}$. That is, $\mathbf{0} = (n-1)\mathbf{S}\boldsymbol{\ell} = (\mathbf{X} - \bar{\mathbf{x}}\mathbf{1}')(\mathbf{X} - \bar{\mathbf{x}}\mathbf{1}')'\boldsymbol{\ell}$. Premultiplying by $\boldsymbol{\ell}'$ yields

$$0 = \boldsymbol{\ell}'(\mathbf{X} - \bar{\mathbf{x}}\mathbf{1}')(\mathbf{X} - \bar{\mathbf{x}}\mathbf{1}')'\boldsymbol{\ell} = L^2_{(\mathbf{X} - \bar{\mathbf{x}}\mathbf{1}')'\boldsymbol{\ell}}$$

and, for the length to equal zero, we must have $(\mathbf{X} - \bar{\mathbf{x}}\mathbf{1}')'\boldsymbol{\ell} = \mathbf{0}$. Thus the columns of $(\mathbf{X} - \bar{\mathbf{x}}\mathbf{1}')'$ or, equivalently, the rows of $(\mathbf{X} - \bar{\mathbf{x}}\mathbf{1}')$ are linearly dependent. ■

Example 3.8

Show that $|\mathbf{S}| = 0$ for

$$\underset{(3\times 3)}{\mathbf{X}} = \begin{bmatrix} 1 & 4 & 4 \\ 2 & 1 & 0 \\ 5 & 6 & 4 \end{bmatrix}$$

and determine the degeneracy.

Here $\bar{\mathbf{x}}' = [3, 1, 5]$ so

$$\mathbf{X} - \bar{\mathbf{x}}\mathbf{1}' = \begin{bmatrix} 1-3 & 4-3 & 4-3 \\ 2-1 & 1-1 & 0-1 \\ 5-5 & 6-5 & 4-5 \end{bmatrix} = \begin{bmatrix} -2 & 1 & 1 \\ 1 & 0 & -1 \\ 0 & 1 & -1 \end{bmatrix}$$

The residual (row) vectors are $\mathbf{e}_1' = [-2, 1, 1]$, $\mathbf{e}_2' = [1, 0, -1]$ and $\mathbf{e}_3' = [0, 1, -1]$. Since $\mathbf{e}_3 = \mathbf{e}_1 + 2\mathbf{e}_2$, there is row degeneracy. (Note there is column degeneracy also.) This means that one of the residual vectors, for example $\mathbf{e}_3$, lies in the plane generated by the other two residual vectors. Consequently, the *three*-dimensional volume is zero. This case is illustrated in Figure 3.7 and verified algebraically below by showing $|\mathbf{S}| = 0$.

Now

$$\underset{(3\times 3)}{\mathbf{S}} = \begin{bmatrix} 3 & -\frac{3}{2} & 0 \\ -\frac{3}{2} & 1 & \frac{1}{2} \\ 0 & \frac{1}{2} & 1 \end{bmatrix}$$

and from Definition 2A.24,

$$|\mathbf{S}| = 3\begin{vmatrix} 1 & \frac{1}{2} \\ \frac{1}{2} & 1 \end{vmatrix}(-1)^2 + (-\tfrac{3}{2})\begin{vmatrix} -\frac{3}{2} & \frac{1}{2} \\ 0 & 1 \end{vmatrix}(-1)^3 + (0)\begin{vmatrix} -\frac{3}{2} & 1 \\ 0 & \frac{1}{2} \end{vmatrix}(-1)^4$$

$$= 3(1 - \tfrac{1}{4}) + (\tfrac{3}{2})(-\tfrac{3}{2} - 0) + 0 = \tfrac{9}{4} - \tfrac{9}{4} = 0 \quad \blacksquare$$

In any statistical analysis, $|\mathbf{S}| = 0$ means that the measurements on some variables should be removed from the study as far as the mathematical computations are concerned. The corresponding reduced data matrix will then lead to a covariance matrix of full rank and a nonzero generalized variance. The question of which measurements to remove in degenerate cases is not easy to answer. When there is a choice, one should retain measurements on a (presumed) causal variable instead of those on a secondary characteristic. We shall return to this subject in our discussion of principal components.

At this point, we settle for delineating some simple conditions for $\mathbf{S}$ to be of full rank or of reduced rank.

Result 3.3. If $n \le p$ that is, (sample size) $\le$ (number of variables), then $|\mathbf{S}| = 0$ for all samples.

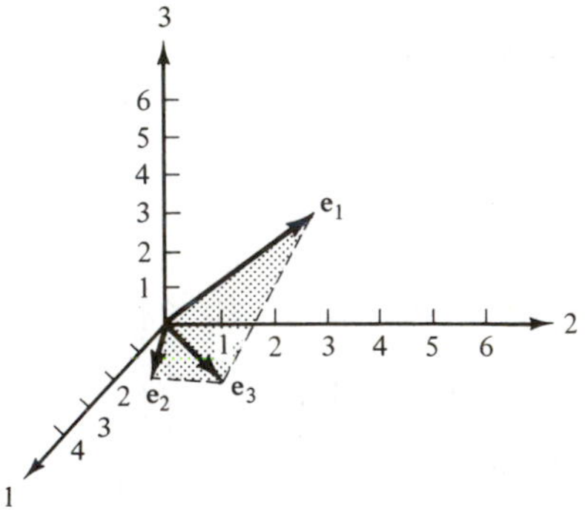

Figure 3.7 A case where the three-dimensional volume is zero ($|\mathbf{S}| = 0$).

Proof. We must show the rank of $\mathbf{S}$ is less than or equal to p and then apply Result 2A.9.

For any fixed sample, the n column vectors in (3-18) sum to the zero vector. The existence of this linear combination means that the rank of $\mathbf{X} - \bar{\mathbf{x}}\mathbf{1}'$ is less than or equal to $n - 1$, which, in turn, is less than or equal to $p - 1$ from the condition $n \leq p$. Since

$$(n-1) \underset{(p\times p)}{\mathbf{S}} = \underset{(p\times n)}{(\mathbf{X} - \bar{\mathbf{x}}\mathbf{1}')} \underset{(n\times p)}{(\mathbf{X} - \bar{\mathbf{x}}\mathbf{1}')'}$$

the kth column of $\mathbf{S}$, $\text{col}_k(\mathbf{S})$, can be written as a linear combination of the columns of $(\mathbf{X} - \bar{\mathbf{x}}\mathbf{1}')$. In particular,

$$\begin{aligned}(n-1)\text{col}_k(\mathbf{S}) &= (\mathbf{X} - \bar{\mathbf{x}}\mathbf{1}')\text{col}_k(\mathbf{X} - \bar{\mathbf{x}}\mathbf{1}')' \\ &= (x_{k1} - \bar{x}_k)\text{col}_1(\mathbf{X} - \bar{\mathbf{x}}\mathbf{1}') + (x_{k2} - \bar{x}_k)\text{col}_2(\mathbf{X} - \bar{\mathbf{x}}\mathbf{1}') \\ &\quad + \cdots + (x_{kn} - \bar{x}_k)\text{col}_n(\mathbf{X} - \bar{\mathbf{x}}\mathbf{1}')\end{aligned}$$

Since the column vectors of $\mathbf{X} - \bar{\mathbf{x}}\mathbf{1}'$ sum to the zero vector, we can write, for example, $\text{col}_1(\mathbf{X} - \bar{\mathbf{x}}\mathbf{1}')$ as the negative of the sum of the remaining column vectors. After substituting for $\text{col}_1(\mathbf{X} - \bar{\mathbf{x}}\mathbf{1}')$ above, we can express $\text{col}_k(\mathbf{S})$ as a linear combination of the at most $n - 1$ linearly independent column vectors $\text{col}_2(\mathbf{X} - \bar{\mathbf{x}}\mathbf{1}'), \ldots, \text{col}_n(\mathbf{X} - \bar{\mathbf{x}}\mathbf{1}')$. The rank of $\mathbf{S}$ is, therefore, less than or equal to $n - 1$, which—as noted at the beginning—is less than or equal to $p - 1$, and $\mathbf{S}$ is singular. This implies, from Result 2A.9, that $|\mathbf{S}| = 0$. ■

Result 3.4. Let the column vectors $\mathbf{x}_1, \mathbf{x}_2, \ldots, \mathbf{x}_n$ of the data matrix $\mathbf{X}$ be realizations of the independent random vectors $\mathbf{X}_1, \mathbf{X}_2, \ldots, \mathbf{X}_n$.

1. If the linear combination $\boldsymbol{\ell}'\mathbf{X}_j$ has positive variance for each constant vector $\boldsymbol{\ell} \neq \mathbf{0}$, then, provided $p < n$, $\mathbf{S}$ has full rank with probability 1 and $|\mathbf{S}| > 0$.
2. If, with probability 1, $\boldsymbol{\ell}'\mathbf{X}_j$ is a constant (for example, c) *for all j*, then $|\mathbf{S}| = 0$.

Proof. (Part 2). If $\boldsymbol{\ell}'\mathbf{X}_j = \ell_1 X_{1j} + \ell_2 X_{2j} + \cdots + \ell_p X_{pj} = c$, the sample mean of this linear combination is $c = \sum_{j=1}^{n} (\ell_1 x_{1j} + \ell_2 x_{2j} + \cdots + \ell_p x_{pj})/n = \ell_1 \bar{x}_1 + \ell_2 \bar{x}_2 + \cdots + \ell_p \bar{x}_p = \boldsymbol{\ell}'\bar{\mathbf{x}}$, and then

$$\begin{aligned}(\mathbf{X} - \bar{\mathbf{x}}\mathbf{1}')'\boldsymbol{\ell} &= \ell_1 \begin{bmatrix} x_{11} - \bar{x}_1 \\ \vdots \\ x_{1n} - \bar{x}_1 \end{bmatrix} + \cdots + \ell_p \begin{bmatrix} x_{p1} - \bar{x}_p \\ \vdots \\ x_{pn} - \bar{x}_p \end{bmatrix} \\ &= \begin{bmatrix} \boldsymbol{\ell}'\mathbf{x}_1 - \boldsymbol{\ell}'\bar{\mathbf{x}} \\ \vdots \\ \boldsymbol{\ell}'\mathbf{x}_n - \boldsymbol{\ell}'\bar{\mathbf{x}} \end{bmatrix} = \begin{bmatrix} c - c \\ \vdots \\ c - c \end{bmatrix} = \mathbf{0}\end{aligned}$$

indicating linear dependence; the conclusion follows from Result 3.2.

The proof of Part (1) is difficult and can be found in [2]. ■

Generalized Variance Determined by |R| *and Its Geometrical Interpretation*

The generalized sample variance is unduly affected by the variability of measurements on a single variable. For example, suppose some s_{ii} is either large or quite small. Geometrically, the corresponding residual vector $\mathbf{e}_i = (\mathbf{y}_i - \bar{x}_i\mathbf{1})$ will be very long or very short and will therefore clearly be an important factor in determining volume. Consequently, it is sometimes useful to scale all the residual vectors so that they all have the same length.

Scaling the residual vectors is equivalent to replacing each original observation x_{ij} by its standardized value $(x_{ij} - \bar{x}_i)/\sqrt{s_{ii}}$. The sample covariance matrix of the standardized variables is then $\mathbf{R}$, the sample correlation matrix of the original variables. (See Exercise 3.11.) We define

$$\left(\begin{array}{c}\text{Generalized sample variance}\\ \text{of the standardized variables}\end{array}\right) = |\mathbf{R}| \tag{3-19}$$

Since the resulting residual vectors

$$\left[(x_{i1} - \bar{x}_i)/\sqrt{s_{ii}},\, (x_{i2} - \bar{x}_i)/\sqrt{s_{ii}},\ldots,(x_{in} - \bar{x}_i)/\sqrt{s_{ii}}\right] = (\mathbf{y}_i - \bar{x}_i\mathbf{1})'/\sqrt{s_{ii}}$$

all have length $\sqrt{n-1}$, the generalized sample variance of the standardized variables will be large when these vectors are nearly perpendicular. It will be small when two or more of these vectors are in almost the same direction. Employing the argument leading to (3-7), it should be evident that the cosine of the angle θ_{ik} between $(\mathbf{y}_i - \bar{x}_i\mathbf{1})/\sqrt{s_{ii}}$ and $(\mathbf{y}_k - \bar{x}_k\mathbf{1})/\sqrt{s_{kk}}$ is the sample correlation coefficient r_{ik}. Therefore, we can make the statement that $|\mathbf{R}|$ is large when all the r_{ik} are nearly zero and it is small when one or more of the r_{ik} are nearly $+1$ or -1.

In summary we have the following result. Let

$$\frac{(\mathbf{y}_i - \bar{x}_i\mathbf{1})}{\sqrt{s_{ii}}} = \begin{bmatrix} \dfrac{x_{i1} - \bar{x}_i}{\sqrt{s_{ii}}} \\ \dfrac{x_{i2} - \bar{x}_i}{\sqrt{s_{ii}}} \\ \vdots \\ \dfrac{x_{in} - \bar{x}_i}{\sqrt{s_{ii}}} \end{bmatrix}, \qquad i = 1, 2, \ldots, p$$

be the residual vectors of the standardized variables. The volume generated in p-space by these vectors can be related to the generalized sample variance. The same steps leading to (3-15) produce

$$\left(\begin{array}{c}\text{Generalized sample variance}\\ \text{of the standardized variables}\end{array}\right) = |\mathbf{R}| = (n-1)^{-p}(\text{volume})^2 \tag{3-20}$$

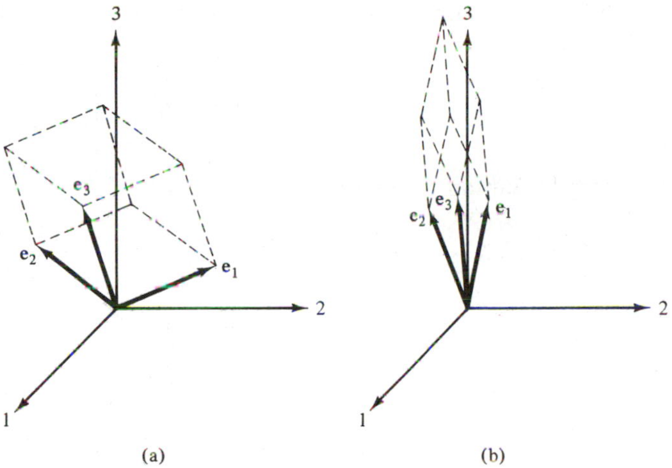

Figure 3.8 The volume generated by equal-length deviation vectors of the standardized variables.

The volume generated by deviation vectors of the standardized variables is illustrated in Figure 3.8 for the deviation vectors graphed in Figure 3.6. A comparison of Figures 3.8 and 3.6 reveals that the influence of the $\mathbf{e}_2$ vector (large variability in x_2) on the squared volume $|\mathbf{S}|$ is much greater than its influence on the squared volume $|\mathbf{R}|$.

The quantities $|\mathbf{S}|$ and $|\mathbf{R}|$ are connected by the relationship

$$|\mathbf{S}| = (s_{11}s_{22}\cdots s_{pp})|\mathbf{R}| \tag{3-21}$$

so

$$\begin{aligned}(n-1)^p|\mathbf{S}| &= (n-1)^p(s_{11}s_{22}\cdots s_{pp})|\mathbf{R}| \\ &= (n-1)s_{11}(n-1)s_{22}\cdots(n-1)s_{pp}|\mathbf{R}|\end{aligned} \tag{3-22}$$

[The proof of (3-21) is considered in Exercise 3.10.]

Interpreting (3-22) in terms of volumes, we see from (3-15) and (3-20) that the squared volume $|\mathbf{R}|$ is proportional to the squared volume $(n-1)^p|\mathbf{S}|$ and that the constant of proportionality is the product of the squared lengths $(n-1)s_{ii}$ of the $\mathbf{e}_i$. Equations (3-21) or (3-22) show, algebraically, how a change in scale measurement for X_1, for example, will alter the relationship between the generalized variances. Since $|\mathbf{R}|$ is based on standardized measurements, it is unaffected by the change in scale. However, the relative value of $|\mathbf{S}|$ will be changed whenever the multiplicative factor s_{11} changes.

Example 3.9

Let us illustrate the relationship in (3-21) for the generalized variances $|\mathbf{S}|$ and $|\mathbf{R}|$ when $p = 3$. Suppose

$$\underset{(3\times 3)}{\mathbf{S}} = \begin{bmatrix} 4 & 3 & 1 \\ 3 & 9 & 2 \\ 1 & 2 & 1 \end{bmatrix}$$

Then $s_{11} = 4$, $s_{22} = 9$, and $s_{33} = 1$. Moreover,

$$\mathbf{R} = \begin{bmatrix} 1 & \frac{1}{2} & \frac{1}{2} \\ \frac{1}{2} & 1 & \frac{2}{3} \\ \frac{1}{2} & \frac{2}{3} & 1 \end{bmatrix}$$

Using Definition 2A.24,

$$|\mathbf{S}| = 4\begin{vmatrix} 9 & 2 \\ 2 & 1 \end{vmatrix}(-1)^2 + 3\begin{vmatrix} 3 & 2 \\ 1 & 1 \end{vmatrix}(-1)^3 + 1\begin{vmatrix} 3 & 9 \\ 1 & 2 \end{vmatrix}(-1)^4$$
$$= 4(9-4) - 3(3-2) + 1(6-9) = 14$$

$$|\mathbf{R}| = 1\begin{vmatrix} 1 & \frac{2}{3} \\ \frac{2}{3} & 1 \end{vmatrix}(-1)^2 + \frac{1}{2}\begin{vmatrix} \frac{1}{2} & \frac{2}{3} \\ \frac{1}{2} & 1 \end{vmatrix}(-1)^3 + \frac{1}{2}\begin{vmatrix} \frac{1}{2} & 1 \\ \frac{1}{2} & \frac{2}{3} \end{vmatrix}(-1)^4$$
$$= \left(1 - \frac{4}{9}\right) - \left(\frac{1}{2}\right)\left(\frac{1}{2} - \frac{1}{3}\right) + \left(\frac{1}{2}\right)\left(\frac{1}{3} - \frac{1}{2}\right) = \frac{7}{18}$$

Given the information above, we see that

$$14 = |\mathbf{S}| = s_{11}s_{22}s_{33}|\mathbf{R}| = (4)(9)(1)\left(\frac{7}{18}\right) = 14 \qquad \text{(check)}$$

■

Another Generalization of Variance

We conclude this discussion by mentioning another generalization of variance. Specifically, we define the *total sample variance* as the sum of the diagonal elements of the sample variance-covariance matrix **S**. Thus

$$\text{Total sample variance} = s_{11} + s_{22} + \cdots + s_{pp} \tag{3-23}$$

Example 3.10

Calculate the total sample variance for the variance-covariance matrices **S** in Examples 3.7 and 3.8.

From Example 3.7,

$$\mathbf{S} = \begin{bmatrix} 14{,}808 & 14{,}213 \\ 14{,}213 & 15{,}538 \end{bmatrix}$$

and

$$\text{Total sample variance} = s_{11} + s_{22} = 14{,}808 + 15{,}538 = 30{,}346$$

From Example 3.8,

$$\mathbf{S} = \begin{bmatrix} 3 & -\frac{3}{2} & 0 \\ -\frac{3}{2} & 1 & \frac{1}{2} \\ 0 & \frac{1}{2} & 1 \end{bmatrix}$$

and

$$\text{Total sample variance} = s_{11} + s_{22} + s_{33} = 3 + 1 + 1 = 5$$

■

Geometrically, the total sample variance is the sum of the squared lengths of the p residual vectors $\mathbf{e}_1 = (\mathbf{y}_1 - \bar{x}_1\mathbf{1}), \ldots, \mathbf{e}_p = (\mathbf{y}_p - \bar{x}_p\mathbf{1})$ divided by $n - 1$. The total sample variance criterion pays no attention to the orientation (correlation structure) of the residual vectors. For instance, it assigns the same values to both sets of residual vectors (a) and (b) in Figure 3.6.

3.5 SAMPLE MEAN, COVARIANCE, AND CORRELATION AS MATRIX OPERATIONS

We have developed geometrical representations of the data matrix $\mathbf{X}$ and the derived descriptive statistics $\bar{\mathbf{x}}$ and $\mathbf{S}$. In addition, it is possible to link algebraically the calculation of $\bar{\mathbf{x}}$ and $\mathbf{S}$ directly to $\mathbf{X}$ using matrix operations. The resulting expressions, which depict the relation between $\bar{\mathbf{x}}$, $\mathbf{S}$, and the full data set $\mathbf{X}$ concisely, are easily programmed on electronic computers.

We have that $\bar{x}_i = (x_{i1} \cdot 1 + x_{i2} \cdot 1 + \cdots + x_{in} \cdot 1)/n = \mathbf{y}_i'\mathbf{1}/n$. Therefore

$$\bar{\mathbf{x}} = \begin{bmatrix} \bar{x}_1 \\ \bar{x}_2 \\ \vdots \\ \bar{x}_p \end{bmatrix} = \begin{bmatrix} \dfrac{\mathbf{y}_1'\mathbf{1}}{n} \\ \dfrac{\mathbf{y}_2'\mathbf{1}}{n} \\ \vdots \\ \dfrac{\mathbf{y}_p'\mathbf{1}}{n} \end{bmatrix} = \frac{1}{n} \begin{bmatrix} x_{11} & x_{12} & \cdots & x_{1n} \\ x_{21} & x_{22} & \cdots & x_{2n} \\ \vdots & \vdots & & \vdots \\ x_{p1} & x_{p2} & \cdots & x_{pn} \end{bmatrix} \begin{bmatrix} 1 \\ 1 \\ \vdots \\ 1 \end{bmatrix}$$

or

$$\bar{\mathbf{x}} = \frac{1}{n}\mathbf{X}\mathbf{1} \tag{3-24}$$

That is, $\bar{\mathbf{x}}$ is calculated from the data matrix by postmultiplying by the vector $\mathbf{1}$ and then multiplying the result by the constant $1/n$.

Next, we create a *$p \times n$ matrix of means* by postmultiplying both sides of (3-24) by $\mathbf{1}'$; that is,

$$\bar{\mathbf{x}}\mathbf{1}' = \frac{1}{n}\mathbf{X}\mathbf{1}\mathbf{1}' = \begin{bmatrix} \bar{x}_1 & \bar{x}_1 & \cdots & \bar{x}_1 \\ \bar{x}_2 & \bar{x}_2 & \cdots & \bar{x}_2 \\ \vdots & \vdots & & \vdots \\ \bar{x}_p & \bar{x}_p & \cdots & \bar{x}_p \end{bmatrix} \tag{3-25}$$

Subtracting this result from $\mathbf{X}$ produces the *$p \times n$ matrix of deviations* (residuals)

$$\mathbf{X} - \frac{1}{n}\mathbf{X}\mathbf{1}\mathbf{1}' = \begin{bmatrix} x_{11} - \bar{x}_1 & x_{12} - \bar{x}_1 & \cdots & x_{1n} - \bar{x}_1 \\ x_{21} - \bar{x}_2 & x_{22} - \bar{x}_2 & \cdots & x_{2n} - \bar{x}_2 \\ \vdots & \vdots & & \vdots \\ x_{p1} - \bar{x}_p & x_{p2} - \bar{x}_p & \cdots & x_{pn} - \bar{x}_p \end{bmatrix} \tag{3-26}$$

Now the matrix $(n-1)\mathbf{S}$ representing sums of squares and cross-products is just the matrix (3-26) times its transpose, or

$$(n-1)\mathbf{S} = \begin{bmatrix} x_{11}-\bar{x}_1 & x_{12}-\bar{x}_1 & \cdots & x_{1n}-\bar{x}_1 \\ x_{21}-\bar{x}_2 & x_{22}-\bar{x}_2 & \cdots & x_{2n}-\bar{x}_2 \\ \vdots & \vdots & & \vdots \\ x_{p1}-\bar{x}_p & x_{p2}-\bar{x}_p & \cdots & x_{pn}-\bar{x}_p \end{bmatrix}$$

$$\times \begin{bmatrix} x_{11}-\bar{x}_1 & x_{21}-\bar{x}_2 & \cdots & x_{p1}-\bar{x}_p \\ x_{12}-\bar{x}_1 & x_{22}-\bar{x}_2 & \cdots & x_{p2}-\bar{x}_p \\ \vdots & \vdots & & \vdots \\ x_{1n}-\bar{x}_1 & x_{2n}-\bar{x}_2 & \cdots & x_{pn}-\bar{x}_p \end{bmatrix}$$

$$= \left(\mathbf{X} - \frac{1}{n}\mathbf{X}\mathbf{1}\mathbf{1}'\right)\left(\mathbf{X} - \frac{1}{n}\mathbf{X}\mathbf{1}\mathbf{1}'\right)' = \mathbf{X}\left(\mathbf{I} - \frac{1}{n}\mathbf{1}\mathbf{1}'\right)\mathbf{X}'$$

since

$$\left(\mathbf{I} - \frac{1}{n}\mathbf{1}\mathbf{1}'\right)\left(\mathbf{I} - \frac{1}{n}\mathbf{1}\mathbf{1}'\right)' = \mathbf{I} - \frac{1}{n}\mathbf{1}\mathbf{1}' - \frac{1}{n}\mathbf{1}\mathbf{1}' + \frac{1}{n^2}\mathbf{1}\mathbf{1}'\mathbf{1}\mathbf{1}' = \mathbf{I} - \frac{1}{n}\mathbf{1}\mathbf{1}'$$

To summarize, the matrix expressions relating $\bar{\mathbf{x}}$ and $\mathbf{S}$ to the data set $\mathbf{X}$ are

$$\bar{\mathbf{x}} = \frac{1}{n}\mathbf{X}\mathbf{1}$$

$$\mathbf{S} = \frac{1}{n-1}\mathbf{X}\left(\mathbf{I} - \frac{1}{n}\mathbf{1}\mathbf{1}'\right)\mathbf{X}' \tag{3-27}$$

The result for $\mathbf{S}_n$ is similar except $1/n$ replaces $1/(n-1)$ as the first factor.

The relations in (3-27) show clearly how matrix operations on the data matrix $\mathbf{X}$ lead to $\bar{\mathbf{x}}$ and $\mathbf{S}$.

Once $\mathbf{S}$ is computed, it can be related to the sample correlation matrix $\mathbf{R}$. The resulting expression can also be "inverted" to relate $\mathbf{R}$ to $\mathbf{S}$. We first define the $p \times p$ *sample standard deviation matrix* $\mathbf{D}^{1/2}$ and compute its inverse, $(\mathbf{D}^{1/2})^{-1} = \mathbf{D}^{-1/2}$. Let

$$\underset{(p\times p)}{\mathbf{D}^{1/2}} = \begin{bmatrix} \sqrt{s_{11}} & 0 & \cdots & 0 \\ 0 & \sqrt{s_{22}} & \cdots & 0 \\ \vdots & \vdots & \ddots & \vdots \\ 0 & 0 & \cdots & \sqrt{s_{pp}} \end{bmatrix} \tag{3-28}$$

Then

$$\underset{(p\times p)}{\mathbf{D}^{-1/2}} = \begin{bmatrix} \frac{1}{\sqrt{s_{11}}} & 0 & \cdots & 0 \\ 0 & \frac{1}{\sqrt{s_{22}}} & \cdots & 0 \\ \vdots & \vdots & \ddots & \vdots \\ 0 & 0 & \cdots & \frac{1}{\sqrt{s_{pp}}} \end{bmatrix}$$

Since

$$\mathbf{S} = \begin{bmatrix} s_{11} & s_{12} & \cdots & s_{1p} \\ \vdots & \vdots & & \vdots \\ s_{1p} & s_{2p} & \cdots & s_{pp} \end{bmatrix}$$

and

$$\mathbf{R} = \begin{bmatrix} \frac{s_{11}}{\sqrt{s_{11}}\sqrt{s_{11}}} & \frac{s_{12}}{\sqrt{s_{11}}\sqrt{s_{22}}} & \cdots & \frac{s_{1p}}{\sqrt{s_{11}}\sqrt{s_{pp}}} \\ \vdots & \vdots & & \vdots \\ \frac{s_{1p}}{\sqrt{s_{11}}\sqrt{s_{pp}}} & \frac{s_{2p}}{\sqrt{s_{22}}\sqrt{s_{pp}}} & \cdots & \frac{s_{pp}}{\sqrt{s_{pp}}\sqrt{s_{pp}}} \end{bmatrix} = \begin{bmatrix} 1 & r_{12} & \cdots & r_{1p} \\ \vdots & \vdots & & \vdots \\ r_{1p} & r_{2p} & \cdots & 1 \end{bmatrix}$$

we have

$$\mathbf{R} = \mathbf{D}^{-1/2}\mathbf{S}\mathbf{D}^{-1/2} \tag{3-29}$$

Postmultiplying and premultiplying both sides of (3-29) by $\mathbf{D}^{1/2}$ and noting that $\mathbf{D}^{-1/2}\mathbf{D}^{1/2} = \mathbf{D}^{1/2}\mathbf{D}^{-1/2} = \mathbf{I}$ gives

$$\mathbf{S} = \mathbf{D}^{1/2}\mathbf{R}\mathbf{D}^{1/2} \tag{3-30}$$

That is, **R** can be obtained from the information in **S**, while **S** can be obtained from $\mathbf{D}^{1/2}$ and **R**. Equations (3-29) and (3-30) are sample analogs of Equations (2-36) and (2-37).

3.6 SAMPLE VALUES OF LINEAR COMBINATIONS OF VARIABLES

We have introduced linear combinations of p variables in Section 2.6. In many multivariate procedures, we are led naturally to consider a linear combination of the form

$$\mathbf{c}'\mathbf{X} = c_1X_1 + c_2X_2 + \cdots + c_pX_p$$

whose observed value on the jth trial is

$$\mathbf{c}'\mathbf{x}_j = c_1 x_{1j} + c_2 x_{2j} + \cdots + c_p x_{pj}, \qquad j = 1, 2, \ldots, n \tag{3-31}$$

The n derived observations in (3-31) have

$$\text{Sample mean} = \frac{(\mathbf{c}'\mathbf{x}_1 + \mathbf{c}'\mathbf{x}_2 + \cdots + \mathbf{c}'\mathbf{x}_n)}{n}$$

$$= \mathbf{c}'(\mathbf{x}_1 + \mathbf{x}_2 + \cdots + \mathbf{x}_n)\frac{1}{n} = \mathbf{c}'\bar{\mathbf{x}} \tag{3-32}$$

Similarly, the sample variance of the derived observations can be obtained. Since $(\mathbf{c}'\mathbf{x}_j - \mathbf{c}'\bar{\mathbf{x}})^2 = (\mathbf{c}'(\mathbf{x}_j - \bar{\mathbf{x}}))^2 = \mathbf{c}'(\mathbf{x}_j - \bar{\mathbf{x}})(\mathbf{x}_j - \bar{\mathbf{x}})'\mathbf{c}$, we have

$$\text{Sample variance} = \frac{(\mathbf{c}'\mathbf{x}_1 - \mathbf{c}'\bar{\mathbf{x}})^2 + (\mathbf{c}'\mathbf{x}_2 - \mathbf{c}'\bar{\mathbf{x}})^2 + \cdots + (\mathbf{c}'\mathbf{x}_n - \mathbf{c}'\bar{\mathbf{x}})^2}{n-1}$$

$$= \frac{\mathbf{c}'(\mathbf{x}_1 - \bar{\mathbf{x}})(\mathbf{x}_1 - \bar{\mathbf{x}})'\mathbf{c} + \mathbf{c}'(\mathbf{x}_2 - \bar{\mathbf{x}})(\mathbf{x}_2 - \bar{\mathbf{x}})'\mathbf{c} + \cdots + \mathbf{c}'(\mathbf{x}_n - \bar{\mathbf{x}})(\mathbf{x}_n - \bar{\mathbf{x}})'\mathbf{c}}{n-1}$$

$$= \mathbf{c}'\left[\frac{(\mathbf{x}_1 - \bar{\mathbf{x}})(\mathbf{x}_1 - \bar{\mathbf{x}})' + (\mathbf{x}_2 - \bar{\mathbf{x}})(\mathbf{x}_2 - \bar{\mathbf{x}})' + \cdots + (\mathbf{x}_n - \bar{\mathbf{x}})(\mathbf{x}_n - \bar{\mathbf{x}})'}{n-1}\right]\mathbf{c}$$

or

$$\text{Sample variance of } \mathbf{c}'\mathbf{X} = \mathbf{c}'\mathbf{S}\mathbf{c} \tag{3-33}$$

Equations (3-32) and (3-33) are sample analogs of (2-43). They correspond to substituting the sample quantities $\bar{\mathbf{x}}$ and $\mathbf{S}$ for the "population" quantities $\boldsymbol{\mu}$ and $\boldsymbol{\Sigma}$, respectively, in (2-43).

Consider a second linear combination

$$\mathbf{b}'\mathbf{X} = b_1 X_1 + b_2 X_2 + \cdots + b_p X_p$$

whose observed value on the jth trial is

$$\mathbf{b}'\mathbf{x}_j = b_1 x_{1j} + b_2 x_{2j} + \cdots + b_p x_{pj}, \qquad j = 1, 2, \ldots, n \tag{3-34}$$

It follows from (3-32) and (3-33) that the sample mean and variance of these derived observations are

$$\text{Sample mean of } \mathbf{b}'\mathbf{X} = \mathbf{b}'\bar{\mathbf{x}}$$

$$\text{Sample variance of } \mathbf{b}'\mathbf{X} = \mathbf{b}'\mathbf{S}\mathbf{b}$$

Moreover, the sample covariance computed from pairs of observations on $\mathbf{b}'\mathbf{X}$ and $\mathbf{c}'\mathbf{X}$ is

Sample covariance =

$$\frac{(\mathbf{b}'\mathbf{x}_1 - \mathbf{b}'\bar{\mathbf{x}})(\mathbf{c}'\mathbf{x}_1 - \mathbf{c}'\bar{\mathbf{x}}) + (\mathbf{b}'\mathbf{x}_2 - \mathbf{b}'\bar{\mathbf{x}})(\mathbf{c}'\mathbf{x}_2 - \mathbf{c}'\bar{\mathbf{x}}) + \cdots + (\mathbf{b}'\mathbf{x}_n - \mathbf{b}'\bar{\mathbf{x}})(\mathbf{c}'\mathbf{x}_n - \mathbf{c}'\bar{\mathbf{x}})}{n-1}$$

$$= \frac{\mathbf{b}'(\mathbf{x}_1 - \bar{\mathbf{x}})(\mathbf{x}_1 - \bar{\mathbf{x}})'\mathbf{c} + \mathbf{b}'(\mathbf{x}_2 - \bar{\mathbf{x}})(\mathbf{x}_2 - \bar{\mathbf{x}})'\mathbf{c} + \cdots + \mathbf{b}'(\mathbf{x}_n - \bar{\mathbf{x}})(\mathbf{x}_n - \bar{\mathbf{x}})'\mathbf{c}}{n-1}$$

$$= \mathbf{b}'\left[\frac{(\mathbf{x}_1 - \bar{\mathbf{x}})(\mathbf{x}_1 - \bar{\mathbf{x}})' + (\mathbf{x}_2 - \bar{\mathbf{x}})(\mathbf{x}_2 - \bar{\mathbf{x}})' + \cdots + (\mathbf{x}_n - \bar{\mathbf{x}})(\mathbf{x}_n - \bar{\mathbf{x}})'}{n-1}\right]\mathbf{c}$$

or

$$\text{Sample covariance of } \mathbf{b'X} \text{ and } \mathbf{c'X} = \mathbf{b'Sc} \tag{3-35}$$

In summary, we have the following result.

Result 3.5. The linear combinations

$$\mathbf{b'X} = b_1X_1 + b_2X_2 + \cdots + b_pX_p$$
$$\mathbf{c'X} = c_1X_1 + c_2X_2 + \cdots + c_pX_p$$

have sample means, variances, and covariances that are related to $\bar{\mathbf{x}}$ and $\mathbf{S}$ by

$$\begin{aligned} \text{Sample mean of } \mathbf{b'X} &= \mathbf{b'\bar{x}} \\ \text{Sample mean of } \mathbf{c'X} &= \mathbf{c'\bar{x}} \\ \text{Sample variance of } \mathbf{b'X} &= \mathbf{b'Sb} \\ \text{Sample variance of } \mathbf{c'X} &= \mathbf{c'Sc} \\ \text{Sample covariance of } \mathbf{b'X} \text{ and } \mathbf{c'X} &= \mathbf{b'Sc} \end{aligned} \tag{3-36}$$

■

Example 3.11

We shall consider two linear combinations and their derived values for the $n = 3$ observations given in Example 3.8 as

$$\mathbf{X} = \begin{bmatrix} x_{11} & x_{12} & x_{13} \\ x_{21} & x_{22} & x_{23} \\ x_{31} & x_{32} & x_{33} \end{bmatrix} = \begin{bmatrix} 1 & 4 & 4 \\ 2 & 1 & 0 \\ 5 & 6 & 4 \end{bmatrix}$$

The means, variances, and covariance will be evaluated directly by (3-36).

Consider the two linear combinations

$$\mathbf{b'X} = [2 \quad 2 \quad -1] \begin{bmatrix} X_1 \\ X_2 \\ X_3 \end{bmatrix} = 2X_1 + 2X_2 - X_3$$

and

$$\mathbf{c'X} = [1 \quad -1 \quad 3] \begin{bmatrix} X_1 \\ X_2 \\ X_3 \end{bmatrix} = X_1 - X_2 + 3X_3$$

Observations on these linear combinations are obtained by replacing X_1, X_2, and X_3 with their observed values. For example, the $n = 3$ observations on $\mathbf{b'X}$ are

$$\mathbf{b'x}_1 = 2x_{11} + 2x_{21} - x_{31} = 2(1) + 2(2) - (5) = 1$$
$$\mathbf{b'x}_2 = 2x_{12} + 2x_{22} - x_{32} = 2(4) + 2(1) - (6) = 4$$
$$\mathbf{b'x}_3 = 2x_{13} + 2x_{23} - x_{33} = 2(4) + 2(0) - (4) = 4$$

The sample mean and variance of these values are, respectively,

$$\text{Sample mean} = \frac{(1 + 4 + 4)}{3} = 3$$

$$\text{Sample variance} = \frac{(1-3)^2 + (4-3)^2 + (4-3)^2}{3-1} = 3$$

In a similar manner, the $n = 3$ observations on $\mathbf{c}'\mathbf{X}$ are

$$\mathbf{c}'\mathbf{x}_1 = 1x_{11} - 1x_{21} + 3x_{31} = 1(1) - 1(2) + 3(5) = 14$$
$$\mathbf{c}'\mathbf{x}_2 = 1(4) - 1(1) + 3(6) = 21$$
$$\mathbf{c}'\mathbf{x}_3 = 1(4) - 1(0) + 3(4) = 16$$

and

$$\text{Sample mean} = \frac{(14 + 21 + 16)}{3} = 17$$

$$\text{Sample variance} = \frac{(14-17)^2 + (21-17)^2 + (16-17)^2}{3-1} = 13$$

Moreover, the sample covariance computed from the pairs of observations $(\mathbf{b}'\mathbf{x}_1, \mathbf{c}'\mathbf{x}_1)$, $(\mathbf{b}'\mathbf{x}_2, \mathbf{c}'\mathbf{x}_2)$, and $(\mathbf{b}'\mathbf{x}_3, \mathbf{c}'\mathbf{x}_3)$ is

Sample covariance

$$= \frac{(1-3)(14-17) + (4-3)(21-17) + (4-3)(16-17)}{3-1} = \frac{9}{2}$$

Alternatively, we use the sample mean vector $\bar{\mathbf{x}}$ and sample covariance matrix $\mathbf{S}$ derived from the original data matrix $\mathbf{X}$ to calculate the sample means, variances, and covariances for the linear combinations. Thus if only the descriptive statistics are of interest, we do not even need to calculate the observations $\mathbf{b}'\mathbf{x}_j$ and $\mathbf{c}'\mathbf{x}_j$.

From Example 3.8,

$$\bar{\mathbf{x}} = \begin{bmatrix} 3 \\ 1 \\ 5 \end{bmatrix} \quad \text{and} \quad \mathbf{S} = \begin{bmatrix} 3 & -\frac{3}{2} & 0 \\ -\frac{3}{2} & 1 & \frac{1}{2} \\ 0 & \frac{1}{2} & 1 \end{bmatrix}$$

Consequently, using (3-36), the two sample means for the derived observations are

$$\text{Sample mean of } \mathbf{b}'\mathbf{X} = \mathbf{b}'\bar{\mathbf{x}} = [2 \quad 2 \quad -1]\begin{bmatrix} 3 \\ 1 \\ 5 \end{bmatrix} = 3 \qquad \text{(check)}$$

$$\text{Sample mean of } \mathbf{c}'\mathbf{X} = \mathbf{c}'\bar{\mathbf{x}} = [1 \quad -1 \quad 3]\begin{bmatrix} 3 \\ 1 \\ 5 \end{bmatrix} = 17 \qquad \text{(check)}$$

Using (3-36) we have

Sample variance of $\mathbf{b}'\mathbf{X} = \mathbf{b}'\mathbf{S}\mathbf{b}$

$$= [2 \quad 2 \quad -1]\begin{bmatrix} 3 & -\frac{3}{2} & 0 \\ -\frac{3}{2} & 1 & \frac{1}{2} \\ 0 & \frac{1}{2} & 1 \end{bmatrix}\begin{bmatrix} 2 \\ 2 \\ -1 \end{bmatrix}$$

$$= [2 \quad 2 \quad -1]\begin{bmatrix} 3 \\ -\frac{3}{2} \\ 0 \end{bmatrix} = 3 \qquad \text{(check)}$$

Sample variance of $\mathbf{c}'\mathbf{X} = \mathbf{c}'\mathbf{S}\mathbf{c}$

$$= [1 \quad -1 \quad 3]\begin{bmatrix} 3 & -\frac{3}{2} & 0 \\ -\frac{3}{2} & 1 & \frac{1}{2} \\ 0 & \frac{1}{2} & 1 \end{bmatrix}\begin{bmatrix} 1 \\ -1 \\ 3 \end{bmatrix}$$

$$= [1 \quad -1 \quad 3]\begin{bmatrix} \frac{9}{2} \\ -1 \\ \frac{5}{2} \end{bmatrix} = 13 \qquad \text{(check)}$$

Sample covariance of $\mathbf{b}'\mathbf{X}$ and $\mathbf{c}'\mathbf{X} = \mathbf{b}'\mathbf{S}\mathbf{c}$

$$= [2 \quad 2 \quad -1]\begin{bmatrix} 3 & -\frac{3}{2} & 0 \\ -\frac{3}{2} & 1 & \frac{1}{2} \\ 0 & \frac{1}{2} & 1 \end{bmatrix}\begin{bmatrix} 1 \\ -1 \\ 3 \end{bmatrix}$$

$$= [2 \quad 2 \quad -1]\begin{bmatrix} \frac{9}{2} \\ -1 \\ \frac{5}{2} \end{bmatrix} = \frac{9}{2} \qquad \text{(check)}$$

As indicated, these last results check with the corresponding sample quantities computed directly from the observations on the linear combinations. ■

The sample mean and covariance relations in Result 3.5 pertain to any number of linear combinations. Consider the q linear combinations

$$a_{i1}X_1 + a_{i2}X_2 + \cdots + a_{ip}X_p, \qquad i = 1, 2, \ldots, q \tag{3-37}$$

These can be expressed in matrix notation as

$$\begin{bmatrix} a_{11}X_1 + a_{12}X_2 + \cdots + a_{1p}X_p \\ a_{21}X_1 + a_{22}X_2 + \cdots + a_{2p}X_p \\ \vdots \qquad \vdots \qquad \vdots \\ a_{q1}X_1 + a_{q2}X_2 + \cdots + a_{qp}X_p \end{bmatrix} = \begin{bmatrix} a_{11} & a_{12} & \cdots & a_{1p} \\ a_{21} & a_{22} & \cdots & a_{2p} \\ \vdots & \vdots & & \vdots \\ a_{q1} & a_{q2} & \cdots & a_{qp} \end{bmatrix}\begin{bmatrix} X_1 \\ X_2 \\ \vdots \\ X_p \end{bmatrix} = \mathbf{AX} \tag{3-38}$$

Taking the ith row of $\mathbf{A}$, $\mathbf{a}_i'$, to be $\mathbf{b}'$ and the kth row of $\mathbf{A}$, $\mathbf{a}_k'$, to be $\mathbf{c}'$, Equations (3-36) imply the ith row of $\mathbf{AX}$ has sample mean $\mathbf{a}_i'\bar{\mathbf{x}}$ and the ith and kth rows of $\mathbf{AX}$ have sample covariance $\mathbf{a}_i'\mathbf{S}\mathbf{a}_k$. Note that $\mathbf{a}_i'\mathbf{S}\mathbf{a}_k$ is the (i, k) element of $\mathbf{ASA}'$.

Result 3.6. The q linear combinations $\mathbf{AX}$ in (3-38) have sample mean vector $\mathbf{A}\bar{\mathbf{x}}$ and sample covariance matrix $\mathbf{ASA}'$. ■

3.7 TREATING AN OBSERVED SAMPLE AS A POPULATION

In certain cases, it is possible to treat an observed sample as a population of values. This is useful for two reasons: (1) It serves as a powerful mechanism for deducing general properties of samples from corresponding population properties; and (2) It provides the relations for calculating population means and covariances from data collected in a complete census—that is, in a situation where the data matrix contains *all* the available information about a subject. For example, such a situation occurs when a car manufacturer, interested in annual automobile sales in a state, collects complete figures from *all* of its dealerships.

Let the n (observed) columns of the $p \times n$ data matrix $\mathbf{X} = [\mathbf{x}_1, \mathbf{x}_2, \ldots, \mathbf{x}_n]$ represent a complete census of the measurements we can obtain on the p variables of interest. Suppose we regard each column vector as being "equally informative" and assign each a weight of $1/n$. This is equivalent to distributing probability in a uniform manner so that observed vector $\mathbf{x}_j$ is assigned probability $1/n$ for each $j = 1, 2, \ldots, n$.

Let $\mathbf{X}$ denote the vector random variable that assumes the value $\mathbf{x}_j$ with probability $1/n$. For this distribution, the expected value is

$$\text{Mean} = E(\mathbf{X}) = \mathbf{x}_1\left(\frac{1}{n}\right) + \mathbf{x}_2\left(\frac{1}{n}\right) + \cdots + \mathbf{x}_n\left(\frac{1}{n}\right) = \bar{\mathbf{x}} \qquad \text{(sample mean)} \tag{3-39}$$

Similarly, the covariance matrix of $\mathbf{X}$ is

$$\begin{aligned}\text{Covariance} &= E(\mathbf{X} - E(\mathbf{X}))(\mathbf{X} - E(\mathbf{X}))' \\ &= E(\mathbf{X} - \bar{\mathbf{x}})(\mathbf{X} - \bar{\mathbf{x}})' \\ &= (\mathbf{x}_1 - \bar{\mathbf{x}})(\mathbf{x}_1 - \bar{\mathbf{x}})'\left(\frac{1}{n}\right) + (\mathbf{x}_2 - \bar{\mathbf{x}})(\mathbf{x}_2 - \bar{\mathbf{x}})'\left(\frac{1}{n}\right) \\ &\quad + \cdots + (\mathbf{x}_n - \bar{\mathbf{x}})(\mathbf{x}_n - \bar{\mathbf{x}})'\left(\frac{1}{n}\right) \\ &= \mathbf{S}_n \qquad \text{(sample covariance)}\end{aligned} \tag{3-40}$$

Equations (3-39) and (3-40) establish the formal relationship between calculating $\bar{\mathbf{x}}$ and $\mathbf{S}_n$ (see (1-4) and (1-8)) as descriptive statistics and treating them as population moments. Result 3.7 summarizes these ideas.

Result 3.7. If the observed values $\mathbf{x}_1, \mathbf{x}_2, \ldots, \mathbf{x}_n$ are treated as the complete set of values for a random variable and each is treated alike and assigned probability

$1/n$, the resulting distribution has

$$\text{Population mean vector} = \bar{\mathbf{x}}$$
$$\text{Population covariance matrix} = \mathbf{S}_n \qquad \blacksquare$$

We conclude by illustrating how sample relations involving $\mathbf{S}_n$ are deduced from known population results by specializing to the current situation in which $\bar{\mathbf{x}}$ is a population mean $\boldsymbol{\mu}$ and $\mathbf{S}_n$ a population covariance matrix $\boldsymbol{\Sigma}$.

Set

$$\mathbf{D}_n^{1/2} = \begin{bmatrix} \sqrt{s_{11}} & 0 & \cdots & 0 \\ 0 & \sqrt{s_{22}} & \cdots & 0 \\ \vdots & \vdots & \ddots & \vdots \\ 0 & 0 & \cdots & \sqrt{s_{pp}} \end{bmatrix}$$

where the $s_{ii} = \frac{1}{n}\sum_{j=1}^{n} (x_{ij} - \bar{x}_i)^2$ are diagonal elements of $\mathbf{S}_n$. Then $\mathbf{D}_n^{1/2}$ replaces $\mathbf{V}^{1/2}$ in (2-37) and $\mathbf{R}$ replaces $\boldsymbol{\rho}$, so

$$\mathbf{D}_n^{-1/2}\mathbf{S}_n\mathbf{D}_n^{-1/2} = \mathbf{R} \quad \text{and} \quad \mathbf{S}_n = \mathbf{D}_n^{1/2}\mathbf{R}\mathbf{D}_n^{1/2} \tag{3-41}$$

EXERCISES

3.1. Given the data matrix

$$\mathbf{X} = \begin{bmatrix} 9 & 5 & 1 \\ 1 & 3 & 2 \end{bmatrix}$$

(a) Graph the scatterplot in $p = 2$ dimensions. Locate the sample mean on your diagram.

(b) Sketch the $n = 3$-dimensional representation of the data and plot the deviation vectors $\mathbf{y}_1 - \bar{x}_1\mathbf{1}$ and $\mathbf{y}_2 - \bar{x}_2\mathbf{1}$.

(c) Sketch the deviation vectors in (b) emanating from the origin. Calculate the lengths of these vectors and the cosine of the angle between them. Relate these quantities to $\mathbf{S}_n$ and $\mathbf{R}$.

3.2. Given the data matrix

$$\mathbf{X} = \begin{bmatrix} 3 & 6 & 3 \\ 4 & -2 & 1 \end{bmatrix}$$

(a) Graph the scatterplot in $p = 2$ dimensions and locate the sample mean on your diagram.

(b) Sketch the $n = 3$-space representation of the data and plot the deviation vectors $\mathbf{y}_1 - \bar{x}_1\mathbf{1}$ and $\mathbf{y}_2 - \bar{x}_2\mathbf{1}$.

(c) Sketch the deviation vectors in (b) emanating from the origin. Calculate their lengths and the cosine of the angle between them. Relate these quantities to $\mathbf{S}_n$ and $\mathbf{R}$.

3.3. Perform the decomposition of $\mathbf{y}_1$ into $\bar{x}_1\mathbf{1}$ and $\mathbf{y}_1 - \bar{x}_1\mathbf{1}$ using the first row of the data matrix in Example 3.8.

3.4. Use the six observations on the variable X_1 from Table 1.1.

(a) Find the projection on $\mathbf{1} = [1, 1, 1, 1, 1, 1]'$.

(b) Calculate the deviation vector $\mathbf{y}_1 - \bar{x}_1\mathbf{1}$. Relate its length to the sample standard deviation.

(c) Graph (to scale) the triangle formed by $\mathbf{y}_1$, $\bar{x}_1\mathbf{1}$ and $\mathbf{y}_1 - \bar{x}_1\mathbf{1}$. Identify the length of each component in your graph.

(d) Repeat Parts a–c for the variable X_2 in the Table 1.1.

(e) Graph (to scale) the two deviation vectors $\mathbf{y}_1 - \bar{x}_1\mathbf{1}$ and $\mathbf{y}_2 - \bar{x}_2\mathbf{1}$. Calculate the value of the angle between them.

3.5. Calculate the generalized sample variance $|\mathbf{S}|$ for (a) the data matrix $\mathbf{X}$ in Exercise 3.1, and (b) the data matrix $\mathbf{X}$ in Exercise 3.2.

3.6. Consider the data matrix

$$\mathbf{X} = \begin{bmatrix} -1 & 2 & 5 \\ 3 & 4 & 2 \\ -2 & 2 & 3 \end{bmatrix}$$

(a) Calculate the matrix of deviations (residuals), $\mathbf{X} - \bar{\mathbf{x}}\mathbf{1}'$. Is this matrix of full rank? Explain.

(b) Determine $\mathbf{S}$ and calculate the generalized sample variance $|\mathbf{S}|$. Interpret the latter geometrically.

(c) Using the results in (b), calculate the total sample variance. [See (3-23).]

3.7. Sketch the solid ellipsoids $(\mathbf{x} - \bar{\mathbf{x}})'\mathbf{S}^{-1}(\mathbf{x} - \bar{\mathbf{x}}) \leq 1$ [see (3-16)] for the three matrices

$$\mathbf{S} = \begin{bmatrix} 5 & 4 \\ 4 & 5 \end{bmatrix}, \quad \mathbf{S} = \begin{bmatrix} 5 & -4 \\ -4 & 5 \end{bmatrix}, \quad \mathbf{S} = \begin{bmatrix} 3 & 0 \\ 0 & 3 \end{bmatrix}$$

(Note that these matrices have the *same* generalized variance $|\mathbf{S}|$.)

3.8. Given

$$\mathbf{S} = \begin{bmatrix} 1 & 0 & 0 \\ 0 & 1 & 0 \\ 0 & 0 & 1 \end{bmatrix} \quad \text{and} \quad \mathbf{S} = \begin{bmatrix} 1 & -\frac{1}{2} & -\frac{1}{2} \\ -\frac{1}{2} & 1 & -\frac{1}{2} \\ -\frac{1}{2} & -\frac{1}{2} & 1 \end{bmatrix}$$

(a) Calculate the total sample variance for each $\mathbf{S}$. Compare the results.

(b) Calculate the generalized sample variance for each $\mathbf{S}$ and compare the results. Comment on the discrepancies, if any, found between Parts a and b.

3.9. Use the sample covariance obtained in Example 3.7 to verify (3-29) and (3-30), which state that $\mathbf{R} = \mathbf{D}^{-1/2}\mathbf{S}\mathbf{D}^{-1/2}$ and $\mathbf{D}^{1/2}\mathbf{R}\mathbf{D}^{1/2} = \mathbf{S}$.

3.10. Show $|\mathbf{S}| = (s_{11}s_{22}\cdots s_{pp})|\mathbf{R}|$.
[*Hint:* From Equation (3-30), $\mathbf{S} = \mathbf{D}^{1/2}\mathbf{R}\mathbf{D}^{1/2}$. Taking determinants gives $|\mathbf{S}| = |\mathbf{D}^{1/2}||\mathbf{R}||\mathbf{D}^{1/2}|$. (See Result 2A.11.) Now examine $|\mathbf{D}^{1/2}|$.]

3.11. Given a data matrix $\mathbf{X}$, and the resulting sample correlation matrix $\mathbf{R}$, consider the standardized observations $(x_{ij} - \bar{x}_i)/\sqrt{s_{ii}}$, $i = 1, 2, \ldots, p$; $j = 1, 2, \ldots, n$. Show that these standardized quantities have sample covariance matrix $\mathbf{R}$.

3.12. Consider the data matrix $\mathbf{X}$ in Exercise 3.1. We have $n = 3$ observations on $p = 2$ variables, X_1 and X_2. Form the linear combinations

$$\mathbf{c}'\mathbf{X} = [-1 \quad 2]\begin{bmatrix} X_1 \\ X_2 \end{bmatrix} = -X_1 + 2X_2$$

$$\mathbf{b}'\mathbf{X} = [2 \quad 3]\begin{bmatrix} X_1 \\ X_2 \end{bmatrix} = 2X_1 + 3X_2$$

(a) Evaluate the sample means, variances, and covariance of $\mathbf{b}'\mathbf{X}$ and $\mathbf{c}'\mathbf{X}$ from first principles. That is, calculate the observed values of $\mathbf{b}'\mathbf{X}$ and $\mathbf{c}'\mathbf{X}$ and then use the sample mean, variance, and covariance formulas.

(b) Calculate the sample means, variances, and covariance of $\mathbf{b}'\mathbf{X}$ and $\mathbf{c}'\mathbf{X}$ using (3-36). Compare the results in (a) and (b).

3.13. Repeat Exercise 3.12 using the data matrix

$$\mathbf{X} = \begin{bmatrix} 1 & 6 & 8 \\ 4 & 2 & 3 \\ 3 & 6 & 3 \end{bmatrix}$$

and the linear combinations

$$\mathbf{b}'\mathbf{X} = [1 \quad 1 \quad 1] \begin{bmatrix} X_1 \\ X_2 \\ X_3 \end{bmatrix}$$

and

$$\mathbf{c}'\mathbf{X} = [1 \quad 2 \quad -3] \begin{bmatrix} X_1 \\ X_2 \\ X_3 \end{bmatrix}$$

3.14. Let $\mathbf{V}$ be a vector random variable with mean vector $E(\mathbf{V}) = \boldsymbol{\mu}_V$ and covariance matrix $E(\mathbf{V} - \boldsymbol{\mu}_V)(\mathbf{V} - \boldsymbol{\mu}_V)' = \boldsymbol{\Sigma}_V$. Show that $E(\mathbf{V}\mathbf{V}') = \boldsymbol{\Sigma}_V + \boldsymbol{\mu}_V\boldsymbol{\mu}_V'$.

3.15. Show that, if $\underset{(p\times 1)}{\mathbf{X}}$ and $\underset{(q\times 1)}{\mathbf{Z}}$ are independent, then each component of $\mathbf{X}$ is independent of each component of $\mathbf{Z}$.
(*Hint:* $P[X_1 \le x_1, X_2 \le x_2, \dots, X_p \le x_p$ and $Z_1 \le z_1, \dots, Z_q \le z_q] = P[X_1 \le x_1, X_2 \le x_2, \dots, X_p \le x_p] \cdot P[Z_1 \le z_1, \dots, Z_q \le z_q]$ by independence. Let $x_2, \dots, x_p$ and $z_2, \dots, z_q$ tend to infinity to obtain $P[X_1 \le x_1$ and $Z_1 \le z_1] = P[X_1 \le x_1] \cdot P[Z_1 \le z_1]$ for all x_1, z_1, so X_1 and Z_1 are independent. Repeat for other pairs.)

REFERENCES

[1] Anderson, T. W., *An Introduction to Multivariate Statistical Analysis*, New York: John Wiley, 1958.

[2] Eaton, M., and M. Perlman, "The Non-singularity of Generalized Sample Covariance Matrices," *Annals of Statistics*, **1** (1973), 710–717.

4

The Multivariate Normal Distribution

4.1 INTRODUCTION

A generalization of the familiar bell-shaped normal density to several dimensions plays a fundamental role in multivariate analysis. In fact, most of the techniques encountered in this book are based on the assumption that the data were generated from a *multivariate* normal distribution. While real data are never *exactly* multivariate normal, the normal density is often a useful approximation to the "true" population distribution.

One advantage of the multivariate normal distribution stems from the fact that it is mathematically tractable and "nice" results can be obtained. This is frequently not the case for other data-generating distributions. Of course, mathematical attractiveness per se is of little use to the practitioner. It turns out, however, that normal distributions are useful in practice for two reasons. First, the normal distribution serves as a bona fide population model in some instances. Secondly, the sampling distributions of many multivariate statistics are approximately normal, regardless of the form of the parent population, because of a *central limit* effect.

To summarize, many real-world problems fall naturally within the framework of normal theory. The importance of the normal distribution rests on its dual role as both population model for certain natural phenomena and approximate sampling distribution for many statistics.

4.2 THE MULTIVARIATE NORMAL DENSITY AND ITS PROPERTIES

The multivariate normal density is a generalization of the univariate normal density to $p \geq 2$ dimensions. Recall the univariate normal distribution, with mean μ and variance σ^2, has the probability density function

$$f(x) = \frac{1}{\sqrt{2\pi\sigma^2}} e^{-\frac{1}{2}\left(\frac{x-\mu}{\sigma}\right)^2} \qquad -\infty < x < \infty \tag{4-1}$$

A plot of this function yields the familiar bell-shaped curve shown in Figure 4.1. Also shown in the figure are approximate areas under the curve within ± 1 standard deviations and ± 2 standard deviations of the mean. These areas represent probabilities and thus, for the normal random variable X,

$$P(\mu - \sigma \leq X \leq \mu + \sigma) \doteq .68$$

$$P(\mu - 2\sigma \leq X \leq \mu + 2\sigma) \doteq .95$$

It is convenient to denote the normal density function with mean μ and variance σ^2 by $N(\mu, \sigma^2)$. Therefore $N(10, 4)$ refers to the function in (4-1) with $\mu = 10$ and $\sigma = 2$. This notation will be extended to the multivariate case later.

The term

$$\left(\frac{x-\mu}{\sigma}\right)^2 = (x-\mu)(\sigma^2)^{-1}(x-\mu) \tag{4-2}$$

in the exponent of the univariate normal density function measures the squared distance from x to μ in standard deviation units. This can be generalized for a $p \times 1$ vector $\mathbf{x}$ of observations on several variables as

$$(\mathbf{x} - \boldsymbol{\mu})'\boldsymbol{\Sigma}^{-1}(\mathbf{x} - \boldsymbol{\mu}) \tag{4-3}$$

The $p \times 1$ vector $\boldsymbol{\mu}$ represents the expected value of the random vector $\mathbf{X}$ and the $p \times p$ matrix $\boldsymbol{\Sigma}$ is its variance-covariance matrix. [See (2-30) and (2-31).] We shall assume the symmetric matrix $\boldsymbol{\Sigma}$ is positive definite, so the expression in (4-3) is the squared generalized distance from $\mathbf{x}$ to $\boldsymbol{\mu}$.

The multivariate normal density is obtained by replacing the univariate distance in (4-2) by the multivariate generalized distance of (4-3) in the density function of (4-1). When this replacement is made, the univariate normalizing constant $(2\pi)^{-1/2}(\sigma^2)^{-1/2}$ must be changed to a more general constant that makes the *volume* under the surface of the multivariate density function unity for any p. This is necessary because in the multivariate case, probabilities are represented by volumes

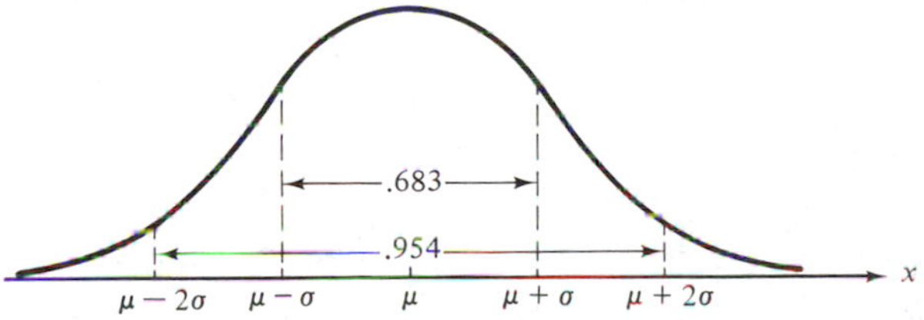

Figure 4.1 A normal density with mean μ and variance σ^2 and selected areas under the curve.

under the surface over regions defined by intervals of the x_i values. It can be shown (see [1]) that this constant is $(2\pi)^{-p/2}|\boldsymbol{\Sigma}|^{-1/2}$, and consequently a p-dimensional normal density for the random vector $\mathbf{X} = [X_1, X_2, \ldots, X_p]'$ has the form

$$f(\mathbf{x}) = \frac{1}{(2\pi)^{p/2}|\boldsymbol{\Sigma}|^{1/2}} e^{-\frac{1}{2}(\mathbf{x}-\boldsymbol{\mu})'\boldsymbol{\Sigma}^{-1}(\mathbf{x}-\boldsymbol{\mu})} \tag{4-4}$$

where $-\infty < x_i < \infty$, $i = 1, 2, \ldots, p$. We shall denote this p-dimensional normal density by $N_p(\boldsymbol{\mu}, \boldsymbol{\Sigma})$, which is analogous to the univariate case.

Example 4.1 (Bivariate Normal Density)

Let us evaluate the $p = 2$ variate normal density in terms of the individual parameters $\mu_1 = E(X_1)$, $\mu_2 = E(X_2)$, $\sigma_{11} = \text{Var}(X_1)$, $\sigma_{22} = \text{Var}(X_2)$, and $\rho_{12} = \sigma_{12}/\sqrt{\sigma_{11}}\sqrt{\sigma_{22}} = \text{Corr}(X_1, X_2)$.

Using Result 2A.8, the inverse of the covariance matrix

$$\boldsymbol{\Sigma} = \begin{bmatrix} \sigma_{11} & \sigma_{12} \\ \sigma_{12} & \sigma_{22} \end{bmatrix}$$

is

$$\boldsymbol{\Sigma}^{-1} = \frac{1}{\sigma_{11}\sigma_{22} - \sigma_{12}^2} \begin{bmatrix} \sigma_{22} & -\sigma_{12} \\ -\sigma_{12} & \sigma_{11} \end{bmatrix}$$

Introducing the correlation coefficient ρ_{12} by writing $\sigma_{12} = \rho_{12}\sqrt{\sigma_{11}}\sqrt{\sigma_{22}}$, we obtain $\sigma_{11}\sigma_{22} - \sigma_{12}^2 = \sigma_{11}\sigma_{22}(1 - \rho_{12}^2)$ and the squared distance becomes

$$(\mathbf{x} - \boldsymbol{\mu})'\boldsymbol{\Sigma}^{-1}(\mathbf{x} - \boldsymbol{\mu})$$

$$= [x_1 - \mu_1, x_2 - \mu_2] \frac{1}{\sigma_{11}\sigma_{22}(1 - \rho_{12}^2)} \begin{bmatrix} \sigma_{22} & -\rho_{12}\sqrt{\sigma_{11}}\sqrt{\sigma_{22}} \\ -\rho_{12}\sqrt{\sigma_{11}}\sqrt{\sigma_{22}} & \sigma_{11} \end{bmatrix} \begin{bmatrix} x_1 - \mu_1 \\ x_2 - \mu_2 \end{bmatrix}$$

$$= \frac{\sigma_{22}(x_1 - \mu_1)^2 + \sigma_{11}(x_2 - \mu_2)^2 - 2\rho_{12}\sqrt{\sigma_{11}}\sqrt{\sigma_{22}}\,(x_1 - \mu_1)(x_2 - \mu_2)}{\sigma_{11}\sigma_{22}(1 - \rho_{12}^2)}$$

$$= \frac{1}{1 - \rho_{12}^2}\left[\left(\frac{x_1 - \mu_1}{\sqrt{\sigma_{11}}}\right)^2 + \left(\frac{x_2 - \mu_2}{\sqrt{\sigma_{22}}}\right)^2 - 2\rho_{12}\left(\frac{x_1 - \mu_1}{\sqrt{\sigma_{11}}}\right)\left(\frac{x_2 - \mu_2}{\sqrt{\sigma_{22}}}\right)\right] \tag{4-5}$$

The last expression is written in terms of the standardized values $(x_1 - \mu_1)/\sqrt{\sigma_{11}}$ and $(x_2 - \mu_2)/\sqrt{\sigma_{22}}$.

Next, since $|\boldsymbol{\Sigma}| = \sigma_{11}\sigma_{22} - \sigma_{12}^2 = \sigma_{11}\sigma_{22}(1 - \rho_{12}^2)$, we can substitute for $\boldsymbol{\Sigma}^{-1}$ and $|\boldsymbol{\Sigma}|$ in (4-4) to get the expression for the bivariate ($p = 2$) normal density involving the individual parameters μ_1, μ_2, σ_{11}, σ_{22}, and ρ_{12}. Thus

$$f(x_1, x_2) = \frac{1}{2\pi\sqrt{\sigma_{11}\sigma_{22}(1 - \rho_{12}^2)}} \tag{4-6}$$

$$\times \exp\left\{-\frac{1}{2(1 - \rho_{12}^2)}\left[\left(\frac{x_1 - \mu_1}{\sqrt{\sigma_{11}}}\right)^2 + \left(\frac{x_2 - \mu_2}{\sqrt{\sigma_{22}}}\right)^2 - 2\rho_{12}\left(\frac{x_1 - \mu_1}{\sqrt{\sigma_{11}}}\right)\left(\frac{x_2 - \mu_2}{\sqrt{\sigma_{22}}}\right)\right]\right\}$$

The expression in (4-6) is somewhat unwieldy and the compact general form in (4-4) is more informative in many ways. On the other hand, the expression in (4-6) is useful for discussing certain properties of the normal distribution. For example, if the random variables X_1 and X_2 are uncorrelated so that $\rho_{12} = 0$, the joint density can be written as the product of two univariate normal densities each of the form of (4-1). That is, $f(x_1, x_2) = f(x_1)f(x_2)$ and X_1 and X_2 are independent. [See (2-28).] This result is true in general. (See Result 4.5.)

Two bivariate distributions with $\sigma_{11} = \sigma_{22}$ are shown in Figure 4.2. In

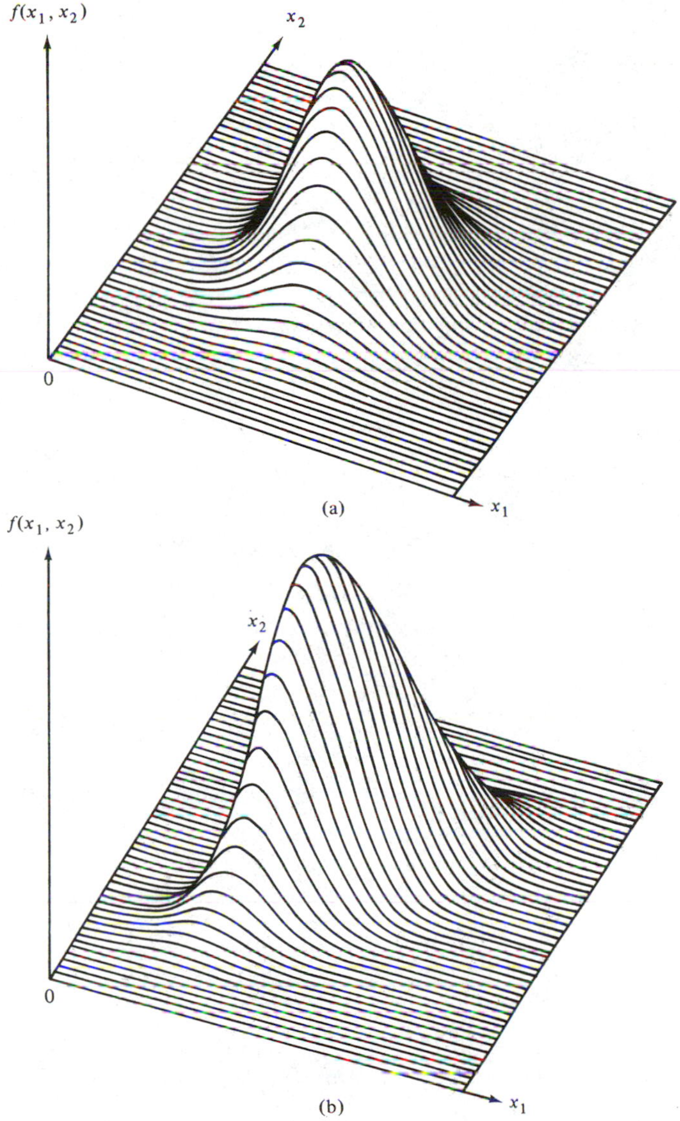

Figure 4.2 Two bivariate normal distributions. (a) $\sigma_{11} = \sigma_{22}$ and $\rho_{12} = 0$ (b) $\sigma_{11} = \sigma_{22}$, $\rho_{12} = .75$.

Figure 4.2(a), X_1 and X_2 are independent ($\rho_{12} = 0$). In Figure 4.2(b), $\rho_{12} = .75$. Notice how the presence of correlation causes the probability to concentrate along a line. ■

From the expression in (4-4) for the density of a p-dimensional normal variable, it should be clear that the paths of $\mathbf{x}$ values yielding a constant height for the density are ellipsoids. That is, the multivariate normal density is constant on surfaces where the distance $(\mathbf{x} - \boldsymbol{\mu})'\boldsymbol{\Sigma}^{-1}(\mathbf{x} - \boldsymbol{\mu})$ is constant. These paths are called *contours*.

Constant probability density contour

$$= \{\text{all } \mathbf{x} \text{ such that } (\mathbf{x} - \boldsymbol{\mu})'\boldsymbol{\Sigma}^{-1}(\mathbf{x} - \boldsymbol{\mu}) = c^2\}$$
$$= \text{surface of an ellipsoid centered at } \boldsymbol{\mu}$$

The axes of each ellipsoid of constant density are in the direction of the eigenvectors of $\boldsymbol{\Sigma}^{-1}$ and their lengths are proportional to the reciprocals of the square roots of the eigenvalues of $\boldsymbol{\Sigma}^{-1}$. Fortunately, we can avoid the calculation of $\boldsymbol{\Sigma}^{-1}$ when determining these axes since these ellipsoids are also determined by the eigenvalues and eigenvectors of $\boldsymbol{\Sigma}$. We state the correspondence formally for later reference.

Result 4.1. If $\boldsymbol{\Sigma}$ is positive definite so that $\boldsymbol{\Sigma}^{-1}$ exists,

$$\boldsymbol{\Sigma}\mathbf{e} = \lambda\mathbf{e} \quad \text{implies} \quad \boldsymbol{\Sigma}^{-1}\mathbf{e} = \left(\frac{1}{\lambda}\right)\mathbf{e}$$

so $(\lambda, \mathbf{e})$ is an eigenvalue-eigenvector pair for $\boldsymbol{\Sigma}$ corresponding to the pair $(1/\lambda, \mathbf{e})$ for $\boldsymbol{\Sigma}^{-1}$. Also, $\boldsymbol{\Sigma}^{-1}$ is positive definite.

Proof. For $\boldsymbol{\Sigma}$ positive definite and $\mathbf{e} \neq \mathbf{0}$ an eigenvector, we have $0 < \mathbf{e}'\boldsymbol{\Sigma}\mathbf{e} = \mathbf{e}'(\boldsymbol{\Sigma}\mathbf{e}) = \mathbf{e}'(\lambda\mathbf{e}) = \lambda\mathbf{e}'\mathbf{e} = \lambda$. Moreover $\mathbf{e} = \boldsymbol{\Sigma}^{-1}(\boldsymbol{\Sigma}\mathbf{e}) = \boldsymbol{\Sigma}^{-1}(\lambda\mathbf{e})$ or $\mathbf{e} = \lambda\boldsymbol{\Sigma}^{-1}\mathbf{e}$ and division by $\lambda > 0$ gives $\boldsymbol{\Sigma}^{-1}\mathbf{e} = (1/\lambda)\mathbf{e}$. Thus $(1/\lambda, \mathbf{e})$ is an eigenvalue-eigenvector pair for $\boldsymbol{\Sigma}^{-1}$. For any $p \times 1\mathbf{x}$,

$$\mathbf{x}'\boldsymbol{\Sigma}^{-1}\mathbf{x} = \mathbf{x}'\left(\sum_{i=1}^{p}(1/\lambda_i)\mathbf{e}_i\mathbf{e}_i'\right)\mathbf{x}$$
$$= \sum_{i=1}^{p}(1/\lambda_i)(\mathbf{x}'\mathbf{e}_i)^2 \geq 0$$

since each term $\lambda_i^{-1}(\mathbf{x}'\mathbf{e}_i)^2$ is nonnegative. Also, $\mathbf{x}'\mathbf{e}_i = 0$ for all i only if $\mathbf{x} = \mathbf{0}$. So $\mathbf{x} \neq \mathbf{0}$ implies $\sum_{i=1}^{p}(1/\lambda_i)(\mathbf{x}'\mathbf{e}_i)^2 > 0$ and $\boldsymbol{\Sigma}^{-1}$ is positive definite. ■

The following summarizes these concepts.

> Contours of constant density for the p-dimensional normal distribution are ellipsoids defined by $\mathbf{x}$ such that
>
> $$(\mathbf{x}-\boldsymbol{\mu})'\boldsymbol{\Sigma}^{-1}(\mathbf{x}-\boldsymbol{\mu}) = c^2 \tag{4-7}$$
>
> These ellipsoids are centered at $\boldsymbol{\mu}$ and have axes $\pm c\sqrt{\lambda_i}\,\mathbf{e}_i$, where $\boldsymbol{\Sigma}\mathbf{e}_i = \lambda_i \mathbf{e}_i$, $i = 1, 2, \ldots, p$.

A contour of constant density for a bivariate normal distribution with $\sigma_{11} = \sigma_{22}$ is obtained in the following example.

Example 4.2 (Contours of the Bivariate Normal Density)

We shall obtain the axes of constant probability density contours for a bivariate normal distribution when $\sigma_{11} = \sigma_{22}$. From (4-7) these axes are given by the eigenvalues and eigenvectors of $\boldsymbol{\Sigma}$. Here $|\boldsymbol{\Sigma} - \lambda\mathbf{I}| = 0$ becomes

$$0 = \begin{vmatrix} \sigma_{11}-\lambda & \sigma_{12} \\ \sigma_{12} & \sigma_{11}-\lambda \end{vmatrix} = (\sigma_{11}-\lambda)^2 - \sigma_{12}^2$$

$$= (\lambda - \sigma_{11} - \sigma_{12})(\lambda - \sigma_{11} + \sigma_{12})$$

Consequently, the eigenvalues are $\lambda_1 = \sigma_{11} + \sigma_{12}$ and $\lambda_2 = \sigma_{11} - \sigma_{12}$. The eigenvector $\mathbf{e}_1$ is determined from

$$\begin{bmatrix} \sigma_{11} & \sigma_{12} \\ \sigma_{12} & \sigma_{11} \end{bmatrix} \begin{bmatrix} e_1 \\ e_2 \end{bmatrix} = (\sigma_{11} + \sigma_{12}) \begin{bmatrix} e_1 \\ e_2 \end{bmatrix}$$

or

$$\sigma_{11}e_1 + \sigma_{12}e_2 = (\sigma_{11} + \sigma_{12})e_1$$
$$\sigma_{12}e_1 + \sigma_{11}e_2 = (\sigma_{11} + \sigma_{12})e_2$$

These equations imply $e_1 = e_2$ and, after normalization, the first eigenvalue-eigenvector pair is

$$\lambda_1 = \sigma_{11} + \sigma_{12}, \qquad \mathbf{e}_1 = \begin{bmatrix} \dfrac{1}{\sqrt{2}} \\ \dfrac{1}{\sqrt{2}} \end{bmatrix}$$

Similarly, $\lambda_2 = \sigma_{11} - \sigma_{12}$ yields the eigenvector $\mathbf{e}_2' = [1/\sqrt{2}, -1/\sqrt{2}]$.

When the covariance σ_{12} (or correlation ρ_{12}) is positive, $\lambda_1 = \sigma_{11} + \sigma_{12}$ is the *largest* eigenvalue and its associated eigenvector, $\mathbf{e}_1' = [1/\sqrt{2}, 1/\sqrt{2}]$ lies along the 45° line through the point $\boldsymbol{\mu}' = [\mu_1, \mu_2]$. This is true for any positive value of the covariance (correlation). Since the axes of the constant density ellipses are given by $\pm c\sqrt{\lambda_1}\,\mathbf{e}_1$ and $\pm c\sqrt{\lambda_2}\,\mathbf{e}_2$ [see (4-7)] and the eigenvectors

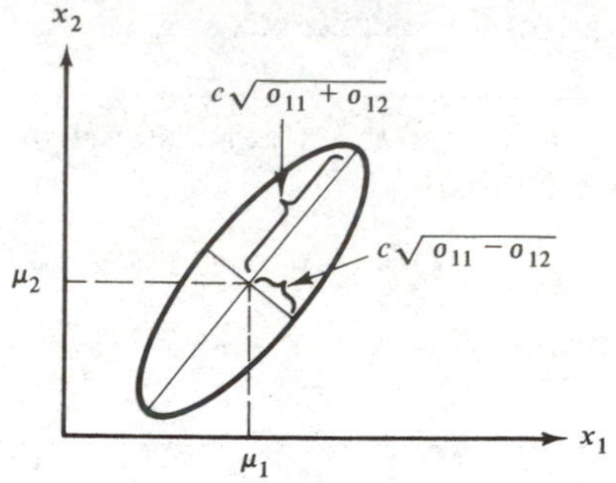

Figure 4.3 A constant density contour for a bivariate normal distribution with $\sigma_{11} = \sigma_{22}$ and $\sigma_{12} > 0$ (or $\rho_{12} > 0$).

each have length 1, the major axis will be associated with the largest eigenvalue. For positively correlated normal random variables then, the *major* axis of the constant density ellipses will be along the 45° line through $\boldsymbol{\mu}$ (see Figure 4.3). When the covariance (correlation) is negative, $\lambda_2 = \sigma_{11} - \sigma_{12}$ will be the largest eigenvalue and the major axes of the constant density ellipses will lie along a line at right angles to the 45° line through $\boldsymbol{\mu}$. (These results are true only for $\sigma_{11} = \sigma_{22}$.)

To summarize, the axes of the ellipses of constant density for a bivariate normal distribution with $\sigma_{11} = \sigma_{22}$ are determined by

$$\pm c\sqrt{\sigma_{11} + \sigma_{12}} \begin{bmatrix} \frac{1}{\sqrt{2}} \\ \frac{1}{\sqrt{2}} \end{bmatrix} \quad \text{and} \quad \pm c\sqrt{\sigma_{11} - \sigma_{12}} \begin{bmatrix} \frac{1}{\sqrt{2}} \\ \frac{-1}{\sqrt{2}} \end{bmatrix}$$ ■

We show in Result 4.7 that the choice $c^2 = \chi_p^2(\alpha)$, where $\chi_p^2(\alpha)$ is the upper (100α)th percentile of a chi-square distribution with p degrees of freedom, leads to contours that contain $(1 - \alpha) \times 100\%$ of the probability. Specifically, the following is true for a p-dimensional normal distribution.

> The solid ellipsoid of $\mathbf{x}$ values satisfying
>
> $$(\mathbf{x} - \boldsymbol{\mu})'\boldsymbol{\Sigma}^{-1}(\mathbf{x} - \boldsymbol{\mu}) \le \chi_p^2(\alpha) \tag{4-8}$$
>
> has probability $1 - \alpha$.

The constant density contours containing 50% and 90% of the probability under the bivariate normal surfaces in Figure 4.2 are pictured in Figure 4.4.

The p-variate normal density in (4-4) has a maximum value when the squared distance of (4-3) is zero; that is, when $\mathbf{x} = \boldsymbol{\mu}$. Thus $\boldsymbol{\mu}$ is the point of maximum density, or *mode*, as well as the expected value of $\mathbf{X}$, or *mean*. The fact that $\boldsymbol{\mu}$ is the mean of the multivariate normal distribution follows from the symmetry exhibited by the constant density contours. These contours are centered, or balanced, at $\boldsymbol{\mu}$.

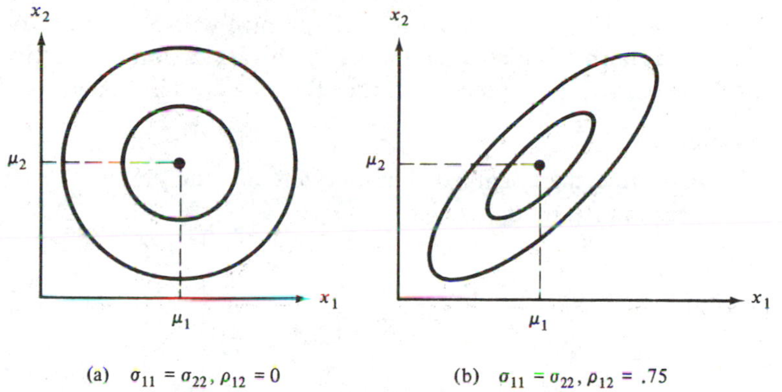

Figure 4.4 The 50% and 90% contours for the bivariate normal distributions in Figure 4.2.

Additional Properties of the Multivariate Normal Distribution

Certain properties of the normal distribution will be needed repeatedly in our explanations of statistical models and methods. These properties make it possible to manipulate normal distributions easily and as we suggested in Section 4.1, are partly responsible for the popularity of the normal distribution. The key properties, which we shall soon discuss in some mathematical detail, can be stated rather simply.

The following are true for a random vector $\mathbf{X}$ having a multivariate normal distribution.

1. Linear combinations of the components of $\mathbf{X}$ are normally distributed.
2. All subsets of the components of $\mathbf{X}$ have a (multivariate) normal distribution.
3. Zero covariance implies that the corresponding components are independently distributed.
4. The conditional distributions of the components are (multivariate) normal.

The statements above are reproduced mathematically in the following results. Many of these results are illustrated with examples. The proofs that are included should help improve your understanding of matrix manipulations and also lead you to an appreciation for the manner in which the results successively build on themselves.

Result 4.2 can be taken as a working definition of the normal distribution. With this in hand, the subsequent properties are almost immediate. Our partial proof of Result 4.2 indicates how the linear combination definition of a normal density relates to the multivariate density in (4-4).

Result 4.2. If $\mathbf{X}$ is distributed as $N_p(\boldsymbol{\mu}, \boldsymbol{\Sigma})$, then any linear combination of variables $\mathbf{a}'\mathbf{X} = a_1X_1 + a_2X_2 + \cdots + a_pX_p$ is distributed as $N(\mathbf{a}'\boldsymbol{\mu}, \mathbf{a}'\boldsymbol{\Sigma}\mathbf{a})$. Also, if $\mathbf{a}'\mathbf{X}$ is distributed as $N(\mathbf{a}'\boldsymbol{\mu}, \mathbf{a}'\boldsymbol{\Sigma}\mathbf{a})$ for every $\mathbf{a}$, then $\mathbf{X}$ must be $N_p(\boldsymbol{\mu}, \boldsymbol{\Sigma})$.

Proof. The expected value and variance of $\mathbf{a}'\mathbf{X}$ follow from (2-43). The fact that $\mathbf{a}'\mathbf{X}$ is normally distributed if $\mathbf{X}$ is multivariate normal is more difficult. You can find a proof in [1]. The second part of Result 4.2 is also demonstrated in [1]. ■

Example 4.3

Consider the linear combination $\mathbf{a}'\mathbf{X}$ of a multivariate normal random vector determined by the choice $\mathbf{a}' = [1, 0, \ldots, 0]$. Since

$$\mathbf{a}'\mathbf{X} = [1, 0, \ldots, 0] \begin{bmatrix} X_1 \\ X_2 \\ \vdots \\ X_p \end{bmatrix} = X_1$$

and

$$\mathbf{a}'\boldsymbol{\mu} = [1, 0, \ldots, 0] \begin{bmatrix} \mu_1 \\ \mu_2 \\ \vdots \\ \mu_p \end{bmatrix} = \mu_1$$

$$\mathbf{a}'\boldsymbol{\Sigma}\mathbf{a} = [1, 0, \ldots, 0] \begin{bmatrix} \sigma_{11} & \sigma_{12} & \cdots & \sigma_{1p} \\ \sigma_{12} & \sigma_{22} & \cdots & \sigma_{2p} \\ \vdots & \vdots & & \vdots \\ \sigma_{1p} & \sigma_{2p} & \cdots & \sigma_{pp} \end{bmatrix} \begin{bmatrix} 1 \\ 0 \\ \vdots \\ 0 \end{bmatrix} = \sigma_{11}$$

it follows from Result 4.2 that X_1 is distributed as $N(\mu_1, \sigma_{11})$. More generally, the marginal distribution of any component X_i of $\mathbf{X}$ is $N(\mu_i, \sigma_{ii})$. ■

The next result considers several linear combinations of a multivariate normal vector $\mathbf{X}$.

Result 4.3. If $\mathbf{X}$ is distributed as $N_p(\boldsymbol{\mu}, \boldsymbol{\Sigma})$, the q linear combinations

$$\underset{(q\times p)}{\mathbf{A}} \underset{(p\times 1)}{\mathbf{X}} = \begin{bmatrix} a_{11}X_1 + \cdots + a_{1p}X_p \\ a_{21}X_1 + \cdots + a_{2p}X_p \\ \vdots \\ a_{q1}X_1 + \cdots + a_{qp}X_p \end{bmatrix}$$

are distributed as $N_q(\mathbf{A}\boldsymbol{\mu}, \mathbf{A}\boldsymbol{\Sigma}\mathbf{A}')$. Also $\underset{(p\times 1)}{\mathbf{X}} + \underset{(p\times 1)}{\mathbf{d}}$, where $\mathbf{d}$ is a vector of constants, is distributed as $N_p(\boldsymbol{\mu} + \mathbf{d}, \boldsymbol{\Sigma})$.

Proof. The expected value $E(\mathbf{AX})$ and the covariance matrix of $\mathbf{AX}$ follow from (2-45). Any linear combination $\mathbf{b}'(\mathbf{AX})$ is a linear combination of $\mathbf{X}$, of the form $\mathbf{a}'\mathbf{X}$ with $\mathbf{a} = \mathbf{A}'\mathbf{b}$. Thus the conclusion concerning $\mathbf{AX}$ follows directly from Result 4.2.

The second part of the result can be obtained by considering $\mathbf{a}'(\mathbf{X} + \mathbf{d}) = \mathbf{a}'\mathbf{X} + (\mathbf{a}'\mathbf{d})$, where $\mathbf{a}'\mathbf{X}$ is distributed as $N(\mathbf{a}'\boldsymbol{\mu}, \mathbf{a}'\boldsymbol{\Sigma}\mathbf{a})$. It is known from the univariate case that adding a constant $\mathbf{a}'\mathbf{d}$ to the random variable $\mathbf{a}'\mathbf{X}$ leaves the variance unchanged and translates the mean to $\mathbf{a}'\boldsymbol{\mu} + \mathbf{a}'\mathbf{d} = \mathbf{a}'(\boldsymbol{\mu} + \mathbf{d})$. Since $\mathbf{a}$ was arbitrary, $\mathbf{X} + \mathbf{d}$ is distributed as $N_p(\boldsymbol{\mu} + \mathbf{d}, \boldsymbol{\Sigma})$. ■

Example 4.4

For $\mathbf{X}$ distributed as $N_3(\boldsymbol{\mu}, \boldsymbol{\Sigma})$, find the distribution of

$$\begin{bmatrix} X_1 - X_2 \\ X_2 - X_3 \end{bmatrix} = \begin{bmatrix} 1 & -1 & 0 \\ 0 & 1 & -1 \end{bmatrix} \begin{bmatrix} X_1 \\ X_2 \\ X_3 \end{bmatrix} = \mathbf{AX}$$

By Result 4.3, the distribution of $\mathbf{AX}$ is multivariate normal with mean

$$\mathbf{A}\boldsymbol{\mu} = \begin{bmatrix} 1 & -1 & 0 \\ 0 & 1 & -1 \end{bmatrix} \begin{bmatrix} \mu_1 \\ \mu_2 \\ \mu_3 \end{bmatrix} = \begin{bmatrix} \mu_1 - \mu_2 \\ \mu_2 - \mu_3 \end{bmatrix}$$

and covariance matrix

$$\mathbf{A\Sigma A}' = \begin{bmatrix} 1 & -1 & 0 \\ 0 & 1 & -1 \end{bmatrix} \begin{bmatrix} \sigma_{11} & \sigma_{12} & \sigma_{13} \\ \sigma_{12} & \sigma_{22} & \sigma_{23} \\ \sigma_{13} & \sigma_{23} & \sigma_{33} \end{bmatrix} \begin{bmatrix} 1 & 0 \\ -1 & 1 \\ 0 & -1 \end{bmatrix}$$

$$= \begin{bmatrix} \sigma_{11} - \sigma_{12} & \sigma_{12} - \sigma_{22} & \sigma_{13} - \sigma_{23} \\ \sigma_{12} - \sigma_{13} & \sigma_{22} - \sigma_{23} & \sigma_{23} - \sigma_{33} \end{bmatrix} \begin{bmatrix} 1 & 0 \\ -1 & 1 \\ 0 & -1 \end{bmatrix}$$

$$= \begin{bmatrix} \sigma_{11} - 2\sigma_{12} + \sigma_{22} & \sigma_{12} + \sigma_{23} - \sigma_{22} - \sigma_{13} \\ \sigma_{12} + \sigma_{23} - \sigma_{22} - \sigma_{13} & \sigma_{22} - 2\sigma_{23} + \sigma_{33} \end{bmatrix}$$

Alternatively, the mean vector $\mathbf{A}\boldsymbol{\mu}$ and covariance matrix $\mathbf{A\Sigma A}'$ may be verified by direct calculation of the means and covariances of the two random variables $Y_1 = X_1 - X_2$ and $Y_2 = X_2 - X_3$. ■

We have mentioned that all subsets of a multivariate normal random vector $\mathbf{X}$ are themselves normally distributed. We state this property formally as Result 4.4.

Result 4.4. All subsets of $\mathbf{X}$ are normally distributed. If we partition $\mathbf{X}$, its mean vector $\boldsymbol{\mu}$, and covariance matrix $\boldsymbol{\Sigma}$ as

$$\underset{(p\times 1)}{\mathbf{X}} = \begin{bmatrix} \underset{(q\times 1)}{\mathbf{X}_1} \\ \text{-----} \\ \underset{((p-q)\times 1)}{\mathbf{X}_2} \end{bmatrix} \qquad \underset{(p\times 1)}{\boldsymbol{\mu}} = \begin{bmatrix} \underset{(q\times 1)}{\boldsymbol{\mu}_1} \\ \text{-----} \\ \underset{((p-q)\times 1)}{\boldsymbol{\mu}_2} \end{bmatrix}$$

and

$$\underset{(p\times p)}{\boldsymbol{\Sigma}} = \left[\begin{array}{c|c} \underset{(q\times q)}{\boldsymbol{\Sigma}_{11}} & \underset{(q\times(p-q))}{\boldsymbol{\Sigma}_{12}} \\ \hline \underset{((p-q)\times q)}{\boldsymbol{\Sigma}_{21}} & \underset{((p-q)\times(p-q))}{\boldsymbol{\Sigma}_{22}} \end{array}\right]$$

then $\mathbf{X}_1$ is distributed as $N_q(\boldsymbol{\mu}_1, \boldsymbol{\Sigma}_{11})$.

Proof. Set $\underset{(q\times p)}{\mathbf{A}} = \left[\underset{(q\times q)}{\mathbf{I}} \;\middle|\; \underset{(q\times(p-q))}{\mathbf{0}}\right]$ in Result 4.3 and the conclusion follows. To apply Result 4.4 to an *arbitrary* subset of the components of $\mathbf{X}$, we simply relabel the subset of interest as $\mathbf{X}_1$ and select the corresponding component means and covariances as $\boldsymbol{\mu}_1$ and $\boldsymbol{\Sigma}_{11}$, respectively. ■

Example 4.5

If $\mathbf{X}$ is distributed as $N_5(\boldsymbol{\mu}, \boldsymbol{\Sigma})$, find the distribution of $\begin{bmatrix} X_2 \\ X_4 \end{bmatrix}$ We set $\mathbf{X}_1 = \begin{bmatrix} X_2 \\ X_4 \end{bmatrix}$, $\boldsymbol{\mu}_1 = \begin{bmatrix} \mu_2 \\ \mu_4 \end{bmatrix}$, $\boldsymbol{\Sigma}_{11} = \begin{bmatrix} \sigma_{22} & \sigma_{24} \\ \sigma_{24} & \sigma_{44} \end{bmatrix}$ and note that with this assignment, $\mathbf{X}$, $\boldsymbol{\mu}$, and $\boldsymbol{\Sigma}$ can be rearranged and partitioned as

$$\mathbf{X} = \begin{bmatrix} X_2 \\ X_4 \\ \hline X_1 \\ X_3 \\ X_5 \end{bmatrix}, \quad \boldsymbol{\mu} = \begin{bmatrix} \mu_2 \\ \mu_4 \\ \hline \mu_1 \\ \mu_3 \\ \mu_5 \end{bmatrix}, \quad \boldsymbol{\Sigma} = \left[\begin{array}{cc|ccc} \sigma_{22} & \sigma_{24} & \sigma_{12} & \sigma_{23} & \sigma_{25} \\ \sigma_{24} & \sigma_{44} & \sigma_{14} & \sigma_{34} & \sigma_{45} \\ \hline \sigma_{12} & \sigma_{14} & \sigma_{11} & \sigma_{13} & \sigma_{15} \\ \sigma_{23} & \sigma_{34} & \sigma_{13} & \sigma_{33} & \sigma_{35} \\ \sigma_{25} & \sigma_{45} & \sigma_{15} & \sigma_{35} & \sigma_{55} \end{array}\right]$$

or

$$\mathbf{X} = \begin{bmatrix} \underset{(2\times 1)}{\mathbf{X}_1} \\ \hline \underset{(3\times 1)}{\mathbf{X}_2} \end{bmatrix}, \quad \boldsymbol{\mu} = \begin{bmatrix} \underset{(2\times 1)}{\boldsymbol{\mu}_1} \\ \hline \underset{(3\times 1)}{\boldsymbol{\mu}_2} \end{bmatrix}, \quad \boldsymbol{\Sigma} = \left[\begin{array}{c|c} \underset{(2\times 2)}{\boldsymbol{\Sigma}_{11}} & \underset{(2\times 3)}{\boldsymbol{\Sigma}_{12}} \\ \hline \underset{(3\times 2)}{\boldsymbol{\Sigma}_{21}} & \underset{(3\times 3)}{\boldsymbol{\Sigma}_{22}} \end{array}\right]$$

Thus, using Result 4.4, for

$$\mathbf{X}_1 = \begin{bmatrix} X_2 \\ X_4 \end{bmatrix}$$

we have the distribution

$$N_2(\boldsymbol{\mu}_1, \boldsymbol{\Sigma}_{11}) = N_2\left(\begin{bmatrix} \mu_2 \\ \mu_4 \end{bmatrix}, \begin{bmatrix} \sigma_{22} & \sigma_{24} \\ \sigma_{24} & \sigma_{44} \end{bmatrix}\right)$$

It is clear from this example that the normal distribution for any subset can be expressed by simply selecting the appropriate means and covariances from the original $\boldsymbol{\mu}$ and $\boldsymbol{\Sigma}$. The formal process of relabeling and partitioning is unnecessary. ■

We are now in a position to state that zero correlation between normal random variables or sets of normal random variables is equivalent to statistical independence.

Result 4.5. (a) If $\underset{(q_1\times 1)}{\mathbf{X}_1}$ and $\underset{(q_2\times 1)}{\mathbf{X}_2}$ are independent, then it is always true that $\text{Cov}(\mathbf{X}_1, \mathbf{X}_2) = \mathbf{0}$, a $q_1 \times q_2$ matrix of zeros.

(b) If $\begin{bmatrix} \mathbf{X}_1 \\ \hline \mathbf{X}_2 \end{bmatrix}$ is $N_{q_1+q_2}\left(\begin{bmatrix} \boldsymbol{\mu}_1 \\ \hline \boldsymbol{\mu}_2 \end{bmatrix}, \begin{bmatrix} \boldsymbol{\Sigma}_{11} & \boldsymbol{\Sigma}_{12} \\ \hline \boldsymbol{\Sigma}_{21} & \boldsymbol{\Sigma}_{22} \end{bmatrix}\right)$, then $\mathbf{X}_1$ and $\mathbf{X}_2$ are independent if and only if $\boldsymbol{\Sigma}_{12} = \mathbf{0}$.

(c) If $\mathbf{X}_1$ and $\mathbf{X}_2$ are independent and are distributed as $N_{q_1}(\boldsymbol{\mu}_1, \boldsymbol{\Sigma}_{11})$ and $N_{q_2}(\boldsymbol{\mu}_2, \boldsymbol{\Sigma}_{22})$, respectively, then $\begin{bmatrix} \mathbf{X}_1 \\ \hline \mathbf{X}_2 \end{bmatrix}$ is multivariate normal $N_{q_1+q_2}\left(\begin{bmatrix} \boldsymbol{\mu}_1 \\ \hline \boldsymbol{\mu}_2 \end{bmatrix}, \begin{bmatrix} \boldsymbol{\Sigma}_{11} & \mathbf{0} \\ \hline \mathbf{0}' & \boldsymbol{\Sigma}_{22} \end{bmatrix}\right)$.

Proof. (See Exercise 4.11 for partial proofs based upon factoring the density function when $\boldsymbol{\Sigma}_{12} = \mathbf{0}$.) ∎

Example 4.6

Let $\underset{(3\times 1)}{\mathbf{X}}$ be $N_3(\boldsymbol{\mu}, \boldsymbol{\Sigma})$ with

$$\boldsymbol{\Sigma} = \begin{bmatrix} 4 & 1 & 0 \\ 1 & 3 & 0 \\ 0 & 0 & 2 \end{bmatrix}$$

Are X_1 and X_2 independent? What about (X_1, X_2) and X_3?

Since X_1 and X_2 have covariance $\sigma_{12} = 1$, they are not independent. However, partitioning $\mathbf{X}$ and $\boldsymbol{\Sigma}$ as

$$\mathbf{X} = \begin{bmatrix} X_1 \\ X_2 \\ \hline X_3 \end{bmatrix}, \qquad \boldsymbol{\Sigma} = \left[\begin{array}{cc|c} 4 & 1 & 0 \\ 1 & 3 & 0 \\ \hline 0 & 0 & 2 \end{array}\right] = \left[\begin{array}{c|c} \underset{(2\times 2)}{\boldsymbol{\Sigma}_{11}} & \underset{(2\times 1)}{\boldsymbol{\Sigma}_{12}} \\ \hline \underset{(1\times 2)}{\boldsymbol{\Sigma}_{21}} & \underset{(1\times 1)}{\boldsymbol{\Sigma}_{22}} \end{array}\right]$$

we see that $\mathbf{X}_1 = \begin{bmatrix} X_1 \\ X_2 \end{bmatrix}$ and X_3 have covariance matrix $\boldsymbol{\Sigma}_{12} = \begin{bmatrix} 0 \\ 0 \end{bmatrix}$. Therefore (X_1, X_2) and X_3 are independent by Result 4.5. This implies X_3 is independent of X_1 and also of X_2. ∎

We pointed out in our discussion of the bivariate normal distribution that $\rho_{12} = 0$ (zero correlation) implied independence because the joint density function [see (4-6)] could then be written as the product of the marginal (normal) densities of X_1 and X_2. This fact, which we encouraged you to verify directly, is simply a special case of Result 4.5 with $q_1 = q_2 = 1$.

Result 4.6. Let $\mathbf{X} = \begin{bmatrix} \mathbf{X}_1 \\ \hline \mathbf{X}_2 \end{bmatrix}$ be distributed as $N_p(\boldsymbol{\mu}, \boldsymbol{\Sigma})$ with $\boldsymbol{\mu} = \begin{bmatrix} \boldsymbol{\mu}_1 \\ \hline \boldsymbol{\mu}_2 \end{bmatrix}$, $\boldsymbol{\Sigma} = \begin{bmatrix} \boldsymbol{\Sigma}_{11} & \boldsymbol{\Sigma}_{12} \\ \hline \boldsymbol{\Sigma}_{21} & \boldsymbol{\Sigma}_{22} \end{bmatrix}$ and $|\boldsymbol{\Sigma}_{22}| > 0$. Then the conditional distribution of $\mathbf{X}_1$, given

$\mathbf{X}_2 = \mathbf{x}_2$, is normal with

$$\text{Mean} = \boldsymbol{\mu}_1 + \boldsymbol{\Sigma}_{12}\boldsymbol{\Sigma}_{22}^{-1}(\mathbf{x}_2 - \boldsymbol{\mu}_2)$$

and

$$\text{Covariance} = \boldsymbol{\Sigma}_{11} - \boldsymbol{\Sigma}_{12}\boldsymbol{\Sigma}_{22}^{-1}\boldsymbol{\Sigma}_{21}$$

Note that the covariance does not depend on the value $\mathbf{x}_2$ of the conditioning variable.

Proof. We shall give an indirect proof. (See Exercise 4.10, which uses the densities directly.) Take

$$\underset{(p\times p)}{\mathbf{A}} = \left[\begin{array}{c|c} \underset{(q\times q)}{\mathbf{I}} & \underset{q\times(p-q)}{-\boldsymbol{\Sigma}_{12}\boldsymbol{\Sigma}_{22}^{-1}} \\ \hline \underset{(p-q)\times q}{\mathbf{0}} & \underset{(p-q)\times(p-q)}{\mathbf{I}} \end{array}\right]$$

so

$$\mathbf{A}(\mathbf{X} - \boldsymbol{\mu}) = \mathbf{A}\left[\begin{array}{c} \mathbf{X}_1 - \boldsymbol{\mu}_1 \\ \hline \mathbf{X}_2 - \boldsymbol{\mu}_2 \end{array}\right] = \left[\begin{array}{c} \mathbf{X}_1 - \boldsymbol{\mu}_1 - \boldsymbol{\Sigma}_{12}\boldsymbol{\Sigma}_{22}^{-1}(\mathbf{X}_2 - \boldsymbol{\mu}_2) \\ \hline \mathbf{X}_2 - \boldsymbol{\mu}_2 \end{array}\right]$$

is jointly normal with covariance matrix $\mathbf{A\Sigma A}'$ given by

$$\left[\begin{array}{c|c} \mathbf{I} & -\boldsymbol{\Sigma}_{12}\boldsymbol{\Sigma}_{22}^{-1} \\ \hline \mathbf{0} & \mathbf{I} \end{array}\right]\left[\begin{array}{c|c} \boldsymbol{\Sigma}_{11} & \boldsymbol{\Sigma}_{12} \\ \hline \boldsymbol{\Sigma}_{21} & \boldsymbol{\Sigma}_{22} \end{array}\right]\left[\begin{array}{c|c} \mathbf{I} & \mathbf{0}' \\ \hline (-\boldsymbol{\Sigma}_{12}\boldsymbol{\Sigma}_{22}^{-1})' & \mathbf{I} \end{array}\right] = \left[\begin{array}{c|c} \boldsymbol{\Sigma}_{11} - \boldsymbol{\Sigma}_{12}\boldsymbol{\Sigma}_{22}^{-1}\boldsymbol{\Sigma}_{21} & \mathbf{0}' \\ \hline \mathbf{0} & \boldsymbol{\Sigma}_{22} \end{array}\right]$$

Since $\mathbf{X}_1 - \boldsymbol{\mu}_1 - \boldsymbol{\Sigma}_{12}\boldsymbol{\Sigma}_{22}^{-1}(\mathbf{X}_2 - \boldsymbol{\mu}_2)$ and $\mathbf{X}_2 - \boldsymbol{\mu}_2$ have zero covariance, they are independent. Moreover $\mathbf{X}_1 - \boldsymbol{\mu}_1 - \boldsymbol{\Sigma}_{12}\boldsymbol{\Sigma}_{22}^{-1}(\mathbf{X}_2 - \boldsymbol{\mu}_2)$ has distribution $N_q(\mathbf{0}, \boldsymbol{\Sigma}_{11} - \boldsymbol{\Sigma}_{12}\boldsymbol{\Sigma}_{22}^{-1}\boldsymbol{\Sigma}_{21})$. Given $\mathbf{X}_2 = \mathbf{x}_2$, $\boldsymbol{\mu}_1 + \boldsymbol{\Sigma}_{12}\boldsymbol{\Sigma}_{22}^{-1}(\mathbf{x}_2 - \boldsymbol{\mu}_2)$ is a constant. Because $\mathbf{X}_1 - \boldsymbol{\mu}_1 - \boldsymbol{\Sigma}_{12}\boldsymbol{\Sigma}_{22}^{-1}(\mathbf{X}_2 - \boldsymbol{\mu}_2)$ and $\mathbf{X}_2 - \boldsymbol{\mu}_2$ are independent, the conditional distribution of $\mathbf{X}_1 - \boldsymbol{\mu}_1 - \boldsymbol{\Sigma}_{12}\boldsymbol{\Sigma}_{22}^{-1}(\mathbf{x}_2 - \boldsymbol{\mu}_2)$ is the same as the unconditional distribution of $\mathbf{X}_1 - \boldsymbol{\mu}_1 - \boldsymbol{\Sigma}_{12}\boldsymbol{\Sigma}_{22}^{-1}(\mathbf{X}_2 - \boldsymbol{\mu}_2)$. Since $\mathbf{X}_1 - \boldsymbol{\mu}_1 - \boldsymbol{\Sigma}_{12}\boldsymbol{\Sigma}_{22}^{-1}(\mathbf{X}_2 - \boldsymbol{\mu}_2)$ is $N_q(\mathbf{0}, \boldsymbol{\Sigma}_{11} - \boldsymbol{\Sigma}_{12}\boldsymbol{\Sigma}_{22}^{-1}\boldsymbol{\Sigma}_{21})$, so is the random vector $\mathbf{X}_1 - \boldsymbol{\mu}_1 - \boldsymbol{\Sigma}_{12}\boldsymbol{\Sigma}_{22}^{-1}(\mathbf{x}_2 - \boldsymbol{\mu}_2)$ when $\mathbf{X}_2$ has the particular value $\mathbf{x}_2$. Equivalently, given $\mathbf{X}_2 = \mathbf{x}_2$, $\mathbf{X}_1$ is distributed as $N_q(\boldsymbol{\mu}_1 + \boldsymbol{\Sigma}_{12}\boldsymbol{\Sigma}_{22}^{-1}(\mathbf{x}_2 - \boldsymbol{\mu}_2), \boldsymbol{\Sigma}_{11} - \boldsymbol{\Sigma}_{12}\boldsymbol{\Sigma}_{22}^{-1}\boldsymbol{\Sigma}_{21})$. ■

Example 4.7 (The Conditional Density of a Bivariate Normal Distribution)

The conditional density of X_1, given $X_2 = x_2$ for any bivariate distribution, is defined by

$$f(x_1 \mid x_2) = \{\text{conditional density of } X_1 \text{ given } X_2 = x_2\} = \frac{f(x_1, x_2)}{f(x_2)}$$

where $f(x_2)$ is the marginal distribution of X_2. If $f(x_1, x_2)$ is the bivariate normal density, show that $f(x_1 \mid x_2)$ is

$$N\left(\mu_1 + \frac{\sigma_{12}}{\sigma_{22}}(x_2 - \mu_2), \sigma_{11} - \frac{\sigma_{12}^2}{\sigma_{22}}\right)$$

Here $\sigma_{11} - \sigma_{12}^2/\sigma_{22} = \sigma_{11}(1 - \rho_{12}^2)$. The two terms involving $x_1 - \mu_1$ in the exponent of the bivariate normal density [see Equation (4-6)] become, apart from the multiplicative constant $-1/2(1 - \rho_{12}^2)$,

$$\frac{(x_1 - \mu_1)^2}{\sigma_{11}} - 2\rho_{12}\frac{(x_1 - \mu_1)(x_2 - \mu_2)}{\sqrt{\sigma_{11}}\sqrt{\sigma_{22}}}$$
$$= \frac{1}{\sigma_{11}}\left[x_1 - \mu_1 - \rho_{12}\frac{\sqrt{\sigma_{11}}}{\sqrt{\sigma_{22}}}(x_2 - \mu_2)\right]^2 - \frac{\rho_{12}^2}{\sigma_{22}}(x_2 - \mu_2)^2$$

Because $\rho_{12} = \sigma_{12}/\sqrt{\sigma_{11}}\sqrt{\sigma_{22}}$, or $\rho_{12}\sqrt{\sigma_{11}}/\sqrt{\sigma_{22}} = \sigma_{12}/\sigma_{22}$, the complete exponent is

$$\frac{-1}{2(1 - \rho_{12}^2)}\left[\frac{(x_1 - \mu_1)^2}{\sigma_{11}} - 2\rho_{12}\frac{(x_1 - \mu_1)(x_2 - \mu_2)}{\sqrt{\sigma_{11}}\sqrt{\sigma_{22}}} + \frac{(x_2 - \mu_2)^2}{\sigma_{22}}\right]$$
$$= \frac{-1}{2\sigma_{11}(1 - \rho_{12}^2)}\left(x_1 - \mu_1 - \rho_{12}\frac{\sqrt{\sigma_{11}}}{\sqrt{\sigma_{22}}}(x_2 - \mu_2)\right)^2$$
$$- \frac{1}{2(1 - \rho_{12}^2)}\left(\frac{1}{\sigma_{22}} - \frac{\rho_{12}^2}{\sigma_{22}}\right)(x_2 - \mu_2)^2$$
$$= \frac{-1}{2\sigma_{11}(1 - \rho_{12}^2)}\left(x_1 - \mu_1 - \frac{\sigma_{12}}{\sigma_{22}}(x_2 - \mu_2)\right)^2 - \frac{1}{2}\frac{(x_2 - \mu_2)^2}{\sigma_{22}}$$

The constant term $2\pi\sqrt{\sigma_{11}\sigma_{22}(1 - \rho_{12}^2)}$ also factors as

$$\sqrt{2\pi}\sqrt{\sigma_{22}} \times \sqrt{2\pi}\sqrt{\sigma_{11}(1 - \rho_{12}^2)}$$

Dividing the joint density of X_1 and X_2 by the marginal density

$$f(x_2) = \frac{1}{\sqrt{2\pi}\sqrt{\sigma_{22}}}e^{-(x_2-\mu_2)^2/2\sigma_{22}}$$

and canceling terms yields the conditional density

$$f(x_1 \mid x_2) = \frac{f(x_1, x_2)}{f(x_2)}$$
$$= \frac{1}{\sqrt{2\pi}\sqrt{\sigma_{11}(1 - \rho_{12}^2)}}e^{-[x_1-\mu_1-(\sigma_{12}/\sigma_{22})(x_2-\mu_2)]^2/2\sigma_{11}(1-\rho_{12}^2)},$$
$$-\infty < x_1 < \infty$$

Thus with our customary notation, the conditional distribution of X_1 given $X_2 = x_2$ is $N(\mu_1 + (\sigma_{12}/\sigma_{22})(x_2 - \mu_2),\ \sigma_{11}(1 - \rho_{12}^2))$. Note that $\mathbf{\Sigma}_{11} - \mathbf{\Sigma}_{12}\mathbf{\Sigma}_{22}^{-1}\mathbf{\Sigma}_{21} = \sigma_{11} - \sigma_{12}^2/\sigma_{22} = \sigma_{11}(1 - \rho_{12}^2)$ and $\mathbf{\Sigma}_{12}\mathbf{\Sigma}_{22}^{-1} = \sigma_{12}/\sigma_{22}$ agreeing with Result 4.6, which we obtained by an indirect method. ■

For the multivariate normal situation it is worth emphasizing the following.

1. All conditional distributions are (multivariate) normal.
2. The conditional mean is of the form

$$\begin{gathered}\mu_1 + \beta_{1,q+1}(x_{q+1} - \mu_{q+1}) + \cdots + \beta_{1,p}(x_p - \mu_p) \\ \vdots \\ \mu_q + \beta_{q,q+1}(x_{q+1} - \mu_{q+1}) + \cdots + \beta_{q,p}(x_p - \mu_p)\end{gathered} \tag{4-9}$$

where the β's are defined by

$$\mathbf{\Sigma}_{12}\mathbf{\Sigma}_{22}^{-1} = \begin{bmatrix} \beta_{1,q+1} & \beta_{1,q+2} & \cdots & \beta_{1,p} \\ \beta_{2,q+1} & \beta_{2,q+2} & \cdots & \beta_{2,p} \\ \vdots & \vdots & & \vdots \\ \beta_{q,q+1} & \beta_{q,q+2} & \cdots & \beta_{q,p} \end{bmatrix}$$

3. The conditional covariance, $\mathbf{\Sigma}_{11} - \mathbf{\Sigma}_{12}\mathbf{\Sigma}_{22}^{-1}\mathbf{\Sigma}_{21}$, does not depend upon the value(s) of the conditioning variable(s).

We conclude this section by presenting two final properties of multivariate normal random vectors. One is concerned with the probability content of the ellipsoids of constant density. The other result discusses the distribution of another form of linear combinations.

The chi-square distribution determines the variability of the sample variance $s^2 = s_{11}$ for samples from a univariate normal population. It also plays a basic role in the multivariate case.

Result 4.7. Let $\mathbf{X}$ be distributed as $N_p(\boldsymbol{\mu}, \mathbf{\Sigma})$ with $|\mathbf{\Sigma}| > 0$. Then:

(a) $(\mathbf{X} - \boldsymbol{\mu})'\mathbf{\Sigma}^{-1}(\mathbf{X} - \boldsymbol{\mu})$ is distributed as χ_p^2, where χ_p^2 denotes the chi-square distribution with p degrees of freedom.

(b) The $N_p(\boldsymbol{\mu}, \mathbf{\Sigma})$ distribution assigns probability $1 - \alpha$ to the solid ellipsoid $\{\mathbf{x} : (\mathbf{x} - \boldsymbol{\mu})'\mathbf{\Sigma}^{-1}(\mathbf{x} - \boldsymbol{\mu}) \le \chi_p^2(\alpha)\}$, where $\chi_p^2(\alpha)$ denotes the upper (100α)th percentile of the χ_p^2 distribution.

Proof. We know χ_p^2 is defined as the distribution of the sum $Z_1^2 + Z_2^2 + \cdots + Z_p^2$ where $Z_1, Z_2, \ldots, Z_p$ are independent $N(0, 1)$ random variables. Next by the spectral decomposition [see Equations (2-16) and (2-21) with $\mathbf{A} = \mathbf{\Sigma}$, and see Result 4.1] $\mathbf{\Sigma}^{-1} = \sum_{i=1}^{p} \frac{1}{\lambda_i}\mathbf{e}_i\mathbf{e}_i'$, where $\mathbf{\Sigma}\mathbf{e}_i = \lambda_i\mathbf{e}_i$ so $\mathbf{\Sigma}^{-1}\mathbf{e}_i = (1/\lambda_i)\mathbf{e}_i$. Consequently, $(\mathbf{X} - \boldsymbol{\mu})'\mathbf{\Sigma}^{-1}(\mathbf{X} - \boldsymbol{\mu}) = \sum_{i=1}^{p} (1/\lambda_i)(\mathbf{X} - \boldsymbol{\mu})'\mathbf{e}_i\mathbf{e}_i'(\mathbf{X} - \boldsymbol{\mu}) = \sum_{i=1}^{p} (1/\lambda_i)(\mathbf{e}_i'(\mathbf{X} - \boldsymbol{\mu}))^2 =$

$\sum_{i=1}^{p} [(1/\sqrt{\lambda_i})\mathbf{e}_i'(\mathbf{X}-\boldsymbol{\mu})]^2 = \sum_{i=1}^{p} Z_i^2$, for instance. Now $\mathbf{Z} = \mathbf{A}(\mathbf{X}-\boldsymbol{\mu})$, where

$$\underset{(p\times 1)}{\mathbf{Z}} = \begin{bmatrix} Z_1 \\ Z_2 \\ \vdots \\ Z_p \end{bmatrix}, \qquad \underset{(p\times p)}{\mathbf{A}} = \begin{bmatrix} \frac{1}{\sqrt{\lambda_1}}\mathbf{e}_1' \\ \frac{1}{\sqrt{\lambda_2}}\mathbf{e}_2' \\ \vdots \\ \frac{1}{\sqrt{\lambda_p}}\mathbf{e}_p' \end{bmatrix}$$

and $\mathbf{X}-\boldsymbol{\mu}$ is distributed as $N_p(\mathbf{0}, \boldsymbol{\Sigma})$. Therefore, by Result 4.3, $\mathbf{Z} = \mathbf{A}(\mathbf{X}-\boldsymbol{\mu})$ is distributed as $N_p(\mathbf{0}, \mathbf{A}\boldsymbol{\Sigma}\mathbf{A}')$, where

$$\underset{(p\times p)}{\mathbf{A}}\,\underset{(p\times p)}{\boldsymbol{\Sigma}}\,\underset{(p\times p)}{\mathbf{A}'} = \begin{bmatrix} \frac{1}{\sqrt{\lambda_1}}\mathbf{e}_1' \\ \frac{1}{\sqrt{\lambda_2}}\mathbf{e}_2' \\ \vdots \\ \frac{1}{\sqrt{\lambda_p}}\mathbf{e}_p' \end{bmatrix} \left[\sum_{i=1}^{p} \lambda_i \mathbf{e}_i \mathbf{e}_i'\right] \left[\frac{1}{\sqrt{\lambda_1}}\mathbf{e}_1 \;\vdots\; \frac{1}{\sqrt{\lambda_2}}\mathbf{e}_2 \;\vdots\; \cdots \;\vdots\; \frac{1}{\sqrt{\lambda_p}}\mathbf{e}_p\right]$$

$$= \begin{bmatrix} \sqrt{\lambda_1}\,\mathbf{e}_1' \\ \sqrt{\lambda_2}\,\mathbf{e}_2' \\ \vdots \\ \sqrt{\lambda_p}\,\mathbf{e}_p' \end{bmatrix} \left[\frac{1}{\sqrt{\lambda_1}}\mathbf{e}_1 \;\vdots\; \frac{1}{\sqrt{\lambda_2}}\mathbf{e}_2 \;\vdots\; \cdots \;\vdots\; \frac{1}{\sqrt{\lambda_p}}\mathbf{e}_p\right] = \mathbf{I}$$

By Result 4.5, $Z_1, Z_2, \ldots, Z_p$ are *independent* standard normal variables and we conclude that $(\mathbf{X}-\boldsymbol{\mu})'\boldsymbol{\Sigma}^{-1}(\mathbf{X}-\boldsymbol{\mu})$ has a χ_p^2-distribution.

For Part b, we note that $P[(\mathbf{X}-\boldsymbol{\mu})'\boldsymbol{\Sigma}^{-1}(\mathbf{X}-\boldsymbol{\mu}) \le c^2]$ is the probability assigned to the ellipsoid $(\mathbf{x}-\boldsymbol{\mu})'\boldsymbol{\Sigma}^{-1}(\mathbf{x}-\boldsymbol{\mu}) \le c^2$ by the density $N_p(\boldsymbol{\mu}, \boldsymbol{\Sigma})$. But from Part a, $P[(\mathbf{X}-\boldsymbol{\mu})'\boldsymbol{\Sigma}^{-1}(\mathbf{X}-\boldsymbol{\mu}) \le \chi_p^2(\alpha)] = 1-\alpha$ and Part b holds. ■

Next consider the linear combination of vector random variables

$$c_1\mathbf{X}_1 + c_2\mathbf{X}_2 + \cdots + c_n\mathbf{X}_n = \underset{(p\times n)}{[\mathbf{X}_1 \;\vdots\; \mathbf{X}_2 \;\vdots\; \cdots \;\vdots\; \mathbf{X}_n]}\;\underset{(n\times 1)}{\mathbf{c}} \tag{4-10}$$

This linear combination differs from the linear combinations considered earlier.

Equation (4-10) defines a $p \times 1$ *vector* random variable that is a linear combination of vectors. Previously we discussed a *single* random variable that could be written as a linear combination of other univariate random variables.

Result 4.8. Let $\mathbf{X}_1, \mathbf{X}_2, \ldots, \mathbf{X}_n$ be mutually independent with $\mathbf{X}_j$ distributed as $N_p(\boldsymbol{\mu}_j, \boldsymbol{\Sigma})$. (Note each $\mathbf{X}_j$ has the *same* covariance matrix $\boldsymbol{\Sigma}$.) Then

$$\mathbf{V}_1 = c_1\mathbf{X}_1 + c_2\mathbf{X}_2 + \cdots + c_n\mathbf{X}_n$$

is distributed as $N_p\left(\sum_{j=1}^{n} c_j\boldsymbol{\mu}_j, \left(\sum_{j=1}^{n} c_j^2\right)\boldsymbol{\Sigma}\right)$. Moreover $\mathbf{V}_1$ and $\mathbf{V}_2 = b_1\mathbf{X}_1 + b_2\mathbf{X}_2 + \cdots + b_n\mathbf{X}_n$ are jointly multivariate normal with covariance matrix

$$\begin{bmatrix} \left(\sum_{j=1}^{n} c_j^2\right)\boldsymbol{\Sigma} & (\mathbf{b}'\mathbf{c})\boldsymbol{\Sigma} \\ (\mathbf{b}'\mathbf{c})\boldsymbol{\Sigma} & \left(\sum_{j=1}^{n} b_j^2\right)\boldsymbol{\Sigma} \end{bmatrix}$$

Consequently, $\mathbf{V}_1$ and $\mathbf{V}_2$ are independent if $\mathbf{b}'\mathbf{c} = \sum_{j=1}^{n} c_j b_j = 0$.

Proof. By Result 4.5(c), the pn component vector

$$[X_{11}, \ldots, X_{p1}, X_{12}, \ldots, X_{p2}, \ldots, X_{pn}] = [\mathbf{X}_1', \mathbf{X}_2', \ldots, \mathbf{X}_n'] = \underset{(1\times np)}{\mathbf{X}'}$$

is multivariate normal. In particular, $\underset{(np\times 1)}{\mathbf{X}}$ is distributed as $N_{np}(\boldsymbol{\mu}, \boldsymbol{\Sigma}_{\mathbf{X}})$ where

$$\underset{(np\times 1)}{\boldsymbol{\mu}} = \begin{bmatrix} \boldsymbol{\mu}_1 \\ \boldsymbol{\mu}_2 \\ \vdots \\ \boldsymbol{\mu}_{np} \end{bmatrix} \quad \text{and} \quad \underset{(np\times np)}{\boldsymbol{\Sigma}_{\mathbf{X}}} = \begin{bmatrix} \boldsymbol{\Sigma} & \mathbf{0} & \cdots & \mathbf{0} \\ \mathbf{0} & \boldsymbol{\Sigma} & \cdots & \mathbf{0} \\ \vdots & \vdots & & \vdots \\ \mathbf{0} & \mathbf{0} & \cdots & \boldsymbol{\Sigma} \end{bmatrix}$$

The choice

$$\underset{(2p\times np)}{\mathbf{A}} = \begin{bmatrix} c_1\mathbf{I} & c_2\mathbf{I} & \cdots & c_n\mathbf{I} \\ b_1\mathbf{I} & b_2\mathbf{I} & \cdots & b_n\mathbf{I} \end{bmatrix}$$

where $\mathbf{I}$ is the $p \times p$ identity matrix, gives

$$\mathbf{AX} = \begin{bmatrix} \sum_{j=1}^{n} c_j\mathbf{X}_j \\ \sum_{j=1}^{n} b_j\mathbf{X}_j \end{bmatrix} = \begin{bmatrix} \mathbf{V}_1 \\ \mathbf{V}_2 \end{bmatrix}$$

and $\mathbf{AX}$ is normal $N_{2p}(\mathbf{A}\boldsymbol{\mu}, \mathbf{A}\boldsymbol{\Sigma}_{\mathbf{X}}\mathbf{A}')$ by Result 4.3. Straightforward block multiplication shows that $\mathbf{A}\boldsymbol{\Sigma}_{\mathbf{X}}\mathbf{A}'$ has the first block diagonal term $[c_1\boldsymbol{\Sigma}, c_2\boldsymbol{\Sigma}, \ldots, c_n\boldsymbol{\Sigma}]$

$\times[c_1\mathbf{I}, c_2\mathbf{I}, \ldots, c_n\mathbf{I}]' = (\sum_{j=1}^{n} c_j^2)\mathbf{\Sigma}$. The off-diagonal term is

$$[c_1\mathbf{\Sigma}, c_2\mathbf{\Sigma}, \ldots, c_n\mathbf{\Sigma}][b_1\mathbf{I}, b_2\mathbf{I}, \ldots, b_n\mathbf{I}]' = \left(\sum_{j=1}^{n} c_j b_j\right)\mathbf{\Sigma}$$

This term is the covariance matrix for $\mathbf{V}_1, \mathbf{V}_2$. Consequently, when $\sum_{j=1}^{n} c_j b_j = \mathbf{b}'\mathbf{c} = 0$, so that $(\sum_{j=1}^{n} c_j b_j)\mathbf{\Sigma} = \underset{(p\times p)}{\mathbf{0}}$, $\mathbf{V}_1$ and $\mathbf{V}_2$ are independent by Result 4.5(b). ■

For sums of the type in (4-10), the property of zero correlation is equivalent to requiring the coefficient vectors **b** and **c** to be perpendicular.

4.3 SAMPLING FROM A MULTIVARIATE NORMAL DISTRIBUTION AND MAXIMUM LIKELIHOOD ESTIMATION

We discussed sampling and simple random samples briefly in Chapter 3. In this section we shall be concerned with samples from a multivariate normal population and in particular with the sampling distribution of $\bar{\mathbf{X}}$ and $\mathbf{S}$.

The Multivariate Normal Likelihood

Let us assume that the $p \times 1$ vectors $\mathbf{X}_1, \mathbf{X}_2, \ldots, \mathbf{X}_n$ represent a random sample from a multivariate normal population with mean vector $\boldsymbol{\mu}$ and covariance matrix $\mathbf{\Sigma}$. Since $\mathbf{X}_1, \mathbf{X}_2, \ldots, \mathbf{X}_n$ are mutually independent and each has distribution $N_p(\boldsymbol{\mu}, \mathbf{\Sigma})$, the joint density function of all the observations is the product of the marginal normal densities. Thus

$$\begin{Bmatrix} \text{Joint density} \\ \text{of } \mathbf{X}_1, \mathbf{X}_2, \ldots, \mathbf{X}_n \end{Bmatrix} = \prod_{j=1}^{n} \left\{ \frac{1}{(2\pi)^{p/2}|\mathbf{\Sigma}|^{1/2}} e^{-\frac{1}{2}(\mathbf{x}_j-\boldsymbol{\mu})'\mathbf{\Sigma}^{-1}(\mathbf{x}_j-\boldsymbol{\mu})} \right\}$$

$$= \frac{1}{(2\pi)^{np/2}} \frac{1}{|\mathbf{\Sigma}|^{n/2}} e^{-\frac{1}{2}\sum_{j=1}^{n}(\mathbf{x}_j-\boldsymbol{\mu})'\mathbf{\Sigma}^{-1}(\mathbf{x}_j-\boldsymbol{\mu})} \qquad (4\text{-}11)$$

When the numerical values of the observations become available, these may be substituted for the $\mathbf{x}_j$ in Equation (4-11). The resulting expression, now considered as a function of $\boldsymbol{\mu}$ and $\mathbf{\Sigma}$ for the fixed set of observations $\mathbf{x}_1, \mathbf{x}_2, \ldots, \mathbf{x}_n$, is called the *likelihood*.

Many good statistical procedures employ values for the population parameters that "best" explain the observed data. One meaning of *best* is to select the parameter values that *maximize* the joint density evaluated at the observations. This technique is called *maximum likelihood estimation* and the maximizing parameter values are called *maximum likelihood estimates*.

At this point, we shall consider maximum likelihood estimation of the parameters $\boldsymbol{\mu}$ and $\boldsymbol{\Sigma}$ for a multivariate normal population. To do so, we take the observations $\mathbf{x}_1, \mathbf{x}_2, \dots, \mathbf{x}_n$ as fixed and consider the joint density of Equation (4-11) at these values. The result is the likelihood function. In order to simplify matters it is necessary to rewrite the likelihood function in another form. We shall need some additional properties for the trace of a square matrix. (The trace of a matrix is the sum of its diagonal elements and its properties are discussed in Definition 2A.28 and Result 2A.12.)

Result 4.9. Let $\mathbf{A}$ be a $k \times k$ symmetric matrix and $\mathbf{x}$ be a $k \times 1$ vector.

(a) $\mathbf{x}'\mathbf{A}\mathbf{x} = \text{tr}(\mathbf{x}'\mathbf{A}\mathbf{x}) = \text{tr}(\mathbf{A}\mathbf{x}\mathbf{x}')$

(b) $\text{tr}(\mathbf{A}) = \sum_{i=1}^{k} \lambda_i$, where the λ_i are the eigenvalues of $\mathbf{A}$.

Proof. For Part a, we note that $\mathbf{x}'\mathbf{A}\mathbf{x}$ is a scalar, so $\mathbf{x}'\mathbf{A}\mathbf{x} = \text{tr}(\mathbf{x}'\mathbf{A}\mathbf{x})$. We pointed out in Result 2A.12 that $\text{tr}(\mathbf{BC}) = \text{tr}(\mathbf{CB})$ for any two matrices $\mathbf{B}$ and $\mathbf{C}$ of dimensions $m \times k$ and $k \times m$, respectively. This follows because $\mathbf{BC}$ has $\sum_{j=1}^{k} b_{ij}c_{ji}$ as its ith diagonal element, so $\text{tr}(\mathbf{BC}) = \sum_{i=1}^{m} \left(\sum_{j=1}^{k} b_{ij}c_{ji} \right)$. Similarly the jth diagonal element of $\mathbf{CB}$ is $\sum_{i=1}^{m} c_{ji}b_{ij}$, so $\text{tr}(\mathbf{CB}) = \sum_{j=1}^{k} \left(\sum_{i=1}^{m} c_{ji}b_{ij} \right) = \sum_{i=1}^{m} \left(\sum_{j=1}^{k} b_{ij}c_{ji} \right) = \text{tr}(\mathbf{BC})$. Let $\mathbf{x}'$ be the matrix $\mathbf{B}$ with $m = 1$ and let $\mathbf{Ax}$ play the role of the matrix $\mathbf{C}$. Then $\text{tr}(\mathbf{x}'(\mathbf{Ax})) = \text{tr}((\mathbf{Ax})\mathbf{x}')$, and the result follows.

Part b is proved by using the spectral decomposition of (2-20) to write $\mathbf{A} = \mathbf{P}'\boldsymbol{\Lambda}\mathbf{P}$, where $\mathbf{PP}' = \mathbf{I}$ and $\boldsymbol{\Lambda}$ is a diagonal matrix with entries $\lambda_1, \lambda_2, \dots, \lambda_k$. Therefore $\text{tr}(\mathbf{A}) = \text{tr}(\mathbf{P}'\boldsymbol{\Lambda}\mathbf{P}) = \text{tr}(\boldsymbol{\Lambda}\mathbf{PP}') = \text{tr}(\boldsymbol{\Lambda}) = \lambda_1 + \lambda_2 + \cdots + \lambda_k$. ■

Now the exponent in the joint density in (4-11) can be simplified. By Result 4.9(a),

$$(\mathbf{x}_j - \boldsymbol{\mu})'\boldsymbol{\Sigma}^{-1}(\mathbf{x}_j - \boldsymbol{\mu}) = \text{tr}\left[(\mathbf{x}_j - \boldsymbol{\mu})'\boldsymbol{\Sigma}^{-1}(\mathbf{x}_j - \boldsymbol{\mu})\right]$$
$$= \text{tr}\left[\boldsymbol{\Sigma}^{-1}(\mathbf{x}_j - \boldsymbol{\mu})(\mathbf{x}_j - \boldsymbol{\mu})'\right] \qquad (4\text{-}12)$$

Next

$$\sum_{j=1}^{n} (\mathbf{x}_j - \boldsymbol{\mu})'\boldsymbol{\Sigma}^{-1}(\mathbf{x}_j - \boldsymbol{\mu}) = \sum_{j=1}^{n} \text{tr}\left[(\mathbf{x}_j - \boldsymbol{\mu})'\boldsymbol{\Sigma}^{-1}(\mathbf{x}_j - \boldsymbol{\mu})\right]$$
$$= \sum_{j=1}^{n} \text{tr}\left[\boldsymbol{\Sigma}^{-1}(\mathbf{x}_j - \boldsymbol{\mu})(\mathbf{x}_j - \boldsymbol{\mu})'\right]$$
$$= \text{tr}\left[\boldsymbol{\Sigma}^{-1}\left(\sum_{j=1}^{n} (\mathbf{x}_j - \boldsymbol{\mu})(\mathbf{x}_j - \boldsymbol{\mu})' \right)\right] \qquad (4\text{-}13)$$

since the trace of a sum of matrices is equal to the sum of the traces of the matrices, according to Result 2A.12(b). We can add and subtract $\bar{\mathbf{x}} = (1/n) \sum_{j=1}^{n} \mathbf{x}_j$ to each term $(\mathbf{x}_j - \boldsymbol{\mu})$ in $\sum_{j=1}^{n} (\mathbf{x}_j - \boldsymbol{\mu})(\mathbf{x}_j - \boldsymbol{\mu})'$ to give

$$\begin{aligned}\sum_{j=1}^{n} (\mathbf{x}_j - \bar{\mathbf{x}} + \bar{\mathbf{x}} - \boldsymbol{\mu})(\mathbf{x}_j - \bar{\mathbf{x}} + \bar{\mathbf{x}} - \boldsymbol{\mu})' \\ &= \sum_{j=1}^{n} (\mathbf{x}_j - \bar{\mathbf{x}})(\mathbf{x}_j - \bar{\mathbf{x}})' + \sum_{j=1}^{n} (\bar{\mathbf{x}} - \boldsymbol{\mu})(\bar{\mathbf{x}} - \boldsymbol{\mu})' \\ &= \sum_{j=1}^{n} (\mathbf{x}_j - \bar{\mathbf{x}})(\mathbf{x}_j - \bar{\mathbf{x}})' + n(\bar{\mathbf{x}} - \boldsymbol{\mu})(\bar{\mathbf{x}} - \boldsymbol{\mu})'\end{aligned} \tag{4-14}$$

because the crossproduct terms, $\sum_{j=1}^{n} (\mathbf{x}_j - \bar{\mathbf{x}})(\bar{\mathbf{x}} - \boldsymbol{\mu})'$ and $\sum_{j=1}^{n} (\bar{\mathbf{x}} - \boldsymbol{\mu})(\mathbf{x}_j - \bar{\mathbf{x}})'$, are both matrices of zeros (see Exercise 4.12). Consequently, using Equations (4-13) and (4-14), the joint density of a random sample from a multivariate normal population can be written

$$\begin{Bmatrix}\text{Joint density of} \\ \mathbf{X}_1, \mathbf{X}_2, \ldots, \mathbf{X}_n\end{Bmatrix} = \frac{1}{(2\pi)^{np/2} |\boldsymbol{\Sigma}|^{n/2}} e^{-\frac{1}{2}\operatorname{tr}[\boldsymbol{\Sigma}^{-1}(\sum_{j=1}^{n} (\mathbf{x}_j - \bar{\mathbf{x}})(\mathbf{x}_j - \bar{\mathbf{x}})' + n(\bar{\mathbf{x}} - \boldsymbol{\mu})(\bar{\mathbf{x}} - \boldsymbol{\mu})')]} \tag{4-15}$$

Substituting the observed values $\mathbf{x}_1, \mathbf{x}_2, \ldots, \mathbf{x}_n$ into the joint density yields the likelihood function. We shall denote this function by $L(\boldsymbol{\mu}, \boldsymbol{\Sigma})$ to stress the fact that it is a function of the (unknown) population parameters $\boldsymbol{\mu}$ and $\boldsymbol{\Sigma}$. Thus when the vectors $\mathbf{x}_j$ contain the specific numbers actually observed, we have

$$L(\boldsymbol{\mu}, \boldsymbol{\Sigma}) = \frac{1}{(2\pi)^{np/2} |\boldsymbol{\Sigma}|^{n/2}} e^{-\frac{1}{2}\operatorname{tr}[\boldsymbol{\Sigma}^{-1}(\sum_{j=1}^{n} (\mathbf{x}_j - \bar{\mathbf{x}})(\mathbf{x}_j - \bar{\mathbf{x}})' + n(\bar{\mathbf{x}} - \boldsymbol{\mu})(\bar{\mathbf{x}} - \boldsymbol{\mu})')]} \tag{4-16}$$

It will be convenient in later sections of this book to express the exponent in the likelihood function (4-16) in different ways. In particular, we shall make use of the identity

$$\begin{aligned}\operatorname{tr}\left[\boldsymbol{\Sigma}^{-1}\left(\sum_{j=1}^{n} (\mathbf{x}_j - \bar{\mathbf{x}})(\mathbf{x}_j - \bar{\mathbf{x}})' + n(\bar{\mathbf{x}} - \boldsymbol{\mu})(\bar{\mathbf{x}} - \boldsymbol{\mu})'\right)\right] \\ &= \operatorname{tr}\left[\boldsymbol{\Sigma}^{-1}\left(\sum_{j=1}^{n} (\mathbf{x}_j - \bar{\mathbf{x}})(\mathbf{x}_j - \bar{\mathbf{x}})'\right)\right] + n \operatorname{tr}\left[\boldsymbol{\Sigma}^{-1}(\bar{\mathbf{x}} - \boldsymbol{\mu})(\bar{\mathbf{x}} - \boldsymbol{\mu})'\right] \\ &= \operatorname{tr}\left[\boldsymbol{\Sigma}^{-1}\left(\sum_{j=1}^{n} (\mathbf{x}_j - \bar{\mathbf{x}})(\mathbf{x}_j - \bar{\mathbf{x}})'\right)\right] + n(\bar{\mathbf{x}} - \boldsymbol{\mu})'\boldsymbol{\Sigma}^{-1}(\bar{\mathbf{x}} - \boldsymbol{\mu})\end{aligned} \tag{4-17}$$

Maximum Likelihood Estimation of $\boldsymbol{\mu}$ and $\boldsymbol{\Sigma}$

The next result will eventually allow us to obtain the maximum likelihood estimators of $\boldsymbol{\mu}$ and $\boldsymbol{\Sigma}$.

Result 4.10. Given a $p \times p$ symmetric positive definite matrix $\mathbf{B}$ and scalar $b > 0$,

$$\frac{1}{|\boldsymbol{\Sigma}|^b} e^{-\frac{1}{2}\operatorname{tr}(\boldsymbol{\Sigma}^{-1}\mathbf{B})} \le \frac{1}{|\mathbf{B}|^b}(2b)^{pb} e^{-bp}$$

for all positive definite $\underset{(p\times p)}{\boldsymbol{\Sigma}}$, with equality holding only for $\boldsymbol{\Sigma} = (1/2b)\mathbf{B}$.

Proof. Let $\mathbf{B}^{1/2}$ be the symmetric square root of $\mathbf{B}$ [see Equation (2-22)], so $\mathbf{B}^{1/2}\mathbf{B}^{1/2} = \mathbf{B}$, $\mathbf{B}^{1/2}\mathbf{B}^{-1/2} = \mathbf{I}$, and $\mathbf{B}^{-1/2}\mathbf{B}^{-1/2} = \mathbf{B}^{-1}$. Then $\operatorname{tr}(\boldsymbol{\Sigma}^{-1}\mathbf{B}) = \operatorname{tr}[(\boldsymbol{\Sigma}^{-1}\mathbf{B}^{1/2})\mathbf{B}^{1/2}] = \operatorname{tr}[\mathbf{B}^{1/2}(\boldsymbol{\Sigma}^{-1}\mathbf{B}^{1/2})]$. Let η be an eigenvalue of $\mathbf{B}^{1/2}\boldsymbol{\Sigma}^{-1}\mathbf{B}^{1/2}$. This matrix is positive definite because $\mathbf{y}'\mathbf{B}^{1/2}\boldsymbol{\Sigma}^{-1}\mathbf{B}^{1/2}\mathbf{y} = (\mathbf{B}^{1/2}\mathbf{y})'\boldsymbol{\Sigma}^{-1}(\mathbf{B}^{1/2}\mathbf{y}) > 0$ if $\mathbf{B}^{1/2}\mathbf{y} \ne \mathbf{0}$ or, equivalently, $\mathbf{y} \ne \mathbf{0}$. Thus the eigenvalues η_i of $\mathbf{B}^{1/2}\boldsymbol{\Sigma}^{-1}\mathbf{B}^{1/2}$ are positive by Exercise 2.17. Result 4.9(b) gives

$$\operatorname{tr}(\boldsymbol{\Sigma}^{-1}\mathbf{B}) = \operatorname{tr}(\mathbf{B}^{1/2}\boldsymbol{\Sigma}^{-1}\mathbf{B}^{1/2}) = \sum_{i=1}^{p} \eta_i$$

and $|\mathbf{B}^{1/2}\boldsymbol{\Sigma}^{-1}\mathbf{B}^{1/2}| = \prod_{i=1}^{p} \eta_i$ by Exercise 2.12. From the properties of determinants in Result 2A.11, we can write

$$|\mathbf{B}^{1/2}\boldsymbol{\Sigma}^{-1}\mathbf{B}^{1/2}| = |\mathbf{B}^{1/2}||\boldsymbol{\Sigma}^{-1}||\mathbf{B}^{1/2}| = |\boldsymbol{\Sigma}^{-1}||\mathbf{B}^{1/2}||\mathbf{B}^{1/2}|$$

$$= |\boldsymbol{\Sigma}^{-1}||\mathbf{B}| = \frac{1}{|\boldsymbol{\Sigma}|}|\mathbf{B}|$$

or

$$\frac{1}{|\boldsymbol{\Sigma}|} = \frac{|\mathbf{B}^{1/2}\boldsymbol{\Sigma}^{-1}\mathbf{B}^{1/2}|}{|\mathbf{B}|} = \frac{\prod_{i=1}^{p} \eta_i}{|\mathbf{B}|}$$

Combining the results for the trace and determinant,

$$\frac{1}{|\boldsymbol{\Sigma}|^b} e^{-\frac{1}{2}\operatorname{tr}[\boldsymbol{\Sigma}^{-1}B]} = \frac{\left(\prod_{i=1}^{p} \eta_i\right)^b}{|\mathbf{B}|^b} e^{-\frac{1}{2}\sum_{i=1}^{p}\eta_i} = \frac{1}{|\mathbf{B}|^b}\prod_{i=1}^{p} \eta_i^b e^{-\frac{1}{2}\eta_i}$$

But the function $\eta^b e^{-\eta/2}$ has a maximum, with respect to η, of $(2b)^b e^{-b}$, occurring at $\eta = 2b$. The choice $\eta_i = 2b$, for each i, therefore gives

$$\frac{1}{|\boldsymbol{\Sigma}|^b} e^{-\frac{1}{2}\operatorname{tr}(\boldsymbol{\Sigma}^{-1}\mathbf{B})} \le \frac{1}{|\mathbf{B}|^b}(2b)^{pb} e^{-bp}$$

The upper bound is uniquely attained when $\boldsymbol{\Sigma} = (1/2b)\mathbf{B}$, since for this choice,

$$\mathbf{B}^{1/2}\boldsymbol{\Sigma}^{-1}\mathbf{B}^{1/2} = \mathbf{B}^{1/2}(2b)\mathbf{B}^{-1}\mathbf{B}^{1/2} = (2b)\underset{(p\times p)}{\mathbf{I}}$$

and

$$\text{tr}[\boldsymbol{\Sigma}^{-1}\mathbf{B}] = \text{tr}[\mathbf{B}^{1/2}\boldsymbol{\Sigma}^{-1}\mathbf{B}^{1/2}] = \text{tr}[(2b)\mathbf{I}] = 2bp$$

Moreover

$$\frac{1}{|\boldsymbol{\Sigma}|} = \frac{|\mathbf{B}^{1/2}\boldsymbol{\Sigma}^{-1}\mathbf{B}^{1/2}|}{|\mathbf{B}|} = \frac{|(2b)\mathbf{I}|}{|\mathbf{B}|} = \frac{(2b)^p}{|\mathbf{B}|}$$

Straightforward substitution for $\text{tr}[\boldsymbol{\Sigma}^{-1}\mathbf{B}]$ and $1/|\boldsymbol{\Sigma}|^b$ yields the bound asserted. ■

The maximum likelihood estimates of $\boldsymbol{\mu}$ and $\boldsymbol{\Sigma}$ are those values—denoted by $\hat{\boldsymbol{\mu}}$ and $\hat{\boldsymbol{\Sigma}}$—that maximize the function $L(\boldsymbol{\mu}, \boldsymbol{\Sigma})$ in (4-16). The estimates $\hat{\boldsymbol{\mu}}$ and $\hat{\boldsymbol{\Sigma}}$ will depend on the observed values $\mathbf{x}_1, \mathbf{x}_2, \ldots, \mathbf{x}_n$ through the summary statistics $\bar{\mathbf{x}}$ and $\mathbf{S}$.

Result 4.11. Let $\mathbf{X}_1, \mathbf{X}_2, \ldots, \mathbf{X}_n$ be a random sample from a normal population with mean $\boldsymbol{\mu}$ and covariance $\boldsymbol{\Sigma}$. Then

$$\hat{\boldsymbol{\mu}} = \bar{\mathbf{X}} \quad \text{and} \quad \hat{\boldsymbol{\Sigma}} = \frac{1}{n}\sum_{j=1}^{n}(\mathbf{X}_j - \bar{\mathbf{X}})(\mathbf{X}_j - \bar{\mathbf{X}})' = \frac{(n-1)}{n}\mathbf{S}$$

are the *maximum likelihood estimators* of $\boldsymbol{\mu}$ and $\boldsymbol{\Sigma}$, respectively. Their observed values, $\bar{\mathbf{x}}$ and $(1/n)\sum_{j=1}^{n}(\mathbf{x}_j - \bar{\mathbf{x}})(\mathbf{x}_j - \bar{\mathbf{x}})'$ are called the *maximum likelihood estimates* of $\boldsymbol{\mu}$ and $\boldsymbol{\Sigma}$.

Proof. The exponent in the likelihood function [see Equation (4-16)], apart from the multiplicative factor $-\frac{1}{2}$, is [see (4-17)]

$$\text{tr}\left[\boldsymbol{\Sigma}^{-1}\left(\sum_{j=1}^{n}(\mathbf{x}_j - \bar{\mathbf{x}})(\mathbf{x}_j - \bar{\mathbf{x}})'\right)\right] + n(\bar{\mathbf{x}} - \boldsymbol{\mu})'\boldsymbol{\Sigma}^{-1}(\bar{\mathbf{x}} - \boldsymbol{\mu})$$

By Result 4.1, $\boldsymbol{\Sigma}^{-1}$ is positive definite, so the distance $(\bar{\mathbf{x}} - \boldsymbol{\mu})'\boldsymbol{\Sigma}^{-1}(\bar{\mathbf{x}} - \boldsymbol{\mu}) > 0$ unless $\boldsymbol{\mu} = \bar{\mathbf{x}}$. Thus the likelihood is maximized with respect to $\boldsymbol{\mu}$ at $\hat{\boldsymbol{\mu}} = \bar{\mathbf{x}}$. It remains to maximize

$$L(\hat{\boldsymbol{\mu}}, \boldsymbol{\Sigma}) = \frac{1}{(2\pi)^{np/2}|\boldsymbol{\Sigma}|^{n/2}} e^{-\frac{1}{2}\text{tr}[\boldsymbol{\Sigma}^{-1}(\sum_{j=1}^{n}(\mathbf{x}_j - \bar{\mathbf{x}})(\mathbf{x}_j - \bar{\mathbf{x}})')]}$$

over $\boldsymbol{\Sigma}$. By Result 4.10 with $b = n/2$ and $\mathbf{B} = \sum_{j=1}^{n}(\mathbf{x}_j - \bar{\mathbf{x}})(\mathbf{x}_j - \bar{\mathbf{x}})'$, the maximum occurs at $\hat{\boldsymbol{\Sigma}} = (1/n)\sum_{j=1}^{n}(\mathbf{x}_j - \bar{\mathbf{x}})(\mathbf{x}_j - \bar{\mathbf{x}})'$ as stated.

The maximum likelihood estimators are random quantities. They are obtained by replacing the observations $\mathbf{x}_1, \mathbf{x}_2, \ldots, \mathbf{x}_n$ in the expressions for $\hat{\boldsymbol{\mu}}$ and $\hat{\boldsymbol{\Sigma}}$ with the corresponding random vectors, $\mathbf{X}_1, \mathbf{X}_2, \ldots, \mathbf{X}_n$. ■

We note that the maximum likelihood estimator $\bar{\mathbf{X}}$ is a random vector and the maximum likelihood estimator $\hat{\mathbf{\Sigma}}$ is a random matrix. The maximum likelihood estimates are their particular values for the given data set. In addition, the maximum of the likelihood is

$$L(\hat{\boldsymbol{\mu}}, \hat{\mathbf{\Sigma}}) = \frac{1}{(2\pi)^{np/2}} e^{-np/2} \frac{1}{|\hat{\mathbf{\Sigma}}|^{n/2}} \qquad (4\text{-}18)$$

or, since $|\hat{\mathbf{\Sigma}}| = [(n-1)/n]^p |\mathbf{S}|$,

$$L(\hat{\boldsymbol{\mu}}, \hat{\mathbf{\Sigma}}) = \text{constant} \times (\text{generalized variance})^{-n/2} \qquad (4\text{-}19)$$

The generalized variance determines the "peakedness" of the likelihood function and, consequently, is a natural measure of variability when the parent population is multivariate normal.

Maximum likelihood estimators possess an *invariance property*. Let $\hat{\boldsymbol{\theta}}$ be the maximum likelihood estimator of $\boldsymbol{\theta}$ and consider estimating the parameter $h(\boldsymbol{\theta})$, which is a function of $\boldsymbol{\theta}$. Then the *maximum likelihood estimate* of

$$\underset{(\text{a function of } \boldsymbol{\theta})}{h(\boldsymbol{\theta})} \quad \text{is given by} \quad \underset{(\text{same function of } \hat{\boldsymbol{\theta}})}{h(\hat{\boldsymbol{\theta}})} \qquad (4\text{-}20)$$

(See [1] and [10].) For example:

1. The maximum likelihood estimator of $\boldsymbol{\mu}'\mathbf{\Sigma}^{-1}\boldsymbol{\mu}$ is $\hat{\boldsymbol{\mu}}'\hat{\mathbf{\Sigma}}^{-1}\hat{\boldsymbol{\mu}}$, where $\hat{\boldsymbol{\mu}} = \bar{\mathbf{X}}$ and $\hat{\mathbf{\Sigma}} = ((n-1)/n)\mathbf{S}$ are the maximum likelihood estimators of $\boldsymbol{\mu}$ and $\mathbf{\Sigma}$, respectively.
2. The maximum likelihood estimator of $\sqrt{\sigma_{ii}}$ is $\sqrt{\hat{\sigma}_{ii}}$, where

$$\hat{\sigma}_{ii} = (1/n) \sum_{j=1}^{n} (X_{ij} - \bar{X}_i)^2$$

is the maximum likelihood estimator of $\sigma_{ii} = \text{Var}(X_i)$.

Sufficient Statistics

From expression (4-15), the joint density depends on the whole set of observations $\mathbf{x}_1, \mathbf{x}_2, \ldots, \mathbf{x}_n$ only through the sample mean, $\bar{\mathbf{x}}$, and the sum of squares and cross products matrix, $\sum_{j=1}^{n} (\mathbf{x}_j - \bar{\mathbf{x}})(\mathbf{x}_j - \bar{\mathbf{x}})' = (n-1)\mathbf{S}$. We express this fact by saying that $\bar{\mathbf{x}}$ and $(n-1)\mathbf{S}$ (or $\mathbf{S}$) are *sufficient statistics*.

Let $\mathbf{X}_1, \mathbf{X}_2, \ldots, \mathbf{X}_n$ be a random sample from a multivariate normal population with mean $\boldsymbol{\mu}$ and covariance $\mathbf{\Sigma}$. Then

$$\bar{\mathbf{X}} \text{ and } \mathbf{S} \text{ are } \textit{sufficient statistics} \qquad (4\text{-}21)$$

The importance of sufficient statistics for normal populations is that all of the sample information in the data matrix $\mathbf{X}$ is contained in $\bar{\mathbf{x}}$ and $\mathbf{S}$, regardless of the sample size n. This generally is not true for nonnormal populations. Since many multivariate techniques begin with sample means and covariances, it is prudent to check on the *adequacy* of the multivariate normal assumption (see Section 4.6). If the data cannot be regarded as multivariate normal, techniques that depend solely on $\bar{\mathbf{x}}$ and $\mathbf{S}$ may be ignoring other useful sample information.

4.4 THE SAMPLING DISTRIBUTION OF $\bar{\mathbf{X}}$ AND S

The tentative assumption that the columns of

$$\mathbf{X} = \begin{bmatrix} X_{11} & X_{12} & \cdots & X_{1n} \\ X_{21} & X_{22} & \cdots & X_{2n} \\ \vdots & \vdots & & \vdots \\ X_{p1} & X_{p2} & \cdots & X_{pn} \end{bmatrix} = [\mathbf{X}_1, \mathbf{X}_2, \ldots, \mathbf{X}_n]$$

constitute a random sample from a normal population, with mean $\boldsymbol{\mu}$ and covariance $\boldsymbol{\Sigma}$, completely determines the sampling distributions of $\bar{\mathbf{X}}$ and $\mathbf{S}$. Here we present the results on the sampling distributions of $\bar{\mathbf{X}}$ and $\mathbf{S}$ by drawing a parallel with the familiar univariate conclusions.

In the univariate case ($p = 1$), we know that $\bar{X}$ is normal with mean $\mu =$ (population mean) and variance

$$\frac{1}{n}\sigma^2 = \frac{\text{population variance}}{\text{sample size}}$$

The result for the multivariate case ($p \geq 2$) is analogous in that $\bar{\mathbf{X}}$ has a normal distribution with mean $\boldsymbol{\mu}$ and covariance matrix $(1/n)\boldsymbol{\Sigma}$.

For the sample variance, we recall that $(n-1)s^2 = \sum_{j=1}^{n} (X_j - \bar{X})^2$ is distributed as σ^2 times a chi-square variable having $n - 1$ degrees of freedom (d.f.). In turn, this chi-square is the distribution of a sum of squared independent standard normal random variables. That is $(n-1)s^2$ is distributed as $\sigma^2(Z_1^2 + \cdots + Z_{n-1}^2) = (\sigma Z_1)^2 + \cdots + (\sigma Z_{n-1})^2$. The individual terms (σZ_i) are independently distributed as $N(0, \sigma^2)$. It is this latter form that is suitably generalized to the basic sampling distribution for the sample covariance matrix.

The sampling distribution of the sample covariance matrix is called the *Wishart distribution*, after its discoverer; it is defined as the sum of independent products of multivariate normal random vectors. Specifically,

$$\begin{aligned} W_m(\cdot \,|\, \boldsymbol{\Sigma}) &= \text{Wishart distribution with } m \text{ d.f.} \\ &= \text{distribution of } \sum_{j=1}^{m} \mathbf{Z}_j \mathbf{Z}_j' \end{aligned} \qquad (4\text{-}22)$$

where the $\mathbf{Z}_j$ are each independently distributed as $N_p(\mathbf{0}, \boldsymbol{\Sigma})$.

We summarize the sampling distribution results below.

> Let $\mathbf{X}_1, \mathbf{X}_2, \ldots, \mathbf{X}_n$ be a random sample of size n from a p-variate *normal* distribution with mean $\boldsymbol{\mu}$ and covariance matrix $\boldsymbol{\Sigma}$. Then:
>
> 1. $\overline{\mathbf{X}}$ is distributed as $N_p(\boldsymbol{\mu}, (1/n)\boldsymbol{\Sigma})$; (4-23)
> 2. $(n-1)\mathbf{S}$ is distributed as a Wishart random matrix with $n-1$ d.f.;
> 3. $\overline{\mathbf{X}}$ and $\mathbf{S}$ are independent.

Because $\boldsymbol{\Sigma}$ is unknown, the distribution of $\overline{\mathbf{X}}$ cannot be used directly to make inferences about $\boldsymbol{\mu}$. However, $\mathbf{S}$ provides independent information about $\boldsymbol{\Sigma}$ and the distribution of $\mathbf{S}$ does not depend on $\boldsymbol{\mu}$. This allows us to construct a statistic for making inferences about $\boldsymbol{\mu}$, as we shall see in Chapter 5.

For the present, we record some further results from multivariable distribution theory. The properties of the Wishart distribution in (4-24) follow directly from its definition as a sum of the independent products, $\mathbf{Z}_j\mathbf{Z}_j'$. Proofs can be found in [1].

Properties of the Wishart Distribution

1. If $\mathbf{A}_1$ is distributed as $W_{m_1}(\mathbf{A}_1 \mid \boldsymbol{\Sigma})$ independently of $\mathbf{A}_2$, which is distributed as $W_{m_2}(\mathbf{A}_2 \mid \boldsymbol{\Sigma})$, then $\mathbf{A}_1 + \mathbf{A}_2$ is distributed as $W_{m_1+m_2}(\mathbf{A}_1 + \mathbf{A}_2 \mid \boldsymbol{\Sigma})$. That is, the degrees of freedom add. (4-24)
2. If $\mathbf{A}$ is distributed as $W_m(\mathbf{A} \mid \boldsymbol{\Sigma})$, then $\mathbf{CAC}'$ is distributed as $W_m(\mathbf{CAC}' \mid \mathbf{C\Sigma C}')$.

Although we do not have any particular need for the probability density function of the Wishart distribution, it may be of some interest to see its rather complicated form. The density does not exist unless the sample size n is greater than the number of variables p. When it does exist, its value at the positive definite matrix $\mathbf{A}$ is

$$w_{n-1}(\mathbf{A} \mid \boldsymbol{\Sigma}) = \frac{|\mathbf{A}|^{\frac{1}{2}(n-p-2)} e^{-\frac{1}{2}\operatorname{tr} \mathbf{A}\boldsymbol{\Sigma}^{-1}}}{2^{\frac{1}{2}p(n-1)} \pi^{p(p-1)/4} |\boldsymbol{\Sigma}|^{(n-1)/2} \prod_{i=1}^{p} \Gamma\left(\frac{1}{2}(n-i)\right)}, \quad \mathbf{A} \text{ positive definite} \tag{4-25}$$

where $\Gamma(\cdot)$ is the gamma function (see [1]).

4.5 LARGE SAMPLE BEHAVIOR OF $\overline{\mathbf{X}}$ AND S

Suppose the quantity X is determined by a large number of independent causes, $V_1, V_2, \ldots, V_n$, where the random variables V_i, representing the causes, have approximately the same variability. If X is the sum

$$X = V_1 + V_2 + \cdots + V_n$$

then the central limit theorem applies and we conclude that X has a distribution that is nearly normal. This is true for virtually any parent distribution of the V_i's, provided n is large enough.

The univariate central limit theorem also tells us that the sampling distribution of the sample mean, $\bar{X}$, for large sample size is nearly normal whatever the form of the underlying population distribution. A similar result holds for many other important univariate statistics.

It turns out that certain multivariate statistics, like $\bar{\mathbf{X}}$ and $\mathbf{S}$, have large sample properties analogous to their univariate counterparts. As the sample size is increased without bound, certain regularities govern the sampling variation in $\bar{\mathbf{X}}$ and $\mathbf{S}$, irrespective of the form of the parent population. Therefore the conclusions presented in this section do not require multivariate normal populations. The only requirements are that the parent population, whatever its form, have a mean $\boldsymbol{\mu}$ and a finite covariance $\boldsymbol{\Sigma}$.

Result 4.12 (Law of large numbers). Let $Y_1, Y_2, \ldots, Y_n$ be independent observations from a population with mean $E(Y_i) = \mu$. Then

$$\bar{Y} = \frac{Y_1 + Y_2 + \cdots + Y_n}{n}$$

converges in probability to μ as n increases without bound. That is, for any prescribed accuracy $\varepsilon > 0$, $P[-\varepsilon < \bar{Y} - \mu < \varepsilon]$ approaches 1 as $n \to \infty$.

Proof. See [8]. ■

As a direct consequence of the law of large numbers, which says each $\bar{X}_i$ converges in probability to μ_i, $i = 1, 2, \ldots, p$,

$$\bar{\mathbf{X}} \text{ converges in probability to } \boldsymbol{\mu} \tag{4-26}$$

Also each sample covariance s_{ik} converges in probability to σ_{ik}, $i, k = 1, 2, \ldots, p$ and

$$\mathbf{S} \ (\text{or } \hat{\boldsymbol{\Sigma}} = \mathbf{S}_n) \text{ converges in probability to } \boldsymbol{\Sigma} \tag{4-27}$$

Statement (4-27) follows from writing

$$\begin{aligned}
(n-1)s_{ik} &= \sum_{j=1}^{n} (X_{ij} - \bar{X}_i)(X_{kj} - \bar{X}_k) \\
&= \sum_{j=1}^{n} (X_{ij} - \mu_i + \mu_i - \bar{X}_i)(X_{kj} - \mu_k + \mu_k - \bar{X}_k) \\
&= \sum_{j=1}^{n} (X_{ij} - \mu_i)(X_{kj} - \mu_k) + n(\bar{X}_i - \mu_i)(\bar{X}_k - \mu_k)
\end{aligned}$$

Letting $Y_j = (X_{ij} - \mu_i)(X_{kj} - \mu_k)$, with $E(Y_j) = \sigma_{ik}$, we see that the first term in s_{ik} converges to σ_{ik} and the second term converges to zero, by applying the law of large numbers.

The practical interpretation of statements (4-26) and (4-27) is that, with high probability, $\bar{\mathbf{X}}$ will be close to $\boldsymbol{\mu}$ and $\mathbf{S}$ will be close to $\boldsymbol{\Sigma}$ whenever the sample size is large. The statement concerning $\bar{\mathbf{X}}$ is made even more precise by a multivariate version of the central limit theorem.

Result 4.13 (The central limit theorem). Let $\mathbf{X}_1, \mathbf{X}_2, \ldots, \mathbf{X}_n$ be independent observations from any population with mean $\boldsymbol{\mu}$ and finite covariance $\boldsymbol{\Sigma}$. Then

$$\sqrt{n}\,(\bar{\mathbf{X}} - \boldsymbol{\mu}) \text{ has an approximate } N_p(\mathbf{0}, \boldsymbol{\Sigma}) \text{ distribution}$$

for large sample sizes. Here n should also be large relative to p.

Proof. See [1]. ■

The approximation provided by the central limit theorem applies to discrete, as well as continuous, multivariate populations. Mathematically, the limit is exact and the approach to normality is often fairly rapid. Moreover, from the results in Section 4.4, we know that $\bar{\mathbf{X}}$ is exactly normally distributed when the underlying population is normal. Thus we would expect the central limit theorem approximation to be quite good for moderate n when the parent population is nearly normal.

As we have seen, when n is large, $\mathbf{S}$ is close to $\boldsymbol{\Sigma}$ with high probability. Consequently, replacing $\boldsymbol{\Sigma}$ by $\mathbf{S}$ in the approximating normal distribution for $\bar{\mathbf{X}}$ will have a negligible effect on subsequent probability calculations.

Result 4.7 can be used to show that $n(\bar{\mathbf{X}} - \boldsymbol{\mu})'\boldsymbol{\Sigma}^{-1}(\bar{\mathbf{X}} - \boldsymbol{\mu})$ has a χ_p^2 distribution when $\bar{\mathbf{X}}$ is distributed as $N_p\left(\boldsymbol{\mu}, \frac{1}{n}\boldsymbol{\Sigma}\right)$ or, equivalently, when $\sqrt{n}\,(\bar{\mathbf{X}} - \boldsymbol{\mu})$ has a $N_p(\mathbf{0}, \boldsymbol{\Sigma})$ distribution. The distribution χ_p^2 is *approximately* the sampling distribution of $n(\bar{\mathbf{X}} - \boldsymbol{\mu})'\boldsymbol{\Sigma}^{-1}(\bar{\mathbf{X}} - \boldsymbol{\mu})$ when $\bar{\mathbf{X}}$ is approximately normally distributed. Replacing $\boldsymbol{\Sigma}^{-1}$ by $\mathbf{S}^{-1}$ does not seriously affect this approximation for n large and much greater than p.

We summarize the major conclusions of this section as follows.

> Let $\mathbf{X}_1, \mathbf{X}_2, \ldots, \mathbf{X}_n$ be independent observations from a population with mean $\boldsymbol{\mu}$ and finite (nonsingular) covariance $\boldsymbol{\Sigma}$. Then
>
> $$\sqrt{n}\,(\bar{\mathbf{X}} - \boldsymbol{\mu}) \text{ is approximately } N_p(\mathbf{0}, \mathbf{S})$$
>
> and (4-28)
>
> $$n(\bar{\mathbf{X}} - \boldsymbol{\mu})'\mathbf{S}^{-1}(\bar{\mathbf{X}} - \boldsymbol{\mu}) \text{ is approximately } \chi_p^2$$
>
> for $n - p$ large.

In the next sections we consider ways of verifying the assumption of normality and methods for transforming nonnormal observations into observations that are approximately normal.

4.6 ASSESSING THE ASSUMPTION OF NORMALITY

As we have pointed out, most of the statistical techniques discussed in subsequent chapters assume that each vector observation $\mathbf{X}_j$ comes from a multivariate normal distribution. On the other hand, in situations where the sample size is large and the techniques depend solely on the behavior of $\overline{\mathbf{X}}$, or distances involving $\overline{\mathbf{X}}$ of the form $n(\overline{\mathbf{X}} - \boldsymbol{\mu})'\mathbf{S}^{-1}(\overline{\mathbf{X}} - \boldsymbol{\mu})$, the assumption of normality for the individual observations is less crucial. But to some degree, the *quality* of inferences made by these methods depends on how closely the true parent population resembles the multivariate normal form. It is imperative, then, that procedures exist for detecting cases where the data exhibit moderate to extreme departures from what is expected under multivariate normality.

We want to answer this question: Do the observations $\mathbf{X}_j$ appear to violate the assumption that they came from a normal population? Based on the properties of normal distributions, we know that all linear combinations of normal variables are normal and the contours of the multivariate normal density are ellipsoids. Therefore we address these questions.

1. Do the marginal distributions of the elements of $\mathbf{X}$ appear to be normal? What about a few linear combinations of the components X_i?
2. Do the scatterplots of pairs of observations on different characteristics give the elliptical appearance expected from normal populations?
3. Are there any "wild" observations that should be checked for accuracy?

It will become clear that our investigations of normality will concentrate on the behavior of the observations in one or two dimensions (for example, marginal distributions and scatterplots). As might be expected, it has proved difficult to construct a "good" overall test of joint normality in more than two dimensions because of the large number of things that can go wrong. To some extent we must pay a price for concentrating on univariate and bivariate examinations of normality. We can never be sure that we have not missed some feature that is only revealed in higher dimensions. (It is possible, for example, to construct a nonnormal bivariate distribution with normal marginals. [See Exercise 4.6.]) Yet many types of nonnormality are often reflected in the marginal distributions and scatterplots. Moreover, for most practical work, one-dimensional and two-dimensional investigations are ordinarily sufficient. Fortunately, pathological data sets that are normal in lower-dimensional representations but nonnormal in higher dimensions have not frequently been detected.

Evaluating the Normality of the Univariate Marginal Distributions

Dot diagrams for smaller n and histograms for $n > 25$, or so, help reveal situations where one tail of a univariate distribution is much longer than the other. If the histogram for a variable X_i appears reasonably symmetric, we can check further by counting the number of observations in certain intervals. A univariate normal distribution assigns probability .683 to the interval $(\mu_i - \sqrt{\sigma_{ii}}, \mu_i + \sqrt{\sigma_{ii}})$ and probability .954 to the interval $(\mu_i - 2\sqrt{\sigma_{ii}}, \mu_i + 2\sqrt{\sigma_{ii}})$. Consequently, with a large sample size n, we expect the observed proportion $\hat{p}_{i1}$ of the observations lying in the interval $(\bar{x}_i - \sqrt{s_{ii}}, \bar{x}_i + \sqrt{s_{ii}})$ to be about .683. Similarly, the observed proportion, $\hat{p}_{i2}$ of the observations in $(\bar{x}_i - 2\sqrt{s_{ii}}, \bar{x}_i + 2\sqrt{s_{ii}})$ should be about .954. Using the normal approximation to the sampling distribution of $\hat{p}_i$ (see [8]), either

$$|\hat{p}_{i1} - .683| > 3\sqrt{\frac{(.683)(.317)}{n}} = \frac{1.396}{\sqrt{n}}$$

or

$$|\hat{p}_{i2} - .954| > 3\sqrt{\frac{(.954)(.046)}{n}} = \frac{.628}{\sqrt{n}} \qquad (4\text{-}29)$$

would indicate departures from an assumed normal distribution for the ith characteristic. When the observed proportions are too small, parent distributions with thicker tails than the normal are suggested.

Plots are always useful devices in any data analysis. Special plots called *Q-Q plots* can be used to assess the assumption of normality. These plots can be performed for the marginal distributions of the sample observations on each variable. They are, in effect, a plot of the sample quantile versus the quantile one would expect to observe if the observations actually were normally distributed. When the points lie very nearly along a straight line, the normality assumption remains tenable. Normality is suspect if the points deviate from a straight line. Moreover, the pattern of the deviations can provide clues about the nature of the nonnormality. Once the reasons for the nonnormality are identified, corrective action is often possible. (See Section 4.7.)

To simplify notation, let $x_1, x_2, \ldots, x_n$ represent n observations on any single characteristic X_i. Let $x_{(1)} \le x_{(2)} \le \cdots \le x_{(n)}$ represent these observations after they are ordered according to magnitude. For example, $x_{(2)}$ is the second smallest observation and $x_{(n)}$ is the largest observation. The $x_{(i)}$'s are the sample quantiles. When the $x_{(i)}$ are distinct, exactly i observations are less than or equal to $x_{(i)}$. (This is theoretically always true when the observations are of the continuous type, which we usually assume.) The proportion i/n of the sample at or to the left of $x_{(i)}$ is often approximated by $(i - \frac{1}{2})/n$ for analytical convenience.[1] A plot of $(i - \frac{1}{2})/n$ versus $x_{(i)}$ gives the empirical distribution function and the *Q-Q* plots provide a measure of the agreement between the pairs $(x_{(i)}, (i - \frac{1}{2})/n)$ and the corresponding theoretical values for a normal distribution.

For a standard normal distribution the quantiles $q_{(i)}$ are defined by the relation

$$P[Z \le q_{(i)}] = \int_{-\infty}^{q_{(i)}} \frac{1}{\sqrt{2\pi}} e^{-z^2/2}\, dz = p_{(i)} \qquad (4\text{-}30)$$

[1] The $\frac{1}{2}$ in the numerator of $(i - \frac{1}{2})/n$ is a "continuity" correction. Some authors (see [5]) have suggested replacing $(i - \frac{1}{2})/n$ by $(i - \frac{3}{8})/(n + \frac{1}{4})$.

(See Table 1 in the appendix.) Here $p_{(i)}$ is the probability of getting a value less than or equal to $q_{(i)}$ in a single drawing from a standard normal population. A comparison of the empirical cumulative distribution function with a normal can be made by setting $p_{(i)} = (i - \frac{1}{2})/n$. The idea is to look at the pairs of quantiles $(q_{(i)}, x_{(i)})$ with the same associated cumulative probability, $(i - \frac{1}{2})/n$. If the empirical and theoretical distribution functions "look very much alike," the pairs $(q_{(i)}, x_{(i)})$ will be approximately linearly related.

Example 4.8

A sample of $n = 10$ observations gives the values in the following table.

Ordered observations $x_{(i)}$	Probability levels $(i - \frac{1}{2})/n$	Standard normal quantiles $q_{(i)}$
−1.00	.05	−1.645
−.10	.15	−1.036
.16	.25	−.674
.41	.35	−.385
.62	.45	−.125
.80	.55	.125
1.26	.65	.385
1.54	.75	.674
1.71	.85	1.036
2.30	.95	1.645

Here, for example, $P[Z \leq .385] = \int_{-\infty}^{.385} \frac{1}{\sqrt{2\pi}} e^{-z^2/2}\, dz = .65$. [See (4-30).]

Let us now construct the Q-Q plot and comment on its appearance. The Q-Q plot for the above data, which is a plot of the ordered data $x_{(i)}$ against the normal quantiles $q_{(i)}$, is shown in Figure 4.5. The pairs of points $(q_{(i)}, x_{(i)})$ lie very nearly along a straight line and we would not reject the notion that these data are normally distributed—particularly with a sample size as small as $n = 10$. ■

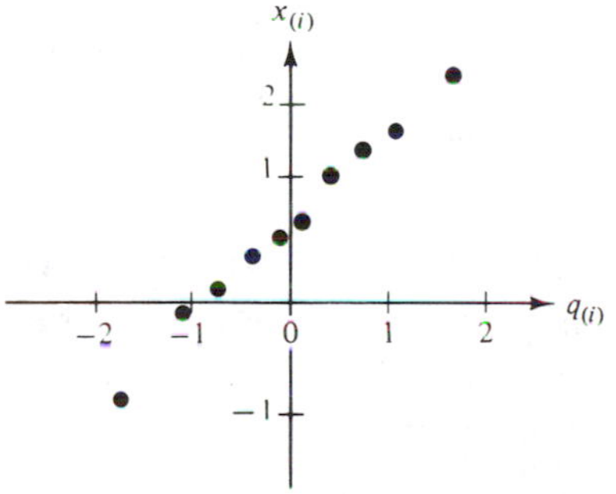

Figure 4.5 A Q-Q plot for the data in Example 4.8.

The calculations required for Q-Q plots are easily programmed for electronic computers. Some collections of statistical programs available commercially are capable of producing Q-Q plots.

The steps leading to a Q-Q plot are:

1. Order the original observations to get $x_{(1)}, x_{(2)}, \ldots, x_{(n)}$ and their corresponding probability values $(1 - \frac{1}{2})/n, (2 - \frac{1}{2})/n, \ldots, (n - \frac{1}{2})/n$;
2. Calculate the standard normal quantiles $q_{(1)}, q_{(2)}, \ldots, q_{(n)}$; and
3. Plot the pairs of observations $(q_{(1)}, x_{(1)}), (q_{(2)}, x_{(2)}), \ldots, (q_{(n)}, x_{(n)})$ and examine the "straightness" of the outcome.

The Q-Q plots are not particularly informative unless the sample size is moderate to large; for instance, $n \geq 20$. There can be quite a bit of variability in the straightness of the Q-Q plot for small samples, even when the observations are known to come from a normal population.

Example 4.9

The quality-control department of a manufacturer of microwave ovens is required by the federal government to monitor the amount of radiation emitted when the doors of the ovens are closed. Observations of the radiation emitted through closed doors of $n = 42$ randomly selected ovens were made. These data are listed in Table 4.1.

In order to determine the probability of exceeding a prespecified tolerance level, a probability distribution for the radiation emitted was needed. Can we regard the observations here as being normally distributed?

TABLE 4.1 RADIATION DATA (DOOR CLOSED)

Oven no.	Radiation	Oven no.	Radiation	Oven no.	Radiation
1	.15	16	.10	31	.10
2	.09	17	.02	32	.20
3	.18	18	.10	33	.11
4	.10	19	.01	34	.30
5	.05	20	.40	35	.02
6	.12	21	.10	36	.20
7	.08	22	.05	37	.20
8	.05	23	.03	38	.30
9	.08	24	.05	39	.30
10	.10	25	.15	40	.40
11	.07	26	.10	41	.30
12	.02	27	.15	42	.05
13	.01	28	.09		
14	.10	29	.08		
15	.10	30	.18		

SOURCE: Data courtesy of J. D. Cryer.

A computer was used to assemble the pairs $(q_{(i)}, x_{(i)})$ and construct a *Q-Q* plot, which is pictured in Figure 4.6. It appears from the plot that the data as a whole are not normally distributed. The points indicated by the circled locations in the figure are outliers, values that are too large relative to the rest of the observations.

For the radiation data, several observations are equal. When this occurs, those observations with like values are associated with the same normal quantile. This quantile is calculated using the average of the ranks, (i), the tied observations would have if they all differed slightly. ■

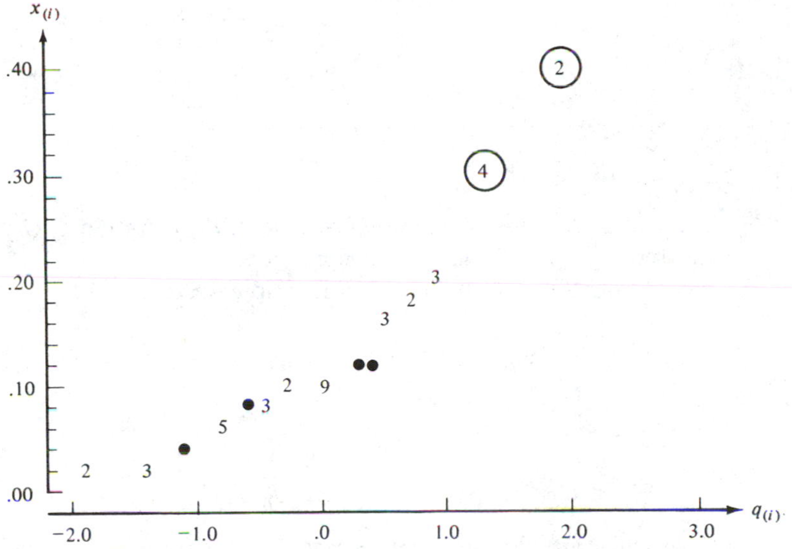

Figure 4.6 A *Q-Q* plot of the radiation data (door closed) from Example 4.9. (The integers in the plot indicate the number of points occupying the same location.)

The straightness of the *Q-Q* plot can be measured by calculating the correlation coefficient of the points in the plot. The correlation coefficient for the *Q-Q* plot is defined by

$$r_Q = \frac{\sum_{i=1}^{n} (x_{(i)} - \bar{x})(q_{(i)} - \bar{q})}{\sqrt{\sum_{i=1}^{n} (x_{(i)} - \bar{x})^2} \sqrt{\sum_{i=1}^{n} (q_{(i)} - \bar{q})^2}} \tag{4-31}$$

and a powerful test of normality can be based on it. (See [5] and [9].) Formally, we reject the hypothesis of normality at level of significance α if r_Q falls *below* the appropriate value in Table 4.2.

TABLE 4.2 CRITICAL POINTS FOR THE *Q-Q* PLOT CORRELATION COEFFICIENT TEST FOR NORMALITY

Sample size	Significance levels α		
n	.01	.05	.10
10	.880	.918	.935
15	.911	.938	.951
20	.929	.950	.960
25	.941	.958	.966
30	.949	.964	.971
40	.960	.972	.977
50	.966	.976	.981
60	.971	.980	.984
75	.976	.984	.987
100	.981	.986	.989
150	.987	.991	.992
200	.990	.993	.994

Example 4.10

Let us calculate the correlation coefficient, r_Q, from the *Q-Q* plot of Example 4.8 (see Figure 4.5) and test for normality.

Using the information from Example 4.8, $\bar{x} = .770$,

$$\sum_{i=1}^{10} (x_{(i)} - \bar{x})q_{(i)} = 8.584, \quad \sum_{i=1}^{10} (x_{(i)} - \bar{x})^2 = 8.472, \text{ and } \sum_{i=1}^{10} q_{(i)}^2 = 8.795$$

Since $\bar{q} = 0$,

$$r_Q = \frac{8.584}{\sqrt{8.472}\sqrt{8.795}} = .994$$

A test of normality at the 10% level of significance is provided by referring $r_Q = .994$ to the entry in Table 4.2 corresponding to $n = 10$ and $\alpha = .10$. This entry is .935. Since $r_Q > .935$, we do not reject the hypothesis of normality. ■

Linear combinations of more than one characteristic can be investigated. Many statisticians suggest plotting

$$\hat{\mathbf{e}}_1'\mathbf{x}_j \quad \text{where} \quad \mathbf{S}\hat{\mathbf{e}}_1 = \hat{\lambda}_1\hat{\mathbf{e}}_1$$

and $\hat{\lambda}_1$ is the largest eigenvalue of $\mathbf{S}$. Here $\mathbf{x}_j' = [x_{1j}, x_{2j}, \ldots, x_{pj}]$ is the jth observation on the p variables $X_1, X_2, \ldots, X_p$. The linear combination $\hat{\mathbf{e}}_p'\mathbf{x}_j$ corresponding to the smallest eigenvalue is also frequently singled out for inspection. See Chapter 8 and [6] for further details.

Evaluating Bivariate Normality

We would like to check on the assumption of normality for all distributions of $2, 3, \ldots, p$ dimensions. However, as we have pointed out, for practical work it is usually sufficient to investigate the univariate and bivariate distributions. We considered univariate marginal distributions earlier. It is now of interest to examine the bivariate case.

In Chapter 1 we described scatterplots for pairs of characteristics. If the observations were generated from a multivariate normal distribution, each bivariate distribution would be normal and the contours of constant density would be ellipses. The scatterplot should conform to this structure by exhibiting an overall pattern that is nearly elliptical.

Moreover, by Result 4.7, the set of bivariate outcomes $\mathbf{x}$ such that

$$(\mathbf{x}-\boldsymbol{\mu})'\boldsymbol{\Sigma}^{-1}(\mathbf{x}-\boldsymbol{\mu}) \le \chi_2^2(.5)$$

has probability .5. Thus we should expect *roughly* the same percentage, 50%, of sample observations to lie in the ellipse

$$\text{all } \mathbf{x} \text{ such that } (\mathbf{x}-\bar{\mathbf{x}})'\mathbf{S}^{-1}(\mathbf{x}-\bar{\mathbf{x}}) \le \chi_2^2(.5)$$

where we have replaced $\boldsymbol{\mu}$ by its estimate $\bar{\mathbf{x}}$ and $\boldsymbol{\Sigma}^{-1}$ by its estimate $\mathbf{S}^{-1}$. If not, the normality assumption is suspect.

Example 4.11

Although not a random sample, data consisting of the pairs of observations (x_1 = assets, x_2 = net income) for the 10 largest U.S. industrial corporations are listed in Exercise 1.4. These data give

$$\bar{\mathbf{x}} = \begin{bmatrix} 19.32 \\ 1.51 \end{bmatrix}, \qquad \mathbf{S} = \begin{bmatrix} 70.41 & 5.87 \\ 5.87 & .97 \end{bmatrix}$$

so

$$\mathbf{S}^{-1} = \frac{1}{(70.41)(.97)-(5.87)^2}\begin{bmatrix} .97 & -5.87 \\ -5.87 & 70.41 \end{bmatrix} = \begin{bmatrix} .029 & -.173 \\ -.173 & 2.080 \end{bmatrix}$$

From Table 3 in the appendix, $\chi_2^2(.5) = 1.39$. Any observation $\mathbf{x}' = [x_1, x_2]$ satisfying

$$\begin{bmatrix} x_1 - 19.32 \\ x_2 - 1.51 \end{bmatrix}' \begin{bmatrix} .029 & -.173 \\ -.173 & 2.080 \end{bmatrix} \begin{bmatrix} x_1 - 19.32 \\ x_2 - 1.51 \end{bmatrix} \le 1.39$$

is on or inside the estimated 50% contour. Otherwise the observation is outside this contour. The first pair of observations in Exercise 1.4 is $[x_1, x_2]' = [26.7, 3.3]'$. In this case

$$\begin{bmatrix} 26.7 - 19.32 \\ 3.3 - 1.51 \end{bmatrix}' \begin{bmatrix} .029 & -.173 \\ -.173 & 2.080 \end{bmatrix} \begin{bmatrix} 26.7 - 19.32 \\ 3.3 - 1.51 \end{bmatrix} = 3.67 > 1.39$$

and this point falls outside the 50% contour. The remaining 9 points have generalized distances from $\bar{\mathbf{x}}$ of 6.32, .07, .81, .69, .34, 3.08, .55, .47 and 2.21, respectively. Since 6 of these distances are less than 1.39, a proportion .60 of the data falls within the 50% contour. If these observations were normally distributed, we would expect about $\frac{1}{2}$, or 5, of the observations to be within this contour. Given our small sample size of 10, we would conclude that this evidence is not sufficient to reject the notion of bivariate normality. ■

Computing the fraction of the points within a contour and subjectively comparing it with the theoretical probability is a useful, but rather rough, procedure. A somewhat more formal method for judging the joint normality of a data set is based on the squared generalized distances

$$d_j^2 = (\mathbf{x}_j - \bar{\mathbf{x}})'\mathbf{S}^{-1}(\mathbf{x}_j - \bar{\mathbf{x}}), \quad j = 1, 2, \ldots, n \tag{4-32}$$

where $\mathbf{x}_1, \mathbf{x}_2, \ldots, \mathbf{x}_n$ are the sample observations. The procedure we are about to describe is not limited to the bivariate case. It can be used for all $p \geq 2$.

When the parent population is multivariate normal and both n and $n - p$ are greater than 25 or 30, each of the squared distances $d_1^2, d_2^2, \ldots, d_n^2$ should behave like a chi-square random variable. [See Result 4.7 and Equations (4-26) and (4-27).] Although these distances are *not* independent or exactly chi-square distributed, it is helpful to plot them as if they were. The resulting plot is called a *chi-square plot*, or *gamma plot* because the chi-square distribution is a special case of the more general gamma distribution. (See [6].)

To construct the chi-square plot:

1. Order the squared distances in (4-32) from smallest to largest as $d_{(1)}^2 \leq d_{(2)}^2 \leq \cdots \leq d_{(n)}^2$.
2. Graph the pairs $(d_{(j)}^2, \chi_p^2((j - \frac{1}{2})/n))$, where $\chi_p^2((j - \frac{1}{2})/n)$ is the $100(j - \frac{1}{2})/n$ percentile of the chi-square distribution with p degrees of freedom.

The plot should resemble a straight line. A systematic curved pattern suggests lack of normality. One or two points far to the right of the line indicate large distances, or outlying observations, that merit further attention.

Example 4.12

Let us construct a chi-square plot of the generalized distances given in Example 4.11. The ordered distances and the corresponding chi-square percentiles for $p = 2$ and $n = 10$ are listed below.

j	$d_{(j)}^2$	$\chi_2^2\left(\frac{j - \frac{1}{2}}{10}\right)$
1	.07	.10
2	.34	.33
3	.47	.58
4	.55	.86
5	.69	1.20
6	.81	1.60
7	2.21	2.10
8	3.08	2.77
9	3.67	3.79
10	6.32	5.99

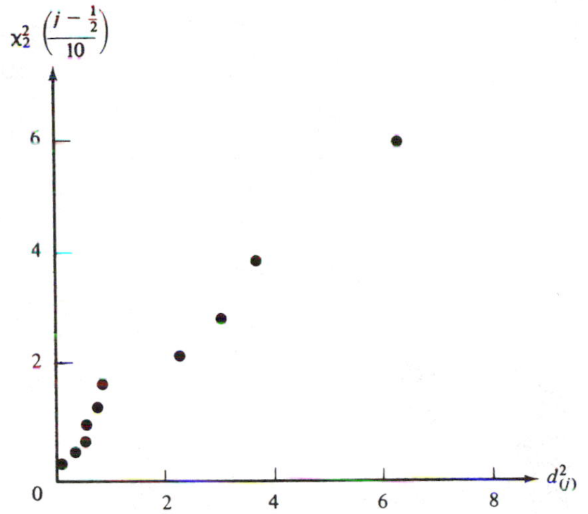

Figure 4.7 A chi-square plot of the ordered distances in Example 4.12.

A graph of the pairs $(d_{(j)}^2, \chi_2^2((j-\frac{1}{2})/10))$ is shown in Figure 4.7.

There is some evidence of a systematic deviation from straightness. In particular, the middle distances appear to be too small relative to the distances expected from bivariate normal populations for samples of size 10. On the other hand, the evidence against normality is slight and, given the small sample, it would seem reasonable to analyze this data as if it were bivariate normal. ■

Example 4.13

The data in Table 4.3 were obtained by taking four different measures of stiffness, x_1, x_2, x_3, and x_4, of each of $n = 30$ boards. The first measurement involves sending a shock wave down the board, the second measurement is determined while vibrating the board, and the last two measurements are obtained from static tests. The standardized measurements

$$z_{ij} = (x_{ij} - \bar{x}_i)/\sqrt{s_{ii}}, \; i = 1, 2, 3, 4; \; j = 1, 2, \ldots, 30$$

and the squared distances $d_j^2 = (\mathbf{x}_j - \bar{\mathbf{x}})'\mathbf{S}^{-1}(\mathbf{x}_j - \bar{\mathbf{x}})$ are also presented in the table.

The marginal distributions appear quite normal (see Exercise 4.25) with the possible exception of the specimen (board) 9. The last column in Table 4.3 reveals that specimen 16 is a multivariate outlier since $\chi_4^2(.005) = 14.86$; yet all of the individual measurements are well within their respective univariate scatters.

Scientists specializing in wood properties conjectured that specimen 9 was unusually clear and therefore very stiff and strong. It would also appear that specimen 16 is a bit unusual, since both dynamic measurements are above average and the two static measurements are low. Unfortunately, it was not possible to investigate this specimen further because the material was no longer available. ■

TABLE 4.3 FOUR MEASUREMENTS OF STIFFNESS

x_1	x_2	x_3	x_4	Obs. no.	z_1	z_2	z_3	z_4	d^2
1889	1651	1561	1778	1	−.1	−.3	.2	.2	.60
2403	2048	2087	2197	2	1.5	.9	1.9	1.5	5.48
2119	1700	1815	2222	3	.7	−.2	1.0	1.5	7.62
1645	1627	1110	1533	4	−.8	−.4	−1.3	−.6	5.21
1976	1916	1614	1883	5	.2	.5	.3	.5	1.40
1712	1712	1439	1546	6	−.6	−.1	−.2	−.6	2.22
1943	1685	1271	1671	7	.1	−.2	−.8	−.2	4.99
2104	1820	1717	1874	8	.6	.2	.7	.5	1.49
2983	2794	2412	2581	9	3.3	3.3	3.0	2.7	12.26
1745	1600	1384	1508	10	−.5	−.5	−.4	−.7	.77
1710	1591	1518	1667	11	−.6	−.5	.0	−.2	1.93
2046	1907	1627	1898	12	.4	.5	.4	.5	.46
1840	1841	1595	1741	13	−.2	.3	.3	.0	2.70
1867	1685	1493	1678	14	−.1	−.2	−.1	−.1	.13
1859	1649	1389	1714	15	−.1	−.3	−.4	−.0	1.08
1954	2149	1180	1281	16	.1	1.3	−1.1	−1.4	16.85
1325	1170	1002	1176	17	−1.8	−1.8	−1.7	−1.7	3.50
1419	1371	1252	1308	18	−1.5	−1.2	−.8	−1.3	3.99
1828	1634	1602	1755	19	−.2	−.4	.3	.1	1.36
1725	1594	1313	1646	20	−.6	−.5	−.6	−.2	1.46
2276	2189	1547	2111	21	1.1	1.4	.1	1.2	9.90
1899	1614	1422	1477	22	−.0	−.4	−.3	−.8	5.06
1633	1513	1290	1516	23	−.8	−.7	−.7	−.6	.80
2061	1867	1646	2037	24	.5	.4	.5	1.0	2.54
1856	1493	1356	1533	25	−.2	−.8	−.5	−.6	4.58
1727	1412	1238	1469	26	−.6	−1.1	−.9	−.8	3.40
2168	1896	1701	1834	27	.8	.5	.6	.3	2.38
1655	1675	1414	1597	28	−.8	−.2	−.3	−.4	3.00
2326	2301	2065	2234	29	1.3	1.7	1.8	1.6	6.28
1490	1382	1214	1284	30	−1.3	−1.2	−1.0	−1.4	2.58

SOURCE: Data courtesy of William Galligan.

We have discussed some rather simple techniques for checking the normality assumption. See [6] for a more complete exposition of methods for assessing normality.

All measures of goodness-of-fit suffer the same serious drawback. When the sample size is small, only the most aberrant behavior will be identified as lack of fit. On the other hand, very large samples invariably produce statistically significant lack of fit. Yet the departure from the specified distribution may be very small and technically unimportant to the inferential conclusions.

4.7 TRANSFORMATIONS TO NEAR NORMALITY

If normality is not a viable assumption, what is the next step? One alternative is to ignore the findings of a normality check and proceed as if the data were normally distributed. This practice is not recommended since, in many instances, it could lead to incorrect conclusions. A second alternative is to make nonnormal data more "normal looking" by considering *transformations* of the data. Normal theory analyses can then be carried out with the suitably transformed data.

Transformations are nothing more than a reexpression of the data in different units. For example when a histogram of positive observations exhibits a long right-hand tail, transforming the observations by taking their logarithms or square roots will often markedly improve the symmetry about the mean and the approximation to a normal distribution. It frequently happens that the new units provide more natural expressions of the characteristics being studied.

Appropriate transformations are suggested by (1) theoretical considerations or (2) the data themselves (or both). It has been shown theoretically that data that are counts can often be made more normal by taking their *square roots*. Similarly, the *logit transformation* applied to proportions and *Fisher's z-transformation* applied to correlation coefficients yield quantities that are approximately normally distributed.

Helpful Transformations to Near Normality

Original Scale	*Transformed Scale*	
1. Counts, y	$\sqrt{y}$	(4-33)
2. Proportions, $\hat{p}$	$\text{logit}(\hat{p}) = \frac{1}{2}\log\left(\frac{\hat{p}}{1-\hat{p}}\right)$	
3. Correlations, r	Fisher's $z(r) = \frac{1}{2}\log\left(\frac{1+r}{1-r}\right)$	

In many instances the choice of a transformation to improve the approximation to normality is not obvious. For such cases it is convenient to let the data suggest a transformation. A useful family of transformations for this purpose is the family of *power transformations*.

Power transformations are defined only for positive variables. However, this is not as restrictive as it seems, because a single constant can be added to each observation in the data set if some of the values are negative.

Let x represent an arbitrary observation. The power family of transformations is indexed by a parameter λ. A given value for λ implies a particular transformation. For example, consider x^λ with $\lambda = -1$. Since $x^{-1} = 1/x$, this choice of λ corresponds to the reciprocal transformation. We can trace the family of transformations as λ ranges from negative to positive powers of x. For $\lambda = 0$, we define $x^0 = \ln x$. A sequence of possible transformations is

$$\underbrace{\ldots, x^{-1} = \frac{1}{x}, x^0 = \ln x, x^{1/4} = \sqrt[4]{x}, x^{1/2} = \sqrt{x}}_{\text{Shrinks large values of } x}, \underbrace{x^2, x^3, \ldots}_{\text{Increases large values of } x}$$

To select a power transformation, an investigator looks at the marginal dot diagram or histogram and decides whether large values have to be "pulled in" or "pushed out" to improve the symmetry about the mean. Trial-and-error calculations with a few of the transformations indicated above should produce an improvement. The final choice should always be examined by a *Q-Q* plot or other checks to see if the tentative normal assumption is satisfactory.

The transformations we have been discussing are data based in the sense that it is only the appearance of the data themselves that influences the choice of an appropriate transformation. There are no external considerations involved, although the transformation actually used is often determined by some mix of information supplied by the data and extra-data factors, such as simplicity, or ease of interpretation.

A convenient analytical method is available for choosing a power transformation. We begin by focusing our attention on the univariate case.

Box and Cox [3] consider the slightly modified family of power transformations

$$x^{(\lambda)} = \begin{cases} \dfrac{x^\lambda - 1}{\lambda} & \lambda \neq 0 \\ \ln x & \lambda = 0 \end{cases} \tag{4-34}$$

which is continuous in λ for $x > 0$. (See [7].) Given the observations $x_1, x_2, \ldots, x_n$, the Box–Cox solution for the choice of an appropriate power λ is the one which *maximizes* the expression

$$\ell(\lambda) = -\frac{n}{2}\ln\left[\frac{1}{n}\sum_{j=1}^{n}\left(x_j^{(\lambda)} - \overline{x^{(\lambda)}}\right)^2\right] + (\lambda - 1)\sum_{j=1}^{n}\ln x_j \tag{4-35}$$

We note that $x_j^{(\lambda)}$ is defined in (4-34) and

$$\overline{x^{(\lambda)}} = \frac{1}{n}\sum_{j=1}^{n} x_j^{(\lambda)} = \frac{1}{n}\sum_{j=1}^{n}\left(\frac{x_j^\lambda - 1}{\lambda}\right) \tag{4-36}$$

is the arithmetic average of the transformed observations. The first term in (4-35) is, apart from a constant, the logarithm of a normal likelihood function, after maximizing it with respect to the population mean and variance parameters.

The calculation of $\ell(\lambda)$ for many values of λ is an easy task for a computer. It is helpful to have a graph of $\ell(\lambda)$ versus λ, as well as a tabular display of the pairs $(\lambda, \ell(\lambda))$, in order to study the behavior near the maximizing value $\hat{\lambda}$. For instance, if either $\lambda = 0$ (logarithm) or $\lambda = \frac{1}{2}$ (square root) is near $\hat{\lambda}$, one of these may be preferred because of its simplicity.

Comment. It is now understood that the transformation obtained by maximizing $\ell(\lambda)$ usually improves the approximation to normality. However, there is no guarantee that even the best choice of λ will produce a transformed set of values that adequately conform to a normal distribution. The outcomes produced by a transformation selected according to (4-35) should always be carefully examined for possible violations of the tentative assumption of normality. This warning applies with equal force to transformations selected by any other technique.

Example 4.14

We gave readings of the microwave radiation emitted through the closed doors of $n = 42$ ovens in Example 4.9. The *Q-Q* plot of this data in Figure 4.6 indicates that the observations deviate from what would be expected if they were normally distributed. Since all the observations are positive, let us perform a power transformation of the data which, we hope, will produce results that are more nearly normal. Restricting our attention to the family of transformations in (4-34), we must find that value of λ maximizing the function $\ell(\lambda)$ in (4-35).

The pairs $(\lambda, \ell(\lambda))$ are listed below for several values of λ.

λ	$\ell(\lambda)$	λ	$\ell(\lambda)$
−1.00	70.52		
−.90	75.65	.40	106.20
−.80	80.46	.50	105.50
−.70	84.94	.60	104.43
−.60	89.06	.70	103.03
−.50	92.79	.80	101.33
−.40	96.10	.90	99.34
−.30	98.97	1.00	97.10
−.20	101.39	1.10	94.64
−.10	103.35	1.20	91.96
.00	104.83	1.30	89.10
.10	105.84	1.40	86.07
.20	106.39	1.50	82.88
.30	106.51		

It is evident from the table above that a value of $\hat{\lambda}$ around .30 maximizes $\ell(\lambda)$. For convenience, we choose $\hat{\lambda} = .25$. The data x_i were reexpressed as

$$x_i^{(1/4)} = \frac{x_i^{1/4} - 1}{\frac{1}{4}} \qquad i = 1, 2, \ldots, 42$$

and a *Q-Q* plot was constructed from the transformed quantities. This plot is shown in Figure 4.8. The quantile pairs fall very close to a straight line, and we would conclude from this evidence that the $x_i^{(1/4)}$ are approximately normally distributed. ■

With multivariate observations, a power transformation must be selected for each of the variables. Let $\lambda_1, \lambda_2, \ldots, \lambda_p$ be the power transformations for the p measured characteristics. Each λ_k can be selected by *maximizing*

$$\ell_k(\lambda) = -\frac{n}{2}\ln\left[\frac{1}{n}\sum_{j=1}^{n}\left(x_{kj}^{(\lambda_k)} - \overline{x_k^{(\lambda_k)}}\right)^2\right] + (\lambda_k - 1)\sum_{j=1}^{n}\ln x_{kj} \tag{4-37}$$

where $x_{k1}, x_{k2}, \ldots, x_{kn}$ are the n observations on the kth variable, $k = 1, 2, \ldots, p$.

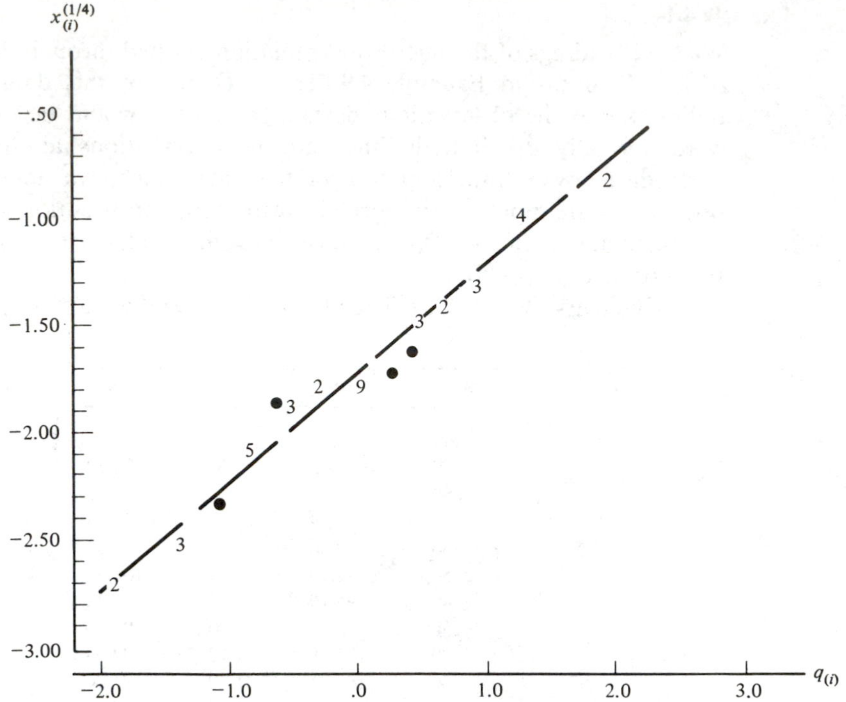

Figure 4.8 A *Q-Q* plot of the transformed radiation data (door closed). (The integers in the plot indicate the number of points occupying the same location.)

Here

$$\overline{x_k^{(\lambda_k)}} = \frac{1}{n}\sum_{j=1}^{n} x_{kj}^{(\lambda_k)} = \frac{1}{n}\sum_{j=1}^{n}\left(\frac{x_{kj}^{\lambda_k} - 1}{\lambda_k}\right) \tag{4-38}$$

is the arithmetic average of the transformed observations. The jth transformed multivariate observation is

$$\mathbf{x}_j^{(\hat{\lambda})} = \begin{bmatrix} \dfrac{x_{1j}^{\hat{\lambda}_1} - 1}{\hat{\lambda}_1} \\ \dfrac{x_{2j}^{\hat{\lambda}_2} - 1}{\hat{\lambda}_2} \\ \vdots \\ \dfrac{x_{pj}^{\hat{\lambda}_p} - 1}{\hat{\lambda}_p} \end{bmatrix}$$

where $\hat{\lambda}_1, \hat{\lambda}_2, \ldots, \hat{\lambda}_p$ are the values that individually maximize (4-37).

The procedure just described is equivalent to making each marginal distribution approximately normal. Although normal marginals are not sufficient to ensure that the joint distribution is normal, in practical applications it may be good enough. If not, we could start with the values $\hat{\lambda}_1, \hat{\lambda}_2, \ldots, \hat{\lambda}_p$ obtained above and iterate toward the set of values $\boldsymbol{\lambda}' = [\lambda_1, \lambda_2, \ldots, \lambda_p]$, which collectively maximize

$$\begin{aligned} \ell(\lambda_1, \lambda_2, \ldots, \lambda_p) = -\frac{n}{2}\ln|\mathbf{S}(\boldsymbol{\lambda})| + (\lambda_1 - 1) \sum_{j=1}^{n} \ln x_{1j} \\ + (\lambda_2 - 1) \sum_{j=1}^{n} \ln x_{2j} \\ + \cdots + (\lambda_p - 1) \sum_{j=1}^{n} \ln x_{pj} \end{aligned} \qquad (4\text{-}39)$$

where $\mathbf{S}(\boldsymbol{\lambda})$ is the sample covariance matrix computed from

$$\mathbf{x}_j^{(\boldsymbol{\lambda})} = \begin{bmatrix} \dfrac{x_{1j}^{\lambda_1} - 1}{\lambda_1} \\ \dfrac{x_{2j}^{\lambda_2} - 1}{\lambda_2} \\ \vdots \\ \dfrac{x_{pj}^{\lambda_p} - 1}{\lambda_p} \end{bmatrix} \qquad j = 1, 2, \ldots, n$$

Maximizing (4-39) is not only substantially more difficult than maximizing the individual expressions in (4-37), it is also unlikely to yield remarkably better results. The selection method based on Equation (4-39) is equivalent to maximizing a multivariate likelihood over $\boldsymbol{\mu}$, $\boldsymbol{\Sigma}$, and $\boldsymbol{\lambda}$, whereas the method based on (4-37) corresponds to maximizing the kth univariate likelihood over μ_k, σ_{kk}, and λ_k. The latter likelihood is generated by pretending there is some λ_k for which the observations $(x_{kj}^{\lambda_k} - 1)/\lambda_k$, $j = 1, 2, \ldots, n$, have a normal distribution. See [3] and [2] for detailed discussions of the univariate and multivariate cases, respectively. (Also, see [7].)

Example 4.15

Radiation measurements were also recorded through the open doors of the $n = 42$ microwave ovens introduced in Example 4.9. The amount of radiation emitted through the open doors of these ovens is listed in Table 4.4.

Following the procedure outlined in Example 4.14, a power transformation for this data was selected by maximizing $\ell(\lambda)$ in (4-35). The approximate maximizing value was $\hat{\lambda} = .30$. Figure 4.9 on page 167 shows Q-Q plots of the untransformed and transformed door-open radiation data. (These data were

TABLE 4.4 RADIATION DATA (DOOR OPEN)

Oven no.	Radiation	Oven no.	Radiation	Oven no.	Radiation
1	.30	16	.20	31	.10
2	.09	17	.04	32	.10
3	.30	18	.10	33	.10
4	.10	19	.01	34	.30
5	.10	20	.60	35	.12
6	.12	21	.12	36	.25
7	.09	22	.10	37	.20
8	.10	23	.05	38	.40
9	.09	24	.05	39	.33
10	.10	25	.15	40	.32
11	.07	26	.30	41	.12
12	.05	27	.15	42	.12
13	.01	28	.09		
14	.45	29	.09		
15	.12	30	.28		

SOURCE: Data courtesy of J. D. Cryer.

actually transformed by taking the fourth root, as in Example 4.14.) It is clear from the figure that the transformed data are more nearly normal, although the normal approximation is not as good as it was for the door-closed data.

Let us denote the door-closed data by $x_{11}, x_{12}, \ldots, x_{1,42}$ and the door-open data by $x_{21}, x_{22}, \ldots, x_{2,42}$. Choosing a power transformation for each set by maximizing the expression in (4-35) is equivalent to maximizing $\ell_k(\lambda)$ in (4-37) with $k = 1, 2$. Thus, using the outcomes from Example 4.14 and the results above, we have $\hat{\lambda}_1 = .30$ and $\hat{\lambda}_2 = .30$. These powers were determined for the *marginal* distributions of x_1 and x_2.

We can consider the *joint* distribution of x_1 and x_2 and simultaneously determine the pair of powers (λ_1, λ_2) that makes this joint distribution approximately bivariate normal. To do this we must maximize $\ell(\lambda_1, \lambda_2)$ in (4-39) with respect to both λ_1 and λ_2.

We have computed $\ell(\lambda_1, \lambda_2)$ over a grid of λ_1, λ_2 values, and some of the outcomes are displayed in Figure 4.10 on page 168. The grid was determined by choosing $0 \le \lambda_1 \le .50$ and $0 \le \lambda_2 \le .50$ and incrementing both λ_1 and λ_2 by .10. We see that the maximum occurs at $\hat{\lambda}_1 = .20$, $\hat{\lambda}_2 = .20$.

The "best" power transformations for the bivariate case do not differ substantially from those obtained by considering each marginal distribution. ■

As we saw in Example 4.15, making each marginal distribution approximately normal is roughly equivalent to addressing the bivariate distribution directly and making it approximately normal. It is generally easier to select appropriate transformations for the marginal distributions than for the joint distributions.

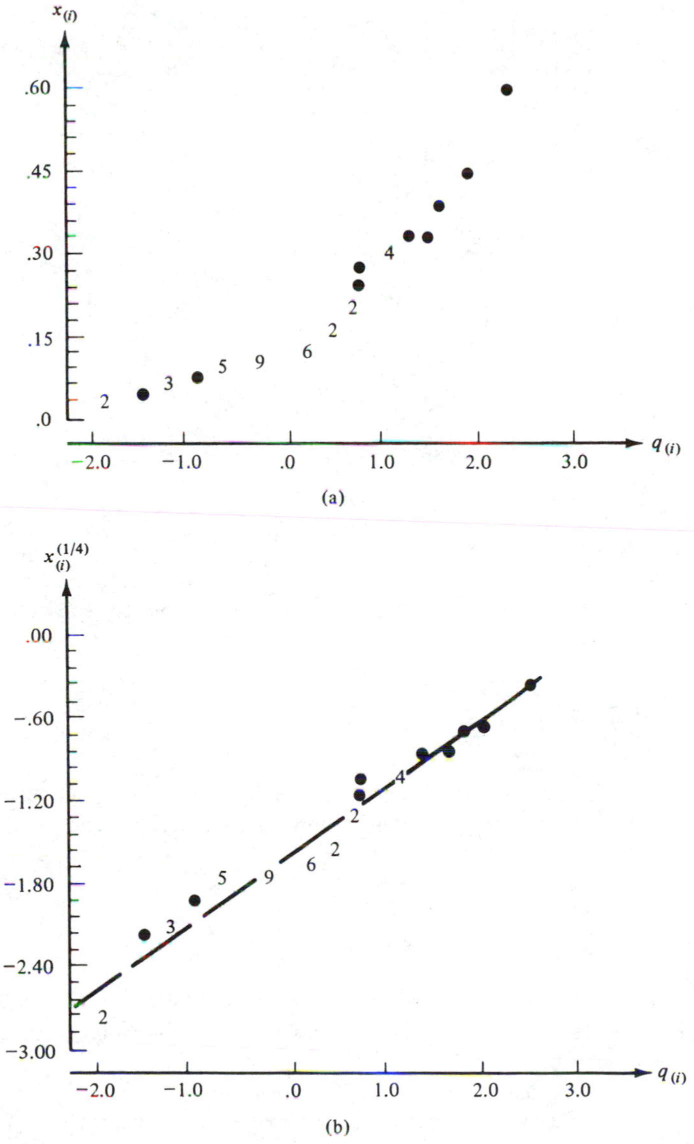

Figure 4.9 *Q-Q* plots of the (a) original and (b) transformed radiation data (door open). (The integers in the plot indicate the number of points occupying the same location.)

λ_2 \ λ_1	0	.10	.20	.30	.40	.50
.50	219.15	220.75	221.72	222.05	221.77	220.92
.40	221.46	222.94	223.68	223.80	223.32	222.27
.30	223.36	224.57	225.10	224.98	224.25	222.97
.20	224.67	225.60	225.83	225.42	224.42	222.91
.10	225.20	225.78	225.68	224.96	223.70	221.97
0	224.76	224.96	224.51	223.49	221.99	220.07

Figure 4.10 Values of $\ell(\lambda_1, \lambda_2)$ for selected pairs of λ_1 and λ_2 (radiation data).

EXERCISES

4.1. Let $\mathbf{X}_1$, $\mathbf{X}_2$, $\mathbf{X}_3$, and $\mathbf{X}_4$ be independent $N_p(\boldsymbol{\mu}, \boldsymbol{\Sigma})$ random vectors.

(a) Find the marginal distributions of the random vectors

$$\mathbf{V}_1 = \tfrac{1}{4}\mathbf{X}_1 - \tfrac{1}{4}\mathbf{X}_2 + \tfrac{1}{4}\mathbf{X}_3 - \tfrac{1}{4}\mathbf{X}_4$$

and

$$\mathbf{V}_2 = \tfrac{1}{4}\mathbf{X}_1 + \tfrac{1}{4}\mathbf{X}_2 - \tfrac{1}{4}\mathbf{X}_3 - \tfrac{1}{4}\mathbf{X}_4$$

(b) Find the joint density of the random vectors $\mathbf{V}_1$ and $\mathbf{V}_2$ defined in (a).

4.2. Consider a bivariate normal population with $\mu_1 = 0$, $\mu_2 = 2$, $\sigma_{11} = 2$, $\sigma_{22} = 1$, and $\rho_{12} = .5$.

(a) Write out the bivariate normal density.

(b) Write out the squared generalized distance expression $(\mathbf{x} - \boldsymbol{\mu})'\boldsymbol{\Sigma}^{-1}(\mathbf{x} - \boldsymbol{\mu})$ as a function of x_1 and x_2.

(c) Determine (and sketch) the constant density contour that contains 50% of the probability.

4.3. Let $\mathbf{X}$ be $N_3(\boldsymbol{\mu}, \boldsymbol{\Sigma})$ with $\boldsymbol{\mu}' = [-3, 1, 4]$ and

$$\boldsymbol{\Sigma} = \begin{bmatrix} 1 & -2 & 0 \\ -2 & 5 & 0 \\ 0 & 0 & 2 \end{bmatrix}$$

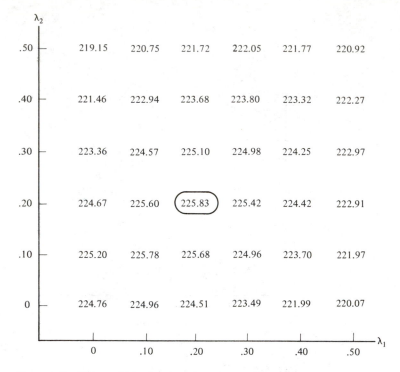

Which of the following random variables are independent? Explain.

(a) X_1 and X_2
(b) X_2 and X_3
(c) (X_1, X_2) and X_3
(d) $\dfrac{X_1 + X_2}{2}$ and X_3
(e) X_2 and $X_2 - \frac{5}{2}X_1 - X_3$

4.4. Let $\mathbf{X}$ be $N_3(\boldsymbol{\mu}, \boldsymbol{\Sigma})$ with $\boldsymbol{\mu}' = [2, -3, 1]$ and

$$\boldsymbol{\Sigma} = \begin{bmatrix} 1 & 1 & 1 \\ 1 & 3 & 2 \\ 1 & 2 & 2 \end{bmatrix}$$

(a) Find the distribution of $3X_1 - 2X_2 + X_3$.
(b) Relabel the variables, if necessary, and find a 2×1 vector $\mathbf{a}$ such that X_2 and $X_2 - \mathbf{a}'\begin{bmatrix} X_1 \\ X_3 \end{bmatrix}$ are independent.

4.5. Specify each of the following.

(a) The conditional distribution of X_1 given $X_2 = x_2$ for the joint distribution in Exercise 4.2.
(b) The conditional distribution of X_2 given $X_1 = x_1$ and $X_3 = x_3$ for the joint distribution in Exercise 4.3.
(c) The conditional distribution of X_3 given $X_1 = x_1$ and $X_2 = x_2$ for the joint distribution in Exercise 4.4.

4.6. (Example of a nonnormal bivariate distribution with normal marginals.) Let X_1 be $N(0, 1)$ and

$$X_2 = \begin{cases} -X_1 & \text{if } -1 \le X_1 \le 1 \\ X_1 & \text{otherwise} \end{cases}$$

Show each of the following.

(a) X_2 also has a $N(0, 1)$ distribution.
(b) X_1 and X_2 do *not* have a bivariate normal distribution.

(*Hint:*

(a) Since X_1 is $N(0, 1)$, $P[-1 < X_1 \le x] = P[-x \le X_1 < 1]$ for any x. When $-1 < x_2 < 1$, $P[X_2 \le x_2] = P[X_2 \le -1] + P[-1 < X_2 \le x_2] = P[X_1 \le -1] + P[-1 < -X_1 \le x_2] = P[X_1 \le -1] + P[-x_2 \le X_1 < 1]$. But $P[-x_2 \le X_1 < 1] = P[-1 < X_1 \le x_2]$ from the symmetry argument in the first line immediately above. Thus $P[X_2 \le x_2] = P[X_1 \le -1] + P[-1 < X_1 \le x_2] = P[X_1 \le x_2]$, which is a standard normal probability.
(b) Consider the linear combination $X_1 - X_2$, which equals zero with probability $P[|X_1| > 1] = .3174$.)

4.7. Show each of the following.

(a)

$$\begin{vmatrix} \mathbf{A} & \mathbf{0} \\ \mathbf{0}' & \mathbf{B} \end{vmatrix} = |\mathbf{A}||\mathbf{B}|$$

(b)

$$\begin{vmatrix} \mathbf{A} & \mathbf{C} \\ \mathbf{0}' & \mathbf{B} \end{vmatrix} = |\mathbf{A}||\mathbf{B}| \quad \text{for} \quad |\mathbf{A}| \ne 0$$

(*Hint:*

(a) $\begin{vmatrix} \mathbf{A} & \mathbf{0} \\ \mathbf{0}' & \mathbf{B} \end{vmatrix} = \begin{vmatrix} \mathbf{A} & \mathbf{0} \\ \mathbf{0}' & \mathbf{I} \end{vmatrix} \begin{vmatrix} \mathbf{I} & \mathbf{0} \\ \mathbf{0}' & \mathbf{B} \end{vmatrix}$. Expanding the determinant $\begin{vmatrix} \mathbf{I} & \mathbf{0} \\ \mathbf{0}' & \mathbf{B} \end{vmatrix}$ by the first row (see Definition 2A.24) gives 1 times a determinant of the same form, with the order of $\mathbf{I}$ reduced by one. This procedure is repeated until $1 \times |\mathbf{B}|$ is obtained. Similarly, expanding the determinant $\begin{vmatrix} \mathbf{A} & \mathbf{0} \\ \mathbf{0}' & \mathbf{I} \end{vmatrix}$ by the last row gives $\begin{vmatrix} \mathbf{A} & \mathbf{0} \\ \mathbf{0}' & \mathbf{I} \end{vmatrix} = |\mathbf{A}|$.

(b) $\begin{vmatrix} \mathbf{A} & \mathbf{C} \\ \mathbf{0}' & \mathbf{B} \end{vmatrix} = \begin{vmatrix} \mathbf{A} & \mathbf{0} \\ \mathbf{0}' & \mathbf{B} \end{vmatrix} \begin{vmatrix} \mathbf{I} & \mathbf{A}^{-1}\mathbf{C} \\ \mathbf{0}' & \mathbf{I} \end{vmatrix}$. But expanding the determinant $\begin{vmatrix} \mathbf{I} & \mathbf{A}^{-1}\mathbf{C} \\ \mathbf{0}' & \mathbf{I} \end{vmatrix}$ by the last row gives $\begin{vmatrix} \mathbf{I} & \mathbf{A}^{-1}\mathbf{C} \\ \mathbf{0}' & \mathbf{I} \end{vmatrix} = 1$. Now use the result in Part (a).)

4.8. Show that, if $\mathbf{A}$ is square,

$$|\mathbf{A}| = |\mathbf{A}_{22}||\mathbf{A}_{11} - \mathbf{A}_{12}\mathbf{A}_{22}^{-1}\mathbf{A}_{21}| \qquad \text{for } |\mathbf{A}_{22}| \neq 0$$

$$= |\mathbf{A}_{11}||\mathbf{A}_{22} - \mathbf{A}_{21}\mathbf{A}_{11}^{-1}\mathbf{A}_{12}| \qquad \text{for } |\mathbf{A}_{11}| \neq 0$$

(*Hint:* Partition $\mathbf{A}$ and verify that

$$\begin{bmatrix} \mathbf{I} & -\mathbf{A}_{12}\mathbf{A}_{22}^{-1} \\ \mathbf{0}' & \mathbf{I} \end{bmatrix} \begin{bmatrix} \mathbf{A}_{11} & \mathbf{A}_{12} \\ \mathbf{A}_{21} & \mathbf{A}_{22} \end{bmatrix} \begin{bmatrix} \mathbf{I} & \mathbf{0} \\ -\mathbf{A}_{22}^{-1}\mathbf{A}_{21} & \mathbf{I} \end{bmatrix} = \begin{bmatrix} \mathbf{A}_{11} - \mathbf{A}_{12}\mathbf{A}_{22}^{-1}\mathbf{A}_{21} & \mathbf{0} \\ \mathbf{0}' & \mathbf{A}_{22} \end{bmatrix}$$

Take determinants on both sides of this equality. Use Exercise 4.7 for the first and third determinants on the left and for the determinant on the right. The second equality for $|\mathbf{A}|$ follows by considering

$$\begin{bmatrix} \mathbf{I} & \mathbf{0} \\ -\mathbf{A}_{21}\mathbf{A}_{11}^{-1} & \mathbf{I} \end{bmatrix} \begin{bmatrix} \mathbf{A}_{11} & \mathbf{A}_{12} \\ \mathbf{A}_{21} & \mathbf{A}_{22} \end{bmatrix} \begin{bmatrix} \mathbf{I} & -\mathbf{A}_{11}^{-1}\mathbf{A}_{12} \\ \mathbf{0}' & \mathbf{I} \end{bmatrix} = \begin{bmatrix} \mathbf{A}_{11} & \mathbf{0} \\ \mathbf{0}' & \mathbf{A}_{22} - \mathbf{A}_{21}\mathbf{A}_{11}^{-1}\mathbf{A}_{12} \end{bmatrix} \quad)$$

4.9. Show that, for $\mathbf{A}$ symmetric,

$$\mathbf{A}^{-1} = \begin{bmatrix} \mathbf{I} & \mathbf{0} \\ -\mathbf{A}_{22}^{-1}\mathbf{A}_{21} & \mathbf{I} \end{bmatrix} \begin{bmatrix} (\mathbf{A}_{11} - \mathbf{A}_{12}\mathbf{A}_{22}^{-1}\mathbf{A}_{21})^{-1} & \mathbf{0} \\ \mathbf{0}' & \mathbf{A}_{22}^{-1} \end{bmatrix} \begin{bmatrix} \mathbf{I} & -\mathbf{A}_{12}\mathbf{A}_{22}^{-1} \\ \mathbf{0}' & \mathbf{I} \end{bmatrix}$$

Thus $(\mathbf{A}_{11} - \mathbf{A}_{12}\mathbf{A}_{22}^{-1}\mathbf{A}_{21})^{-1}$ is the upper left-hand block of $\mathbf{A}^{-1}$.

(*Hint:* Premultiply the expression in the hint to Exercise 4.8 by $\begin{bmatrix} \mathbf{I} & -\mathbf{A}_{12}\mathbf{A}_{22}^{-1} \\ \mathbf{0}' & \mathbf{I} \end{bmatrix}^{-1}$ and postmultiply by $\begin{bmatrix} \mathbf{I} & \mathbf{0} \\ -\mathbf{A}_{22}^{-1}\mathbf{A}_{21} & \mathbf{I} \end{bmatrix}^{-1}$. Take inverses of the resulting expression.)

4.10. Show the following if $|\boldsymbol{\Sigma}| \neq 0$.

(a) Check that $|\boldsymbol{\Sigma}| = |\boldsymbol{\Sigma}_{22}||\boldsymbol{\Sigma}_{11} - \boldsymbol{\Sigma}_{12}\boldsymbol{\Sigma}_{22}^{-1}\boldsymbol{\Sigma}_{21}|$. (Note $|\boldsymbol{\Sigma}|$ can be factored into the product of contributions from the marginal and conditional distributions.)

(b) Check that

$$\begin{aligned}(\mathbf{x} - \boldsymbol{\mu})'\boldsymbol{\Sigma}^{-1}(\mathbf{x} - \boldsymbol{\mu}) &= [\mathbf{x}_1 - \boldsymbol{\mu}_1 - \boldsymbol{\Sigma}_{12}\boldsymbol{\Sigma}_{22}^{-1}(\mathbf{x}_2 - \boldsymbol{\mu}_2)]' \\ &\quad \times (\boldsymbol{\Sigma}_{11} - \boldsymbol{\Sigma}_{12}\boldsymbol{\Sigma}_{22}^{-1}\boldsymbol{\Sigma}_{21})^{-1} \\ &\quad \times [\mathbf{x}_1 - \boldsymbol{\mu}_1 - \boldsymbol{\Sigma}_{12}\boldsymbol{\Sigma}_{22}^{-1}(\mathbf{x}_2 - \boldsymbol{\mu}_2)] \\ &\quad + (\mathbf{x}_2 - \boldsymbol{\mu}_2)'\boldsymbol{\Sigma}_{22}^{-1}(\mathbf{x}_2 - \boldsymbol{\mu}_2)\end{aligned}$$

(Thus the joint density exponent can be written as the sum of two terms corresponding to contributions from the conditional and marginal distributions.)

(c) Given the results in Parts a and b, identify the marginal distribution of $\mathbf{X}_2$ and the conditional distribution of $\mathbf{X}_1 | \mathbf{X}_2 = \mathbf{x}_2$.

(*Hint:*
(a) Apply Exercise 4.8.
(b) Note from Exercise 4.9 we can write $(\mathbf{x}-\boldsymbol{\mu})'\boldsymbol{\Sigma}^{-1}(\mathbf{x}-\boldsymbol{\mu})$ as

$$\begin{bmatrix} \mathbf{x}_1-\boldsymbol{\mu}_1 \\ \mathbf{x}_2-\boldsymbol{\mu}_2 \end{bmatrix}' \begin{bmatrix} \mathbf{I} & \mathbf{0} \\ -\boldsymbol{\Sigma}_{22}^{-1}\boldsymbol{\Sigma}_{21} & \mathbf{I} \end{bmatrix} \begin{bmatrix} (\boldsymbol{\Sigma}_{11}-\boldsymbol{\Sigma}_{12}\boldsymbol{\Sigma}_{22}^{-1}\boldsymbol{\Sigma}_{21})^{-1} & \mathbf{0} \\ \mathbf{0}' & \boldsymbol{\Sigma}_{22}^{-1} \end{bmatrix}$$

$$\times \begin{bmatrix} \mathbf{I} & -\boldsymbol{\Sigma}_{12}\boldsymbol{\Sigma}_{22}^{-1} \\ \mathbf{0}' & \mathbf{I} \end{bmatrix} \begin{bmatrix} \mathbf{x}_1-\boldsymbol{\mu}_1 \\ \mathbf{x}_2-\boldsymbol{\mu}_2 \end{bmatrix}$$

Grouping the product so that

$$\begin{bmatrix} \mathbf{I} & -\boldsymbol{\Sigma}_{12}\boldsymbol{\Sigma}_{22}^{-1} \\ \mathbf{0}' & \mathbf{I} \end{bmatrix} \begin{bmatrix} \mathbf{x}_1-\boldsymbol{\mu}_1 \\ \mathbf{x}_2-\boldsymbol{\mu}_2 \end{bmatrix} = \begin{bmatrix} \mathbf{x}_1-\boldsymbol{\mu}_1-\boldsymbol{\Sigma}_{12}\boldsymbol{\Sigma}_{22}^{-1}(\mathbf{x}_2-\boldsymbol{\mu}_2) \\ \mathbf{x}_2-\boldsymbol{\mu}_2 \end{bmatrix}$$

the result follows.)

4.11. If $\mathbf{X}$ is distributed as $N_p(\boldsymbol{\mu}, \boldsymbol{\Sigma})$ with $|\boldsymbol{\Sigma}| \neq 0$, show that the joint density can be written as the product of marginal densities for

$$\underset{(q\times 1)}{\mathbf{X}_1} \quad \text{and} \quad \underset{((p-q)\times 1)}{\mathbf{X}_2} \quad \text{if } \boldsymbol{\Sigma}_{12} = \underset{((p-q)\times q)}{\mathbf{0}}.$$

(*Hint:* Show by block multiplication that

$$\begin{bmatrix} \boldsymbol{\Sigma}_{11}^{-1} & \mathbf{0} \\ \mathbf{0}' & \boldsymbol{\Sigma}_{22}^{-1} \end{bmatrix} \quad \text{is the inverse of} \quad \boldsymbol{\Sigma} = \begin{bmatrix} \boldsymbol{\Sigma}_{11} & \mathbf{0} \\ \mathbf{0}' & \boldsymbol{\Sigma}_{22} \end{bmatrix}$$

Then write

$$(\mathbf{x}-\boldsymbol{\mu})'\boldsymbol{\Sigma}^{-1}(\mathbf{x}-\boldsymbol{\mu}) = [\,(\mathbf{x}_1-\boldsymbol{\mu}_1)',\ (\mathbf{x}_2-\boldsymbol{\mu}_2)'\,]$$

$$\times \begin{bmatrix} \boldsymbol{\Sigma}_{11}^{-1} & \mathbf{0} \\ \mathbf{0}' & \boldsymbol{\Sigma}_{22}^{-1} \end{bmatrix} \begin{bmatrix} \mathbf{x}_1-\boldsymbol{\mu}_1 \\ \mathbf{x}_2-\boldsymbol{\mu}_2 \end{bmatrix}$$

$$= (\mathbf{x}_1-\boldsymbol{\mu}_1)'\boldsymbol{\Sigma}_{11}^{-1}(\mathbf{x}_1-\boldsymbol{\mu}_1) + (\mathbf{x}_2-\boldsymbol{\mu}_2)'\boldsymbol{\Sigma}_{22}^{-1}(\mathbf{x}_2-\boldsymbol{\mu}_2)$$

Note that $|\boldsymbol{\Sigma}| = |\boldsymbol{\Sigma}_{11}||\boldsymbol{\Sigma}_{22}|$ from Exercise 4.7(a). Now factor the joint density.)

4.12. Show $\sum_{j=1}^{n} (\mathbf{x}_j - \bar{\mathbf{x}})(\bar{\mathbf{x}} - \boldsymbol{\mu})'$ and $\sum_{j=1}^{n} (\bar{\mathbf{x}} - \boldsymbol{\mu})(\mathbf{x}_j - \bar{\mathbf{x}})'$ are both $p \times p$ matrices of zeros. Here $\mathbf{x}_j' = [x_{1j}, x_{2j}, \ldots, x_{pj}]$, $j = 1, 2, \ldots, n$, and

$$\bar{\mathbf{x}} = \frac{1}{n}\sum_{j=1}^{n} \mathbf{x}_j$$

4.13. Find the maximum likelihood estimates of the 2×1 mean vector $\boldsymbol{\mu}$ and the 2×2 covariance matrix $\boldsymbol{\Sigma}$ based on the random sample

$$\mathbf{X} = \begin{bmatrix} 3 & 4 & 5 & 4 \\ 6 & 4 & 7 & 7 \end{bmatrix}$$

from a bivariate normal population.

4.14. Let $\mathbf{X}_1, \mathbf{X}_2, \ldots, \mathbf{X}_{20}$ be a random sample of size $n = 20$ from a $N_6(\boldsymbol{\mu}, \boldsymbol{\Sigma})$ population. Specify each of the following completely.

(a) The distribution of $(\mathbf{X}_1 - \boldsymbol{\mu})'\boldsymbol{\Sigma}^{-1}(\mathbf{X}_1 - \boldsymbol{\mu})$.

(b) The distributions of $\overline{\mathbf{X}}$ and $\sqrt{n}(\overline{\mathbf{X}} - \boldsymbol{\mu})$.

(c) The distribution of $(n - 1)\mathbf{S}$.

4.15. For the random variables $\mathbf{X}_1, \mathbf{X}_2, \ldots, \mathbf{X}_{20}$ in Exercise 4.14, specify the distribution of $\mathbf{B}(19\mathbf{S})\mathbf{B}'$ in each case.

(a) $\mathbf{B} = \begin{bmatrix} 1 & -\frac{1}{2} & -\frac{1}{2} & 0 & 0 & 0 \\ 0 & 0 & 0 & -\frac{1}{2} & -\frac{1}{2} & 1 \end{bmatrix}$

(b) $\mathbf{B} = \begin{bmatrix} 1 & 0 & 0 & 0 & 0 & 0 \\ 0 & 0 & 1 & 0 & 0 & 0 \end{bmatrix}$

4.16. Let $\mathbf{X}_1, \mathbf{X}_2, \ldots, \mathbf{X}_{75}$ be a random sample from a population distribution with mean $\boldsymbol{\mu}$ and covariance matrix $\boldsymbol{\Sigma}$. What is the approximate distribution of each of the following?

(a) $\overline{\mathbf{X}}$

(b) $n(\overline{\mathbf{X}} - \boldsymbol{\mu})'\mathbf{S}^{-1}(\overline{\mathbf{X}} - \boldsymbol{\mu})$

4.17. Consider the annual rates of return (including dividends) on the Dow Jones Industrial Average for the years 1963–1972. These data, multiplied by 100, are: 20.6, 18.7, 14.2, −15.7, 19.0, 7.7, −11.6, 8.8, 9.8, and 18.2. Use these 10 observations to answer the following.

(a) Construct a *Q-Q* plot. Do the data seem to be normally distributed? Explain.

(b) Carry out a test of normality based on the correlation coefficient r_Q. [See (4-31).] Let the significance level be $\alpha = .10$.

4.18. Exercise 1.2 gives the age x_1, measured in years, as well as the selling price x_2, measured in thousands of dollars, for $n = 10$ used cars. These data are reproduced below.

x_1	3	5	5	7	7	7	8	9	10	11
x_2	2.30	1.90	1.00	.70	.30	1.00	1.05	.45	.70	.30

(a) Use the results of Exercise 1.2 to calculate the squared generalized distances $(\mathbf{x}_j - \overline{\mathbf{x}})'\mathbf{S}^{-1}(\mathbf{x}_j - \overline{\mathbf{x}})$, $j = 1, 2, \ldots, 10$, where $\mathbf{x}_j' = [x_{1j}, x_{2j}]$.

(b) Using the distances in Part a, determine the proportion of the observations falling within the estimated 50% probability contour of a bivariate normal distribution.

(c) Order the distances in Part a and construct a chi-square plot.

(d) Given the results in Parts b and c, are these data approximately bivariate normal? Explain.

4.19. Consider the radiation data (door closed) in Example 4.9. Construct a *Q-Q* plot for the natural logarithms of these data. [Note the natural-logarithm transformation corresponds to the value $\lambda = 0$ in (4-34).] Do the natural logarithms appear to be normally distributed? Compare your results with Figure 4.8. Does the choice $\lambda = \frac{1}{4}$ or $\lambda = 0$ make much difference in this case?

The following exercises may require a computer.

4.20. Consider the air-pollution data given in Table 1.2. Construct a *Q-Q* plot for the solar radiation measurements and carry out a test for normality based on the correlation coefficient r_Q (see (4-31)). Let $\alpha = .05$ and use the entry corresponding to $n = 40$ in Table 4.2.

4.21. Given the air pollution data in Table 1.2, examine the pairs $x_5 = NO_2$ and $x_6 = O_3$ for bivariate normality.

(a) Calculate generalized distances $(\mathbf{x}_j - \bar{\mathbf{x}})'\mathbf{S}^{-1}(\mathbf{x}_j - \bar{\mathbf{x}})$, $j = 1, 2, \ldots, 42$, where $\mathbf{x}'_j = [x_{5j}, x_{6j}]$.

(b) Determine the proportion of observations $\mathbf{x}'_j = [x_{5j}, x_{6j}]$, $j = 1, 2, \ldots, 42$, falling within the approximate 50% probability contour of a bivariate normal distribution.

(c) Construct a chi-square plot of the ordered distances in Part a.

4.22. Consider the used-car data in Exercise 4.18.

(a) Determine the power transformation $\hat{\lambda}_1$ that makes the x_1 values approximately normal. Construct a *Q-Q* plot for the transformed data.

(b) Determine the power transformation $\hat{\lambda}_2$ that makes the x_2 values approximately normal. Construct a *Q-Q* plot for the transformed data.

(c) Determine the power transformations $\hat{\boldsymbol{\lambda}}' = [\hat{\lambda}_1, \hat{\lambda}_2]$ that make the $[x_1, x_2]$ values jointly normal using (4-39). Compare the results with those obtained in Parts a and b.

4.23. Examine the marginal normality of the observations on variables $X_1, X_2, \ldots, X_5$ for the multiple-sclerosis data in Table 1.3. Treat the non–multiple-sclerosis and multiple-sclerosis groups separately. Use whatever methodology, including transformations, you feel is appropriate.

4.24. Examine the marginal normality of the observations on variables $X_1, X_2, \ldots, X_6$ for the radiotherapy data in Table 1.4. Use whatever methodology, including transformations, you feel is appropriate.

4.25. Examine the marginal and bivariate normality of the observations on variables X_1, X_2, X_3, and X_4 for the data in Table 4.3.

REFERENCES

[1] Anderson, T. W., *An Introduction to Multivariate Statistical Analysis*, New York: John Wiley, 1958.

[2] Andrews, D. F., R. Gnanadesikan, and J. L. Warner, "Transformations of Multivariate Data," *Biometrics*, **27**, no. 4 (1971), 825–840.

[3] Box, G. E. P., and D. R. Cox, "An Analysis of Transformations," (with discussion), *Journal of the Royal Statistical Society* (*B*), **26**, no. 2 (1964), 211–252.

[4] Daniel, C., and F. S. Wood, *Fitting Equations to Data* (2nd ed.), New York: John Wiley, 1980.

[5] Filliben, J. J., "The Probability Plot Correlation Coefficient Test for Normality," *Technometrics*, **17**, no. 1 (1975), 111–117.

[6] Gnanadesikan, R., *Methods for Statistical Data Analysis of Multivariate Observations*, New York: John Wiley, 1977.

[7] Hernandez, F., and R. A. Johnson, "The Large-Sample Behavior of Transformations to Normality," *Journal of the American Statistical Association*, **75**, no. 372 (1980), 855–861.

[8] Hogg, R. V., and A. T. Craig, *Introduction to Mathematical Statistics* (3rd ed.), New York: Macmillan, 1970.

[9] Shapiro, S. S., and M. B. Wilk, "An Analysis of Variance Test for Normality (Complete Samples)," *Biometrika*, **52**, no. 4 (1965), 591–611.

[10] Zehna, P. "Invariance of Maximum Likelihood Estimators," *Annals of Mathematical Statistics*, **37**, no. 3 (1966), 744.

Part II

Inferences about Multivariate Means and Linear Models

5

Inferences about a Mean Vector

5.1 INTRODUCTION

This chapter is the first of the methodological sections of the book. We shall now use the concepts and results in Chapters 1 through 4 to develop techniques for analyzing data. A large part of any analysis is concerned with *inference*; that is, reaching valid conclusions on the basis of sample information.

At this point we shall concentrate on inferences about a population mean vector and its component parts. Although we introduce statistical inference through initial discussions of tests of hypotheses, our ultimate aim is to present a full statistical analysis of the component means based on simultaneous confidence statements.

One of the central messages of multivariate analysis is that p correlated variables must be analyzed jointly. This principle is exemplified by the methods presented in this chapter.

5.2 THE PLAUSIBILITY OF μ_0 AS A VALUE FOR A NORMAL POPULATION MEAN

Let us start by recalling the univariate theory for determining if a specific value, μ_0, is a plausible value for the population mean μ. From the point of view of hypothesis testing, this problem can be formulated as a *test* of the competing *hypotheses*

$$H_0: \mu = \mu_0 \quad \text{against} \quad H_1: \mu \neq \mu_0$$

Here H_0 is the null hypothesis and H_1 is the (two-sided) alternative hypothesis. If $X_1, X_2, \ldots, X_n$ denote a random sample from a normal population, the appropriate test statistic is

$$t = \frac{(\bar{X} - \mu_0)}{s/\sqrt{n}}, \qquad \text{where } \bar{X} = \frac{1}{n}\sum_{j=1}^{n} X_j \quad \text{and} \quad s^2 = \frac{1}{n-1}\sum_{j=1}^{n}(X_j - \bar{X})^2$$

This test statistic has a student's t-distribution with $n - 1$ d.f. We reject H_0, that μ_0 is a plausible value of μ, if the observed $|t|$ exceeds a specified percentage point of a t-distribution with $n - 1$ d.f.

Rejecting H_0 when $|t|$ is large is equivalent to rejecting H_0 if its square,

$$t^2 = \frac{(\bar{X} - \mu_0)^2}{s^2/n} = n(\bar{X} - \mu_0)(s^2)^{-1}(\bar{X} - \mu_0) \tag{5-1}$$

is large. The variable t^2 in (5-1) is the squared distance from the sample mean $\bar{X}$ to the test value μ_0. The units of distance are expressed in terms of $s/\sqrt{n}$ or estimated standard deviations of $\bar{X}$. Once $\bar{X}$ and s^2 are observed, the test becomes: Reject H_0 in favor of H_1, at significance level α, if

$$n(\bar{x} - \mu_0)(s^2)^{-1}(\bar{x} - \mu_0) > t_{n-1}^2(\alpha/2) \tag{5-2}$$

where $t_{n-1}(\alpha/2)$ denotes the upper (100α)th percentile of the t-distribution with $n - 1$ d.f.

If H_0 is not rejected, we conclude μ_0 is a plausible value for the normal population mean. Are there other values of μ which are also consistent with the data? The answer is Yes! In fact there is always a *set* of plausible values for a normal population mean. From the well-known correspondence between acceptance regions for tests of H_0: $\mu = \mu_0$ versus H_1: $\mu \neq \mu_0$ and confidence intervals for μ we have

$$\{\text{Do not reject } H_0: \mu = \mu_0 \text{ at level } \alpha\} \quad \text{or} \quad \left|\frac{\bar{x} - \mu_0}{s/\sqrt{n}}\right| \leq t_{n-1}(\alpha/2)$$

is equivalent to

$$\left\{\mu_0 \text{ lies in the } 100(1-\alpha)\% \text{ confidence interval } \bar{x} \pm t_{n-1}(\alpha/2)\frac{s}{\sqrt{n}}\right\}$$

or

$$\bar{x} - t_{n-1}(\alpha/2)\frac{s}{\sqrt{n}} \leq \mu_0 \leq \bar{x} + t_{n-1}(\alpha/2)\frac{s}{\sqrt{n}} \tag{5-3}$$

The confidence interval consists of all those values μ_0 that would not be rejected by the test of H_0: $\mu = \mu_0$.

Before the sample is selected, the $100(1 - \alpha)\%$ confidence interval in (5-3) is a *random interval* because the endpoints depend upon the random variables, $\bar{X}$ and s.

The probability that the interval contains μ is $1 - \alpha$; among large numbers of such independent intervals, $100\,(1 - \alpha)\%$ of them will contain μ.

Consider now the problem of determining if a given $p \times 1$ vector $\boldsymbol{\mu}_0$ is a plausible value for the mean of a multivariate normal distribution. We shall proceed by analogy to the univariate development just presented.

A natural generalization of the squared distance in (5-1) is its multivariate analog

$$T^2 = (\bar{\mathbf{X}} - \boldsymbol{\mu}_0)'\left(\frac{\mathbf{S}}{n}\right)^{-1}(\bar{\mathbf{X}} - \boldsymbol{\mu}_0) = n(\bar{\mathbf{X}} - \boldsymbol{\mu}_0)'\mathbf{S}^{-1}(\bar{\mathbf{X}} - \boldsymbol{\mu}_0) \tag{5-4}$$

where

$$\underset{(p\times 1)}{\bar{\mathbf{X}}} = \frac{1}{n}\sum_{j=1}^{n}\mathbf{X}_j, \qquad \underset{(p\times p)}{\mathbf{S}} = \frac{1}{n-1}\sum_{j=1}^{n}(\mathbf{X}_j - \bar{\mathbf{X}})(\mathbf{X}_j - \bar{\mathbf{X}})', \quad \text{and} \quad \underset{(p\times 1)}{\boldsymbol{\mu}_0} = \begin{bmatrix} \mu_{10} \\ \mu_{20} \\ \vdots \\ \mu_{p0} \end{bmatrix}$$

The statistic T^2 is called *Hotelling's* T^2 in honor of Harold Hotelling, a pioneer in multivariate analysis who first obtained its sampling distribution. Here $(1/n)\mathbf{S}$ is the estimated covariance matrix of $\bar{\mathbf{X}}$ (see Result 3.1).

If the observed generalized distance T^2 is too large—that is, $\bar{\mathbf{x}}$ is "too far" from $\boldsymbol{\mu}_0$—the hypothesis $H_0: \boldsymbol{\mu} = \boldsymbol{\mu}_0$ is rejected. It turns out that special tables of T^2 percentage points are not required for formal tests of hypotheses. This is true because

$$T^2 \quad \text{is distributed as} \quad \frac{(n-1)p}{(n-p)}F_{p,\,n-p} \tag{5-5}$$

where $F_{p,\,n-p}$ denotes a random variable with an F-distribution with p and $n - p$ d.f.

To summarize, we have the following.

Let $\mathbf{X}_1, \mathbf{X}_2, \ldots, \mathbf{X}_n$ be a random sample from a $N_p(\boldsymbol{\mu}, \boldsymbol{\Sigma})$ population. Then with $\bar{\mathbf{X}} = \frac{1}{n}\sum_{j=1}^{n}\mathbf{X}_j$ and $\mathbf{S} = \frac{1}{(n-1)}\sum_{j=1}^{n}(\mathbf{X}_j - \bar{\mathbf{X}})(\mathbf{X}_j - \bar{\mathbf{X}})'$,

$$\begin{aligned}\alpha &= P\left[T^2 > \frac{(n-1)p}{(n-p)}F_{p,\,n-p}(\alpha)\right] \\ &= P\left[n(\bar{\mathbf{X}} - \boldsymbol{\mu})'\mathbf{S}^{-1}(\bar{\mathbf{X}} - \boldsymbol{\mu}) > \frac{(n-1)p}{(n-p)}F_{p,\,n-p}(\alpha)\right]\end{aligned} \tag{5-6}$$

whatever the true $\boldsymbol{\mu}$ and $\boldsymbol{\Sigma}$. Here $F_{p,\,n-p}(\alpha)$ is the upper (100α)th percentile of the $F_{p,\,n-p}$ distribution.

Statement (5-6) leads immediately to a test of the hypothesis $H_0: \boldsymbol{\mu} = \boldsymbol{\mu}_0$ versus $H_1: \boldsymbol{\mu} \neq \boldsymbol{\mu}_0$. At the α level of significance, reject H_0 in favor of H_1 if

$$T^2 = n(\bar{\mathbf{x}} - \boldsymbol{\mu}_0)'\mathbf{S}^{-1}(\bar{\mathbf{x}} - \boldsymbol{\mu}_0) > \frac{(n-1)p}{(n-p)} F_{p,n-p}(\alpha) \tag{5-7}$$

It is informative to discuss the nature of the T^2-distribution briefly and its correspondence with the univariate test statistic. In Section 4.4, we described the manner in which the Wishart distribution generalizes the chi-square distribution. We can write

$$T^2 = \sqrt{n}\,(\bar{\mathbf{X}} - \boldsymbol{\mu}_0)' \left(\frac{\sum_{j=1}^{n} (\mathbf{X}_j - \bar{\mathbf{X}})(\mathbf{X}_j - \bar{\mathbf{X}})'}{n-1} \right)^{-1} \sqrt{n}\,(\bar{\mathbf{X}} - \boldsymbol{\mu}_0)$$

which is of the form

$$\begin{pmatrix} \text{multivariate normal} \\ \text{random vector} \end{pmatrix}' \left(\frac{\text{Wishart random matrix}}{\text{d.f.}} \right)^{-1} \begin{pmatrix} \text{multivariate normal} \\ \text{random vector} \end{pmatrix}$$

This is analogous to

$$t^2 = \sqrt{n}\,(\bar{X} - \mu_0)(s^2)^{-1}\sqrt{n}\,(\bar{X} - \mu_0)$$

or

$$\begin{pmatrix} \text{normal} \\ \text{random variable} \end{pmatrix} \left(\frac{\text{(scaled) chi-square random variable}}{\text{d.f.}} \right)^{-1} \begin{pmatrix} \text{normal} \\ \text{random variable} \end{pmatrix}$$

for the univariate case. Since the multivariate normal and Wishart random variables are independently distributed [see (4-23)], their joint distribution is the product of the marginal normal and Wishart distributions. Using calculus, the distribution of T^2 as given above can be derived from this joint distribution.

It is rare, in multivariate situations, to be content with a test of $H_0: \boldsymbol{\mu} = \boldsymbol{\mu}_0$, where all of the mean vector components are specified under the null hypothesis. Ordinarily it is preferable to find regions of $\boldsymbol{\mu}$ values that are plausible in light of the observed data. We shall return to this issue in Section 5.4.

Example 5.1

Let the data matrix for a random sample of size $n = 3$ from a bivariate normal population be

$$\mathbf{X} = \begin{bmatrix} 6 & 10 & 8 \\ 9 & 6 & 3 \end{bmatrix}$$

Evaluate the observed T^2 for $\boldsymbol{\mu}_0' = [9, 5]$. What is the sampling distribution of

T^2 in this case? We find

$$\bar{\mathbf{x}} = \begin{bmatrix} \bar{x}_1 \\ \bar{x}_2 \end{bmatrix} = \begin{bmatrix} \dfrac{6+10+8}{3} \\ \dfrac{9+6+3}{3} \end{bmatrix} = \begin{bmatrix} 8 \\ 6 \end{bmatrix}$$

and

$$s_{11} = \frac{(6-8)^2 + (10-8)^2 + (8-8)^2}{2} = 4$$

$$s_{12} = \frac{(6-8)(9-6) + (10-8)(6-6) + (8-8)(3-6)}{2} = -3$$

$$s_{22} = \frac{(9-6)^2 + (6-6)^2 + (3-6)^2}{2} = 9$$

so

$$\mathbf{S} = \begin{bmatrix} 4 & -3 \\ -3 & 9 \end{bmatrix}$$

Thus

$$\mathbf{S}^{-1} = \frac{1}{(4)(9) - (-3)(-3)} \begin{bmatrix} 9 & 3 \\ 3 & 4 \end{bmatrix} = \begin{bmatrix} 1/3 & 1/9 \\ 1/9 & 4/27 \end{bmatrix}$$

and, from (5-4),

$$T^2 = 3[8-9, 6-5] \begin{bmatrix} 1/3 & 1/9 \\ 1/9 & 4/27 \end{bmatrix} \begin{bmatrix} 8-9 \\ 6-5 \end{bmatrix} = 3[-1, 1] \begin{bmatrix} -2/9 \\ 1/27 \end{bmatrix} = 7/9$$

Before the sample is selected, T^2 has the distribution of a

$$\frac{(3-1)2}{(3-2)} F_{2,3-2} = 4F_{2,1}$$

random variable. ■

The next example illustrates a test of the hypothesis H_0: $\boldsymbol{\mu} = \boldsymbol{\mu}_0$ using data collected as part of a search for new diagnostic techniques at the University of Wisconsin Medical School.

Example 5.2

Perspiration from 20 healthy females was analyzed. Three components, X_1 = sweat rate, X_2 = sodium content, and X_3 = potassium content, were measured and the results, which we call the *sweat data*, are presented in Table 5.1.

Test the hypothesis H_0: $\boldsymbol{\mu}' = [4, 50, 10]$ against H_1: $\boldsymbol{\mu}' \neq [4, 50, 10]$ at level of significance $\alpha = .10$.

Computer calculations provide

$$\bar{\mathbf{x}} = \begin{bmatrix} 4.640 \\ 45.400 \\ 9.965 \end{bmatrix}, \qquad \mathbf{S} = \begin{bmatrix} 2.879 & 10.002 & -1.810 \\ 10.002 & 199.798 & -5.627 \\ -1.810 & -5.627 & 3.628 \end{bmatrix}$$

TABLE 5.1 SWEAT DATA

Individual	X_1 (Sweat rate)	X_2 (Sodium)	X_3 (Potassium)
1	3.7	48.5	9.3
2	5.7	65.1	8.0
3	3.8	47.2	10.9
4	3.2	53.2	12.0
5	3.1	55.5	9.7
6	4.6	36.1	7.9
7	2.4	24.8	14.0
8	7.2	33.1	7.6
9	6.7	47.4	8.5
10	5.4	54.1	11.3
11	3.9	36.9	12.7
12	4.5	58.8	12.3
13	3.5	27.8	9.8
14	4.5	40.2	8.4
15	1.5	13.5	10.1
16	8.5	56.4	7.1
17	4.5	71.6	8.2
18	6.5	52.8	10.9
19	4.1	44.1	11.2
20	5.5	40.9	9.4

SOURCE: Courtesy of Dr. Gerald Bargman.

and

$$\mathbf{S}^{-1} = \begin{bmatrix} .586 & -.022 & .258 \\ -.022 & .006 & -.002 \\ .258 & -.002 & .402 \end{bmatrix}$$

We evaluate

$$T^2 = 20[4.640 - 4, 45.400 - 50, 9.965 - 10]$$

$$\times \begin{bmatrix} .586 & -.022 & .258 \\ -.022 & .006 & -.002 \\ .258 & -.002 & .402 \end{bmatrix} \begin{bmatrix} 4.640 - 4 \\ 45.400 - 50 \\ 9.965 - 10 \end{bmatrix}$$

$$= 20[.640, -4.600, -.035] \begin{bmatrix} .467 \\ -.042 \\ .160 \end{bmatrix} = 9.74$$

Comparing the observed $T^2 = 9.74$ with the critical value

$$\frac{(n-1)p}{(n-p)} F_{p,n-p}(.10) = \frac{19(3)}{17} F_{3,17}(.10) = 3.353(2.44) = 8.18$$

we see that $T^2 = 9.74 > 8.18$, and consequently we reject H_0 at the 10% level of significance.

We note that H_0 will be rejected if one or more of the component means, or some combination of means, differs too much from the hypothesized values

[4, 50, 10]. At this point, we have no idea which of these hypothesized values may not be supported by the data.

We have assumed the sweat data are multivariate normal. The *Q-Q* plots constructed from the marginal distributions of X_1, X_2, and X_3 all approximate straight lines. Moreover, scatter plots for pairs of observations have elliptical shapes and we concluded that the normality assumption was realistic in this case (see Exercise 5.4). ■

One feature of the T^2-statistic is that it is invariant (unchanged) under changes in the units of measurements for **X** of the form

$$\underset{(p\times 1)}{\mathbf{Y}} = \underset{(p\times p)}{\mathbf{C}} \underset{(p\times 1)}{\mathbf{X}} + \underset{(p\times 1)}{\mathbf{d}}, \qquad \mathbf{C} \text{ nonsingular} \tag{5-8}$$

A transformation of the observations of this kind arises when a constant b_i is subtracted from the ith variable to form $X_i - b_i$ and the result is multiplied by a constant $a_i > 0$ to get $a_i(X_i - b_i)$. Premultiplication of the *centered* and *scaled* quantities $a_i(X_i - b_i)$ by any nonsingular matrix will yield Equation (5-8). As an example, the operations involved in changing X_i to $a_i(X_i - b_i)$ correspond exactly to the process of converting temperature from a Fahrenheit to a Celsius reading.

Given observations $\mathbf{x}_1, \mathbf{x}_2, \ldots, \mathbf{x}_n$ and the transformation in (5-8), it immediately follows from Result 3.6 that

$$\bar{\mathbf{y}} = \mathbf{C}\bar{\mathbf{x}} + \mathbf{d} \quad \text{and} \quad \mathbf{S}_{\mathbf{y}} = \frac{1}{n-1}\sum_{j=1}^{n}(\mathbf{y}_j - \bar{\mathbf{y}})(\mathbf{y}_j - \bar{\mathbf{y}})' = \mathbf{CSC}'$$

Moreover, by (2-24) and (2-45),

$$\boldsymbol{\mu}_{\mathbf{Y}} = E(\mathbf{Y}) = E(\mathbf{CX} + \mathbf{d}) = E(\mathbf{CX}) + E(\mathbf{d}) = \mathbf{C}\boldsymbol{\mu} + \mathbf{d}$$

Therefore T^2 computed with the **y**'s and a hypothesized value $\boldsymbol{\mu}_{\mathbf{Y},0} = \mathbf{C}\boldsymbol{\mu}_0 + \mathbf{d}$ is

$$\begin{aligned}
T^2 &= n(\bar{\mathbf{y}} - \boldsymbol{\mu}_{\mathbf{Y},0})'\mathbf{S}_{\mathbf{y}}^{-1}(\bar{\mathbf{y}} - \boldsymbol{\mu}_{\mathbf{Y},0}) \\
&= n(\mathbf{C}(\bar{\mathbf{x}} - \boldsymbol{\mu}_0))'(\mathbf{CSC}')^{-1}(\mathbf{C}(\bar{\mathbf{x}} - \boldsymbol{\mu}_0)) \\
&= n(\bar{\mathbf{x}} - \boldsymbol{\mu}_0)'\mathbf{C}'(\mathbf{CSC}')^{-1}\mathbf{C}(\bar{\mathbf{x}} - \boldsymbol{\mu}_0) \\
&= n(\bar{\mathbf{x}} - \boldsymbol{\mu}_0)'\mathbf{C}'(\mathbf{C}')^{-1}\mathbf{S}^{-1}\mathbf{C}^{-1}\mathbf{C}(\bar{\mathbf{x}} - \boldsymbol{\mu}_0) = n(\bar{\mathbf{x}} - \boldsymbol{\mu}_0)'\mathbf{S}^{-1}(\bar{\mathbf{x}} - \boldsymbol{\mu}_0)
\end{aligned}$$

The last expression is recognized as the value of T^2 computed with the **x**'s.

5.3 HOTELLING'S T^2 AND LIKELIHOOD RATIO TESTS

We introduced the T^2-statistic by analogy with the univariate squared distance, t^2. There is a general principle for constructing test procedures called the *likelihood ratio method*, and the T^2-statistic can be derived as the likelihood ratio test of $H_0: \boldsymbol{\mu} = \boldsymbol{\mu}_0$. The general theory of likelihood ratio tests is beyond the scope of this book. (See [1] for a treatment of this topic.) Likelihood ratio tests have several optimal properties for reasonably large samples, and they are particularly convenient for hypotheses formulated in terms of multivariate normal parameters.

We know from (4-18) that the maximum of the multivariate normal likelihood as $\boldsymbol{\mu}$ and $\boldsymbol{\Sigma}$ are varied over their possible values is given by

$$\max_{\boldsymbol{\mu}, \boldsymbol{\Sigma}} L(\boldsymbol{\mu}, \boldsymbol{\Sigma}) = \frac{1}{(2\pi)^{np/2} |\hat{\boldsymbol{\Sigma}}|^{n/2}} e^{-np/2} \tag{5-9}$$

where

$$\hat{\boldsymbol{\Sigma}} = \frac{1}{n} \sum_{j=1}^{n} (\mathbf{x}_j - \bar{\mathbf{x}})(\mathbf{x}_j - \bar{\mathbf{x}})' \quad \text{and} \quad \hat{\boldsymbol{\mu}} = \bar{\mathbf{x}} = \frac{1}{n} \sum_{j=1}^{n} \mathbf{x}_j$$

are the maximum likelihood estimates. Recall that the maximum likelihood estimates $\hat{\boldsymbol{\mu}}$ and $\hat{\boldsymbol{\Sigma}}$ are those choices for $\boldsymbol{\mu}$ and $\boldsymbol{\Sigma}$ that best explain the observed values of the random sample.

Under the hypothesis $H_0: \boldsymbol{\mu} = \boldsymbol{\mu}_0$, the normal likelihood specializes to

$$L(\boldsymbol{\mu}_0, \boldsymbol{\Sigma}) = \frac{1}{(2\pi)^{np/2} |\boldsymbol{\Sigma}|^{n/2}} e^{-\frac{1}{2} \sum_{j=1}^{n} (\mathbf{x}_j - \boldsymbol{\mu}_0)' \boldsymbol{\Sigma}^{-1} (\mathbf{x}_j - \boldsymbol{\mu}_0)} \tag{5-10}$$

The mean $\boldsymbol{\mu}_0$ is now fixed, but $\boldsymbol{\Sigma}$ can be varied to find the value that is "most likely" to have led, with $\boldsymbol{\mu}_0$ fixed, to the observed sample. This value is obtained by maximizing $L(\boldsymbol{\mu}_0, \boldsymbol{\Sigma})$ with respect to $\boldsymbol{\Sigma}$.

Following the steps in (4-13), the exponent in $L(\boldsymbol{\mu}_0, \boldsymbol{\Sigma})$ may be written as

$$\begin{aligned} -\frac{1}{2} \sum_{j=1}^{n} (\mathbf{x}_j - \boldsymbol{\mu}_0)' \boldsymbol{\Sigma}^{-1} (\mathbf{x}_j - \boldsymbol{\mu}_0) &= -\frac{1}{2} \sum_{j=1}^{n} \operatorname{tr}\left[\boldsymbol{\Sigma}^{-1} (\mathbf{x}_j - \boldsymbol{\mu}_0)(\mathbf{x}_j - \boldsymbol{\mu}_0)'\right] \\ &= -\frac{1}{2} \operatorname{tr}\left[\boldsymbol{\Sigma}^{-1} \left(\sum_{j=1}^{n} (\mathbf{x}_j - \boldsymbol{\mu}_0)(\mathbf{x}_j - \boldsymbol{\mu}_0)' \right)\right] \end{aligned}$$

Applying Result 4.10 with $\mathbf{B} = \sum_{j=1}^{n} (\mathbf{x}_j - \boldsymbol{\mu}_0)(\mathbf{x}_j - \boldsymbol{\mu}_0)'$ and $b = n/2$, we have

$$\max_{\boldsymbol{\Sigma}} L(\boldsymbol{\mu}_0, \boldsymbol{\Sigma}) = \frac{1}{(2\pi)^{np/2} |\hat{\boldsymbol{\Sigma}}_0|^{n/2}} e^{-np/2} \tag{5-11}$$

$$\text{with} \quad \hat{\boldsymbol{\Sigma}}_0 = \frac{1}{n} \sum_{j=1}^{n} (\mathbf{x}_j - \boldsymbol{\mu}_0)(\mathbf{x}_j - \boldsymbol{\mu}_0)'$$

To determine whether $\boldsymbol{\mu}_0$ is plausible value for $\boldsymbol{\mu}$, the maximum of $L(\boldsymbol{\mu}_0, \boldsymbol{\Sigma})$ is compared with the unrestricted maximum of $L(\boldsymbol{\mu}, \boldsymbol{\Sigma})$. The resulting ratio is called the *likelihood ratio statistic*.

Using Equations (5-9) and (5-11),

$$\text{Likelihood ratio} = \Lambda = \frac{\max_{\boldsymbol{\Sigma}} L(\boldsymbol{\mu}_0, \boldsymbol{\Sigma})}{\max_{\boldsymbol{\mu}, \boldsymbol{\Sigma}} L(\boldsymbol{\mu}, \boldsymbol{\Sigma})} = \left(\frac{|\hat{\boldsymbol{\Sigma}}|}{|\hat{\boldsymbol{\Sigma}}_0|} \right)^{n/2} \tag{5-12}$$

The equivalent statistic $\Lambda^{2/n} = |\hat{\boldsymbol{\Sigma}}|/|\hat{\boldsymbol{\Sigma}}_0|$ is called *Wilks' lambda*. If the observed value of this likelihood ratio is too small, the hypothesis $H_0: \boldsymbol{\mu} = \boldsymbol{\mu}_0$ is unlikely to be true and is, therefore, rejected. Specifically, the likelihood ratio test of $H_0: \boldsymbol{\mu} = \boldsymbol{\mu}_0$ against $H_1: \boldsymbol{\mu} \neq \boldsymbol{\mu}_0$ rejects H_0 if

$$\Lambda = \left(\frac{|\hat{\boldsymbol{\Sigma}}|}{|\hat{\boldsymbol{\Sigma}}_0|} \right)^{n/2} = \left(\frac{\left| \sum_{j=1}^{n} (\mathbf{x}_j - \bar{\mathbf{x}})(\mathbf{x}_j - \bar{\mathbf{x}})' \right|}{\left| \sum_{j=1}^{n} (\mathbf{x}_j - \boldsymbol{\mu}_0)(\mathbf{x}_j - \boldsymbol{\mu}_0)' \right|} \right)^{n/2} < c_\alpha \qquad (5\text{-}13)$$

where c_α is the lower (100α)th percentile of the distribution of Λ. (Note that the likelihood-ratio-test statistic is a power of the ratio of generalized variances.) Fortunately, because of the following relation between T^2 and Λ, we do not need the distribution of the latter to carry out the test.

Result 5.1. Let $\mathbf{X}_1, \mathbf{X}_2, \ldots, \mathbf{X}_n$ be a random sample from a $N_p(\boldsymbol{\mu}, \boldsymbol{\Sigma})$ population. Then the test in (5-7) based on T^2 is equivalent to the likelihood ratio test of $H_0: \boldsymbol{\mu} = \boldsymbol{\mu}_0$ versus $H_1: \boldsymbol{\mu} \neq \boldsymbol{\mu}_0$ because $\Lambda^{2/n} = \left(1 + \dfrac{T^2}{(n-1)}\right)^{-1}$.

Proof. Let the $(p+1) \times (p+1)$ matrix $\mathbf{A}$ be defined by

$$\mathbf{A} = \left[\begin{array}{c|c} \sum_{j=1}^{n} (\mathbf{x}_j - \bar{\mathbf{x}})(\mathbf{x}_j - \bar{\mathbf{x}})' & \sqrt{n}\,(\bar{\mathbf{x}} - \boldsymbol{\mu}_0) \\ \hline \sqrt{n}\,(\bar{\mathbf{x}} - \boldsymbol{\mu}_0)' & -1 \end{array} \right] = \left[\begin{array}{c|c} \mathbf{A}_{11} & \mathbf{A}_{12} \\ \hline \mathbf{A}_{21} & \mathbf{A}_{22} \end{array} \right]$$

By Exercise 4.8, $|\mathbf{A}| = |\mathbf{A}_{22}||\mathbf{A}_{11} - \mathbf{A}_{12}\mathbf{A}_{22}^{-1}\mathbf{A}_{21}| = |\mathbf{A}_{11}||\mathbf{A}_{22} - \mathbf{A}_{21}\mathbf{A}_{11}^{-1}\mathbf{A}_{12}|$, from which we obtain

$$(-1)\left| \sum_{j=1}^{n} (\mathbf{x}_j - \bar{\mathbf{x}})(\mathbf{x}_j - \bar{\mathbf{x}})' + n(\bar{\mathbf{x}} - \boldsymbol{\mu}_0)(\bar{\mathbf{x}} - \boldsymbol{\mu}_0)' \right|$$

$$= \left| \sum_{j=1}^{n} (\mathbf{x}_j - \bar{\mathbf{x}})(\mathbf{x}_j - \bar{\mathbf{x}})' \right| \left| -1 - n(\bar{\mathbf{x}} - \boldsymbol{\mu}_0)' \left(\sum_{j=1}^{n} (\mathbf{x}_j - \bar{\mathbf{x}})(\mathbf{x}_j - \bar{\mathbf{x}})' \right)^{-1} (\bar{\mathbf{x}} - \boldsymbol{\mu}_0) \right|$$

Since, by (4-14) page 143,

$$\sum_{j=1}^{n} (\mathbf{x}_j - \boldsymbol{\mu}_0)(\mathbf{x}_j - \boldsymbol{\mu}_0)' = \sum_{j=1}^{n} (\mathbf{x}_j - \bar{\mathbf{x}} + \bar{\mathbf{x}} - \boldsymbol{\mu}_0)(\mathbf{x}_j - \bar{\mathbf{x}} + \bar{\mathbf{x}} - \boldsymbol{\mu}_0)'$$

$$= \sum_{j=1}^{n} (\mathbf{x}_j - \bar{\mathbf{x}})(\mathbf{x}_j - \bar{\mathbf{x}})' + n(\bar{\mathbf{x}} - \boldsymbol{\mu}_0)(\bar{\mathbf{x}} - \boldsymbol{\mu}_0)',$$

the equality above involving determinants can be written

$$(-1)\left| \sum_{j=1}^{n} (\mathbf{x}_j - \boldsymbol{\mu}_0)(\mathbf{x}_j - \boldsymbol{\mu}_0)' \right| = \left| \sum_{j=1}^{n} (\mathbf{x}_j - \bar{\mathbf{x}})(\mathbf{x}_j - \bar{\mathbf{x}})' \right| (-1)\left(1 + \frac{T^2}{(n-1)}\right)$$

or

$$|n\hat{\boldsymbol{\Sigma}}_0| = |n\hat{\boldsymbol{\Sigma}}|\left(1 + \frac{T^2}{(n-1)}\right)$$

Thus

$$\Lambda^{2/n} = \frac{|\hat{\boldsymbol{\Sigma}}|}{|\hat{\boldsymbol{\Sigma}}_0|} = \left(1 + \frac{T^2}{(n-1)}\right)^{-1} \tag{5-14}$$

Here H_0 is rejected for small values of $\Lambda^{2/n}$ or, equivalently, large values of T^2. The critical values of T^2 are determined by (5-6). ■

Incidentally, Relation (5-14) shows that T^2 may be calculated from two determinants, thus avoiding the computation of $\mathbf{S}^{-1}$. Solving (5-14) for T^2, we have

$$T^2 = \frac{(n-1)|\hat{\boldsymbol{\Sigma}}_0|}{|\hat{\boldsymbol{\Sigma}}|} - (n-1) = \frac{(n-1)\left|\sum_{j=1}^{n}(\mathbf{x}_j - \boldsymbol{\mu}_0)(\mathbf{x}_j - \boldsymbol{\mu}_0)'\right|}{\left|\sum_{j=1}^{n}(\mathbf{x}_j - \bar{\mathbf{x}})(\mathbf{x}_j - \bar{\mathbf{x}})'\right|} - (n-1) \tag{5-15}$$

Likelihood ratio tests are common in multivariate analysis. Their optimal large sample properties hold in very general contexts, as we shall indicate shortly. They are well suited for the testing situations considered in this book. Likelihood ratio methods yield test statistics that reduce to the familiar *F*- and *t*-statistics in univariate situations.

Generalized Likelihood Ratio Method

We shall now consider the general likelihood ratio method. Let $\boldsymbol{\theta}$ be a vector consisting of all the *unknown* population parameters, and let $L(\boldsymbol{\theta})$ be the likelihood function obtained by evaluating the joint density of $\mathbf{X}_1, \mathbf{X}_2, \ldots, \mathbf{X}_n$ at their observed values $\mathbf{x}_1, \mathbf{x}_2, \ldots, \mathbf{x}_n$. The parameter vector $\boldsymbol{\theta}$ takes its value in the parameter set $\boldsymbol{\Theta}$. For example, in the p-dimensional multivariate normal case, $\boldsymbol{\theta}' = [\mu_1, \ldots, \mu_p, \sigma_{11}, \ldots, \sigma_{1p}, \sigma_{22}, \ldots, \sigma_{2p}, \ldots, \sigma_{p-1,p}, \sigma_{pp}]$ and $\boldsymbol{\Theta}$ consists of the union of the p-dimensional space where $-\infty < \mu_1 < \infty, \ldots, -\infty < \mu_p < \infty$ and the $[p(p+1)/2]$-dimensional space of variances and covariances such that $\boldsymbol{\Sigma}$ is positive definite. Therefore $\boldsymbol{\Theta}$ has dimension $\nu = p + p(p+1)/2$. Under the null hypothesis $H_0: \boldsymbol{\theta} = \boldsymbol{\theta}_0$, $\boldsymbol{\theta}$ is restricted to lie in a subset, $\boldsymbol{\Theta}_0$ of $\boldsymbol{\Theta}$. For the multivariate normal situation with $\boldsymbol{\mu} = \boldsymbol{\mu}_0$ and $\boldsymbol{\Sigma}$ unspecified, $\boldsymbol{\Theta}_0 = \{\mu_1 = \mu_{10}, \mu_2 = \mu_{20}, \ldots, \mu_p = \mu_{p0}; \sigma_{11}, \ldots, \sigma_{1p}, \sigma_{22}, \ldots, \sigma_{2p}, \ldots, \sigma_{p-1,p}, \sigma_{pp}$ with $\boldsymbol{\Sigma}$ positive definite$\}$, so $\boldsymbol{\Theta}_0$ has dimension $\nu_0 = 0 + p(p+1)/2 = p(p+1)/2$.

A likelihood ratio test of $H_0: \boldsymbol{\theta} \in \boldsymbol{\Theta}_0$ rejects H_0 in favor of $H_1: \boldsymbol{\theta} \notin \boldsymbol{\Theta}_0$ if

$$\Lambda = \frac{\max\limits_{\boldsymbol{\theta} \in \boldsymbol{\Theta}_0} L(\boldsymbol{\theta})}{\max\limits_{\boldsymbol{\theta} \in \boldsymbol{\Theta}} L(\boldsymbol{\theta})} < c \tag{5-16}$$

where c is a suitably chosen constant. Intuitively, we reject H_0 if the maximum of the likelihood obtained by allowing $\boldsymbol{\theta}$ to vary over the set $\boldsymbol{\Theta}_0$ is much smaller than the maximum of the likelihood obtained by varying $\boldsymbol{\theta}$ over all values in $\boldsymbol{\Theta}$. When the maximum in the numerator of expression (5-16) is much smaller than the maximum in the denominator, $\boldsymbol{\Theta}_0$ does not contain plausible values for $\boldsymbol{\theta}$.

In each application of the likelihood ratio method, we must obtain the sampling distribution of the likelihood ratio test statistic Λ. Then c can be selected to produce a test with a specified significance level α. However when the sample size is large and certain regularity conditions are satisfied, the sampling distribution of $-2\ln\Lambda$ is well approximated by a chi-square distribution. This attractive feature accounts, in part, for the popularity of likelihood ratio procedures.

Result 5.2. When the sample size n is large,

$$-2\ln\Lambda = -2\ln\left(\frac{\max\limits_{\boldsymbol{\theta}\in\boldsymbol{\Theta}_0} L(\boldsymbol{\theta})}{\max\limits_{\boldsymbol{\theta}\in\boldsymbol{\Theta}} L(\boldsymbol{\theta})}\right) \text{ is, approximately, a } \chi^2_{\nu-\nu_0}$$

random variable. Here the degrees of freedom are $\nu - \nu_0 =$ (dimension of $\boldsymbol{\Theta}$) $-$ (dimension of $\boldsymbol{\Theta}_0$). ■

Statistical tests are compared on the basis of *power*, which is defined as the curve or surface whose height is $P[\text{test rejects } H_0 \mid \boldsymbol{\theta}]$ evaluated at each parameter vector $\boldsymbol{\theta}$. Power measures the ability of a test to reject H_0 when it is not true. In the rare situation where $\boldsymbol{\theta} = \boldsymbol{\theta}_0$ is completely specified under H_0, and the alternative H_1, consists of the single specified value $\boldsymbol{\theta} = \boldsymbol{\theta}_1$, the likelihood ratio test has the highest power among all tests with the same significance level $\alpha = P[\text{test rejects } H_0 \mid \boldsymbol{\theta} = \boldsymbol{\theta}_0]$. In many single parameter cases ($\boldsymbol{\theta}$ has one component), the likelihood ratio test is uniformly most powerful against all alternatives to one side of $H_0: \theta = \theta_0$. In other cases, this property holds approximately for large samples.

We shall not give the technical details required for discussing the optimal properties of likelihood ratio tests in the multivariate situation. The general import of these properties, for our purposes, is that they have the highest possible (average) power when the sample size is large. A discussion of the power of the test based on Hotelling's T^2 (which corresponds to a likelihood ratio test) is contained in Supplement 5B.

5.4 CONFIDENCE REGIONS AND SIMULTANEOUS COMPARISONS OF COMPONENT MEANS

To obtain our primary method for making inferences from a sample, we need to extend the concept of a univariate confidence interval to a multivariate *confidence region*. Let $\boldsymbol{\theta}$ be a vector of unknown population parameters and $\boldsymbol{\Theta}$ the set of all possible values for $\boldsymbol{\theta}$. A confidence region is a region of likely $\boldsymbol{\theta}$ values. This region is determined by the data and, for the moment, we shall denote it by $R(\mathbf{X})$, where $\mathbf{X} = [\mathbf{X}_1, \mathbf{X}_2, \ldots, \mathbf{X}_n]$ is the data matrix.

The region $R(\mathbf{X})$ is said to be a *100(1 − α)% confidence region* if, before the sample is selected,

$$P[R(\mathbf{X}) \text{ will cover the true } \boldsymbol{\theta}] = 1 - \alpha \tag{5-17}$$

This probability is calculated under the true, but unknown, value of $\boldsymbol{\theta}$.

The confidence region for the mean $\boldsymbol{\mu}$ of a p-dimensional normal population is available from (5-6). Before the sample is selected,

$$P\left[n(\overline{\mathbf{X}} - \boldsymbol{\mu})'\mathbf{S}^{-1}(\overline{\mathbf{X}} - \boldsymbol{\mu}) \le \frac{(n-1)p}{(n-p)} F_{p,n-p}(\alpha)\right] = 1 - \alpha$$

whatever the values of the unknown $\boldsymbol{\mu}$ and $\boldsymbol{\Sigma}$. In words, $\overline{\mathbf{X}}$ will be within

$$[(n-1)pF_{p,n-p}(\alpha)/(n-p)]^{1/2}$$

of $\boldsymbol{\mu}$, with probability $1 - \alpha$, provided distance is defined in terms of $(\mathbf{S}/n)^{-1}$. For a particular sample, $\bar{\mathbf{x}}$ and $\mathbf{S}$ can be computed and the inequality $n(\bar{\mathbf{x}} - \boldsymbol{\mu})'\mathbf{S}^{-1}(\bar{\mathbf{x}} - \boldsymbol{\mu}) \le (n-1)pF_{p,n-p}(\alpha)/(n-p)$ will define a region, $R(\mathbf{X})$, within the space of all possible parameter values. In this case, the region will be an ellipsoid centered at $\bar{\mathbf{x}}$. This ellipsoid is the 100(1 − α)% confidence region for $\boldsymbol{\mu}$.

A *100(1 − α)% confidence region* for the mean of a p-dimensional normal distribution is the set determined by all $\boldsymbol{\mu}$ such that

$$n(\bar{\mathbf{x}} - \boldsymbol{\mu})'\mathbf{S}^{-1}(\bar{\mathbf{x}} - \boldsymbol{\mu}) \le \frac{p(n-1)}{(n-p)} F_{p,n-p}(\alpha) \tag{5-18}$$

where $\bar{\mathbf{x}} = \frac{1}{n}\sum_{j=1}^{n} \mathbf{x}_j$, $\mathbf{S} = \frac{1}{(n-1)}\sum_{j=1}^{n} (\mathbf{x}_j - \bar{\mathbf{x}})(\mathbf{x}_j - \bar{\mathbf{x}})'$ and $\mathbf{x}_1, \mathbf{x}_2, \ldots, \mathbf{x}_n$ are the sample observations.

To determine whether any $\boldsymbol{\mu}_0$ falls in the confidence region (is a plausible value for $\boldsymbol{\mu}$), we need compute the generalized squared distance $n(\bar{\mathbf{x}} - \boldsymbol{\mu}_0)'\mathbf{S}^{-1}(\bar{\mathbf{x}} - \boldsymbol{\mu}_0)$ and compare it with $[p(n-1)/(n-p)]F_{p,n-p}(\alpha)$. If the squared distance is larger than $[p(n-1)/(n-p)]F_{p,n-p}(\alpha)$, $\boldsymbol{\mu}_0$ is not in the confidence region. Since this is analogous to testing $H_0: \boldsymbol{\mu} = \boldsymbol{\mu}_0$ versus $H_1: \boldsymbol{\mu} \ne \boldsymbol{\mu}_0$ [see (5-7)], we see that the confidence region of (5-18) consists of all $\boldsymbol{\mu}_0$ vectors for which the T^2-test would *not* reject H_0 in favor of H_1 at significance level α.

For $p \ge 4$, we cannot graph the joint confidence region for $\boldsymbol{\mu}$. However, we can calculate the axes of the confidence ellipsoid and their relative lengths. These are determined from the eigenvalues λ_i and eigenvectors $\mathbf{e}_i$ of $\mathbf{S}$. As in (4-7), the direction and lengths of the axes of

$$n(\bar{\mathbf{x}} - \boldsymbol{\mu})'\mathbf{S}^{-1}(\bar{\mathbf{x}} - \boldsymbol{\mu}) \le c^2 = p(n-1)F_{p,n-p}(\alpha)/(n-p)$$

are determined by going

$$\sqrt{\lambda_i}\, c/\sqrt{n} = \sqrt{\lambda_i}\sqrt{p(n-1)F_{p,n-p}(\alpha)/n(n-p)}$$

units along the eigenvectors $\mathbf{e}_i$. Beginning at the center $\bar{\mathbf{x}}$, the axes of the confidence ellipsoid are

$$\pm\sqrt{\lambda_i}\sqrt{\frac{p(n-1)}{n(n-p)}F_{p,n-p}(\alpha)}\,\mathbf{e}_i \qquad \text{where } \mathbf{S}\mathbf{e}_i = \lambda_i\mathbf{e}_i, \quad i = 1, 2, \ldots, p \tag{5-19}$$

The ratios of the λ_i's will help identify relative amounts of elongation along pairs of axes.

Example 5.3

The data for radiation from microwave ovens were introduced in Examples 4.9 and 4.15. Let

$$x_1 = \sqrt[4]{\text{measured radiation with door closed}}$$

and

$$x_2 = \sqrt[4]{\text{measured radiation with door open}}$$

For the $n = 42$ pairs of transformed observations, we find

$$\bar{\mathbf{x}} = \begin{bmatrix} .564 \\ .603 \end{bmatrix}, \qquad \mathbf{S} = \begin{bmatrix} .0144 & .0117 \\ .0117 & .0146 \end{bmatrix}, \qquad \mathbf{S}^{-1} = \begin{bmatrix} 203.018 & -163.391 \\ -163.391 & 200.228 \end{bmatrix}$$

The eigenvalue and eigenvector pairs for $\mathbf{S}$ are

$$\lambda_1 = .026, \qquad \mathbf{e}_1' = [.704, .710]$$
$$\lambda_2 = .002, \qquad \mathbf{e}_2' = [-.710, .704]$$

The 95% confidence ellipse for $\boldsymbol{\mu}$ consists of all values (μ_1, μ_2) satisfying

$$42[.564 - \mu_1, .603 - \mu_2]\begin{bmatrix} 203.018 & -163.391 \\ -163.391 & 200.228 \end{bmatrix}\begin{bmatrix} .564 - \mu_1 \\ .603 - \mu_2 \end{bmatrix} \le \frac{2\ (41\ )}{40}F_{2,40}(.05)$$

or, since $F_{2,40}(.05) = 3.23$,

$$42(203.018)(.564 - \mu_1)^2 + 42(200.228)(.603 - \mu_2)^2 - 84(163.391)(.564 - \mu_1)(.603 - \mu_2) \le 6.62$$

To see if $\boldsymbol{\mu}' = [.562, .589]$ is in the confidence region, we compute

$$42(203.018)(.564 - .562)^2 + 42(200.228)(.603 - .589)^2 - 84(163.391)(.564 - .562)(.603 - .589) = 1.30 \le 6.62$$

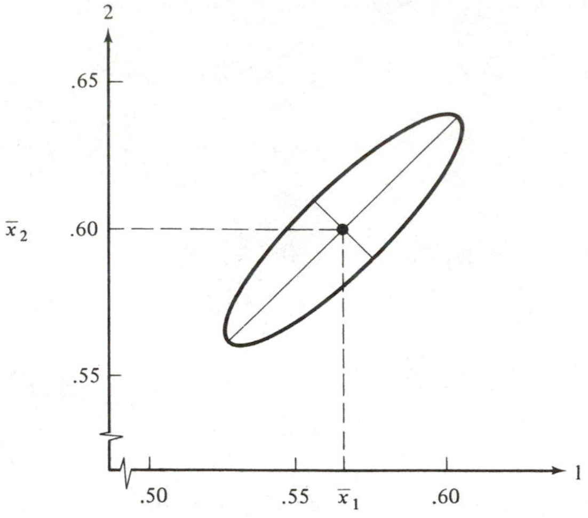

Figure 5.1 A 95% confidence ellipse for $\boldsymbol{\mu}$ based on the microwave-radiation data.

and conclude it is in the region. Equivalently, a test of $H_0: \boldsymbol{\mu} = \begin{bmatrix} .562 \\ .589 \end{bmatrix}$ would not be rejected in favor of $H_1: \boldsymbol{\mu} \neq \begin{bmatrix} .562 \\ .589 \end{bmatrix}$ at the $\alpha = .05$ level of significance.

The joint confidence ellipsoid is plotted in Figure 5.1. The center is at $\bar{\mathbf{x}}' = [.564, .603]$ and the half-lengths of the major and minor axes are given by

$$\sqrt{\lambda_1}\sqrt{\frac{p(n-1)}{n(n-p)}F_{p,n-p}(\alpha)} = \sqrt{.026}\sqrt{\frac{2(41)}{42(40)}(3.23)} = .064$$

and

$$\sqrt{\lambda_2}\sqrt{\frac{p(n-1)}{n(n-p)}F_{p,n-p}(\alpha)} = \sqrt{.002}\sqrt{\frac{2(41)}{42(40)}(3.23)} = .018$$

respectively. The axes lie along $\mathbf{e}_1' = [.704, .710]$ and $\mathbf{e}_2' = [-.710, .704]$ when these vectors are plotted with $\bar{\mathbf{x}}$ as the origin. An indication of the elongation of the confidence ellipse is provided by the ratio of lengths of the major and minor axes. This ratio is

$$\frac{2\sqrt{\lambda_1}\sqrt{\dfrac{p(n-1)}{n(n-p)}F_{p,n-p}(\alpha)}}{2\sqrt{\lambda_2}\sqrt{\dfrac{p(n-1)}{n(n-p)}F_{p,n-p}(\alpha)}} = \frac{\sqrt{\lambda_1}}{\sqrt{\lambda_2}} = \frac{.161}{.045} = 3.6$$

The length of the major axis is 3.6 times the length of the minor axis. ■

Simultaneous Confidence Statements

While the confidence region $n(\bar{\mathbf{x}} - \boldsymbol{\mu})'\mathbf{S}^{-1}(\bar{\mathbf{x}} - \boldsymbol{\mu}) \leq c^2$, c a constant, correctly assesses the joint knowledge concerning plausible values for $\boldsymbol{\mu}$, any summary of conclusions ordinarily includes confidence statements about the individual component means. In so doing, we adopt the attitude that all of the separate confidence statements should hold *simultaneously* with a specified high probability. It is the guarantee of a specified probability against *any* statement being incorrect that motivates the term *simultaneous confidence intervals.* We begin by considering simultaneous confidence statements, which are intimately related to the joint confidence region based on the T^2-statistic.

Let $\mathbf{X}$ have a $N_p(\boldsymbol{\mu}, \boldsymbol{\Sigma})$ distribution and form the linear combination

$$Z = \ell_1 X_1 + \ell_2 X_2 + \cdots + \ell_p X_p = \boldsymbol{\ell}'\mathbf{X}$$

From (2-43),

$$\mu_Z = E(Z) = \boldsymbol{\ell}'\boldsymbol{\mu}$$

and

$$\sigma_Z^2 = \text{Var}(Z) = \boldsymbol{\ell}'\boldsymbol{\Sigma}\boldsymbol{\ell}$$

Moreover, by Result 4.2, Z has a $N(\boldsymbol{\ell}'\boldsymbol{\mu}, \boldsymbol{\ell}'\boldsymbol{\Sigma}\boldsymbol{\ell})$ distribution. If a random sample $\mathbf{X}_1, \mathbf{X}_2, \ldots, \mathbf{X}_n$ from the $N_p(\boldsymbol{\mu}, \boldsymbol{\Sigma})$ population is available, a corresponding sample of Z's can be created by taking linear combinations. Thus

$$Z_j = \ell_1 X_{1j} + \ell_2 X_{2j} + \cdots + \ell_p X_{pj} = \boldsymbol{\ell}'\mathbf{X}_j, \qquad j = 1, 2, \ldots, n$$

The sample mean and variance of $z_1, z_2, \ldots, z_n$ are, by (3-36),

$$\bar{z} = \boldsymbol{\ell}'\bar{\mathbf{x}}$$

and

$$s_z^2 = \boldsymbol{\ell}'\mathbf{S}\boldsymbol{\ell}$$

where $\bar{\mathbf{x}}$ and $\mathbf{S}$ are the sample mean vector and covariance matrix of the $\mathbf{x}_j$'s, respectively.

Simultaneous confidence intervals can be developed from consideration of confidence intervals for $\boldsymbol{\ell}'\boldsymbol{\mu}$ for various choices of $\boldsymbol{\ell}$. The argument proceeds as follows.

For $\boldsymbol{\ell}$ *fixed* and σ_Z^2 unknown, a $100(1 - \alpha)\%$ confidence interval for $\mu_Z = \boldsymbol{\ell}'\boldsymbol{\mu}$ is based on student's t-ratio

$$t = \frac{\bar{z} - \mu_Z}{s_z/\sqrt{n}} = \frac{\sqrt{n}\,(\boldsymbol{\ell}'\bar{\mathbf{x}} - \boldsymbol{\ell}'\boldsymbol{\mu})}{\sqrt{\boldsymbol{\ell}'\mathbf{S}\boldsymbol{\ell}}} \tag{5-20}$$

and leads to the statement

$$\bar{z} - t_{n-1}(\alpha/2)\frac{s_z}{\sqrt{n}} \le \mu_Z \le \bar{z} + t_{n-1}(\alpha/2)\frac{s_z}{\sqrt{n}}$$

or

$$\boldsymbol{\ell}'\bar{\mathbf{x}} - t_{n-1}(\alpha/2)\frac{\sqrt{\boldsymbol{\ell}'\mathbf{S}\boldsymbol{\ell}}}{\sqrt{n}} \le \boldsymbol{\ell}'\boldsymbol{\mu} \le \boldsymbol{\ell}'\bar{\mathbf{x}} + t_{n-1}(\alpha/2)\frac{\sqrt{\boldsymbol{\ell}'\mathbf{S}\boldsymbol{\ell}}}{\sqrt{n}} \tag{5-21}$$

where $t_{n-1}(\alpha/2)$ is the upper $100(\alpha/2)$th percentile of a t-distribution with $n - 1$ d.f.

Inequality (5-21) can be interpreted as a statement about the components of the mean vector $\boldsymbol{\mu}$. For example, with $\boldsymbol{\ell}' = [1, 0, \ldots, 0]$, $\boldsymbol{\ell}'\boldsymbol{\mu} = \mu_1$ and (5-21) becomes the usual confidence interval for a normal population mean. (Note, in this case, $\boldsymbol{\ell}'\mathbf{S}\boldsymbol{\ell} = s_{11}$.) Clearly, we could make several confidence statements about the components of $\boldsymbol{\mu}$, each with associated confidence coefficient $1 - \alpha$, by choosing different coefficient vectors $\boldsymbol{\ell}$. However, the confidence associated with all of the statements taken together is *not* $1 - \alpha$.

Intuitively, it would be desirable to associate a "collective" confidence coefficient of $1 - \alpha$ with the confidence intervals that can be generated by all choices of $\boldsymbol{\ell}$. A price must be paid for the convenience of a large simultaneous confidence coefficient. The price is in the form of intervals that are wider (less precise) than the interval of (5-21) for a specific choice of $\boldsymbol{\ell}$.

Given a data set $\mathbf{x}_1, \mathbf{x}_2, \ldots, \mathbf{x}_n$ and a particular $\boldsymbol{\ell}$, the confidence interval in (5-21) is that set of $\boldsymbol{\ell}'\boldsymbol{\mu}$ values for which

$$|t| = \left| \frac{\sqrt{n}\,(\boldsymbol{\ell}'\bar{\mathbf{x}} - \boldsymbol{\ell}'\boldsymbol{\mu})}{\sqrt{\boldsymbol{\ell}'\mathbf{S}\boldsymbol{\ell}}} \right| \le t_{n-1}(\alpha/2)$$

or, equivalently,

$$t^2 = \frac{n(\boldsymbol{\ell}'\bar{\mathbf{x}} - \boldsymbol{\ell}'\boldsymbol{\mu})^2}{\boldsymbol{\ell}'\mathbf{S}\boldsymbol{\ell}} = \frac{n(\boldsymbol{\ell}'(\bar{\mathbf{x}} - \boldsymbol{\mu}))^2}{\boldsymbol{\ell}'\mathbf{S}\boldsymbol{\ell}} \le t_{n-1}^2(\alpha/2) \qquad (5\text{-}22)$$

A simultaneous confidence region is given by the set of $\boldsymbol{\ell}'\boldsymbol{\mu}$ values such that t^2 is relatively small for *all* choices of $\boldsymbol{\ell}$. It seems reasonable to expect that the constant $t_{n-1}^2(\alpha/2)$ in (5-22) will be replaced by a larger value, c^2, when statements are developed for many choices of $\boldsymbol{\ell}$.

Considering the values of $\boldsymbol{\ell}$ for which $t^2 \le c^2$, we are naturally led to the determination of

$$\max_{\boldsymbol{\ell}} t^2 = \max_{\boldsymbol{\ell}} \frac{n(\boldsymbol{\ell}'(\bar{\mathbf{x}} - \boldsymbol{\mu}))^2}{\boldsymbol{\ell}'\mathbf{S}\boldsymbol{\ell}}$$

Using the maximization lemma in (2-50) with $\mathbf{x} = \boldsymbol{\ell}$, $\mathbf{d} = (\bar{\mathbf{x}} - \boldsymbol{\mu})$, and $\mathbf{B} = \mathbf{S}$,

$$\max_{\boldsymbol{\ell}} \frac{n(\boldsymbol{\ell}'(\bar{\mathbf{x}} - \boldsymbol{\mu}))^2}{\boldsymbol{\ell}'\mathbf{S}\boldsymbol{\ell}} = n\left[\max_{\boldsymbol{\ell}} \frac{(\boldsymbol{\ell}'(\bar{\mathbf{x}} - \boldsymbol{\mu}))^2}{\boldsymbol{\ell}'\mathbf{S}\boldsymbol{\ell}}\right] = n(\bar{\mathbf{x}} - \boldsymbol{\mu})'\mathbf{S}^{-1}(\bar{\mathbf{x}} - \boldsymbol{\mu}) = T^2 \qquad (5\text{-}23)$$

with the maximum occurring for $\boldsymbol{\ell}$ proportional to $\mathbf{S}^{-1}(\bar{\mathbf{x}} - \boldsymbol{\mu})$.

Result 5.3. Let $\mathbf{X}_1, \mathbf{X}_2, \ldots, \mathbf{X}_n$ be a random sample from a $N_p(\boldsymbol{\mu}, \boldsymbol{\Sigma})$ population with $\boldsymbol{\Sigma}$ positive definite. Then, simultaneously for all $\boldsymbol{\ell}$, the interval

$$\left(\boldsymbol{\ell}'\bar{\mathbf{X}} - \sqrt{\frac{p(n-1)}{n(n-p)}F_{p,n-p}(\alpha)\boldsymbol{\ell}'\mathbf{S}\boldsymbol{\ell}}\,,\ \boldsymbol{\ell}'\bar{\mathbf{X}} + \sqrt{\frac{p(n-1)}{n(n-p)}F_{p,n-p}(\alpha)\boldsymbol{\ell}'\mathbf{S}\boldsymbol{\ell}}\right)$$

will contain $\boldsymbol{\ell}'\boldsymbol{\mu}$ with probability $1 - \alpha$.

Proof. From (5-23),

$$T^2 = n(\bar{\mathbf{x}} - \boldsymbol{\mu})'\mathbf{S}^{-1}(\bar{\mathbf{x}} - \boldsymbol{\mu}) \le c^2 \quad \text{implies} \quad \frac{n(\boldsymbol{\ell}'\bar{\mathbf{x}} - \boldsymbol{\ell}'\boldsymbol{\mu})^2}{\boldsymbol{\ell}'\mathbf{S}\boldsymbol{\ell}} \le c^2$$

for every $\boldsymbol{\ell}$, or

$$\boldsymbol{\ell}'\bar{\mathbf{x}} - c\sqrt{\frac{\boldsymbol{\ell}'\mathbf{S}\boldsymbol{\ell}}{n}} \le \boldsymbol{\ell}'\boldsymbol{\mu} \le \boldsymbol{\ell}'\bar{\mathbf{x}} + c\sqrt{\frac{\boldsymbol{\ell}'\mathbf{S}\boldsymbol{\ell}}{n}}$$

for every $\boldsymbol{\ell}$. Choosing $c^2 = p(n-1)F_{p,n-p}(\alpha)/(n-p)$ [see (5-6)] gives intervals that will contain $\boldsymbol{\ell}'\boldsymbol{\mu}$ for all $\boldsymbol{\ell}$, with probability $1 - \alpha = P[T^2 \le c^2]$. ■

It is convenient to refer to the simultaneous intervals of Result 5.3 as *T^2-intervals*, since the coverage probability is determined by the distribution of T^2. The successive choices $\boldsymbol{\ell}' = [1, 0, \ldots, 0]$, $\boldsymbol{\ell}' = [0, 1, \ldots, 0]$, and so on through $\boldsymbol{\ell}' = [0, 0, \ldots, 1]$ for the T^2-intervals allow us to conclude

$$\bar{x}_1 - \sqrt{\frac{p(n-1)}{(n-p)}F_{p,n-p}(\alpha)}\sqrt{\frac{s_{11}}{n}} < \mu_1 < \bar{x}_1 + \sqrt{\frac{p(n-1)}{(n-p)}F_{p,n-p}(\alpha)}\sqrt{\frac{s_{11}}{n}}$$

$$\bar{x}_2 - \sqrt{\frac{p(n-1)}{(n-p)}F_{p,n-p}(\alpha)}\sqrt{\frac{s_{22}}{n}} < \mu_2 < \bar{x}_2 + \sqrt{\frac{p(n-1)}{(n-p)}F_{p,n-p}(\alpha)}\sqrt{\frac{s_{22}}{n}}$$

$$\vdots \qquad\qquad \vdots \qquad\qquad \vdots$$

$$\bar{x}_p - \sqrt{\frac{p(n-1)}{(n-p)}F_{p,n-p}(\alpha)}\sqrt{\frac{s_{pp}}{n}} < \mu_p < \bar{x}_p + \sqrt{\frac{p(n-1)}{(n-p)}F_{p,n-p}(\alpha)}\sqrt{\frac{s_{pp}}{n}} \tag{5-24}$$

all hold simultaneously with confidence coefficient $1 - \alpha$. Note that, without modifying the coefficient $1 - \alpha$, we can make statements about the differences $\mu_i - \mu_k$ corresponding to $\boldsymbol{\ell}' = [0, \ldots, \ell_i, 0, \ldots, \ell_k, \ldots, 0]$ where $\ell_i = 1$ and $\ell_k = -1$. In this case $\boldsymbol{\ell}'\mathbf{S}\boldsymbol{\ell} = s_{ii} - 2s_{ik} + s_{kk}$ and we have the statement

$$\bar{x}_i - \bar{x}_k - \sqrt{\frac{p(n-1)}{(n-p)}F_{p,n-p}(\alpha)}\sqrt{\frac{s_{ii} - 2s_{ik} + s_{kk}}{n}} < \mu_i - \mu_k$$

$$< \bar{x}_i - \bar{x}_k + \sqrt{\frac{p(n-1)}{(n-p)}F_{p,n-p}(\alpha)}\sqrt{\frac{s_{ii} - 2s_{ik} + s_{kk}}{n}} \tag{5-25}$$

The simultaneous T^2 confidence intervals are ideal for "data snooping." The confidence coefficient $1 - \alpha$ remains unchanged for any choice of $\boldsymbol{\ell}$, so linear combinations of the components μ_i that merit inspection *based upon an examination of the data* can be estimated.

Example 5.4

The scores obtained by $n = 87$ college students on the College Level Examination Program (CLEP) subtest X_1 and the College Qualification Test (CQT) subtests X_2 and X_3 are given in Table 5.2 for X_1 = social science and history,

TABLE 5.2 COLLEGE TEST DATA

Individual	X_1 (Social science and history)	X_2 (Verbal)	X_3 (Science)	Individual	X_1 (Social science and history)	X_2 (Verbal)	X_3 (Science)
1	468	41	26	45	494	41	24
2	428	39	26	46	541	47	25
3	514	53	21	47	362	36	17
4	547	67	33	48	408	28	17
5	614	61	27	49	594	68	23
6	501	67	29	50	501	25	26
7	421	46	22	51	687	75	33
8	527	50	23	52	633	52	31
9	527	55	19	53	647	67	29
10	620	72	32	54	647	65	34
11	587	63	31	55	614	59	25
12	541	59	19	56	633	65	28
13	561	53	26	57	448	55	24
14	468	62	20	58	408	51	19
15	614	65	28	59	441	35	22
16	527	48	21	60	435	60	20
17	507	32	27	61	501	54	21
18	580	64	21	62	507	42	24
19	507	59	21	63	620	71	36
20	521	54	23	64	415	52	20
21	574	52	25	65	554	69	30
22	587	64	31	66	348	28	18
23	488	51	27	67	468	49	25
24	488	62	18	68	507	54	26
25	587	56	26	69	527	47	31
26	421	38	16	70	527	47	26
27	481	52	26	71	435	50	28
28	428	40	19	72	660	70	25
29	640	65	25	73	733	73	33
30	574	61	28	74	507	45	28
31	547	64	27	75	527	62	29
32	580	64	28	76	428	37	19
33	494	53	26	77	481	48	23
34	554	51	21	78	507	61	19
35	647	58	23	79	527	66	23
36	507	65	23	80	488	41	28
37	454	52	28	81	607	69	28
38	427	57	21	82	561	59	34
39	521	66	26	83	614	70	23
40	468	57	14	84	527	49	30
41	587	55	30	85	474	41	16
42	507	61	31	86	441	47	26
43	574	54	31	87	607	67	32
44	507	53	23				

SOURCE: Data courtesy of Richard W. Johnson.

X_2 = verbal, and X_3 = science. These data give

$$\bar{\mathbf{x}} = \begin{bmatrix} 527.74 \\ 54.69 \\ 25.13 \end{bmatrix} \quad \text{and} \quad \mathbf{S} = \begin{bmatrix} 5691.34 & 600.51 & 217.25 \\ 600.51 & 126.05 & 23.37 \\ 217.25 & 23.37 & 23.11 \end{bmatrix}$$

Let us compute the 95% simultaneous confidence intervals for μ_1, μ_2, and μ_3.

Here

$$\frac{p(n-1)}{n-p} F_{p,n-p}(\alpha) = \frac{3(87-1)}{(87-3)} F_{3,84}(.05) = \frac{3(86)}{84}(2.7) = 8.29$$

and we obtain the simultaneous confidence statements [see (5-24)]

$$527.74 - \sqrt{8.29}\sqrt{\frac{5691.34}{87}} \le \mu_1 \le 527.74 + \sqrt{8.29}\sqrt{\frac{5691.34}{87}}$$

or

$$504.45 \le \mu_1 \le 551.03$$

$$54.69 - \sqrt{8.29}\sqrt{\frac{126.05}{87}} \le \mu_2 \le 54.69 + \sqrt{8.29}\sqrt{\frac{126.05}{87}}$$

or

$$51.22 \le \mu_2 \le 58.16$$

$$25.13 - \sqrt{8.29}\sqrt{\frac{23.11}{87}} \le \mu_3 \le 25.13 + \sqrt{8.29}\sqrt{\frac{23.11}{87}}$$

or

$$23.65 \le \mu_3 \le 26.61$$

With the possible exception of the verbal scores, the marginal *Q-Q* plots and two-dimensional scatter plots do not reveal any serious departures from normality for the college qualification test data (see Exercise 5.13). Moreover, the sample size is large enough to justify the methodology even though the data are not quite normally distributed (see Section 5.6). ■

A Comparison of Simultaneous Confidence Intervals with One-at-a-Time Intervals

An alternate, though somewhat misguided, approach to construction of confidence intervals is to consider the components μ_i one at a time, as suggested by (5-21) with $\ell' = [0, \ldots, 0, \ell_i, 0, \ldots, 0]$ where $\ell_i = 1$. This approach ignores the covariance structure of the p variables and leads to the intervals

$$\begin{aligned} \bar{x}_1 - t_{n-1}(\alpha/2)\sqrt{\frac{s_{11}}{n}} &\le \mu_1 \le \bar{x}_1 + t_{n-1}(\alpha/2)\sqrt{\frac{s_{11}}{n}} \\ \bar{x}_2 - t_{n-1}(\alpha/2)\sqrt{\frac{s_{22}}{n}} &\le \mu_2 \le \bar{x}_2 + t_{n-1}(\alpha/2)\sqrt{\frac{s_{22}}{n}} \\ &\vdots \\ \bar{x}_p - t_{n-1}(\alpha/2)\sqrt{\frac{s_{pp}}{n}} &\le \mu_p \le \bar{x}_p + t_{n-1}(\alpha/2)\sqrt{\frac{s_{pp}}{n}} \end{aligned} \tag{5-26}$$

Although prior to sampling, the ith interval above has probability $1 - \alpha$ of covering μ_i, we do not know what to assert, in general, about the probability of *all* intervals containing their respective μ_i's. As we have pointed out, this probability is not $1 - \alpha$.

To shed some light on this problem, consider the special case where the observations have a joint normal distribution and

$$\mathbf{\Sigma} = \begin{bmatrix} \sigma_{11} & 0 & \cdots & 0 \\ 0 & \sigma_{22} & \cdots & 0 \\ \vdots & \vdots & \ddots & \vdots \\ 0 & 0 & \cdots & \sigma_{pp} \end{bmatrix}$$

Since the observations on the first variable are independent of those on the second variable, and so on, the product rule for independent events can be applied and, before the sample is selected,

$$P\left[\text{all } t\text{-intervals in (5-26) contain the } \mu_i\text{'s}\right] = (1-\alpha)(1-\alpha)\cdots(1-\alpha) = (1-\alpha)^p$$

If $1 - \alpha = .95$ and $p = 6$, this probability is $(.95)^6 = .74$.

To guarantee a probability of $1 - \alpha$ that all of the statements about component means hold simultaneously, the individual intervals must be wider than the separate t-intervals; just how much larger depends on both p and n, as well as on $1 - \alpha$.

For $1 - \alpha = .95$, $n = 15$, and $p = 4$, the multipliers of $\sqrt{s_{ii}/n}$ in (5-24) and (5-26) are

$$\sqrt{\frac{p(n-1)}{(n-p)} F_{p,n-p}(.05)} = \sqrt{\frac{4(14)}{11}(3.36)} = 4.14$$

and $t_{n-1}(.025) = 2.145$, respectively. Consequently, in this case the simultaneous intervals are $100(4.14 - 2.145)/2.145 = 93\%$ wider than those derived from the one-at-a-time t method.

TABLE 5.3 CRITICAL DISTANCE MULTIPLIERS FOR ONE-AT-A-TIME t-INTERVALS AND T^2-INTERVALS FOR SELECTED n AND p $(1 - \alpha = .95)$

		$\sqrt{\frac{(n-1)p}{(n-p)} F_{p,n-p}(.05)}$	
n	$t_{n-1}(.025)$	$p = 4$	$p = 10$
15	2.145	4.14	11.52
25	2.064	3.60	6.39
50	2.010	3.31	5.05
100	1.970	3.19	4.61
∞	1.960	3.08	4.28

Table 5.3 gives some critical distance multipliers for one-at-a-time t-intervals computed according to (5-21) and the corresponding simultaneous T^2-intervals. In general, the width of the T^2-intervals, relative to the t-intervals, increases for fixed n as p increases and decreases for fixed p as n increases.

The Bonferroni Method of Multiple Comparisons

Often attention is restricted to a small number of individual confidence statements. In these situations it is possible to do better than the simultaneous intervals of Result 5.3. If the number m of specified component means μ_i, or linear combinations $\boldsymbol{\ell}'\boldsymbol{\mu} = \ell_1\mu_1 + \ell_2\mu_2 + \cdots + \ell_p\mu_p$, is small, simultaneous confidence intervals can be developed that are shorter (more precise) than the simultaneous T^2-intervals. The alternative method for multiple comparisons is called the *Bonferroni method*, because it is developed from a probability inequality carrying that name.

Suppose, prior to the collection of data, confidence statements about m linear combinations $\boldsymbol{\ell}_1'\boldsymbol{\mu}, \boldsymbol{\ell}_2'\boldsymbol{\mu}, \ldots, \boldsymbol{\ell}_m'\boldsymbol{\mu}$ are required. Let C_i denote a confidence statement about the value of $\boldsymbol{\ell}_i'\boldsymbol{\mu}$ with $P[C_i \text{ true}] = 1 - \alpha_i$, $i = 1, 2, \ldots, m$. Now (see Exercise 5.6)

$$\begin{aligned} P[\textit{all } C_i \text{ true}] &= 1 - P[\text{at least one } C_i \text{ false}] \\ &\geq 1 - \sum_{i=1}^{m} P(C_i \text{ false}) = 1 - \sum_{i=1}^{m} (1 - P(C_i \text{ true})) \\ &= 1 - (\alpha_1 + \alpha_2 + \cdots + \alpha_m) \end{aligned} \tag{5-27}$$

Inequality (5-27), a special case of the Bonferroni inequality, allows an investigator to control the overall error rate $\alpha_1 + \alpha_2 + \cdots + \alpha_m$, regardless of the correlation structure behind the confidence statements. There is also the flexibility of controlling the error rate for a group of important statements and balancing it by another choice for the less-important statements.

Let us develop simultaneous interval estimates for the restricted set consisting of the components μ_i of $\boldsymbol{\mu}$. Lacking information on the relative importance of these components, we consider the individual t-intervals

$$\bar{x}_i \pm t_{n-1}\left(\frac{\alpha_i}{2}\right)\sqrt{\frac{s_{ii}}{n}} \qquad i = 1, 2, \ldots, m$$

with $\alpha_i = \alpha/m$. Since $P[\bar{X}_i \pm t_{n-1}(\alpha/2m)\sqrt{s_{ii}/n}$ contains $\mu_i] = 1 - \alpha/m$, $i = 1, 2, \ldots, m$, we have from (5-27)

$$\begin{aligned} P\left[\bar{X}_i \pm t_{n-1}\left(\frac{\alpha}{2m}\right)\sqrt{\frac{s_{ii}}{n}} \text{ contain } \mu_i \text{ for all } i\right] &\geq 1 - \underbrace{\left(\frac{\alpha}{m} + \frac{\alpha}{m} + \cdots + \frac{\alpha}{m}\right)}_{m \text{ terms}} \\ &= 1 - \alpha \end{aligned} \tag{5-28}$$

Therefore, with an overall confidence level greater than or equal to $1 - \alpha$, we can make the $m = p$ statements:

$$\begin{aligned}
\bar{x}_1 - t_{n-1}\left(\frac{\alpha}{2p}\right)\sqrt{\frac{s_{11}}{n}} &\le \mu_1 \le \bar{x}_1 + t_{n-1}\left(\frac{\alpha}{2p}\right)\sqrt{\frac{s_{11}}{n}} \\
\bar{x}_2 - t_{n-1}\left(\frac{\alpha}{2p}\right)\sqrt{\frac{s_{22}}{n}} &\le \mu_2 \le \bar{x}_2 + t_{n-1}\left(\frac{\alpha}{2p}\right)\sqrt{\frac{s_{22}}{n}} \\
&\vdots \\
\bar{x}_p - t_{n-1}\left(\frac{\alpha}{2p}\right)\sqrt{\frac{s_{pp}}{n}} &\le \mu_p \le \bar{x}_p + t_{n-1}\left(\frac{\alpha}{2p}\right)\sqrt{\frac{s_{pp}}{n}}
\end{aligned} \tag{5-29}$$

The statements in (5-29) can be compared with those in (5-24). The percentage point $t_{n-1}(\alpha/2p)$ replaces $\sqrt{(n-1)pF_{p,n-p}(\alpha)/(n-p)}$ but otherwise the intervals are of the same structure.

Example 5.5

Let us return to the sweat data introduced in Example 5.2 (see Table 5.1). We shall obtain the simultaneous 95% Bonferroni confidence intervals for μ_1, μ_2 and μ_3 corresponding to the choice $\alpha_i = .05/3$, $i = 1, 2, 3$.

We make use of the results in Example 5.2, noting that $n = 20$ and $t_{19}(.05/2(3)) = t_{19}(.0083) = 2.625$, to get

$$\bar{x}_1 \pm t_{19}(.0083)\sqrt{\frac{s_{11}}{n}} = 4.64 \pm 2.625\sqrt{\frac{2.879}{20}} \quad \text{or} \quad 3.64 \le \mu_1 \le 5.64$$

$$\bar{x}_2 \pm t_{19}(.0083)\sqrt{\frac{s_{22}}{n}} = 45.4 \pm 2.625\sqrt{\frac{199.798}{20}} \quad \text{or} \quad 37.10 \le \mu_2 \le 53.70$$

$$\bar{x}_3 \pm t_{19}(.0083)\sqrt{\frac{s_{33}}{n}} = 9.965 \pm 2.625\sqrt{\frac{3.628}{20}} \quad \text{or} \quad 8.85 \le \mu_3 \le 11.08 \quad \blacksquare$$

The Bonferroni intervals for linear combinations $\boldsymbol{\ell}'\boldsymbol{\mu}$ and the analogous T^2-intervals (recall Result 5.3) have the same general form:

$$\boldsymbol{\ell}'\bar{\mathbf{x}} \pm (\text{critical value})\sqrt{\frac{\boldsymbol{\ell}'\mathbf{S}\boldsymbol{\ell}}{n}}$$

Consequently, in every instance where $\alpha_i = \alpha/m$,

$$\frac{\text{Length of Bonferroni interval}}{\text{Length of } T^2\text{-interval}} = \frac{t_{n-1}(\alpha/2m)}{\sqrt{\dfrac{p(n-1)}{n-p}F_{p,n-p}(\alpha)}} \tag{5-30}$$

which does not depend on the random quantities $\bar{\mathbf{X}}$ and $\mathbf{S}$. As we have pointed out, for a small number m of specified parametric functions $\boldsymbol{\ell}'\boldsymbol{\mu}$, the Bonferroni intervals will always be shorter. How much shorter is indicated in Table 5.4 for selected n and p.

We see from Table 5.4 that the Bonferroni method provides shorter intervals when $m = p$. However, because the T^2-procedure has an associated $1 - \alpha$ confidence coefficient for *all* statements about linear combinations of $\boldsymbol{\mu}$, including those

TABLE 5.4 (LENGTH OF BONFERRONI INTERVAL)/(LENGTH OF T^2-INTERVAL) FOR $1 - \alpha = .95$ AND $\alpha_i = .05/m$

	$m = p$		
n	2	4	10
15	.88	.69	.29
25	.90	.75	.48
50	.91	.78	.58
100	.91	.80	.62
∞	.91	.81	.66

suggested by the data, we tend to use these intervals in our summaries. Cases where the shorter Bonferroni intervals merit consideration *require* a general agreement beforehand to restrict attention to a few specified parametric functions.

5.5 SIMULTANEOUS CONFIDENCE ELLIPSES FOR PAIRS OF MEAN COMPONENTS

The confidence ellipsoid (5-18) for the mean vector $\boldsymbol{\mu}$ is difficult to visualize if $p \geq 4$. It is often informative to investigate the components of $\boldsymbol{\mu}$ in pairs. This is convenient because pairs of components can be investigated in two dimensions and visual displays of plausible values are readily available, as we shall see shortly.

Confidence statements for pairs of mean components follow as special cases from the more general problem of determining a joint confidence ellipse for the two linear combinations

$$\boldsymbol{\ell}_1'\boldsymbol{\mu} = \ell_{11}\mu_1 + \ell_{21}\mu_2 + \cdots + \ell_{p1}\mu_p$$

and

$$\boldsymbol{\ell}_2'\boldsymbol{\mu} = \ell_{12}\mu_1 + \ell_{22}\mu_2 + \cdots + \ell_{p2}\mu_p$$

In matrix notation we set

$$\underset{(p\times 2)}{\mathbf{L}} = [\boldsymbol{\ell}_1, \boldsymbol{\ell}_2] \quad \text{so that} \quad \mathbf{L}'\boldsymbol{\mu} = \begin{bmatrix} \boldsymbol{\ell}_1' \\ \boldsymbol{\ell}_2' \end{bmatrix} \boldsymbol{\mu} = \begin{bmatrix} \boldsymbol{\ell}_1'\boldsymbol{\mu} \\ \boldsymbol{\ell}_2'\boldsymbol{\mu} \end{bmatrix} \tag{5-31}$$

If $\mathbf{X}_1, \mathbf{X}_2, \ldots, \mathbf{X}_n$ is a random sample from a $N_p(\boldsymbol{\mu}, \boldsymbol{\Sigma})$ population, then Result 4.3 implies that the bivariate observations $\mathbf{L}'\mathbf{X}_j = \begin{bmatrix} \boldsymbol{\ell}_1'\mathbf{X}_j \\ \boldsymbol{\ell}_2'\mathbf{X}_j \end{bmatrix}$ are normal with

$$E(\mathbf{L}'\mathbf{X}_j) = \mathbf{L}'\boldsymbol{\mu} = \begin{bmatrix} \boldsymbol{\ell}_1'\boldsymbol{\mu} \\ \boldsymbol{\ell}_2'\boldsymbol{\mu} \end{bmatrix}$$

and

$$\text{Cov}(\mathbf{L}'\mathbf{X}_j) = \mathbf{L}'\boldsymbol{\Sigma}\mathbf{L}$$

Also, the $\mathbf{L}'\mathbf{X}_j$, $j = 1, 2, \ldots, n$, have sample mean $\mathbf{L}'\overline{\mathbf{X}}$ and sample covariance $\mathbf{L}'\mathbf{S}\mathbf{L}$ where $\mathbf{S} = [1/(n-1)] \sum_{j=1}^{n} (\mathbf{X}_j - \overline{\mathbf{X}})(\mathbf{X}_j - \overline{\mathbf{X}})'$. (See Result 3.6.)

From Result 5C.2, $\mathbf{L}'\boldsymbol{\mu}$ lies in the two-dimensional ellipse

$$n(\mathbf{L}'\overline{\mathbf{x}} - \mathbf{L}'\boldsymbol{\mu})'(\mathbf{L}'\mathbf{S}\mathbf{L})^{-1}(\mathbf{L}'\overline{\mathbf{x}} - \mathbf{L}'\boldsymbol{\mu}) \leq \frac{(n-1)p}{(n-p)} F_{p,n-p}(\alpha) \tag{5-32}$$

if and only if $\boldsymbol{\mu}$ lies in the p-dimensional confidence ellipsoid [see (5-18)]

$$n(\overline{\mathbf{x}} - \boldsymbol{\mu})'\mathbf{S}^{-1}(\overline{\mathbf{x}} - \boldsymbol{\mu}) \leq \frac{(n-1)p}{(n-p)} F_{p,n-p}(\alpha)$$

The critical (squared) distance, $(n-1)pF_{p,n-p}(\alpha)/(n-p)$, is the same for both the ellipse of (5-32) and the ellipsoid (5-18). The matrices in the quadratic forms on the left-hand side of the inequalities are the inverses of the sample covariance computed from $\mathbf{L}\mathbf{X}_j$ and $\mathbf{X}_j$, respectively. These covariances, when divided by the sample size n, become the estimated covariances of $\mathbf{L}'\overline{\mathbf{X}}$ and $\overline{\mathbf{X}}$.

Ruling out the case $\ell_1 = (\text{constant})\ell_2$, which is essentially a single linear combination situation, we obtain Result 5.4.

Result 5.4. Let $\boldsymbol{\mu}$ belong to the p-dimensional $100(1-\alpha)\%$ confidence ellipsoid $n(\overline{\mathbf{x}} - \boldsymbol{\mu})'\mathbf{S}^{-1}(\overline{\mathbf{x}} - \boldsymbol{\mu}) \leq (n-1)pF_{p,n-p}(\alpha)/(n-p)$. Then for every $p \times 2$ matrix $\mathbf{L}$ of rank two, $\mathbf{L}'\boldsymbol{\mu}$ belongs to the $100(1-\alpha)\%$ confidence ellipse

$$n(\mathbf{L}'\overline{\mathbf{x}} - \mathbf{L}'\boldsymbol{\mu})'(\mathbf{L}'\mathbf{S}\mathbf{L})^{-1}(\mathbf{L}'\overline{\mathbf{x}} - \mathbf{L}'\boldsymbol{\mu}) \leq \frac{(n-1)p}{(n-p)} F_{p,n-p}(\alpha)$$

Consequently (μ_i, μ_k) belongs to the ellipse

$$n\begin{bmatrix} \bar{x}_i - \mu_i \\ \bar{x}_k - \mu_k \end{bmatrix}' \begin{bmatrix} s_{ii} & s_{ik} \\ s_{ik} & s_{kk} \end{bmatrix}^{-1} \begin{bmatrix} \bar{x}_i - \mu_i \\ \bar{x}_k - \mu_k \end{bmatrix} \leq \frac{(n-1)p}{(n-p)} F_{p,n-p}(\alpha) \tag{5-33}$$

for all $i \neq k$, with confidence coefficient $1 - \alpha$.

Proof. Set $\mathbf{A} = \sum_{j=1}^{n} (\mathbf{x}_j - \overline{\mathbf{x}})(\mathbf{x}_j - \overline{\mathbf{x}})'/(n-1)$, $c^2 = \dfrac{(n-1)p}{(n-p)} F_{p,n-p}(\alpha)$ and $\mathbf{z} = \boldsymbol{\mu} - \overline{\mathbf{x}}$ in Result 5C.2. For pairs of mean components, set $\ell_1' = [0, \ldots, 0, 1, 0, \ldots, 0]$, with 1 in the ith position, and $\ell_2' = [0, \ldots, 0, 1, 0, \ldots, 0]$, with 1 in the kth position. ■

Thus, in addition to the single component T^2-intervals of Result 5.3, we can also include statements about all $\frac{1}{2}p(p-1)$ pairs of component means in a $100(1-\alpha)\%$ confidence summary.

Graphing the confidence ellipses allows one to visually distinguish between the three situations pictured in Figure 5.2. All three cases give the same simultaneous intervals for the individual means μ_i and μ_k.[1] However, more precise inferences can

[1]Result 5C.3 shows that the individual simultaneous intervals are *shadows* of the p-dimensional confidence ellipsoid for $\boldsymbol{\mu}$. These shadows may be obtained by projecting the p-dimensional ellipsoid onto two dimensions and then taking the shadows of these ellipses on each coordinate axis.

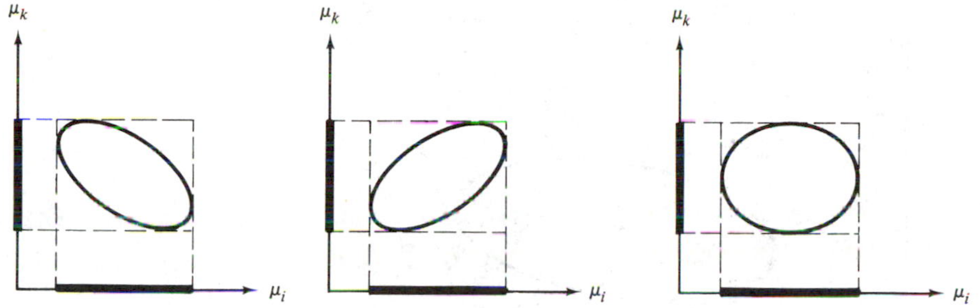

Figure 5.2 Three simultaneous confidence ellipses yielding the same individual intervals for μ_i and μ_k.

be obtained by considering (μ_i, μ_k) jointly in terms of the confidence ellipses. This precision increases as the absolute value of the correlation between X_i and X_j increases, in which case the ellipses become more attenuated.

Example 5.6

We shall obtain the pairwise simultaneous 95% confidence ellipses for the test scores introduced in Example 5.4. Recall $n = 87$ and

$$\bar{\mathbf{x}} = \begin{bmatrix} 527.74 \\ 54.69 \\ 25.13 \end{bmatrix}; \qquad \mathbf{S} = \begin{bmatrix} 5691.34 & 600.51 & 217.25 \\ 600.51 & 126.05 & 23.37 \\ 217.25 & 23.37 & 23.11 \end{bmatrix}$$

In Example 5.4 we calculated $p(n-1)F_{p,\,n-p}(.05)/(n-p) = 3(86)(2.7)/84 = 8.29$. Using (5-33) we obtain the simultaneous 95% statements

$$87\begin{bmatrix} 527.74 - \mu_1 \\ 54.69 - \mu_2 \end{bmatrix}'\begin{bmatrix} 5691.34 & 600.51 \\ 600.51 & 126.05 \end{bmatrix}^{-1}\begin{bmatrix} 527.74 - \mu_1 \\ 54.69 - \mu_2 \end{bmatrix} \le 8.29$$

$$87\begin{bmatrix} 527.74 - \mu_1 \\ 25.13 - \mu_3 \end{bmatrix}'\begin{bmatrix} 5691.34 & 217.25 \\ 217.25 & 23.11 \end{bmatrix}^{-1}\begin{bmatrix} 527.74 - \mu_1 \\ 25.13 - \mu_3 \end{bmatrix} \le 8.29$$

$$87\begin{bmatrix} 54.69 - \mu_2 \\ 25.13 - \mu_3 \end{bmatrix}'\begin{bmatrix} 126.05 & 23.37 \\ 23.37 & 23.11 \end{bmatrix}^{-1}\begin{bmatrix} 54.69 - \mu_2 \\ 25.13 - \mu_3 \end{bmatrix} \le 8.29$$

The sets of inequalities above define the confidence ellipses for pairs of component means. These confidence ellipses can be determined from the eigenvalues and eigenvectors of the appropriate covariance matrix.

Consider the pair (μ_2, μ_3). The confidence ellipse is centered at $[\bar{x}_2, \bar{x}_3]' = [54.69, 25.13]'$ and the major and minor axes are determined by the eigenvectors $\mathbf{e}_1$ and $\mathbf{e}_2$ of the covariance matrix

$$\begin{bmatrix} s_{22} & s_{23} \\ s_{23} & s_{33} \end{bmatrix} = \begin{bmatrix} 126.05 & 23.37 \\ 23.37 & 23.11 \end{bmatrix}$$

The eigenvalues of this matrix and $p(n-1)F_{p,\,n-p}(.05)/(n-p) = 8.29$ determine the lengths of these axes. Straightforward calculations produce the

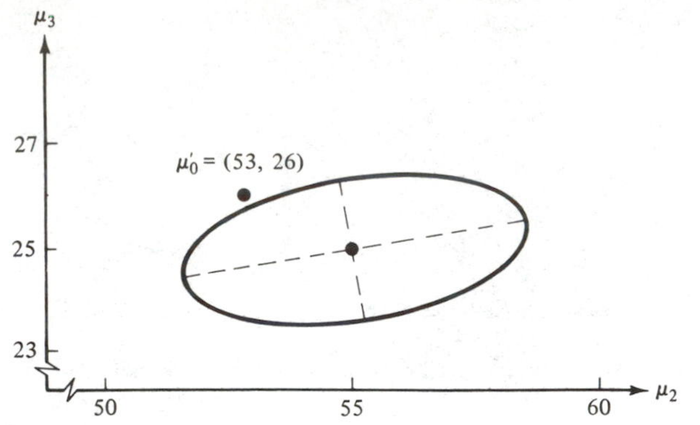

Figure 5.3 A simultaneous 95% confidence ellipse for μ_2, μ_3 (college test data).

pairs

$$\lambda_1 = 131.1, \qquad \mathbf{e}_1 = \begin{bmatrix} .98 \\ .21 \end{bmatrix}$$

$$\lambda_2 = 18.1, \qquad \mathbf{e}_2 = \begin{bmatrix} -.21 \\ .98 \end{bmatrix}$$

Given the preceding information and (5-33), the half-lengths of the major and minor axes are

$$\sqrt{\lambda_1}\sqrt{p(n-1)F_{p,n-p}(.05)/(n-p)n} = \sqrt{(131.1)(8.29)/87} = 3.53$$

and

$$\sqrt{\lambda_2}\sqrt{p(n-1)F_{p,n-p}(.05)/(n-p)n} = \sqrt{(18.1)(8.29)/87} = 1.31$$

respectively. These axes fall along directions $\mathbf{e}_1$ and $\mathbf{e}_2$.

The simultaneous confidence ellipse for (μ_2, μ_3) is plotted in Figure 5.3. It is clear from the figure, for example, that jointly $\mu_{20} = 53$ and $\mu_{30} = 26$ are not plausible values for the population average verbal and science scores since the point with these coordinates falls outside of the ellipse. Note, however, that $\mu_{20} = 53$ and $\mu_{30} = 26$ fall within the simultaneous individual T^2-intervals formed by taking the shadows of the ellipse on each coordinate axis. ■

5.6 LARGE SAMPLE INFERENCES ABOUT A POPULATION MEAN VECTOR

When the sample size is large, tests of hypotheses and confidence regions for $\boldsymbol{\mu}$ can be constructed without the assumption of a normal population. As illustrated more fully by the examples in Section 5.7, for large n, we are able to make inferences about the population mean even though the parent distribution is discrete. In fact, serious departures from a normal population can be overcome by large sample sizes. Both tests of hypotheses and simultaneous confidence statements will then possess (approximately) their nominal levels.

The advantages associated with large samples may be partially offset by a loss in sample information caused by using only the summary statistics $\bar{\mathbf{x}}$ and $\mathbf{S}$. On the other hand, since $(\bar{\mathbf{x}}, \mathbf{S})$ is a sufficient summary for *normal* populations [see (4-21)], the closer the underlying population is to multivariate normal, the more efficiently the sample information will be utilized in making inferences.

All large sample inferences about $\boldsymbol{\mu}$ are based on a χ^2-distribution. From (4-28), we know $(\bar{\mathbf{X}} - \boldsymbol{\mu})'(\mathbf{S}/n)^{-1}(\bar{\mathbf{X}} - \boldsymbol{\mu}) = n(\bar{\mathbf{X}} - \boldsymbol{\mu})'\mathbf{S}^{-1}(\bar{\mathbf{X}} - \boldsymbol{\mu})$ is approximately χ^2 with p d.f. and thus

$$P\left(n(\bar{\mathbf{X}} - \boldsymbol{\mu})'\mathbf{S}^{-1}(\bar{\mathbf{X}} - \boldsymbol{\mu}) \leq \chi_p^2(\alpha)\right) \doteq 1 - \alpha \tag{5-34}$$

where $\chi_p^2(\alpha)$ is the upper (100α)th percentile of the χ_p^2-distribution.

Equation (5-34) immediately leads to large sample tests of hypotheses and simultaneous confidence regions. These procedures are summarized in Results 5.5 and 5.6.

Result 5.5. Let $\mathbf{X}_1, \mathbf{X}_2, \ldots, \mathbf{X}_n$ be a random sample from a population with mean $\boldsymbol{\mu}$ and positive definite covariance matrix $\boldsymbol{\Sigma}$. When $n - p$ is large, the hypothesis H_0: $\boldsymbol{\mu} = \boldsymbol{\mu}_0$ is rejected in favor of $H_1 : \boldsymbol{\mu} \neq \boldsymbol{\mu}_0$, at a level of significance approximately α, if

$$n(\bar{\mathbf{x}} - \boldsymbol{\mu}_0)'\mathbf{S}^{-1}(\bar{\mathbf{x}} - \boldsymbol{\mu}_0) > \chi_p^2(\alpha)$$

Here $\chi_p^2(\alpha)$ is the upper (100α)th percentile of a chi-square distribution with p d.f. ■

Comparing the test in Result 5.5 with the corresponding *normal theory* test in (5-7), we see that the test statistics have the same structure, but the critical values are different. A closer examination however, reveals that both tests yield essentially the same result in situations where the χ^2-test of Result 5.5 is appropriate. This follows directly from the fact that $(n-1)pF_{p,n-p}(\alpha)/(n-p)$ and $\chi_p^2(\alpha)$ are approximately equal for n large relative to p (see Tables 3 and 4 in the appendix).

Result 5.6. Let $\mathbf{X}_1, \mathbf{X}_2, \ldots, \mathbf{X}_n$ be a random sample from a population with mean $\boldsymbol{\mu}$ and positive definite covariance $\boldsymbol{\Sigma}$. If $n - p$ is large,

$$\boldsymbol{\ell}'\bar{\mathbf{X}} \pm \sqrt{\chi_p^2(\alpha)}\,\sqrt{(\boldsymbol{\ell}'\mathbf{S}\boldsymbol{\ell}/n)}$$

will contain $\boldsymbol{\ell}'\boldsymbol{\mu}$, for every $\boldsymbol{\ell}$ with probability approximately $1 - \alpha$. Consequently, we can make the $100(1 - \alpha)\%$ simultaneous confidence statements

$$\bar{x}_1 \pm \sqrt{\chi_p^2(\alpha)}\sqrt{\frac{s_{11}}{n}} \quad \text{contains} \quad \mu_1$$

$$\bar{x}_2 \pm \sqrt{\chi_p^2(\alpha)}\sqrt{\frac{s_{22}}{n}} \quad \text{contains} \quad \mu_2$$

$$\vdots \qquad\qquad\qquad \vdots$$

$$\bar{x}_p \pm \sqrt{\chi_p^2(\alpha)}\sqrt{\frac{s_{pp}}{n}} \quad \text{contains} \quad \mu_p$$

and, in addition, for all pairs (μ_i, μ_k), $i, k = 1, 2, \ldots, p$, the sample mean centered ellipses

$$n[\bar{x}_i - \mu_i, \bar{x}_k - \mu_k]\begin{bmatrix} s_{ii} & s_{ik} \\ s_{ik} & s_{kk} \end{bmatrix}^{-1}\begin{bmatrix} \bar{x}_i - \mu_i \\ \bar{x}_k - \mu_k \end{bmatrix} \le \chi_p^2(\alpha) \quad \text{contain} \quad (\mu_i, \mu_k)$$

Proof. The first part follows from Result 5C.1, with $c^2 = \chi_p^2(\alpha)$. The probability level is a consequence of (5-34). The statements for the μ_i are obtained by the special choices $\ell' = [0, \ldots, 0, \ell_i, 0, \ldots, 0]$, where $\ell_i = 1$, $i = 1, 2, \ldots, p$. The ellipsoids for pairs of means follow from Result 5C.2 with $c^2 = \chi_p^2(\alpha)$. The overall confidence level of approximately $1 - \alpha$ for all statements is, once again, a result of the large-sample distribution theory summarized in (5-34). ■

It is good statistical practice to subject these large-sample inference procedures to the same checks required of the normal-theory methods. Although small to moderate departures from normality do not cause any difficulties for n large, *extreme* deviations could cause problems. Specifically, the true error rate may be far removed from the nominal level α. If, on the basis of Q-Q plots and other investigative devices, outliers and other forms of extreme departures are indicated, appropriate corrective actions, including transformations, are desirable. In some instances Results 5.5 and 5.6 are only useful for *very* large samples.

The next example allows us to illustrate the construction of large sample simultaneous statements for single mean components.

Example 5.7

A music educator tested thousands of Finnish students on native musical ability in order to set national norms in Finland. Summary statistics for part of the data set are given in Table 5.5. These statistics are based on a sample of $n = 96$ Finnish twelfth graders.

Let us construct 90% simultaneous confidence intervals for the individual mean components μ_i, $i = 1, 2, \ldots, 7$.

From Result 5.6, simultaneous 90% confidence limits are given by $\bar{x}_i \pm \sqrt{\chi_7^2(.10)}\sqrt{\frac{s_{ii}}{n}}$, $i = 1, 2, \ldots, 7$, where $\chi_7^2(.10) = 12.02$. Thus, with approxi-

TABLE 5.5 MUSICAL APTITUDE PROFILE MEANS AND STANDARD DEVIATIONS FOR 96 TWELFTH-GRADE FINNISH STUDENTS PARTICIPATING IN A STANDARDIZATION PROGRAM

	Raw score	
	Mean ($\bar{x}_i$)	Standard deviation ($\sqrt{s_{ii}}$)
X_1 = melody	28.1	5.76
X_2 = harmony	26.6	5.85
X_3 = tempo	35.4	3.82
X_4 = meter	34.2	5.12
X_5 = phrasing	23.6	3.76
X_6 = balance	22.0	3.93
X_7 = style	22.7	4.03

SOURCE: Data courtesy of V. Sell.

mately 90% confidence,

$$28.1 \pm \sqrt{12.02}\,\frac{5.76}{\sqrt{96}} \quad \text{contains } \mu_1 \quad \text{or} \quad 26.06 \leq \mu_1 \leq 30.14$$

$$26.6 \pm \sqrt{12.02}\,\frac{5.85}{\sqrt{96}} \quad \text{contains } \mu_2 \quad \text{or} \quad 24.53 \leq \mu_2 \leq 28.67$$

$$35.4 \pm \sqrt{12.02}\,\frac{3.82}{\sqrt{96}} \quad \text{contains } \mu_3 \quad \text{or} \quad 34.05 \leq \mu_3 \leq 36.75$$

$$34.2 \pm \sqrt{12.02}\,\frac{5.12}{\sqrt{96}} \quad \text{contains } \mu_4 \quad \text{or} \quad 32.39 \leq \mu_4 \leq 36.01$$

$$23.6 \pm \sqrt{12.02}\,\frac{3.76}{\sqrt{96}} \quad \text{contains } \mu_5 \quad \text{or} \quad 22.27 \leq \mu_5 \leq 24.93$$

$$22.0 \pm \sqrt{12.02}\,\frac{3.93}{\sqrt{96}} \quad \text{contains } \mu_6 \quad \text{or} \quad 20.61 \leq \mu_6 \leq 23.39$$

$$22.7 \pm \sqrt{12.02}\,\frac{4.03}{\sqrt{96}} \quad \text{contains } \mu_7 \quad \text{or} \quad 21.27 \leq \mu_7 \leq 24.13$$

Based, perhaps, upon thousands of American students, the investigator could hypothesize the musical aptitude profile scores to be

$$\boldsymbol{\mu}_0' = [31, 27, 34, 31, 23, 22, 22]$$

We see from the simultaneous statements above that the melody, tempo, and meter components of $\boldsymbol{\mu}_0$ do not appear to be plausible values for the corresponding means of Finnish scores. ■

5.7 LARGE SAMPLE INFERENCES ABOUT PROPORTIONS

Frequently, some or all of the population characteristics of interest are in the form of *attributes*. Each individual in the population may then be described in terms of the attributes they possess. For convenience, attributes are usually numerically coded with respect to their presence or absence. As an example, a student at the University of Wisconsin may or may not possess the attribute *registered to vote*. If we let the variable X_1 pertain to voter registration, then we can distinguish between the presence or absence of this attribute by defining

$$X_1 = \begin{cases} 1 & \text{if student } \textit{is} \text{ registered to vote} \\ 0 & \text{if student } \textit{is not} \text{ registered to vote} \end{cases}$$

In this way we can assign numerical values to qualitative characteristics.

When attributes are numerically coded as 0–1 variables, a random sample from the population of interest results in statistics that consist of the *counts* of the number of sample items that have each distinct set of characteristics. If the sample counts are large, our previously derived methods for simultaneous confidence statements can be easily adapted to situations involving proportions.

We consider the situation where an individual with a particular combination of attributes can be classified into one of $q + 1$ mutually exclusive and exhaustive categories. The corresponding probabilities are denoted by $p_1, p_2, \ldots, p_q, p_{q+1}$. Since the categories include all possibilities, we take $p_{q+1} = 1 - (p_1 + p_2 + \cdots + p_q)$. An individual from category k will be assigned the $((q + 1) \times 1)$ vector value $[0, \ldots, 0, 1, 0, \ldots, 0]'$ with 1 in the kth position.

Example 5.8

As part of a larger marketing research project, a consultant for the Bank of Shorewood wants to know the proportion of savers that use its facilities as their primary savings bank. The consultant would also like to know the proportions of savers who use the three major competitors, Bank B, Bank C and Bank D. Each individual contacted in a survey responded to the question:

Which bank is your primary savings bank?

Response:	Bank of Shorewood	Bank B	Bank C	Bank D	Another Bank	No Savings

(The people with no savings will be ignored in the comparison of savers so there are 5 categories.) Let the population proportions be

$$\begin{aligned} p_1 &= \text{proportion of savers at Bank of Shorewood} \\ p_2 &= \text{proportion of savers at Bank B} \\ p_3 &= \text{proportion of savers at Bank C} \\ p_4 &= \text{proportion of savers at Bank D} \\ 1 - (p_1 + p_2 + p_3 + p_4) &= \text{proportion of savers at other banks} \end{aligned}$$

A saver at Bank B would have the observation vector $[0, 1, 0, 0, 0]'$. ■

The probability distribution for an observation from the population of individuals in $q + 1$ mutually exclusive and exhaustive categories is known as the *multinomial distribution*. It has the following structure

$$\begin{array}{lccccccc} \text{Category:} & 1 & 2 & \cdots & k & \cdots & q & q+1 \\ \text{Outcome (value):} & \begin{bmatrix} 1 \\ 0 \\ 0 \\ \vdots \\ \vdots \\ 0 \end{bmatrix} & \begin{bmatrix} 0 \\ 1 \\ 0 \\ \vdots \\ \vdots \\ 0 \end{bmatrix} & \cdots & \begin{bmatrix} 0 \\ \vdots \\ 0 \\ 1 \\ 0 \\ \vdots \\ 0 \end{bmatrix} & \cdots & \begin{bmatrix} 0 \\ 0 \\ 0 \\ \vdots \\ 0 \\ 1 \\ 0 \end{bmatrix} & \begin{bmatrix} 0 \\ 0 \\ 0 \\ \vdots \\ 0 \\ 1 \end{bmatrix} \\ \begin{array}{l}\text{Probability} \\ \text{(proportion):}\end{array} & p_1 & p_2 & \cdots & p_k & \cdots & p_q & p_{q+1} = 1 - \sum_{i=1}^{q} p_i \end{array} \qquad (5\text{-}35)$$

Let $\mathbf{X}_j$, $j = 1, 2, \ldots, n$, be a random sample of size n from the multinomial distribution. The kth component, X_{kj}, of $\mathbf{X}_j$ is 1 if the observation (individual) is from category k and 0 otherwise. Since X_{kj} will be 1 with probability p_k,

$$\mu_k = E(X_{kj}) = 1(p_k) + 0(1 - p_k) = p_k$$

$$\sigma_{kk} = \text{Var}(X_{kj}) = E(X_{kj}^2) - E^2(X_{kj}) = 1^2(p_k) + 0^2(1 - p_k) - p_k^2$$
$$= p_k(1 - p_k) \qquad (5\text{-}36)$$

The covariance structure for the components of the $\mathbf{X}_j$'s can be derived in the usual way. Since X_{ij} and X_{kj}, $i \neq k$, cannot both be 1 for the jth individual, the covariances are all negative. In particular,

$$\sigma_{ik} = \text{Cov}(X_{ij}, X_{kj}) = E(X_{ij}X_{kj}) - E(X_{ij})E(X_{kj})$$
$$= E(0) - p_i p_k = -p_i p_k \qquad (5\text{-}37)$$

The random sample $\mathbf{X}_1, \mathbf{X}_2, \ldots, \mathbf{X}_n$ can be converted to a sample proportion vector, which, given the nature of the observations (5-35), is a sample mean vector. Thus

$$\hat{\mathbf{p}} = \begin{bmatrix} \hat{p}_1 \\ \hat{p}_2 \\ \vdots \\ \hat{p}_{q+1} \end{bmatrix} = \frac{1}{n}\sum_{j=1}^{n} \mathbf{X}_j \quad \text{with} \quad E(\hat{\mathbf{p}}) = \mathbf{p} = \begin{bmatrix} p_1 \\ p_2 \\ \vdots \\ p_{q+1} \end{bmatrix}$$

and

$$\text{Cov}(\hat{\mathbf{p}}) = \frac{1}{n}\text{Cov}(\mathbf{X}_j) = \frac{1}{n}\boldsymbol{\Sigma} = \frac{1}{n}\begin{bmatrix} \sigma_{11} & \sigma_{12} & \cdots & \sigma_{1,q+1} \\ \sigma_{12} & \sigma_{22} & \cdots & \sigma_{2,q+1} \\ \vdots & \vdots & & \vdots \\ \sigma_{1,q+1} & \sigma_{2,q+1} & \cdots & \sigma_{q+1,q+1} \end{bmatrix}$$
$$(5\text{-}38)$$

For large n, the approximate sampling distribution of $\hat{\mathbf{p}}$ is provided by the central limit theorem. We have

$$\sqrt{n}\,(\hat{\mathbf{p}} - \mathbf{p}) \quad \text{is approximately} \quad N(\mathbf{0}, \boldsymbol{\Sigma}) \qquad (5\text{-}39)$$

where the elements of $\boldsymbol{\Sigma}$ are defined in (5-36) and (5-37). The approximation in (5-39) remains valid when σ_{kk} is estimated by $\hat{\sigma}_{kk} = \hat{p}_k(1 - \hat{p}_k)$ and σ_{ik} is estimated by $\hat{\sigma}_{ik} = -\hat{p}_i \hat{p}_k$, $i \neq k$.

Since each individual must belong to exactly one category, $X_{q+1,j} = 1 - (X_{1j} + X_{2j} + \cdots + X_{qj})$ and $\hat{p}_{q+1} = 1 - (\hat{p}_1 + \hat{p}_2 + \cdots + \hat{p}_q)$ and, as a result, $\hat{\boldsymbol{\Sigma}}$ has rank q. The usual inverse of $\hat{\boldsymbol{\Sigma}}$ does not exist, but it is still possible to develop simultaneous $100(1 - \alpha)\%$ confidence intervals for all linear combinations $\boldsymbol{\ell}'\mathbf{p}$.

Result 5.7. Let $\mathbf{X}_1, \mathbf{X}_2, \ldots, \mathbf{X}_n$ be a random sample from a $q + 1$ category multinomial distribution with $P[X_{kj} = 1] = p_k$, $k = 1, 2, \ldots, q + 1$, $j = 1, 2, \ldots, n$.

Approximate simultaneous $100(1-\alpha)\%$ confidence regions for all linear combinations $\boldsymbol{\ell}'\mathbf{p} = \ell_1 p_1 + \ell_2 p_2 + \cdots + \ell_{q+1} p_{q+1}$ are given by the observed values of

$$\boldsymbol{\ell}'\hat{\mathbf{p}} \pm \sqrt{\chi_q^2(\alpha)}\sqrt{\frac{\boldsymbol{\ell}'\hat{\boldsymbol{\Sigma}}\boldsymbol{\ell}}{n}}$$

provided $n - q$ is large. Here $\hat{\mathbf{p}} = (1/n)\sum_{j=1}^{n} \mathbf{X}_j$, and $\hat{\boldsymbol{\Sigma}} = \{\hat{\sigma}_{ik}\}$ is a $(q+1) \times (q+1)$ matrix with $\hat{\sigma}_{kk} = \hat{p}_k(1-\hat{p}_k)$ and $\hat{\sigma}_{ik} = -\hat{p}_i\hat{p}_k$, $i \neq k$. Also, $\chi_q^2(\alpha)$ is the upper (100α)th percentile of the chi-square distribution with q d.f. ■

In Result 5.7, the requirement that $n - q$ is large is interpreted to mean $n\hat{p}_k$ is about 20 or more for each category.

Example 5.9

In Example 5.8 we discussed a banking study conducted by a market researcher. A sample of $n = 355$ people, with savings accounts, produced the counts shown below when asked to indicate their primary savings banks.

Bank (category)	Bank of Shorewood	Bank B	Bank C	Bank D	Another bank	
Observed number	105	119	56	25	50	Total $n = 355$
Population proportion	p_1	p_2	p_3	p_4	$p_5 = 1 - (p_1 + p_2 + p_3 + p_4)$	
Observed sample proportion	$\hat{p}_1 = \frac{105}{355} = .30$	$\hat{p}_2 = .33$	$\hat{p}_3 = .16$	$\hat{p}_4 = .07$	$\hat{p}_5 = .14$	

With $\chi_4^2(.05) = 9.49$, simultaneous 95% confidence intervals for $p_1, p_2, \ldots, p_5$ are given by

$$p_1: \hat{p}_1 \pm \sqrt{\chi_4^2(.05)}\sqrt{\frac{\hat{p}_1(1-\hat{p}_1)}{n}} = .30 \pm \sqrt{9.49\left(\frac{.30(.70)}{355}\right)} \quad \text{or} \quad .23 \leq p_1 \leq .37$$

$$p_2: \hat{p}_2 \pm \sqrt{\chi_4^2(.05)}\sqrt{\frac{\hat{p}_2(1-\hat{p}_2)}{n}} = .33 \pm \sqrt{9.49\left(\frac{.33(.67)}{355}\right)} \quad \text{or} \quad .25 \leq p_2 \leq .41$$

$$p_3: \hat{p}_3 \pm \sqrt{\chi_4^2(.05)}\sqrt{\frac{\hat{p}_3(1-\hat{p}_3)}{n}} = .16 \pm \sqrt{9.49\left(\frac{.16(.84)}{355}\right)} \quad \text{or} \quad .10 \leq p_3 \leq .22$$

$$p_4: \hat{p}_4 \pm \sqrt{\chi_4^2(.05)}\sqrt{\frac{\hat{p}_4(1-\hat{p}_4)}{n}} = .07 \pm \sqrt{9.49\left(\frac{.07(.93)}{355}\right)} \quad \text{or} \quad .03 \leq p_4 \leq .11$$

$$p_5: \hat{p}_5 \pm \sqrt{\chi_4^2(.05)}\sqrt{\frac{\hat{p}_5(1-\hat{p}_5)}{n}} = .14 \pm \sqrt{9.49\left(\frac{.14(.86)}{355}\right)} \quad \text{or} \quad .08 \leq p_5 \leq .20$$

The intervals above follow from Result 5.7 by choosing

$$\boldsymbol{\ell}' = [0, \ldots, 0, \ell_i, 0, \ldots, 0]$$

with $\ell_i = 1$, $i = 1, 2, \ldots, 5$. The choice $\ell' = [1, -1, 0, 0, 0]$ gives $\ell'\mathbf{p} = (p_1 - p_2)$ and allows a comparison of the Bank of Shorewood with its major competitor. In this case, $\ell'\hat{\Sigma}\ell = \hat{\sigma}_{11} - 2\hat{\sigma}_{12} + \hat{\sigma}_{22} = \hat{p}_1(1 - \hat{p}_1) - 2(-\hat{p}_1\hat{p}_2) + \hat{p}_2(1 - \hat{p}_2)$ and a simultaneous 95% confidence interval for $p_1 - p_2$ is

$$(\hat{p}_1 - \hat{p}_2) \pm \sqrt{\chi_4^2(.05)}\sqrt{\frac{\hat{p}_1(1 - \hat{p}_1) + 2\hat{p}_1\hat{p}_2 + \hat{p}_2(1 - \hat{p}_2)}{n}}$$

$$= (.30 - .33) \pm \sqrt{9.49\left[\frac{.30(.70) + 2(.30)(.33) + .33(.67)}{355}\right]}$$

$$= -.03 \pm .13$$

or

$$-.16 \le p_1 - p_2 \le .10$$

Since zero is in this last interval, we conclude, with 95% confidence, that there is no difference between the population proportions of savers who use the Bank of Shorewood and Bank B. Moreover, the overall confidence level remains at 95% even though we first looked at the data to identify the bank with the largest market share in the sample before selecting it for comparison. ■

We have only touched on the possibilities for the analysis of categorical data. A complete discussion of categorical data analysis based upon parsimonious log linear models is available in [2].

5.8 INFERENCES ABOUT MEAN VECTORS WHEN SOME OBSERVATIONS ARE MISSING

Often some components of a vector observation are unavailable. This may occur because of a breakdown in the recording equipment or because of the unwillingness of a respondent to answer a particular item on a survey questionnaire. The best way to handle incomplete observations, or missing values, depends, to a large extent, on the experimental context. If the pattern of missing values is closely tied to the value of the response, such as people with extremely high incomes who refuse to respond in a salary survey, subsequent inferences may be seriously biased. To date, no statistical techniques have been developed for these cases. However, we are able to treat situations where data are missing at random—that is, cases in which the chance mechanism responsible for the missing values is *not* influenced by the values of the variables.

A general approach for computing maximum likelihood estimates from incomplete data is given by Dempster and others in [4]. Their technique, called the *EM algorithm*, consists of an iterative calculation involving two steps. We call them the *prediction* and *estimation* steps.

1. *Prediction step* Given some estimate $\tilde{\boldsymbol{\theta}}$ of the unknown parameters, predict the contribution of any missing observation to the (complete data) sufficient statistics.

2. *Estimation step* Use the predicted sufficient statistics to compute a revised estimate of the parameters.

The calculation cycles from one step to the other until the revised estimates do not differ appreciably from the one obtained in the previous iteration.

When the observations $\mathbf{X}_1, \mathbf{X}_2, \ldots, \mathbf{X}_n$ are a random sample from a p-variate normal population, the prediction-estimation algorithm is based on the complete data sufficient statistics [see (4-21)]

$$\mathbf{T}_1 = \sum_{j=1}^{n} \mathbf{X}_j = n\overline{\mathbf{X}}$$

$$\mathbf{T}_2 = \sum_{j=1}^{n} \mathbf{X}_j\mathbf{X}_j' = (n-1)\mathbf{S} + n\overline{\mathbf{X}}\overline{\mathbf{X}}'$$

In this case, the algorithm proceeds as follows. We assume the population mean and variance—$\boldsymbol{\mu}$ and $\boldsymbol{\Sigma}$, respectively—are unknown and must be estimated.

Prediction step. For each vector $\mathbf{x}_j$ with missing values, let $\mathbf{x}_j^{(1)}$ denote the missing components and $\mathbf{x}_j^{(2)}$ denote those components available. Thus $\mathbf{x}_j' = [\mathbf{x}_j^{(1)\prime}, \mathbf{x}_j^{(2)\prime}]'$.

Given estimates $\tilde{\boldsymbol{\mu}}$ and $\tilde{\boldsymbol{\Sigma}}$ from the estimation step, use the mean of the conditional normal distribution of $\mathbf{x}^{(1)}$, given $\mathbf{x}^{(2)}$, to estimate the missing values. That is,[2]

$$\tilde{\mathbf{x}}_j^{(1)} = E\left(\mathbf{X}_j^{(1)} \mid \mathbf{x}_j^{(2)}; \tilde{\boldsymbol{\mu}}, \tilde{\boldsymbol{\Sigma}}\right) = \tilde{\boldsymbol{\mu}}^{(1)} + \tilde{\boldsymbol{\Sigma}}_{12}\tilde{\boldsymbol{\Sigma}}_{22}^{-1}\left(\mathbf{x}_j^{(2)} - \tilde{\boldsymbol{\mu}}^{(2)}\right) \qquad (5\text{-}40)$$

estimates the contribution of $\mathbf{x}_j^{(1)}$ to $\mathbf{T}_1$.

Next, the predicted contribution of $\mathbf{x}_j^{(1)}$ to $\mathbf{T}_2$ is

$$\widetilde{\mathbf{x}_j^{(1)}\mathbf{x}_j^{(1)\prime}} = E\left[\mathbf{X}_j^{(1)}\mathbf{X}_j^{(1)\prime} \mid \mathbf{x}_j^{(2)}; \tilde{\boldsymbol{\mu}}, \tilde{\boldsymbol{\Sigma}}\right) = \tilde{\boldsymbol{\Sigma}}_{11} - \tilde{\boldsymbol{\Sigma}}_{12}\tilde{\boldsymbol{\Sigma}}_{22}^{-1}\tilde{\boldsymbol{\Sigma}}_{21} + \tilde{\mathbf{x}}_j^{(1)}\tilde{\mathbf{x}}_j^{(1)\prime} \qquad (5\text{-}41)$$

and

$$\widetilde{\mathbf{x}_j^{(1)}\mathbf{x}_j^{(2)\prime}} = E\left(\mathbf{X}_j^{(1)}\mathbf{X}_j^{(2)\prime} \mid \mathbf{x}_j^{(2)}; \tilde{\boldsymbol{\mu}}, \tilde{\boldsymbol{\Sigma}}\right) = \tilde{\mathbf{x}}_j^{(1)}\mathbf{x}_j^{(2)\prime}$$

The contributions in (5-40) and (5-41) are summed over all $\mathbf{x}_j$ with missing components. The results are combined with the sample data to yield $\tilde{\mathbf{T}}_1$ and $\tilde{\mathbf{T}}_2$.

Estimation step. Compute the revised maximum likelihood estimates (see Result 4.11),

$$\tilde{\boldsymbol{\mu}} = \frac{\tilde{\mathbf{T}}_1}{n}, \qquad \tilde{\boldsymbol{\Sigma}} = \frac{1}{n}\tilde{\mathbf{T}}_2 - \tilde{\boldsymbol{\mu}}\tilde{\boldsymbol{\mu}}' \qquad (5\text{-}42)$$

We illustrate the computational aspects of the prediction-estimation algorithm in Example 5.10.

[2]If all the components $\mathbf{x}_j$ are missing, set $\tilde{\mathbf{x}}_j = \tilde{\boldsymbol{\mu}}$ and $\tilde{\mathbf{x}}_j\tilde{\mathbf{x}}_j' = \tilde{\boldsymbol{\Sigma}} + \tilde{\boldsymbol{\mu}}\tilde{\boldsymbol{\mu}}'$.

Example 5.10

Estimate the normal population mean $\boldsymbol{\mu}$ and covariance $\boldsymbol{\Sigma}$ using the incomplete data set

$$\mathbf{X} = \begin{bmatrix} - & 7 & 5 & - \\ 0 & 2 & 1 & - \\ 3 & 6 & 2 & 5 \end{bmatrix}$$

Here $p = 3$, $n = 4$, and parts of observation vectors $\mathbf{x}_1$ and $\mathbf{x}_4$ are missing.

We obtain the initial sample averages

$$\tilde{\mu}_1 = \frac{7 + 5}{2} = 6, \qquad \tilde{\mu}_2 = \frac{0 + 2 + 1}{3} = 1, \qquad \tilde{\mu}_3 = \frac{3 + 6 + 2 + 5}{4} = 4$$

from the available observations. Substituting these averages for any missing values, so that $\tilde{x}_{11} = 6$, for example, we can obtain initial covariance estimates. We shall construct these estimates using the divisor n because the algorithm eventually produces the maximum likelihood estimate $\hat{\boldsymbol{\Sigma}}$. Thus

$$\tilde{\sigma}_{11} = \frac{(6-6)^2 + (7-6)^2 + (5-6)^2 + (6-6)^2}{4} = \frac{1}{2},$$

$$\tilde{\sigma}_{22} = \frac{1}{2}, \qquad \tilde{\sigma}_{33} = \frac{5}{2}$$

$$\tilde{\sigma}_{12} = \frac{(6-6)(0-1) + (7-6)(2-1) + (5-6)(1-1) + (6-6)(1-1)}{4}$$

$$= \frac{1}{4},$$

$$\tilde{\sigma}_{23} = \frac{3}{4}, \qquad \tilde{\sigma}_{13} = 1$$

The prediction step consists of using the initial estimates $\tilde{\boldsymbol{\mu}}$ and $\tilde{\boldsymbol{\Sigma}}$ to predict the contributions of the missing values to the sufficient statistics $\mathbf{T}_1$ and $\mathbf{T}_2$. [See (5-40) and (5-41).]

The first component of $\mathbf{x}_1$ is missing, so we partition $\tilde{\boldsymbol{\mu}}$ and $\tilde{\boldsymbol{\Sigma}}$ as

$$\tilde{\boldsymbol{\mu}} = \begin{bmatrix} \tilde{\mu}_1 \\ \hline \tilde{\mu}_2 \\ \tilde{\mu}_3 \end{bmatrix} = \begin{bmatrix} \tilde{\boldsymbol{\mu}}^{(1)} \\ \hline \tilde{\boldsymbol{\mu}}^{(2)} \end{bmatrix}, \qquad \tilde{\boldsymbol{\Sigma}} = \left[\begin{array}{c|cc} \tilde{\sigma}_{11} & \tilde{\sigma}_{12} & \tilde{\sigma}_{13} \\ \hline \tilde{\sigma}_{12} & \tilde{\sigma}_{22} & \tilde{\sigma}_{23} \\ \tilde{\sigma}_{13} & \tilde{\sigma}_{23} & \tilde{\sigma}_{33} \end{array}\right] = \left[\begin{array}{c|c} \tilde{\boldsymbol{\Sigma}}_{11} & \tilde{\boldsymbol{\Sigma}}_{12} \\ \hline \tilde{\boldsymbol{\Sigma}}_{21} & \tilde{\boldsymbol{\Sigma}}_{22} \end{array}\right]$$

and predict

$$\tilde{x}_{11} = \tilde{\mu}_1 + \tilde{\boldsymbol{\Sigma}}_{12}\tilde{\boldsymbol{\Sigma}}_{22}^{-1}\begin{bmatrix} x_{21} - \tilde{\mu}_2 \\ x_{31} - \tilde{\mu}_3 \end{bmatrix} = 6 + [\tfrac{1}{4}, 1]\begin{bmatrix} \frac{1}{2} & \frac{3}{4} \\ \frac{3}{4} & \frac{5}{2} \end{bmatrix}^{-1}\begin{bmatrix} 0 - 1 \\ 3 - 4 \end{bmatrix} = 5.73$$

$$\widetilde{x_{11}^2} = \tilde{\sigma}_{11} - \tilde{\boldsymbol{\Sigma}}_{12}\tilde{\boldsymbol{\Sigma}}_{22}^{-1}\tilde{\boldsymbol{\Sigma}}_{21} + \tilde{x}_{11}^2 = \tfrac{1}{2} - [\tfrac{1}{4}, 1]\begin{bmatrix} \frac{1}{2} & \frac{3}{4} \\ \frac{3}{4} & \frac{5}{2} \end{bmatrix}^{-1}\begin{bmatrix} \frac{1}{4} \\ 1 \end{bmatrix} + (5.73)^2 = 32.99$$

$$\widetilde{x_{11}[x_{21}, x_{31}]} = \tilde{x}_{11}[x_{21}, x_{31}] = 5.73[0, 3] = [0, 17.18]$$

For the two missing components of $\mathbf{x}_4$, we partition $\tilde{\boldsymbol{\mu}}$ and $\tilde{\boldsymbol{\Sigma}}$ as

$$\tilde{\boldsymbol{\mu}} = \begin{bmatrix} \tilde{\mu}_1 \\ \tilde{\mu}_2 \\ \hline \tilde{\mu}_3 \end{bmatrix} = \begin{bmatrix} \tilde{\boldsymbol{\mu}}^{(1)} \\ \hline \tilde{\mu}^{(2)} \end{bmatrix}, \qquad \tilde{\boldsymbol{\Sigma}} = \left[\begin{array}{cc|c} \sigma_{11} & \sigma_{12} & \sigma_{13} \\ \sigma_{12} & \sigma_{22} & \sigma_{23} \\ \hline \sigma_{13} & \sigma_{23} & \sigma_{33} \end{array}\right] = \left[\begin{array}{c|c} \tilde{\boldsymbol{\Sigma}}_{11} & \tilde{\boldsymbol{\Sigma}}_{12} \\ \hline \tilde{\boldsymbol{\Sigma}}_{21} & \tilde{\boldsymbol{\Sigma}}_{22} \end{array}\right]$$

and predict

$$\widetilde{\begin{bmatrix} x_{14} \\ x_{24} \end{bmatrix}} = E\left(\begin{bmatrix} X_{14} \\ X_{24} \end{bmatrix} \middle| x_{34} = 5; \tilde{\boldsymbol{\mu}}, \tilde{\boldsymbol{\Sigma}}\right) = \begin{bmatrix} \tilde{\mu}_1 \\ \tilde{\mu}_2 \end{bmatrix} + \tilde{\boldsymbol{\Sigma}}_{12}\tilde{\boldsymbol{\Sigma}}_{22}^{-1}(x_{34} - \tilde{\mu}_3)$$

$$= \begin{bmatrix} 6 \\ 1 \end{bmatrix} + \begin{bmatrix} 1 \\ \frac{3}{4} \end{bmatrix}\left(\tfrac{5}{2}\right)^{-1}(5 - 4) = \begin{bmatrix} 6.4 \\ 1.3 \end{bmatrix}$$

for the contribution to $\mathbf{T}_1$. Also, from (5-41),

$$\begin{bmatrix} \widetilde{x_{14}^2} & \widetilde{x_{14}x_{24}} \\ \widetilde{x_{14}x_{24}} & \widetilde{x_{24}^2} \end{bmatrix} = E\left(\begin{bmatrix} X_{14}^2 & X_{14}X_{24} \\ X_{14}X_{24} & X_{24}^2 \end{bmatrix} \middle| x_{34} = 5; \tilde{\boldsymbol{\mu}}, \tilde{\boldsymbol{\Sigma}}\right)$$

$$= \begin{bmatrix} \frac{1}{2} & \frac{1}{4} \\ \frac{1}{4} & \frac{1}{2} \end{bmatrix} - \begin{bmatrix} 1 \\ \frac{3}{4} \end{bmatrix}\left(\tfrac{5}{2}\right)^{-1}\begin{bmatrix} 1 & \frac{3}{4} \end{bmatrix} + \begin{bmatrix} 6.4 \\ 1.3 \end{bmatrix}\begin{bmatrix} 6.4 & 1.3 \end{bmatrix}$$

$$= \begin{bmatrix} 41.06 & 8.27 \\ 8.27 & 1.97 \end{bmatrix}$$

and

$$\widetilde{\begin{bmatrix} x_{14} \\ x_{24} \end{bmatrix}(x_{34})} = E\left(\begin{bmatrix} X_{14}X_{34} \\ X_{24}X_{34} \end{bmatrix} \middle| x_{34} = 5; \tilde{\boldsymbol{\mu}}, \tilde{\boldsymbol{\Sigma}}\right) = \widetilde{\begin{bmatrix} x_{14} \\ x_{24} \end{bmatrix}}(x_{34}) = \begin{bmatrix} 6.4 \\ 1.3 \end{bmatrix}(5) = \begin{bmatrix} 32.0 \\ 6.5 \end{bmatrix}$$

are the contributions to $\mathbf{T}_2$. Thus the predicted complete data sufficient statistics are

$$\tilde{\mathbf{T}}_1 = \begin{bmatrix} \tilde{x}_{11} + x_{12} + x_{13} + \tilde{x}_{14} \\ x_{21} + x_{22} + x_{23} + \tilde{x}_{24} \\ x_{31} + x_{32} + x_{33} + x_{34} \end{bmatrix} = \begin{bmatrix} 5.73 + 7 + 5 + 6.4 \\ 0 + 2 + 1 + 1.3 \\ 3 + 6 + 2 + 5 \end{bmatrix} = \begin{bmatrix} 24.13 \\ 4.30 \\ 16.00 \end{bmatrix}$$

$$\tilde{\mathbf{T}}_2 =$$

$$\begin{bmatrix} \widetilde{x_{11}^2} + x_{12}^2 + x_{13}^2 + \widetilde{x_{14}^2} & & \\ \widetilde{x_{11}x_{21}} + x_{12}x_{22} + x_{13}x_{23} + \widetilde{x_{14}x_{24}} & x_{21}^2 + x_{22}^2 + x_{23}^2 + \widetilde{x_{24}^2} & \\ \widetilde{x_{11}x_{31}} + x_{12}x_{32} + x_{13}x_{33} + \widetilde{x_{14}x_{34}} & x_{21}x_{31} + x_{22}x_{32} + x_{23}x_{33} + \widetilde{x_{24}x_{34}} & x_{31}^2 + x_{32}^2 + x_{33}^2 + x_{34}^2 \end{bmatrix}$$

$$= \begin{bmatrix} 32.99 + 7^2 + 5^2 + 41.06 & & \\ 0 + 7(2) + 5(1) + 8.27 & 0^2 + 2^2 + 1^2 + 1.97 & \\ 17.18 + 7(6) + 5(2) + 32 & 0(3) + 2(6) + 1(2) + 6.5 & 3^2 + 6^2 + 2^2 + 5^2 \end{bmatrix}$$

$$= \begin{bmatrix} 148.05 & 27.27 & 101.18 \\ 27.27 & 6.97 & 20.50 \\ 101.18 & 20.50 & 74.00 \end{bmatrix}$$

This completes one prediction step.

The next estimation step, using (5-42), provides the revised estimates

$$\tilde{\boldsymbol{\mu}} = \frac{1}{n}\tilde{\mathbf{T}}_1 = \tfrac{1}{4}\begin{bmatrix} 24.13 \\ 4.30 \\ 16.00 \end{bmatrix} = \begin{bmatrix} 6.03 \\ 1.08 \\ 4.00 \end{bmatrix}$$

$$\tilde{\boldsymbol{\Sigma}} = \frac{1}{n}\tilde{\mathbf{T}}_2 - \tilde{\boldsymbol{\mu}}\tilde{\boldsymbol{\mu}}'$$

$$= \tfrac{1}{4}\begin{bmatrix} 148.05 & 27.27 & 101.18 \\ 27.27 & 6.97 & 20.50 \\ 101.18 & 20.50 & 74.00 \end{bmatrix} - \begin{bmatrix} 6.03 \\ 1.08 \\ 4.00 \end{bmatrix}\begin{bmatrix} 6.03 & 1.08 & 4.00 \end{bmatrix}$$

$$= \begin{bmatrix} .65 & .31 & 1.18 \\ .31 & .58 & .81 \\ 1.18 & .81 & 2.50 \end{bmatrix}$$

Note that $\tilde{\sigma}_{11} = .65$ and $\tilde{\sigma}_{22} = .58$ are larger than the corresponding initial estimates obtained by replacing the missing observations on the first and second variables by the sample means of the remaining values. The third variance estimate $\tilde{\sigma}_{33}$ remains unchanged, because it is not affected by the missing components.

The iteration between the prediction and estimation steps continues until the elements of $\tilde{\boldsymbol{\mu}}$ and $\tilde{\boldsymbol{\Sigma}}$ remain essentially unchanged. Calculations of this sort are easily handled with a computer. ■

Once final estimates $\hat{\boldsymbol{\mu}}$ and $\hat{\boldsymbol{\Sigma}}$ are obtained and relatively few missing components occur in $\mathbf{X}$, it seems reasonable to treat

$$\text{all } \boldsymbol{\mu} \text{ such that } n(\hat{\boldsymbol{\mu}} - \boldsymbol{\mu})'\hat{\boldsymbol{\Sigma}}^{-1}(\hat{\boldsymbol{\mu}} - \boldsymbol{\mu}) \le \chi_p^2(\alpha) \qquad (5\text{-}43)$$

as an approximate $100(1 - \alpha)\%$ confidence ellipsoid. The simultaneous confidence statements would then follow as in Section 5.6, but with $\bar{\mathbf{x}}$ replaced by $\hat{\boldsymbol{\mu}}$ and $\mathbf{S}$ replaced by $\hat{\boldsymbol{\Sigma}}$.

Caution. The prediction-estimation algorithm that we discussed is developed on the basis that component observations are missing at random. If missing values are related to the response levels, then handling missing values as suggested above may introduce serious biases into the estimation procedures. Typically, missing values *are* related to the responses being measured. Consequently, we must be dubious of any computational scheme that fills in values as if they were lost at random. When more than a few values are missing, it is imperative that the investigator search for the systematic causes that created them.

We have concentrated, in this section, on missing components of multivariate normal data vectors. Users of sample survey techniques know that the presence or absence of a sensitive attribute cannot always be recorded. Thus the individual cannot be assigned to a particular multinomial category. A formal procedure for estimating the population proportions associated with each multinomial category when some of the observations are incomplete is available in [5] and [4].

Supplement 5A: The Sample Geometry of T^2

Two geometrical interpretations, based upon the representation of the data as p vectors in n dimensions, will be given for Hotelling's T^2-statistic. We begin by considering the single variable situation for which the data matrix $\mathbf{X}$ has one row.

Geometry for the student's *t*-statistic. The student's t-statistic, or its square,

$$t^2 = \frac{n(\bar{x} - \mu_0)^2}{\sum_{j=1}^{n} (x_j - \bar{x})^2/(n-1)}$$

is used for testing H_0: $\mu = \mu_0$ against H_1: $\mu \neq \mu_0$ when the observations come from a $N(\mu, \sigma^2)$ population. Large values of t^2 call for the rejection of H_0.

It is convenient, for our geometrical representation, to consider the special case[3] H_0: $\mu = 0$ versus H_1: $\mu \neq 0$. The test statistic is then

$$t^2 = \frac{n\bar{x}^2}{\sum_{j=1}^{n} (x_j - \bar{x})^2/(n-1)} = \frac{(\bar{x}\mathbf{1})'(\bar{x}\mathbf{1})}{(\mathbf{y} - \bar{x}\mathbf{1})'(\mathbf{y} - \bar{x}\mathbf{1})/(n-1)} = \frac{(n-1)L^2_{\bar{x}\mathbf{1}}}{L^2_{\mathbf{y}-\bar{x}\mathbf{1}}} \tag{5A-1}$$

where

$$\mathbf{y}' = [x_1, x_2, \ldots, x_n]$$

$$L^2_{\bar{x}\mathbf{1}} = \left(\begin{array}{l}\text{squared length of projection of } \mathbf{y} \text{ on the} \\ \text{equiangular vector } \mathbf{1}' = [1, 1, \ldots, 1]\end{array}\right)$$

$$L^2_{\mathbf{y}-\bar{x}\mathbf{1}} = \left(\begin{array}{l}\text{squared length of vector between } \mathbf{y} \text{ and} \\ \text{its projection } \bar{x}\mathbf{1}\end{array}\right)$$

If θ is the angle between the data vector $\mathbf{y}$ and the equiangular vector $\mathbf{1}$, then

$$\cot^2(\theta) = \frac{L^2_{\bar{x}\mathbf{1}}}{L^2_{\mathbf{y}-\bar{x}\mathbf{1}}} = \frac{t^2}{(n-1)} \tag{5A-2}$$

The situation is pictured in Figure 5.4. Large values of t^2 correspond to small values of the angle θ.

The geometrical arguments above allow us to digress briefly and explain the $n - 1$ degrees of freedom associated with the sample variance

$$s^2 = \sum_{j=1}^{n} (x_j - \bar{x})^2/(n-1)$$

[3] The general case, H_0: $\mu = \mu_0$ versus H_1: $\mu \neq \mu_0$, can be handled by replacing the observations x_j by $x_j - \mu_0$ in the subsequent development.

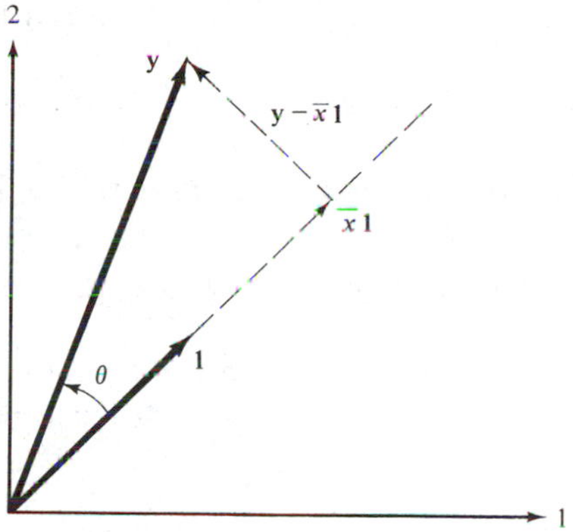

Figure 5.4 The geometry of student's t-statistic: $t^2/(n-1) = \cot^2(\theta)$.

The vector of deviations from the sample mean

$$(\mathbf{y} - \bar{x}\mathbf{1})' = [x_1 - \bar{x}, x_2 - \bar{x}, \ldots, x_n - \bar{x}]$$

satisfies $(\mathbf{y} - \bar{x}\mathbf{1})'\mathbf{1} = \sum_{j=1}^{n} (x_j - \bar{x}) = 0$. Consequently $(\mathbf{y} - \bar{x}\mathbf{1})$ is perpendicular to the direction of $\mathbf{1}$ for every possible set of observations $x_1, x_2, \ldots, x_n$. The deviation vector has the "freedom" only to lie in the $(n-1)$-dimensional portion of the space that is perpendicular to the equiangular line. Borrowing terminology from physics, the vector $\mathbf{y} - \bar{x}\mathbf{1}$ (or its squared length $\sum_{j=1}^{n} (x_j - \bar{x})^2$), is said to have $n-1$ degrees of freedom.

Geometry for Hotelling's T^2. For a p-dimensional normal population, a test of H_0: $\boldsymbol{\mu} = \boldsymbol{\mu}_0$ versus H_1: $\boldsymbol{\mu} \neq \boldsymbol{\mu}_0$ is based on the statistic

$$T^2 = n(\bar{\mathbf{x}} - \boldsymbol{\mu}_0)'\mathbf{S}^{-1}(\bar{\mathbf{x}} - \boldsymbol{\mu}_0)$$

Large values of T^2 are evidence against H_0.

Again, for convenience, we consider the special case H_0: $\boldsymbol{\mu} = \mathbf{0}$ against H_1: $\boldsymbol{\mu} \neq 0$. For this situation, the test statistic is $T^2 = n\bar{\mathbf{x}}'\mathbf{S}^{-1}\bar{\mathbf{x}}$. Let $\mathbf{y}_1' = [x_{11}, x_{12}, \ldots, x_{1n}]$, $\mathbf{y}_2' = [x_{21}, x_{22}, \ldots, x_{2n}], \ldots, \mathbf{y}_p' = [x_{p1}, x_{p2}, \ldots, x_{pn}]$. The linear combinations

$$\ell_1\mathbf{y}_1 + \ell_2\mathbf{y}_2 + \cdots + \ell_p\mathbf{y}_p = [\boldsymbol{\ell}'\mathbf{x}_1, \boldsymbol{\ell}'\mathbf{x}_2, \ldots, \boldsymbol{\ell}'\mathbf{x}_n]' = \mathbf{z}$$

lie in the p-dimensional hyperplane formed by the vectors $\mathbf{y}_1, \mathbf{y}_2, \ldots, \mathbf{y}_p$. For fixed $\boldsymbol{\ell}$, the univariate observations $z_1 = \boldsymbol{\ell}'\mathbf{x}_1, z_2 = \boldsymbol{\ell}'\mathbf{x}_2, \ldots, z_n = \boldsymbol{\ell}'\mathbf{x}_n$ have sample mean $\boldsymbol{\ell}'\bar{\mathbf{x}}$ and sample covariance $\boldsymbol{\ell}'\mathbf{S}\boldsymbol{\ell}$. Thus under the null hypothesis $\mu_z = 0$ and, using (5A-1) and (5A-2),

$$t^2 = \frac{n(\boldsymbol{\ell}'\bar{\mathbf{x}})^2}{\boldsymbol{\ell}'\mathbf{S}\boldsymbol{\ell}} = \frac{n\bar{z}^2}{\sum_{j=1}^{n} (z_j - \bar{z})^2/(n-1)} = \frac{(\bar{z}\mathbf{1})'(\bar{z}\mathbf{1})}{(\mathbf{z} - \bar{z}\mathbf{1})'(\mathbf{z} - \bar{z}\mathbf{1})/(n-1)}$$

$$= (n-1)\cot^2(\theta)$$

where θ is the angle between $\mathbf{z}' = [z_1,\ldots,z_n] = [\ell'\mathbf{x}_1, \ell'\mathbf{x}_2,\ldots,\ell'\mathbf{x}_n]$ and the equiangular vector $\mathbf{1}$.

Rejecting H_0: $\boldsymbol{\mu} = \mathbf{0}$ on the basis of T^2 corresponds to varying ℓ, or the vector in the hyperplane determined by $\mathbf{y}_1, \mathbf{y}_2,\ldots,\mathbf{y}_p$, to maximize t^2. [See (5-23).] Maximizing t^2, or $\cot^2(\theta)$, is equivalent to choosing ℓ to minimize the angle θ. To summarize,

$$\frac{T^2}{(n-1)} = \cot^2(\theta) \tag{5A-3}$$

where θ is the (smallest) angle between the plane formed from the vectors $\mathbf{y}_1, \mathbf{y}_2,\ldots,\mathbf{y}_p$ and the equiangular line in the direction $\mathbf{1}$. We have illustrated the geometry in Figure 5.5 for $p = 2$.

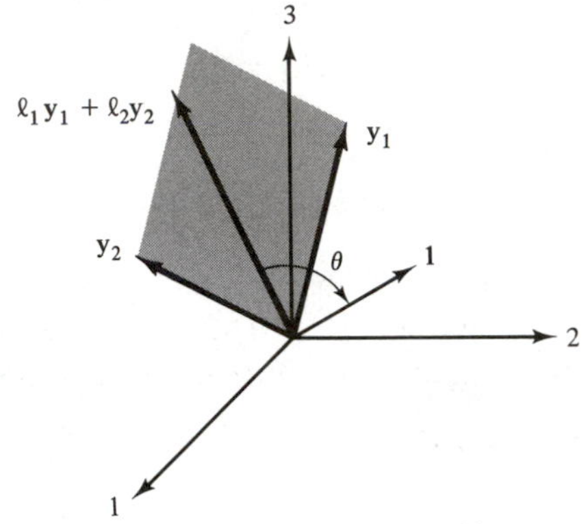

Figure 5.5 The geometry of T^2 for $p = 2$: $T^2/(n-1) = \cot^2(\theta)$.

A second geometrical interpretation of T^2 is based on the relation in (5-15). Using this relation, a representation of T^2 is given by

$$\frac{T^2}{(n-1)} = \frac{|\hat{\mathbf{\Sigma}}_0|}{|\hat{\mathbf{\Sigma}}|} - 1 = \frac{\text{squared volume generated by the } \mathbf{y}_i}{\text{squared volume generated by the } \mathbf{y}_i - \bar{x}_i\mathbf{1}} - 1$$

If H_0 is true, the volume in the numerator will tend to be approximately equal to the volume in the denominator, and T^2 will be small. On the other hand, if H_0 is false, the volume in the denominator will be smaller than the volume in the numerator and T^2 will be large. (See Figure 3.6 for large and small volumes generated by vectors of the form $\mathbf{y}_i - \bar{x}_i\mathbf{1}$.)

Supplement 5B: The Power of the T^2 Test and the Importance of Additional Variables

The power of the T^2-test for H_0: $\boldsymbol{\mu} = \boldsymbol{\mu}_0$ against H_1: $\boldsymbol{\mu} \neq \boldsymbol{\mu}_0$ is given by

$$\text{Power} = P\left[n(\overline{\mathbf{X}} - \boldsymbol{\mu}_0)'\mathbf{S}^{-1}(\overline{\mathbf{X}} - \boldsymbol{\mu}_0) > \frac{p(n-1)}{(n-p)} F_{p,n-p}(\alpha) \mid \text{true mean is } \boldsymbol{\mu}\right]$$

$$= P\left[\text{noncentral } F_{p,n-p} > F_{p,n-p}(\alpha) \mid \Delta^2 = (\boldsymbol{\mu} - \boldsymbol{\mu}_0)'\boldsymbol{\Sigma}^{-1}(\boldsymbol{\mu} - \boldsymbol{\mu}_0)\right] \tag{5B-1}$$

where $\delta^2 = n\Delta^2 = n(\boldsymbol{\mu} - \boldsymbol{\mu}_0)'\boldsymbol{\Sigma}^{-1}(\boldsymbol{\mu} - \boldsymbol{\mu}_0)$ is the *noncentrality* parameter and $F_{p,n-p}(\alpha)$ is the upper (100α)th percentile of the usual F-distribution with p and $n - p$ d.f. For a fixed sample size, the power depends only on the squared distance, $(\boldsymbol{\mu} - \boldsymbol{\mu}_0)'\boldsymbol{\Sigma}^{-1}(\boldsymbol{\mu} - \boldsymbol{\mu}_0) = \Delta^2$, from $\boldsymbol{\mu}$ to its hypothesized value $\boldsymbol{\mu}_0$. Moreover, from the discussion following (5-8), δ^2 (or Δ^2) is unchanged if $\mathbf{X}$ is transformed to $\mathbf{CX} + \mathbf{d}$. That is, power also does not depend on the choice of measurement units.

When the number of variables, p, is fixed, the power in (5B-1) will always increase with an increase in δ^2. If it is possible to measure r additional variables, however, the question arises as to whether these "extra" variables will provide important information. One way of judging the contribution of additional variables is to examine the effect of these variables on the power of the T^2-test. Surprisingly, the information contributed by additional variables will not necessarily enhance the ability of the T^2-statistic to detect true alternative hypotheses. A *decrease* in power occurs in those cases where a small increase in the noncentrality parameter δ^2 is more than offset by a loss in the degrees of freedom of the denominator $n - p$ to $n - p - r$ and a gain in the degrees of freedom of the numerator from p to $p + r$.

We illustrate these remarks in the next example.

Example 5B.1

Consider testing H_0: $\boldsymbol{\mu} = \mathbf{0}$ when, with $|\rho| < 1$,

$$\boldsymbol{\Sigma} = \left[\begin{array}{cc|cc} 1 & .2 & \rho & 0 \\ .2 & 1 & 0 & \rho \\ \hline \rho & 0 & 1 & .1 \\ 0 & \rho & .1 & 1 \end{array}\right] = \left[\begin{array}{c|c} \boldsymbol{\Sigma}_{11} & \boldsymbol{\Sigma}_{12} \\ \hline \boldsymbol{\Sigma}_{21} & \boldsymbol{\Sigma}_{22} \end{array}\right]$$

and a sample of size $n = 26$ is available from a multivariate normal population.

Power depends on the particular value of $\boldsymbol{\mu}$ under the alternative hypothesis. We consider the choice, $\boldsymbol{\mu}' = [\mu, \mu, \mu, \mu]$. The noncentrality parameter is

$$\delta_4^2 = n\boldsymbol{\mu}'\boldsymbol{\Sigma}^{-1}\boldsymbol{\mu}$$

When only the first two variables are used, their means $\boldsymbol{\mu}_1' = [\mu, \mu]$, under the alternative, produce the noncentrality parameter

$$\delta_2^2 = n\boldsymbol{\mu}_1'\boldsymbol{\Sigma}_{11}^{-1}\boldsymbol{\mu}_1$$

Although it can be shown that $\delta_4^2 - \delta_2^2 \geq 0$ for all alternatives, Table 5B.1 indicates that the power of the T^2-test with four variables can be *smaller* than the corresponding power with two variables. We note that this phenomenon seems to occur for our alternative when the correlation ρ between the two original variables and the additional variables is high. For negative ρ, in the $\boldsymbol{\mu}' = [\mu, \mu, \mu, \mu]$ situation, the power of the test with all four variables is considerably higher than that with two variables. ■

TABLE 5B.1 POWER OF THE T^2-TEST (SIGNIFICANCE LEVEL = α = .05)

Alternative μ	Two variables	Four variables $\rho = .7$	$\rho = .8$	$\rho = .9$
.2	.181	.160	.154	.148
.3	.364	.330	.314	.299
.4	.592	.561	.536	.514
.5	.796	.779	.754	.731

SOURCE: Computer calculations courtesy of J. Klotz.

See [3] and [7] for more technical discussions of the power of the T^2-test.

Supplement 5C: Simultaneous Confidence Intervals and Ellipses as Shadows of the p-Dimensional Ellipsoids

We begin by establishing the general result concerning the projection (shadow) of an ellipsoid onto a line.

Result 5C.1. Let the constant $c^2 > 0$ and positive definite $p \times p$ matrix $\mathbf{A}$ determine the ellipsoid $\{\mathbf{z} : \mathbf{z}'\mathbf{A}^{-1}\mathbf{z} \le c^2\}$. For a given vector $\boldsymbol{\ell} \neq \mathbf{0}$, and $\mathbf{z}$ belonging to the ellipsoid, the

$$\begin{pmatrix} \text{Projection (shadow) of} \\ \{\mathbf{z}'\mathbf{A}^{-1}\mathbf{z} \le c^2\} \text{ on } \boldsymbol{\ell} \end{pmatrix} = c\frac{\sqrt{\boldsymbol{\ell}'\mathbf{A}\boldsymbol{\ell}}}{\boldsymbol{\ell}'\boldsymbol{\ell}}\boldsymbol{\ell}$$

which extends from $\mathbf{0}$ along $\boldsymbol{\ell}$ with length $c\sqrt{\boldsymbol{\ell}'\mathbf{A}\boldsymbol{\ell}/\boldsymbol{\ell}'\boldsymbol{\ell}}$. When $\boldsymbol{\ell}$ is a unit vector, the shadow extends $c\sqrt{\boldsymbol{\ell}'\mathbf{A}\boldsymbol{\ell}}$ units, so $|\mathbf{z}'\boldsymbol{\ell}| \le c\sqrt{\boldsymbol{\ell}'\mathbf{A}\boldsymbol{\ell}}$.

Proof. By Definition 2A.12, the projection of any $\mathbf{z}$ on $\boldsymbol{\ell}$ is given by $(\mathbf{z}'\boldsymbol{\ell})\boldsymbol{\ell}/\boldsymbol{\ell}'\boldsymbol{\ell}$. Its squared length is $(\mathbf{z}'\boldsymbol{\ell})^2/\boldsymbol{\ell}'\boldsymbol{\ell}$. We want to maximize this shadow over all $\mathbf{z}$ with $\mathbf{z}'\mathbf{A}^{-1}\mathbf{z} \le c^2$. The extended Cauchy–Schwarz inequality in (2-49) states that $(\mathbf{b}'\mathbf{d})^2 \le (\mathbf{b}'\mathbf{B}\mathbf{b})(\mathbf{d}'\mathbf{B}^{-1}\mathbf{d})$, with equality when $\mathbf{b} = k\mathbf{B}^{-1}\mathbf{d}$. Setting $\mathbf{b} = \mathbf{z}$, $\mathbf{d} = \boldsymbol{\ell}$, and $\mathbf{B} = \mathbf{A}^{-1}$,

$$(\boldsymbol{\ell}'\boldsymbol{\ell})(\text{length of projection})^2 = (\mathbf{z}'\boldsymbol{\ell})^2 \le (\mathbf{z}'\mathbf{A}^{-1}\mathbf{z})(\boldsymbol{\ell}'\mathbf{A}\boldsymbol{\ell})$$
$$\le c^2\boldsymbol{\ell}'\mathbf{A}\boldsymbol{\ell}, \qquad \text{all } \mathbf{z}\text{: } \mathbf{z}'\mathbf{A}^{-1}\mathbf{z} \le c^2$$

The choice $\mathbf{z} = c\mathbf{A}\boldsymbol{\ell}/\sqrt{\boldsymbol{\ell}'\mathbf{A}\boldsymbol{\ell}}$ yields equalities and thus gives the maximum shadow, besides belonging to the boundary of the ellipsoid. That is, $\mathbf{z}'\mathbf{A}^{-1}\mathbf{z} = c^2\boldsymbol{\ell}'\mathbf{A}\boldsymbol{\ell}/\boldsymbol{\ell}'\mathbf{A}\boldsymbol{\ell} = c^2$ for this $\mathbf{z}$ that provides the longest shadow. Consequently, the projection of the ellipsoid on $\boldsymbol{\ell}$ is $c\sqrt{\boldsymbol{\ell}'\mathbf{A}\boldsymbol{\ell}}\,\boldsymbol{\ell}/\boldsymbol{\ell}'\boldsymbol{\ell}$ and its length is $c\sqrt{\boldsymbol{\ell}'\mathbf{A}\boldsymbol{\ell}/\boldsymbol{\ell}'\boldsymbol{\ell}}$. With the unit vector $\mathbf{e}_{\boldsymbol{\ell}} = \boldsymbol{\ell}/\sqrt{\boldsymbol{\ell}'\boldsymbol{\ell}}$, the projection extends

$$\sqrt{c^2\mathbf{e}_{\boldsymbol{\ell}}'\mathbf{A}\mathbf{e}_{\boldsymbol{\ell}}} = \frac{c}{\sqrt{\boldsymbol{\ell}'\boldsymbol{\ell}}}\sqrt{\boldsymbol{\ell}'\mathbf{A}\boldsymbol{\ell}} \quad \text{units along } \boldsymbol{\ell} \qquad \blacksquare$$

Result 5C.2. Suppose the ellipsoid $\{\mathbf{z} : \mathbf{z}'\mathbf{A}^{-1}\mathbf{z} \le c^2\}$ is given and that $\mathbf{L} = [\boldsymbol{\ell}_1 \mid \boldsymbol{\ell}_2]$ is arbitrary but of rank two.

$$\begin{Bmatrix} \mathbf{z} \text{ in the ellipsoid} \\ \text{based on } \mathbf{A}^{-1} \text{ and } c^2 \end{Bmatrix} \quad \text{implies that} \quad \begin{Bmatrix} \text{for all } \mathbf{L}, \mathbf{L}'\mathbf{z} \text{ is in the ellipsoid} \\ \text{based on } (\mathbf{L}'\mathbf{A}\mathbf{L})^{-1} \text{ and } c^2 \end{Bmatrix}$$

or

$$\mathbf{z}'\mathbf{A}^{-1}\mathbf{z} \le c^2 \quad \text{implies that} \quad (\mathbf{L}'\mathbf{z})'(\mathbf{L}'\mathbf{A}\mathbf{L})^{-1}(\mathbf{L}'\mathbf{z}) \le c^2 \qquad \text{for all } \mathbf{L}$$

Proof. We first establish a basic inequality. Set $\mathbf{B} = \mathbf{A}^{1/2}\mathbf{L}(\mathbf{L}'\mathbf{A}\mathbf{L})^{-1}\mathbf{L}'\mathbf{A}^{1/2}$, where $\mathbf{A} = \mathbf{A}^{1/2}\mathbf{A}^{1/2}$. Note that $\mathbf{B} = \mathbf{B}'$ and $\mathbf{B}^2 = \mathbf{B}$, so $(\mathbf{I} - \mathbf{B})\mathbf{B}' = \mathbf{B} - \mathbf{B}^2 = \mathbf{0}$.

Next, using $\mathbf{A}^{-1} = \mathbf{A}^{-1/2}\mathbf{A}^{-1/2}$, write $\mathbf{z}'\mathbf{A}^{-1}\mathbf{z} = (\mathbf{A}^{-1/2}\mathbf{z})'(\mathbf{A}^{-1/2}\mathbf{z})$ and $\mathbf{A}^{-1/2}\mathbf{z} = \mathbf{B}\mathbf{A}^{-1/2}\mathbf{z} + (\mathbf{I}-\mathbf{B})\mathbf{A}^{-1/2}\mathbf{z}$. Then

$$\begin{aligned}
\mathbf{z}'\mathbf{A}^{-1}\mathbf{z} &= (\mathbf{A}^{-1/2}\mathbf{z})'(\mathbf{A}^{-1/2}\mathbf{z}) \\
&= \big(\mathbf{B}\mathbf{A}^{-1/2}\mathbf{z} + (\mathbf{I}-\mathbf{B})\mathbf{A}^{-1/2}\mathbf{z}\big)'\big(\mathbf{B}\mathbf{A}^{-1/2}\mathbf{z} + (\mathbf{I}-\mathbf{B})\mathbf{A}^{-1/2}\mathbf{z}\big) \\
&= (\mathbf{B}\mathbf{A}^{-1/2}\mathbf{z})'(\mathbf{B}\mathbf{A}^{-1/2}\mathbf{z}) + \big((\mathbf{I}-\mathbf{B})\mathbf{A}^{-1/2}\mathbf{z}\big)'\big((\mathbf{I}-\mathbf{B})\mathbf{A}^{-1/2}\mathbf{z}\big) \\
&\geq \mathbf{z}'\mathbf{A}^{-1/2}\mathbf{B}'\mathbf{B}\mathbf{A}^{-1/2}\mathbf{z} = \mathbf{z}'\mathbf{A}^{-1/2}\mathbf{B}\mathbf{A}^{-1/2}\mathbf{z} = \mathbf{z}'\mathbf{L}(\mathbf{L}'\mathbf{A}\mathbf{L})^{-1}\mathbf{L}'\mathbf{z}.
\end{aligned} \tag{5C-1}$$

Since $\mathbf{z}'\mathbf{A}^{-1}\mathbf{z} \leq c^2$ and $\mathbf{L}$ was arbitrary, the result follows. ■

Our next result establishes the two-dimensional confidence ellipse as a projection of the p-dimensional ellipsoid (see Figure 5.6).

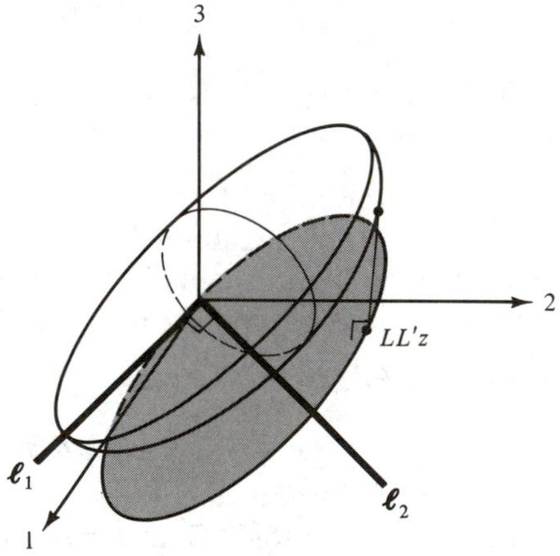

Figure 5.6 The shadow of the ellipsoid $\mathbf{z}'\mathbf{A}^{-1}\mathbf{z} \leq c^2$ on the ℓ_1, ℓ_2 plane is an ellipse.

Projection on a plane is simplest when the two vectors ℓ_1 and ℓ_2 determining the plane are first converted to perpendicular vectors of unit length. (See Result 2A.3.)

Result 5C.3. Given the ellipsoid $\{\mathbf{z} : \mathbf{z}'\mathbf{A}^{-1}\mathbf{z} \leq c^2\}$ and two perpendicular unit vectors ℓ_1 and ℓ_2, the projection (or shadow) of $\{\mathbf{z}'\mathbf{A}^{-1}\mathbf{z} \leq c^2\}$ on the ℓ_1, ℓ_2 plane results in the two-dimensional ellipse $\{(\mathbf{L}'\mathbf{z})'(\mathbf{L}'\mathbf{A}\mathbf{L})^{-1}(\mathbf{L}'\mathbf{z}) \leq c^2\}$, where $\mathbf{L} = [\ell_1 \mid \ell_2]$.

Proof. By Result 2A.3, the projection of a vector $\mathbf{z}$ on the ℓ_1, ℓ_2 plane is

$$(\ell_1'\mathbf{z})\ell_1 + (\ell_2'\mathbf{z})\ell_2 = [\ell_1 \mid \ell_2]\begin{bmatrix}\ell_1'\mathbf{z} \\ \ell_2'\mathbf{z}\end{bmatrix} = \mathbf{L}\mathbf{L}'\mathbf{z}$$

The projection of the ellipsoid $\{\mathbf{z} : \mathbf{z}'\mathbf{A}^{-1}\mathbf{z} \leq c^2\}$ consists of all $\mathbf{L}\mathbf{L}'\mathbf{z}$ with $\mathbf{z}'\mathbf{A}^{-1}\mathbf{z} \leq c^2$. Consider the two coordinates $\mathbf{L}'\mathbf{z}$ of the projection $\mathbf{L}(\mathbf{L}'\mathbf{z})$. Let $\mathbf{z}$ belong to the set

$\{\mathbf{z} : \mathbf{z}'\mathbf{A}^{-1}\mathbf{z} \le c^2\}$ so $\mathbf{L}\mathbf{L}'\mathbf{z}$ belongs to the shadow of the ellipsoid. By Result 5C.2,

$$(\mathbf{L}'\mathbf{z})'(\mathbf{L}'\mathbf{A}\mathbf{L})^{-1}(\mathbf{L}'\mathbf{z}) \le c^2$$

so the ellipse $\{(\mathbf{L}'\mathbf{z})'(\mathbf{L}'\mathbf{A}\mathbf{L})^{-1}(\mathbf{L}'\mathbf{z}) \le c^2\}$ contains the coefficient vectors for the shadow of the ellipsoid.

Let $\mathbf{L}\mathbf{a}$ be a vector in the $\boldsymbol{\ell}_1$, $\boldsymbol{\ell}_2$ plane whose coefficients, $\mathbf{a}$, belong to the ellipse $\{\mathbf{a}'(\mathbf{L}'\mathbf{A}\mathbf{L})^{-1}\mathbf{a} \le c^2\}$. Setting $\mathbf{z} = \mathbf{A}\mathbf{L}(\mathbf{L}'\mathbf{A}\mathbf{L})^{-1}\mathbf{a}$, it follows that

$$\mathbf{L}'\mathbf{z} = \mathbf{L}'\mathbf{A}\mathbf{L}(\mathbf{L}'\mathbf{A}\mathbf{L})^{-1}\mathbf{a} = \mathbf{a}$$

and

$$\mathbf{z}'\mathbf{A}^{-1}\mathbf{z} = \mathbf{a}'(\mathbf{L}'\mathbf{A}\mathbf{L})^{-1}\mathbf{L}'\mathbf{A}\mathbf{A}^{-1}\mathbf{A}\mathbf{L}(\mathbf{L}'\mathbf{A}\mathbf{L})^{-1}\mathbf{a} = \mathbf{a}'(\mathbf{L}'\mathbf{A}\mathbf{L})^{-1}\mathbf{a} \le c^2$$

Thus $\mathbf{L}'\mathbf{z}$ belongs to the coefficient vector ellipse and $\mathbf{z}$ belongs to the ellipsoid $\mathbf{z}'\mathbf{A}^{-1}\mathbf{z} \le c^2$. Consequently, the ellipse contains only coefficient vectors from the projection of $\{\mathbf{z} : \mathbf{z}'\mathbf{A}^{-1}\mathbf{z} \le c^2\}$ onto the $\boldsymbol{\ell}_1$, $\boldsymbol{\ell}_2$ plane. ■

Remark. Projecting the ellipsoid $\mathbf{z}'\mathbf{A}^{-1}\mathbf{z} \le c^2$ first to the $\boldsymbol{\ell}_1$, $\boldsymbol{\ell}_2$ plane and then to the line $\boldsymbol{\ell}_1$ is the same as projecting it directly to the line determined by $\boldsymbol{\ell}_1$. In the context of confidence ellipsoids, the shadows of the two-dimensional ellipses give the single component intervals.

Remark. Results 5C.2 and 5C.3 remain valid if $\mathbf{L} = [\boldsymbol{\ell}_1, \ldots, \boldsymbol{\ell}_q]$ consists of $2 < q \le p$ linearly independent columns.

EXERCISES

5.1. **(a)** Evaluate T^2, for testing H_0: $\boldsymbol{\mu}' = [7, 11]$, using the data

$$\mathbf{X} = \begin{bmatrix} 2 & 8 & 6 & 8 \\ 12 & 9 & 9 & 10 \end{bmatrix}$$

(b) Specify the distribution of T^2 for the situation in (a).

(c) Using (a) and (b), test H_0 at the $\alpha = .05$ level. What conclusion is reached?

5.2. Using the data in Example 5.1, verify that T^2 remains unchanged if each observation $\mathbf{x}_j$, $j = 1, 2, 3$, is replaced by $\mathbf{C}\mathbf{x}_j$ where

$$\mathbf{C} = \begin{bmatrix} 1 & -1 \\ 1 & 1 \end{bmatrix}$$

Note that the observations

$$\mathbf{C}\mathbf{x}_j = \begin{bmatrix} x_{1j} - x_{2j} \\ x_{1j} + x_{2j} \end{bmatrix}$$

yield the data matrix

$$\begin{bmatrix} (6-9) & (10-6) & (8-3) \\ (6+9) & (10+6) & (8+3) \end{bmatrix}$$

5.3. **(a)** Use expression (5-15) to evaluate T^2 for the data in Exercise 5.1.

(b) Use the data in Exercise 5.1 to evaluate Λ in (5-13). Also evaluate Wilk's lambda.

5.4. Use the sweat data in Table 5.1 (see Example 5.2).

(a) Determine the axes of the 90% confidence ellipsoid for $\boldsymbol{\mu}$. Determine the lengths of these axes.

(b) Construct *Q-Q* plots for the observations on sweat rate, sodium content, and potassium content, respectively. Construct the three possible scatter plots for pairs of observations. Does the multivariate normal assumption seem justified in this case? Comment.

5.5. The quantities $\bar{\mathbf{x}}$, $\mathbf{S}$, and $\mathbf{S}^{-1}$ are given in Example 5.3 for the transformed microwave-radiation data. Conduct a test of the null hypothesis H_0: $\boldsymbol{\mu}' = [.55, .60]$ at the $\alpha = .05$ level of significance. Is your result consistent with the 95% confidence ellipse for $\boldsymbol{\mu}$ pictured in Figure 5.1? Explain.

5.6. Verify the Bonferroni inequality in (5-27) for $m = 3$. (*Hint:* A Venn diagram for the three events C_1, C_2, and C_3 may help.)

5.7. Use the sweat data in Table 5.1 (see Example 5.2).

(a) Find simultaneous 95% confidence intervals for μ_1, μ_2, and μ_3 using Result 5.3. Compare the outcomes with the Bonferroni intervals in Example 5.5. Comment.

(b) Determine and sketch the three simultaneous confidence ellipses for the pairs $[\mu_i, \mu_k]'$ using (5-33).

5.8. From (5-23) we know that T^2 is equal to the largest squared univariate t-value constructed from the linear combination $\boldsymbol{\ell}'\mathbf{x}_j$ with $\boldsymbol{\ell} = \mathbf{S}^{-1}(\bar{\mathbf{x}} - \boldsymbol{\mu}_0)$. Using the results in Example 5.3 and the H_0 in Exercise 5.5, evaluate $\boldsymbol{\ell}$ for the transformed microwave-radiation data. Verify that the t^2-value computed with this $\boldsymbol{\ell}$ is equal to T^2 in Exercise 5.5.

5.9. A physical anthropologist performed a mineral analysis of 9 ancient Peruvian hairs. The results for the chromium (x_1) and strontium (x_2) levels, in parts per million (p.p.m.), were as follows.

x_1 (Cr)	.48	40.53	2.19	.55	.74	.66	.93	.37	.22
x_2 (Sr)	12.57	73.68	11.13	20.03	20.29	.78	4.64	.43	1.08

SOURCE: Benfer and others, "Mineral Analysis of Ancient Peruvian Hair," *American Journal of Physical Anthropology*, **48**, no. 3 (1978), 277–282.

It is known that low levels (less than or equal to .100 p.p.m.) of chromium suggest the presence of diabetes, while strontium is an indication of animal protein intake.

(a) Construct and plot a 90% joint confidence ellipse for the population mean vector $\boldsymbol{\mu}' = [\mu_1, \mu_2]$ assuming these nine Peruvian hairs represent a random sample from individuals belonging to a particular ancient Peruvian culture.

(b) Obtain the individual simultaneous 90% confidence intervals for μ_1 and μ_2 by "projecting" the ellipse constructed in Part a on each coordinate axis. (Alternatively, we could use Result 5.3.) Does it appear as if this Peruvian culture has a mean strontium level of 10? That is, are any of the points (μ_1 arbitrary, 10) in the confidence regions? Is $[.30, 10]'$ a plausible value for $\boldsymbol{\mu}$? Discuss.

(c) Do these data appear to be bivariate normal? Discuss with reference to *Q-Q* plots and a scatter diagram. If the data are *not* bivariate normal, what implications does this have for the results in Parts a and b?

(d) Repeat the analysis with the obvious "outlying" observation removed. Do the inferences change? Comment.

5.10. Given the data with missing components,

$$\mathbf{X} = \begin{bmatrix} 3 & 4 & - & 5 \\ 6 & 4 & 8 & - \\ 0 & 3 & 3 & - \end{bmatrix}$$

use the prediction-estimation algorithm of Section 5.8 to estimate $\boldsymbol{\mu}$ and $\boldsymbol{\Sigma}$. Determine the initial estimates and iterate to find the *first* revised estimates.

5.11. In order to assess the prevalence of a drug problem among high school students in a particular city, a random sample of 200 students from the city's five high schools were surveyed. One of the survey questions and the corresponding responses are given below.

What is your typical weekly marijuana usage?

	Category		
	None	Moderate (1–3 joints)	Heavy (4 or more joints)
Number of responses	117	62	21

Construct 95% simultaneous confidence intervals for the three proportions p_1, p_2 and $p_3 = 1 - (p_1 + p_2)$.

5.12. Determine the approximate distribution of $-n\ln(|\hat{\boldsymbol{\Sigma}}|/|\hat{\boldsymbol{\Sigma}}_0|)$ for the sweat data in Table 5.1 (see Result 5.2).

The following exercises may require a computer.

5.13. Use the college test data in Table 5.2 (see Example 5.4).

(a) Test the null hypothesis H_0: $\boldsymbol{\mu}' = [500, 50, 30]$ versus H_1: $\boldsymbol{\mu}' \neq [500, 50, 30]$ at the $\alpha = .05$ level of significance. Suppose $[500, 50, 30]'$ represent average scores for thousands of college students over the last 10 years. Is there reason to believe the current group of students taking these tests is scoring differently? Explain.

(b) Determine the lengths and directions for the axes of the 95% confidence ellipsoid for $\boldsymbol{\mu}$.

(c) Construct *Q-Q* plots from the marginal distributions of social science and history, verbal, and science scores. Also construct the three possible scatter diagrams from the pairs of observations on different variables. Do these data appear to be normally distributed? Discuss.

5.14. Measurements on X_1 = stiffness and X_2 = bending strength for a sample of $n = 30$ pieces of a particular grade of lumber are given in Table 5.6. The units are pounds/(inches)^2.

Using the data in Table 5.6 on page 224:

(a) Construct and sketch a 95% confidence ellipse for the pair $[\mu_1, \mu_2]'$, where $\mu_1 = E(X_1)$ and $\mu_2 = E(X_2)$.

(b) Suppose $\mu_{10} = 2000$ and $\mu_{20} = 10{,}000$ represent "typical" values for stiffness and bending strength respectively. Given the result in (a), are the data in Table 5.6 consistent with these values? Explain.

(c) Is the bivariate normal distribution a viable population model? Explain with reference to *Q-Q* plots and a scatter diagram.

5.15. A wildlife ecologist measured x_1 = tail length (in millimeters) and x_2 = wing length (in millimeters) for a sample of $n = 45$ female hook-billed kites. These data are displayed in Table 5.7 on the next page.

TABLE 5.6 LUMBER DATA

x_1 (Stiffness: modulus of elasticity)	x_2 (Bending strength)	x_1 (Stiffness: modulus of elasticity)	x_2 (Bending strength)
1,232	4,175	1,712	7,749
1,115	6,652	1,932	6,818
2,205	7,612	1,820	9,307
1,897	10,914	1,900	6,457
1,932	10,850	2,426	10,102
1,612	7,627	1,558	7,414
1,598	6,954	1,470	7,556
1,804	8,365	1,858	7,833
1,752	9,469	1,587	8,309
2,067	6,410	2,208	9,559
2,365	10,327	1,487	6,255
1,646	7,320	2,206	10,723
1,579	8,196	2,332	5,430
1,880	9,709	2,540	12,090
1,773	10,370	2,322	10,072

SOURCE: Data courtesy of United States Forest Products Laboratory.

TABLE 5.7 BIRD DATA

x_1 (Tail length)	x_2 (Wing length)	x_1 (Tail length)	x_2 (Wing length)	x_1 (Tail length)	x_2 (Wing length)
191	284	186	266	173	271
197	285	197	285	194	280
208	288	201	295	198	300
180	273	190	282	180	272
180	275	209	305	190	292
188	280	187	285	191	286
210	283	207	297	196	285
196	288	178	268	207	286
191	271	202	271	209	303
179	257	205	285	179	261
208	289	190	280	186	262
202	285	189	277	174	245
200	272	211	310	181	250
192	282	216	305	189	262
199	280	189	274	188	258

SOURCE: Data courtesy of S. Temple.

Using the data in Table 5.7:

(a) Find and sketch the 95% confidence ellipse for the population means μ_1 and μ_2. Suppose it is known that $\mu_1 = 190$ mm and $\mu_2 = 275$ mm for *male* hook-billed kites. Are these plausible values for the mean tail length and mean wing length for the female birds? Explain.

(b) Construct the simultaneous 95% T^2-intervals for μ_1 and μ_2 and the 95% Bonferroni intervals for μ_1 and μ_2. Compare the two sets of intervals. What advantage, if any, do the T^2-intervals have over the Bonferroni intervals?

(c) Is the bivariate normal distribution a viable population model? Explain with reference to Q-Q plots and a scatter diagram.

REFERENCES

[1] Bickel, P. J., and K. A. Doksum, *Mathematical Statistics: Basic Ideas and Selected Topics*, San Francisco: Holden-Day, 1977.

[2] Bishop, Y. M. M., S. E. Feinberg, and P. W. Holland, *Discrete Multivariate Analysis: Theory and Practice*, Cambridge, Mass.: The MIT Press, 1975.

[3] DasGupta, S., and M. Perlman, "Power of the Non-central F-test: Effect of Additional Variables on Hotelling's T^2-test," *Journal of the American Statistical Association*, **69**, no. 345 (1974), 174–180.

[4] Dempster, A. P., N. M. Laird, and D. B. Rubin, "Maximum Likelihood from Incomplete Data via the EM Algorithm (with Discussion)," *Journal of the Royal Statistical Society* (B), **39**, no. 1 (1977), 1–38.

[5] Hartley, H. O., "Maximum Likelihood Estimation from Incomplete Data," *Biometrics*, **14** (1958), 174–194.

[6] Hartley, H. O., and R. R. Hocking, "The Analysis of Incomplete Data," *Biometrics*, **27** (1971), 783–808.

[7] Rao, C. R., *Linear Statistical Inference and Its Applications* (2nd ed.), New York: John Wiley, 1973.

6

Comparisons of Several Multivariate Means

6.1 INTRODUCTION

The ideas developed in Chapter 5 can be extended to handle problems involving the comparison of several mean vectors. The theory is a little more complicated and rests on an assumption of multivariate normal distributions or large sample sizes. Similarly, the notation becomes a bit cumbersome. To circumvent these problems, we shall often review univariate procedures for comparing several means and then generalize to the corresponding multivariate cases by analogy. The numerical examples we present will help cement the concepts.

Because comparisons of means frequently (and should) emanate from designed experiments, we take the opportunity to discuss some of the tenets of good experimental practice. A *repeated measures* design, useful in behavioral studies, is explicitly considered.

We begin by considering *pairs* of mean vectors. In later sections we discuss several comparisons among mean vectors arranged according to levels of treatments. The corresponding test statistics depend upon a partitioning of the total variation into pieces of variation attributable to the treatment sources and error. This partitioning is known as the *multivariate analysis of variance* (MANOVA).

6.2 PAIRED COMPARISONS AND A REPEATED MEASURES DESIGN

Paired Comparisons

Measurements are often recorded under different sets of experimental conditions to see if the responses differ significantly over these sets. For example, the efficacy of a new drug or of a saturation advertising campaign may be determined by comparing measurements before the "treatment" (drug or advertising) with those after the treatment. In other situations, *two or more* treatments can be administered to the same or similar experimental units and responses can be compared to assess the effects of the treatments.

One rational approach to comparing two treatments, or the presence and absence of a single treatment, is to assign both treatments to the *same* or *identical* units (individuals, stores, plots of land and so forth). The paired responses may then be analyzed by computing their differences, thereby eliminating much of the influence of extraneous unit-to-unit variation.

In the single response (univariate) case, let X_{1j} denote the response to treatment 1 (or the response before treatment) and let X_{2j} denote the response to treatment 2 (or the response after treatment) for the jth trial. That is, (X_{1j}, X_{2j}) are measurements recorded on the jth unit or jth pair of like units. By design, the n differences

$$D_j = X_{1j} - X_{2j}, \qquad j = 1, 2, \ldots, n \tag{6-1}$$

should reflect only the differential effects of the treatments.

Assuming the differences, D_j, in (6-1) represent independent observations from an $N(\delta, \sigma_d^2)$ distribution, the variable

$$t = \frac{\bar{D} - \delta}{s_d/\sqrt{n}} \tag{6-2}$$

where

$$\bar{D} = \frac{1}{n}\sum_{j=1}^{n} D_j \qquad \text{and} \qquad s_d^2 = \frac{1}{n-1}\sum_{j=1}^{n}\left(D_j - \bar{D}\right)^2 \tag{6-3}$$

has a t-distribution with $n - 1$ d.f. Consequently, an α-level test of

$$H_0 : \delta = 0 \qquad \text{(zero mean difference for treatments)}$$

versus

$$H_1 : \delta \neq 0$$

may be conducted by comparing $|t|$ with $t_{n-1}(\alpha/2)$—the upper $(100\alpha/2)$th percentile of a t-distribution with $n - 1$ d.f. A $100(1 - \alpha)\%$ confidence interval for the mean difference $\delta = E(X_{1j} - X_{2j})$ is provided by the statement

$$\bar{d} - t_{n-1}(\alpha/2)\frac{s_d}{\sqrt{n}} \leq \delta \leq \bar{d} + t_{n-1}(\alpha/2)\frac{s_d}{\sqrt{n}} \tag{6-4}$$

(for example, see [7]).

Additional notation is required for the multivariate extension of the paired comparison procedure. It is necessary to distinguish between p responses, 2 treatments, and n experimental units. We label the p responses within the *jth unit* as

$$
\begin{aligned}
X_{11j} &= \text{variable 1 under treatment 1} \\
X_{12j} &= \text{variable 2 under treatment 1} \\
&\vdots \\
X_{1pj} &= \text{variable } p \text{ under treatment 1} \\
\hline
X_{21j} &= \text{variable 1 under treatment 2} \\
X_{22j} &= \text{variable 2 under treatment 2} \\
&\vdots \\
X_{2pj} &= \text{variable } p \text{ under treatment 2}
\end{aligned}
$$

and the p paired difference random variables become

$$
\begin{aligned}
D_{1j} &= X_{11j} - X_{21j} \\
D_{2j} &= X_{12j} - X_{22j} \\
&\vdots \\
D_{pj} &= X_{1pj} - X_{2pj}
\end{aligned}
\qquad (6\text{-}5)
$$

Let $\mathbf{D}_j' = [D_{1j}, D_{2j}, \ldots, D_{pj}]$ and assume, for $j = 1, 2, \ldots, n$, that

$$
E(\mathbf{D}_j) = \boldsymbol{\delta} = \begin{bmatrix} \delta_1 \\ \delta_2 \\ \vdots \\ \delta_p \end{bmatrix} \quad \text{and} \quad \text{Cov}(\mathbf{D}_j) = \boldsymbol{\Sigma}_d \qquad (6\text{-}6)
$$

If, in addition, $\mathbf{D}_1, \mathbf{D}_2, \ldots, \mathbf{D}_n$ are independent $N_p(\boldsymbol{\delta}, \boldsymbol{\Sigma}_d)$ random vectors, inferences about the vector of mean differences $\boldsymbol{\delta}$ can be based upon a T^2-statistic.

Specifically,

$$
T^2 = n(\overline{\mathbf{D}} - \boldsymbol{\delta})'\mathbf{S}_d^{-1}(\overline{\mathbf{D}} - \boldsymbol{\delta}) \qquad (6\text{-}7)
$$

where

$$
\overline{\mathbf{D}} = \frac{1}{n} \sum_{j=1}^{n} \mathbf{D}_j \quad \text{and} \quad \mathbf{S}_d = \frac{1}{n-1} \sum_{j=1}^{n} (\mathbf{D}_j - \overline{\mathbf{D}})(\mathbf{D}_j - \overline{\mathbf{D}})' \qquad (6\text{-}8)
$$

Result 6.1. Let the differences $\mathbf{D}_1, \mathbf{D}_2, \ldots, \mathbf{D}_n$ be a random sample from an $N_p(\boldsymbol{\delta}, \boldsymbol{\Sigma}_d)$ population. Then

$$
T^2 = n(\overline{\mathbf{D}} - \boldsymbol{\delta})'\mathbf{S}_d^{-1}(\overline{\mathbf{D}} - \boldsymbol{\delta})
$$

is distributed as an $[(n-1)p/(n-p)]F_{p,n-p}$ random variable whatever the true $\boldsymbol{\delta}$ and $\boldsymbol{\Sigma}_d$.

If n and $n - p$ are both large, T^2 is approximately distributed as a χ_p^2 random variable regardless of the form of the underlying population of differences.

Proof. The exact distribution of T^2 is a restatement of the summary in (5-6), with vectors of differences for the observation vectors. The approximate distribution of T^2, for n and $n - p$ large, follows from (4-28). ■

The condition $\boldsymbol{\delta} = \mathbf{0}$ is equivalent to "no average difference between the two treatments." For the ith variable, $\delta_i > 0$ implies treatment 2 is higher, on average, than treatment 1. In general, inferences about $\boldsymbol{\delta}$ can be made using Result 6.1.

Given the observed differences $\mathbf{d}'_j = [d_{1j}, d_{2j}, \ldots, d_{pj}]$, $j = 1, 2, \ldots, n$, corresponding to the random variables in (6-5), an *α-level test of $H_0 : \boldsymbol{\delta} = \mathbf{0}$ versus $H_1 : \boldsymbol{\delta} \neq \mathbf{0}$* for an $N_p(\boldsymbol{\delta}, \boldsymbol{\Sigma}_d)$ population rejects H_0 if the observed

$$T^2 = n\bar{\mathbf{d}}'\mathbf{S}_d^{-1}\bar{\mathbf{d}} > [(n-1)p/(n-p)]F_{p,n-p}(\alpha)$$

where $F_{p,n-p}(\alpha)$ is the upper (100α)th percentile of an F-distribution with p and $n - p$ d.f. Here $\bar{\mathbf{d}}$ and $\mathbf{S}_d$ are given by (6-8).

A $100(1 - \alpha)$% *confidence region for* $\boldsymbol{\delta}$ consists of all $\boldsymbol{\delta}$ such that

$$(\bar{\mathbf{d}} - \boldsymbol{\delta})'\mathbf{S}_d^{-1}(\bar{\mathbf{d}} - \boldsymbol{\delta}) \leq \frac{(n-1)p}{n(n-p)}F_{p,n-p}(\alpha) \qquad \text{(6-9)}$$

Also, $100(1 - \alpha)$% *simultaneous confidence intervals for the individual mean differences* δ_i are given by

$$\delta_i : \bar{d}_i \pm \sqrt{\frac{(n-1)p}{(n-p)}F_{p,n-p}(\alpha)}\sqrt{\frac{s_{d_i}^2}{n}} \qquad \text{(6-10)}$$

where $\bar{d}_i$ is the ith element of $\bar{\mathbf{d}}$ and $s_{d_i}^2$ is the ith diagonal element of $\mathbf{S}_d$.

For $n - p$ large, $[(n-1)p/(n-p)]F_{p,n-p}(\alpha) \doteq \chi_p^2(\alpha)$ and normality need not be assumed.

Example 6.1

Municipal wastewater treatment plants are required by law to monitor their discharges into rivers and streams on a regular basis. Concern about the reliability of data from one of these self-monitoring programs led to a study in which samples of effluent were divided and sent to two laboratories for testing. One-half of each sample was sent to the Wisconsin State Laboratory of Hygiene and one-half was sent to a private commercial laboratory routinely used in the monitoring program. Measurements of biochemical oxygen demand (BOD) and suspended solids (SS) were obtained, for $n = 11$ sample splits, from the two laboratories. The data are displayed in Table 6.1.

TABLE 6.1 EFFLUENT DATA

Sample j	Commercial lab x_{11j}(BOD)	x_{12j}(SS)	State lab of hygiene x_{21j}(BOD)	x_{22j}(SS)
1	6	27	25	15
2	6	23	28	13
3	18	64	36	22
4	8	44	35	29
5	11	30	15	31
6	34	75	44	64
7	28	26	42	30
8	71	124	54	64
9	43	54	34	56
10	33	30	29	20
11	20	14	39	21

SOURCE: Data courtesy of S. Weber.

Do the two laboratories' chemical analyses agree? If differences exist, what is their nature?

The T^2-statistic for testing $H_0: \boldsymbol{\delta}' = [\delta_1, \delta_2] = [0, 0]$ is constructed from the differences of paired observations:

$d_{1j} = x_{11j} - x_{21j}$	−19	−22	−18	−27	−4	−10	−14	17	9	4	−19
$d_{2j} = x_{12j} - x_{22j}$	12	10	42	15	−1	11	−4	60	−2	10	−7

Here

$$\bar{\mathbf{d}} = \begin{bmatrix} \bar{d}_1 \\ \bar{d}_2 \end{bmatrix} = \begin{bmatrix} -9.36 \\ 13.27 \end{bmatrix}, \qquad \mathbf{S}_d = \begin{bmatrix} 199.26 & 88.38 \\ 88.38 & 418.61 \end{bmatrix}$$

and

$$T^2 = 11[-9.36, \quad 13.27]\begin{bmatrix} .0055 & -.0012 \\ -.0012 & .0026 \end{bmatrix}\begin{bmatrix} -9.36 \\ 13.27 \end{bmatrix} = 13.6$$

Taking $\alpha = .05$, we find $[p(n-1)/(n-p)]F_{p,n-p}(.05) = [2(10)/9]F_{2,9}(.05) = 9.47$. Since $T^2 = 13.6 > 9.47$, we reject H_0 and conclude there is a nonzero mean difference between the measurements of the two laboratories. It appears, from inspection of the data, that the commercial lab tends to produce lower BOD measurements and higher SS measurements than the State Lab of Hygiene. The 95% simultaneous confidence intervals for the mean differences δ_1 and δ_2 can be computed using (6-10). These intervals are

$$\delta_1: \quad \bar{d}_1 \pm \sqrt{\frac{(n-1)p}{(n-p)}F_{p,n-p}(\alpha)}\sqrt{\frac{s_{d_1}^2}{n}} = -9.36 \pm \sqrt{9.47}\sqrt{\frac{199.26}{11}}$$
$$\text{or} \quad (-22.46, 3.74)$$

$$\delta_2: \quad 13.27 \pm \sqrt{9.47}\sqrt{\frac{418.61}{11}} \quad \text{or} \quad (-5.71, 32.25)$$

The 95% *simultaneous confidence intervals* include zero, yet the hypothesis H_0: $\boldsymbol{\delta} = \mathbf{0}$ was rejected at the 5% level. What are we to conclude?

The evidence points towards real differences. The point $\boldsymbol{\delta} = \mathbf{0}$ falls outside the 95% *confidence region* for $\boldsymbol{\delta}$ (see Exercise 6.1), and this result is consistent with the T^2-test. The 95% simultaneous confidence coefficient applies to the *entire* set of intervals that could be constructed for all possible linear combinations of the form $\ell_1\delta_1 + \ell_2\delta_2$. The particular intervals corresponding to the choices ($\ell_1 = 1, \ell_2 = 0$) and ($\ell_1 = 0, \ell_2 = 1$) contain zero. Other choices of ℓ_1 and ℓ_2 will produce simultaneous intervals that do *not* contain zero. (If the hypothesis H_0: $\boldsymbol{\delta} = \mathbf{0}$ was not rejected, then *all* simultaneous intervals would include zero.)

Our analysis assumed a normal distribution for the $\mathbf{D}_j$. In fact, the situation is further complicated by the presence of one or, possibly, two outliers. ■

The experimenter in Example 6.1 actually divided a sample by first shaking it and then pouring it rapidly back and forth into two bottles for chemical analysis. This was prudent because a simple division of the sample into two pieces obtained by pouring the top half into one bottle and the remainder into another bottle might result in more suspended solids in the lower half due to settling. The two laboratories would then not be working with the same, or like, experimental units and the conclusions would not pertain to laboratory competence, measuring techniques, and so forth.

Whenever an investigator can control the assignment of treatments to experimental units, an appropriate pairing of units and a randomized assignment of treatments can enhance the statistical analysis. Differences, if any, between supposedly identical units must be identified and the most-alike units paired. Further, a random assignment of treatment 1 to one unit and treatment 2 to the other unit will help eliminate the systematic effects of uncontrolled sources of variation. Randomization can be implemented by flipping a coin to determine if the first unit in a pair receives treatment 1 (heads) or treatment 2 (tails). The remaining treatment is then

Experimental Design for Paired Comparisons

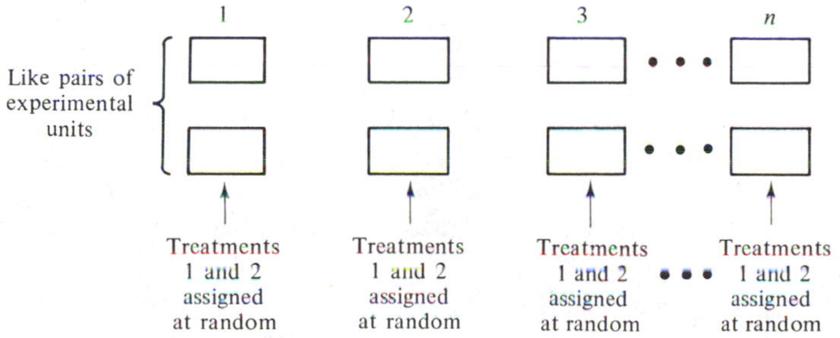

(6 - 10)

assigned to the other unit. A separate independent randomization is conducted for each pair.

We conclude our discussion of paired comparisons by noting that $\bar{\mathbf{d}}$ and $\mathbf{S}_d$, and hence T^2, may be calculated from the full sample quantities $\bar{\mathbf{x}}$ and $\mathbf{S}$. Here $\bar{\mathbf{x}}$ is the $2p \times 1$ vector of sample averages for the p variables on the two treatments given by

$$\bar{\mathbf{x}}' = [\bar{x}_{11}, \bar{x}_{12}, \ldots, \bar{x}_{1p}, \bar{x}_{21}, \bar{x}_{22}, \ldots, \bar{x}_{2p}] \tag{6-11}$$

and $\mathbf{S}$ is the $2p \times 2p$ matrix of sample variances and covariances arranged as

$$\mathbf{S} = \begin{bmatrix} \underset{(p\times p)}{\mathbf{S}_{11}} & \underset{(p\times p)}{\mathbf{S}_{12}} \\ \underset{(p\times p)}{\mathbf{S}_{21}} & \underset{(p\times p)}{\mathbf{S}_{22}} \end{bmatrix} \tag{6-12}$$

The matrix $\mathbf{S}_{11}$ contains the sample variances and covariances for the p variables on treatment 1. Similarly, $\mathbf{S}_{22}$ contains the sample variances and covariances computed for the p variables on treatment 2. Finally, $\mathbf{S}_{12} = \mathbf{S}'_{21}$ are the matrices of sample covariances computed from observations on pairs of treatment 1 and treatment 2 variables.

Defining the matrix

$$\underset{(p\times 2p)}{\mathbf{C}} = \left[\begin{array}{cccc|cccc} 1 & 0 & \cdots & 0 & -1 & 0 & \cdots & 0 \\ 0 & 1 & \cdots & 0 & 0 & -1 & \cdots & 0 \\ \vdots & \vdots & \ddots & \vdots & \vdots & \vdots & \ddots & \vdots \\ 0 & 0 & \cdots & 1 & 0 & 0 & \cdots & -1 \end{array}\right] \tag{6-13}$$

$\uparrow$ $(p+1)$st column

it can be verified (see Exercise 6.6) that

$$\mathbf{d}_j = \mathbf{C}\mathbf{x}_j, \qquad j = 1, 2, \ldots, n$$

$$\bar{\mathbf{d}} = \mathbf{C}\bar{\mathbf{x}} \quad \text{and} \quad \mathbf{S}_d = \mathbf{C}\mathbf{S}\mathbf{C}' \tag{6-14}$$

Thus

$$T^2 = n\bar{\mathbf{x}}'\mathbf{C}'[\mathbf{C}\mathbf{S}\mathbf{C}']^{-1}\mathbf{C}\bar{\mathbf{x}} \tag{6-15}$$

and it is not necessary first to calculate the differences $\mathbf{d}_1, \mathbf{d}_2, \ldots, \mathbf{d}_n$. On the other hand, it is wise to calculate these differences in order to check normality and the random sample assumption.

Each row $\mathbf{c}'_i$ of the matrix $\mathbf{C}$ in (6-13) is a *contrast vector* because its elements sum to zero. Attention is usually centered on contrasts when comparing treatments.

Each contrast is perpendicular to the vector $\mathbf{1}' = [1, 1, \ldots, 1]$ since $\mathbf{c}_i'\mathbf{1} = 0$. The component $\mathbf{1}'\mathbf{x}_j$, representing the overall treatment sum, is ignored by the test statistic T^2 presented in this section.

A Repeated-Measures Design for Comparing Treatments

Another generalization of the univariate paired t-statistic arises in situations where q treatments are compared with respect to a *single* response variable. Each subject or experimental unit receives each treatment once over successive periods of time. The jth observation is

$$\mathbf{X}_j = \begin{bmatrix} X_{1j} \\ X_{2j} \\ \vdots \\ X_{qj} \end{bmatrix}, \qquad j = 1, 2, \ldots, n$$

where X_{ij} is the response to the ith treatment on the jth unit. The name *repeated measures* stems from the fact that all treatments are administered to each unit.

For comparative purposes, we consider contrasts of the components of $\boldsymbol{\mu} = E(\mathbf{X}_j)$. These could be

$$\begin{bmatrix} \mu_1 - \mu_2 \\ \mu_1 - \mu_3 \\ \vdots \\ \mu_1 - \mu_q \end{bmatrix} = \begin{bmatrix} 1 & -1 & 0 & \cdots & 0 \\ 1 & 0 & -1 & \cdots & 0 \\ \vdots & \vdots & \vdots & & \vdots \\ 1 & 0 & 0 & \cdots & -1 \end{bmatrix} \begin{bmatrix} \mu_1 \\ \mu_2 \\ \vdots \\ \mu_q \end{bmatrix} = \mathbf{C}_1\boldsymbol{\mu}$$

or

$$\begin{bmatrix} \mu_2 - \mu_1 \\ \mu_3 - \mu_2 \\ \vdots \\ \mu_q - \mu_{q-1} \end{bmatrix} = \begin{bmatrix} -1 & 1 & 0 & \cdots & 0 & 0 \\ 0 & -1 & 1 & \cdots & 0 & 0 \\ \vdots & \vdots & \vdots & & \vdots & \vdots \\ 0 & 0 & 0 & \cdots & -1 & 1 \end{bmatrix} \begin{bmatrix} \mu_1 \\ \mu_2 \\ \vdots \\ \mu_q \end{bmatrix} = \mathbf{C}_2\boldsymbol{\mu}$$

Both $\mathbf{C}_1$ and $\mathbf{C}_2$ are called *contrast matrices* because their $q - 1$ rows are linearly independent and each is a contrast vector. The nature of the design eliminates much of the influence of unit-to-unit variation on treatment comparisons. Of course, the experimenter should randomize the order in which the treatments are presented to each subject.

When the treatment means are equal, $\mathbf{C}_1\boldsymbol{\mu} = \mathbf{C}_2\boldsymbol{\mu} = \mathbf{0}$. In general, the hypothesis that there are no differences in treatments (equal treatment means) becomes $\mathbf{C}\boldsymbol{\mu} = \mathbf{0}$ for any choice of the contrast matrix $\mathbf{C}$.

Test for Equality of Treatments in a Repeated-Measures Design

Consider an $N_q(\boldsymbol{\mu}, \boldsymbol{\Sigma})$ population and let $\mathbf{C}$ be a contrast matrix. An α-level test of $H_0: \mathbf{C}\boldsymbol{\mu} = \mathbf{0}$ (equal treatment means) versus $H_1: \mathbf{C}\boldsymbol{\mu} \neq \mathbf{0}$ is: Reject H_0 if

$$T^2 = n(\mathbf{C}\bar{\mathbf{x}})'(\mathbf{CSC}')^{-1}\mathbf{C}\bar{\mathbf{x}} > \frac{(n-1)(q-1)}{(n-q+1)} F_{q-1,\, n-q+1}(\alpha) \qquad \text{(6-16)}$$

where $F_{q-1,\, n-q+1}(\alpha)$ is the upper (100α)th percentile of an F-distribution with $q-1$ and $n-q+1$ d.f. Here $\bar{\mathbf{x}}$ and $\mathbf{S}$ are the sample mean vector and covariance matrix defined by

$$\bar{\mathbf{x}} = \frac{1}{n} \sum_{j=1}^{n} \mathbf{x}_j \quad \text{and} \quad \mathbf{S} = \frac{1}{n-1} \sum_{j=1}^{n} (\mathbf{x}_j - \bar{\mathbf{x}})(\mathbf{x}_j - \bar{\mathbf{x}})'$$

It can be shown that T^2 in (6-16) does not depend on the particular choice of $\mathbf{C}$.[1]

A confidence region for contrasts $\mathbf{C}\boldsymbol{\mu}$, with $\boldsymbol{\mu}$ the mean of a normal population, is determined by the set of all $\mathbf{C}\boldsymbol{\mu}$ such that

$$(\mathbf{C}\bar{\mathbf{x}} - \mathbf{C}\boldsymbol{\mu})'(\mathbf{CSC}')^{-1}(\mathbf{C}\bar{\mathbf{x}} - \mathbf{C}\boldsymbol{\mu}) \leq \frac{(n-1)(q-1)}{(n-q+1)} F_{q-1,\, n-q+1}(\alpha) \qquad \text{(6-17)}$$

where $\bar{\mathbf{x}}$ and $\mathbf{S}$ are defined in (6-16). Consequently, simultaneous $100(1-\alpha)\%$ confidence intervals for single contrasts $\mathbf{c}'\boldsymbol{\mu}$ for any contrast vectors of interest are given by (see Result 5C.1)

$$\mathbf{c}'\boldsymbol{\mu}: \quad \mathbf{c}'\bar{\mathbf{x}} \pm \sqrt{\frac{(n-1)(q-1)}{(n-q+1)} F_{q-1,\, n-q+1}(\alpha)} \sqrt{\frac{\mathbf{c}'\mathbf{S}\mathbf{c}}{n}} \qquad \text{(6-18)}$$

Example 6.2

Improved anesthetics are often developed by first studying their effects on animals. In one study, 19 dogs were initially given the drug pentobarbitol. Each dog was then administered carbon dioxide (CO_2) at each of 2 pressure levels. Next, halothane (H) was added and the administration of CO_2 was repeated. The response, milliseconds between heartbeats, was measured for the 4 treatment combinations.

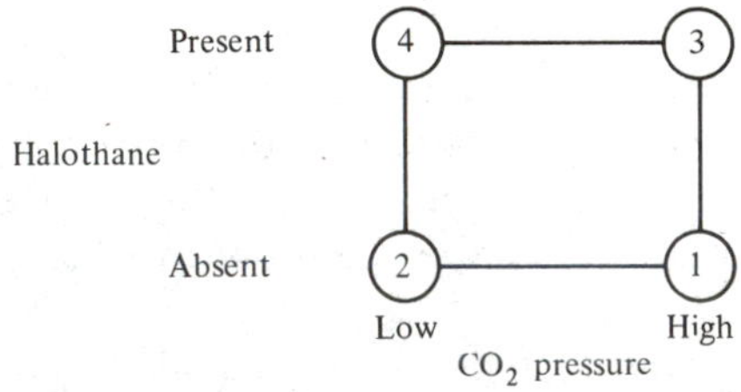

[1]Any pair of contrast matrices $\mathbf{C}_1$ and $\mathbf{C}_2$ must be related by $\mathbf{C}_1 = \mathbf{BC}_2$, with $\mathbf{B}$ nonsingular. This follows because each $\mathbf{C}$ has the largest possible number, $q-1$, of linearly independent rows all perpendicular to the vector $\mathbf{1}$. Then $(\mathbf{BC}_2)'(\mathbf{BC}_2\mathbf{SC}_2'\mathbf{B}')^{-1}(\mathbf{BC}_2) = \mathbf{C}_2'\mathbf{B}'(\mathbf{B}')^{-1}(\mathbf{C}_2\mathbf{SC}_2')^{-1}\mathbf{B}^{-1}\mathbf{BC}_2 = \mathbf{C}_2'(\mathbf{C}_2\mathbf{SC}_2')^{-1}\mathbf{C}_2$, so T^2 computed with $\mathbf{C}_2$ or $\mathbf{C}_1 = \mathbf{BC}_2$ gives the same result.

TABLE 6.2 SLEEPING-DOG DATA

Dog	Treatment 1	2	3	4
1	426	609	556	600
2	253	236	392	395
3	359	433	349	357
4	432	431	522	600
5	405	426	513	513
6	324	438	507	539
7	310	312	410	456
8	326	326	350	504
9	375	447	547	548
10	286	286	403	422
11	349	382	473	497
12	429	410	488	547
13	348	377	447	514
14	412	473	472	446
15	347	326	455	468
16	434	458	637	524
17	364	367	432	469
18	420	395	508	531
19	397	556	645	625

SOURCE: Data courtesy of Dr. J. Atlee.

Table 6.2 contains the four measurements for each of the 19 dogs.

Treatment 1 = high CO_2 pressure without H

Treatment 2 = low CO_2 pressure without H

Treatment 3 = high CO_2 pressure with H

Treatment 4 = low CO_2 pressure with H

We shall analyze the anesthetizing effects of CO_2 pressure and halothane from this repeated-measures design.

There are three treatment contrasts that might be of interest in this experiment. Let μ_1, μ_2, μ_3, and μ_4 correspond to the mean responses for treatments 1, 2, 3 and 4, respectively. Then

$$(\mu_3 + \mu_4) - (\mu_1 + \mu_2) = \begin{pmatrix} \text{Halothane contrast representing the difference} \\ \text{between the presence and absence of halothane} \end{pmatrix}$$

$$(\mu_1 + \mu_3) - (\mu_2 + \mu_4) = \begin{pmatrix} \text{CO}_2 \text{ contrast representing the difference} \\ \text{between high and low CO}_2 \text{ pressure} \end{pmatrix}$$

$$(\mu_1 + \mu_4) - (\mu_2 + \mu_3) = \begin{pmatrix} \text{Contrast representing the influence} \\ \text{of halothane on CO}_2 \text{ pressure differences} \\ \text{(H–CO}_2 \text{ pressure ``interaction'')} \end{pmatrix}$$

With $\boldsymbol{\mu}' = [\mu_1, \mu_2, \mu_3, \mu_4]$, the contrast matrix $\mathbf{C}$ is

$$\mathbf{C} = \begin{bmatrix} -1 & -1 & 1 & 1 \\ 1 & -1 & 1 & -1 \\ 1 & -1 & -1 & 1 \end{bmatrix}$$

The data (see Table 6.2) give

$$\bar{\mathbf{x}} = \begin{bmatrix} 368.21 \\ 404.63 \\ 479.26 \\ 502.89 \end{bmatrix} \quad \text{and} \quad \mathbf{S} = \begin{bmatrix} 2819.29 & & & \\ 3568.42 & 7963.14 & & \\ 2943.49 & 5303.98 & 6851.32 & \\ 2295.35 & 4065.44 & 4499.63 & 4878.99 \end{bmatrix}$$

It may be verified that

$$\mathbf{C}\bar{\mathbf{x}} = \begin{bmatrix} 209.31 \\ -60.05 \\ -12.79 \end{bmatrix}; \qquad \mathbf{CSC}' = \begin{bmatrix} 9432.32 & 1098.92 & 927.62 \\ 1098.92 & 5195.84 & 914.54 \\ 927.62 & 914.54 & 7557.44 \end{bmatrix}$$

and

$$T^2 = n(\mathbf{C}\bar{\mathbf{x}})'(\mathbf{CSC}')^{-1}(\mathbf{C}\bar{\mathbf{x}}) = 19(6.11) = 116.$$

With $\alpha = .05$,

$$\frac{(n-1)(q-1)}{(n-q+1)} F_{q-1,\, n-q+1}(\alpha) = \frac{18(3)}{16} F_{3,16}(.05) = \frac{18(3)}{16}(3.24) = 10.94$$

Using (6-16), $T^2 = 116 > 10.94$ and we reject $H_0\colon \mathbf{C}\boldsymbol{\mu} = \mathbf{0}$ (no treatment effects). To see which of the contrasts are responsible for the rejection of H_0, we construct 95% simultaneous confidence intervals for these contrasts. From (6-18), the contrast

$$\mathbf{c}_1'\boldsymbol{\mu} = (\mu_3 + \mu_4) - (\mu_1 + \mu_2) = \text{halothane influence}$$

is estimated by the interval

$$(\bar{x}_3 + \bar{x}_4) - (\bar{x}_1 + \bar{x}_2) \pm \sqrt{\frac{18(3)}{16} F_{3,16}(.05)} \sqrt{\frac{\mathbf{c}_1'\mathbf{S}\mathbf{c}_1}{19}}$$

$$= 209.31 \pm \sqrt{10.94} \sqrt{\frac{9432.32}{19}}$$

$$= 209.31 \pm 73.70$$

where $\mathbf{c}_1'$ is the first row of $\mathbf{C}$. Similarly, the remaining contrasts are estimated by

CO_2 pressure influence $= (\mu_1 + \mu_3) - (\mu_2 + \mu_4)$:

$$-60.05 \pm \sqrt{10.94} \sqrt{\frac{5195.84}{19}} = -60.05 \pm 54.70$$

H–CO_2 pressure "interaction" $= (\mu_1 + \mu_4) - (\mu_2 + \mu_3)$:

$$-12.79 \pm \sqrt{10.94} \sqrt{\frac{7557.44}{19}} = -12.79 \pm 65.97$$

The first confidence interval implies there is a halothane effect. The presence of halothane produces longer times between heartbeats. This occurs at both levels of CO_2 pressure, since the H–CO_2 pressure interaction contrast,

$(\mu_1 + \mu_4) - (\mu_2 + \mu_3)$, is not significantly different from zero (see the third confidence interval). The second confidence interval indicates there is an effect due to CO_2 pressure. The *lower* CO_2 pressure produces longer times between heartbeats.

Some caution must be exercised in our interpretation of the results because the trials with halothane must necessarily follow those without. The apparent H-effect may be due to a time trend. (Ideally, the time order of *all* treatments should be determined at random.) ■

The test in (6-16) is appropriate when the covariance matrix, $\text{Cov}(\mathbf{X}) = \mathbf{\Sigma}$, cannot be assumed to have any special structure. If it is reasonable to assume that $\mathbf{\Sigma}$ has a particular structure, tests designed with this structure in mind have higher power than the one in (6-16). (For $\mathbf{\Sigma}$ with the equal correlation structure (8-14), see a discussion of the "randomized block" design in [11] or [18].)

6.3 COMPARING MEAN VECTORS FROM TWO POPULATIONS

A T^2-statistic for testing the equality of vector means from two multivariate populations can be developed by analogy with the univariate procedure (see [7] for a discussion of the univariate case). This T^2-statistic is appropriate for comparing responses from one set of experimental settings (population 1) with independent responses from another set of experimental settings (population 2). This can be done without explicitly controlling for unit-to-unit variability, as in the paired-comparison case.

If possible, the experimental units should be randomly assigned to the sets of experimental conditions. Randomization will, to some extent, mitigate the effect of unit-to-unit variability in a subsequent comparison of treatments. Although some precision is lost relative to paired comparisons, the inferences in the two population case are, ordinarily, applicable to a more general collection of experimental units simply because unit homogeneity is not required.

Consider a random sample of size n_1 from population 1 and a sample of size n_2 from population 2. The observations on p variables can be arranged as:

Sample	Summary statistics	
(Population 1) $\mathbf{x}_{11}, \mathbf{x}_{12}, \ldots, \mathbf{x}_{1n_1}$	$\bar{\mathbf{x}}_1 = \frac{1}{n_1}\sum_{j=1}^{n_1} \mathbf{x}_{1j}$	$\mathbf{S}_1 = \frac{1}{n_1 - 1}\sum_{j=1}^{n_1} (\mathbf{x}_{1j} - \bar{\mathbf{x}}_1)(\mathbf{x}_{1j} - \bar{\mathbf{x}}_1)'$
(Population 2) $\mathbf{x}_{21}, \mathbf{x}_{22}, \ldots, \mathbf{x}_{2n_2}$	$\bar{\mathbf{x}}_2 = \frac{1}{n_2}\sum_{j=1}^{n_2} \mathbf{x}_{2j}$	$\mathbf{S}_2 = \frac{1}{n_2 - 1}\sum_{j=1}^{n_2} (\mathbf{x}_{2j} - \bar{\mathbf{x}}_2)(\mathbf{x}_{2j} - \bar{\mathbf{x}}_2)'$

In this notation, the first subscript—1 or 2—denotes the population.

We want to make inferences about (mean vector of population 1)—(mean vector of population 2) $= \boldsymbol{\mu}_1 - \boldsymbol{\mu}_2$. For instance, we shall want to answer the question, Is $\boldsymbol{\mu}_1 = \boldsymbol{\mu}_2$ (or, equivalently, $\boldsymbol{\mu}_1 - \boldsymbol{\mu}_2 = \mathbf{0}$)? Also, if $\boldsymbol{\mu}_1 - \boldsymbol{\mu}_2 \neq \mathbf{0}$, which component means, if any, are different?

With a few tentative assumptions we are able to provide answers to these questions.

Assumptions concerning the Structure of the Data

1. The sample $\mathbf{X}_{11}, \mathbf{X}_{12}, \ldots, \mathbf{X}_{1n_1}$ is a random sample of size n_1 from a p-variate population with mean vector $\boldsymbol{\mu}_1$ and covariance matrix $\boldsymbol{\Sigma}_1$.
2. The sample $\mathbf{X}_{21}, \mathbf{X}_{22}, \ldots, \mathbf{X}_{2n_2}$ is a random sample of size n_2 from a p-variate population with mean vector $\boldsymbol{\mu}_2$ and covariance matrix $\boldsymbol{\Sigma}_2$. (6-19)
3. Also, $\mathbf{X}_{11}, \mathbf{X}_{12}, \ldots, \mathbf{X}_{1n_1}$ are independent of $\mathbf{X}_{21}, \mathbf{X}_{22}, \ldots, \mathbf{X}_{2n_2}$

We shall see later that, for large samples, this structure is sufficient for making inferences about the $p \times 1$ vector $\boldsymbol{\mu}_1 - \boldsymbol{\mu}_2$. However, when the sample sizes n_1 and n_2 are small, more assumptions are needed.

Further Assumptions when n_1 and n_2 Are Small

1. Both populations are multivariate normal.
2. Also, $\boldsymbol{\Sigma}_1 = \boldsymbol{\Sigma}_2$ (same covariance matrix). (6-20)

The second assumption, $\boldsymbol{\Sigma}_1 = \boldsymbol{\Sigma}_2$, is much stronger than its univariate counterpart. Here we are assuming that several pairs of variances and covariances are nearly equal.

When $\boldsymbol{\Sigma}_1 = \boldsymbol{\Sigma}_2 = \boldsymbol{\Sigma}$, $\sum_{j=1}^{n_1} (\mathbf{x}_{1j} - \bar{\mathbf{x}}_1)(\mathbf{x}_{1j} - \bar{\mathbf{x}}_1)'$ is an estimate of $(n_1 - 1)\boldsymbol{\Sigma}$ and $\sum_{j=1}^{n_2} (\mathbf{x}_{2j} - \bar{\mathbf{x}}_2)(\mathbf{x}_{2j} - \bar{\mathbf{x}}_2)'$ is an estimate of $(n_2 - 1)\boldsymbol{\Sigma}$. Consequently, we can pool the information in both samples in order to estimate the common covariance $\boldsymbol{\Sigma}$.

We set

$$\mathbf{S}_{\text{pooled}} = \frac{\sum_{j=1}^{n_1} (\mathbf{x}_{1j} - \bar{\mathbf{x}}_1)(\mathbf{x}_{1j} - \bar{\mathbf{x}}_1)' + \sum_{j=1}^{n_2} (\mathbf{x}_{2j} - \bar{\mathbf{x}}_2)(\mathbf{x}_{2j} - \bar{\mathbf{x}}_2)'}{n_1 + n_2 - 2}$$

$$= \frac{(n_1 - 1)\mathbf{S}_1 + (n_2 - 1)\mathbf{S}_2}{n_1 + n_2 - 2} \qquad (6\text{-}21)$$

Since $\sum_{j=1}^{n_1} (\mathbf{x}_{1j} - \bar{\mathbf{x}}_1)(\mathbf{x}_{1j} - \bar{\mathbf{x}}_1)'$ has $n_1 - 1$ d.f. and $\sum_{j=1}^{n_2} (\mathbf{x}_{2j} - \bar{\mathbf{x}}_2)(\mathbf{x}_{2j} - \bar{\mathbf{x}}_2)'$ has $n_2 - 1$ d.f., the divisor $(n_1 - 1) + (n_2 - 1)$ in (6-21) is obtained by combining the two component degrees of freedom. [See (4-24).] Additional support for the pooling procedure comes from consideration of the likelihood. (See Exercise 6.8.)

To test the hypothesis that $\boldsymbol{\mu}_1 - \boldsymbol{\mu}_2 = \boldsymbol{\delta}_0$, a specified vector, we consider the squared statistical distance from $\bar{\mathbf{x}}_1 - \bar{\mathbf{x}}_2$ to $\boldsymbol{\delta}_0$. Now

$$E(\bar{\mathbf{X}}_1 - \bar{\mathbf{X}}_2) = E(\bar{\mathbf{X}}_1) - E(\bar{\mathbf{X}}_2) = \boldsymbol{\mu}_1 - \boldsymbol{\mu}_2$$

Since the independence assumption in (6-19) implies $\bar{\mathbf{X}}_1$ and $\bar{\mathbf{X}}_2$ are independent and thus $\text{Cov}(\bar{\mathbf{X}}_1, \bar{\mathbf{X}}_2) = \mathbf{0}$ (see Result 4.5), by (3-9),

$$\text{Cov}(\bar{\mathbf{X}}_1 - \bar{\mathbf{X}}_2) = \text{Cov}(\bar{\mathbf{X}}_1) + \text{Cov}(\bar{\mathbf{X}}_2) = \frac{1}{n_1}\boldsymbol{\Sigma} + \frac{1}{n_2}\boldsymbol{\Sigma} = \left(\frac{1}{n_1} + \frac{1}{n_2}\right)\boldsymbol{\Sigma} \tag{6-22}$$

Because $\mathbf{S}_{\text{pooled}}$ estimates $\boldsymbol{\Sigma}$, we see that

$$\left(\frac{1}{n_1} + \frac{1}{n_2}\right)\mathbf{S}_{\text{pooled}}$$

is an estimator of $\text{Cov}(\bar{\mathbf{X}}_1 - \bar{\mathbf{X}}_2)$.

The likelihood ratio test of

$$H_0: \boldsymbol{\mu}_1 - \boldsymbol{\mu}_2 = \boldsymbol{\delta}_0$$

is based on the squared statistical distance, T^2, and is given by (see [1]): Reject H_0 if

$$T^2 = (\bar{\mathbf{x}}_1 - \bar{\mathbf{x}}_2 - \boldsymbol{\delta}_0)'\left[\left(\frac{1}{n_1} + \frac{1}{n_2}\right)\mathbf{S}_{\text{pooled}}\right]^{-1}(\bar{\mathbf{x}}_1 - \bar{\mathbf{x}}_2 - \boldsymbol{\delta}_0) > c^2 \tag{6-23}$$

where the critical distance, c^2, is determined from the distribution of the two-sample T^2-statistic.

Result 6.2. When $\mathbf{X}_{11}, \mathbf{X}_{12}, \ldots, \mathbf{X}_{1n_1}$ is a random sample of size n_1 from $N_p(\boldsymbol{\mu}_1, \boldsymbol{\Sigma})$ and $\mathbf{X}_{21}, \mathbf{X}_{22}, \ldots, \mathbf{X}_{2n_2}$ is an independent random sample of size n_2 from $N_p(\boldsymbol{\mu}_2, \boldsymbol{\Sigma})$, then

$$T^2 = [\bar{\mathbf{X}}_1 - \bar{\mathbf{X}}_2 - (\boldsymbol{\mu}_1 - \boldsymbol{\mu}_2)]'\left[\left(\frac{1}{n_1} + \frac{1}{n_2}\right)\mathbf{S}_{\text{pooled}}\right]^{-1}[\bar{\mathbf{X}}_1 - \bar{\mathbf{X}}_2 - (\boldsymbol{\mu}_1 - \boldsymbol{\mu}_2)]$$

is distributed as

$$\frac{(n_1 + n_2 - 2)p}{(n_1 + n_2 - p - 1)}F_{p,\, n_1+n_2-p-1}$$

Consequently,

$$P\left[(\bar{\mathbf{X}}_1 - \bar{\mathbf{X}}_2 - (\boldsymbol{\mu}_1 - \boldsymbol{\mu}_2))'\left[\left(\frac{1}{n_1} + \frac{1}{n_2}\right)\mathbf{S}_{\text{pooled}}\right]^{-1}(\bar{\mathbf{X}}_1 - \bar{\mathbf{X}}_2 - (\boldsymbol{\mu}_1 - \boldsymbol{\mu}_2)) \le c^2\right] = 1 - \alpha \tag{6-24}$$

where

$$c^2 = \frac{(n_1 + n_2 - 2)p}{(n_1 + n_2 - p - 1)}F_{p,\, n_1+n_2-p-1}(\alpha)$$

Proof. We first note that

$$\bar{\mathbf{X}}_1 - \bar{\mathbf{X}}_2 = \frac{1}{n_1}\mathbf{X}_{11} + \frac{1}{n_1}\mathbf{X}_{12} + \cdots + \frac{1}{n_1}\mathbf{X}_{1n_1} - \frac{1}{n_2}\mathbf{X}_{21} - \frac{1}{n_2}\mathbf{X}_{22} - \cdots - \frac{1}{n_2}\mathbf{X}_{2n_2}$$

is

$$N_p\left(\boldsymbol{\mu}_1 - \boldsymbol{\mu}_2, \left(\frac{1}{n_1} + \frac{1}{n_2}\right)\boldsymbol{\Sigma}\right)$$

by Result 4.8, with $c_1 = c_2 = \cdots = c_{n_1} = 1/n_1$ and $c_{n_1+1} = c_{n_1+2} = \cdots = c_{n_1+n_2} = 1/n_2$. According to (4-23),

$$(n_1 - 1)\mathbf{S}_1 \text{ is distributed as } W_{n_1-1}(\cdot\,|\,\boldsymbol{\Sigma}) \text{ and } (n_2 - 1)\mathbf{S}_2 \text{ as } W_{n_2-1}(\cdot\,|\,\boldsymbol{\Sigma})$$

By assumption, the $\mathbf{X}_{1j}$'s and the $\mathbf{X}_{2j}$'s are independent, so $(n_1 - 1)\mathbf{S}_1$ and $(n_2 - 1)\mathbf{S}_2$ are also independent. Using (4-24), $(n_1 - 1)\mathbf{S}_1 + (n_2 - 1)\mathbf{S}_2$ is then distributed as $W_{n_1+n_2-2}(\cdot\,|\,\boldsymbol{\Sigma})$. Therefore

$$T^2 = \left(\frac{1}{n_1} + \frac{1}{n_2}\right)^{-1/2}(\bar{\mathbf{X}}_1 - \bar{\mathbf{X}}_2 - (\boldsymbol{\mu}_1 - \boldsymbol{\mu}_2))'\mathbf{S}_{\text{pooled}}^{-1}$$

$$\times\left(\frac{1}{n_1} + \frac{1}{n_2}\right)^{-1/2}(\bar{\mathbf{X}}_1 - \bar{\mathbf{X}}_2 - (\boldsymbol{\mu}_1 - \boldsymbol{\mu}_2))$$

$$= \begin{pmatrix}\text{multivariate normal}\\ \text{random vector}\end{pmatrix}'\left(\frac{\text{Wishart random matrix}}{\text{degrees of freedom}}\right)^{-1}\begin{pmatrix}\text{multivariate normal}\\ \text{random vector}\end{pmatrix}$$

$$= N_p(\mathbf{0}, \boldsymbol{\Sigma})'\left[\frac{W_{n_1+n_2-2}(\cdot\,|\,\boldsymbol{\Sigma})}{n_1 + n_2 - 2}\right]^{-1}N_p(\mathbf{0}, \boldsymbol{\Sigma})$$

which is the T^2-distribution specified in (5-5), with n replaced by $n_1 + n_2 - 1$. ■

We are primarily interested in confidence regions for $\boldsymbol{\mu}_1 - \boldsymbol{\mu}_2$. From (6-24) we conclude that all $\boldsymbol{\mu}_1 - \boldsymbol{\mu}_2$ "close to" $\bar{\mathbf{x}}_1 - \bar{\mathbf{x}}_2$ constitute the confidence region. The confidence region is an ellipsoid centered at the observed difference $\bar{\mathbf{x}}_1 - \bar{\mathbf{x}}_2$, whose axes are determined by the eigenvalues and eigenvectors of $\mathbf{S}_{\text{pooled}}$ (or $\mathbf{S}_{\text{pooled}}^{-1}$).

Example 6.3

Fifty bars of soap are manufactured in each of two ways. Two characteristics X_1 = lather and X_2 = mildness are measured. The summary statistics for bars produced by methods 1 and 2 are

$$\bar{\mathbf{x}}_1 = \begin{bmatrix} 8.3 \\ 4.1 \end{bmatrix}, \qquad \mathbf{S}_1 = \begin{bmatrix} 2 & 1 \\ 1 & 6 \end{bmatrix}$$

$$\bar{\mathbf{x}}_2 = \begin{bmatrix} 10.2 \\ 3.9 \end{bmatrix}, \qquad \mathbf{S}_2 = \begin{bmatrix} 2 & 1 \\ 1 & 4 \end{bmatrix}$$

Obtain a 95% confidence region for $\boldsymbol{\mu}_1 - \boldsymbol{\mu}_2$.

We first note that $\mathbf{S}_1$ and $\mathbf{S}_2$ are approximately equal so that it is reasonable to pool. Hence

$$\mathbf{S}_{\text{pooled}} = \frac{(50-1)\mathbf{S}_1 + (50-1)\mathbf{S}_2}{50+50-2} = \begin{bmatrix} 2 & 1 \\ 1 & 5 \end{bmatrix}$$

Also

$$\bar{\mathbf{x}}_1 - \bar{\mathbf{x}}_2 = \begin{bmatrix} -1.9 \\ .2 \end{bmatrix}$$

so the confidence ellipse is centered at $[-1.9, .2]'$. The eigenvalues and eigenvectors of $\mathbf{S}_{\text{pooled}}$ are obtained from the equation

$$0 = |\mathbf{S}_{\text{pooled}} - \lambda\mathbf{I}| = \begin{vmatrix} 2-\lambda & 1 \\ 1 & 5-\lambda \end{vmatrix} = \lambda^2 - 7\lambda + 9$$

so $\lambda = (7 \pm \sqrt{49-36})/2$. Consequently, $\lambda_1 = 5.303$ and $\lambda_2 = 1.697$, and the corresponding eigenvectors, $\mathbf{e}_1$ and $\mathbf{e}_2$, determined from

$$(\mathbf{S}_{\text{pooled}})\mathbf{e}_i = \lambda_i \mathbf{e}_i \qquad i = 1, 2$$

are

$$\mathbf{e}_1 = \begin{bmatrix} .290 \\ .957 \end{bmatrix} \quad \text{and} \quad \mathbf{e}_2 = \begin{bmatrix} .957 \\ -.290 \end{bmatrix}$$

By Result 6.2,

$$\left(\frac{1}{n_1} + \frac{1}{n_2}\right)c^2 = \left(\frac{1}{50} + \frac{1}{50}\right)\frac{(98)(2)}{(97)} F_{2,97}(.05) = .25$$

since $F_{2,97}(.05) = 3.1$. The confidence ellipse extends

$$\sqrt{\lambda_i}\sqrt{\left(\frac{1}{n_1} + \frac{1}{n_2}\right)c^2} = \sqrt{\lambda_i}\sqrt{.25}$$

units along the eigenvector $\mathbf{e}_i$, or 1.15 units in the $\mathbf{e}_1$ direction and .65 units in the $\mathbf{e}_2$ direction. The 95% confidence ellipse is shown in Figure 6.1. Clearly

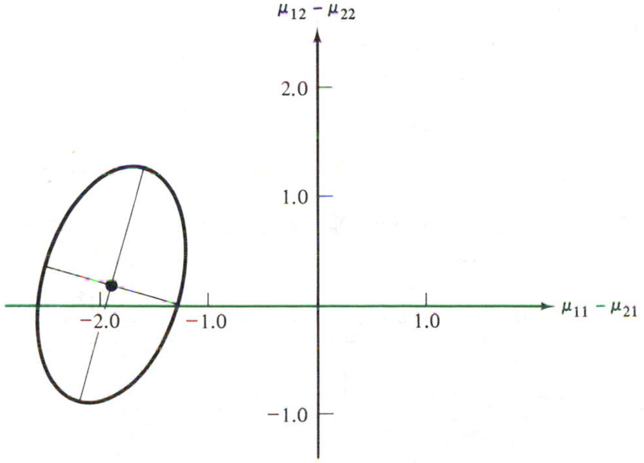

Figure 6.1 95% confidence ellipse for $\mu_1 - \mu_2$.

$\boldsymbol{\mu}_1 - \boldsymbol{\mu}_2 = \mathbf{0}$ is not in the ellipse, and we conclude the two methods of manufacturing soap produce different results. It appears as if the two processes give bars of soap with about the same mildness (X_2), but those from the second process have more lather (X_1). ■

Simultaneous Confidence Intervals

It is possible to derive simultaneous confidence intervals for the components of the vector $\boldsymbol{\mu}_1 - \boldsymbol{\mu}_2$. These confidence intervals are developed from a consideration of all possible linear combinations of the differences in the mean vectors. It is assumed that the parent multivariate populations are normal with a common covariance $\boldsymbol{\Sigma}$.

Result 6.3. Let $c^2 = [(n_1 + n_2 - 2)p/(n_1 + n_2 - p - 1)]F_{p,\, n_1+n_2-p-1}(\alpha)$. With probability $1 - \alpha$,

$$\boldsymbol{\ell}'(\overline{\mathbf{X}}_1 - \overline{\mathbf{X}}_2) \pm c\sqrt{\boldsymbol{\ell}'\left(\frac{1}{n_1} + \frac{1}{n_2}\right)\mathbf{S}_{\text{pooled}}\boldsymbol{\ell}}$$

will cover $\boldsymbol{\ell}'(\boldsymbol{\mu}_1 - \boldsymbol{\mu}_2)$ for all $\boldsymbol{\ell}$. In particular $\mu_{1i} - \mu_{2i}$ will be covered by

$$(\overline{X}_{1i} - \overline{X}_{2i}) \pm c\sqrt{\left(\frac{1}{n_1} + \frac{1}{n_2}\right)s_{ii,\text{pooled}}} \qquad \text{for } i = 1, 2, \ldots, p$$

Proof. Consider univariate linear combinations of the observations

$$\mathbf{X}_{11}, \mathbf{X}_{12}, \ldots, \mathbf{X}_{1n_1} \qquad \text{and} \qquad \mathbf{X}_{21}, \mathbf{X}_{22}, \ldots, \mathbf{X}_{2n_2}$$

given by $\boldsymbol{\ell}'\mathbf{X}_{1j} = \ell_1 X_{11j} + \ell_2 X_{12j} + \cdots + \ell_p X_{1pj}$ and $\boldsymbol{\ell}'\mathbf{X}_{2j} = \ell_1 X_{21j} + \ell_2 X_{22j} + \cdots + \ell_p X_{2pj}$. These linear combinations have sample means and covariances: $\boldsymbol{\ell}'\overline{\mathbf{X}}_1$, $\boldsymbol{\ell}'\mathbf{S}_1\boldsymbol{\ell}$ and $\boldsymbol{\ell}'\overline{\mathbf{X}}_2$, $\boldsymbol{\ell}'\mathbf{S}_2\boldsymbol{\ell}$, respectively, where $\overline{\mathbf{X}}_1$, $\mathbf{S}_1$, and $\overline{\mathbf{X}}_2$, $\mathbf{S}_2$ are the mean and covariance statistics for the two original samples (see Result 3.5). When both parent populations have the same covariance, $s^2_{1,\ell} = \boldsymbol{\ell}'\mathbf{S}_1\boldsymbol{\ell}$ and $s^2_{2,\ell} = \boldsymbol{\ell}'\mathbf{S}_2\boldsymbol{\ell}$ are both estimators of $\boldsymbol{\ell}'\boldsymbol{\Sigma}\boldsymbol{\ell}$, the common population variance of the linear combinations $\boldsymbol{\ell}'\mathbf{X}_1$ and $\boldsymbol{\ell}'\mathbf{X}_2$. Pooling these estimators we obtain

$$\begin{aligned} s^2_{\ell,\text{pooled}} &= \frac{(n_1 - 1)s^2_{1,\ell} + (n_2 - 1)s^2_{2,\ell}}{(n_1 + n_2 - 2)} \\ &= \boldsymbol{\ell}'\left[\frac{(n_1 - 1)\mathbf{S}_1 + (n_2 - 1)\mathbf{S}_2}{(n_1 + n_2 - 2)}\right]\boldsymbol{\ell} \\ &= \boldsymbol{\ell}'\mathbf{S}_{\text{pooled}}\boldsymbol{\ell} \end{aligned} \tag{6-25}$$

To test $H_0: \boldsymbol{\ell}'(\boldsymbol{\mu}_1 - \boldsymbol{\mu}_2) = \boldsymbol{\ell}'\boldsymbol{\delta}_0$, on the basis of the $\boldsymbol{\ell}'\mathbf{X}_{1j}$ and $\boldsymbol{\ell}'\mathbf{X}_{2j}$, we can form the square of the univariate two-sample t-statistic

$$t^2_\ell = \frac{\left[\boldsymbol{\ell}'(\overline{\mathbf{X}}_1 - \overline{\mathbf{X}}_2) - \boldsymbol{\ell}'(\boldsymbol{\mu}_1 - \boldsymbol{\mu}_2)\right]^2}{\left(\frac{1}{n_1} + \frac{1}{n_2}\right)s^2_{\ell,\text{pooled}}} = \frac{\left[\boldsymbol{\ell}'(\overline{\mathbf{X}}_1 - \overline{\mathbf{X}}_2 - (\boldsymbol{\mu}_1 - \boldsymbol{\mu}_2))\right]^2}{\boldsymbol{\ell}'\left(\frac{1}{n_1} + \frac{1}{n_2}\right)\mathbf{S}_{\text{pooled}}\boldsymbol{\ell}} \tag{6-26}$$

According to the maximization lemma in (2-50) with $\mathbf{d} = [\bar{\mathbf{X}}_1 - \bar{\mathbf{X}}_2 - (\boldsymbol{\mu}_1 - \boldsymbol{\mu}_2)]$ and $\mathbf{B} = (1/n_1 + 1/n_2)\mathbf{S}_{\text{pooled}}$,

$$t_{\ell}^2 \le (\bar{\mathbf{X}}_1 - \bar{\mathbf{X}}_2 - (\boldsymbol{\mu}_1 - \boldsymbol{\mu}_2))'\left[\left(\frac{1}{n_1} + \frac{1}{n_2}\right)\mathbf{S}_{\text{pooled}}\right]^{-1}(\bar{\mathbf{X}}_1 - \bar{\mathbf{X}}_2 - (\boldsymbol{\mu}_1 - \boldsymbol{\mu}_2))$$
$$= T^2$$

for all $\boldsymbol{\ell} \neq \mathbf{0}$. Thus

$$(1-\alpha) = P[T^2 \le c^2] = P\left[t_{\ell}^2 \le c^2, \quad \text{for all } \boldsymbol{\ell}\right]$$
$$= P\left[\left|\boldsymbol{\ell}'(\bar{\mathbf{X}}_1 - \bar{\mathbf{X}}_2) - \boldsymbol{\ell}'(\boldsymbol{\mu}_1 - \boldsymbol{\mu}_2)\right| \le c\sqrt{\boldsymbol{\ell}'\left(\frac{1}{n_1} + \frac{1}{n_2}\right)\mathbf{S}_{\text{pooled}}\boldsymbol{\ell}}, \quad \text{for all } \boldsymbol{\ell}\right]$$

where c^2 is selected according to Result 6.2. ∎

Remark. For testing $H_0: \boldsymbol{\mu}_1 - \boldsymbol{\mu}_2 = \mathbf{0}$, the linear combination $\hat{\boldsymbol{\ell}}'(\bar{\mathbf{x}}_1 - \bar{\mathbf{x}}_2)$, with coefficient vector $\hat{\boldsymbol{\ell}} \propto \mathbf{S}_{\text{pooled}}^{-1}(\bar{\mathbf{x}}_1 - \bar{\mathbf{x}}_2)$, quantifies the largest population difference. That is, if T^2 rejects H_0, then $\hat{\boldsymbol{\ell}}'(\bar{\mathbf{x}}_1 - \bar{\mathbf{x}}_2)$ has a nonzero mean. Frequently we try to interpret the components of this linear combination both for subject matter and statistical importance.

Example 6.4

Samples of sizes $n_1 = 45$ and $n_2 = 55$ were taken of Wisconsin homeowners with and without air conditioning, respectively. (Data courtesy of Statistical Laboratory, University of Wisconsin). Two measurements of electrical usage (in kilowatt hours) were considered. The first is a measure of total *on*-peak consumption (X_1) during July 1977 and the second is a measure of total *off*-peak consumption (X_2) during July 1977. The resulting summary statistics are

$$\bar{\mathbf{x}}_1 = \begin{bmatrix} 204.4 \\ 556.6 \end{bmatrix}, \quad \mathbf{S}_1 = \begin{bmatrix} 13825.3 & 23823.4 \\ 23823.4 & 73107.4 \end{bmatrix}, \quad n_1 = 45$$
$$\bar{\mathbf{x}}_2 = \begin{bmatrix} 130.0 \\ 355.0 \end{bmatrix}, \quad \mathbf{S}_2 = \begin{bmatrix} 8632.0 & 19616.7 \\ 19616.7 & 55964.5 \end{bmatrix}, \quad n_2 = 55$$

(The off-peak consumption is higher than the on-peak consumption because there are more off-peak hours in a month.)

Let us find 95% simultaneous confidence intervals for the differences in the mean components.

Although there appears to be somewhat of a discrepancy in the sample variances, for illustrative purposes we proceed to a calculation of the pooled sample covariance matrix. Here

$$\mathbf{S}_{\text{pooled}} = \frac{(n_1 - 1)\mathbf{S}_1 + (n_2 - 1)\mathbf{S}_2}{n_1 + n_2 - 2} = \begin{bmatrix} 10963.7 & 21505.5 \\ 21505.5 & 63661.3 \end{bmatrix}$$

and

$$c^2 = \frac{(n_1 + n_2 - 2)p}{n_1 + n_2 - p - 1}F_{p,\, n_1+n_2-p-1}(\alpha) = \frac{98(2)}{97}F_{2,97}(.05) = (2.02)(3.1) = 6.26$$

With $\boldsymbol{\mu}_1' - \boldsymbol{\mu}_2' = [\mu_{11} - \mu_{21}, \mu_{12} - \mu_{22}]$, the 95% simultaneous confidence intervals for the population differences are

$$\mu_{11} - \mu_{21}: \quad (204.4 - 130.0) \pm \sqrt{6.26}\sqrt{\left(\frac{1}{45} + \frac{1}{55}\right)10963.7}$$

or

$$21.7 \le \mu_{11} - \mu_{21} \le 127.1$$

$$\mu_{12} - \mu_{22}: \quad (556.6 - 355.0) \pm \sqrt{6.26}\sqrt{\left(\frac{1}{45} + \frac{1}{55}\right)63661.3}$$

or

$$74.7 \le \mu_{12} - \mu_{22} \le 328.5$$

We conclude there is a difference in electrical consumption between those with air conditioning and those without. This difference is evident in both on-peak and off-peak consumption.

The 95% confidence ellipse for $\boldsymbol{\mu}_1 - \boldsymbol{\mu}_2$ is determined from the eigenvalue-eigenvector pairs $\lambda_1 = 71323.5$, $\mathbf{e}_1' = [.336, .942]$ and $\lambda_2 = 3301.5$, $\mathbf{e}_2' = [.942, -.336]$. Since

$$\sqrt{\lambda_1}\sqrt{\left(\frac{1}{n_1} + \frac{1}{n_2}\right)c^2} = \sqrt{71323.5}\sqrt{\left(\frac{1}{45} + \frac{1}{55}\right)6.26} = 134.3$$

and

$$\sqrt{\lambda_2}\sqrt{\left(\frac{1}{n_1} + \frac{1}{n_2}\right)c^2} = \sqrt{3301.5}\sqrt{\left(\frac{1}{45} + \frac{1}{55}\right)6.26} = 28.9$$

we obtain the 95% confidence ellipse for $\boldsymbol{\mu}_1 - \boldsymbol{\mu}_2$ sketched in Figure 6.2. Because the confidence ellipse for the difference in means does not cover $\mathbf{0}' = [0, 0]$, the T^2-statistic will reject $H_0: \boldsymbol{\mu}_1 - \boldsymbol{\mu}_2 = \mathbf{0}$ at the 5% level. The coefficient vector for the linear combination most responsible for rejection is proportional to $\mathbf{S}_{\text{pooled}}^{-1}(\bar{\mathbf{x}}_1 - \bar{\mathbf{x}}_2)$. (See Exercise 6.4.) ■

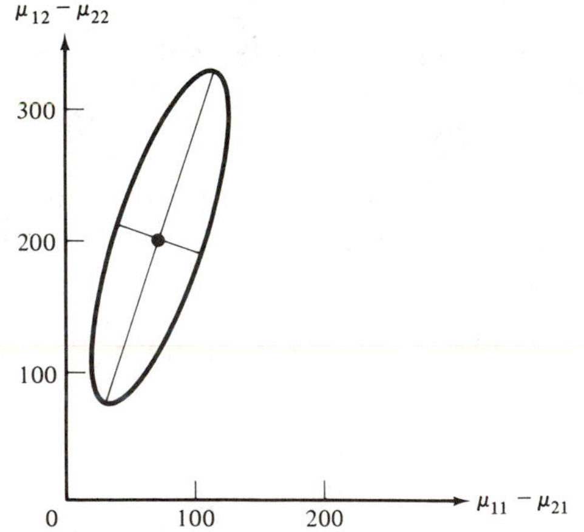

Figure 6.2 95% confidence ellipse for $\boldsymbol{\mu}_1' - \boldsymbol{\mu}_2' = (\mu_{11} - \mu_{21}, \mu_{12} - \mu_{22})$.

The Two-Sample Situation when $\Sigma_1 \neq \Sigma_2$

When $\boldsymbol{\Sigma}_1 \neq \boldsymbol{\Sigma}_2$, we are unable to find a "distance" measure like T^2, whose distribution does not depend on the unknowns $\boldsymbol{\Sigma}_1$ and $\boldsymbol{\Sigma}_2$. Bartlett's test [2] is used to test equality of $\boldsymbol{\Sigma}_1$ and $\boldsymbol{\Sigma}_2$ in terms of generalized variances. Unfortunately, the conclusions can be seriously misleading when the populations are nonnormal. Nonnormality and unequal covariances cannot be separated with Bartlett's test.

We suggest, without much factual support, that any discrepancy of the order $\sigma_{1,ii} = 4\sigma_{2,ii}$, or vice versa, is probably serious. This is true in the univariate case. The size of the discrepancies that are critical in the multivariate situation probably depends, to a large extent, on the number of variables, p.

A transformation may improve things when the marginal variances are quite different. However, for n_1 and n_2 large, we can avoid the complexities due to unequal covariance matrices.

Result 6.4. Let the sample sizes be such that $n_1 - p$ and $n_2 - p$ are large. An approximate $100(1-\alpha)\%$ confidence ellipsoid for $\boldsymbol{\mu}_1 - \boldsymbol{\mu}_2$ is given by all $\boldsymbol{\mu}_1 - \boldsymbol{\mu}_2$ satisfying

$$[\bar{\mathbf{x}}_1 - \bar{\mathbf{x}}_2 - (\boldsymbol{\mu}_1 - \boldsymbol{\mu}_2)]' \left[\frac{1}{n_1}\mathbf{S}_1 + \frac{1}{n_2}\mathbf{S}_2\right]^{-1} [\bar{\mathbf{x}}_1 - \bar{\mathbf{x}}_2 - (\boldsymbol{\mu}_1 - \boldsymbol{\mu}_2)] \leq \chi_p^2(\alpha)$$

where $\chi_p^2(\alpha)$ is the upper (100α)th percentile of a chi-square distribution with p d.f. Also, $100(1-\alpha)\%$ simultaneous confidence intervals for all linear combinations $\boldsymbol{\ell}'(\boldsymbol{\mu}_1 - \boldsymbol{\mu}_2)$ are provided by

$$\boldsymbol{\ell}'(\boldsymbol{\mu}_1 - \boldsymbol{\mu}_2) \quad \text{belongs to} \quad \boldsymbol{\ell}'(\bar{\mathbf{x}}_1 - \bar{\mathbf{x}}_2) \pm \sqrt{\chi_p^2(\alpha)} \sqrt{\boldsymbol{\ell}'\left(\frac{1}{n_1}\mathbf{S}_1 + \frac{1}{n_2}\mathbf{S}_2\right)\boldsymbol{\ell}}$$

Proof. From (6-22) and (3-9),

$$E(\bar{\mathbf{X}}_1 - \bar{\mathbf{X}}_2) = \boldsymbol{\mu}_1 - \boldsymbol{\mu}_2$$

and

$$\text{Cov}(\bar{\mathbf{X}}_1 - \bar{\mathbf{X}}_2) = \text{Cov}(\bar{\mathbf{X}}_1) + \text{Cov}(\bar{\mathbf{X}}_2) = \frac{1}{n_1}\boldsymbol{\Sigma}_1 + \frac{1}{n_2}\boldsymbol{\Sigma}_2$$

By the central limit theorem, $\bar{\mathbf{X}}_1 - \bar{\mathbf{X}}_2$ is nearly $N_p[\boldsymbol{\mu}_1 - \boldsymbol{\mu}_2, (1/n_1)\boldsymbol{\Sigma}_1 + (1/n_2)\boldsymbol{\Sigma}_2]$. If $\boldsymbol{\Sigma}_1$ and $\boldsymbol{\Sigma}_2$ were known, the squared statistical distance from $\bar{\mathbf{X}}_1 - \bar{\mathbf{X}}_2$ to $\boldsymbol{\mu}_1 - \boldsymbol{\mu}_2$ would be

$$[\bar{\mathbf{X}}_1 - \bar{\mathbf{X}}_2 - (\boldsymbol{\mu}_1 - \boldsymbol{\mu}_2)]' \left(\frac{1}{n_1}\boldsymbol{\Sigma}_1 + \frac{1}{n_2}\boldsymbol{\Sigma}_2\right)^{-1} [\bar{\mathbf{X}}_1 - \bar{\mathbf{X}}_2 - (\boldsymbol{\mu}_1 - \boldsymbol{\mu}_2)]$$

This squared distance has an approximate χ_p^2-distribution by Result 4.7. When n_1 and n_2 are large, with high probability, $\mathbf{S}_1$ will be close to $\boldsymbol{\Sigma}_1$ and $\mathbf{S}_2$ will be close to $\boldsymbol{\Sigma}_2$. Consequently, the approximation holds with $\mathbf{S}_1$ and $\mathbf{S}_2$ in place of $\boldsymbol{\Sigma}_1$ and $\boldsymbol{\Sigma}_2$ respectively.

The results concerning the simultaneous confidence intervals follow from Result 5C.1. ■

Remark. If $n_1 = n_2 = n$, $(n-1)/(n+n-2) = 1/2$, so

$$\frac{1}{n_1}\mathbf{S}_1 + \frac{1}{n_2}\mathbf{S}_2 = \frac{1}{n}(\mathbf{S}_1 + \mathbf{S}_2) = \frac{(n-1)\mathbf{S}_1 + (n-1)\mathbf{S}_2}{n+n-2}\left(\frac{1}{n} + \frac{1}{n}\right)$$

$$= \mathbf{S}_{\text{pooled}}\left(\frac{1}{n} + \frac{1}{n}\right)$$

With equal sample sizes, the large sample procedure is essentially the same as the procedure based on the pooled covariance matrix (see Result 6.2). In one dimension it is well known that the effect of unequal variances is least when $n_1 = n_2$ and greatest when n_1 is much less than n_2, or vice versa.

Example 6.5

We shall analyze the electrical-consumption data discussed in Example 6.4 using the large-sample approach. We first calculate

$$\frac{1}{n_1}\mathbf{S}_1 + \frac{1}{n_2}\mathbf{S}_2 = \frac{1}{45}\begin{bmatrix} 13825.3 & 23823.4 \\ 23823.4 & 73107.4 \end{bmatrix} + \frac{1}{55}\begin{bmatrix} 8632.0 & 19616.7 \\ 19616.7 & 55964.5 \end{bmatrix}$$

$$= \begin{bmatrix} 464.17 & 886.08 \\ 886.08 & 2642.15 \end{bmatrix}$$

The 95% simultaneous confidence intervals for the linear combinations

$$\boldsymbol{\ell}'(\boldsymbol{\mu}_1 - \boldsymbol{\mu}_2) = [1, 0]\begin{bmatrix} \mu_{11} - \mu_{21} \\ \mu_{12} - \mu_{22} \end{bmatrix} = \mu_{11} - \mu_{21}$$

and

$$\boldsymbol{\ell}'(\boldsymbol{\mu}_1 - \boldsymbol{\mu}_2) = [0, 1]\begin{bmatrix} \mu_{11} - \mu_{21} \\ \mu_{12} - \mu_{22} \end{bmatrix} = \mu_{12} - \mu_{22}$$

are (see Result 6.4)

$$\mu_{11} - \mu_{21}: \quad 74.4 \pm \sqrt{5.99}\,\sqrt{464.17} \qquad \text{or} \quad (21.7, 127.1)$$

$$\mu_{12} - \mu_{22}: \quad 201.6 \pm \sqrt{5.99}\,\sqrt{2642.15} \qquad \text{or} \quad (75.8, 327.4)$$

Notice that these intervals differ negligibly from the intervals in Example 6.4, where the pooling procedure was employed. The statistic T^2 for testing $H_0\colon \boldsymbol{\mu}_1 - \boldsymbol{\mu}_2 = \mathbf{0}$ is

$$T^2 = [\bar{\mathbf{x}}_1 - \bar{\mathbf{x}}_2]'\left[\frac{1}{n_1}\mathbf{S}_1 + \frac{1}{n_2}\mathbf{S}_2\right]^{-1}[\bar{\mathbf{x}}_1 - \bar{\mathbf{x}}_2]$$

$$= \begin{bmatrix} 204.4 - 130.0 \\ 556.6 - 355.0 \end{bmatrix}'\begin{bmatrix} 464.17 & 886.08 \\ 886.08 & 2642.15 \end{bmatrix}^{-1}\begin{bmatrix} 204.4 - 130.0 \\ 556.6 - 355.0 \end{bmatrix}$$

$$= [74.4 \quad 201.6](10^{-4})\begin{bmatrix} 59.874 & -20.080 \\ -20.080 & 10.519 \end{bmatrix}\begin{bmatrix} 74.4 \\ 201.6 \end{bmatrix} = 15.66$$

For $\alpha = .05$, the critical value is $\chi_2^2(.05) = 5.99$ and, since $T^2 = 15.66 > \chi_2^2(.05) = 5.99$, we reject H_0.

The most critical linear combination leading to the rejection of H_0 has coefficient vector

$$\hat{\ell} \propto \left(\frac{1}{n_1}\mathbf{S}_1 + \frac{1}{n_2}\mathbf{S}_2\right)^{-1}(\bar{\mathbf{x}}_1 - \bar{\mathbf{x}}_2) = (10^{-4})\begin{bmatrix} 59.874 & -20.080 \\ -20.080 & 10.519 \end{bmatrix}\begin{bmatrix} 74.4 \\ 201.6 \end{bmatrix}$$

$$= \begin{bmatrix} .041 \\ .063 \end{bmatrix}$$

The difference in *off*-peak electrical consumption, between those with air conditioning and those without, contributes more (.063 versus .041) than the corresponding difference in *on*-peak consumption to the rejection of H_0: $\boldsymbol{\mu}_1 - \boldsymbol{\mu}_2 = \mathbf{0}$. ■

6.4 COMPARISON OF SEVERAL MULTIVARIATE POPULATION MEANS (ONE-WAY MANOVA)

Often, more than two populations need to be compared. Random samples, collected from each of g populations, are arranged as

$$\begin{aligned} &\text{Population } 1: \quad \mathbf{X}_{11}, \mathbf{X}_{12}, \ldots, \mathbf{X}_{1n_1} \\ &\text{Population } 2: \quad \mathbf{X}_{21}, \mathbf{X}_{22}, \ldots, \mathbf{X}_{2n_2} \\ &\quad \vdots \qquad\qquad\qquad \vdots \\ &\text{Population } g: \quad \mathbf{X}_{g1}, \mathbf{X}_{g2}, \ldots, \mathbf{X}_{gn_g} \end{aligned} \tag{6-27}$$

MANOVA is used first to investigate whether the population mean vectors are the same and, if not, which mean components differ significantly.

Assumptions concerning the Structure of the Data

1. $\mathbf{X}_{\ell 1}, \mathbf{X}_{\ell 2}, \ldots, \mathbf{X}_{\ell n_\ell}$ is a random sample of size n_ℓ from a population with mean $\boldsymbol{\mu}_\ell$, $\ell = 1, 2, \ldots, g$. The random samples from different populations are independent.
2. All populations have a common covariance matrix $\boldsymbol{\Sigma}$.
3. Each population is multivariate normal.

Condition 3 can be relaxed by appealing to the central limit theorem (Result 4.13) when the sample sizes, n_ℓ, are large.

A review of the univariate analysis of variance (ANOVA) will facilitate our discussion of the multivariate assumptions and solution methods.

A Summary of Univariate ANOVA

In the univariate situation, the assumptions become: $X_{\ell 1}, X_{\ell 2}, \ldots, X_{\ell n_\ell}$ is a random sample from an $N(\mu_\ell, \sigma^2)$ population, $\ell = 1, 2, \ldots, g$, and the random samples are independent. Although the null hypothesis of equality of means could be formulated as $\mu_1 = \mu_2 = \cdots = \mu_g$, it is customary to regard μ_ℓ as the sum of an overall mean component, such as μ, and a component due to the specific population. For instance, we can write $\mu_\ell = \mu + (\mu_\ell - \mu)$ or $\mu_\ell = \mu + \tau_\ell$.

Populations usually correspond to different sets of experimental conditions, and therefore it is convenient to investigate the deviations, τ_ℓ, associated with the ℓth population (treatment).

The *reparameterization*

$$\underset{\left(\begin{array}{c}\ell\text{th population}\\ \text{mean}\end{array}\right)}{\mu_\ell} = \underset{\left(\begin{array}{c}\text{overall}\\ \text{mean}\end{array}\right)}{\mu} + \underset{\left(\begin{array}{c}\ell\text{th population}\\ \text{(treatment) effect}\end{array}\right)}{\tau_\ell} \tag{6-28}$$

leads to a restatement of the hypothesis of equality of means. The null hypothesis becomes

$$H_0\colon \tau_1 = \tau_2 = \cdots = \tau_g = 0$$

The response $X_{\ell j}$, distributed as $N(\mu + \tau_\ell, \sigma^2)$, can be expressed in the suggestive form

$$X_{\ell j} = \underset{\text{(overall mean)}}{\mu} + \underset{\left(\begin{array}{c}\text{treatment}\\ \text{effect}\end{array}\right)}{\tau_\ell} + \underset{\left(\begin{array}{c}\text{random}\\ \text{error}\end{array}\right)}{e_{\ell j}} \tag{6-29}$$

where the $e_{\ell j}$ are independent $N(0, \sigma^2)$ random variables. To uniquely define the model parameters and their least squares estimates it is customary to impose the constraint $\sum_{\ell=1}^{g} n_\ell \tau_\ell = 0$.

Motivated by the decomposition in (6-29), the analysis of variance is based upon an analogous decomposition of the observations,

$$\underset{\text{(observation)}}{x_{\ell j}} = \underset{\left(\begin{array}{c}\text{overall}\\ \text{sample mean}\end{array}\right)}{\bar{x}} + \underset{\left(\begin{array}{c}\text{estimated}\\ \text{treatment effect}\end{array}\right)}{(\bar{x}_\ell - \bar{x})} + \underset{\text{(residual)}}{(x_{\ell j} - \bar{x}_\ell)} \tag{6-30}$$

where $\bar{x}$ is an estimate of μ, $\hat{\tau}_\ell = (\bar{x}_\ell - \bar{x})$ is an estimate of τ_ℓ, and $(x_{\ell j} - \bar{x}_\ell)$ is an estimate of the error $e_{\ell j}$.

Example 6.6

Consider the following independent samples.

Population 1:	9, 6, 9
Population 2:	0, 2
Population 3:	3, 1, 2

Since, for example, $\bar{x}_3 = (3 + 1 + 2)/3 = 2$ and $\bar{x} = (9 + 6 + 9 + 0 + 2 + 3 + 1 + 2)/8 = 4$, we find

$$\begin{aligned} 3 = x_{31} &= \bar{x} + (\bar{x}_3 - \bar{x}) + (x_{31} - \bar{x}_3) \\ &= 4 + (2 - 4) + (3 - 2) \\ &= 4 + (-2) + 1 \end{aligned}$$

Repeating this operation for each observation, we obtain the arrays

$$\begin{pmatrix} 9 & 6 & 9 \\ 0 & 2 & \\ 3 & 1 & 2 \end{pmatrix} = \begin{pmatrix} 4 & 4 & 4 \\ 4 & 4 & \\ 4 & 4 & 4 \end{pmatrix} + \begin{pmatrix} 4 & 4 & 4 \\ -3 & -3 & \\ -2 & -2 & -2 \end{pmatrix} + \begin{pmatrix} 1 & -2 & 1 \\ -1 & 1 & \\ 1 & -1 & 0 \end{pmatrix}$$

$$\begin{array}{cccc} \text{observation} = & \text{mean} & + \text{ treatment effect} & + \text{ residual} \\ (x_{\ell j}) & (\bar{x}) & (\bar{x}_\ell - \bar{x}) & (x_{\ell j} - \bar{x}_\ell) \end{array}$$

The question of equality of means is answered by assessing whether the contribution of the treatment array is large relative to the residuals. (Our estimates $\hat{\tau}_\ell = \bar{x}_\ell - \bar{x}$ of τ_ℓ always satisfy $\sum_{\ell=1}^{g} n_\ell \hat{\tau}_\ell = 0$. Under H_0, each $\hat{\tau}_\ell$ is an estimate of zero.) If the treatment contribution is large, H_0 should be rejected. The size of an array is quantified by stringing the rows of the array out into a vector and calculating its squared length. This quantity is called the *sum of squares* (SS). For the observations, we construct the vector $\mathbf{y}' = [9, 6, 9, 0, 2, 3, 1, 2]$. Its squared length is:

$$SS_{obs} = 9^2 + 6^2 + 9^2 + 0^2 + 2^2 + 3^2 + 1^2 + 2^2 = 216$$

Similarly,

$$\begin{aligned} SS_{mean} &= 4^2 + 4^2 + 4^2 + 4^2 + 4^2 + 4^2 + 4^2 + 4^2 = 8(4^2) = 128 \\ SS_{tr} &= 4^2 + 4^2 + 4^2 + (-3)^2 + (-3)^2 + (-2)^2 + (-2)^2 + (-2)^2 \\ &= 3(4^2) + 2(-3)^2 + 3(-2)^2 = 78 \end{aligned}$$

and the residual sum of squares is

$$SS_{res} = 1^2 + (-2)^2 + 1^2 + (-1)^2 + 1^2 + 1^2 + (-1)^2 + 0^2 = 10$$

The sums of squares satisfy the same decomposition, (6-30), as the observations. Consequently,

$$SS_{obs} = SS_{mean} + SS_{tr} + SS_{res}$$

or $216 = 128 + 78 + 10$. The breakup into sums of squares apportions variability in the combined samples into mean, treatment, and residual (error) components. An analysis of variance proceeds by comparing the relative sizes of SS_{tr} and SS_{res}. If H_0 is true, variances computed from SS_{tr} and SS_{res} should be approximately equal. ■

The sum of squares decomposition illustrated numerically in Example 6.6 is so basic that the algebraic equivalent will now be developed.

Subtracting $\bar{x}$ from both sides of (6-30) and squaring gives

$$(x_{\ell j} - \bar{x})^2 = (\bar{x}_\ell - \bar{x})^2 + (x_{\ell j} - \bar{x}_\ell)^2 + 2(\bar{x}_\ell - \bar{x})(x_{\ell j} - \bar{x}_\ell)$$

We can sum both sides over j, note that $\sum_{j=1}^{n_\ell} (x_{\ell j} - \bar{x}_\ell) = 0$, and obtain

$$\sum_{j=1}^{n_\ell} (x_{\ell j} - \bar{x})^2 = n_\ell(\bar{x}_\ell - \bar{x})^2 + \sum_{j=1}^{n_\ell} (x_{\ell j} - \bar{x}_\ell)^2$$

Next, summing both sides over ℓ we get

$$\sum_{\ell=1}^{g}\sum_{j=1}^{n_\ell} (x_{\ell j} - \bar{x})^2 = \sum_{\ell=1}^{g} n_\ell(\bar{x}_\ell - \bar{x})^2 + \sum_{\ell=1}^{g}\sum_{j=1}^{n_\ell} (x_{\ell j} - \bar{x}_\ell)^2 \qquad (6\text{-}31)$$

$$\begin{pmatrix} SS_{cor} \\ \text{total (corrected) SS} \end{pmatrix} = \begin{pmatrix} SS_{tr} \\ \text{between (sample) SS} \end{pmatrix} + \begin{pmatrix} SS_{res} \\ \text{within (samples) SS} \end{pmatrix}$$

or

$$\sum_{\ell=1}^{g}\sum_{j=1}^{n_\ell} x_{\ell j}^2 = (n_1 + n_2 + \cdots + n_g)\bar{x}^2 + \sum_{\ell=1}^{g} n_\ell(\bar{x}_\ell - \bar{x})^2 + \sum_{\ell=1}^{g}\sum_{j=1}^{n_\ell} (x_{\ell j} - \bar{x}_\ell)^2$$

$$(SS_{obs}) \quad = \quad (SS_{mean}) \quad + \quad (SS_{tr}) \quad + \quad (SS_{res}) \qquad (6\text{-}32)$$

In the course of establishing (6-32), we have verified that the arrays representing the mean, treatment effects, and residuals are *orthogonal*. That is, these arrays considered as vectors are perpendicular whatever the observation vector $\mathbf{y}' = [x_{11}, \ldots, x_{1n_1}, x_{21}, \ldots, x_{2n_2}, \ldots, x_{gn_g}]$. Consequently we could obtain SS_{res} by subtraction, without having to calculate the individual residuals, because $SS_{res} = SS_{obs} - SS_{mean} - SS_{tr}$. However, this is false economy because plots of the residuals provide checks on the model assumptions.

The vector representations of the arrays involved in the decomposition (6-30) also have geometric interpretations that provide the degrees of freedom. For an arbitrary set of observations $[x_{11}, \ldots, x_{1n_1}, x_{21}, \ldots, x_{2n_2}, \ldots, x_{gn_g}] = \mathbf{y}'$: the observation vector, $\mathbf{y}$, can lie anywhere in $n = n_1 + n_2 + \cdots + n_g$ dimensions; the mean vector, $\bar{x}\mathbf{1} = [\bar{x}, \ldots, \bar{x}]'$, must lie along the equiangular line of $\mathbf{1}$; and the treatment effect vector

$$(\bar{x}_1 - \bar{x})\begin{bmatrix} 1 \\ \vdots \\ 1 \\ 0 \\ \vdots \\ 0 \end{bmatrix}\begin{matrix} \left.\vphantom{\begin{matrix}1\\ \vdots\\ 1\end{matrix}}\right\} n_1 \\ \\ \\ \\ \end{matrix} + (\bar{x}_2 - \bar{x})\begin{bmatrix} 0 \\ \vdots \\ 0 \\ 1 \\ \vdots \\ 1 \\ 0 \\ \vdots \\ 0 \end{bmatrix}\begin{matrix} \\ \\ \\ \left.\vphantom{\begin{matrix}1\\ \vdots\\ 1\end{matrix}}\right\} n_2 \\ \\ \\ \end{matrix} + \cdots + (\bar{x}_g - \bar{x})\begin{bmatrix} 0 \\ \vdots \\ 0 \\ 1 \\ \vdots \\ 1 \end{bmatrix}\begin{matrix} \\ \\ \\ \left.\vphantom{\begin{matrix}1\\ \vdots\\ 1\end{matrix}}\right\} n_g \end{matrix}$$

$$= (\bar{x}_1 - \bar{x})\mathbf{u}_1 + (\bar{x}_2 - \bar{x})\mathbf{u}_2 + \cdots + (\bar{x}_g - \bar{x})\mathbf{u}_g$$

lies in the hyperplane of linear combinations of the g vectors $\mathbf{u}_1, \mathbf{u}_2, \ldots, \mathbf{u}_g$. Since $\mathbf{1} = \mathbf{u}_1 + \mathbf{u}_2 + \cdots + \mathbf{u}_g$, the mean vector also lies in this hyperplane and it is *always* perpendicular to the treatment vector (see Exercise 6.7). Thus the mean vector has the freedom to lie anywhere along the one-dimensional equiangular line and the treatment vector has the freedom to lie anywhere in the other $g - 1$ dimensions. The residual vector, $\hat{\mathbf{e}} = \mathbf{y} - (\bar{x}\mathbf{1}) - [(\bar{x}_1 - \bar{x})\mathbf{u}_1 + \cdots + (\bar{x}_g - \bar{x})\mathbf{u}_g]$ is perpendicular to the mean vector and treatment effect vector and has the freedom to lie anywhere in the subspace of dimension $n - (g - 1) - 1 = n - g$, that is perpendicular to their hyperplane.

To summarize, we attribute 1 d.f. to SS_{mean}, $g - 1$ d.f. to SS_{tr} and $n - g = (n_1 + n_2 + \cdots + n_g) - g$ d.f. to SS_{res}. The total number of degrees of freedom is $n = n_1 + n_2 + \cdots + n_g$. Alternatively, by appealing to the univariate distribution theory, these are the degrees of freedom for the chi-square distributions associated with the corresponding sums of squares.

The calculations of the sums of squares and the associated degrees of freedom are conveniently summarized by an ANOVA table.

ANOVA TABLE FOR COMPARING UNIVARIATE POPULATION MEANS

Source of variation	Sum of squares (SS)	Degrees of freedom (d.f.)
Treatments	$SS_{\text{tr}} = \sum_{\ell=1}^{g} n_\ell(\bar{x}_\ell - \bar{x})^2$	$g - 1$
Residual (Error)	$SS_{\text{res}} = \sum_{\ell=1}^{g} \sum_{j=1}^{n_\ell} (x_{\ell j} - \bar{x}_\ell)^2$	$\sum_{\ell=1}^{g} n_\ell - g$
Total (corrected for the mean)	$SS_{\text{cor}} = \sum_{\ell=1}^{g} \sum_{j=1}^{n_\ell} (x_{\ell j} - \bar{x})^2$	$\sum_{\ell=1}^{g} n_\ell - 1$

The usual F-test rejects $H_0: \tau_1 = \tau_2 = \cdots = \tau_g = 0$; at level α, if

$$F = \frac{SS_{\text{tr}}/(g-1)}{SS_{\text{res}}\Big/\left(\sum_{\ell=1}^{g} n_\ell - g\right)} > F_{g-1, \Sigma n_\ell - g}(\alpha)$$

where $F_{g-1, \Sigma n_\ell - g}(\alpha)$ is the upper (100α)th percentile of the F-distribution with $g - 1$ and $\Sigma n_\ell - g$ degrees of freedom. This is equivalent to rejecting H_0 for large values of $SS_{\text{tr}}/SS_{\text{res}}$ or for large values of $1 + SS_{\text{tr}}/SS_{\text{res}}$. The statistic appropriate for a multivariate generalization rejects H_0 for *small* values of the reciprocal

$$\frac{1}{1 + SS_{\text{tr}}/SS_{\text{res}}} = \frac{SS_{\text{res}}}{SS_{\text{res}} + SS_{\text{tr}}} \tag{6-33}$$

Example 6.7

Using the information in Example 6.6, we have the following ANOVA table.

Source of variation	Sum of squares	Degrees of freedom
Treatments	$SS_{tr} = 78$	$g - 1 = 3 - 1 = 2$
Residual	$SS_{res} = 10$	$\sum_{\ell=1}^{g} n_\ell - g = (3 + 2 + 3) - 3 = 5$
Total (corrected)	$SS_{cor} = 88$	$\sum_{\ell=1}^{g} n_\ell - 1 = 7$

Consequently,

$$F = \frac{SS_{tr}/(g-1)}{SS_{res}/(\Sigma n_\ell - g)} = \frac{78/2}{10/5} = 19.5$$

Since $F = 19.5 > F_{2,5}(.01) = 13.27$, we reject H_0: $\tau_1 = \tau_2 = \tau_3 = 0$ (no treatment effect) at the 1% level. ■

Multivariate Analysis of Variance (MANOVA)

Paralleling the univariate reparameterization, the model becomes the following.

> **MANOVA Model for Comparing g Population Mean Vectors**
>
> $$\mathbf{X}_{\ell j} = \boldsymbol{\mu} + \boldsymbol{\tau}_\ell + \mathbf{e}_{\ell j}, \qquad j = 1, 2, \ldots, n_\ell \quad \text{and} \quad \ell = 1, 2, \ldots, g \qquad \text{(6-34)}$$
>
> where $\mathbf{e}_{\ell j}$ are independent $N_p(\mathbf{0}, \boldsymbol{\Sigma})$ variables. Here the parameter vector $\boldsymbol{\mu}$ is an overall mean (level) and $\boldsymbol{\tau}_\ell$ represents the ℓth treatment effect with $\sum_{\ell=1}^{g} n_\ell \boldsymbol{\tau}_\ell = \mathbf{0}$.

According to the model in (6-34), *each component* of the observation vector $\mathbf{X}_{\ell j}$ satisfies the univariate model (6-29). The errors for the components of $\mathbf{X}_{\ell j}$ are correlated, but the covariance matrix $\boldsymbol{\Sigma}$ is the same for all populations.

A vector of observations may be decomposed as suggested by the model. Thus

$$\underset{\text{(observation)}}{\mathbf{x}_{\ell j}} = \underset{\left(\substack{\text{overall sample}\\ \text{mean, } \hat{\boldsymbol{\mu}}}\right)}{\bar{\mathbf{x}}} + \underset{\left(\substack{\text{estimated}\\ \text{treatment}\\ \text{effect, } \hat{\boldsymbol{\tau}}_\ell}\right)}{(\bar{\mathbf{x}}_\ell - \bar{\mathbf{x}})} + \underset{\left(\substack{\text{residual,}\\ \hat{\mathbf{e}}_{\ell j}}\right)}{(\mathbf{x}_{\ell j} - \bar{\mathbf{x}}_\ell)} \qquad \text{(6-35)}$$

The decomposition in (6-35) leads to the multivariate analog of the univariate sum of squares breakup in (6-31). First we note that the crossproduct

$$(\mathbf{x}_{\ell j} - \bar{\mathbf{x}})(\mathbf{x}_{\ell j} - \bar{\mathbf{x}})'$$

can be written as

$$\begin{aligned}(\mathbf{x}_{\ell j} - \bar{\mathbf{x}})(\mathbf{x}_{\ell j} - \bar{\mathbf{x}})' &= \left[(\mathbf{x}_{\ell j} - \bar{\mathbf{x}}_\ell) + (\bar{\mathbf{x}}_\ell - \bar{\mathbf{x}})\right]\left[(\mathbf{x}_{\ell j} - \bar{\mathbf{x}}_\ell) + (\bar{\mathbf{x}}_\ell - \bar{\mathbf{x}})\right]' \\ &= (\mathbf{x}_{\ell j} - \bar{\mathbf{x}}_\ell)(\mathbf{x}_{\ell j} - \bar{\mathbf{x}}_\ell)' + (\mathbf{x}_{\ell j} - \bar{\mathbf{x}}_\ell)(\bar{\mathbf{x}}_\ell - \bar{\mathbf{x}})' \\ &\quad + (\bar{\mathbf{x}}_\ell - \bar{\mathbf{x}})(\mathbf{x}_{\ell j} - \bar{\mathbf{x}}_\ell)' + (\bar{\mathbf{x}}_\ell - \bar{\mathbf{x}})(\bar{\mathbf{x}}_\ell - \bar{\mathbf{x}})'\end{aligned}$$

The sum over j of the middle two terms is the zero matrix because $\sum_{j=1}^{n_\ell} (\mathbf{x}_{\ell j} - \bar{\mathbf{x}}_\ell) = \mathbf{0}$. Next, since $\sum_{\ell=1}^{g} n_\ell(\bar{\mathbf{x}}_\ell - \bar{\mathbf{x}}) = \mathbf{0}$, summing the crossproduct over ℓ and j yields

$$\sum_{\ell=1}^{g}\sum_{j=1}^{n_\ell} (\mathbf{x}_{\ell j} - \bar{\mathbf{x}})(\mathbf{x}_{\ell j} - \bar{\mathbf{x}})' = \sum_{\ell=1}^{g} n_\ell(\bar{\mathbf{x}}_\ell - \bar{\mathbf{x}})(\bar{\mathbf{x}}_\ell - \bar{\mathbf{x}})' + \sum_{\ell=1}^{g}\sum_{j=1}^{n_\ell} (\mathbf{x}_{\ell j} - \bar{\mathbf{x}}_\ell)(\mathbf{x}_{\ell j} - \bar{\mathbf{x}}_\ell)'$$

$$\begin{pmatrix}\text{total (Corrected) sum} \\ \text{of squares and cross-} \\ \text{products}\end{pmatrix} \quad \begin{pmatrix}\text{treatment (}\underline{\text{B}}\text{etween)} \\ \text{sum of squares and} \\ \text{crossproducts}\end{pmatrix} \quad \begin{pmatrix}\text{residual (}\underline{\text{W}}\text{ithin) sum} \\ \text{of squares and cross-} \\ \text{products}\end{pmatrix} \tag{6-36}$$

The *within* sum of squares and crossproducts matrix can be expressed as

$$\begin{aligned}\mathbf{W} &= \sum_{\ell=1}^{g}\sum_{j=1}^{n_\ell} (\mathbf{x}_{\ell j} - \bar{\mathbf{x}}_\ell)(\mathbf{x}_{\ell j} - \bar{\mathbf{x}}_\ell)' \\ &= (n_1 - 1)\mathbf{S}_1 + (n_2 - 1)\mathbf{S}_2 + \cdots + (n_g - 1)\mathbf{S}_g\end{aligned} \tag{6-37}$$

where $\mathbf{S}_\ell$ is the sample covariance matrix for the ℓth sample. This matrix is a generalization of the $(n_1 + n_2 - 2)\mathbf{S}_{\text{pooled}}$ matrix encountered in the two-sample case. It plays a dominant role in testing for the presence of treatment effects.

Analogous to the univariate result, the hypothesis of no treatment effects

$$H_0\colon \boldsymbol{\tau}_1 = \boldsymbol{\tau}_2 = \cdots = \boldsymbol{\tau}_g = \mathbf{0}$$

is tested by considering the relative sizes of the treatment and residual sums of squares and crossproducts. Equivalently, we may consider the relative sizes of the residual and total (corrected) sum of squares and crossproducts. Formally, we summarize the calculations leading to the test statistic in a MANOVA table. This table is exactly the same form, component by component, as the ANOVA table, except squares of scalars are replaced by their vector counterparts. For example, $(\bar{x}_\ell - \bar{x})^2$ becomes $(\bar{\mathbf{x}}_\ell - \bar{\mathbf{x}})(\bar{\mathbf{x}}_\ell - \bar{\mathbf{x}})'$. The degrees of freedom correspond to the univariate geometry and also to some multivariate distribution theory involving Wishart densities. (See [1].)

MANOVA TABLE FOR COMPARING POPULATION MEAN VECTORS

Source of variation	Matrix of sum of squares and crossproducts (SSP)	Degrees of freedom (d.f.)
Treatment	$\mathbf{B} = \sum_{\ell=1}^{g} n_\ell(\bar{\mathbf{x}}_\ell - \bar{\mathbf{x}})(\bar{\mathbf{x}}_\ell - \bar{\mathbf{x}})'$	$g - 1$
Residual (Error)	$\mathbf{W} = \sum_{\ell=1}^{g} \sum_{j=1}^{n_\ell} (\mathbf{x}_{\ell j} - \bar{\mathbf{x}}_\ell)(\mathbf{x}_{\ell j} - \bar{\mathbf{x}}_\ell)'$	$\sum_{\ell=1}^{g} n_\ell - g$
Total (corrected for the mean)	$\mathbf{B} + \mathbf{W} = \sum_{\ell=1}^{g} \sum_{j=1}^{n_\ell} (\mathbf{x}_{\ell j} - \bar{\mathbf{x}})(\mathbf{x}_{\ell j} - \bar{\mathbf{x}})'$	$\sum_{\ell=1}^{g} n_\ell - 1$

One test of $H_0: \boldsymbol{\tau}_1 = \boldsymbol{\tau}_2 = \cdots = \boldsymbol{\tau}_g = \mathbf{0}$ involves generalized variances. We reject H_0 if the ratio of generalized variances

$$\Lambda^* = \frac{|\mathbf{W}|}{|\mathbf{B} + \mathbf{W}|} = \frac{\left| \sum_{\ell=1}^{g} \sum_{j=1}^{n_\ell} (\mathbf{x}_{\ell j} - \bar{\mathbf{x}}_\ell)(\mathbf{x}_{\ell j} - \bar{\mathbf{x}}_\ell)' \right|}{\left| \sum_{\ell=1}^{g} \sum_{j=1}^{n_\ell} (\mathbf{x}_{\ell j} - \bar{\mathbf{x}})(\mathbf{x}_{\ell j} - \bar{\mathbf{x}})' \right|} \tag{6-38}$$

is too small. The quantity $\Lambda^* = |\mathbf{W}|/|\mathbf{B} + \mathbf{W}|$, proposed originally by Wilks (see [20]), corresponds to the equivalent form (6-33) of the F-test of H_0: no-treatment effects in the univariate case. *Wilks' lambda* has the virtue of being convenient and related to the likelihood ratio criterion. The exact distribution of Λ^* can be derived for the special cases listed in Table 6.3. For other cases and large sample sizes, a modification of Λ^* due to Bartlett (see [3]) can be used to test H_0.

TABLE 6.3 DISTRIBUTION OF WILK'S LAMBDA, $\Lambda^* = |\mathbf{W}|/|\mathbf{B} + \mathbf{W}|$

No. of variables	No. of groups	Sampling distribution for multivariate normal data
$p = 1$	$g \geq 2$	$\left(\frac{\Sigma n_\ell - g}{g - 1}\right)\left(\frac{1 - \Lambda^*}{\Lambda^*}\right) \sim F_{g-1, \Sigma n_\ell - g}$
$p = 2$	$g \geq 2$	$\left(\frac{\Sigma n_\ell - g - 1}{g - 1}\right)\left(\frac{1 - \sqrt{\Lambda^*}}{\sqrt{\Lambda^*}}\right) \sim F_{2(g-1), 2(\Sigma n_\ell - g - 1)}$
$p \geq 1$	$g = 2$	$\left(\frac{\Sigma n_\ell - p - 1}{p}\right)\left(\frac{1 - \Lambda^*}{\Lambda^*}\right) \sim F_{p, \Sigma n_\ell - p - 1}$
$p \geq 1$	$g = 3$	$\left(\frac{\Sigma n_\ell - p - 2}{p}\right)\left(\frac{1 - \sqrt{\Lambda^*}}{\sqrt{\Lambda^*}}\right) \sim F_{2p, 2(\Sigma n_\ell - p - 2)}$

Bartlett (see [3]) has shown that, if H_0 is true and $\Sigma n_\ell = n$ is large,

$$-\left(n - 1 - \frac{(p+g)}{2}\right)\ln \Lambda^* = -\left(n - 1 - \frac{(p+g)}{2}\right)\ln\left(\frac{|\mathbf{W}|}{|\mathbf{B}+\mathbf{W}|}\right) \tag{6-39}$$

has approximately a chi-square distribution with $p(g-1)$ d.f. Consequently, for Σn_ℓ large, we reject H_0 at significance level α if

$$-\left(n - 1 - \frac{(p+g)}{2}\right)\ln\left(\frac{|\mathbf{W}|}{|\mathbf{B}+\mathbf{W}|}\right) > \chi^2_{p(g-1)}(\alpha) \tag{6-40}$$

where $\chi^2_{p(g-1)}(\alpha)$ is the upper (100α)th percentile of a chi-square distribution with $p(g-1)$ d.f.

Example 6.8

Suppose an additional variable is observed along with the variable introduced in Example 6.6. The sample sizes are $n_1 = 3$, $n_2 = 2$, and $n_3 = 3$. Arranging the observation pairs $\mathbf{x}_{\ell j}$ in rows, the data are

$$\begin{pmatrix} \begin{bmatrix}9\\3\end{bmatrix} & \begin{bmatrix}6\\2\end{bmatrix} & \begin{bmatrix}9\\7\end{bmatrix} \\ \begin{bmatrix}0\\4\end{bmatrix} & \begin{bmatrix}2\\0\end{bmatrix} & \\ \begin{bmatrix}3\\8\end{bmatrix} & \begin{bmatrix}1\\9\end{bmatrix} & \begin{bmatrix}2\\7\end{bmatrix} \end{pmatrix} \quad \text{with}$$

$$\bar{\mathbf{x}}_1 = \begin{bmatrix}8\\4\end{bmatrix}, \quad \bar{\mathbf{x}}_2 = \begin{bmatrix}1\\2\end{bmatrix}, \quad \bar{\mathbf{x}}_3 = \begin{bmatrix}2\\8\end{bmatrix}, \quad \bar{\mathbf{x}} = \begin{bmatrix}4\\5\end{bmatrix}$$

We have already expressed the observations on the first variable as the sum of an overall mean, treatment effect, and residual in our discussion of univariate ANOVA. We found

$$\underset{\text{(observation)}}{\begin{pmatrix}9 & 6 & 9\\ 0 & 2 & \\ 3 & 1 & 2\end{pmatrix}} = \underset{\text{(mean)}}{\begin{pmatrix}4 & 4 & 4\\ 4 & 4 & \\ 4 & 4 & 4\end{pmatrix}} + \underset{\left(\substack{\text{treatment}\\\text{effect}}\right)}{\begin{pmatrix}4 & 4 & 4\\ -3 & -3 & \\ -2 & -2 & -2\end{pmatrix}} + \underset{\text{(residual)}}{\begin{pmatrix}1 & -2 & 1\\ -1 & 1 & \\ 1 & -1 & 0\end{pmatrix}}$$

and

$$SS_{obs} = SS_{mean} + SS_{tr} + SS_{res}$$
$$216 = 128 + 78 + 10$$
$$\text{Total SS (corrected)} = SS_{obs} - SS_{mean} = 216 - 128 = 88$$

Repeating this operation for the observations on the second variable, we have

$$\underset{\text{(observation)}}{\begin{pmatrix}3 & 2 & 7\\ 4 & 0 & \\ 8 & 9 & 7\end{pmatrix}} = \underset{\text{(mean)}}{\begin{pmatrix}5 & 5 & 5\\ 5 & 5 & \\ 5 & 5 & 5\end{pmatrix}} + \underset{\left(\substack{\text{treatment}\\\text{effect}}\right)}{\begin{pmatrix}-1 & -1 & -1\\ -3 & -3 & \\ 3 & 3 & 3\end{pmatrix}} + \underset{\text{(residual)}}{\begin{pmatrix}-1 & -2 & 3\\ 2 & -2 & \\ 0 & 1 & -1\end{pmatrix}}$$

and

$$SS_{obs} = SS_{mean} + SS_{tr} + SS_{res}$$
$$272 = 200 + 48 + 24$$
$$\text{Total SS (corrected)} = SS_{obs} - SS_{mean} = 272 - 200 = 72.$$

These two single-component analyses must be augmented with the sum of entry by entry *crossproducts* in order to complete the entries in the MANOVA table. Proceeding row by row in the arrays for the two variables, we obtain the cross-product contributions:

Mean: $4(5) + 4(5) + \cdots + 4(5) = 8(4)(5) = 160$

Treatment: $3(4)(-1) + 2(-3)(-3) + 3(-2)(3) = -12$

Residual: $1(-1) + (-2)(-2) + 1(3) + (-1)(2) + \cdots + 0(-1) = 1$

Total: $9(3) + 6(2) + 9(7) + 0(4) + \cdots + 2(7) = 149$

$$\text{Total (corrected) crossproduct} = \text{total crossproduct} - \text{mean crossproduct} = 149 - 160 = -11$$

Thus the MANOVA table takes the following form

Source of variation	Matrix of sum of squares and crossproducts	Degrees of freedom
Treatment	$\begin{bmatrix} 78 & -12 \\ -12 & 48 \end{bmatrix}$	$3 - 1 = 2$
Residual	$\begin{bmatrix} 10 & 1 \\ 1 & 24 \end{bmatrix}$	$3 + 2 + 3 - 3 = 5$
Total (corrected)	$\begin{bmatrix} 88 & -11 \\ -11 & 72 \end{bmatrix}$	7

Equation (6-36) is verified by noting

$$\begin{bmatrix} 88 & -11 \\ -11 & 72 \end{bmatrix} = \begin{bmatrix} 78 & -12 \\ -12 & 48 \end{bmatrix} + \begin{bmatrix} 10 & 1 \\ 1 & 24 \end{bmatrix}$$

Using (6-38),

$$\Lambda^* = \frac{|\mathbf{W}|}{|\mathbf{B} + \mathbf{W}|} = \frac{\begin{vmatrix} 10 & 1 \\ 1 & 24 \end{vmatrix}}{\begin{vmatrix} 88 & -11 \\ -11 & 72 \end{vmatrix}} = \frac{10(24) - (1)^2}{88(72) - (-11)^2} = \frac{239}{6,215} = .0385$$

Since $p = 2$ and $g = 3$, Table 6.3 indicates that an exact test (assuming normality and equal group covariance matrices) of $H_0: \boldsymbol{\tau}_1 = \boldsymbol{\tau}_2 = \boldsymbol{\tau}_3 = \mathbf{0}$ (no treatment effects) versus H_1: at least one $\boldsymbol{\tau}_\ell \neq \mathbf{0}$ is available. To carry out the test, we compare the test statistic

$$\left(\frac{1-\sqrt{\Lambda^*}}{\sqrt{\Lambda^*}}\right)\frac{(\Sigma n_\ell - g - 1)}{(g-1)} = \left(\frac{1-\sqrt{.0385}}{\sqrt{.0385}}\right)\left(\frac{8-3-1}{3-1}\right) = 8.19$$

with a percentage point of an F-distribution having $\nu_1 = 2(g-1) = 4$ and $\nu_2 = 2(\Sigma n_\ell - g - 1) = 8$ d.f. Since $8.19 > F_{4,8}(.01) = 7.01$, we reject H_0 at the $\alpha = .01$ level and conclude treatment differences exist. ■

When the number of variables, p, is large, the MANOVA table is usually not constructed. Still, it is good practice to have the computer print the matrices **B** and **W** so that especially large entries can be located. Also, the residual vectors

$$\hat{\mathbf{e}}_{\ell j} = \mathbf{x}_{\ell j} - \bar{\mathbf{x}}_\ell$$

should be examined for normality and the presence of outliers using the techniques discussed in Section 4.6 of Chapter 4.

Example 6.9

The Wisconsin Department of Health and Social Services reimburses nursing homes in the state for the services provided. The department develops a set of formula rates for each facility, based on factors such as level of care, mean wage rate, and average wage rate in the state.

Nursing homes can be classified on the basis of ownership (private party, nonprofit organization, and government) and certification (skilled nursing facility (SNF), intermediate care facility (ICF), or combination (SNF & ICF)).

One purpose of a recent study was to investigate the effects of ownership or certification (or both) on costs. Four costs, computed on a per-patient-day basis and measured in hours per patient day, were selected for analysis: X_1 = cost of nursing labor, X_2 = cost of dietary labor, X_3 = cost of plant operation and maintenance labor, and X_4 = cost of housekeeping and laundry labor. A total of $n = 516$ observations on each of the $p = 4$ cost variables were initially separated according to ownership. Summary statistics for each of the $g = 3$ groups are given below.

Group	Number of observations	Sample mean vectors
$\ell = 1$ (Private)	$n_1 = 271$	
$\ell = 2$ (Nonprofit)	$n_2 = 138$	$\bar{\mathbf{x}}_1 = \begin{bmatrix} 2.066 \\ .480 \\ .082 \\ .360 \end{bmatrix}$; $\bar{\mathbf{x}}_2 = \begin{bmatrix} 2.167 \\ .596 \\ .124 \\ .418 \end{bmatrix}$; $\bar{\mathbf{x}}_3 = \begin{bmatrix} 2.273 \\ .521 \\ .125 \\ .383 \end{bmatrix}$
$\ell = 3$ (Government)	$n_3 = 107$	
	$\sum_{\ell=1}^{3} n_\ell = 516$	

Sample covariance matrices

$$\mathbf{S}_1 = \begin{bmatrix} .291 & & & \\ -.001 & .011 & & \\ .002 & .000 & .001 & \\ .010 & .003 & .000 & .010 \end{bmatrix}; \quad \mathbf{S}_2 = \begin{bmatrix} .561 & & & \\ .011 & .025 & & \\ .001 & .004 & .005 & \\ .037 & .007 & .002 & .019 \end{bmatrix};$$

$$\mathbf{S}_3 = \begin{bmatrix} .261 & & & \\ .030 & .017 & & \\ .003 & -.000 & .004 & \\ .018 & .006 & .001 & .013 \end{bmatrix}$$

SOURCE: Data courtesy of State of Wisconsin Department of Health and Social Services.

Since the $\mathbf{S}_\ell$'s seem to be reasonably compatible,[2] they were pooled [see (6-37)] to obtain

$$\mathbf{W} = (n_1 - 1)\mathbf{S}_1 + (n_2 - 1)\mathbf{S}_2 + (n_3 - 1)\mathbf{S}_3$$

$$= \begin{bmatrix} 182.962 & & & \\ 4.408 & 8.200 & & \\ 1.695 & .633 & 1.484 & \\ 9.581 & 2.428 & .394 & 6.538 \end{bmatrix}$$

Also

$$\bar{\mathbf{x}} = \frac{n_1\bar{\mathbf{x}}_1 + n_2\bar{\mathbf{x}}_2 + n_3\bar{\mathbf{x}}_3}{n_1 + n_2 + n_3} = \begin{bmatrix} 2.136 \\ .519 \\ .102 \\ .380 \end{bmatrix}$$

and

$$\mathbf{B} = \sum_{\ell=1}^{3} n_\ell(\bar{\mathbf{x}}_\ell - \bar{\mathbf{x}})(\bar{\mathbf{x}}_\ell - \bar{\mathbf{x}})' = \begin{bmatrix} 3.475 & & & \\ 1.111 & 1.225 & & \\ .821 & .453 & .235 & \\ .584 & .610 & .230 & .304 \end{bmatrix}$$

To test H_0: $\boldsymbol{\tau}_1 = \boldsymbol{\tau}_2 = \boldsymbol{\tau}_3$ (no ownership effects or, equivalently, no difference in average costs among the three types of owners, private, nonprofit and government), we can use the result in Table 6.3 for $g = 3$.

Computer-based calculations give

$$\Lambda^* = \frac{|\mathbf{W}|}{|\mathbf{B} + \mathbf{W}|} = .7714$$

and

$$\left(\frac{\Sigma n_\ell - p - 2}{p}\right)\left(\frac{1 - \sqrt{\Lambda^*}}{\sqrt{\Lambda^*}}\right) = \left(\frac{516 - 4 - 2}{4}\right)\left(\frac{1 - \sqrt{.7714}}{\sqrt{.7714}}\right) = 17.67$$

[2] However, a normal theory test of H_0: $\boldsymbol{\Sigma}_1 = \boldsymbol{\Sigma}_2 = \boldsymbol{\Sigma}_3$ would reject H_0 at any reasonable significance level because of the large sample sizes.

Let $\alpha = .01$ so that $F_{2(4),2(510)}(.01) \doteq \chi_8^2(.01)/8 = 2.51$. Since $17.67 > F_{8,1020}(.01) \doteq 2.51$, we reject H_0 at the 1% level and conclude average costs differ depending on type of ownership.

It is informative to compare the results based on the "exact" test above with those obtained using the large-sample procedure summarized in Equations (6-39) and (6-40). For the present example, $\Sigma n_\ell = n = 516$ is large and H_0 can be tested at the $\alpha = .01$ level by comparing

$$-(n - 1 - (p + g)/2)\ln\left(\frac{|\mathbf{W}|}{|\mathbf{B} + \mathbf{W}|}\right) = -511.5\ln(.7714) = 132.76$$

with $\chi^2_{p(g-1)}(.01) = \chi_8^2(.01) = 20.09$. Since $132.76 > \chi_8^2(.01) = 20.09$, we reject H_0 at the 1% level. This result is consistent with the result based on the F-statistic above. ■

6.5 SIMULTANEOUS CONFIDENCE INTERVALS FOR TREATMENT EFFECTS

When the hypothesis of equal treatment effects is rejected, those effects, or linear combinations of effects, that led to the rejection of the hypothesis are of interest. Roy's approach [17] to the testing of multivariate statistical hypotheses can be used to construct simultaneous confidence intervals for the differences $\boldsymbol{\tau}_\ell - \boldsymbol{\tau}_k$ (or $\boldsymbol{\mu}_\ell - \boldsymbol{\mu}_k$), $\ell \neq k$.

The technique will be expressed in terms of the linear combinations $\mathbf{a}'\boldsymbol{\tau}_\ell$. The ith component of $\boldsymbol{\tau}_\ell$ is examined by choosing $\mathbf{a}' = [0,\ldots,0,1,0,\ldots,0]$ with 1 in the ith position. Treatment differences can be investigated by forming *contrasts* with the linear combinations $\mathbf{a}'\boldsymbol{\tau}_\ell$. For example, let $\mathbf{c}' = [c_1, c_2,\ldots,c_g]$ with $\sum_{\ell=1}^{g} c_\ell = 0$. With $c_1 = 1$, $c_2 = -1$, $c_3 = \cdots = c_g = 0$ and $\mathbf{a}' = [1,0,\ldots,0]$, the contrast

$$c_1\mathbf{a}'\boldsymbol{\tau}_1 + c_2\mathbf{a}'\boldsymbol{\tau}_2 + \cdots + c_g\mathbf{a}'\boldsymbol{\tau}_g = c_1\mathbf{a}'\boldsymbol{\mu}_1 + c_2\mathbf{a}'\boldsymbol{\mu}_2 + \cdots + c_g\mathbf{a}'\boldsymbol{\mu}_g \tag{6-41}$$

becomes $\tau_{11} - \tau_{21}$; the difference in the effects of treatments 1 and 2 on the first variable.

Result 6.5. Let $\sum_{\ell=1}^{g} n_\ell - g - p > 1$ for the model in (6-34). Then with probability $(1 - \alpha)$,

$$\sum_{\ell=1}^{g} c_\ell\mathbf{a}'\boldsymbol{\tau}_\ell \quad \text{belongs to} \quad \sum_{\ell=1}^{g} c_\ell\mathbf{a}'\hat{\boldsymbol{\tau}}_\ell \pm \sqrt{\eta_\alpha}\sqrt{\mathbf{a}'\mathbf{W}\mathbf{a}\left(\sum_{\ell=1}^{g}\frac{c_\ell^2}{n_\ell}\right)}$$

for all $\mathbf{a}$ and all constants c_ℓ such that $\sum_{\ell=1}^{g} c_\ell = 0$. Here η_α is the upper (100α)th percentile of the distribution of the largest root of $|\tilde{\mathbf{B}} - \eta\mathbf{W}| = 0$, where $\tilde{\mathbf{B}}$ and $\mathbf{W}$ are independent Wishart matrices.

Proof. Since $\boldsymbol{\mu}_\ell = \boldsymbol{\mu} + \boldsymbol{\tau}_\ell$, $\bar{\boldsymbol{\tau}} = \sum_{\ell=1}^{g} n_\ell \boldsymbol{\tau}_\ell \Big/ \sum_{\ell=1}^{g} n_\ell = \mathbf{0}$, and $\bar{\boldsymbol{\mu}} = \sum_{\ell=1}^{g} n_\ell \boldsymbol{\mu}_\ell \Big/ \sum_{\ell=1}^{g} n_\ell = \boldsymbol{\mu}$, so $\overline{\mathbf{X}}_\ell - \overline{\mathbf{X}}$ has expected value $\boldsymbol{\mu}_\ell - \bar{\boldsymbol{\mu}} = \boldsymbol{\tau}_\ell$. Set $\tilde{\mathbf{B}} = \sum_{\ell=1}^{g} n_\ell(\overline{\mathbf{X}}_\ell - \overline{\mathbf{X}} - \boldsymbol{\tau}_\ell)(\overline{\mathbf{X}}_\ell - \overline{\mathbf{X}} - \boldsymbol{\tau}_\ell)'$. The property $\sum_{\ell=1}^{g} c_\ell = 0$ allows us to write

$$\sum_{\ell=1}^{g} c_\ell \boldsymbol{\mu}_\ell = \sum_{\ell=1}^{g} c_\ell(\boldsymbol{\mu}_\ell - \bar{\boldsymbol{\mu}}) = \sum_{\ell=1}^{g} c_\ell \boldsymbol{\tau}_\ell$$

and

$$\sum_{\ell=1}^{g} c_\ell \overline{\mathbf{X}}_\ell = \sum_{\ell=1}^{g} c_\ell(\overline{\mathbf{X}}_\ell - \overline{\mathbf{X}}) = \sum_{\ell=1}^{g} c_\ell \hat{\boldsymbol{\tau}}_\ell$$

Therefore

$$\sum_{\ell=1}^{g} c_\ell \overline{\mathbf{X}}_\ell - \sum_{\ell=1}^{g} c_\ell \boldsymbol{\tau}_\ell = \sum_{\ell=1}^{g} c_\ell(\overline{\mathbf{X}}_\ell - \overline{\mathbf{X}} - \boldsymbol{\tau}_\ell)$$

Consider a fixed linear combination $\mathbf{a}'(\overline{\mathbf{X}}_\ell - \overline{\mathbf{X}} - \boldsymbol{\tau}_\ell)$. By the Cauchy–Schwarz inequality in (2-48) with $\mathbf{b}' = [n_1^{-1/2}c_1, n_2^{-1/2}c_2, \ldots, n_g^{-1/2}c_g]$ and $\mathbf{d}' = [n_1^{1/2}\mathbf{a}'(\overline{\mathbf{X}}_1 - \overline{\mathbf{X}} - \boldsymbol{\tau}_1), n_2^{1/2}\mathbf{a}'(\overline{\mathbf{X}}_2 - \overline{\mathbf{X}} - \boldsymbol{\tau}_2), \ldots, n_g^{1/2}\mathbf{a}'(\overline{\mathbf{X}}_g - \overline{\mathbf{X}} - \boldsymbol{\tau}_g)]$,

$$\begin{aligned}
\left|\sum_{\ell=1}^{g} c_\ell \mathbf{a}'\overline{\mathbf{X}}_\ell - \sum_{\ell=1}^{g} c_\ell \mathbf{a}'\boldsymbol{\tau}_\ell\right|^2 &= \left|\sum_{\ell=1}^{g} c_\ell \mathbf{a}'(\overline{\mathbf{X}}_\ell - \overline{\mathbf{X}} - \boldsymbol{\tau}_\ell)\right|^2 \\
&= \left|\sum_{\ell=1}^{g} c_\ell n_\ell^{-1/2} n_\ell^{1/2} \mathbf{a}'(\overline{\mathbf{X}}_\ell - \overline{\mathbf{X}} - \boldsymbol{\tau}_\ell)\right|^2 \\
&\le \left(\sum_{\ell=1}^{g} \frac{c_\ell^2}{n_\ell}\right) \sum_{\ell=1}^{g} n_\ell \big(\mathbf{a}'(\overline{\mathbf{X}}_\ell - \overline{\mathbf{X}} - \boldsymbol{\tau}_\ell)\big)^2 \\
&= \left(\sum_{\ell=1}^{g} \frac{c_\ell^2}{n_\ell}\right) \sum_{\ell=1}^{g} n_\ell \mathbf{a}'(\overline{\mathbf{X}}_\ell - \overline{\mathbf{X}} - \boldsymbol{\tau}_\ell)(\overline{\mathbf{X}}_\ell - \overline{\mathbf{X}} - \boldsymbol{\tau}_\ell)'\mathbf{a} \\
&= \left(\sum_{\ell=1}^{g} \frac{c_\ell^2}{n_\ell}\right) \mathbf{a}'\tilde{\mathbf{B}}\mathbf{a} \qquad \text{(6-42)}
\end{aligned}$$

Equality can be attained by an appropriate choice of $c_1, c_2, \ldots, c_g$. The bound in (6-42), for a fixed $\mathbf{a}$, yields

$$\frac{\left|\sum_{\ell=1}^{g} c_\ell \mathbf{a}'\overline{\mathbf{X}}_\ell - \sum_{\ell=1}^{g} c_\ell \mathbf{a}'\boldsymbol{\tau}_\ell\right|^2}{\mathbf{a}'\mathbf{W}\mathbf{a}} \le \left(\sum_{\ell=1}^{g} \frac{c_\ell^2}{n_\ell}\right) \frac{\mathbf{a}'\tilde{\mathbf{B}}\mathbf{a}}{\mathbf{a}'\mathbf{W}\mathbf{a}} \qquad \text{(6-43)}$$

Let $\mathbf{W}^{1/2}$ be the symmetric square root matrix [see (2-22)] such that $\mathbf{W}^{1/2}\mathbf{W}^{1/2} = \mathbf{W}$ and $(\mathbf{W}^{1/2})^{-1} = \mathbf{W}^{-1/2}$. Setting $\mathbf{v} = \mathbf{W}^{1/2}\mathbf{a}$, we obtain

$$\begin{aligned}
\mathbf{a}'\tilde{\mathbf{B}}\mathbf{a}/\mathbf{a}'\mathbf{W}\mathbf{a} &= \mathbf{a}'\mathbf{W}^{1/2}\mathbf{W}^{-1/2}\tilde{\mathbf{B}}\mathbf{W}^{-1/2}\mathbf{W}^{1/2}\mathbf{a}/\mathbf{a}'\mathbf{W}^{1/2}\mathbf{W}^{1/2}\mathbf{a} \\
&= \mathbf{v}'(\mathbf{W}^{-1/2}\tilde{\mathbf{B}}\mathbf{W}^{-1/2})\mathbf{v}/\mathbf{v}'\mathbf{v}
\end{aligned}$$

By (2-51),

$$\frac{\mathbf{a}'\tilde{\mathbf{B}}\mathbf{a}}{\mathbf{a}'\mathbf{W}\mathbf{a}} = \frac{\mathbf{v}'(\mathbf{W}^{-1/2}\tilde{\mathbf{B}}\mathbf{W}^{-1/2})\mathbf{v}}{\mathbf{v}'\mathbf{v}} \le \eta_1, \qquad \text{for all } \mathbf{a} \ne \mathbf{0}$$

where η_1 is the *largest* eigenvalue of $\mathbf{W}^{-1/2}\tilde{\mathbf{B}}\mathbf{W}^{-1/2}$. Thus, taking square roots in (6-43),

$$\left|\sum_{\ell=1}^{g} c_\ell \mathbf{a}'\overline{\mathbf{X}}_\ell - \sum_{\ell=1}^{g} c_\ell \mathbf{a}'\boldsymbol{\tau}_\ell\right| = \left|\sum_{\ell=1}^{g} c_\ell \mathbf{a}'\hat{\boldsymbol{\tau}}_\ell - \sum_{\ell=1}^{g} c_\ell \mathbf{a}'\boldsymbol{\tau}_\ell\right|$$

$$\le \sqrt{\eta_1}\sqrt{\left(\sum_{\ell=1}^{g}\frac{c_\ell^2}{n_\ell}\right)\mathbf{a}'\mathbf{W}\mathbf{a}}$$

with $|\mathbf{W}^{-1/2}\tilde{\mathbf{B}}\mathbf{W}^{-1/2} - \eta_1\mathbf{I}| = 0$. By the properties of determinants (see Result 2A.11),

$$\begin{aligned} 0 &= |\mathbf{W}^{1/2}||\mathbf{W}^{-1/2}\tilde{\mathbf{B}}\mathbf{W}^{-1/2} - \eta\mathbf{I}||\mathbf{W}^{1/2}| \\ &= |\mathbf{W}^{1/2}\mathbf{W}^{-1/2}\tilde{\mathbf{B}}\mathbf{W}^{-1/2}\mathbf{W}^{1/2} - \eta\mathbf{W}^{1/2}\mathbf{W}^{1/2}| \\ &= |\tilde{\mathbf{B}} - \eta\mathbf{W}| \end{aligned}$$

so that the maximum eigenvalue of $\mathbf{W}^{-1/2}\tilde{\mathbf{B}}\mathbf{W}^{-1/2}$ is also the largest root of $|\tilde{\mathbf{B}} - \eta\mathbf{W}| = 0$.

Because $\tilde{\mathbf{B}}$ and $\mathbf{W}$ are random, we must obtain the upper (100α)th percentile of the distribution of η_1 for the simultaneous confidence procedure. From (4-23), $\sum_{j=1}^{n_\ell}(\mathbf{X}_j - \overline{\mathbf{X}}_\ell)(\mathbf{X}_j - \overline{\mathbf{X}}_\ell)'$ is distributed according to the Wishart distribution, $W_{n_\ell - 1}(\cdot|\boldsymbol{\Sigma})$. By the additive property (4-24) of Wishart random matrices, the sum $\mathbf{W} = \sum_{\ell=1}^{g}\sum_{j=1}^{n_\ell}(\mathbf{X}_j - \overline{\mathbf{X}}_\ell)(\mathbf{X}_j - \overline{\mathbf{X}}_\ell)'$ is distributed as $W_{\Sigma n_\ell - g}(\cdot|\boldsymbol{\Sigma})$. (With the condition $\sum_{\ell=1}^{g}(n_\ell - 1) - p - 1 > 0$, $\mathbf{W}$ has a density function that places probability 1 on matrices that are positive definite. Consequently, $\mathbf{W}^{1/2}$ and $\mathbf{W}^{-1/2}$ exist.) It can be shown (see [1]) that $\tilde{\mathbf{B}}$ is distributed independently of $\mathbf{W}$ as a Wishart random matrix with degrees of freedom equal to $\text{rank}(\tilde{\mathbf{B}}) = \min(p, g-1)$. Thus η_1 is the largest root satisfying

$$0 = |\mathbf{W}^{-1/2}\tilde{\mathbf{B}}\mathbf{W}^{-1/2} - \eta\mathbf{I}| = |\tilde{\mathbf{B}} - \eta\mathbf{W}| \tag{6-44}$$

for the independent Wishart matrices $\tilde{\mathbf{B}}$ and $\mathbf{W}$. ■

Comment. To implement Result 6.5, we need percentiles of the distribution of the largest root, η_1, of $0 = |\tilde{\mathbf{B}} - \eta\mathbf{W}|$ for the two Wishart matrices $\tilde{\mathbf{B}}$ and $\mathbf{W}$. Existing tables of percentiles (due to Pillai [16] and Heck [10]) are available for the related roots θ of $0 = |\tilde{\mathbf{B}} - \theta(\mathbf{W} + \tilde{\mathbf{B}})|$. Since $0 = |\tilde{\mathbf{B}} - \theta(\mathbf{W} + \tilde{\mathbf{B}})| = |(1-\theta)\tilde{\mathbf{B}} - \theta\mathbf{W}| = (1-\theta)^p|\tilde{\mathbf{B}} - (\theta/(1-\theta))\mathbf{W}|$ (see Result 2A.11(f)), the upper (100α)th percentile of η_1, η_α, is related to the same percentile for θ, θ_α, by $\eta_\alpha = \theta_\alpha/(1-\theta_\alpha)$. (See [6] for tables.) The distribution of the largest eigenvalue does not depend on the unknown $\boldsymbol{\Sigma}$. It depends only on the degrees of freedom $\sum_{\ell=1}^{g} (n_\ell - 1) - p - 1$, the number of groups, g, and the number of variables, p. Selected percentiles are tabled for the parameters $\tilde{s} = \min(p, g-1)$, $\tilde{m} = (|g - p - 1| - 1)/2$ and $\tilde{n} = \left(\sum_{\ell=1}^{g} (n_\ell - 1) - p - 1 \right) \Big/ 2$.

We shall illustrate the construction of simultaneous interval estimates for the contrasts in treatment means using the nursing-home data introduced in Example 6.9.

Example 6.10

We saw in Example 6.9 that average costs for nursing homes differ, depending on the type of ownership. We can use Result 6.5 to estimate the magnitudes of the cost differences. A comparison of the cost variable X_3, costs of plant operation and maintenance labor, between privately owned nursing homes and government-owned nursing homes can be made by estimating the contrast; $\sum_{\ell=1}^{3} c_\ell \mathbf{a}'\boldsymbol{\tau}_\ell = \sum_{\ell=1}^{3} c_\ell \mathbf{a}'\boldsymbol{\mu}_\ell$ with the particular choices $c_1 = 1$, $c_2 = 0$, $c_3 = -1$, and $\mathbf{a}' = [0, 0, 1, 0]$. Thus

$$\sum_{\ell=1}^{3} c_\ell \mathbf{a}'\boldsymbol{\tau}_\ell = 1[0 \quad 0 \quad 1 \quad 0] \begin{bmatrix} \tau_{11} \\ \tau_{12} \\ \tau_{13} \\ \tau_{14} \end{bmatrix} - 1[0 \quad 0 \quad 1 \quad 0] \begin{bmatrix} \tau_{31} \\ \tau_{32} \\ \tau_{33} \\ \tau_{34} \end{bmatrix} = \tau_{13} - \tau_{33}$$

Using (6-35) and the information in Example 6.9,

$$\hat{\boldsymbol{\tau}}_1 = (\bar{\mathbf{x}}_1 - \bar{\mathbf{x}}) = \begin{bmatrix} -.070 \\ -.039 \\ -.020 \\ -.020 \end{bmatrix}, \qquad \hat{\boldsymbol{\tau}}_3 = (\bar{\mathbf{x}}_3 - \bar{\mathbf{x}}) = \begin{bmatrix} .137 \\ .002 \\ .023 \\ .003 \end{bmatrix}$$

$$\mathbf{W} = \begin{bmatrix} 182.962 & & & \\ 4.408 & 8.200 & & \\ 1.695 & .633 & 1.484 & \\ 9.581 & 2.428 & .394 & 6.538 \end{bmatrix}$$

Consequently,

$$\sum_{\ell=1}^{3} c_\ell \mathbf{a}'\hat{\boldsymbol{\tau}}_\ell = \hat{\tau}_{13} - \hat{\tau}_{33} = -.020 - .023 = -.043$$

and

$$\mathbf{a}'\mathbf{W}\mathbf{a}\left(\sum_{\ell=1}^{3} \frac{c_\ell^2}{n_\ell} \right) = 1.484\left(\frac{1^2}{271} + \frac{(-1)^2}{107} \right) = .0193$$

Now $\tilde{s} = \min(g-1, p) = \min(2, 4) = 2$, $\tilde{m} = (|g - p - 1| - 1)/2 = (|3 - 5 - 1| - 1)/2 = \frac{1}{2}$ and $\tilde{n} = (\Sigma(n_\ell - 1) - p - 1)/2 = (513 - 4 - 1)/2 = 254$.

From Heck's charts in [6], we find $\eta_{.05} = \theta_{.05}/(1 - \theta_{.05}) \doteq .025/(1 - .025) = .026$. The 95% simultaneous confidence statement is

$$\tau_{13} - \tau_{33} \quad \text{belongs to} \quad \hat{\tau}_{13} - \hat{\tau}_{33} \pm \sqrt{\eta_{.05}} \sqrt{\mathbf{a}'\mathbf{W}\mathbf{a}\left(\sum_{\ell=1}^{3} \frac{c_\ell^2}{n_\ell}\right)}$$

$$= -.043 \pm \sqrt{.026}\sqrt{.0193}$$

$$= -.043 \pm .022, \quad \text{or} \quad (-.065, -.021)$$

We conclude that the average maintenance and labor cost for government-owned nursing homes is higher by .021 to .065 hours per patient day than for privately owned nursing homes. With the same 95% confidence,

$$\tau_{13} - \tau_{23} \text{ belongs to the interval } (-.063, -.021)$$

and

$$\tau_{23} - \tau_{33} \text{ belongs to the interval } (-.026, .024)$$

Thus a difference in this cost exists between private and nonprofit nursing homes, but no difference is observed between nonprofit and government nursing homes. ■

6.6 PROFILE ANALYSIS

Profile analysis pertains to situations where a battery of p treatments (tests, questions, and so forth) are administered to two or more groups of subjects. It is assumed that the responses for the different groups are independent of one another, but all responses must be expressed in similar units. Ordinarily we might pose the question, Are the population mean vectors the same? In profile analysis, the question of equality of mean vectors is divided into several specific possibilities.

Consider the population means $\boldsymbol{\mu}_1' = [\mu_{11}, \mu_{12}, \mu_{13}, \mu_{14}]$ representing the average responses to four treatments for the first group. A plot of these means connected by straight lines is shown in Figure 6.3. This broken-line graph is the *profile* for population 1. Profiles can be constructed for each population (group). We shall concentrate on two groups.

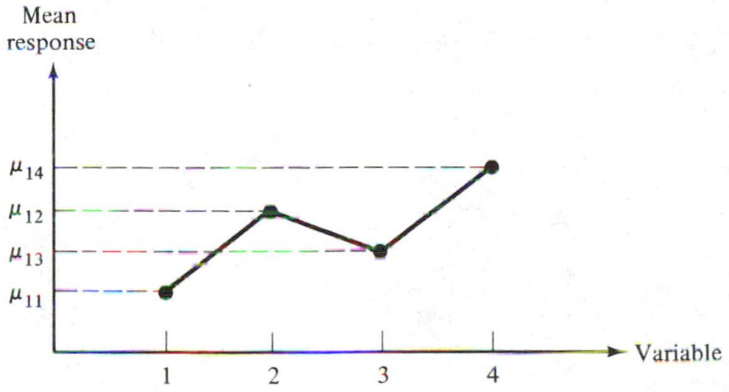

Figure 6.3 The population profile, $p = 4$.

Let $\boldsymbol{\mu}_1' = [\mu_{11}, \mu_{12}, \ldots, \mu_{1p}]$ and $\boldsymbol{\mu}_2' = [\mu_{21}, \mu_{22}, \ldots, \mu_{2p}]$ be the mean responses to p treatments for populations 1 and 2, respectively. The hypothesis $H_0\colon \boldsymbol{\mu}_1 = \boldsymbol{\mu}_2$ implies the treatments have the same (average) effect on the two populations. In terms of the population profiles, we can formulate the question of equality in a stagewise fashion.

1. Are the profiles parallel?
   Equivalently: Is $H_{01}\colon \mu_{1i} - \mu_{1i-1} = \mu_{2i} - \mu_{2i-1}$, $i = 2, 3, \ldots, p$, acceptable?
2. Assuming the profiles *are* parallel, are the profiles coincident?
   Equivalently: Is $H_{02}\colon \mu_{1i} = \mu_{2i}$, $i = 1, 2, \ldots, p$, acceptable?
3. Assuming the profiles *are* coincident, are the profiles level? That is, are all the means equal to the same constant?
   Equivalently: Is $H_{03}\colon \mu_{11} = \mu_{12} = \cdots = \mu_{1p} = \mu_{21} = \mu_{22} = \cdots = \mu_{2p}$ acceptable?

The null hypothesis in stage 1 can be written

$$H_{01}\colon \mathbf{C}\boldsymbol{\mu}_1 = \mathbf{C}\boldsymbol{\mu}_2$$

where $\mathbf{C}$ is the contrast matrix

$$\underset{((p-1)\times p)}{\mathbf{C}} = \begin{bmatrix} -1 & 1 & 0 & 0 & \cdots & 0 & 0 \\ 0 & -1 & 1 & 0 & \cdots & 0 & 0 \\ \vdots & \vdots & \vdots & \vdots & & \vdots & \vdots \\ 0 & 0 & 0 & 0 & \cdots & -1 & 1 \end{bmatrix} \tag{6-45}$$

For independent samples of sizes n_1 and n_2 from the two populations, the null hypothesis can be tested by constructing the transformed observations

$$\mathbf{C}\mathbf{x}_{1j}, \qquad j = 1, 2, \ldots, n_1$$

and

$$\mathbf{C}\mathbf{x}_{2j}, \qquad j = 1, 2, \ldots, n_2$$

These have sample mean vectors $\mathbf{C}\bar{\mathbf{x}}_1$ and $\mathbf{C}\bar{\mathbf{x}}_2$, respectively, and pooled covariance matrix, $\mathbf{C}\mathbf{S}_{\text{pooled}}\mathbf{C}'$.

Since the two sets of transformed observations have $N_{p-1}(\mathbf{C}\boldsymbol{\mu}_1, \mathbf{C}\boldsymbol{\Sigma}\mathbf{C}')$ and $N_{p-1}(\mathbf{C}\boldsymbol{\mu}_2, \mathbf{C}\boldsymbol{\Sigma}\mathbf{C}')$ distributions, respectively, an application of Result 6.2 provides a test for parallel profiles.

Test for Parallel Profiles for Two Normal Populations

Reject $H_{01}\colon \mathbf{C}\boldsymbol{\mu}_1 = \mathbf{C}\boldsymbol{\mu}_2$ (parallel profiles) at level α if

$$T^2 = (\bar{\mathbf{x}}_1 - \bar{\mathbf{x}}_2)'\mathbf{C}'\left[\left(\frac{1}{n_1} + \frac{1}{n_2}\right)\mathbf{C}\mathbf{S}_{\text{pooled}}\mathbf{C}'\right]^{-1}\mathbf{C}(\bar{\mathbf{x}}_1 - \bar{\mathbf{x}}_2) > c^2 \tag{6-46}$$

where

$$c^2 = \frac{(n_1 + n_2 - 2)(p-1)}{n_1 + n_2 - p} F_{p-1, n_1+n_2-p}(\alpha)$$

When the profiles are parallel, the first is either above the second, $\mu_{1i} > \mu_{2i}$, all i, or vice versa. Under this condition, the profiles will be coincident only if the total heights $\mu_{11} + \mu_{12} + \cdots + \mu_{1p} = \mathbf{1}'\boldsymbol{\mu}_1$ and $\mu_{21} + \mu_{22} + \cdots + \mu_{2p} = \mathbf{1}'\boldsymbol{\mu}_2$ are equal. Therefore the null hypothesis at stage 2 can be written in the equivalent form

$$H_{02}: \quad \mathbf{1}'\boldsymbol{\mu}_1 = \mathbf{1}'\boldsymbol{\mu}_2$$

We can then test H_{02} with the usual two-sample t-statistic based on the univariate observations $\mathbf{1}'\mathbf{x}_{1j}, j = 1, 2, \ldots, n_1$, and $\mathbf{1}'\mathbf{x}_{2j}, j = 1, 2, \ldots, n_2$.

> **Test for Coincident Profiles, Given that Profiles Are Parallel**
>
> For two normal populations: Reject H_{02}: $\mathbf{1}'\boldsymbol{\mu}_1 = \mathbf{1}'\boldsymbol{\mu}_2$ (profiles coincident) at level α if
>
> $$T^2 = \mathbf{1}'(\bar{\mathbf{x}}_1 - \bar{\mathbf{x}}_2)\left[\left(\frac{1}{n_1} + \frac{1}{n_2}\right)\mathbf{1}'\mathbf{S}_{\text{pooled}}\mathbf{1}\right]^{-1}\mathbf{1}'(\bar{\mathbf{x}}_1 - \bar{\mathbf{x}}_2) \qquad \text{(6-47)}$$
>
> $$= \left(\frac{\mathbf{1}'(\bar{\mathbf{x}}_1 - \bar{\mathbf{x}}_2)}{\sqrt{\left(\frac{1}{n_1} + \frac{1}{n_2}\right)\mathbf{1}'\mathbf{S}_{\text{pooled}}\mathbf{1}}}\right)^2 > t^2_{n_1+n_2-2}\left(\frac{\alpha}{2}\right) = F_{1,\, n_1+n_2-2}(\alpha)$$

For coincident profiles, $\mathbf{x}_{11}, \mathbf{x}_{12}, \ldots, \mathbf{x}_{1n_1}$ and $\mathbf{x}_{21}, \mathbf{x}_{22}, \ldots, \mathbf{x}_{2n_2}$ are all observations from the same normal population. The next step is to see if all variables have the same mean, so that the common profile is level.

The null hypothesis at stage 3 can be written as

$$H_{03}: \quad \mathbf{C}(\boldsymbol{\mu}_1 + \boldsymbol{\mu}_2) = \mathbf{0}$$

where $\mathbf{C}$ is given by (6-45). When H_{01} and H_{02} are tenable, the common mean vector is estimated by

$$\bar{\mathbf{x}} = \frac{\sum_{j=1}^{n_1} \mathbf{x}_{1j} + \sum_{j=1}^{n_2} \mathbf{x}_{2j}}{n_1 + n_2} = \frac{n_1}{(n_1 + n_2)}\bar{\mathbf{x}}_1 + \frac{n_2}{(n_1 + n_2)}\bar{\mathbf{x}}_2$$

Consequently, we have the following test.

> **Test for Level Profiles, Given that Profiles Are Coincident**
>
> For two normal populations: Reject H_{03}: $\mathbf{C}(\boldsymbol{\mu}_1 + \boldsymbol{\mu}_2) = \mathbf{0}$ (profiles level) at level α if
>
> $$(n_1 + n_2)\bar{\mathbf{x}}'\mathbf{C}'[\mathbf{C}\mathbf{S}_{\text{pooled}}\mathbf{C}']^{-1}\mathbf{C}\bar{\mathbf{x}} > F_{p-1,\, n_1+n_2-p}(\alpha) \qquad \text{(6-48)}$$

Example 6.11

As part of a larger study of love and marriage, a sociologist surveyed adults with respect to their marriage "contributions" and "outcomes" and their levels of "passionate" and "companionate" love. (Data courtesy of E. Hatfield.)

Recently married males and females were asked to respond to these questions, using the 8-point scale in the figure below.

1. All things considered, how would you describe *your contributions* to the marriage?
2. All things considered, how would you describe *your outcomes* from the marriage?

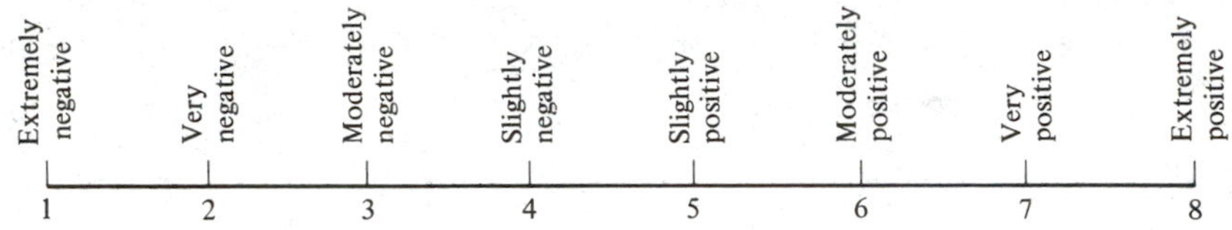

Subjects were also asked to respond to the following questions, using the 5-point scale shown.

3. What is the level of *passionate* love that you feel for your partner?
4. What is the level of *companionate* love that you feel for your partner?

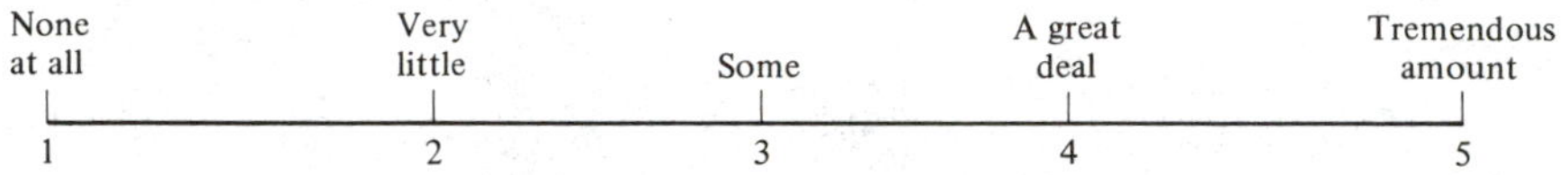

Let

$$x_1 = \text{an 8-point scale response to Question 1}$$
$$x_2 = \text{an 8-point scale response to Question 2}$$
$$x_3 = \text{a 5-point scale response to Question 3}$$
$$x_4 = \text{a 5-point scale response to Question 4}$$

and the two populations be defined as

$$\text{Population 1} = \text{married men}$$
$$\text{Population 2} = \text{married women}$$

The population means are the average responses to the $p = 4$ questions for the populations of males and females. Assuming a common covariance matrix $\mathbf{\Sigma}$, it is of interest to see if the profiles of males and females are the same.

A sample of $n_1 = 30$ males and $n_2 = 30$ females gave the sample mean vectors:

$$\bar{\mathbf{x}}_1 = \begin{bmatrix} 6.833 \\ 7.033 \\ 3.967 \\ 4.700 \end{bmatrix}, \qquad \bar{\mathbf{x}}_2 = \begin{bmatrix} 6.633 \\ 7.000 \\ 4.000 \\ 4.533 \end{bmatrix}$$
$$\text{(males)} \qquad\qquad \text{(females)}$$

and pooled covariance matrix

$$\mathbf{S}_{\text{pooled}} = \begin{bmatrix} .606 & .262 & .066 & .161 \\ .262 & .637 & .173 & .143 \\ .066 & .173 & .810 & .029 \\ .161 & .143 & .029 & .306 \end{bmatrix}$$

The sample mean vectors are plotted as sample profiles in Figure 6.4.

Since the sample sizes are reasonably large, we shall use the normal theory methodology, even though the data, which are integers, are clearly nonnormal. To test for parallelism (H_{01}: $\mathbf{C}\boldsymbol{\mu}_1 = \mathbf{C}\boldsymbol{\mu}_2$), we compute

$$\mathbf{C}\mathbf{S}_{\text{pooled}}\mathbf{C}' = \begin{bmatrix} -1 & 1 & 0 & 0 \\ 0 & -1 & 1 & 0 \\ 0 & 0 & -1 & 1 \end{bmatrix} \mathbf{S}_{\text{pooled}} \begin{bmatrix} -1 & 0 & 0 \\ 1 & -1 & 0 \\ 0 & 1 & -1 \\ 0 & 0 & 1 \end{bmatrix}$$

$$= \begin{bmatrix} .719 & -.268 & -.125 \\ -.268 & 1.101 & -.751 \\ -.125 & -.751 & 1.058 \end{bmatrix}$$

and

$$\mathbf{C}(\bar{\mathbf{x}}_1 - \bar{\mathbf{x}}_2) = \begin{bmatrix} -1 & 1 & 0 & 0 \\ 0 & -1 & 1 & 0 \\ 0 & 0 & -1 & 1 \end{bmatrix} \begin{bmatrix} .200 \\ .033 \\ -.033 \\ .167 \end{bmatrix} = \begin{bmatrix} -.167 \\ -.066 \\ .200 \end{bmatrix}$$

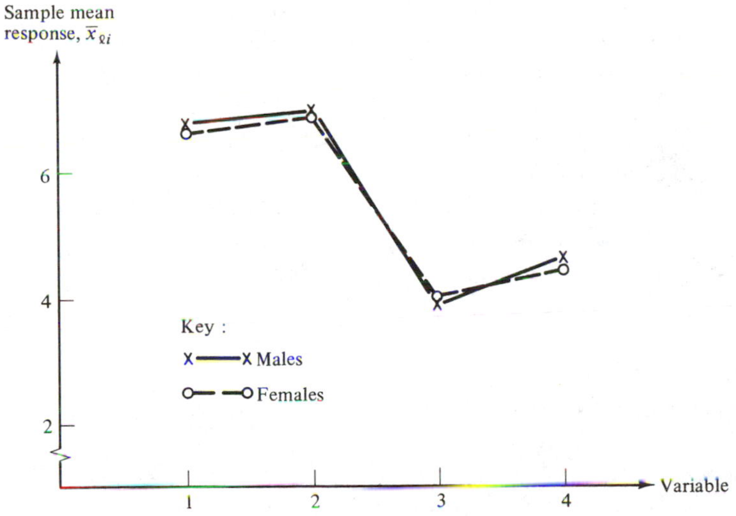

Figure 6.4 Sample profiles for marriage–love responses.

Thus

$$T^2 =$$

$$[-.167, -.066, .200]\left(\tfrac{1}{30} + \tfrac{1}{30}\right)^{-1}\begin{bmatrix} .719 & -.268 & -.125 \\ -.268 & 1.101 & -.751 \\ -.125 & -.751 & 1.058 \end{bmatrix}^{-1}\begin{bmatrix} -.167 \\ -.066 \\ .200 \end{bmatrix}$$

$$= 15(.067) = 1.005$$

Moreover, with $\alpha = .05$, $c^2 = [(30 + 30 - 2)(4 - 1)/(30 + 30 - 4)]F_{3,56}(.05) = 3.11(2.8) = 8.7$. Since $T^2 = 1.005 < 8.7$, we conclude the hypothesis of parallel profiles for men and women is tenable. Given the plot in Figure 6.4, this finding is not surprising.

Assuming the profiles are parallel, we can test for *coincident* profiles. To test H_{02}: $\mathbf{1}'\boldsymbol{\mu}_1 = \mathbf{1}'\boldsymbol{\mu}_2$ (profiles coincident), we need:

$$\text{Sum of elements in } (\bar{\mathbf{x}}_1 - \bar{\mathbf{x}}_2) = \mathbf{1}'(\bar{\mathbf{x}}_1 - \bar{\mathbf{x}}_2) = .367$$

$$\text{Sum of elements in } \mathbf{S}_{\text{pooled}} = \mathbf{1}'\mathbf{S}_{\text{pooled}}\mathbf{1} = 4.027$$

Using (6-47),

$$T^2 = \left(\frac{.367}{\sqrt{\left(\frac{1}{30} + \frac{1}{30}\right)4.027}}\right)^2 = .708$$

With $\alpha = .05$, $F_{1,58}(.05) = 4.0$, and $T^2 = .708 < F_{1,58}(.05) = 4.0$, we cannot reject the hypothesis that the profiles are coincident. That is, the responses of men and women to the four questions posed appear to be the same.

We could now test for level profiles; however, it does not make sense to carry out this test for our example since Questions 1 and 2 were measured on a 1–8 scale, while Questions 3 and 4 were measured on a 1–5 scale. The incompatability of measurement scales makes the test for level profiles meaningless and illustrates the need for similar measurements in order to carry out a complete profile analysis. ■

When the sample sizes are small, a profile analysis will depend on the normality assumption. This assumption can be checked, using methods discussed in Chapter 4, with the original observations $\mathbf{x}_{\ell j}$ or the contrast observations $\mathbf{C}\mathbf{x}_{\ell j}$.

The analysis of profiles for several populations proceeds in much the same fashion as that for two populations. In fact, the general measures of comparison are analogous to those just discussed (see [13]).

6.7 TWO-WAY MULTIVARIATE ANALYSIS OF VARIANCE

Following our approach to the one-way MANOVA, we shall briefly review the analysis for a *univariate* two-way fixed-effects model and then simply generalize to the multivariate case by analogy.

Univariate Two-Way Fixed-Effects Model with Interaction

We assume that measurements are recorded at various levels of two factors. In some cases, these experimental conditions represent levels of a single treatment arranged within several blocks. The particular experimental design employed will not concern us in this book. (See [9] and [11] for discussions of experimental design.) We shall, however, assume that observations at different combinations of experimental conditions are independent of one another.

Let the two sets of experimental conditions be the levels of, for instance, factor 1 and factor 2, respectively.[3] Suppose there are g levels of factor 1, b levels of factor 2, and n independent observations can be observed at each of the gb combinations of levels. Denoting the rth observation at level ℓ of factor 1 and level k of factor 2 by $X_{\ell kr}$, the univariate two-way model is

$$X_{\ell kr} = \mu + \tau_\ell + \beta_k + \gamma_{\ell k} + e_{\ell kr},$$
$$\ell = 1,2,\ldots,g, \quad k = 1,2,\ldots,b, \quad r = 1,2,\ldots,n \tag{6-49}$$

where $\sum_{\ell=1}^{g} \tau_\ell = \sum_{k=1}^{b} \beta_k = \sum_{\ell=1}^{g} \gamma_{\ell k} = \sum_{k=1}^{b} \gamma_{\ell k} = 0$ and $e_{\ell kr}$ are independent $N(0, \sigma^2)$ random variables. Here μ represents an overall level, τ_ℓ represents the fixed effect of factor 1, β_k represents the fixed effect of factor 2, and $\gamma_{\ell k}$ is the interaction between factor 1 and factor 2. The expected response at the ℓth level of factor 1 and the kth level of factor 2 is thus

$$E(X_{\ell kr}) = \mu + \tau_\ell + \beta_k + \gamma_{\ell k}, \tag{6-50}$$

$$\begin{pmatrix}\text{mean}\\ \text{response}\end{pmatrix} = \begin{pmatrix}\text{overall}\\ \text{level}\end{pmatrix} + \begin{pmatrix}\text{effect of}\\ \text{factor 1}\end{pmatrix} + \begin{pmatrix}\text{effect of}\\ \text{factor 2}\end{pmatrix} + \begin{pmatrix}\text{factor 1–factor 2}\\ \text{interaction}\end{pmatrix}$$

$$\ell = 1,2,\ldots, g, \quad k = 1,2,\ldots, b$$

The presence of interaction, $\gamma_{k\ell}$, implies the factor effects are not additive and complicates the interpretation of the results. Figure 6.5(a) and (b) show expected responses as a function of the factor levels with and without interaction, respectively.

In a manner analogous to (6-49), each observation can be decomposed as

$$x_{\ell kr} = \bar{x} + (\bar{x}_{\ell\cdot} - \bar{x}) + (\bar{x}_{\cdot k} - \bar{x}) + (\bar{x}_{\ell k} - \bar{x}_{\ell\cdot} - \bar{x}_{\cdot k} + \bar{x}) + (x_{\ell kr} - \bar{x}_{\ell k}) \tag{6-51}$$

where $\bar{x}$ is the overall average, $\bar{x}_{\ell\cdot}$ is the average for the ℓth level of factor 1, $\bar{x}_{\cdot k}$ is the average for the kth level of factor 2 and $\bar{x}_{\ell k}$ is the average for the ℓth level of factor 1 *and* the kth level of factor 2. Squaring and summing the deviations

[3] The use of the term *factor* to indicate an experimental condition is convenient. The factors discussed here should not be confused with the unobservable factors considered in Chapter 9 in the context of factor analysis.

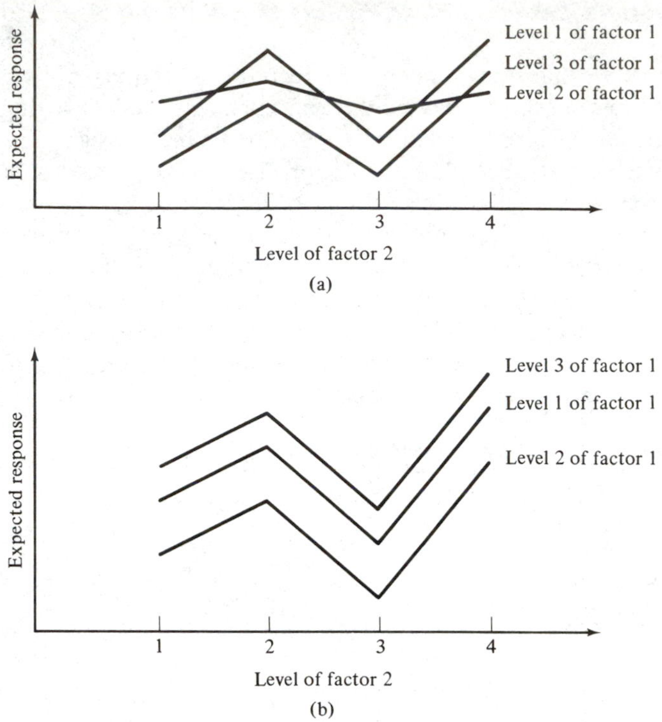

Figure 6.5 Curves for expected responses (a) with interaction (b) without interaction.

$(x_{\ell kr} - \bar{x})$ gives

$$\sum_{\ell=1}^{g} \sum_{k=1}^{b} \sum_{r=1}^{n} (x_{\ell kr} - \bar{x})^2 = \sum_{\ell=1}^{g} bn(\bar{x}_{\ell\cdot} - \bar{x})^2 + \sum_{k=1}^{b} gn(\bar{x}_{\cdot k} - \bar{x})^2 + \sum_{\ell=1}^{g} \sum_{k=1}^{b} n(\bar{x}_{\ell k} - \bar{x}_{\ell\cdot} - \bar{x}_{\cdot k} + \bar{x})^2 + \sum_{\ell=1}^{g} \sum_{k=1}^{b} \sum_{r=1}^{n} (x_{\ell kr} - \bar{x}_{\ell k})^2 \qquad (6\text{-}52)$$

or

$$SS_{\text{cor}} = SS_{\text{fac 1}} + SS_{\text{fac 2}} + SS_{\text{int}} + SS_{\text{res}}$$

The corresponding degrees of freedom associated with the sums of squares in the breakup in (6-52) are

$$gbn - 1 = (g - 1) + (b - 1) + (g - 1)(b - 1) + gb(n - 1) \qquad (6\text{-}53)$$

The analysis of variance table takes the following form.

ANOVA Table for Comparing Effects of Two Factors and Their Interaction

Source of variation	Sum of squares (SS)	Degrees of freedom (d.f.)
Factor 1	$SS_{\text{fac }1} = \sum_{\ell=1}^{g} bn(\bar{x}_{\ell\cdot} - \bar{x})^2$	$g - 1$
Factor 2	$SS_{\text{fac }2} = \sum_{k=1}^{b} gn(\bar{x}_{\cdot k} - \bar{x})^2$	$b - 1$
Interaction	$SS_{\text{int}} = \sum_{\ell=1}^{g} \sum_{k=1}^{b} n(\bar{x}_{\ell k} - \bar{x}_{\ell\cdot} - \bar{x}_{\cdot k} + \bar{x})^2$	$(g-1)(b-1)$
Residual (Error)	$SS_{\text{res}} = \sum_{\ell=1}^{g} \sum_{k=1}^{b} \sum_{r=1}^{n} (x_{\ell kr} - \bar{x}_{\ell k})^2$	$gb(n-1)$
Total (corrected)	$SS_{\text{cor}} = \sum_{\ell=1}^{g} \sum_{k=1}^{b} \sum_{r=1}^{n} (x_{\ell kr} - \bar{x})^2$	$gbn - 1$

The F-ratios of the mean squares, $SS_{\text{fac }1}/(g-1)$, $SS_{\text{fac }2}/(b-1)$, and $SS_{\text{int}}/(g-1)(b-1)$ to the mean square, $SS_{\text{res}}/(gb(n-1))$ can be used to test for the effects of factor 1, factor 2, and factor 1–factor 2 interaction, respectively. (See [7] for a discussion of univariate two-way analysis of variance.)

Multivariate Two-Way Fixed-Effects Model with Interaction

Proceeding by analogy, the two-way fixed-effects model for a *vector* response consisting of p components is (see (6-49))

$$\mathbf{X}_{\ell kr} = \boldsymbol{\mu} + \boldsymbol{\tau}_\ell + \boldsymbol{\beta}_k + \boldsymbol{\gamma}_{\ell k} + \mathbf{e}_{\ell kr},$$
$$\ell = 1,2,\ldots,g, \quad k = 1,2,\ldots,b, \quad r = 1,2,\ldots,n \tag{6-54}$$

where $\sum_{\ell=1}^{g} \boldsymbol{\tau}_\ell = \sum_{k=1}^{b} \boldsymbol{\beta}_k = \sum_{\ell=1}^{g} \boldsymbol{\gamma}_{\ell k} = \sum_{k=1}^{b} \boldsymbol{\gamma}_{\ell k} = \mathbf{0}$. The vectors are all of order $p \times 1$ and $\mathbf{e}_{\ell kr}$ is assumed to be an $N_p(\mathbf{0}, \boldsymbol{\Sigma})$ random vector. Thus the responses consist of p measurements replicated n times at each of the possible combinations of levels of factors 1 and 2.

Following (6-51) the observation vectors $\mathbf{x}_{\ell kr}$ can be decomposed as

$$\mathbf{x}_{\ell kr} = \bar{\mathbf{x}} + (\bar{\mathbf{x}}_{\ell\cdot} - \bar{\mathbf{x}}) + (\bar{\mathbf{x}}_{\cdot k} - \bar{\mathbf{x}}) + (\bar{\mathbf{x}}_{\ell k} - \bar{\mathbf{x}}_{\ell\cdot} - \bar{\mathbf{x}}_{\cdot k} + \bar{\mathbf{x}}) + (\mathbf{x}_{\ell kr} - \bar{\mathbf{x}}_{\ell k}) \tag{6-55}$$

where $\bar{\mathbf{x}}$ is the overall average of the observation vectors, $\bar{\mathbf{x}}_{\ell\cdot}$ is the average of the observation vectors at the ℓth level of factor 1, $\bar{\mathbf{x}}_{\cdot k}$ is the average of the observation vectors at the kth level of factor 2 and $\bar{\mathbf{x}}_{\ell k}$ is the average of the observation vectors at the ℓth level of factor 1 *and* the kth level of factor 2.

Straightforward generalizations of (6-52) and (6-53) give the breakups of the sum of squares and crossproducts and degrees of freedom:

$$\sum_{\ell=1}^{g}\sum_{k=1}^{b}\sum_{r=1}^{n}(\mathbf{x}_{\ell kr}-\bar{\mathbf{x}})(\mathbf{x}_{\ell kr}-\bar{\mathbf{x}})' = \sum_{\ell=1}^{g} bn(\bar{\mathbf{x}}_{\ell\cdot}-\bar{\mathbf{x}})(\bar{\mathbf{x}}_{\ell\cdot}-\bar{\mathbf{x}})'$$

$$+\sum_{k=1}^{b} gn(\bar{\mathbf{x}}_{\cdot k}-\bar{\mathbf{x}})(\bar{\mathbf{x}}_{\cdot k}-\bar{\mathbf{x}})'$$

$$+\sum_{\ell=1}^{g}\sum_{k=1}^{b} n(\bar{\mathbf{x}}_{\ell k}-\bar{\mathbf{x}}_{\ell\cdot}-\bar{\mathbf{x}}_{\cdot k}+\bar{\mathbf{x}})(\bar{\mathbf{x}}_{\ell k}-\bar{\mathbf{x}}_{\ell\cdot}-\bar{\mathbf{x}}_{\cdot k}+\bar{\mathbf{x}})'$$

$$+\sum_{\ell=1}^{g}\sum_{k=1}^{b}\sum_{r=1}^{n}(\mathbf{x}_{\ell kr}-\bar{\mathbf{x}}_{\ell k})(\mathbf{x}_{\ell kr}-\bar{\mathbf{x}}_{\ell k})' \qquad (6\text{-}56)$$

$$gbn-1=(g-1)+(b-1)+(g-1)(b-1)+gb(n-1) \qquad (6\text{-}57)$$

Again, the generalization from the univariate to the multivariate analysis consists simply of replacing a scalar like $(\bar{x}_{\ell\cdot}-\bar{x})^2$ with the corresponding matrix $(\bar{\mathbf{x}}_{\ell\cdot}-\bar{\mathbf{x}})(\bar{\mathbf{x}}_{\ell\cdot}-\bar{\mathbf{x}})'$.

The MANOVA table is the following.

MANOVA Table for Comparing Factors and Their Interaction

Source of variation	Matrix of sum of squares and crossproducts (SSP)	Degrees of freedom (d.f.)
Factor 1	$\text{SSP}_{\text{fac 1}}=\sum_{\ell=1}^{g} bn(\bar{\mathbf{x}}_{\ell\cdot}-\bar{\mathbf{x}})(\bar{\mathbf{x}}_{\ell\cdot}-\bar{\mathbf{x}})'$	$g-1$
Factor 2	$\text{SSP}_{\text{fac 2}}=\sum_{k=1}^{b} gn(\bar{\mathbf{x}}_{\cdot k}-\bar{\mathbf{x}})(\bar{\mathbf{x}}_{\cdot k}-\bar{\mathbf{x}})'$	$b-1$
Interaction	$\text{SSP}_{\text{int}}=\sum_{\ell=1}^{g}\sum_{k=1}^{b} n(\bar{\mathbf{x}}_{\ell k}-\bar{\mathbf{x}}_{\ell\cdot}-\bar{\mathbf{x}}_{\cdot k}+\bar{\mathbf{x}})(\bar{\mathbf{x}}_{\ell k}-\bar{\mathbf{x}}_{\ell\cdot}-\bar{\mathbf{x}}_{\cdot k}+\bar{\mathbf{x}})'$	$(g-1)(b-1)$
Residual (Error)	$\text{SSP}_{\text{res}}=\sum_{\ell=1}^{g}\sum_{k=1}^{b}\sum_{r=1}^{n}(\mathbf{x}_{\ell kr}-\bar{\mathbf{x}}_{\ell k})(\mathbf{x}_{\ell kr}-\bar{\mathbf{x}}_{\ell k})'$	$gb(n-1)$
Total (corrected)	$\text{SSP}_{\text{cor}}=\sum_{\ell=1}^{g}\sum_{k=1}^{b}\sum_{r=1}^{n}(\mathbf{x}_{\ell kr}-\bar{\mathbf{x}})(\mathbf{x}_{\ell kr}-\bar{\mathbf{x}})'$	$gbn-1$

A test (the likelihood ratio test)[4] of

$$H_0:\quad \boldsymbol{\gamma}_{11}=\boldsymbol{\gamma}_{12}=\cdots=\boldsymbol{\gamma}_{gb}=\mathbf{0} \qquad \text{(no interaction effects)} \qquad (6\text{-}58)$$

[4] The likelihood test procedures require $p \le gb(n-1)$ so that SSP_{res} will be positive definite (with probability 1).

versus

$$H_1: \quad \text{At least one } \gamma_{\ell k} \neq \mathbf{0}$$

is conducted by rejecting H_0 for small values of the ratio

$$\Lambda^* = \frac{|\,\text{SSP}_{\text{res}}\,|}{|\,\text{SSP}_{\text{int}} + \text{SSP}_{\text{res}}\,|} \tag{6-59}$$

For large samples, Wilks' lambda, Λ^*, can be referred to a chi-square percentile. Using Bartlett's multiplier (see [5]) to improve the chi-square approximation: Reject H_0: $\gamma_{11} = \gamma_{12} = \cdots = \gamma_{gb} = \mathbf{0}$ at the α level if

$$-\left[gb(n-1) - \frac{p+1-(g-1)(b-1)}{2}\right]\ln \Lambda^* > \chi^2_{(g-1)(b-1)p}(\alpha) \tag{6-60}$$

where Λ^* is given by (6-59) and $\chi^2_{(g-1)(b-1)p}(\alpha)$ is the upper (100α)th percentile of a chi-square distribution with $(g-1)(b-1)p$ d.f.

Ordinarily, the test for interaction is carried out before the tests for main factor effects. If interaction effects exist, the factor effects do not have a clear interpretation. From a practical standpoint, it is not advisable to proceed with the additional multivariate tests. Instead p univariate two-way analyses of variance (one for each variable) are often conducted to see if the interaction appears in some responses but not others. Those responses without interaction may be interpreted in terms of additive factor 1 and 2 effects, provided the latter effects exist.

Assuming there is no interaction, we test for factor 1 and factor 2 main effects. First, consider the hypotheses: H_0: $\boldsymbol{\tau}_1 = \boldsymbol{\tau}_2 = \cdots = \boldsymbol{\tau}_g = \mathbf{0}$ and H_1: at least one $\boldsymbol{\tau}_\ell \neq \mathbf{0}$. These hypotheses specify *no* factor 1 effects and *some* factor 1 effects, respectively. Let

$$\Lambda^* = \frac{|\,\text{SSP}_{\text{res}}\,|}{|\,\text{SSP}_{\text{fac 1}} + \text{SSP}_{\text{res}}\,|} \tag{6-61}$$

so that small values of Λ^* are consistent with H_1. Using Bartlett's correction, the likelihood ratio test is: Reject H_0: $\boldsymbol{\tau}_1 = \boldsymbol{\tau}_2 = \cdots = \boldsymbol{\tau}_g = \mathbf{0}$ (no factor 1 effects) at level α if

$$-\left[gb(n-1) - \frac{p+1-(g-1)}{2}\right]\ln \Lambda^* > \chi^2_{(g-1)p}(\alpha) \tag{6-62}$$

where Λ^* is given by (6-61) and $\chi^2_{(g-1)p}(\alpha)$ is the upper (100α)th percentile of a chi-square distribution with $(g-1)p$ d.f.

In a similar manner, factor 2 effects are tested by considering H_0: $\boldsymbol{\beta}_1 = \boldsymbol{\beta}_2 = \cdots = \boldsymbol{\beta}_b = \mathbf{0}$ and H_1: at least one $\boldsymbol{\beta}_k \neq \mathbf{0}$. Small values of

$$\Lambda^* = \frac{|\,\text{SSP}_{\text{res}}\,|}{|\,\text{SSP}_{\text{fac 2}} + \text{SSP}_{\text{res}}\,|} \tag{6-63}$$

are consistent with H_1. Once again, for large samples and using Bartlett's correction: Reject H_0: $\boldsymbol{\beta}_1 = \boldsymbol{\beta}_2 = \cdots = \boldsymbol{\beta}_b = \mathbf{0}$ (no factor 2 effects) at level α if

$$-\left[gb(n-1) - \frac{p+1-(b-1)}{2}\right]\ln \Lambda^* > \chi^2_{(b-1)p}(\alpha) \tag{6-64}$$

where Λ^* is given by (6-63) and $\chi^2_{(b-1)p}(\alpha)$ is the upper (100α)th percentile of a chi-square distribution with $(b-1)p$ degrees of freedom.

Simultaneous confidence intervals for contrasts in the model parameters can provide insights into the nature of the factor effects. Results comparable to Result 6.5 are available for the two-way model. We take the position that interaction effects are negligible and, therefore, concentrate on contrasts in the factor 1 and factor 2 main effects.

Let $\sum_{\ell=1}^{g} c_\ell \mathbf{a}'\boldsymbol{\tau}_\ell$ and $\sum_{k=1}^{b} c_k \mathbf{a}'\boldsymbol{\beta}_k$ be contrasts in linear compounds of the factor 1 effects and factor 2 effects, respectively. The $100(1-\alpha)\%$ simultaneous confidence intervals for $\sum_{\ell=1}^{g} c_\ell \mathbf{a}'\boldsymbol{\tau}_\ell$ are

$$\sum_{\ell=1}^{g} c_\ell \mathbf{a}'\boldsymbol{\tau}_\ell \quad \text{belongs to} \quad \sum_{\ell=1}^{g} c_\ell \mathbf{a}'\hat{\boldsymbol{\tau}}_\ell \pm \sqrt{\eta_\alpha}\sqrt{\mathbf{a}'\mathbf{E}\mathbf{a}\left(\frac{\sum_{\ell=1}^{g} c_\ell^2}{bn}\right)} \tag{6-65}$$

where $\hat{\boldsymbol{\tau}}_\ell = (\bar{\mathbf{x}}_{\ell\cdot} - \bar{\mathbf{x}})$, $\mathbf{E} = \mathrm{SSP}_{\mathrm{res}}$, $\eta_\alpha = \theta_\alpha/(1-\theta_\alpha)$ and θ_α is the upper (100α)th percentile of the largest root distribution with parameters $\tilde{s} = \min(p, g-1)$, $\tilde{m} = [|g-p-1|-1]/2$, and $\tilde{n} = [gb(n-1)-p-1]/2$.

The $100(1-\alpha)\%$ simultaneous confidence intervals for $\sum_{k=1}^{b} c_k \mathbf{a}'\boldsymbol{\beta}_k$ are

$$\sum_{k=1}^{b} c_k \mathbf{a}'\boldsymbol{\beta}_k \quad \text{belongs to} \quad \sum_{k=1}^{b} c_k \mathbf{a}'\hat{\boldsymbol{\beta}}_k \pm \sqrt{\eta_\alpha}\sqrt{\mathbf{a}'\mathbf{E}\mathbf{a}\left(\frac{\sum_{k=1}^{b} c_k^2}{gn}\right)} \tag{6-66}$$

where $\hat{\boldsymbol{\beta}}_k = (\bar{\mathbf{x}}_{\cdot k} - \bar{\mathbf{x}})$, $\mathbf{E} = \mathrm{SSP}_{\mathrm{res}}$, $\eta_\alpha = \theta_\alpha/(1-\theta_\alpha)$ and θ_α is the upper (100α)th percentile of the largest root distribution with parameters $\tilde{s} = \min(p, b-1)$, $\tilde{m} = [|b-p-1|-1]/2$, and $\tilde{n} = [gb(n-1)-p-1]/2$.

Comment. We have considered the multivariate two-way model with replications. That is, the model allows for n replications of the responses at each combination of factor levels. This enables us to examine the "interaction" of the factors. If only one observation vector is available at each combination of factor levels, the two-way model does not allow for the possibility of a general interaction term, $\gamma_{\ell k}$. The corresponding MANOVA table includes only factor 1, factor 2, and residual sources of variation as components of the total variation (see Exercise 6.9).

Example 6.12

The optimum conditions for extruding plastic film have been examined using a technique called Evolutionary Operation (see [8]). In the course of this study three responses—X_1 = tear resistance, X_2 = gloss, and X_3 = opacity—were measured at two levels of the factors, *rate of extrusion* and *amount of an*

TABLE 6.4 PLASTIC FILM DATA

x_1 = tear resistance, x_2 = gloss, and x_3 = opacity

		Factor 2: Amount of additive					
		Low (1.0%)			High (1.5%)		
		x_1	x_2	x_3	x_1	x_2	x_3
Factor 1: Change in rate of extrusion	Low (−10%)	[6.5	9.5	4.4]	[6.9	9.1	5.7]
		[6.2	9.9	6.4]	[7.2	10.0	2.0]
		[5.8	9.6	3.0]	[6.9	9.9	3.9]
		[6.5	9.6	4.1]	[6.1	9.5	1.9]
		[6.5	9.2	0.8]	[6.3	9.4	5.7]
		x_1	x_2	x_3	x_1	x_2	x_3
	High (10%)	[6.7	9.1	2.8]	[7.1	9.2	8.4]
		[6.6	9.3	4.1]	[7.0	8.8	5.2]
		[7.2	8.3	3.8]	[7.2	9.7	6.9]
		[7.1	8.4	1.6]	[7.5	10.1	2.7]
		[6.8	8.5	3.4]	[7.6	9.2	1.9]

additive. The measurements were repeated $n = 5$ times at each combination of the factor levels. The data are displayed in Table 6.4.

The matrices of the appropriate sum of squares and crossproducts were calculated, leading to the following MANOVA table:

Source of variation	SSP	d.f.
Factor 1: Change in rate of extrusion	$\begin{bmatrix} 1.7405 & -1.5045 & .8555 \\ & 1.3005 & -.7395 \\ & & .4205 \end{bmatrix}$	1
Factor 2: Amount of additive	$\begin{bmatrix} .7605 & .6825 & 1.9305 \\ & .6125 & 1.7325 \\ & & 4.9005 \end{bmatrix}$	1
Interaction	$\begin{bmatrix} .0005 & .0165 & .0445 \\ & .5445 & 1.4685 \\ & & 3.9605 \end{bmatrix}$	1
Residual	$\begin{bmatrix} 1.7640 & .0200 & -3.0700 \\ & 2.6280 & -.5520 \\ & & 64.9240 \end{bmatrix}$	16
Total (corrected)	$\begin{bmatrix} 4.2655 & -.7855 & -.2395 \\ & 5.0855 & 1.9095 \\ & & 74.2055 \end{bmatrix}$	19

To test for interaction we compute

$$\Lambda^* = \frac{|\,\mathrm{SSP_{res}}\,|}{|\,\mathrm{SSP_{int}} + \mathrm{SSP_{res}}\,|} = \frac{275.7098}{354.7906} = .7771$$

For $(g-1)(b-1) = 1$,

$$F = \left(\frac{1-\Lambda^*}{\Lambda^*}\right)\frac{(gb(n-1)-p+1)/2}{(|\,(g-1)(b-1)-p\,|+1)/2}$$

has an exact F-distribution with $\nu_1 = |\,(g-1)(b-1)-p\,|+1$ and $\nu_2 = gb(n-1)-p+1$ d.f. (see [1]). For our example,

$$F = \left(\frac{1-.7771}{.7771}\right)\frac{(2(2)(4)-3+1)/2}{(|\,1(1)-3\,|+1)/2} = 1.34$$

$$\nu_1 = (|\,1(1)-3\,|+1) = 3$$

$$\nu_2 = (2(2)(4)-3+1) = 14$$

and $F_{3,14}(.05) = 3.34$. Since $F = 1.34 < F_{3,14}(.05) = 3.34$, we do not reject the hypothesis H_0: $\gamma_{11} = \gamma_{12} = \gamma_{21} = \gamma_{22} = \mathbf{0}$ (no interaction effects). (Note the approximate chi-square statistic for this test is $-[2(2)(4) - (3 + 1 - 1(1))/2]\ln(.7771) = 3.66$, from (6-60). Since $\chi_3^2(.05) = 7.81$, we would reach the same conclusion as provided by the exact F-test.)

To test for factor 1 and factor 2 effects, we require

$$\Lambda_1^* = \frac{|\,\mathrm{SSP_{res}}\,|}{|\,\mathrm{SSP_{fac\,1}} + \mathrm{SSP_{res}}\,|} = \frac{275.7098}{722.0212} = .3819$$

and

$$\Lambda_2^* = \frac{|\,\mathrm{SSP_{res}}\,|}{|\,\mathrm{SSP_{fac\,2}} + \mathrm{SSP_{res}}\,|} = \frac{275.7098}{527.1347} = .5230$$

For both $g - 1 = 1$ and $b - 1 = 1$,

$$F_1 = \left(\frac{1-\Lambda_1^*}{\Lambda_1^*}\right)\frac{(gb(n-1)-p+1)/2}{(|\,(g-1)-p\,|+1)/2}$$

and

$$F_2 = \left(\frac{1-\Lambda_2^*}{\Lambda_2^*}\right)\frac{(gb(n-1)-p+1)/2}{(|\,(b-1)-p\,|+1)/2}$$

have F-distributions with degrees of freedom $\nu_1 = (|\,(g-1)-p\,|+1)$, $\nu_2 = (gb(n-1)-p+1)$ and $\nu_1 = (|\,(b-1)-p\,|+1)$, $\nu_2 = (gb(n-1)-p+1)$, respectively (see [1]). In our case,

$$F_1 = \left(\frac{1-.3819}{.3819}\right)\frac{(16-3+1)/2}{(|\,1-3\,|+1)/2} = 7.55$$

$$F_2 = \left(\frac{1-.5230}{.5230}\right)\frac{(16-3+1)/2}{(|\,1-3\,|+1)/2} = 4.26$$

and

$$\nu_1 = |1-3|+1 = 3 \qquad \nu_2 = (16-3+1) = 14$$

From above, $F_{3,14}(.05) = 3.34$. We have $F_1 = 7.55 > F_{3,14}(.05) = 3.34$, and therefore we reject H_0: $\boldsymbol{\tau}_1 = \boldsymbol{\tau}_2 = \mathbf{0}$ (no factor 1 effects) at the 5% level. Similarly, $F_2 = 4.26 > F_{3,14}(.05) = 3.34$, and we reject H_0: $\boldsymbol{\beta}_1 = \boldsymbol{\beta}_2 = \mathbf{0}$ (no factor 2 effects) at the 5% level. We conclude that both the *change in rate of extrusion* and the *amount of additive* affect the responses, and they do so in an additive manner.

The *nature* of the effects of factors 1 and 2 on the responses is explored in Exercise 6.11. In that exercise, simultaneous confidence intervals for contrasts in the components of $\boldsymbol{\tau}_\ell$ and $\boldsymbol{\beta}_k$ are considered. ■

EXERCISES

6.1. Construct and sketch a joint 95% confidence region for the mean difference vector $\boldsymbol{\delta}$ using the effluent data and results in Example 6.1. Note that the point $\boldsymbol{\delta} = \mathbf{0}$ falls outside the 95% contour. Is this result consistent with the test of H_0: $\boldsymbol{\delta} = \mathbf{0}$ considered in Example 6.1? Explain.

6.2. A researcher considered three indices measuring severity of heart attacks. The values of these indices for $n = 40$ heart attack patients arriving at a hospital emergency room produced the summary statistics

$$\bar{\mathbf{x}} = \begin{bmatrix} 46.1 \\ 57.3 \\ 50.4 \end{bmatrix} \quad \text{and} \quad \mathbf{S} = \begin{bmatrix} 101.3 & 63.0 & 71.0 \\ 63.0 & 80.2 & 55.6 \\ 71.0 & 55.6 & 97.4 \end{bmatrix}$$

(a) All three indices are evaluated for each patient. Test for the equality of mean indices using (6-16) with $\alpha = .05$.

(b) Judge the differences in pairs of mean indices using 95% simultaneous confidence intervals [see (6-18)].

6.3. Use the data for treatments 2 and 3 in Exercise 6.5.

(a) Calculate $\mathbf{S}_{\text{pooled}}$.

(b) Test H_0: $\boldsymbol{\mu}_2 - \boldsymbol{\mu}_3 = \mathbf{0}$ employing a two-sample approach with $\alpha = .01$.

(c) Construct 99% simultaneous confidence intervals for the differences $\mu_{2i} - \mu_{3i}$, $i = 1, 2$.

6.4. Using the summary statistics for the electricity-demand data given in Example 6.4, compute T^2 and test the hypothesis H_0: $\boldsymbol{\mu}_1 - \boldsymbol{\mu}_2 = \mathbf{0}$, assuming $\boldsymbol{\Sigma}_1 = \boldsymbol{\Sigma}_2$. Set $\alpha = .05$. Also determine the linear combination of mean components most responsible for the rejection of H_0.

6.5. Observations on two responses are collected for three treatments. The observation vectors $\begin{bmatrix} x_1 \\ x_2 \end{bmatrix}$ are

$$\text{Treatment 1:} \quad \begin{bmatrix} 6 \\ 7 \end{bmatrix}, \quad \begin{bmatrix} 5 \\ 9 \end{bmatrix}, \quad \begin{bmatrix} 8 \\ 6 \end{bmatrix}, \quad \begin{bmatrix} 4 \\ 9 \end{bmatrix}, \quad \begin{bmatrix} 7 \\ 9 \end{bmatrix}$$

$$\text{Treatment 2:} \quad \begin{bmatrix} 3 \\ 3 \end{bmatrix}, \quad \begin{bmatrix} 1 \\ 6 \end{bmatrix}, \quad \begin{bmatrix} 2 \\ 3 \end{bmatrix}$$

$$\text{Treatment 3:} \quad \begin{bmatrix} 2 \\ 3 \end{bmatrix}, \quad \begin{bmatrix} 5 \\ 1 \end{bmatrix}, \quad \begin{bmatrix} 3 \\ 1 \end{bmatrix}, \quad \begin{bmatrix} 2 \\ 3 \end{bmatrix}$$

(a) Break up the observations into mean, treatment, and residual components as in (6-35). Construct the corresponding arrays for each variable (see Example 6.8).

(b) Using the information in Part a, construct the one-way MANOVA table.

(c) Evaluate Wilk's lambda, Λ^*, and use Table 6.3 to test for treatment effects. Set $\alpha = .01$. Repeat the test using the chi-square approximation with Bartlett's correction [see (6-39)]. Compare the conclusions.

6.6. Using the contrast matrix $\mathbf{C}$ in (6-13), verify the relationships $\mathbf{d}_j = \mathbf{C}\mathbf{x}_j$; $\bar{\mathbf{d}} = \mathbf{C}\bar{\mathbf{x}}$, and $\mathbf{S}_d = \mathbf{C}\mathbf{S}\mathbf{C}'$ in (6-14).

6.7. Consider the univariate one-way decomposition of the observation $x_{\ell j}$ given by (6-30). Show that the mean vector $\bar{x}\mathbf{1}$ is always perpendicular to the treatment effect vector $(\bar{x}_1 - \bar{x})\mathbf{u}_1 + (\bar{x}_2 - \bar{x})\mathbf{u}_2 + \cdots + (\bar{x}_g - \bar{x})\mathbf{u}_g$ where

$$\mathbf{u}_1 = \begin{bmatrix} 1 \\ \vdots \\ 1 \\ 0 \\ \vdots \\ 0 \end{bmatrix} \begin{matrix} \left.\vphantom{\begin{matrix}1\\ \vdots \\ 1\end{matrix}}\right\} n_1 \\ \vphantom{\begin{matrix}0\\ \vdots \\ 0\end{matrix}} \end{matrix}, \quad \mathbf{u}_2 = \begin{bmatrix} 0 \\ \vdots \\ 0 \\ 1 \\ \vdots \\ 1 \\ 0 \\ \vdots \\ 0 \end{bmatrix} \begin{matrix} \vphantom{\begin{matrix}0\\ \vdots \\ 0\end{matrix}} \\ \left.\vphantom{\begin{matrix}1\\ \vdots \\ 1\end{matrix}}\right\} n_2 \\ \vphantom{\begin{matrix}0\\ \vdots \\ 0\end{matrix}} \end{matrix}, \quad \ldots, \quad \mathbf{u}_g = \begin{bmatrix} 0 \\ \vdots \\ 0 \\ 1 \\ \vdots \\ 1 \end{bmatrix} \begin{matrix} \vphantom{\begin{matrix}0\\ \vdots \\ 0\end{matrix}} \\ \left.\vphantom{\begin{matrix}1\\ \vdots \\ 1\end{matrix}}\right\} n_g \end{matrix}$$

6.8. A likelihood argument provides additional support for pooling the two independent sample covariance matrices to estimate a common covariance matrix in the case of two normal populations. Give the likelihood function, $L(\boldsymbol{\mu}_1, \boldsymbol{\mu}_2, \boldsymbol{\Sigma})$, for two independent samples of sizes n_1 and n_2 from $N_p(\boldsymbol{\mu}_1, \boldsymbol{\Sigma})$ and $N_p(\boldsymbol{\mu}_2, \boldsymbol{\Sigma})$ populations, respectively. Show that this likelihood is maximized by the choices $\hat{\boldsymbol{\mu}}_1 = \bar{\mathbf{x}}_1$, $\hat{\boldsymbol{\mu}}_2 = \bar{\mathbf{x}}_2$ and

$$\hat{\boldsymbol{\Sigma}} = \frac{1}{n_1 + n_2}[(n_1 - 1)\mathbf{S}_1 + (n_2 - 1)\mathbf{S}_2] = \left(\frac{n_1 + n_2 - 2}{n_1 + n_2}\right)\mathbf{S}_{\text{pooled}}$$

[*Hint:* Use (4-16) and the maximization Result 4.10.]

6.9. (Two-way MANOVA without replications.) Consider the observations on two responses, x_1 and x_2, displayed in the form of a two-way table below. (Note there is a *single* observation vector at each combination of factor levels.)

		Factor 2			
		Level 1	Level 2	Level 3	Level 4
	Level 1	$\begin{bmatrix} 6 \\ 8 \end{bmatrix}$	$\begin{bmatrix} 4 \\ 6 \end{bmatrix}$	$\begin{bmatrix} 8 \\ 12 \end{bmatrix}$	$\begin{bmatrix} 2 \\ 6 \end{bmatrix}$
Factor 1:	Level 2	$\begin{bmatrix} 3 \\ 8 \end{bmatrix}$	$\begin{bmatrix} -3 \\ 2 \end{bmatrix}$	$\begin{bmatrix} 4 \\ 3 \end{bmatrix}$	$\begin{bmatrix} -4 \\ 3 \end{bmatrix}$
	Level 3	$\begin{bmatrix} -3 \\ 2 \end{bmatrix}$	$\begin{bmatrix} -4 \\ -5 \end{bmatrix}$	$\begin{bmatrix} 3 \\ -3 \end{bmatrix}$	$\begin{bmatrix} -4 \\ -6 \end{bmatrix}$

With no replications, the two-way MANOVA model is

$$\mathbf{X}_{\ell k} = \boldsymbol{\mu} + \boldsymbol{\tau}_\ell + \boldsymbol{\beta}_k + \mathbf{e}_{\ell k}; \qquad \sum_{\ell=1}^{g} \boldsymbol{\tau}_\ell = \sum_{k=1}^{b} \boldsymbol{\beta}_k = \mathbf{0}$$

where $\mathbf{e}_{\ell k}$ are independent $N_p(\mathbf{0}, \boldsymbol{\Sigma})$ random vectors.

(a) Decompose the observations for each of the two variables as

$$x_{\ell k} = \bar{x} + (\bar{x}_{\ell \cdot} - \bar{x}) + (\bar{x}_{\cdot k} - \bar{x}) + (x_{\ell k} - \bar{x}_{\ell \cdot} - \bar{x}_{\cdot k} + \bar{x})$$

similar to the arrays in Example 6.8.
For each response, this decomposition will result in several 3×4 matrices. Here $\bar{x}$ is the overall average, $\bar{x}_{\ell \cdot}$ is the average for the ℓth level of factor 1 and $\bar{x}_{\cdot k}$ is the average at the kth level of factor 2.

(b) Regard the rows of the matrices in Part a to be strung out in a single "long" vector and compute the sums of squares

$$SS_{tot} = SS_{mean} + SS_{fac\,1} + SS_{fac\,2} + SS_{res}$$

and sums of crossproducts

$$SCP_{tot} = SCP_{mean} + SCP_{fac\,1} + SCP_{fac\,2} + SCP_{res}$$

Consequently, obtain the matrices SSP_{cor}, $SSP_{fac\,1}$, $SSP_{fac\,2}$, and SSP_{res} with degrees of freedom $gb - 1$, $g - 1$, $b - 1$, and $(g - 1)(b - 1)$, respectively.

(c) Summarize the calculations in Part b in a MANOVA table. (*Hint:* This MANOVA table is consistent with the two-way MANOVA table for comparing factors and their interactions where $n = 1$. Note with $n = 1$, SSP_{res} in the general two-way MANOVA table is a zero matrix with zero degrees of freedom. The matrix of interaction sum of squares and crossproducts now becomes the *residual* sum of squares and crossproducts matrix.)

(d) Given the summary in Part c, test for factor 1 and factor 2 main effects at the $\alpha = .05$ level.
[*Hint:* Use the results in (6-62) and (6-64) with $gb(n - 1)$ replaced by $(g - 1)(b - 1)$.]
Note: The tests require $p \le (g - 1)(b - 1)$ so that SSP_{res} will be positive definite (with probability 1).

6.10. A *replicate* of the experiment in Exercise 6.9 yields the data:

		Factor 2 Level 1	Factor 2 Level 2	Factor 2 Level 3	Factor 2 Level 4
	Level 1	$\begin{bmatrix} 14 \\ 8 \end{bmatrix}$	$\begin{bmatrix} 6 \\ 2 \end{bmatrix}$	$\begin{bmatrix} 8 \\ 2 \end{bmatrix}$	$\begin{bmatrix} 16 \\ -4 \end{bmatrix}$
Factor 1:	Level 2	$\begin{bmatrix} 1 \\ 6 \end{bmatrix}$	$\begin{bmatrix} 5 \\ 12 \end{bmatrix}$	$\begin{bmatrix} 0 \\ 15 \end{bmatrix}$	$\begin{bmatrix} 2 \\ 7 \end{bmatrix}$
	Level 3	$\begin{bmatrix} 3 \\ -2 \end{bmatrix}$	$\begin{bmatrix} -2 \\ 7 \end{bmatrix}$	$\begin{bmatrix} -11 \\ 1 \end{bmatrix}$	$\begin{bmatrix} -6 \\ 6 \end{bmatrix}$

(a) Use these data to decompose each of the two measurements in the observation vector as

$$x_{\ell k} = \bar{x} + (\bar{x}_{\ell \cdot} - \bar{x}) + (\bar{x}_{\cdot k} - \bar{x}) + (x_{\ell k} - \bar{x}_{\ell \cdot} - \bar{x}_{\cdot k} + \bar{x})$$

where $\bar{x}$ is the overall average, $\bar{x}_{\ell\cdot}$ is the average for the ℓth level of factor 1 and $\bar{x}_{\cdot k}$ is the average for the kth level of factor 2. Form the corresponding arrays for each of the two responses.

(b) Combine the data above with that in Exercise 6.9 and carry out the necessary calculations to complete the general two-way MANOVA table.

(c) Given the results in Part b, test for interactions, and if the interactions do not exist, test for factor 1 and factor 2 main effects. Use the likelihood ratio test with $\alpha = .05$.

(d) If main effects but no interactions exist, examine the nature of the main effects by constructing simultaneous 95% confidence intervals for contrasts in the components of the factor effect parameters.

6.11. Refer to Example 6.12.

(a) Carry out approximate chi-square (likelihood ratio) tests for the factor 1 and factor 2 effects. Set $\alpha = .05$. Compare these results with the results for the exact F-tests given in the example. Explain any differences.

(b) Using (6-65) and $\eta_{.05} = (\frac{3}{14})F_{3,14}(.05) = .72$, construct simultaneous 95% confidence intervals for contrasts in the factor 1 effect parameters for *pairs* of the three responses. Interpret these intervals. Repeat these calculations for factor 2 effect parameters.

The following exercises may require the use of a computer.

6.12. Four measures of the response *stiffness* on each of 30 boards are listed in Table 4.3 (see Example 4.13). The measures, on a given board, are repeated in the sense that they were made one after another. Assuming the measures of stiffness arise from 4 treatments, test for the equality of treatments in a *repeated measures design* context. Set $\alpha = .05$. Construct a 95% (simultaneous) confidence interval for a contrast in the mean levels representing a comparison of the dynamic measurements with the static measurements.

6.13. Jolicoeur and Mosimann [12] studied the relationship of size and shape for painted turtles. Table 6.5 contains their measurements on the carapaces of 24 female and 24 male turtles.

(a) Test for equality of the two population mean vectors using $\alpha = .05$.

(b) If the hypothesis in Part a is rejected, find the linear combination of mean components most responsible for rejecting H_0.

(c) Find simultaneous confidence intervals for the component mean differences.

(*Hint:* You may wish to consider logarithmic transformations of the observations.)

6.14. In the first phase of a study of the cost of transporting milk from farms to dairy plants, a survey was taken of firms engaged in milk transportation. Cost data on X_1 = fuel, X_2 = repair, and X_3 = capital, all measured on a per-mile basis, are presented in Table 6.6 on page 282 for $n_1 = 36$ gasoline and $n_2 = 23$ diesel trucks.

(a) Test for differences in the mean cost vectors. Set $\alpha = .01$.

(b) If the hypothesis of equal cost vectors is rejected in Part a, find the linear combination of mean components most responsible for the rejection.

(c) Construct 99% confidence intervals for the pairs of mean components. Which costs, if any, appear to be quite different?

(d) Comment on the validity of the assumptions used in your analysis.

6.15. The tail lengths in millimeters (x_1) and wing lengths in millimeters (x_2) for 45 *male* hook-billed kites are given in Table 6.7 on page 283. Similar measurements for female hook-billed kites were given in Table 5.7.

(a) Plot the male hook-billed kite data as a scatter diagram and (visually) check for outliers.

TABLE 6.5 CARAPACE MEASUREMENTS (IN MILLIMETERS) FOR PAINTED TURTLES

Female			Male		
Length (x_1)	Width (x_2)	Height (x_3)	Length (x_1)	Width (x_2)	Height (x_3)
98	81	38	93	74	37
103	84	38	94	78	35
103	86	42	96	80	35
105	86	42	101	84	39
109	88	44	102	85	38
123	92	50	103	81	37
123	95	46	104	83	39
133	99	51	106	83	39
133	102	51	107	82	38
133	102	51	112	89	40
134	100	48	113	88	40
136	102	49	114	86	40
138	98	51	116	90	43
138	99	51	117	90	41
141	105	53	117	91	41
147	108	57	119	93	41
149	107	55	120	89	40
153	107	56	120	93	44
155	115	63	121	95	42
155	117	60	125	93	45
158	115	62	127	96	45
159	118	63	128	95	45
162	124	61	131	95	46
177	132	67	135	106	47

(b) Test for equality of mean vectors for the populations of male and female hook-billed kites. Set $\alpha = .05$. If $H_0\colon \boldsymbol{\mu}_1 - \boldsymbol{\mu}_2 = \mathbf{0}$ is rejected, find the linear combination most responsible for the rejection of H_0. (You may want to eliminate any outliers found in Part a for the male hook-billed kite data before conducting this test.)

(c) Determine the 95% confidence region for $\boldsymbol{\mu}_1 - \boldsymbol{\mu}_2$ and 95% simultaneous confidence intervals for the components of $\boldsymbol{\mu}_1 - \boldsymbol{\mu}_2$.

(d) Are male or female birds generally larger?

6.16. Using Moody's bond ratings, samples of 20 Aa (middle high quality) corporate bonds and 20 Baa (top medium quality) corporate bonds were selected. For each of the corresponding companies, the ratios

X_1 = current ratio (a measure of short-term liquidity)

X_2 = times long-term interest rate (a measure of interest coverage)

X_3 = debt to equity ratio (a measure of financial risk or leverage)

X_4 = rate of return on equity (a measure of profitability)

were recorded. The summary statistics are as follows.

TABLE 6.6 MILK TRANSPORTATION-COST DATA

Gasoline trucks			Diesel trucks		
x_1	x_2	x_3	x_1	x_2	x_3
16.44	12.43	11.23	8.50	12.26	9.11
7.19	2.70	3.92	7.42	5.13	17.15
9.92	1.35	9.75	10.28	3.32	11.23
4.24	5.78	7.78	10.16	14.72	5.99
11.20	5.05	10.67	12.79	4.17	29.28
14.25	5.78	9.88	9.60	12.72	11.00
13.50	10.98	10.60	6.47	8.89	19.00
13.32	14.27	9.45	11.35	9.95	14.53
29.11	15.09	3.28	9.15	2.94	13.68
12.68	7.61	10.23	9.70	5.06	20.84
7.51	5.80	8.13	9.77	17.86	35.18
9.90	3.63	9.13	11.61	11.75	17.00
10.25	5.07	10.17	9.09	13.25	20.66
11.11	6.15	7.61	8.53	10.14	17.45
12.17	14.26	14.39	8.29	6.22	16.38
10.24	2.59	6.09	15.90	12.90	19.09
10.18	6.05	12.14	11.94	5.69	14.77
8.88	2.70	12.23	9.54	16.77	22.66
12.34	7.73	11.68	10.43	17.65	10.66
8.51	14.02	12.01	10.87	21.52	28.47
26.16	17.44	16.89	7.13	13.22	19.44
12.95	8.24	7.18	11.88	12.18	21.20
16.93	13.37	17.59	12.03	9.22	23.09
14.70	10.78	14.58			
10.32	5.16	17.00			
8.98	4.49	4.26			
9.70	11.59	6.83			
12.72	8.63	5.59			
9.49	2.16	6.23			
8.22	7.95	6.72			
13.70	11.22	4.91			
8.21	9.85	8.17			
15.86	11.42	13.06			
9.18	9.18	9.49			
12.49	4.67	11.94			
17.32	6.86	4.44			

SOURCE: Data courtesy of M. Keaton.

Aa bond companies: $n_1 = 20$, $\bar{\mathbf{x}}_1 = [2.287, 12.600, .347, 14.830]'$, and

$$\mathbf{S}_1 = \begin{bmatrix} .459 & .254 & -.026 & -.244 \\ .254 & 27.465 & -.589 & -.267 \\ -.026 & -.589 & .030 & .102 \\ -.244 & -.267 & .102 & 6.854 \end{bmatrix}$$

TABLE 6.7 MALE HOOK-BILLED KITE DATA

x_1 (Tail length)	x_2 (Wing length)	x_1 (Tail length)	x_2 (Wing length)	x_1 (Tail length)	x_2 (Wing length)
180	278	185	282	284	277
186	277	195	285	176	281
206	308	183	276	185	287
184	290	202	308	191	295
177	273	177	254	177	267
177	284	177	268	197	310
176	267	170	260	199	299
200	281	186	274	190	273
191	287	177	272	180	278
193	271	178	266	189	280
212	302	192	281	194	290
181	254	204	276	186	287
195	297	191	290	191	286
187	281	178	265	187	288
190	284	177	275	186	275

SOURCE: Data courtesy of S. Temple.

Baa bond companies: $n_1 = 20$, $\bar{\mathbf{x}}_2 = [2.404, 7.155, .524, 12.840]'$,

$$\mathbf{S}_2 = \begin{bmatrix} .944 & -.089 & .002 & -.719 \\ -.089 & 16.432 & -.400 & 19.044 \\ .002 & -.400 & .024 & -.094 \\ -.719 & 19.044 & -.094 & 61.854 \end{bmatrix}$$

and

$$\mathbf{S}_{\text{pooled}} = \begin{bmatrix} .701 & .083 & -.012 & -.481 \\ .083 & 21.949 & -.494 & 9.388 \\ -.012 & -.494 & .027 & .004 \\ -.481 & 9.388 & .004 & 34.354 \end{bmatrix}$$

(a) Does pooling appear reasonable here? Comment on the pooling procedure in this case.

(b) Are the financial characteristics of firms with Aa bonds different from those with Baa bonds? Using the pooled covariance matrix, test for the equality of mean vectors. Set $\alpha = .05$.

(c) Calculate the linear combination of mean components most responsible for rejecting H_0: $\boldsymbol{\mu}_1 - \boldsymbol{\mu}_2 = \mathbf{0}$ in Part b.

(d) Bond rating companies are interested in a company's ability to satisfy its outstanding debt obligations as they mature.
Does it appear as if one or more of the financial ratios above might be useful in helping to classify a bond as "high" or "medium" quality? Explain.

6.17. Researchers interested in assessing pulmonary function in nonpathological populations asked subjects to run on a treadmill until exhaustion. Samples of air were collected at definite intervals and the gas contents analyzed. The results on 4 measures of oxygen consumption for 25 males and 25 females are given in Table 6.8. The variables were

$$X_1 = \text{resting volume } O_2 \text{ (L/min)}$$

$$X_2 = \text{resting volume } O_2 \text{ (mL/kg/min)}$$

$$X_3 = \text{maximum volume } O_2 \text{ (L/min)}$$

$$X_4 = \text{maximum volume } O_2 \text{ (mL/kg/min)}$$

(a) Look for sex differences by testing for equality of group means. Use $\alpha = .05$. If you reject H_0: $\boldsymbol{\mu}_1 - \boldsymbol{\mu}_2 = \mathbf{0}$, find the linear combination most responsible.
(b) Construct the 95% simultaneous confidence intervals for each $\mu_{1i} - \mu_{2i}$, $i = 1, 2, 3, 4$.
(c) The data in Table 6.8 were collected from graduate-student volunteers and thus they do not represent a random sample. Comment on the possible implications of this information.

6.18. Construct a one-way MANOVA of the iris data in Table 10.4. Construct 95% simultaneous confidence intervals for differences in mean components for the two responses for each pair of populations. Comment on the validity of the assumption $\boldsymbol{\Sigma}_1 = \boldsymbol{\Sigma}_2 = \boldsymbol{\Sigma}_3$.

6.19. Construct a one-way MANOVA of the crude-oil data listed in Table 10.6. Construct 95% simultaneous confidence intervals to determine which mean components differ among the populations. (You may want to consider transformations of the data to make them more closely conform to the usual MANOVA assumptions.)

6.20. A project was designed to investigate how consumers in Green Bay, Wisconsin, would react to an electrical time-of-use pricing scheme. The cost of electricity during peak periods for some customers was set at eight times the cost of electricity during off-peak hours. Hourly consumption (in kilowatt-hours) was measured on a hot summer day in July and this consumption was compared, for both the test group and the control group, with baseline consumption measured on a similar day before the experimental rates began. The responses,

$$\log(\text{current consumption}) - \log(\text{baseline consumption})$$

for the hours ending 9 A.M., 11 A.M. (a peak hour), 1 P.M., and 3 P.M. (a peak hour) produced the following summary statistics.

Test group: $n_1 = 28$, $\bar{\mathbf{x}}_1 = [.153, -.231, -.322, -.339]'$

Control group: $n_2 = 58$, $\bar{\mathbf{x}}_2 = [.151, .180, .256, .275]'$

and

$$\mathbf{S}_{\text{pooled}} = \begin{bmatrix} .804 & .355 & .228 & .232 \\ .355 & .722 & .233 & .199 \\ .228 & .233 & .592 & .239 \\ .232 & .199 & .239 & .479 \end{bmatrix}$$

SOURCE: Data courtesy of Statistical Laboratory, University of Wisconsin.

Perform a profile analysis. Does time-of-use pricing seem to make a difference in electrical consumption? What is the nature of this difference, if any? Comment. (Use a significance level of $\alpha = .05$ for any statistical tests.)

6.21. In the study of love and marriage of Example 6.11, a sample of husbands and a sample of wives were asked to respond to these questions:

1. What is the level of passionate love you feel for your partner?

TABLE 6.8 OXYGEN-CONSUMPTION DATA

Males				Females			
x_1 Resting O_2 (L/min)	x_2 Resting O_2 (mL/kg/min)	x_3 Maximum O_2 (L/min)	x_4 Maximum O_2 (mL/kg/min)	x_1 Resting O_2 (L/min)	x_2 Resting O_2 (mL/kg/min)	x_3 Maximum O_2 (L/min)	x_4 Maximum O_2 (mL/kg/min)
0.34	3.71	2.87	30.87	0.29	5.04	1.93	33.85
0.39	5.08	3.38	43.85	0.28	3.95	2.51	35.82
0.48	5.13	4.13	44.51	0.31	4.88	2.31	36.40
0.31	3.95	3.60	46.00	0.30	5.97	1.90	37.87
0.36	5.51	3.11	47.02	0.28	4.57	2.32	38.30
0.33	4.07	3.95	48.50	0.11	1.74	2.49	39.19
0.43	4.77	4.39	48.75	0.25	4.66	2.12	39.21
0.48	6.69	3.50	48.86	0.26	5.28	1.98	39.94
0.21	3.71	2.82	48.92	0.39	7.32	2.25	42.41
0.32	4.35	3.59	48.38	0.37	6.22	1.71	28.97
0.54	7.89	3.47	50.56	0.31	4.20	2.76	37.80
0.32	5.37	3.07	51.15	0.35	5.10	2.10	31.10
0.40	4.95	4.43	55.34	0.29	4.46	2.50	38.30
0.31	4.97	3.56	56.67	0.33	5.60	3.06	51.80
0.44	6.68	3.86	58.49	0.18	2.80	2.40	37.60
0.32	4.80	3.31	49.99	0.28	4.01	2.58	36.78
0.50	6.43	3.29	42.25	0.44	6.69	3.05	46.16
0.36	5.99	3.10	51.70	0.22	4.55	1.85	38.95
0.48	6.30	4.80	63.30	0.34	5.73	2.43	40.60
0.40	6.00	3.06	46.23	0.30	5.12	2.58	43.69
0.42	6.04	3.85	55.08	0.31	4.77	1.97	30.40
0.55	6.45	5.00	58.80	0.27	5.16	2.03	39.46
0.50	5.55	5.23	57.46	0.66	11.05	2.32	39.34
0.34	4.27	4.00	50.35	0.37	5.23	2.48	34.86
0.40	4.58	2.82	32.48	0.35	5.37	2.25	35.07

SOURCE: Data courtesy of S. Rokicki.

2. What is the level of passionate love that your partner feels for you?
3. What is the level of companionate love that you feel for your partner?
4. What is the level of companionate love that your partner feels for you?
The responses were recorded on a 5-point scale.

None at all	Very little	Some	A great deal	A tremendous amount
1	2	3	4	5

Thirty couples gave the responses in Table 6.9, where X_1 = a 5-point scale response to Question 1, X_2 = a 5-point scale response to Question 2, X_3 = a 5-point scale response to Question 3, and X_4 = a 5-point scale response to Question 4.

(a) Plot the mean vectors for husbands and wives as sample profiles.

(b) Is the husband-profile parallel to the wife-profile? Test for parallel profiles with $\alpha = .05$. If the profiles appear to be parallel, test for coincident profiles at the same

TABLE 6.9 SPOUSE DATA

Husband rating wife				Wife rating husband			
x_1	x_2	x_3	x_4	x_1	x_2	x_3	x_4
2	3	5	5	4	4	5	5
5	5	4	4	4	5	5	5
4	5	5	5	4	4	5	5
4	3	4	4	4	5	5	5
3	3	5	5	4	4	5	5
3	3	4	5	3	3	4	4
3	4	4	4	4	3	5	4
4	4	5	5	3	4	5	5
4	5	5	5	4	4	5	4
4	4	3	3	3	4	4	4
4	4	5	5	4	5	5	5
5	5	4	4	5	5	5	5
4	4	4	4	4	4	5	5
4	3	5	5	4	4	4	4
4	4	5	5	4	4	5	5
3	3	4	5	3	4	4	4
4	5	4	4	5	5	5	5
5	5	5	5	4	5	4	4
5	5	4	4	3	4	4	4
4	4	4	4	5	3	4	4
4	4	4	4	5	3	4	4
4	4	4	4	4	5	4	4
3	4	5	5	2	5	5	5
5	3	5	5	3	4	5	5
5	5	3	3	4	3	5	5
3	3	4	4	4	4	4	4
4	4	4	4	4	4	5	5
3	3	5	5	3	4	4	4
4	4	3	3	4	4	5	4
4	4	5	5	4	4	5	5

SOURCE: Data courtesy of E. Hatfield.

level of significance. Finally, if the profiles are coincident, test for level profiles with $\alpha = .05$. What conclusion(s) can be drawn from this analysis?

6.22. Two species of biting flies (genus: *Leptoconops*) are so similar morphologically, that for many years they were thought to be the same. Biological differences such as sex ratios of emerging flies and biting habits were found to exist. Does the taxonomic data listed in Table 6.10 indicate any difference in the two species *L. carteri* and *L. torrens*? Test for the equality of the two population mean vectors using $\alpha = .05$. If the hypotheses of equal mean vectors is rejected, determine the mean components (or linear combinations of mean components) most responsible for rejecting H_0. Justify your use of normal-theory methods for these data.

TABLE 6.10 BITING FLY DATA

	x_1 (wing length)	x_2 (wing width)	x_3 (third palp length)	x_4 (third palp width)	x_5 (fourth palp length)	x_6 (length antennal segment 12)	x_7 (length antennal segment 13)
	85	41	31	13	25	9	8
	87	38	32	14	22	13	13
	94	44	36	15	27	8	9
	92	43	32	17	28	9	9
	96	43	35	14	26	10	10
	91	44	36	12	24	9	9
	90	42	36	16	26	9	9
	92	43	36	17	26	9	9
	91	41	36	14	23	9	9
	87	38	35	11	24	9	10
	97	45	39	17	27	9	10
	89	38	36	13	22	9	9
	94	45	37	13	26	9	9
	96	44	37	14	24	9	10
	104	49	35	14	21	10	10
	94	41	31	17	26	10	9
	99	44	31	18	28	10	9
	94	38	32	13	22	9	9
L. torrens	94	43	37	16	26	9	10
	93	43	38	14	28	10	10
	95	44	37	18	27	10	10
	95	45	39	13	27	10	10
	96	39	37	12	26	8	8
	103	46	34	18	26	10	10
	108	44	37	14	25	11	11
	106	47	38	15	26	10	10
	105	46	34	14	31	10	11
	103	44	34	15	23	10	10
	100	41	35	14	24	10	10
	109	44	36	13	27	11	10
	104	45	36	15	30	10	10
	95	40	35	14	23	9	10
	104	44	34	15	29	9	10
	90	40	37	12	22	9	10
	104	46	37	14	30	10	10

TABLE 6.10 *(continued)*

	x_1 (wing length)	x_2 (wing width)	x_3 (third palp length)	x_4 (third palp width)	x_5 (fourth palp length)	x_6 (length antennal segment 12)	x_7 (length antennal segment 13)
	86	19	37	11	25	9	9
	94	40	38	14	31	6	7
	103	48	39	14	33	10	10
	82	41	35	12	25	9	8
	103	43	42	15	32	9	9
	101	43	40	15	25	9	9
	103	45	44	14	29	11	11
	100	43	40	18	31	11	10
	99	41	42	15	31	10	10
	100	44	43	16	34	10	10
	112	47	44	16	38	12	11
	99	48	37	14	32	10	9
	98	45	41	19	31	9	8
	101	46	42	14	24	11	10
	99	45	37	13	28	10	9
	103	47	44	15	20	8	9
L. carteri	98	40	38	12	32	9	8
	101	46	36	14	28	10	10
	101	46	40	17	32	9	9
	98	47	39	15	33	10	10
	99	45	42	15	32	10	9
	102	45	44	15	30	10	10
	97	45	37	15	32	10	9
	96	39	40	14	20	9	9
	89	39	33	12	20	9	8
	99	42	38	14	33	9	9
	110	45	41	17	36	9	10
	99	44	35	16	31	10	10
	103	43	38	14	32	10	10
	95	46	36	15	31	8	8
	101	47	38	14	37	11	11
	103	47	40	15	32	11	11
	99	43	37	14	23	11	10
	105	50	40	16	33	12	11
	99	47	39	14	34	7	7

SOURCE: Data courtesy of William Atchley.

REFERENCES

[1] Anderson, T. W., *An Introduction to Multivariate Statistical Analysis*, New York: John Wiley, 1958.

[2] Bartlett, M. S., "Properties of Sufficiency and Statistical Tests," *Proceedings of the Royal Society of London (A)*, **160** (1937), 268–282.

[3] Bartlett, M. S., "Further Aspects of the Theory of Multiple Regression," *Proceedings of the Cambridge Philosophical Society*, **34** (1938), 33–40.

[4] Bartlett, M. S., "Multivariate Analysis," *Journal of the Royal Statistical Society Supplement (B)*, **9** (1947), 176–197.

[5] Bartlett, M. S., "A Note on the Multiplying Factors for Various χ^2 Approximations," *Journal of the Royal Statistical Society (B)*, **16** (1954), 296–298.

[6] Beyer, W. H. (ed.), *Handbook of Tables for Probability and Statistics* (2nd ed.), Cleveland, Ohio: CRC Press, 1968.

[7] Bhattacharyya, G. K., and R. A. Johnson, *Statistical Concepts and Methods*, New York: John Wiley, 1977.

[8] Box, G. E. P., and N. R. Draper, *Evolutionary Operation: A Statistical Method for Process Improvement*, New York: John Wiley, 1969.

[9] Box, G. E. P., W. G. Hunter, and J. S. Hunter, *Statistics for Experimenters*, New York: John Wiley, 1978.

[10] Heck, D. L., "Charts of Some Upper Percentage Points of the Distribution of the Largest Characteristic Root," *Annals of Mathematical Statistics*, **31** (1960), 625–642.

[11] John, P. W. M., *Statistical Design and Analysis of Experiments*, New York: Macmillan, 1971.

[12] Jolicoeur, P., and J. E. Mosimann, "Size and Shape Variation in the Painted Turtle: A Principal Component Analysis," *Growth*, **24** (1960), 339–354.

[13] Morrison, D. F., *Multivariate Statistical Methods* (2nd ed.), New York: McGraw-Hill, 1976.

[14] Pearson, E. S., and H. O. Hartley, eds., *Biometrika Tables for Statisticians*, volume II, England: Cambridge University Press, 1972.

[15] Pillai, K. C. S., "On the Distribution of the Largest Characteristic Root of a Matrix in Multivariate Analysis," *Biometrika*, **52** (1965), 405–414.

[16] Pillai, K. C. S., "Upper Percentage Points of the Largest Root of a Matrix in Multivariate Analysis," *Biometrika*, **54** (1967), 189–193.

[17] Roy, S. N., *Some Aspects of Multivariate Analysis*, New York: John Wiley, 1957.

[18] Scheffé, H., *The Analysis of Variance*, New York: John Wiley, 1959.

[19] Timm, N. H., *Multivariate Analysis with Applications in Education and Psychology*, Monterey, California: Brooks/Cole, 1975.

[20] Wilks, S. S., "Certain Generalizations in the Analysis of Variance," *Biometrika*, **24** (1932), 471–494.

7

Multivariate Linear Regression Models

7.1 INTRODUCTION

Regression analysis is the statistical methodology for predicting values of one or more *response* (dependent) variables from a collection of *predictor* (independent) variable values. It can also be used for assessing the effects of the predictor variables on the responses. Unfortunately, the name *regression*, culled from the title of the first paper on the subject by F. Galton [10] in no way reflects either the importance or breadth of application of this methodology.

In this chapter we first discuss the multiple regression model for the prediction of a *single* response. This model is then generalized to handle the prediction of *several* dependent variables. Our treatment must necessarily be somewhat terse as a vast literature exists on this subject. (If you are interested in pursuing regression analysis, see the following books, in ascending order of difficulty: Neter and Wasserman [15], Draper and Smith [7], Seber [18], and Goldberger [11].) Our abbreviated treatment highlights the regression assumptions and their consequences, alternative formulations of the regression model, and the general applicability of regression techniques to seemingly different situations.

7.2 THE CLASSICAL LINEAR REGRESSION MODEL

Let $z_1, z_2, \ldots, z_r$ be r predictor variables thought to be related to a response variable Y. For example, with $r = 4$, we might have

$$Y = \text{current market value of home}$$

and

$$z_1 = \text{square feet of living area}$$

$$z_2 = \text{location (indicator for zone of city)}$$

$$z_3 = \text{appraised value last year}$$

$$z_4 = \text{quality of construction (price per square foot)}$$

The classical linear regression model states that Y is composed of a mean, which depends in a linear fashion on the z_i's, and random error, ε, which accounts for measurement error and the effects of other variables not explicitly considered in the model. The values of the predictor variables recorded from the experiment or set by the investigator are treated as *fixed*. The error (and hence the response) is viewed as a random variable whose behavior is characterized by a set of distributional assumptions.

Specifically, the linear regression model with a single response, takes the form

$$Y = \beta_0 + \beta_1 z_1 + \cdots + \beta_r z_r + \varepsilon$$

$$[\text{Response}] = [\text{mean (depending on } z_1, z_2, \ldots, z_r)] + [\text{error}]$$

The term *linear* refers to the fact that the mean is a linear function of the unknown parameters $\beta_0, \beta_1, \ldots, \beta_r$. The predictor variables may or may not enter the model as first-order terms.

With n independent observations on Y and the associated values of z_i, the complete model becomes

$$\begin{aligned} Y_1 &= \beta_0 + \beta_1 z_{11} + \beta_2 z_{12} + \cdots + \beta_r z_{1r} + \varepsilon_1 \\ Y_2 &= \beta_0 + \beta_1 z_{21} + \beta_2 z_{22} + \cdots + \beta_r z_{2r} + \varepsilon_2 \\ &\vdots \qquad\qquad \vdots \\ Y_n &= \beta_0 + \beta_1 z_{n1} + \beta_2 z_{n2} + \cdots + \beta_r z_{nr} + \varepsilon_n \end{aligned} \tag{7-1}$$

where the error terms are assumed to have the properties:

1. $E(\varepsilon_j) = 0$;
2. $\text{Var}(\varepsilon_j) = \sigma^2$ (constant); and (7-2)
3. $\text{Cov}(\varepsilon_j, \varepsilon_k) = 0, \quad j \neq k.$

In matrix notation, (7-1) becomes

$$\begin{bmatrix} Y_1 \\ Y_2 \\ \vdots \\ Y_n \end{bmatrix} = \begin{bmatrix} 1 & z_{11} & z_{12} & \cdots & z_{1r} \\ 1 & z_{21} & z_{22} & \cdots & z_{2r} \\ \vdots & \vdots & \vdots & & \vdots \\ 1 & z_{n1} & z_{n2} & \cdots & z_{nr} \end{bmatrix} \begin{bmatrix} \beta_0 \\ \beta_1 \\ \vdots \\ \beta_r \end{bmatrix} + \begin{bmatrix} \varepsilon_1 \\ \varepsilon_2 \\ \vdots \\ \varepsilon_n \end{bmatrix}$$

or

$$\underset{(n\times 1)}{\mathbf{Y}} = \underset{(n\times(r+1))}{\mathbf{Z}} \; \underset{((r+1)\times 1)}{\boldsymbol{\beta}} + \underset{(n\times 1)}{\boldsymbol{\varepsilon}}$$

and the specifications in (7-2) become:

1. $E(\boldsymbol{\varepsilon}) = \mathbf{0}$; and
2. $\text{Cov}(\boldsymbol{\varepsilon}) = E(\boldsymbol{\varepsilon}\boldsymbol{\varepsilon}') = \sigma^2\mathbf{I}$.

Note that a one in the first column of the *design matrix* $\mathbf{Z}$ is the multiplier of the constant term β_0. It is customary to introduce the artificial variable $z_{j0} = 1$ so

$$\beta_0 + \beta_1 z_{j1} + \cdots + \beta_r z_{jr} = \beta_0 z_{j0} + \beta_1 z_{j1} + \cdots + \beta_r z_{jr}$$

Each column of $\mathbf{Z}$ consists of the n values of the corresponding predictor variable, while the jth row of $\mathbf{Z}$ contains the values for all predictor variables on the jth trial.

Classical Linear Regression Model

$$\underset{(n\times 1)}{\mathbf{Y}} = \underset{(n\times(r+1))}{\mathbf{Z}} \; \underset{((r+1)\times 1)}{\boldsymbol{\beta}} + \underset{(n\times 1)}{\boldsymbol{\varepsilon}},$$

$$E(\boldsymbol{\varepsilon}) = \underset{(n\times 1)}{\mathbf{0}}, \text{ and } \text{Cov}(\boldsymbol{\varepsilon}) = \sigma^2 \underset{(n\times n)}{\mathbf{I}}, \qquad (7\text{-}3)$$

where $\boldsymbol{\beta}$ and σ^2 are unknown parameters and the design matrix $\mathbf{Z}$ has jth row $[z_{j0}, z_{j1}, \ldots, z_{jr}]$.

Although the error-term assumptions in (7-2) are very modest, we shall later need to add the assumption of joint normality for making confidence statements and testing hypotheses.

We now provide some examples of the linear regression model.

Example 7.1

Determine the linear regression model for fitting a straight line

$$\text{mean response} = E(Y) = \beta_0 + \beta_1 z_1$$

to the data

z_1	0	1	2	3	4
y	1	4	3	8	9

Before the responses $\mathbf{Y} = [Y_1, Y_2, \ldots, Y_5]'$ are observed, the errors $\boldsymbol{\varepsilon} = [\varepsilon_1, \varepsilon_2, \ldots, \varepsilon_5]'$ are random and we can write

$$\mathbf{Y} = \mathbf{Z}\boldsymbol{\beta} + \boldsymbol{\varepsilon}$$

where

$$\mathbf{Y} = \begin{bmatrix} Y_1 \\ Y_2 \\ \vdots \\ Y_5 \end{bmatrix}; \quad \mathbf{Z} = \begin{bmatrix} 1 & z_{11} \\ 1 & z_{21} \\ \vdots & \vdots \\ 1 & z_{51} \end{bmatrix}; \quad \boldsymbol{\beta} = \begin{bmatrix} \beta_0 \\ \beta_1 \end{bmatrix}; \quad \boldsymbol{\varepsilon} = \begin{bmatrix} \varepsilon_1 \\ \varepsilon_2 \\ \vdots \\ \varepsilon_5 \end{bmatrix}$$

The data for this model are contained in the observed response vector, **y**, and the design matrix, **Z**. Here

$$\mathbf{y} = \begin{bmatrix} 1 \\ 4 \\ 3 \\ 8 \\ 9 \end{bmatrix}; \qquad \mathbf{Z} = \begin{bmatrix} 1 & 0 \\ 1 & 1 \\ 1 & 2 \\ 1 & 3 \\ 1 & 4 \end{bmatrix}$$

Note that we can handle a quadratic expression for the mean response by introducing the term $\beta_2 z_2$, with $z_2 = z_1^2$. The linear regression model for the jth trial in this latter case is

$$Y_j = \beta_0 + \beta_1 z_{j1} + \beta_2 z_{j2} + \varepsilon_j$$

or

$$Y_j = \beta_0 + \beta_1 z_{j1} + \beta_2 z_{j1}^2 + \varepsilon_j$$ ■

Example 7.2

Determine the design matrix if the linear regression model is applied to the one-way ANOVA situation in Example 6.6.

We create so-called *dummy* variables to handle the three population means: $\mu_1 = \mu + \tau_1$, $\mu_2 = \mu + \tau_2$, and $\mu_3 = \mu + \tau_3$. We set

$$z_1 = \begin{cases} 1 & \text{if the observation is from population 1} \\ 0 & \text{otherwise} \end{cases} \qquad z_2 = \begin{cases} 1 & \text{if the observation is from population 2} \\ 0 & \text{otherwise} \end{cases}$$

$$z_3 = \begin{cases} 1 & \text{if the observation is from population 3} \\ 0 & \text{otherwise} \end{cases}$$

and $\beta_0 = \mu$, $\beta_1 = \tau_1$, $\beta_2 = \tau_2$, $\beta_3 = \tau_3$. Then

$$Y_j = \beta_0 + \beta_1 z_{j1} + \beta_2 z_{j2} + \beta_3 z_{j3} + \varepsilon_j, \qquad j = 1, 2, \ldots, 8$$

where we stack the observations from the three populations in sequence. That is, we place each column of observations under the preceeding column to obtain the observed response vector and design matrix

$$\underset{(8\times 1)}{\mathbf{Y}} = \begin{bmatrix} 9 \\ 6 \\ 9 \\ 0 \\ 2 \\ 3 \\ 1 \\ 2 \end{bmatrix}; \qquad \underset{(8\times 4)}{\mathbf{Z}} = \begin{bmatrix} 1 & 1 & 0 & 0 \\ 1 & 1 & 0 & 0 \\ 1 & 1 & 0 & 0 \\ 1 & 0 & 1 & 0 \\ 1 & 0 & 1 & 0 \\ 1 & 0 & 0 & 1 \\ 1 & 0 & 0 & 1 \\ 1 & 0 & 0 & 1 \end{bmatrix}$$ ■

The construction of dummy variables, as in Example 7.2, allows the whole of analysis of variance to be treated within the multiple linear regression framework.

7.3 LEAST SQUARES ESTIMATION

One of the objectives of regression analysis is to develop an equation that will allow the investigator to predict the response for given values of the predictor variables. Thus it is necessary to "fit" the model in (7-3) to the observed y_j corresponding to the known values $1, z_{j1}, \ldots, z_{jr}$. That is, we must determine the values for the *regression coefficients* $\boldsymbol{\beta}$ and the *error variance* σ^2 consistent with the available data.

Let $\mathbf{b}$ be trial values for $\boldsymbol{\beta}$. Consider the difference $y_j - b_0 - b_1 z_{j1} - \cdots - b_r z_{jr}$ between the observed response, y_j, and the value $b_0 + b_1 z_{j1} + \cdots + b_r z_{jr}$ that would be expected if $\mathbf{b}$ was the "true" parameter vector. Typically, the differences $y_j - b_0 - b_1 z_{j1} - \cdots - b_r z_{jr}$ will not be zero because the response fluctuates (in a manner characterized by the error term assumptions) about its expected value. The *method of least squares* selects $\mathbf{b}$ to minimize the sum of squared differences

$$\begin{aligned} S(\mathbf{b}) &= \sum_{j=1}^{n} (y_j - b_0 - b_1 z_{j1} - \cdots - b_r z_{jr})^2 \\ &= (\mathbf{y} - \mathbf{Z}\mathbf{b})'(\mathbf{y} - \mathbf{Z}\mathbf{b}) \end{aligned} \tag{7-4}$$

The coefficients $\mathbf{b}$ chosen by the least squares criterion are called *least squares estimates* of the regression parameters $\boldsymbol{\beta}$. They will henceforth be denoted by $\hat{\boldsymbol{\beta}}$ to emphasize their role as estimates of $\boldsymbol{\beta}$.

The coefficients $\hat{\boldsymbol{\beta}}$ are consistent with the data in the sense that they produce estimated (fitted) mean responses, $\hat{\beta}_0 + \hat{\beta}_1 z_{j1} + \cdots + \hat{\beta}_r z_{jr}$, whose sum of squared differences from the observed y_j is as small as possible. The deviations

$$\hat{\varepsilon}_j = y_j - \hat{\beta}_0 - \hat{\beta}_1 z_{j1} - \cdots - \hat{\beta}_r z_{jr}, \qquad j = 1, 2, \ldots, n \tag{7-5}$$

are called *residuals*. The vector of residuals $\hat{\boldsymbol{\varepsilon}} = \mathbf{y} - \mathbf{Z}\hat{\boldsymbol{\beta}}$ contains the information about the remaining unknown parameter σ^2 (see Result 7.2).

Result 7.1. Let $\mathbf{Z}$ have full rank $r + 1 \le n$.[1] The least squares estimate of $\boldsymbol{\beta}$ in (7-3) is given by

$$\hat{\boldsymbol{\beta}} = (\mathbf{Z}'\mathbf{Z})^{-1}\mathbf{Z}'\mathbf{y}$$

Let $\hat{\mathbf{y}} = \mathbf{Z}\hat{\boldsymbol{\beta}}$ denote the *fitted values* of $\mathbf{y}$. The *residuals*

$$\hat{\boldsymbol{\varepsilon}} = \mathbf{y} - \hat{\mathbf{y}} = \left[\mathbf{I} - \mathbf{Z}(\mathbf{Z}'\mathbf{Z})^{-1}\mathbf{Z}'\right]\mathbf{y}$$

satisfy $\mathbf{Z}'\hat{\boldsymbol{\varepsilon}} = \mathbf{0}$ and $\hat{\mathbf{y}}'\hat{\boldsymbol{\varepsilon}} = 0$. Also, the *residual sum of squares* =

$$\sum_{j=1}^{n} (y_j - \hat{\beta}_0 - \hat{\beta}_1 z_{j1} - \cdots - \hat{\beta}_r z_{jr})^2 = \hat{\boldsymbol{\varepsilon}}'\hat{\boldsymbol{\varepsilon}} = \mathbf{y}'\left[\mathbf{I} - \mathbf{Z}(\mathbf{Z}'\mathbf{Z})^{-1}\mathbf{Z}'\right]\mathbf{y} = \mathbf{y}'\mathbf{y} - \mathbf{y}'\mathbf{Z}\hat{\boldsymbol{\beta}}$$

[1] If $\mathbf{Z}$ is not of full rank, $(\mathbf{Z}'\mathbf{Z})^{-1}$ is replaced by $(\mathbf{Z}'\mathbf{Z})^{-}$, a *generalized inverse* of $\mathbf{Z}'\mathbf{Z}$ (see Exercise 7.6).

Proof. Let $\hat{\boldsymbol{\beta}} = (\mathbf{Z}'\mathbf{Z})^{-1}\mathbf{Z}'\mathbf{y}$ as asserted. Then $\hat{\boldsymbol{\varepsilon}} = \mathbf{y} - \hat{\mathbf{y}} = \mathbf{y} - \mathbf{Z}\hat{\boldsymbol{\beta}} = [\mathbf{I} - \mathbf{Z}(\mathbf{Z}'\mathbf{Z})^{-1}\mathbf{Z}']\mathbf{y}$. The matrix $[\mathbf{I} - \mathbf{Z}(\mathbf{Z}'\mathbf{Z})^{-1}\mathbf{Z}']$ satisfies:

1. $[\mathbf{I} - \mathbf{Z}(\mathbf{Z}'\mathbf{Z})^{-1}\mathbf{Z}']' = [\mathbf{I} - \mathbf{Z}(\mathbf{Z}'\mathbf{Z})^{-1}\mathbf{Z}']$ (symmetric);
2. $$\begin{aligned}&[\mathbf{I} - \mathbf{Z}(\mathbf{Z}'\mathbf{Z})^{-1}\mathbf{Z}'][\mathbf{I} - \mathbf{Z}(\mathbf{Z}'\mathbf{Z})^{-1}\mathbf{Z}']\\ &= \mathbf{I} - 2\mathbf{Z}(\mathbf{Z}'\mathbf{Z})^{-1}\mathbf{Z}' + \mathbf{Z}(\mathbf{Z}'\mathbf{Z})^{-1}\mathbf{Z}'\mathbf{Z}(\mathbf{Z}'\mathbf{Z})^{-1}\mathbf{Z}' \\ &= [\mathbf{I} - \mathbf{Z}(\mathbf{Z}'\mathbf{Z})^{-1}\mathbf{Z}'] \quad \text{(idempotent);}\end{aligned} \tag{7-6}$$
3. $\mathbf{Z}'[\mathbf{I} - \mathbf{Z}(\mathbf{Z}'\mathbf{Z})^{-1}\mathbf{Z}'] = \mathbf{Z}' - \mathbf{Z}' = \mathbf{0}$.

Consequently, $\mathbf{Z}'\hat{\boldsymbol{\varepsilon}} = \mathbf{Z}'(\mathbf{y} - \hat{\mathbf{y}}) = \mathbf{Z}'[\mathbf{I} - \mathbf{Z}(\mathbf{Z}'\mathbf{Z})^{-1}\mathbf{Z}']\mathbf{y} = \mathbf{0}$, so $\hat{\mathbf{y}}'\hat{\boldsymbol{\varepsilon}} = \hat{\boldsymbol{\beta}}'\mathbf{Z}'\hat{\boldsymbol{\varepsilon}} = 0$. Also, $\hat{\boldsymbol{\varepsilon}}'\hat{\boldsymbol{\varepsilon}} = \mathbf{y}'[\mathbf{I} - \mathbf{Z}(\mathbf{Z}'\mathbf{Z})^{-1}\mathbf{Z}'][\mathbf{I} - \mathbf{Z}(\mathbf{Z}'\mathbf{Z})^{-1}\mathbf{Z}']\mathbf{y} = \mathbf{y}'[\mathbf{I} - \mathbf{Z}(\mathbf{Z}'\mathbf{Z})^{-1}\mathbf{Z}']\mathbf{y} = \mathbf{y}'\mathbf{y} - \mathbf{y}'\mathbf{Z}\hat{\boldsymbol{\beta}}$. To verify the expression for $\hat{\boldsymbol{\beta}}$, we write

$$\mathbf{y} - \mathbf{Z}\mathbf{b} = \mathbf{y} - \mathbf{Z}\hat{\boldsymbol{\beta}} + \mathbf{Z}\hat{\boldsymbol{\beta}} - \mathbf{Z}\mathbf{b} = \mathbf{y} - \mathbf{Z}\hat{\boldsymbol{\beta}} + \mathbf{Z}(\hat{\boldsymbol{\beta}} - \mathbf{b})$$

so

$$\begin{aligned} S(\mathbf{b}) &= (\mathbf{y} - \mathbf{Z}\mathbf{b})'(\mathbf{y} - \mathbf{Z}\mathbf{b}) \\ &= (\mathbf{y} - \mathbf{Z}\hat{\boldsymbol{\beta}})'(\mathbf{y} - \mathbf{Z}\hat{\boldsymbol{\beta}}) + (\hat{\boldsymbol{\beta}} - \mathbf{b})'\mathbf{Z}'\mathbf{Z}(\hat{\boldsymbol{\beta}} - \mathbf{b}) + 2(\mathbf{y} - \mathbf{Z}\hat{\boldsymbol{\beta}})'\mathbf{Z}(\hat{\boldsymbol{\beta}} - \mathbf{b}) \\ &= (\mathbf{y} - \mathbf{Z}\hat{\boldsymbol{\beta}})'(\mathbf{y} - \mathbf{Z}\hat{\boldsymbol{\beta}}) + (\hat{\boldsymbol{\beta}} - \mathbf{b})'\mathbf{Z}'\mathbf{Z}(\hat{\boldsymbol{\beta}} - \mathbf{b}) \end{aligned}$$

since $(\mathbf{y} - \mathbf{Z}\hat{\boldsymbol{\beta}})'\mathbf{Z} = \hat{\boldsymbol{\varepsilon}}'\mathbf{Z} = \mathbf{0}'$, as above. The first term in $S(\mathbf{b})$ does not depend on $\mathbf{b}$ and the second is the squared length of $\mathbf{Z}(\hat{\boldsymbol{\beta}} - \mathbf{b})$. Because $\mathbf{Z}$ has full rank, $\mathbf{Z}(\hat{\boldsymbol{\beta}} - \mathbf{b}) \neq \mathbf{0}$ if $\hat{\boldsymbol{\beta}} \neq \mathbf{b}$, so the minimum sum of squares is unique and occurs for $\mathbf{b} = \hat{\boldsymbol{\beta}} = (\mathbf{Z}'\mathbf{Z})^{-1}\mathbf{Z}'\mathbf{y}$. Note that $(\mathbf{Z}'\mathbf{Z})^{-1}$ exists since $\mathbf{Z}'\mathbf{Z}$ has rank $r + 1 \leq n$. (If $\mathbf{Z}'\mathbf{Z}$ is not of full rank, $\mathbf{Z}'\mathbf{Z}\mathbf{a} = \mathbf{0}$ for some $\mathbf{a} \neq \mathbf{0}$, but then $\mathbf{a}'\mathbf{Z}'\mathbf{Z}\mathbf{a} = 0$ or $\mathbf{Z}\mathbf{a} = \mathbf{0}$, which contradicts $\mathbf{Z}$ having full rank $r + 1$.) ■

Result 7.1 shows how the least squares estimator $\hat{\boldsymbol{\beta}}$ and the residuals $\hat{\boldsymbol{\varepsilon}}$ can be obtained from the design matrix $\mathbf{Z}$ and responses $\mathbf{y}$ by simple matrix operations.

Example 7.3

Calculate the least squares estimates $\hat{\boldsymbol{\beta}}$, the residuals $\hat{\boldsymbol{\varepsilon}}$, and the residual sum of squares for a straight-line model

$$Y_j = \beta_0 + \beta_1 z_{j1} + \varepsilon_j$$

fit to the data

z_1	0	1	2	3	4
y	1	4	3	8	9

We have

$\mathbf{Z}'$	$\mathbf{y}$	$\mathbf{Z}'\mathbf{Z}$	$(\mathbf{Z}'\mathbf{Z})^{-1}$	$\mathbf{Z}'\mathbf{y}$
$\begin{bmatrix} 1 & 1 & 1 & 1 & 1 \\ 0 & 1 & 2 & 3 & 4 \end{bmatrix}$	$\begin{bmatrix} 1 \\ 4 \\ 3 \\ 8 \\ 9 \end{bmatrix}$	$\begin{bmatrix} 5 & 10 \\ 10 & 30 \end{bmatrix}$	$\begin{bmatrix} .6 & -.2 \\ -.2 & .1 \end{bmatrix}$	$\begin{bmatrix} 25 \\ 70 \end{bmatrix}$

Consequently,

$$\hat{\boldsymbol{\beta}} = \begin{bmatrix} \hat{\beta}_0 \\ \hat{\beta}_1 \end{bmatrix} = (\mathbf{Z}'\mathbf{Z})^{-1}\mathbf{Z}'\mathbf{y} = \begin{bmatrix} .6 & -.2 \\ -.2 & .1 \end{bmatrix} \begin{bmatrix} 25 \\ 70 \end{bmatrix} = \begin{bmatrix} 1 \\ 2 \end{bmatrix}$$

and the fitted equation is

$$\hat{y} = 1 + 2z$$

The vector of fitted (predicted) values is

$$\hat{\mathbf{y}} = \mathbf{Z}\hat{\boldsymbol{\beta}} = \begin{bmatrix} 1 & 0 \\ 1 & 1 \\ 1 & 2 \\ 1 & 3 \\ 1 & 4 \end{bmatrix} \begin{bmatrix} 1 \\ 2 \end{bmatrix} = \begin{bmatrix} 1 \\ 3 \\ 5 \\ 7 \\ 9 \end{bmatrix} \quad \text{so} \quad \hat{\boldsymbol{\varepsilon}} = \mathbf{y} - \hat{\mathbf{y}} = \begin{bmatrix} 1 \\ 4 \\ 3 \\ 8 \\ 9 \end{bmatrix} - \begin{bmatrix} 1 \\ 3 \\ 5 \\ 7 \\ 9 \end{bmatrix} = \begin{bmatrix} 0 \\ 1 \\ -2 \\ 1 \\ 0 \end{bmatrix}$$

The residual sum of squares is

$$\hat{\boldsymbol{\varepsilon}}'\hat{\boldsymbol{\varepsilon}} = [0 \quad 1 \quad -2 \quad 1 \quad 0] \begin{bmatrix} 0 \\ 1 \\ -2 \\ 1 \\ 0 \end{bmatrix} = 0^2 + 1^2 + (-2)^2 + 1^2 + 0^2 = 6 \quad \blacksquare$$

Sum of Squares Decomposition

According to Result 7.1, $\hat{\mathbf{y}}'\hat{\boldsymbol{\varepsilon}} = 0$ so the total response sum of squares $\mathbf{y}'\mathbf{y} = \sum_{j=1}^{n} y_j^2$ satisfies

$$\mathbf{y}'\mathbf{y} = (\hat{\mathbf{y}} + \mathbf{y} - \hat{\mathbf{y}})'(\hat{\mathbf{y}} + \mathbf{y} - \hat{\mathbf{y}}) = (\hat{\mathbf{y}} + \hat{\boldsymbol{\varepsilon}})'(\hat{\mathbf{y}} + \hat{\boldsymbol{\varepsilon}}) = \hat{\mathbf{y}}'\hat{\mathbf{y}} + \hat{\boldsymbol{\varepsilon}}'\hat{\boldsymbol{\varepsilon}} \qquad (7\text{-}7)$$

Since the first column of $\mathbf{Z}$ is $\mathbf{1}$, the condition $\mathbf{Z}'\hat{\boldsymbol{\varepsilon}} = \mathbf{0}$ includes the requirement $0 = \mathbf{1}'\hat{\boldsymbol{\varepsilon}} = \sum_{j=1}^{n} \hat{\varepsilon}_j = \sum_{j=1}^{n} y_j - \sum_{j=1}^{n} \hat{y}_j$, or $\bar{y} = \bar{\hat{y}}$. Subtracting $n\bar{y}^2 = n\bar{\hat{y}}^2$ from both sides of the decomposition in (7-7), we obtain the basic decomposition of the sum of squares about the mean

$$\mathbf{y}'\mathbf{y} - n\bar{y}^2 = \hat{\mathbf{y}}'\hat{\mathbf{y}} - n\bar{\hat{y}}^2 + \hat{\boldsymbol{\varepsilon}}'\hat{\boldsymbol{\varepsilon}}$$

or

$$\sum_{j=1}^{n} (y_j - \bar{y})^2 = \sum_{j=1}^{n} (\hat{y}_j - \bar{y})^2 + \sum_{j=1}^{n} \hat{\varepsilon}_j^2 \qquad (7\text{-}8)$$

$$\begin{pmatrix} \text{total sum} \\ \text{of squares} \\ \text{about mean} \end{pmatrix} = \begin{pmatrix} \text{regression} \\ \text{sum of} \\ \text{squares} \end{pmatrix} + \begin{pmatrix} \text{residual (error)} \\ \text{sum of squares} \end{pmatrix}$$

The sum of squares breakup above suggests the quality of the model fit can be measured by the *coefficient of determination*

$$R^2 = 1 - \frac{\sum_{j=1}^{n} \hat{\varepsilon}_j^2}{\sum_{j=1}^{n} (y_j - \bar{y})^2} = \frac{\sum_{j=1}^{n} (\hat{y}_j - \bar{y})^2}{\sum_{j=1}^{n} (y_j - \bar{y})^2} \tag{7-9}$$

The quantity R^2 gives the proportion of the total variation in the y_j's "explained" by, or attributable to, the predictor variables $z_1, z_2, \ldots, z_r$. Here R^2 (or the *multiple correlation coefficient* $R = +\sqrt{R^2}$) equals 1 if the fitted equation passes through all the data points so $\hat{\varepsilon}_j = 0$ for all j. At the other extreme, R^2 is 0 if $\hat{\beta}_0 = \bar{y}$ and $\hat{\beta}_1 = \hat{\beta}_2 = \cdots = \hat{\beta}_r = 0$. In this case the predictor variables $z_1, z_2, \ldots, z_r$ have no influence on the response.

Geometry of Least Squares

A geometrical interpretation of the least squares technique highlights the nature of the concept. According to the classical linear regression model

$$\text{Mean response vector} = E(\mathbf{Y}) = \mathbf{Z}\boldsymbol{\beta} = \beta_0 \begin{bmatrix} 1 \\ 1 \\ \vdots \\ 1 \end{bmatrix} + \beta_1 \begin{bmatrix} z_{11} \\ z_{21} \\ \vdots \\ z_{n1} \end{bmatrix} + \cdots + \beta_r \begin{bmatrix} z_{1r} \\ z_{2r} \\ \vdots \\ z_{nr} \end{bmatrix}$$

Thus $E(\mathbf{Y})$ is a linear combination of the columns of $\mathbf{Z}$. As $\boldsymbol{\beta}$ varies, $\mathbf{Z}\boldsymbol{\beta}$ spans the model plane of all linear combinations. Usually, the observation vector $\mathbf{y}$ will not lie in the model plane because of the random error $\boldsymbol{\varepsilon}$; that is, $\mathbf{y}$ is not (exactly) a linear combination of the columns of $\mathbf{Z}$. Recall

$$\underset{\left(\begin{array}{c}\text{response}\\ \text{vector}\end{array}\right)}{\mathbf{Y}} = \underset{\left(\begin{array}{c}\text{vector}\\ \text{in model}\\ \text{plane}\end{array}\right)}{\mathbf{Z}\boldsymbol{\beta}} + \underset{\left(\begin{array}{c}\text{error}\\ \text{vector}\end{array}\right)}{\boldsymbol{\varepsilon}}$$

Once the observations become available, the least squares solution is derived from the deviation vector

$$\mathbf{y} - \mathbf{Zb} = (\text{observation vector}) - (\text{vector in model plane})$$

The squared length $(\mathbf{y} - \mathbf{Zb})'(\mathbf{y} - \mathbf{Zb})$ is the sum of squares $S(\mathbf{b})$. As illustrated in Figure 7.1, $S(\mathbf{b})$ is as small as possible when $\mathbf{b}$ is selected such that $\mathbf{Zb}$ is the point in the model plane closest to $\mathbf{y}$. The point closest to $\mathbf{y}$ occurs at the tip of the perpendicular projection of $\mathbf{y}$ on the plane. That is, for the choice $\mathbf{b} = \hat{\boldsymbol{\beta}}, \hat{\mathbf{y}} = \mathbf{Z}\hat{\boldsymbol{\beta}}$ is the projection of $\mathbf{y}$ on the plane consisting of all linear combinations of the columns of $\mathbf{Z}$. The residual vector $\hat{\boldsymbol{\varepsilon}} = \mathbf{y} - \hat{\mathbf{y}}$ is perpendicular to this plane. This geometry holds even when $\mathbf{Z}$ is not of full rank.

When $\mathbf{Z}$ has full rank, the projection operation is expressed analytically as multiplication by the matrix $\mathbf{Z}(\mathbf{Z}'\mathbf{Z})^{-1}\mathbf{Z}'$. To see this, we use the spectral decomposi-

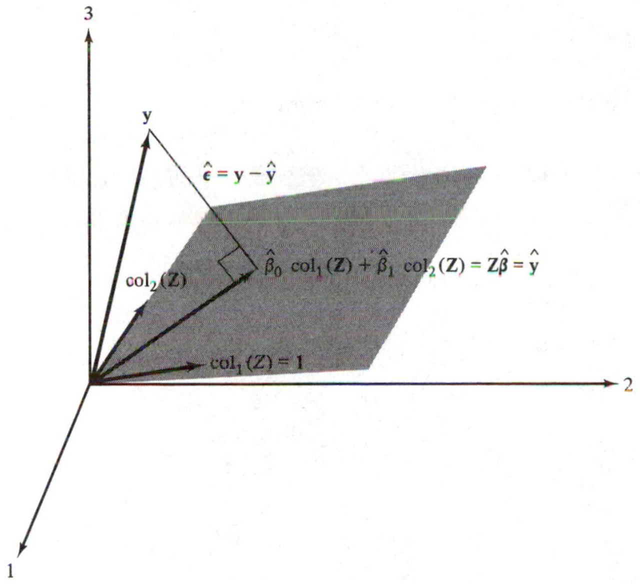

Figure 7.1 Least squares as a projection for $n = 3$, $r = 1$.

tion (2-16) to write

$$\mathbf{Z}'\mathbf{Z} = \lambda_1 \mathbf{e}_1\mathbf{e}_1' + \lambda_2\mathbf{e}_2\mathbf{e}_2' + \cdots + \lambda_{r+1}\mathbf{e}_{r+1}\mathbf{e}_{r+1}'$$

where $\lambda_1 \geq \lambda_2 \geq \cdots \geq \lambda_{r+1} > 0$ are the eigenvalues of $\mathbf{Z}'\mathbf{Z}$ and $\mathbf{e}_1, \mathbf{e}_2, \ldots, \mathbf{e}_{r+1}$ are the corresponding eigenvectors. If $\mathbf{Z}$ is of full rank,

$$(\mathbf{Z}'\mathbf{Z})^{-1} = \frac{1}{\lambda_1}\mathbf{e}_1\mathbf{e}_1' + \frac{1}{\lambda_2}\mathbf{e}_2\mathbf{e}_2' + \cdots + \frac{1}{\lambda_{r+1}}\mathbf{e}_{r+1}\mathbf{e}_{r+1}'$$

Consider $\mathbf{q}_i = \lambda_i^{-1/2}\mathbf{Z}\mathbf{e}_i$, which is a linear combination of the columns of $\mathbf{Z}$. Then $\mathbf{q}_i'\mathbf{q}_k = \lambda_i^{-1/2}\lambda_k^{-1/2}\mathbf{e}_i'\mathbf{Z}'\mathbf{Z}\mathbf{e}_k = \lambda_i^{-1/2}\lambda_k^{-1/2}\mathbf{e}_i'\lambda_k\mathbf{e}_k = 0$ if $i \neq k$ or 1 if $i = k$. That is, the $r + 1$ vectors $\mathbf{q}_i$ are mutually perpendicular and have unit length. Their linear combinations span the space of all linear combinations of the columns of $\mathbf{Z}$. Moreover,

$$\mathbf{Z}(\mathbf{Z}'\mathbf{Z})^{-1}\mathbf{Z}' = \sum_{i=1}^{r+1} \lambda_i^{-1}\mathbf{Z}\mathbf{e}_i\mathbf{e}_i'\mathbf{Z}' = \sum_{i=1}^{r+1} \mathbf{q}_i\mathbf{q}_i'$$

According to Result 2A.2 and Definition 2A.12, the projection of $\mathbf{y}$ on a linear combination of $\{\mathbf{q}_1, \mathbf{q}_2, \ldots, \mathbf{q}_{r+1}\}$ is $\sum_{i=1}^{r+1} (\mathbf{q}_i'\mathbf{y})\mathbf{q}_i = \left(\sum_{i=1}^{r+1} \mathbf{q}_i\mathbf{q}_i'\right)\mathbf{y} = \mathbf{Z}(\mathbf{Z}'\mathbf{Z})^{-1}\mathbf{Z}'\mathbf{y} = \mathbf{Z}\hat{\boldsymbol{\beta}}$. Thus multiplication by $\mathbf{Z}(\mathbf{Z}'\mathbf{Z})^{-1}\mathbf{Z}'$ projects a vector on the space spanned by the columns of $\mathbf{Z}$.[2]

[2]If $\mathbf{Z}$ is not of full rank, we can use the *generalized inverse* $(\mathbf{Z}'\mathbf{Z})^{-} = \sum_{i=1}^{r_1+1} \lambda_i^{-1}\mathbf{e}_i\mathbf{e}_i'$, where $\lambda_1 \geq \lambda_2 \geq \cdots \geq \lambda_{r_1+1} > 0 = \lambda_{r_1+2} = \cdots = \lambda_{r+1}$, as described in Exercise 7.6. Then $\mathbf{Z}(\mathbf{Z}'\mathbf{Z})^{-}\mathbf{Z}' = \sum_{i=1}^{r_1+1} \mathbf{q}_i\mathbf{q}_i'$ has rank $r_1 + 1$ and generates the unique projection of $\mathbf{y}$ on the space spanned by the linearly independent columns of $\mathbf{Z}$. This is true for any choice of the generalized inverse (see [18]).

Similarly, $[\mathbf{I} - \mathbf{Z}(\mathbf{Z}'\mathbf{Z})^{-1}\mathbf{Z}']$ is the matrix for the projection of $\mathbf{y}$ on the plane perpendicular to the plane spanned by the columns of $\mathbf{Z}$.

Sampling Properties of Classical Least Squares Estimators

The least squares estimator $\hat{\boldsymbol{\beta}}$ and the residuals $\hat{\boldsymbol{\varepsilon}}$ have the sampling properties detailed in the next result.

Result 7.2. Under the general linear regression model in (7-3), the least squares estimator $\hat{\boldsymbol{\beta}} = (\mathbf{Z}'\mathbf{Z})^{-1}\mathbf{Z}'\mathbf{Y}$ has

$$E(\hat{\boldsymbol{\beta}}) = \boldsymbol{\beta} \quad \text{and} \quad \text{Cov}(\hat{\boldsymbol{\beta}}) = \sigma^2(\mathbf{Z}'\mathbf{Z})^{-1}$$

The residuals $\hat{\boldsymbol{\varepsilon}}$ have the properties

$$E(\hat{\boldsymbol{\varepsilon}}) = \mathbf{0} \quad \text{and} \quad \text{Cov}(\hat{\boldsymbol{\varepsilon}}) = \sigma^2\left[\mathbf{I} - \mathbf{Z}(\mathbf{Z}'\mathbf{Z})^{-1}\mathbf{Z}'\right]$$

Also $E(\hat{\boldsymbol{\varepsilon}}'\hat{\boldsymbol{\varepsilon}}) = (n - r - 1)\sigma^2$, so defining

$$s^2 = \frac{\hat{\boldsymbol{\varepsilon}}'\hat{\boldsymbol{\varepsilon}}}{n - (r+1)} = \frac{\mathbf{Y}'\left[\mathbf{I} - \mathbf{Z}(\mathbf{Z}'\mathbf{Z})^{-1}\mathbf{Z}'\right]\mathbf{Y}}{n - r - 1}$$

we have

$$E(s^2) = \sigma^2$$

Moreover, $\hat{\boldsymbol{\beta}}$ and $\hat{\boldsymbol{\varepsilon}}$ are uncorrelated.

Proof. Before the response $\mathbf{Y} = \mathbf{Z}\boldsymbol{\beta} + \boldsymbol{\varepsilon}$ is observed, it is a random vector. Now

$$\begin{aligned} \hat{\boldsymbol{\beta}} &= (\mathbf{Z}'\mathbf{Z})^{-1}\mathbf{Z}'\mathbf{Y} = (\mathbf{Z}'\mathbf{Z})^{-1}\mathbf{Z}'(\mathbf{Z}\boldsymbol{\beta} + \boldsymbol{\varepsilon}) = \boldsymbol{\beta} + (\mathbf{Z}'\mathbf{Z})^{-1}\mathbf{Z}'\boldsymbol{\varepsilon} \\ \hat{\boldsymbol{\varepsilon}} &= \left[\mathbf{I} - \mathbf{Z}(\mathbf{Z}'\mathbf{Z})^{-1}\mathbf{Z}'\right]\mathbf{Y} \qquad\qquad (7\text{-}10) \\ &= \left[\mathbf{I} - \mathbf{Z}(\mathbf{Z}'\mathbf{Z})^{-1}\mathbf{Z}'\right][\mathbf{Z}\boldsymbol{\beta} + \boldsymbol{\varepsilon}] = \left[\mathbf{I} - \mathbf{Z}(\mathbf{Z}'\mathbf{Z})^{-1}\mathbf{Z}'\right]\boldsymbol{\varepsilon} \end{aligned}$$

since $[\mathbf{I} - \mathbf{Z}(\mathbf{Z}'\mathbf{Z})^{-1}\mathbf{Z}']\mathbf{Z} = \mathbf{Z} - \mathbf{Z} = \mathbf{0}$. From (2-24) and (2-45),

$$\begin{aligned} E(\hat{\boldsymbol{\beta}}) &= \boldsymbol{\beta} + (\mathbf{Z}'\mathbf{Z})^{-1}\mathbf{Z}'E(\boldsymbol{\varepsilon}) = \boldsymbol{\beta} \\ \text{Cov}(\hat{\boldsymbol{\beta}}) &= (\mathbf{Z}'\mathbf{Z})^{-1}\mathbf{Z}'\,\text{Cov}(\boldsymbol{\varepsilon})\mathbf{Z}(\mathbf{Z}'\mathbf{Z})^{-1} = \sigma^2(\mathbf{Z}'\mathbf{Z})^{-1}\mathbf{Z}'\mathbf{Z}(\mathbf{Z}'\mathbf{Z})^{-1} \\ &= \sigma^2(\mathbf{Z}'\mathbf{Z})^{-1} \\ E(\hat{\boldsymbol{\varepsilon}}) &= \left[\mathbf{I} - \mathbf{Z}(\mathbf{Z}'\mathbf{Z})^{-1}\mathbf{Z}'\right]E(\boldsymbol{\varepsilon}) = \mathbf{0} \\ \text{Cov}(\hat{\boldsymbol{\varepsilon}}) &= \left[\mathbf{I} - \mathbf{Z}(\mathbf{Z}'\mathbf{Z})^{-1}\mathbf{Z}'\right]\text{Cov}(\boldsymbol{\varepsilon})\left[\mathbf{I} - \mathbf{Z}(\mathbf{Z}'\mathbf{Z})^{-1}\mathbf{Z}'\right]' = \sigma^2\left[\mathbf{I} - \mathbf{Z}(\mathbf{Z}'\mathbf{Z})^{-1}\mathbf{Z}'\right] \end{aligned}$$

where the last equality follows from (7-6). Also,

$$\begin{aligned} \text{Cov}(\hat{\boldsymbol{\beta}}, \hat{\boldsymbol{\varepsilon}}) &= E\left[(\hat{\boldsymbol{\beta}} - \boldsymbol{\beta})\hat{\boldsymbol{\varepsilon}}'\right] = (\mathbf{Z}'\mathbf{Z})^{-1}\mathbf{Z}'E(\boldsymbol{\varepsilon}\boldsymbol{\varepsilon}')\left[\mathbf{I} - \mathbf{Z}(\mathbf{Z}'\mathbf{Z})^{-1}\mathbf{Z}'\right] \\ &= \sigma^2(\mathbf{Z}'\mathbf{Z})^{-1}\mathbf{Z}'\left[\mathbf{I} - \mathbf{Z}(\mathbf{Z}'\mathbf{Z})^{-1}\mathbf{Z}'\right] = \mathbf{0} \end{aligned}$$

because $\mathbf{Z}'[\mathbf{I} - \mathbf{Z}(\mathbf{Z}'\mathbf{Z})^{-1}\mathbf{Z}'] = \mathbf{0}$. From (7-10), (7-6), and Result 4.9,

$$\begin{aligned}
\hat{\boldsymbol{\varepsilon}}'\hat{\boldsymbol{\varepsilon}} &= \boldsymbol{\varepsilon}'\left[\mathbf{I} - \mathbf{Z}(\mathbf{Z}'\mathbf{Z})^{-1}\mathbf{Z}'\right]\left[\mathbf{I} - \mathbf{Z}(\mathbf{Z}'\mathbf{Z})^{-1}\mathbf{Z}'\right]\boldsymbol{\varepsilon} \\
&= \boldsymbol{\varepsilon}'\left[\mathbf{I} - \mathbf{Z}(\mathbf{Z}'\mathbf{Z})^{-1}\mathbf{Z}'\right]\boldsymbol{\varepsilon} \\
&= \operatorname{tr}\left[\boldsymbol{\varepsilon}'\left(\mathbf{I} - \mathbf{Z}(\mathbf{Z}'\mathbf{Z})^{-1}\mathbf{Z}'\right)\boldsymbol{\varepsilon}\right] \\
&= \operatorname{tr}\left(\left[\mathbf{I} - \mathbf{Z}(\mathbf{Z}'\mathbf{Z})^{-1}\mathbf{Z}'\right]\boldsymbol{\varepsilon}\boldsymbol{\varepsilon}'\right)
\end{aligned}$$

Now for an arbitrary $n \times n$ random matrix $\mathbf{W}$,

$$\begin{aligned}
E(\operatorname{tr}(\mathbf{W})) &= E(\mathbf{W}_{11} + \mathbf{W}_{22} + \cdots + \mathbf{W}_{nn}) \\
&= E(\mathbf{W}_{11}) + E(\mathbf{W}_{22}) + \cdots + E(\mathbf{W}_{nn}) = \operatorname{tr}\left[E(\mathbf{W})\right].
\end{aligned}$$

Thus, using Result 2A.12,

$$\begin{aligned}
E(\hat{\boldsymbol{\varepsilon}}'\hat{\boldsymbol{\varepsilon}}) &= \operatorname{tr}\left(\left[\mathbf{I} - \mathbf{Z}(\mathbf{Z}'\mathbf{Z})^{-1}\mathbf{Z}'\right]E(\boldsymbol{\varepsilon}\boldsymbol{\varepsilon}')\right) \\
&= \sigma^2 \operatorname{tr}\left[\mathbf{I} - \mathbf{Z}(\mathbf{Z}'\mathbf{Z})^{-1}\mathbf{Z}'\right] \\
&= \sigma^2 \operatorname{tr}(\mathbf{I}) - \sigma^2 \operatorname{tr}\left[\mathbf{Z}(\mathbf{Z}'\mathbf{Z})^{-1}\mathbf{Z}'\right] \\
&= \sigma^2 n - \sigma^2 \operatorname{tr}\left[(\mathbf{Z}'\mathbf{Z})^{-1}\mathbf{Z}'\mathbf{Z}\right] \\
&= n\sigma^2 - \sigma^2 \operatorname{tr}\left[\underset{(r+1)\times(r+1)}{\mathbf{I}}\right] \\
&= \sigma^2(n - r - 1)
\end{aligned}$$

and the result for $s^2 = \hat{\boldsymbol{\varepsilon}}'\hat{\boldsymbol{\varepsilon}}/(n - r - 1)$ follows. ■

The least squares estimator $\hat{\boldsymbol{\beta}}$ possesses a minimum variance property that was first established by Gauss. This result concerns "best" estimators of linear parametric functions of the form $\mathbf{c}'\boldsymbol{\beta} = c_0\beta_0 + c_1\beta_1 + \cdots + c_r\beta_r$ for any $\mathbf{c}$.

Result 7.3 (Gauss[3] least squares theorem). Let $\mathbf{Y} = \mathbf{Z}\boldsymbol{\beta} + \boldsymbol{\varepsilon}$, where $E(\boldsymbol{\varepsilon}) = \mathbf{0}$, $\operatorname{Cov}(\boldsymbol{\varepsilon}) = \sigma^2\mathbf{I}$, and $\mathbf{Z}$ has full rank $r + 1$. For any $\mathbf{c}$, the estimator

$$\mathbf{c}'\hat{\boldsymbol{\beta}} = c_0\hat{\beta}_0 + c_1\hat{\beta}_1 + \cdots + c_r\hat{\beta}_r$$

of $\mathbf{c}'\boldsymbol{\beta}$ has the smallest possible variance among all linear estimators of the form

$$\mathbf{a}'\mathbf{Y} = a_1Y_1 + a_2Y_2 + \cdots + a_nY_n$$

which are unbiased for $\mathbf{c}'\boldsymbol{\beta}$.

Proof. For any fixed $\mathbf{c}$, let $\mathbf{a}'\mathbf{Y}$ be any unbiased estimator of $\mathbf{c}'\boldsymbol{\beta}$. Then $E(\mathbf{a}'\mathbf{Y}) = \mathbf{c}'\boldsymbol{\beta}$, whatever the value of $\boldsymbol{\beta}$. Also, by assumption, $E(\mathbf{a}'\mathbf{Y}) = E(\mathbf{a}'\mathbf{Z}\boldsymbol{\beta} + \mathbf{a}'\boldsymbol{\varepsilon}) = \mathbf{a}'\mathbf{Z}\boldsymbol{\beta}$. Equating the two expected value expressions, $\mathbf{a}'\mathbf{Z}\boldsymbol{\beta} = \mathbf{c}'\boldsymbol{\beta}$ or $(\mathbf{c}' - \mathbf{a}'\mathbf{Z})\boldsymbol{\beta} = 0$ for all $\boldsymbol{\beta}$, including the choice $\boldsymbol{\beta} = (\mathbf{c}' - \mathbf{a}'\mathbf{Z})'$. This implies that $\mathbf{c}' = \mathbf{a}'\mathbf{Z}$ for any unbiased estimator.

[3] Much later, Markov proved a less-general result, which misled many writers to attach his name to this theorem.

Now $\mathbf{c}'\hat{\boldsymbol{\beta}} = \mathbf{c}'(\mathbf{Z}'\mathbf{Z})^{-1}\mathbf{Z}'\mathbf{Y} = \mathbf{a}^{*\prime}\mathbf{Y}$ with $\mathbf{a}^* = \mathbf{Z}(\mathbf{Z}'\mathbf{Z})^{-1}\mathbf{c}$. Moreover, from Result 7.2, $E(\hat{\boldsymbol{\beta}}) = \boldsymbol{\beta}$, so $\mathbf{c}'\hat{\boldsymbol{\beta}} = \mathbf{a}^{*\prime}\mathbf{Y}$ is an unbiased estimator of $\mathbf{c}'\boldsymbol{\beta}$. For any $\mathbf{a}$, satisfying the unbiased requirement $\mathbf{c}' = \mathbf{a}'\mathbf{Z}$,

$$\begin{aligned}\text{Var}(\mathbf{a}'\mathbf{Y}) &= \text{Var}(\mathbf{a}'\mathbf{Z}\boldsymbol{\beta} + \mathbf{a}'\boldsymbol{\varepsilon}) = \text{Var}(\mathbf{a}'\boldsymbol{\varepsilon}) = \mathbf{a}'\mathbf{I}\sigma^2\mathbf{a} \\ &= \sigma^2(\mathbf{a} - \mathbf{a}^* + \mathbf{a}^*)'(\mathbf{a} - \mathbf{a}^* + \mathbf{a}^*) = \sigma^2\left[(\mathbf{a} - \mathbf{a}^*)'(\mathbf{a} - \mathbf{a}^*) + \mathbf{a}^{*\prime}\mathbf{a}^*\right]\end{aligned}$$

since $(\mathbf{a} - \mathbf{a}^*)'\mathbf{a}^* = (\mathbf{a} - \mathbf{a}^*)'\mathbf{Z}(\mathbf{Z}'\mathbf{Z})^{-1}\mathbf{c} = 0$ from the condition $(\mathbf{a} - \mathbf{a}^*)'\mathbf{Z} = \mathbf{a}'\mathbf{Z} - \mathbf{a}^{*\prime}\mathbf{Z} = \mathbf{c}' - \mathbf{c}' = \mathbf{0}'$. Because $\mathbf{a}^*$ is fixed and $(\mathbf{a} - \mathbf{a}^*)'(\mathbf{a} - \mathbf{a}^*)$ is positive unless $\mathbf{a} = \mathbf{a}^*$, $\text{Var}(\mathbf{a}'\mathbf{Y})$ is minimized by the choice $\mathbf{a}^{*\prime}\mathbf{Y} = \mathbf{c}'(\mathbf{Z}'\mathbf{Z})^{-1}\mathbf{Z}'\mathbf{Y} = \mathbf{c}'\hat{\boldsymbol{\beta}}$. ■

This powerful result states that substitution of $\hat{\boldsymbol{\beta}}$ for $\boldsymbol{\beta}$ leads to the best estimator of $\mathbf{c}'\boldsymbol{\beta}$ for any $\mathbf{c}$ of interest. In statistical terminology, the estimator $\mathbf{c}'\hat{\boldsymbol{\beta}}$ is called the *best (minimum variance) linear unbiased estimator* (B.L.U.E.) of $\mathbf{c}'\boldsymbol{\beta}$.

7.4 INFERENCES ABOUT THE REGRESSION MODEL

We describe inferential procedures based on the classical linear regression model in (7-3) with the additional (tentative) assumption that the errors $\boldsymbol{\varepsilon}$ have a normal distribution. Methods for checking the general adequacy of the model are considered in Section 7.6.

Inferences Concerning the Regression Parameters

Before we can assess the importance of particular variables in the *regression function*

$$E(Y) = \beta_0 + \beta_1 z_1 + \cdots + \beta_r z_r \tag{7-11}$$

we must determine the sampling distributions of $\hat{\boldsymbol{\beta}}$ and the residual sum of squares, $\hat{\boldsymbol{\varepsilon}}'\hat{\boldsymbol{\varepsilon}}$. To do so we shall assume the errors, $\boldsymbol{\varepsilon}$, have a normal distribution.

Result 7.4. Let $\mathbf{Y} = \mathbf{Z}\boldsymbol{\beta} + \boldsymbol{\varepsilon}$, where $\mathbf{Z}$ has full rank $r + 1$ and $\boldsymbol{\varepsilon}$ is distributed as $N_n(\mathbf{0}, \sigma^2\mathbf{I})$. The maximum likelihood estimator of $\boldsymbol{\beta}$ is the same as the least squares estimator $\hat{\boldsymbol{\beta}}$. Moreover,

$$\hat{\boldsymbol{\beta}} = (\mathbf{Z}'\mathbf{Z})^{-1}\mathbf{Z}'\mathbf{Y} \quad \text{is distributed as} \quad N_{r+1}\left(\boldsymbol{\beta}, \sigma^2(\mathbf{Z}'\mathbf{Z})^{-1}\right)$$

and is distributed independently of the residuals $\hat{\boldsymbol{\varepsilon}} = \mathbf{Y} - \mathbf{Z}\hat{\boldsymbol{\beta}}$. Further,

$$n\hat{\sigma}^2 = \hat{\boldsymbol{\varepsilon}}'\hat{\boldsymbol{\varepsilon}} \quad \text{is distributed as} \quad \sigma^2\chi^2_{n-r-1}$$

Proof. Given the data and the normal assumption for the errors, the likelihood function for $\boldsymbol{\beta}, \sigma^2$ is

$$\begin{aligned}L(\boldsymbol{\beta}, \sigma^2) &= \prod_{j=1}^{n} \frac{1}{\sqrt{2\pi}\,\sigma} e^{-\varepsilon_j^2/2\sigma^2} = \frac{1}{(2\pi)^{n/2}\sigma^n} e^{-\boldsymbol{\varepsilon}'\boldsymbol{\varepsilon}/2\sigma^2} \\ &= \frac{1}{(2\pi)^{n/2}\sigma^n} e^{-(\mathbf{y}-\mathbf{Z}\boldsymbol{\beta})'(\mathbf{y}-\mathbf{Z}\boldsymbol{\beta})/2\sigma^2}\end{aligned}$$

For a fixed value σ^2, the likelihood is maximized by minimizing $(\mathbf{y} - \mathbf{Z}\boldsymbol{\beta})'(\mathbf{y} - \mathbf{Z}\boldsymbol{\beta})$. But this minimization yields the least squares estimate $\hat{\boldsymbol{\beta}} = (\mathbf{Z}'\mathbf{Z})^{-1}\mathbf{Z}'\mathbf{y}$, which does not depend upon σ^2. Therefore, under the normal assumption, the maximum likelihood and least squares approaches provide the same estimator $\hat{\boldsymbol{\beta}}$. Next maximizing $L(\hat{\boldsymbol{\beta}}, \sigma^2)$ over σ^2 [see (4-18)] gives

$$L(\hat{\boldsymbol{\beta}}, \hat{\sigma}^2) = \frac{1}{(2\pi)^{n/2}(\hat{\sigma}^2)^{n/2}} e^{-n/2} \quad \text{where} \quad \hat{\sigma}^2 = \frac{(\mathbf{y} - \mathbf{Z}\hat{\boldsymbol{\beta}})'(\mathbf{y} - \mathbf{Z}\hat{\boldsymbol{\beta}})}{n} \tag{7-12}$$

From (7-10) we can express $\hat{\boldsymbol{\beta}}$ and $\hat{\boldsymbol{\varepsilon}}$ as linear combinations of the normal variables $\boldsymbol{\varepsilon}$. Specifically,

$$\begin{bmatrix} \hat{\boldsymbol{\beta}} \\ \hline \hat{\boldsymbol{\varepsilon}} \end{bmatrix} = \begin{bmatrix} \boldsymbol{\beta} + (\mathbf{Z}'\mathbf{Z})^{-1}\mathbf{Z}'\boldsymbol{\varepsilon} \\ \hline [\mathbf{I} - \mathbf{Z}(\mathbf{Z}'\mathbf{Z})^{-1}\mathbf{Z}']\boldsymbol{\varepsilon} \end{bmatrix} = \begin{bmatrix} \boldsymbol{\beta} \\ \hline \mathbf{0} \end{bmatrix} + \begin{bmatrix} (\mathbf{Z}'\mathbf{Z})^{-1}\mathbf{Z}' \\ \hline \mathbf{I} - \mathbf{Z}(\mathbf{Z}'\mathbf{Z})^{-1}\mathbf{Z}' \end{bmatrix} \boldsymbol{\varepsilon} = \boldsymbol{\alpha} + \mathbf{A}\boldsymbol{\varepsilon}$$

Because $\mathbf{Z}$ is fixed, Result 4.3 implies the joint normality of $\hat{\boldsymbol{\beta}}$ and $\hat{\boldsymbol{\varepsilon}}$. Their mean vectors and covariance matrices were obtained in Result 7.2. Again, using (7-6),

$$\text{Cov}\left(\begin{bmatrix} \hat{\boldsymbol{\beta}} \\ \hline \hat{\boldsymbol{\varepsilon}} \end{bmatrix}\right) = \mathbf{A}\,\text{Cov}(\boldsymbol{\varepsilon})\mathbf{A}' = \sigma^2 \left[\begin{array}{c|c} (\mathbf{Z}'\mathbf{Z})^{-1} & \mathbf{0} \\ \hline \mathbf{0}' & \mathbf{I} - \mathbf{Z}(\mathbf{Z}'\mathbf{Z})^{-1}\mathbf{Z}' \end{array}\right]$$

Since $\text{Cov}(\hat{\boldsymbol{\beta}}, \hat{\boldsymbol{\varepsilon}}) = \mathbf{0}$ for the normal random vectors $\hat{\boldsymbol{\beta}}$ and $\hat{\boldsymbol{\varepsilon}}$, they are independent. (See Result 4.5.)

Next, let $(\lambda, \mathbf{e})$ be any eigenvalue-eigenvector pair for $\mathbf{I} - \mathbf{Z}(\mathbf{Z}'\mathbf{Z})^{-1}\mathbf{Z}'$. Then, by (7-6), $[\mathbf{I} - \mathbf{Z}(\mathbf{Z}'\mathbf{Z})^{-1}\mathbf{Z}']^2 = [\mathbf{I} - \mathbf{Z}(\mathbf{Z}'\mathbf{Z})^{-1}\mathbf{Z}']$ so

$$\lambda\mathbf{e} = [\mathbf{I} - \mathbf{Z}(\mathbf{Z}'\mathbf{Z})^{-1}\mathbf{Z}']\mathbf{e} = [\mathbf{I} - \mathbf{Z}(\mathbf{Z}'\mathbf{Z})^{-1}\mathbf{Z}']^2\mathbf{e} = \lambda[\mathbf{I} - \mathbf{Z}(\mathbf{Z}'\mathbf{Z})^{-1}\mathbf{Z}']\mathbf{e} = \lambda^2\mathbf{e}$$

That is, $\lambda^2 = 0$ or 1. Now $\text{tr}[\mathbf{I} - \mathbf{Z}(\mathbf{Z}'\mathbf{Z})^{-1}\mathbf{Z}'] = n - r - 1$ (see the proof of Result 7.2) and, from Result 4.9, $\text{tr}[\mathbf{I} - \mathbf{Z}(\mathbf{Z}'\mathbf{Z})^{-1}\mathbf{Z}'] = \lambda_1 + \lambda_2 + \cdots + \lambda_n$, where $\lambda_1 \geq \lambda_2 \geq \cdots \geq \lambda_n$ are the eigenvalues of $[\mathbf{I} - \mathbf{Z}(\mathbf{Z}'\mathbf{Z})^{-1}\mathbf{Z}']$. Consequently, exactly $n - r - 1$ values of λ_i equal one and the rest are zero. It then follows from the spectral decomposition that

$$[\mathbf{I} - \mathbf{Z}(\mathbf{Z}'\mathbf{Z})^{-1}\mathbf{Z}'] = \mathbf{e}_1\mathbf{e}_1' + \mathbf{e}_2\mathbf{e}_2' + \cdots + \mathbf{e}_{n-r-1}\mathbf{e}_{n-r-1}' \tag{7-13}$$

where $\mathbf{e}_1, \mathbf{e}_2, \ldots, \mathbf{e}_{n-r-1}$ are the normalized eigenvectors associated with the eigenvalues $\lambda_1 = \lambda_2 = \cdots = \lambda_{n-r-1} = 1$. Let

$$\mathbf{V} = \begin{bmatrix} V_1 \\ V_2 \\ \vdots \\ V_{n-r-1} \end{bmatrix} = \begin{bmatrix} \mathbf{e}_1' \\ \hline \mathbf{e}_2' \\ \hline \vdots \\ \hline \mathbf{e}_{n-r-1}' \end{bmatrix} \boldsymbol{\varepsilon}$$

Then $\mathbf{V}$ is normal with mean vector $\mathbf{0}$ and

$$\text{Cov}(V_i, V_k) = \begin{cases} \mathbf{e}_i'\sigma^2\mathbf{I}\mathbf{e}_k = \sigma^2\mathbf{e}_i'\mathbf{e}_k = \sigma^2 & i = k \\ 0 & \text{otherwise} \end{cases}$$

That is, the V_i are independent $N(0, \sigma^2)$ and by (7-10),

$$n\hat{\sigma}^2 = \hat{\boldsymbol{\varepsilon}}'\hat{\boldsymbol{\varepsilon}} = \boldsymbol{\varepsilon}'\left[\mathbf{I} - \mathbf{Z}(\mathbf{Z}'\mathbf{Z})^{-1}\mathbf{Z}'\right]\boldsymbol{\varepsilon} = \sigma^2\left(V_1^2 + V_2^2 + \cdots + V_{n-r-1}^2\right)$$

is distributed as $\sigma^2\chi^2_{n-r-1}$. ■

A confidence ellipsoid for $\boldsymbol{\beta}$ is easily constructed. It is expressed in terms of the estimated covariance matrix $s^2(\mathbf{Z}'\mathbf{Z})^{-1}$, where $s^2 = \hat{\boldsymbol{\varepsilon}}'\hat{\boldsymbol{\varepsilon}}/(n - r - 1)$.

Result 7.5. Let $\mathbf{Y} = \mathbf{Z}\boldsymbol{\beta} + \boldsymbol{\varepsilon}$, where $\mathbf{Z}$ has full rank $r + 1$ and $\boldsymbol{\varepsilon}$ is $N_n(\mathbf{0}, \sigma^2\mathbf{I})$. A $100(1 - \alpha)\%$ confidence region for $\boldsymbol{\beta}$ is given by

$$(\boldsymbol{\beta} - \hat{\boldsymbol{\beta}})'\mathbf{Z}'\mathbf{Z}(\boldsymbol{\beta} - \hat{\boldsymbol{\beta}}) \leq (r + 1)s^2F_{r+1,\,n-r-1}(\alpha)$$

where $F_{r+1,\,n-r-1}(\alpha)$ is the upper (100α)th percentile of an F-distribution with $r + 1$ and $n - r - 1$ d.f.

Also, *simultaneous* $100(1 - \alpha)\%$ confidence intervals for the β_i are given by

$$\hat{\beta}_i \pm \sqrt{\widehat{\operatorname{Var}}(\hat{\beta}_i)}\,\sqrt{(r + 1)F_{r+1,\,n-r-1}(\alpha)}\,, \qquad i = 0, 1, \ldots, r$$

where $\widehat{\operatorname{Var}}(\hat{\beta}_i)$ is the diagonal element of $s^2(\mathbf{Z}'\mathbf{Z})^{-1}$ corresponding to $\hat{\beta}_i$.

Proof. Consider the symmetric square-root matrix $(\mathbf{Z}'\mathbf{Z})^{1/2}$ [see (2-22)]. Set $\mathbf{V} = (\mathbf{Z}'\mathbf{Z})^{1/2}(\hat{\boldsymbol{\beta}} - \boldsymbol{\beta})$ and note that $E(\mathbf{V}) = \mathbf{0}$,

$$\operatorname{Cov}(\mathbf{V}) = (\mathbf{Z}'\mathbf{Z})^{1/2}\operatorname{Cov}(\hat{\boldsymbol{\beta}})(\mathbf{Z}'\mathbf{Z})^{1/2} = \sigma^2(\mathbf{Z}'\mathbf{Z})^{1/2}(\mathbf{Z}'\mathbf{Z})^{-1}(\mathbf{Z}'\mathbf{Z})^{1/2} = \sigma^2\mathbf{I}$$

and $\mathbf{V}$ is normally distributed since it consists of linear combinations of the $\hat{\beta}_i$'s. Therefore $\mathbf{V}'\mathbf{V} = (\hat{\boldsymbol{\beta}} - \boldsymbol{\beta})'(\mathbf{Z}'\mathbf{Z})^{1/2}(\mathbf{Z}'\mathbf{Z})^{1/2}(\hat{\boldsymbol{\beta}} - \boldsymbol{\beta}) = (\hat{\boldsymbol{\beta}} - \boldsymbol{\beta})'(\mathbf{Z}'\mathbf{Z})(\hat{\boldsymbol{\beta}} - \boldsymbol{\beta})$ is distributed as $\sigma^2\chi^2_{r+1}$. By Result 7.4, $(n - r - 1)s^2 = \hat{\boldsymbol{\varepsilon}}'\hat{\boldsymbol{\varepsilon}}$ is distributed as $\sigma^2\chi^2_{n-r-1}$, independent of $\hat{\boldsymbol{\beta}}$ and, hence, of $\mathbf{V}$. Consequently, $[\chi^2_{r+1}/(r + 1)]/[\chi^2_{n-r-1}/(n - r - 1)] = [\mathbf{V}'\mathbf{V}/(r + 1)]/s^2$ has an $F_{r+1,\,n-r-1}$ distribution and the confidence ellipsoid follows. Projecting this ellipsoid for $(\hat{\boldsymbol{\beta}} - \boldsymbol{\beta})$ using Result 5C.1 with $\mathbf{A}^{-1} = \mathbf{Z}'\mathbf{Z}$, $c^2 = (r + 1)F_{r+1,\,n-r-1}(\alpha)$, and $\boldsymbol{\ell} = [0, \ldots, 0, 1, 0, \ldots, 0]'$ yields $|\beta_i - \hat{\beta}_i| \leq \sqrt{(r + 1)F_{r+1,\,n-r-1}(\alpha)}\,\sqrt{\widehat{\operatorname{Var}}(\hat{\beta}_i)}$, where $\widehat{\operatorname{Var}}(\hat{\beta}_i)$ is the diagonal element of $s^2(\mathbf{Z}'\mathbf{Z})^{-1}$ corresponding to $\hat{\beta}_i$. ■

The confidence ellipsoid is centered at the maximum likelihood estimate $\hat{\boldsymbol{\beta}}$, and its orientation and size are determined by the eigenvalues and eigenvectors of $\mathbf{Z}'\mathbf{Z}$. If an eigenvalue is nearly zero, the confidence ellipsoid will be very long in the direction of the corresponding eigenvector.

Practitioners often ignore the "simultaneous" confidence property of the interval estimates in Result 7.5. They replace $(r + 1)F_{r+1,\,n-r-1}(\alpha)$ with the one-at-a-time t value, $t_{n-r-1}(\alpha/2)$, and use the intervals

$$\hat{\beta}_i \pm t_{n-r-1}\left(\frac{\alpha}{2}\right)\sqrt{\widehat{\operatorname{Var}}(\hat{\beta}_i)} \tag{7-14}$$

when searching for important predictor variables.

Example 7.4

The assessment data in Table 7.1 were gathered from 20 homes in a Milwaukee, Wisconsin, neighborhood. Fit the regression model

$$Y_j = \beta_0 + \beta_1 z_{j1} + \beta_2 z_{j2} + \varepsilon_j$$

where z_1 = total dwelling size (in hundreds of square feet), z_2 = assessed value (in thousands of dollars) and Y = selling price (in dollars), to these data using the method of least squares. A computer calculation yields

$$(\mathbf{Z}'\mathbf{Z})^{-1} = \begin{bmatrix} 1.9961 & & \\ -.0896 & .0512 & \\ -.0115 & -.0172 & .0067 \end{bmatrix} \quad \text{and} \quad \hat{\boldsymbol{\beta}} = (\mathbf{Z}'\mathbf{Z})^{-1}\mathbf{Z}'\mathbf{y} = \begin{bmatrix} 11{,}870.2 \\ 2{,}634.4 \\ 45.2 \end{bmatrix}$$

Thus the fitted equation is

$$\hat{y} = \underset{(4906)}{11{,}870.2} + \underset{(785)}{2{,}634.4} z_1 + \underset{(285)}{45.2} z_2$$

with $s = 3473$. The numbers in parentheses are the estimated standard deviations of the least squares coefficients. Also, $R^2 = .834$, indicating the data exhibit a strong regression relationship. If the residuals, $\hat{\boldsymbol{\varepsilon}}$, pass the diagnostic checks described in Section 7.6, the fitted equation could be used to predict the selling price of another house in the neighborhood, from its size and assessed

TABLE 7.1 REAL-ESTATE DATA

z_1 Total dwelling size (100 square feet)	z_2 Assessed value ($1000)	Y Selling price (dollars)
15.31	37.3	54,800
15.20	43.8	54,000
16.25	45.4	52,900
14.33	37.0	50,000
14.57	43.8	54,900
17.33	43.2	56,000
14.48	40.2	52,000
14.91	37.7	53,500
15.25	36.4	54,500
13.89	35.6	53,500
15.18	42.6	51,500
14.44	43.4	51,000
14.87	40.2	58,900
18.63	47.2	66,500
15.20	37.1	48,000
25.76	69.6	82,000
19.05	48.6	64,000
15.37	40.1	49,000
18.06	46.3	68,000
16.35	45.8	56,000

value. We note that a 95% confidence interval for β_2 [see (7-14)] is given by

$$\hat{\beta}_2 \pm t_{17}(.025)\sqrt{\widehat{\text{Var}}(\hat{\beta}_2)} = 45.2 \pm 2.110(285)$$

or

$$(-556, 647)$$

Since the confidence interval includes $\beta_2 = 0$, the variable z_2 might be dropped from the regression model and the analysis repeated with the single predictor variable z_1. Given dwelling size, assessed value seems to add little to the prediction of selling price. ■

Likelihood Ratio Tests for the Regression Parameters

Part of regression analysis is concerned with assessing the effects of particular predictor variables on the response variable. One null hypothesis of interest states that certain of the z_i's do not influence the response, Y. These predictors will be labeled $z_{q+1}, z_{q+2}, \ldots, z_r$. The statement that $z_{q+1}, z_{q+2}, \ldots, z_r$ do not influence Y translates into the statistical hypothesis

$$H_0\colon \beta_{q+1} = \beta_{q+2} = \cdots = \beta_r = 0 \quad \text{or} \quad H_0\colon \boldsymbol{\beta}_{(2)} = \mathbf{0} \qquad (7\text{-}15)$$

where $\boldsymbol{\beta}_{(2)} = [\beta_{q+1}, \beta_{q+2}, \ldots, \beta_r]'$. Setting

$$\mathbf{Z} = \left[\underset{n\times(q+1)}{\mathbf{Z}_1} \;\vdots\; \underset{n\times(r-q)}{\mathbf{Z}_2}\right], \qquad \boldsymbol{\beta} = \begin{bmatrix} \underset{((q+1)\times 1)}{\boldsymbol{\beta}_{(1)}} \\ \hline \underset{((r-q)\times 1)}{\boldsymbol{\beta}_{(2)}} \end{bmatrix}$$

the general linear model can be expressed as

$$\mathbf{Y} = \mathbf{Z}\boldsymbol{\beta} + \boldsymbol{\varepsilon} = [\mathbf{Z}_1 \;\vdots\; \mathbf{Z}_2]\begin{bmatrix} \boldsymbol{\beta}_{(1)} \\ \hline \boldsymbol{\beta}_{(2)} \end{bmatrix} + \boldsymbol{\varepsilon} = \mathbf{Z}_1\boldsymbol{\beta}_{(1)} + \mathbf{Z}_2\boldsymbol{\beta}_{(2)} + \boldsymbol{\varepsilon}$$

Under the null hypothesis $H_0\colon \boldsymbol{\beta}_{(2)} = \mathbf{0}$, $\mathbf{Y} = \mathbf{Z}_1\boldsymbol{\beta}_{(1)} + \boldsymbol{\varepsilon}$. The likelihood ratio test of H_0 is based on the

$$\begin{aligned}\text{Extra sum of squares} &= SS_{\text{res}}(\mathbf{Z}_1) - SS_{\text{res}}(\mathbf{Z}) \\ &= (\mathbf{y} - \mathbf{Z}_1\hat{\boldsymbol{\beta}}_{(1)})'(\mathbf{y} - \mathbf{Z}_1\hat{\boldsymbol{\beta}}_{(1)}) - (\mathbf{y} - \mathbf{Z}\hat{\boldsymbol{\beta}})'(\mathbf{y} - \mathbf{Z}\hat{\boldsymbol{\beta}})\end{aligned} \qquad (7\text{-}16)$$

where $\hat{\boldsymbol{\beta}}_{(1)} = (\mathbf{Z}_1'\mathbf{Z}_1)^{-1}\mathbf{Z}_1'\mathbf{y}$.

Result 7.6. Let $\mathbf{Z}$ have full rank $r + 1$ and $\boldsymbol{\varepsilon}$ be distributed as $N_n(\mathbf{0}, \sigma^2\mathbf{I})$. The likelihood ratio test of $H_0\colon \boldsymbol{\beta}_{(2)} = \mathbf{0}$ is equivalent to a test of H_0 based on the extra sum of squares in (7-16) and $s^2 = (\mathbf{y} - \mathbf{Z}\hat{\boldsymbol{\beta}})'(\mathbf{y} - \mathbf{Z}\hat{\boldsymbol{\beta}})/(n - r - 1)$. In particular, the likelihood ratio test rejects H_0 if

$$\frac{(SS_{\text{res}}(\mathbf{Z}_1) - SS_{\text{res}}(\mathbf{Z}))/(r - q)}{s^2} > F_{r-q,\, n-r-1}(\alpha)$$

where $F_{r-q,\,n-r-1}(\alpha)$ is the upper (100α)th percentile of an F-distribution with $r-q$ and $n-r-1$ d.f.

Proof. Given the data and the normal assumption, the likelihood associated with the parameters $\boldsymbol{\beta}$ and σ^2 is

$$L(\boldsymbol{\beta}, \sigma^2) = \frac{1}{(2\pi)^{n/2}\sigma^n} e^{-(\mathbf{y}-\mathbf{Z}\boldsymbol{\beta})'(\mathbf{y}-\mathbf{Z}\boldsymbol{\beta})/2\sigma^2} \le \frac{1}{(2\pi)^{n/2}\hat{\sigma}^n} e^{-n/2}$$

with the maximum occurring at $\hat{\boldsymbol{\beta}} = (\mathbf{Z}'\mathbf{Z})^{-1}\mathbf{Z}'\mathbf{y}$ and $\hat{\sigma}^2 = (\mathbf{y} - \mathbf{Z}\hat{\boldsymbol{\beta}})'(\mathbf{y} - \mathbf{Z}\hat{\boldsymbol{\beta}})/n$. Under the restriction of the null hypothesis, $\mathbf{Y} = \mathbf{Z}_1\boldsymbol{\beta}_{(1)} + \boldsymbol{\varepsilon}$ and

$$\max_{\boldsymbol{\beta}_{(1)},\,\sigma^2} L(\boldsymbol{\beta}_{(1)}, \sigma^2) = \frac{1}{(2\pi)^{n/2}\hat{\sigma}_1^n} e^{-n/2}$$

where the maximum occurs at $\hat{\boldsymbol{\beta}}_{(1)} = (\mathbf{Z}_1'\mathbf{Z}_1)^{-1}\mathbf{Z}_1'\mathbf{y}$ and

$$\hat{\sigma}_1^2 = \left(\mathbf{y} - \mathbf{Z}_1\hat{\boldsymbol{\beta}}_{(1)}\right)'\left(\mathbf{y} - \mathbf{Z}_1\hat{\boldsymbol{\beta}}_{(1)}\right)/n$$

Rejecting $H_0\colon \boldsymbol{\beta}_{(2)} = \mathbf{0}$ for small values of the likelihood ratio

$$\frac{\displaystyle\max_{\boldsymbol{\beta}_{(1)},\,\sigma^2} L(\boldsymbol{\beta}_{(1)}, \sigma^2)}{\displaystyle\max_{\boldsymbol{\beta},\,\sigma^2} L(\boldsymbol{\beta}, \sigma^2)} = \left(\frac{\hat{\sigma}_1^2}{\hat{\sigma}^2}\right)^{-n/2} = \left(\frac{\hat{\sigma}^2 + \hat{\sigma}_1^2 - \hat{\sigma}^2}{\hat{\sigma}^2}\right)^{-n/2} = \left(1 + \frac{\hat{\sigma}_1^2 - \hat{\sigma}^2}{\hat{\sigma}^2}\right)^{-n/2}$$

is equivalent to rejecting H_0 for large values of $(\hat{\sigma}_1^2 - \hat{\sigma}^2)/\hat{\sigma}^2$ or its scaled version

$$\frac{n(\hat{\sigma}_1^2 - \hat{\sigma}^2)/(r-q)}{n\hat{\sigma}^2/(n-r-1)} = \frac{(SS_{\text{res}}(Z_1) - SS_{\text{res}}(Z))/(r-q)}{s^2} = F$$

The F-ratio above has an F-distribution with $r-q$ and $n-r-1$ d.f. (see [18] or Result 7.11 with $m = 1$). ■

Comment. The likelihood ratio test is implemented as follows. To test whether all coefficients in a subset are zero, fit the model with and without the terms corresponding to these coefficients. The improvement in the residual sum of squares (the extra sum of squares) is compared to the residual sum of squares for the full model via the F-ratio. The same procedure applies even in analysis of variance situations where $\mathbf{Z}$ is not of full rank.[4]

More generally, it is possible to formulate null hypotheses concerning $r - q$ linear combinations of $\boldsymbol{\beta}$. Let the $(r-q) \times (r+1)$ matrix $\mathbf{C}$ have full rank and consider

$$H_0\colon \mathbf{C}\boldsymbol{\beta} = \mathbf{0}$$

(This null hypothesis reduces to the previous choice when $\mathbf{C} = \left[\mathbf{0} \;\middle|\; \underset{(r-q)\times(r-q)}{\mathbf{I}}\right]$.)

Under the full model, $\mathbf{C}\hat{\boldsymbol{\beta}}$ is distributed as $N_{r-q}(\mathbf{C}\boldsymbol{\beta}, \sigma^2\mathbf{C}(\mathbf{Z}'\mathbf{Z})^{-1}\mathbf{C}')$. We reject $H_0\colon \mathbf{C}\boldsymbol{\beta} = \mathbf{0}$ at level α if $\mathbf{0}$ does *not* lie in the $100(1-\alpha)\%$ confidence ellipsoid for

[4] In situations where $\mathbf{Z}$ is not of full rank, rank($\mathbf{Z}$) replaces $r + 1$ and rank($\mathbf{Z}_1$) replaces $q + 1$ in Result 7.6.

$\mathbf{C}\boldsymbol{\beta}$. Equivalently, we reject $H_0\colon \mathbf{C}\boldsymbol{\beta} = \mathbf{0}$ if

$$\frac{(\mathbf{C}\hat{\boldsymbol{\beta}})'\left(\mathbf{C}(\mathbf{Z}'\mathbf{Z})^{-1}\mathbf{C}'\right)^{-1}(\mathbf{C}\hat{\boldsymbol{\beta}})}{s^2} > (r-q)F_{r-q,\,n-r-1}(\alpha) \qquad (7\text{-}17)$$

where $s^2 = (\mathbf{y} - \mathbf{Z}\hat{\boldsymbol{\beta}})'(\mathbf{y} - \mathbf{Z}\hat{\boldsymbol{\beta}})/(n-r-1)$ and $F_{r-q,\,n-r-1}(\alpha)$ is the upper (100α)th percentile of an F-distribution with $r-q$ and $n-r-1$ d.f. The test in (7-17) is the likelihood ratio test and the numerator in the F-ratio is the extra residual sum of squares incurred by fitting the model subject to the restriction $\mathbf{C}\boldsymbol{\beta} = \mathbf{0}$ (see [20]).

The next example illustrates how unbalanced experimental designs are easily handled by the general theory described above.

Example 7.5

Male and female patrons rated the service in three establishments (locations) of a large restaurant chain. The service ratings were converted into an index. Table 7.2 contains the data for $n = 18$ customers. Each data point in the table is categorized according to location (1, 2, or 3) and sex (male = 0 and female = 1). This categorization has the format of a two-way table with unequal numbers of observations per cell. For instance, the combination location 1 and male has 5 responses, while the combination location 2 and female has 2 responses. Introducing three dummy variables to account for location and two dummy variables to account for sex, a regression model linking the service index, Y, to location, sex, and their "interaction" can be developed using the design matrix

$$\mathbf{Z} = \begin{array}{c} \\ \\ \end{array}\overset{\begin{array}{cccc}\text{constant} & \text{location} & \text{sex} & \text{interaction}\end{array}}{\left[\begin{array}{c|ccc|cc|cccccc}
1 & 1 & 0 & 0 & 1 & 0 & 1 & 0 & 0 & 0 & 0 & 0 \\
1 & 1 & 0 & 0 & 1 & 0 & 1 & 0 & 0 & 0 & 0 & 0 \\
1 & 1 & 0 & 0 & 1 & 0 & 1 & 0 & 0 & 0 & 0 & 0 \\
1 & 1 & 0 & 0 & 1 & 0 & 1 & 0 & 0 & 0 & 0 & 0 \\
1 & 1 & 0 & 0 & 1 & 0 & 1 & 0 & 0 & 0 & 0 & 0 \\
1 & 1 & 0 & 0 & 0 & 1 & 0 & 1 & 0 & 0 & 0 & 0 \\
1 & 1 & 0 & 0 & 0 & 1 & 0 & 1 & 0 & 0 & 0 & 0 \\
1 & 0 & 1 & 0 & 1 & 0 & 0 & 0 & 1 & 0 & 0 & 0 \\
1 & 0 & 1 & 0 & 1 & 0 & 0 & 0 & 1 & 0 & 0 & 0 \\
1 & 0 & 1 & 0 & 1 & 0 & 0 & 0 & 1 & 0 & 0 & 0 \\
1 & 0 & 1 & 0 & 1 & 0 & 0 & 0 & 1 & 0 & 0 & 0 \\
1 & 0 & 1 & 0 & 1 & 0 & 0 & 0 & 1 & 0 & 0 & 0 \\
1 & 0 & 1 & 0 & 0 & 1 & 0 & 0 & 0 & 1 & 0 & 0 \\
1 & 0 & 1 & 0 & 0 & 1 & 0 & 0 & 0 & 1 & 0 & 0 \\
1 & 0 & 0 & 1 & 1 & 0 & 0 & 0 & 0 & 0 & 1 & 0 \\
1 & 0 & 0 & 1 & 1 & 0 & 0 & 0 & 0 & 0 & 1 & 0 \\
1 & 0 & 0 & 1 & 0 & 1 & 0 & 0 & 0 & 0 & 0 & 1 \\
1 & 0 & 0 & 1 & 0 & 1 & 0 & 0 & 0 & 0 & 0 & 1
\end{array}\right]}
\begin{array}{l}
\left.\begin{array}{c}\\ \\ \\ \\ \\ \end{array}\right\} \text{5 responses} \\
\left.\begin{array}{c}\\ \\ \end{array}\right\} \text{2 responses} \\
\left.\begin{array}{c}\\ \\ \\ \\ \\ \end{array}\right\} \text{5 responses} \\
\left.\begin{array}{c}\\ \\ \end{array}\right\} \text{2 responses} \\
\left.\begin{array}{c}\\ \\ \end{array}\right\} \text{2 responses} \\
\left.\begin{array}{c}\\ \\ \end{array}\right\} \text{2 responses}
\end{array}$$

TABLE 7.2 RESTAURANT-SERVICE DATA

Location	Sex	Service (Y)
1	0	15.2
1	0	21.2
1	0	27.3
1	0	21.2
1	0	21.2
1	1	36.4
1	1	92.4
2	0	27.3
2	0	15.2
2	0	9.1
2	0	18.2
2	0	50.0
2	1	44.0
2	1	63.6
3	0	15.2
3	0	30.3
3	1	36.4
3	1	40.9

The coefficient vector, $\boldsymbol{\beta}$, can be set out as

$$\boldsymbol{\beta} = [\beta_0, \beta_1, \beta_2, \beta_3, \tau_1, \tau_2, \gamma_{11}, \gamma_{12}, \gamma_{21}, \gamma_{22}, \gamma_{31}, \gamma_{32}]'$$

where the β_i's ($i > 0$) represent the effects of the locations on the determination of service, the τ_i's represent the effects of sex on the service index and the γ_{ik}'s represent the location-sex interaction effects.

The design matrix $\mathbf{Z}$ is not of full rank. (For instance, column 1 equals the sum of columns 2 through 4 or columns 5–6.) In fact, rank($\mathbf{Z}$) = 6.

For the complete model, results from a computer program give

$$\text{SS}_{\text{res}}(\mathbf{Z}) = 2977.4$$

and $n - \text{rank}(\mathbf{Z}) = 18 - 6 = 12$.

The model without the interaction terms has the design matrix $\mathbf{Z}_1$ consisting of the first six columns of $\mathbf{Z}$. We find

$$\text{SS}_{\text{res}}(\mathbf{Z}_1) = 3419.1$$

with $n - \text{rank}(\mathbf{Z}_1) = 18 - 4 = 14$. To test H_0: $\gamma_{11} = \gamma_{12} = \cdots = \gamma_{31} = \gamma_{32} = 0$ (no location-sex interaction), we compute

$$F = \frac{(\text{SS}_{\text{res}}(\mathbf{Z}_1) - \text{SS}_{\text{res}}(\mathbf{Z}))/(6-4)}{s^2} = \frac{(\text{SS}_{\text{res}}(\mathbf{Z}_1) - \text{SS}_{\text{res}}(\mathbf{Z}))/2}{\text{SS}_{\text{res}}(\mathbf{Z})/12}$$

$$= \frac{(3419.1 - 2977.4)/2}{2977.4/12} = .89$$

The F-ratio may be compared with an appropriate percentage point of an F-distribution with 2 and 12 d.f. This F-ratio is not significant for any reasonable significance level α. Consequently, we conclude the service index does not depend upon any location-sex interaction and these terms can be dropped from the model.

Using the extra sum of squares approach, it may be verified that there is no difference between locations (no location effect), but that sex is significant: That is, males and females do not give the same ratings to service.

In analysis-of-variance situations where the cell counts are unequal, the variation in the response attributable to different predictor variables and their interactions cannot usually be separated into independent amounts. To evaluate the relative influences of the predictors on the response in this case, it is necessary to fit the model with and without the terms in question and compute the appropriate F-test statistics. ■

7.5 INFERENCES FROM THE ESTIMATED REGRESSION FUNCTION

Once an investigator is satisfied with the fitted regression model, it can be used to solve two prediction problems. Let $\mathbf{z}_0 = [1, z_{01}, \ldots, z_{0r}]'$ be selected values for the predictor variables. Then $\mathbf{z}_0$ and $\hat{\boldsymbol{\beta}}$ can be used (1) to estimate the regression function, $\beta_0 + \beta_1 z_{01} + \cdots + \beta_r z_{0r}$, at $\mathbf{z}_0$, and (2) to estimate the value of the response, Y, at $\mathbf{z}_0$.

Estimating the Regression Function at $\mathbf{z}_0$

Let Y_0 denote the value of the response when the predictor variables have values $\mathbf{z}_0 = [1, z_{01}, \ldots, z_{0r}]'$. According to the model in (7-3), the expected value of Y_0 is

$$E(Y_0 \mid \mathbf{z}_0) = \beta_0 + \beta_1 z_{01} + \cdots + \beta_r z_{0r} = \mathbf{z}_0'\boldsymbol{\beta} \tag{7-18}$$

Its least squares estimate is $\mathbf{z}_0'\hat{\boldsymbol{\beta}}$.

Result 7.7. For the linear regression model in (7-3), $\mathbf{z}_0'\hat{\boldsymbol{\beta}}$ is the unbiased linear estimator of $E(Y_0 \mid \mathbf{z}_0)$ with minimum variance, $\text{Var}(\mathbf{z}_0'\hat{\boldsymbol{\beta}}) = \mathbf{z}_0'(\mathbf{Z}'\mathbf{Z})^{-1}\mathbf{z}_0\sigma^2$. If the errors, $\boldsymbol{\varepsilon}$, are normally distributed, then a $100(1-\alpha)\%$ confidence interval for $E(Y_0 \mid \mathbf{z}_0) = \mathbf{z}_0'\boldsymbol{\beta}$ is provided by

$$\mathbf{z}_0'\hat{\boldsymbol{\beta}} \pm t_{n-r-1}\left(\frac{\alpha}{2}\right)\sqrt{(\mathbf{z}_0'(\mathbf{Z}'\mathbf{Z})^{-1}\mathbf{z}_0)s^2}$$

where $t_{n-r-1}(\alpha/2)$ is the upper $100(\alpha/2)$th percentile of a t-distribution with $n-r-1$ d.f.

Proof. For a fixed $\mathbf{z}_0$, $\mathbf{z}_0'\boldsymbol{\beta}$ is just a linear combination of the β_i's, so Result 7.3 applies. Also $\text{Var}(\mathbf{z}_0'\hat{\boldsymbol{\beta}}) = \mathbf{z}_0'\,\text{Cov}(\hat{\boldsymbol{\beta}})\mathbf{z}_0 = \mathbf{z}_0'(\mathbf{Z}'\mathbf{Z})^{-1}\mathbf{z}_0\sigma^2$ since $\text{Cov}(\hat{\boldsymbol{\beta}}) = \sigma^2(\mathbf{Z}'\mathbf{Z})^{-1}$ by Result 7.2. Under the further assumption that $\boldsymbol{\varepsilon}$ is normally distributed, Result 7.4 asserts that $\hat{\boldsymbol{\beta}}$ is $N_{r+1}(\boldsymbol{\beta}, \sigma^2(\mathbf{Z}'\mathbf{Z})^{-1})$ independently of s^2/σ^2, which is distributed as $\chi^2_{n-r-1}/(n-r-1)$. Consequently, the linear combination $\mathbf{z}_0'\hat{\boldsymbol{\beta}}$ is

$N(\mathbf{z}_0'\boldsymbol{\beta}, \sigma^2\mathbf{z}_0'(\mathbf{Z}'\mathbf{Z})^{-1}\mathbf{z}_0)$ and

$$\frac{(\mathbf{z}_0'\hat{\boldsymbol{\beta}} - \mathbf{z}_0'\boldsymbol{\beta})\Big/\sqrt{\sigma^2\mathbf{z}_0'(\mathbf{Z}'\mathbf{Z})^{-1}\mathbf{z}_0}}{\sqrt{s^2/\sigma^2}} = \frac{(\mathbf{z}_0'\hat{\boldsymbol{\beta}} - \mathbf{z}_0'\boldsymbol{\beta})}{\sqrt{s^2\left(\mathbf{z}_0'(\mathbf{Z}'\mathbf{Z})^{-1}\mathbf{z}_0\right)}}$$

is t_{n-r-1}. The confidence interval follows. ■

Forecasting a New Observation at $\mathbf{z}_0$

Prediction of a new observation, such as Y_0, at $\mathbf{z}_0 = [1, z_{01}, \ldots, z_{0r}]'$ is more uncertain than estimating the *expected value* of Y_0. According to the regression model of (7-3)

$$Y_0 = \mathbf{z}_0'\boldsymbol{\beta} + \varepsilon_0$$

or

$$(\text{new response } Y_0) = (\text{expected value of } Y_0 \text{ at } \mathbf{z}_0) + (\text{new error})$$

where ε_0 is distributed as $N(0, \sigma^2)$ and is independent of $\boldsymbol{\varepsilon}$ and, hence, of $\hat{\boldsymbol{\beta}}$ and s^2. The errors $\boldsymbol{\varepsilon}$ influence the estimators $\hat{\boldsymbol{\beta}}$ and s^2 through the responses $\mathbf{Y}$, but ε_0 does not.

Result 7.8. Given the linear regression model of (7-3), a new observation Y_0 has the *unbiased predictor*

$$\mathbf{z}_0'\hat{\boldsymbol{\beta}} = \hat{\beta}_0 + \hat{\beta}_1 z_{01} + \cdots + \hat{\beta}_r z_{0r}$$

The variance of the *forecast error*, $Y_0 - \mathbf{z}_0'\hat{\boldsymbol{\beta}}$, is

$$\text{Var}\left[Y_0 - \mathbf{z}_0'\hat{\boldsymbol{\beta}}\right] = \sigma^2\left(1 + \mathbf{z}_0'(\mathbf{Z}'\mathbf{Z})^{-1}\mathbf{z}_0\right)$$

When the errors $\boldsymbol{\varepsilon}$ have a normal distribution, a $100(1-\alpha)\%$ *prediction interval* for Y_0 is given by

$$\mathbf{z}_0'\hat{\boldsymbol{\beta}} \pm t_{n-r-1}\left(\frac{\alpha}{2}\right)\sqrt{s^2\left(1 + \mathbf{z}_0'(\mathbf{Z}'\mathbf{Z})^{-1}\mathbf{z}_0\right)}$$

where $t_{n-r-1}(\alpha/2)$ is the upper $100(\alpha/2)$th percentile of a t-distribution with $n - r - 1$ degrees of freedom.

Proof. We forecast Y_0 by $\mathbf{z}_0'\hat{\boldsymbol{\beta}}$, which estimates $E(Y_0 \mid \mathbf{z}_0)$. By Result 7.7 $\mathbf{z}_0'\hat{\boldsymbol{\beta}}$ has $E(\mathbf{z}_0'\hat{\boldsymbol{\beta}}) = \mathbf{z}_0'\boldsymbol{\beta}$ and $\text{Var}(\mathbf{z}_0'\hat{\boldsymbol{\beta}}) = \mathbf{z}_0'(\mathbf{Z}'\mathbf{Z})^{-1}\mathbf{z}_0\sigma^2$. The forecast error is then $Y_0 - \mathbf{z}_0'\hat{\boldsymbol{\beta}} = \mathbf{z}_0'\boldsymbol{\beta} + \varepsilon_0 - \mathbf{z}_0'\hat{\boldsymbol{\beta}} = \varepsilon_0 + \mathbf{z}_0'(\boldsymbol{\beta} - \hat{\boldsymbol{\beta}})$. Thus $E(Y_0 - \mathbf{z}_0'\hat{\boldsymbol{\beta}}) = E(\varepsilon_0) + E(\mathbf{z}_0'(\boldsymbol{\beta} - \hat{\boldsymbol{\beta}})) = 0$ so the predictor is unbiased. Since ε_0 and $\hat{\boldsymbol{\beta}}$ are independent, $\text{Var}(Y_0 - \mathbf{z}_0'\hat{\boldsymbol{\beta}}) = \text{Var}(\varepsilon_0) + \text{Var}(\mathbf{z}_0'\hat{\boldsymbol{\beta}}) = \sigma^2 + \mathbf{z}_0'(\mathbf{Z}'\mathbf{Z})^{-1}\mathbf{z}_0\sigma^2 = \sigma^2(1 + \mathbf{z}_0'(\mathbf{Z}'\mathbf{Z})^{-1}\mathbf{z}_0)$. If it is further assumed that $\boldsymbol{\varepsilon}$ has a normal distribution, $\hat{\boldsymbol{\beta}}$ is normally distributed and so is the linear combination $Y_0 - \mathbf{z}_0'\hat{\boldsymbol{\beta}}$. Consequently, $(Y_0 - \mathbf{z}_0'\hat{\boldsymbol{\beta}})/\sqrt{\sigma^2\left(1 + \mathbf{z}_0'(\mathbf{Z}'\mathbf{Z})^{-1}\mathbf{z}_0\right)}$ is distributed as $N(0, 1)$. Dividing this ratio by $\sqrt{s^2/\sigma^2}$, which is distributed as $\sqrt{\chi^2_{n-r-1}/(n-r-1)}$, we obtain

$$\frac{(Y_0 - \mathbf{z}_0'\hat{\boldsymbol{\beta}})}{\sqrt{s^2(1 + \mathbf{z}_0'(\mathbf{Z}'\mathbf{Z})^{-1}\mathbf{z}_0)}}$$

is t_{n-r-1}. The prediction interval follows immediately. ∎

The prediction interval for Y_0 is wider than the confidence interval for estimating the value of the regression function $E(Y_0 \mid \mathbf{z}_0) = \mathbf{z}_0'\boldsymbol{\beta}$. The additional uncertainty in forecasting Y_0, which is represented by the extra term s^2 in the factor $s^2(1 + \mathbf{z}_0'(\mathbf{Z}'\mathbf{Z})^{-1}\mathbf{z}_0)$, comes from the presence of the unknown error term ε_0.

Example 7.6

Companies considering the purchase of a computer must first assess their future needs in order to determine the proper equipment. A computer scientist collected data from seven similar company sites so that a forecast equation of computer-hardware requirements for inventory management could be developed. The data are given in Table 7.3 for

$$z_1 = \text{customer orders (in thousands)}$$

$$z_2 = \text{add-delete item count (in thousands)}$$

$$Y = \text{CPU (central processing unit) time (in hours)}$$

Construct a 95% confidence interval for the mean CPU time, $E(Y_0 \mid \mathbf{z}_0) = \beta_0 + \beta_1 z_{01} + \beta_2 z_{02}$ at $\mathbf{z}_0 = [1, 130, 7.5]'$. Also find a 95% prediction interval for a new facility's CPU requirement corresponding to the same $\mathbf{z}_0$.

A computer program provides the estimated regression function

$$\hat{y} = 8.42 + 1.08z_1 + .42z_2$$

$$(\mathbf{Z}'\mathbf{Z})^{-1} = \begin{bmatrix} 8.17969 & & \\ -.06411 & .00052 & \\ .08831 & -.00107 & .01440 \end{bmatrix}$$

TABLE 7.3 COMPUTER DATA

z_1 (Orders)	z_2 (Add-delete items)	Y (CPU time)
123.5	2.108	141.5
146.1	9.213	168.9
133.9	1.905	154.8
128.5	.815	146.5
151.5	1.061	172.8
136.2	8.603	160.1
92.0	1.125	108.5

SOURCE: Data taken from H. P. Artis, "Forecasting Computer Requirements: A Forecaster's Dilemma," Piscataway, N. J.: Bell Laboratories, 1979.

and $s = 1.204$. Consequently

$$\mathbf{z}_0'\hat{\boldsymbol{\beta}} = 8.42 + 1.08(130) + .42(7.5) = 151.97$$

and $s\sqrt{\mathbf{z}_0'(\mathbf{Z}'\mathbf{Z})^{-1}\mathbf{z}_0} = 1.204(.58928) = .71$. We have $t_4(.025) = 2.776$, so the 95% confidence interval for the mean CPU time at $\mathbf{z}_0$ is

$$\mathbf{z}_0'\hat{\boldsymbol{\beta}} \pm t_4(.025)s\sqrt{\mathbf{z}_0'(\mathbf{Z}'\mathbf{Z})^{-1}\mathbf{z}_0} = 151.97 \pm 2.776(.71)$$

or (150.00, 153.94).

Since $s\sqrt{1 + \mathbf{z}_0'(\mathbf{Z}'\mathbf{Z})^{-1}\mathbf{z}_0} = (1.204)(1.16071) = 1.40$, a 95% prediction interval for the CPU time at a new facility with conditions $\mathbf{z}_0$ is

$$\mathbf{z}_0'\hat{\boldsymbol{\beta}} \pm t_4(.025)s\sqrt{1 + \mathbf{z}_0'(\mathbf{Z}'\mathbf{Z})^{-1}\mathbf{z}_0} = 151.97 \pm 2.776(1.40)$$

or (148.08, 155.86). ■

7.6 MODEL CHECKING AND OTHER ASPECTS OF REGRESSION

Does the Model Fit?

Assuming the model is "correct," we have used the estimated regression function to make inferences. Of course, it is imperative to examine the adequacy of the model *before* the estimated function becomes a permanent part of the decision-making apparatus.

All of the sample information on lack of fit is contained in the residuals

$$\begin{aligned}
\hat{\varepsilon}_1 &= y_1 - \hat{\beta}_0 - \hat{\beta}_1 z_{11} - \cdots - \hat{\beta}_r z_{1r} \\
\hat{\varepsilon}_2 &= y_2 - \hat{\beta}_0 - \hat{\beta}_1 z_{21} - \cdots - \hat{\beta}_r z_{2r} \\
&\vdots \\
\hat{\varepsilon}_n &= y_n - \hat{\beta}_0 - \hat{\beta}_1 z_{n1} - \cdots - \hat{\beta}_r z_{nr}
\end{aligned}$$

or

$$\hat{\boldsymbol{\varepsilon}} = \left[\mathbf{I} - \mathbf{Z}(\mathbf{Z}'\mathbf{Z})^{-1}\mathbf{Z}'\right]\mathbf{y} \tag{7-19}$$

If the model is valid, each residual $\hat{\varepsilon}_j$ is an estimate of the error ε_j, which is assumed to be a normal random variable with mean zero and variance σ^2. Although the residuals $\hat{\boldsymbol{\varepsilon}}$ have expected value $\mathbf{0}$, their covariance matrix $\sigma^2[\mathbf{I} - \mathbf{Z}(\mathbf{Z}'\mathbf{Z})^{-1}\mathbf{Z}']$ is not diagonal. Residuals have unequal variances and nonzero correlations. Fortunately the correlations are often small and the variances are nearly equal. Consequently, we expect the $\hat{\varepsilon}_j$ to look, approximately, like independent drawings from a normal distribution with a constant variance.

Residuals should be plotted in various ways to detect possible anomalies. For general diagnostic purposes these are the best graphs.

1. *Plot the residuals*, $\hat{\varepsilon}_j$, *against the predicted values*, $\hat{y}_j = \hat{\beta}_0 + \hat{\beta}_1 z_{j1} + \cdots + \hat{\beta}_r z_{jr}$. Departures from the model assumptions are typically indicated by two types of phenomena.
   (a) *A dependence of the residuals on the predicted value*. This is illustrated in Figure 7.2(a). The expected value $E(\mathbf{Y} \mid \mathbf{Z}) = \mathbf{Z}\boldsymbol{\beta}$ is incorrect and more terms must be included or transformations must be attempted, or both.
   (b) *The variance is not constant*. The pattern of residuals may be funnel-shaped, as in Figure 7.2(b), so that there is large variability for large $\hat{y}$ and small variability for small $\hat{y}$. If this is the case, the variance of the error is not constant and transformations or a weighted least squares approach (or both) are required (see Exercise 7.3).
   In Figure 7.2(d) the residuals form a horizontal band. This is ideal and indicates equal variances and no dependence on $\hat{y}$.
2. *Plot the residuals*, $\hat{\varepsilon}_j$, *against a predictor variable, such as* z_1, *or products of predictor variables, like* z_1^2 *or* $z_1 z_2$. A systematic pattern in these plots suggests the need for more terms in the model. This situation is illustrated in Figure 7.2(c).
3. *Q-Q plots and histograms*. Do the errors appear to be normally distributed? To answer this question the residuals, $\hat{\varepsilon}_j$, can be examined using the techniques discussed in Section 4.6. The Q-Q plots, histograms, and dot diagrams help to detect the presence of unusual observations or severe departures from normal-

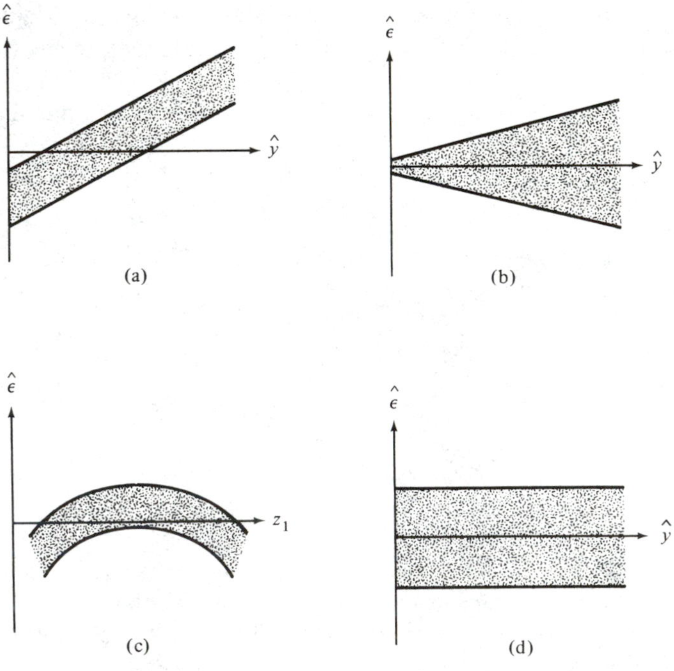

Figure 7.2 Residual plots.

ity that may require special attention in the analysis. If n is large, minor departures from normality will not greatly affect inferences about $\boldsymbol{\beta}$.

4. *Plot the residuals versus time.* The assumption of independence is crucial, but hard to check. If the data are naturally chronological, a plot of the residuals versus time may reveal a systematic pattern (a plot of the positions of the residuals in space may also reveal associations among the errors). For instance, residuals that increase over time indicate a strong positive dependence. A statistical test of independence can be constructed from the first autocorrelation,

$$r_1 = \frac{\sum_{j=2}^{n} \hat{\varepsilon}_j \hat{\varepsilon}_{j-1}}{\sum_{j=1}^{n} \hat{\varepsilon}_j^2} \tag{7-20}$$

of residuals from adjacent time periods. A popular test based on the statistic $\sum_{j=2}^{n} (\hat{\varepsilon}_j - \hat{\varepsilon}_{j-1})^2 / \sum_{j=1}^{n} \hat{\varepsilon}_j^2 \doteq 2(1 - r_1)$ is called the *Durbin–Watson test*. (See [14] or [7] for a description of this test and tables of critical values.)

Example 7.7

Three residual plots for the computer data discussed in Example 7.6 are shown in Figure 7.3. The sample size $n = 7$ is really too small to allow definitive judgments; however, it appears as if the regression assumptions are tenable. ■

If several observations of the response are available for the *same* values of the predictor variables, then a formal test for lack of fit can be carried out (see [7] for a discussion of the pure error lack-of-fit test).

If, after the diagnostic checks, no serious violations of the assumptions are detected, we can make inferences about $\boldsymbol{\beta}$ and future Y values with some assurance that we will not be misled.

Additional Problems in Linear Regression

We shall briefly discuss several important aspects of regression that deserve and receive extensive treatments in texts devoted to regression analysis (see [7], [18], and [5]).

Selecting predictor variables from a large set. In practice, it is often difficult to formulate an appropriate regression function immediately. Which predictor variables should be included? What form should the regression function take?

When the list of possible predictor variables is very large, all the variables cannot be included in the regression function. Techniques and computer programs

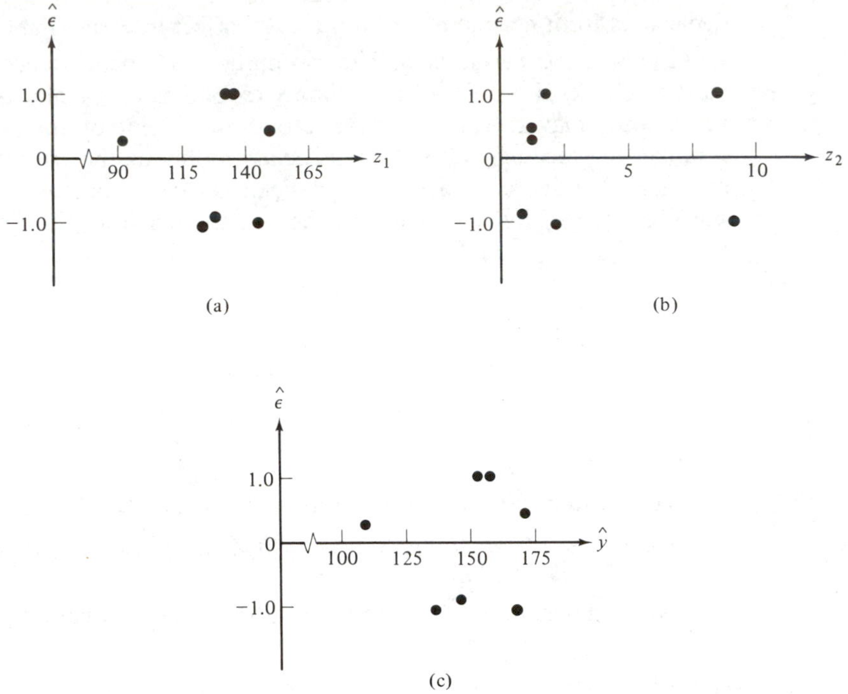

Figure 7.3 Residual plots for computer data.

designed to select the "best" subset of predictors are now readily available. The good ones try all subsets: z_1 alone, z_2 alone, $\ldots, z_1$ and $z_2, \ldots$. The best choice is decided by examining some criterion quantity like R^2 (see (7-9)). However, R^2 always increases with the inclusion of additional predictor variables. Although this problem can be circumvented by using the adjusted R^2, $\bar{R}^2 = 1 - (1 - R^2)(n - 1)/(n - r - 1)$, a better statistic for variable selection seems to be Mallow's C_p statistic (see [7]),

$$C_p = \left(\frac{\begin{array}{c}\text{(residual sum of squares for subset model with}\\ p \text{ parameters, including an intercept)}\end{array}}{\text{(residual variance for full model)}} \right) - (n - 2p)$$

A plot of the pairs (p, C_p), one for each subset of predictors, will indicate models that forecast the observed responses well. Good models typically have (p, C_p) coordinates near the 45° line. In Figure 7.4, we have circled the point corresponding to the "best" subset of predictor variables.

If the list of predictor variables is very long, cost considerations limit the number of models that can be examined. Another approach, called *stepwise regression* (see [7]), attempts to select important predictors without considering all the possibilities. The procedure can be described by listing the basic steps (algorithm) involved in the computations.

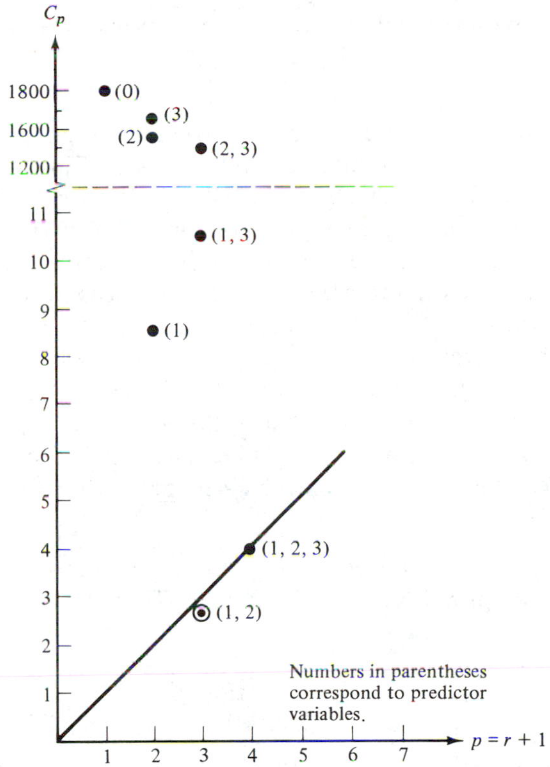

Figure 7.4 C_p plot for computer data with three predictor variables (z_1 = orders, z_2 = add–delete count, z_3 = number of items; see Example 7.6 and original source).

Step 1. All possible *simple* linear regressions are considered. The predictor variable that explains the largest significant proportion of the variation in Y (has the largest correlation with the response) is the first variable to enter the regression function.

Step 2. The next variable to enter is the one (out of those not yet included) that makes the largest significant contribution to the regression sum of squares. The significance of the contribution is determined by an F-test (see Result 7.6). The value of the F-statistic that must be exceeded before the contribution of a variable is deemed significant is often called the *F to enter*.

Step 3. Once an additional variable has been included in the equation, the individual contributions to the regression sum of squares of the other variables already in the equation are checked for significance using F-tests. If the F-statistic is less than the one (called the *F to remove*) corresponding to a prescribed significance level, the variable is deleted from the regression function.

Step 4. Steps 2 and 3 are repeated until all possible additions are nonsignificant and all possible deletions are significant. At this point the selection stops.

Because of the step-by-step procedure, there is no guarantee that this approach will select, for example, the best three variables for prediction. The (automatic) selection methods are not capable of indicating when transformations of variables are useful either.

Colinearity. If $\mathbf{Z}$ is not of full rank, some linear combination, such as $\mathbf{Za}$, must equal $\mathbf{0}$. In this situation the columns are said to be *colinear*. This implies that $\mathbf{Z}'\mathbf{Z}$ does not have an inverse. For most regression situations it is unlikely that $\mathbf{Za} = \mathbf{0}$ exactly. Yet, if linear combinations of the columns of $\mathbf{Z}$ exist which are nearly $\mathbf{0}$, the calculation of $(\mathbf{Z}'\mathbf{Z})^{-1}$ is numerically unstable. Typically, the diagonal entries of $(\mathbf{Z}'\mathbf{Z})^{-1}$ will be large. This yields large estimated variances for the $\hat{\beta}_i$'s and it is then difficult to detect the "significant" regression coefficients β_i. The problems caused by colinearity can be overcome somewhat by (1) deleting one of a pair of predictor variables that are *strongly* correlated or (2) relating the response Y to the *principal components* of the predictor variables; that is, the rows $\mathbf{Z}_j$ of $\mathbf{Z}$ are treated as a sample and the first few principal components are calculated as described in Section 8.3. The response Y is then regressed on these new predictor variables.

Bias caused by a misspecified model. Suppose some important predictor variables are omitted from the proposed regression model. That is, the true model has $\mathbf{Z} = [\mathbf{Z}_1 \,\vdots\, \mathbf{Z}_2]$ with rank $r + 1$ and

$$\underset{(n\times 1)}{\mathbf{Y}} = \left[\underset{(n\times(q+1))}{\mathbf{Z}_1} \;\vdots\; \underset{(n\times(r-q))}{\mathbf{Z}_2}\right] \begin{bmatrix} \underset{((q+1)\times 1)}{\boldsymbol{\beta}_{(1)}} \\ \hline \underset{((r-q)\times 1)}{\boldsymbol{\beta}_{(2)}} \end{bmatrix} + \underset{(n\times 1)}{\boldsymbol{\varepsilon}}$$

$$= \mathbf{Z}_1\boldsymbol{\beta}_{(1)} + \mathbf{Z}_2\boldsymbol{\beta}_{(2)} + \boldsymbol{\varepsilon} \qquad (7\text{-}21)$$

where $E(\boldsymbol{\varepsilon}) = \mathbf{0}$ and $\text{Var}(\boldsymbol{\varepsilon}) = \sigma^2\mathbf{I}$. However, the investigator unknowingly fits a model using only the first q predictors by minimizing the error sum of squares $(\mathbf{Y} - \mathbf{Z}_1\boldsymbol{\beta}_{(1)})'(\mathbf{Y} - \mathbf{Z}_1\boldsymbol{\beta}_{(1)})$. The least squares estimator of $\boldsymbol{\beta}_{(1)}$ is $\hat{\boldsymbol{\beta}}_{(1)} = (\mathbf{Z}_1'\mathbf{Z}_1)^{-1}\mathbf{Z}_1'\mathbf{Y}$. Then, unlike the situation when the model is correct,

$$E(\hat{\boldsymbol{\beta}}_{(1)}) = (\mathbf{Z}_1'\mathbf{Z}_1)^{-1}\mathbf{Z}_1'E(\mathbf{Y}) = (\mathbf{Z}_1'\mathbf{Z}_1)^{-1}\mathbf{Z}_1'(\mathbf{Z}_1\boldsymbol{\beta}_{(1)} + \mathbf{Z}_2\boldsymbol{\beta}_{(2)} + E(\boldsymbol{\varepsilon}))$$

$$= \boldsymbol{\beta}_{(1)} + (\mathbf{Z}_1'\mathbf{Z}_1)^{-1}\mathbf{Z}_1'\mathbf{Z}_2\boldsymbol{\beta}_{(2)} \qquad (7\text{-}22)$$

That is, $\hat{\boldsymbol{\beta}}_{(1)}$ is a biased estimator of $\boldsymbol{\beta}_{(1)}$ unless the columns of $\mathbf{Z}_1$ are perpendicular to those of $\mathbf{Z}_2$ (that is, $\mathbf{Z}_1'\mathbf{Z}_2 = \mathbf{0}$). If important variables appear to be missing from the model, the least squares estimates $\hat{\boldsymbol{\beta}}_{(1)}$ may be misleading.

7.7 MULTIVARIATE MULTIPLE REGRESSION

In this section we consider the problem of modeling the relationship between m responses, $Y_1, Y_2, \ldots, Y_m$ and a single set of predictor variables, $z_1, z_2, \ldots, z_r$. Each response is assumed to follow its own regression model so that

$$\begin{aligned} Y_1 &= \beta_{01} + \beta_{11}z_1 + \cdots + \beta_{r1}z_r + \varepsilon_1 \\ Y_2 &= \beta_{02} + \beta_{12}z_1 + \cdots + \beta_{r2}z_r + \varepsilon_2 \\ &\vdots \qquad\qquad \vdots \\ Y_m &= \beta_{0m} + \beta_{1m}z_1 + \cdots + \beta_{rm}z_r + \varepsilon_m \end{aligned} \tag{7-23}$$

The error term $\boldsymbol{\varepsilon} = [\varepsilon_1, \varepsilon_2, \ldots, \varepsilon_m]'$ has $E(\boldsymbol{\varepsilon}) = \mathbf{0}$ and $\text{Var}(\boldsymbol{\varepsilon}) = \boldsymbol{\Sigma}$. Thus the error terms associated with different responses may be correlated.

To establish notation conforming to the classical linear regression model, let $[z_{j0}, z_{j1}, \ldots, z_{jr}]$ denote the values of the predictor variables for the jth trial, let $\mathbf{Y}_j = [Y_{j1}, Y_{j2}, \ldots, Y_{jm}]'$ be the responses, and let $\boldsymbol{\varepsilon}_j = [\varepsilon_{j1}, \varepsilon_{j2}, \ldots, \varepsilon_{jm}]'$ be the errors. In matrix notation the design matrix

$$\underset{(n\times(r+1))}{\mathbf{Z}} = \begin{bmatrix} z_{10} & z_{11} & \cdots & z_{1r} \\ z_{20} & z_{21} & \cdots & z_{2r} \\ \vdots & \vdots & & \vdots \\ z_{n0} & z_{n1} & \cdots & z_{nr} \end{bmatrix}$$

is the same as that for the single response regression model [see (7-3)]. The other matrix quantities have multivariate counterparts. Set

$$\underset{(n\times m)}{\mathbf{Y}} = \begin{bmatrix} Y_{11} & Y_{12} & \cdots & Y_{1m} \\ Y_{21} & Y_{22} & \cdots & Y_{2m} \\ \vdots & \vdots & & \vdots \\ Y_{n1} & Y_{n2} & \cdots & Y_{nm} \end{bmatrix} = \left[\mathbf{Y}_{(1)} \mid \mathbf{Y}_{(2)} \mid \cdots \mid \mathbf{Y}_{(m)}\right]$$

$$\underset{((r+1)\times m)}{\boldsymbol{\beta}} = \begin{bmatrix} \beta_{01} & \beta_{02} & \cdots & \beta_{0m} \\ \beta_{11} & \beta_{12} & \cdots & \beta_{1m} \\ \vdots & \vdots & & \vdots \\ \beta_{r1} & \beta_{r2} & \cdots & \beta_{rm} \end{bmatrix} = \left[\boldsymbol{\beta}_{(1)} \mid \boldsymbol{\beta}_{(2)} \mid \cdots \mid \boldsymbol{\beta}_{(m)}\right]$$

and

$$\underset{(n\times m)}{\boldsymbol{\varepsilon}} = \begin{bmatrix} \varepsilon_{11} & \varepsilon_{12} & \cdots & \varepsilon_{1m} \\ \varepsilon_{21} & \varepsilon_{22} & \cdots & \varepsilon_{2m} \\ \vdots & \vdots & & \vdots \\ \varepsilon_{n1} & \varepsilon_{n2} & \cdots & \varepsilon_{nm} \end{bmatrix} = \left[\boldsymbol{\varepsilon}_{(1)} \mid \boldsymbol{\varepsilon}_{(2)} \mid \cdots \mid \boldsymbol{\varepsilon}_{(m)}\right]$$

$$= \begin{bmatrix} \boldsymbol{\varepsilon}_1' \\ \hline \boldsymbol{\varepsilon}_2' \\ \hline \vdots \\ \hline \boldsymbol{\varepsilon}_n' \end{bmatrix}$$

The *multivariate linear regression model* is

$$\underset{(n\times m)}{\mathbf{Y}} = \underset{(n\times(r+1))}{\mathbf{Z}} \underset{((r+1)\times m)}{\boldsymbol{\beta}} + \underset{(n\times m)}{\boldsymbol{\varepsilon}}$$

with (7-24)

$$E(\boldsymbol{\varepsilon}_{(i)}) = \mathbf{0}; \qquad \text{Cov}(\boldsymbol{\varepsilon}_{(i)}, \boldsymbol{\varepsilon}_{(k)}) = \sigma_{ik}\mathbf{I} \qquad i, k = 1, 2, \ldots, m$$

The m observations on the jth trial have covariance matrix $\boldsymbol{\Sigma} = \{\sigma_{ik}\}$, but observations from different trials are uncorrelated. Here $\boldsymbol{\beta}$ and σ_{ik} are unknown parameters and the design matrix $\mathbf{Z}$ has jth row $[z_{j0}, z_{j1}, \ldots, z_{jr}]$.

Simply stated, the ith response $\mathbf{Y}_{(i)}$ follows the linear regression model

$$\mathbf{Y}_{(i)} = \mathbf{Z}\boldsymbol{\beta}_{(i)} + \boldsymbol{\varepsilon}_{(i)}, \qquad i = 1, 2, \ldots, m \tag{7-25}$$

with $\text{Cov}(\boldsymbol{\varepsilon}_{(i)}) = \sigma_{ii}\mathbf{I}$. However, the errors for *different* responses on the *same* trial can be correlated.

Given the outcomes $\mathbf{Y}$ and the values of the predictor variables $\mathbf{Z}$, we determine the least squares estimates $\hat{\boldsymbol{\beta}}_{(i)}$ exclusively from the observations, $\mathbf{Y}_{(i)}$, on the ith response. Conforming to the single-response solution, we take

$$\hat{\boldsymbol{\beta}}_{(i)} = (\mathbf{Z}'\mathbf{Z})^{-1}\mathbf{Z}'\mathbf{Y}_{(i)} \tag{7-26}$$

Collecting these univariate least squares estimates, we obtain

$$\hat{\boldsymbol{\beta}} = \left[\hat{\boldsymbol{\beta}}_{(1)} \mid \hat{\boldsymbol{\beta}}_{(2)} \mid \cdots \mid \hat{\boldsymbol{\beta}}_{(m)}\right] = (\mathbf{Z}'\mathbf{Z})^{-1}\mathbf{Z}'\left[\mathbf{Y}_{(1)} \mid \mathbf{Y}_{(2)} \mid \cdots \mid \mathbf{Y}_{(m)}\right]$$

or

$$\hat{\boldsymbol{\beta}} = (\mathbf{Z}'\mathbf{Z})^{-1}\mathbf{Z}'\mathbf{Y} \tag{7-27}$$

For any choice of parameters $\mathbf{B} = \left[\mathbf{b}_{(1)} \mid \mathbf{b}_{(2)} \mid \cdots \mid \mathbf{b}_{(m)}\right]$, the matrix of errors is $\mathbf{Y} - \mathbf{ZB}$. The error sum of squares and crossproducts matrix is

$$(\mathbf{Y} - \mathbf{ZB})'(\mathbf{Y} - \mathbf{ZB})$$

$$= \begin{bmatrix} (\mathbf{Y}_{(1)} - \mathbf{Zb}_{(1)})'(\mathbf{Y}_{(1)} - \mathbf{Zb}_{(1)}) & \cdots & (\mathbf{Y}_{(1)} - \mathbf{Zb}_{(1)})'(\mathbf{Y}_{(m)} - \mathbf{Zb}_{(m)}) \\ \vdots & & \vdots \\ (\mathbf{Y}_{(m)} - \mathbf{Zb}_{(m)})'(\mathbf{Y}_{(1)} - \mathbf{Zb}_{(1)}) & \cdots & (\mathbf{Y}_{(m)} - \mathbf{Zb}_{(m)})'(\mathbf{Y}_{(m)} - \mathbf{Zb}_{(m)}) \end{bmatrix} \tag{7-28}$$

The selection $\mathbf{b}_{(i)} = \hat{\boldsymbol{\beta}}_{(i)}$ minimizes the ith diagonal sum of squares $(\mathbf{Y}_{(i)} - \mathbf{Zb}_{(i)})'(\mathbf{Y}_{(i)} - \mathbf{Zb}_{(i)})$. Consequently, $\text{tr}[(\mathbf{Y} - \mathbf{ZB})'(\mathbf{Y} - \mathbf{ZB})]$ is minimized by the choice $\mathbf{B} = \hat{\boldsymbol{\beta}}$. It can also be shown that the generalized variance $|(\mathbf{Y} - \mathbf{ZB})'(\mathbf{Y} - \mathbf{ZB})|$ is minimized by the least squares estimates $\hat{\boldsymbol{\beta}}$. (See Exercise 7.10 for an additional generalized sum of squares property.)

Using the least squares estimates $\hat{\boldsymbol{\beta}}$, we can form the matrices of

$$\begin{aligned} &\text{Predicted values:} \quad \hat{\mathbf{Y}} = \mathbf{Z}\hat{\boldsymbol{\beta}} = \mathbf{Z}(\mathbf{Z}'\mathbf{Z})^{-1}\mathbf{Z}'\mathbf{Y} \\ &\text{Residuals:} \quad \hat{\boldsymbol{\varepsilon}} = \mathbf{Y} - \hat{\mathbf{Y}} = \left[\mathbf{I} - \mathbf{Z}(\mathbf{Z}'\mathbf{Z})^{-1}\mathbf{Z}'\right]\mathbf{Y}. \end{aligned} \tag{7-29}$$

The orthogonality conditions among the residuals, predicted values, and columns of $\mathbf{Z}$, which hold in classical linear regression, hold in multivariate multiple regression. They follow from $\mathbf{Z}'[\mathbf{I} - \mathbf{Z}(\mathbf{Z}'\mathbf{Z})^{-1}\mathbf{Z}'] = \mathbf{Z}' - \mathbf{Z}' = \mathbf{0}$. Specifically,

$$\mathbf{Z}'\hat{\boldsymbol{\varepsilon}} = \mathbf{Z}'\left[\mathbf{I} - \mathbf{Z}(\mathbf{Z}'\mathbf{Z})^{-1}\mathbf{Z}'\right]\mathbf{Y} = \mathbf{0} \tag{7-30}$$

so the residuals $\hat{\boldsymbol{\varepsilon}}_{(i)}$ are perpendicular to the columns of $\mathbf{Z}$. Also

$$\hat{\mathbf{Y}}'\hat{\boldsymbol{\varepsilon}} = \hat{\boldsymbol{\beta}}'\mathbf{Z}'\left[\mathbf{I} - \mathbf{Z}(\mathbf{Z}'\mathbf{Z})^{-1}\mathbf{Z}'\right]\mathbf{Y} = \mathbf{0} \tag{7-31}$$

confirming that the predicted values $\hat{\mathbf{Y}}_{(i)}$ are perpendicular to all residual vectors $\hat{\boldsymbol{\varepsilon}}_{(k)}$. Because $\mathbf{Y} = \hat{\mathbf{Y}} + \hat{\boldsymbol{\varepsilon}}$,

$$\mathbf{Y}'\mathbf{Y} = (\hat{\mathbf{Y}} + \hat{\boldsymbol{\varepsilon}})'(\hat{\mathbf{Y}} + \hat{\boldsymbol{\varepsilon}}) = \hat{\mathbf{Y}}'\hat{\mathbf{Y}} + \hat{\boldsymbol{\varepsilon}}'\hat{\boldsymbol{\varepsilon}} + \mathbf{0} + \mathbf{0}'$$

or

$$\begin{matrix} \mathbf{Y}'\mathbf{Y} & = & \hat{\mathbf{Y}}'\hat{\mathbf{Y}} & + & \hat{\boldsymbol{\varepsilon}}'\hat{\boldsymbol{\varepsilon}} \\ \begin{pmatrix}\text{total sum of squares}\\ \text{and crossproducts}\end{pmatrix} & = & \begin{pmatrix}\text{predicted sum of squares}\\ \text{and crossproducts}\end{pmatrix} & + & \begin{pmatrix}\text{residual (error) sum}\\ \text{of squares and}\\ \text{crossproducts}\end{pmatrix} \end{matrix} \tag{7-32}$$

The residual sum of squares and crossproducts can also be written as

$$\hat{\boldsymbol{\varepsilon}}'\hat{\boldsymbol{\varepsilon}} = \mathbf{Y}'\mathbf{Y} - \hat{\mathbf{Y}}'\hat{\mathbf{Y}} = \mathbf{Y}'\mathbf{Y} - \hat{\boldsymbol{\beta}}'\mathbf{Z}'\mathbf{Z}\hat{\boldsymbol{\beta}} \tag{7-33}$$

Example 7.8

To illustrate the calculations of $\hat{\boldsymbol{\beta}}$, $\hat{\mathbf{Y}}$ and $\hat{\boldsymbol{\varepsilon}}$, we fit a straight-line regression model

$$Y_{j1} = \beta_{01} + \beta_{11}z_{j1} + \varepsilon_{j1}$$
$$Y_{j2} = \beta_{02} + \beta_{12}z_{j1} + \varepsilon_{j2}, \qquad j = 1, 2, \ldots, 5$$

to two responses Y_1 and Y_2 using the data in Example 7.3. These data augmented by observations on an additional response are

z_1	0	1	2	3	4
y_1	1	4	3	8	9
y_2	−1	−1	2	3	2

The design matrix $\mathbf{Z}$ remains unchanged from the single-response problem. We find

$$\mathbf{Z}' = \begin{bmatrix} 1 & 1 & 1 & 1 & 1 \\ 0 & 1 & 2 & 3 & 4 \end{bmatrix} \qquad (\mathbf{Z}'\mathbf{Z})^{-1} = \begin{bmatrix} .6 & -.2 \\ -.2 & .1 \end{bmatrix}$$

and

$$\mathbf{Z}'\mathbf{y}_{(2)} = \begin{bmatrix} 1 & 1 & 1 & 1 & 1 \\ 0 & 1 & 2 & 3 & 4 \end{bmatrix} \begin{bmatrix} -1 \\ -1 \\ 2 \\ 3 \\ 2 \end{bmatrix} = \begin{bmatrix} 5 \\ 20 \end{bmatrix}$$

so

$$\hat{\boldsymbol{\beta}}_{(2)} = (\mathbf{Z}'\mathbf{Z})^{-1}\mathbf{Z}'\mathbf{y}_{(2)} = \begin{bmatrix} .6 & -.2 \\ -.2 & .1 \end{bmatrix} \begin{bmatrix} 5 \\ 20 \end{bmatrix} = \begin{bmatrix} -1 \\ 1 \end{bmatrix}$$

From Example 7.3,

$$\hat{\boldsymbol{\beta}}_{(1)} = (\mathbf{Z}'\mathbf{Z})^{-1}\mathbf{Z}'\mathbf{y}_{(1)} = \begin{bmatrix} 1 \\ 2 \end{bmatrix}$$

Hence

$$\hat{\boldsymbol{\beta}} = \left[\hat{\boldsymbol{\beta}}_{(1)} \mid \hat{\boldsymbol{\beta}}_{(2)}\right] = \begin{bmatrix} 1 & -1 \\ 2 & 1 \end{bmatrix} = (\mathbf{Z}'\mathbf{Z})^{-1}\mathbf{Z}'\left[\mathbf{y}_{(1)} \mid \mathbf{y}_{(2)}\right]$$

The fitted values are generated from $\hat{y}_1 = 1 + 2z_1$ and $\hat{y}_2 = -1 + z_1$. Collectively,

$$\hat{\mathbf{Y}} = \mathbf{Z}\hat{\boldsymbol{\beta}} = \begin{bmatrix} 1 & 0 \\ 1 & 1 \\ 1 & 2 \\ 1 & 3 \\ 1 & 4 \end{bmatrix} \begin{bmatrix} 1 & -1 \\ 2 & 1 \end{bmatrix} = \begin{bmatrix} 1 & -1 \\ 3 & 0 \\ 5 & 1 \\ 7 & 2 \\ 9 & 3 \end{bmatrix}$$

and

$$\hat{\boldsymbol{\varepsilon}} = \mathbf{Y} - \hat{\mathbf{Y}} = \begin{bmatrix} 0 & 1 & -2 & 1 & 0 \\ 0 & -1 & 1 & 1 & -1 \end{bmatrix}'$$

Note

$$\hat{\boldsymbol{\varepsilon}}'\hat{\mathbf{Y}} = \begin{bmatrix} 0 & 1 & -2 & 1 & 0 \\ 0 & -1 & 1 & 1 & -1 \end{bmatrix} \begin{bmatrix} 1 & -1 \\ 3 & 0 \\ 5 & 1 \\ 7 & 2 \\ 9 & 3 \end{bmatrix} = \begin{bmatrix} 0 & 0 \\ 0 & 0 \end{bmatrix}$$

Since

$$\mathbf{Y}'\mathbf{Y} = \begin{bmatrix} 1 & 4 & 3 & 8 & 9 \\ -1 & -1 & 2 & 3 & 2 \end{bmatrix} \begin{bmatrix} 1 & -1 \\ 4 & -1 \\ 3 & 2 \\ 8 & 3 \\ 9 & 2 \end{bmatrix} = \begin{bmatrix} 171 & 43 \\ 43 & 19 \end{bmatrix}$$

$$\hat{\mathbf{Y}}'\hat{\mathbf{Y}} = \begin{bmatrix} 165 & 45 \\ 45 & 15 \end{bmatrix} \quad \text{and} \quad \hat{\boldsymbol{\varepsilon}}'\hat{\boldsymbol{\varepsilon}} = \begin{bmatrix} 6 & -2 \\ -2 & 4 \end{bmatrix}$$

the sum of squares and crossproducts breakup

$$\mathbf{Y}'\mathbf{Y} = \hat{\mathbf{Y}}'\hat{\mathbf{Y}} + \hat{\boldsymbol{\varepsilon}}'\hat{\boldsymbol{\varepsilon}}$$

is easily verified. ■

Result 7.9. For the least squares estimator $\hat{\boldsymbol{\beta}} = [\hat{\boldsymbol{\beta}}_{(1)} \mid \hat{\boldsymbol{\beta}}_{(2)} \mid \cdots \mid \hat{\boldsymbol{\beta}}_{(m)}]$ determined under the multivariate multiple regression model (7-24) with full rank($\mathbf{Z}$) $= r + 1 < n$,

$$E\left(\hat{\boldsymbol{\beta}}_{(i)}\right) = \boldsymbol{\beta}_{(i)} \quad \text{or} \quad E\left(\hat{\boldsymbol{\beta}}\right) = \boldsymbol{\beta}$$

and

$$\text{Cov}(\hat{\boldsymbol{\beta}}_{(i)}, \hat{\boldsymbol{\beta}}_{(k)}) = \sigma_{ik}(\mathbf{Z}'\mathbf{Z})^{-1}, \qquad i, k = 1, 2, \ldots, r+1$$

The residuals $\hat{\boldsymbol{\varepsilon}} = [\hat{\boldsymbol{\varepsilon}}_{(1)} \mid \hat{\boldsymbol{\varepsilon}}_{(2)} \mid \cdots \mid \hat{\boldsymbol{\varepsilon}}_{(m)}] = \mathbf{Y} - \mathbf{Z}\hat{\boldsymbol{\beta}}$ satisfy $E(\hat{\boldsymbol{\varepsilon}}_{(i)}) = \mathbf{0}$ and $E(\hat{\boldsymbol{\varepsilon}}'_{(i)}\hat{\boldsymbol{\varepsilon}}_{(k)}) = (n - r - 1)\sigma_{ik}$ so

$$E(\hat{\boldsymbol{\varepsilon}}) = \mathbf{0} \quad \text{and} \quad E\left(\frac{\hat{\boldsymbol{\varepsilon}}'\hat{\boldsymbol{\varepsilon}}}{(n-r-1)}\right) = \boldsymbol{\Sigma}$$

Also, $\hat{\boldsymbol{\varepsilon}}$ and $\hat{\boldsymbol{\beta}}$ are uncorrelated.

Proof. The ith response follows the multiple regression model

$$\mathbf{Y}_{(i)} = \mathbf{Z}\boldsymbol{\beta}_{(i)} + \boldsymbol{\varepsilon}_{(i)}, \; E(\boldsymbol{\varepsilon}_{(i)}) = \mathbf{0}, \text{ and } E(\boldsymbol{\varepsilon}_{(i)}\boldsymbol{\varepsilon}'_{(i)}) = \sigma_{ii}\mathbf{I}$$

Also, as in (7-10),

$$\hat{\boldsymbol{\beta}}_{(i)} - \boldsymbol{\beta}_{(i)} = (\mathbf{Z}'\mathbf{Z})^{-1}\mathbf{Z}'\mathbf{Y}_{(i)} - \boldsymbol{\beta}_{(i)} = (\mathbf{Z}'\mathbf{Z})^{-1}\mathbf{Z}'\boldsymbol{\varepsilon}_{(i)} \tag{7-34}$$

and

$$\hat{\boldsymbol{\varepsilon}}_{(i)} = \mathbf{Y}_{(i)} - \hat{\mathbf{Y}}_{(i)} = [\mathbf{I} - \mathbf{Z}(\mathbf{Z}'\mathbf{Z})^{-1}\mathbf{Z}']\mathbf{Y}_{(i)} = [\mathbf{I} - \mathbf{Z}(\mathbf{Z}'\mathbf{Z})^{-1}\mathbf{Z}']\boldsymbol{\varepsilon}_{(i)}$$

so $E(\hat{\boldsymbol{\beta}}_{(i)}) = \boldsymbol{\beta}_{(i)}$ and $E(\hat{\boldsymbol{\varepsilon}}_{(i)}) = \mathbf{0}$.

Next

$$\begin{aligned}
\text{Cov}(\hat{\boldsymbol{\beta}}_{(i)}, \hat{\boldsymbol{\beta}}_{(k)}) &= E(\hat{\boldsymbol{\beta}}_{(i)} - \boldsymbol{\beta}_{(i)})(\hat{\boldsymbol{\beta}}_{(k)} - \boldsymbol{\beta}_{(k)})' \\
&= (\mathbf{Z}'\mathbf{Z})^{-1}\mathbf{Z}'E(\boldsymbol{\varepsilon}_{(i)}\boldsymbol{\varepsilon}'_{(k)})\mathbf{Z}(\mathbf{Z}'\mathbf{Z})^{-1} = \sigma_{ik}(\mathbf{Z}'\mathbf{Z})^{-1}
\end{aligned}$$

Using Result 4.9 and the proof of Result 7.2, with $\mathbf{U}$ any random vector and $\mathbf{A}$ a fixed matrix, $E[\mathbf{U}'\mathbf{A}\mathbf{U}] = E[\text{tr}(\mathbf{A}\mathbf{U}\mathbf{U}')] = \text{tr}[\mathbf{A}\,E(\mathbf{U}\mathbf{U}')]$. Consequently,

$$\begin{aligned}
E(\hat{\boldsymbol{\varepsilon}}'_{(i)}\hat{\boldsymbol{\varepsilon}}_{(k)}) &= E(\boldsymbol{\varepsilon}'_{(i)}(\mathbf{I} - \mathbf{Z}(\mathbf{Z}'\mathbf{Z})^{-1}\mathbf{Z}')\boldsymbol{\varepsilon}_{(k)}) = \text{tr}[(\mathbf{I} - \mathbf{Z}(\mathbf{Z}'\mathbf{Z})^{-1}\mathbf{Z}')\sigma_{ik}\mathbf{I}] \\
&= \sigma_{ik}\text{tr}[(\mathbf{I} - \mathbf{Z}(\mathbf{Z}'\mathbf{Z})^{-1}\mathbf{Z}')] = \sigma_{ik}(n - r - 1)
\end{aligned}$$

as in the proof of Result 7.2. Dividing each entry $\hat{\boldsymbol{\varepsilon}}'_{(i)}\hat{\boldsymbol{\varepsilon}}_{(k)}$ of $\hat{\boldsymbol{\varepsilon}}'\hat{\boldsymbol{\varepsilon}}$ by $n - r - 1$, we obtain the unbiased estimator of $\boldsymbol{\Sigma}$. Finally,

$$\begin{aligned}
\text{Cov}(\hat{\boldsymbol{\beta}}_{(i)}, \hat{\boldsymbol{\varepsilon}}_{(k)}) &= E[(\mathbf{Z}'\mathbf{Z})^{-1}\mathbf{Z}'\boldsymbol{\varepsilon}_{(i)}\boldsymbol{\varepsilon}'_{(k)}(\mathbf{I} - \mathbf{Z}(\mathbf{Z}'\mathbf{Z})^{-1}\mathbf{Z}')] \\
&= (\mathbf{Z}'\mathbf{Z})^{-1}\mathbf{Z}'E(\boldsymbol{\varepsilon}_{(i)}\boldsymbol{\varepsilon}'_{(k)})(\mathbf{I} - \mathbf{Z}(\mathbf{Z}'\mathbf{Z})^{-1}\mathbf{Z}') \\
&= (\mathbf{Z}'\mathbf{Z})^{-1}\mathbf{Z}'\sigma_{ik}\mathbf{I}(\mathbf{I} - \mathbf{Z}(\mathbf{Z}'\mathbf{Z})^{-1}\mathbf{Z}') \\
&= \sigma_{ik}((\mathbf{Z}'\mathbf{Z})^{-1}\mathbf{Z}' - (\mathbf{Z}'\mathbf{Z})^{-1}\mathbf{Z}') = \mathbf{0}
\end{aligned}$$

so each element of $\hat{\boldsymbol{\beta}}$ is uncorrelated with each element of $\hat{\boldsymbol{\varepsilon}}$. ■

The mean vectors and covariance matrices determined in Result 7.9 enable us to obtain the sampling properties of the least squares predictors.

We first consider the problem of estimating the mean vector when the predictor variables have the values $\mathbf{z}_0 = [1, z_{01}, \ldots, z_{0r}]'$. The mean of the ith response variable is $\mathbf{z}_0'\boldsymbol{\beta}_{(i)}$, and this is estimated by $\mathbf{z}_0'\hat{\boldsymbol{\beta}}_{(i)}$, the ith component of the fitted regression relationship. Collectively, the *estimated linear response*

$$\mathbf{z}_0'\hat{\boldsymbol{\beta}} = \left[\mathbf{z}_0'\hat{\boldsymbol{\beta}}_{(1)} \mid \mathbf{z}_0'\hat{\boldsymbol{\beta}}_{(2)} \mid \cdots \mid \mathbf{z}_0'\hat{\boldsymbol{\beta}}_{(m)}\right] \tag{7-35}$$

is an unbiased estimator of $\mathbf{z}_0'\boldsymbol{\beta}$ since $E(\mathbf{z}_0'\hat{\boldsymbol{\beta}}_{(i)}) = \mathbf{z}_0'E(\hat{\boldsymbol{\beta}}_{(i)}) = \mathbf{z}_0'\boldsymbol{\beta}_{(i)}$ for each component. Using the covariance matrix for $\hat{\boldsymbol{\beta}}_{(i)}$ and $\hat{\boldsymbol{\beta}}_{(k)}$, the estimation errors, $\mathbf{z}_0'\boldsymbol{\beta}_{(i)} - \mathbf{z}_0'\hat{\boldsymbol{\beta}}_{(i)}$, have covariances

$$\begin{aligned} E\left[\mathbf{z}_0'\left(\boldsymbol{\beta}_{(i)} - \hat{\boldsymbol{\beta}}_{(i)}\right)\left(\boldsymbol{\beta}_{(k)} - \hat{\boldsymbol{\beta}}_{(k)}\right)'\mathbf{z}_0\right] &= \mathbf{z}_0'\left(E\left(\boldsymbol{\beta}_{(i)} - \hat{\boldsymbol{\beta}}_{(i)}\right)\left(\boldsymbol{\beta}_{(k)} - \hat{\boldsymbol{\beta}}_{(k)}\right)'\right)\mathbf{z}_0 \\ &= \sigma_{ik}\mathbf{z}_0'(\mathbf{Z}'\mathbf{Z})^{-1}\mathbf{z}_0 \end{aligned} \tag{7-36}$$

The related problem is that of forecasting a new observation vector, $\mathbf{Y}_0 = [Y_{01}, Y_{02}, \ldots, Y_{0m}]'$ at $\mathbf{z}_0$. According to the regression model, $Y_{0i} = \mathbf{z}_0'\boldsymbol{\beta}_{(i)} + \varepsilon_{0i}$ where the "new" error $\boldsymbol{\varepsilon}_0 = [\varepsilon_{01}, \varepsilon_{02}, \ldots, \varepsilon_{0m}]'$ is independent of the errors $\boldsymbol{\varepsilon}$ and satisfies $E(\varepsilon_{0i}) = 0$ and $E(\varepsilon_{0i}\varepsilon_{0k}) = \sigma_{ik}$. The *forecast error* for the ith component of $\mathbf{Y}_0$ is

$$\begin{aligned} Y_{0i} - \mathbf{z}_0'\hat{\boldsymbol{\beta}}_{(i)} &= Y_{0i} - \mathbf{z}_0'\boldsymbol{\beta}_{(i)} + \mathbf{z}_0'\boldsymbol{\beta}_{(i)} - \mathbf{z}_0'\hat{\boldsymbol{\beta}}_{(i)} \\ &= \varepsilon_{0i} - \mathbf{z}_0'\left(\hat{\boldsymbol{\beta}}_{(i)} - \boldsymbol{\beta}_{(i)}\right) \end{aligned}$$

so $E(Y_{0i} - \mathbf{z}_0'\hat{\boldsymbol{\beta}}_{(i)}) = E(\varepsilon_{0i}) - \mathbf{z}_0'E(\hat{\boldsymbol{\beta}}_{(i)} - \boldsymbol{\beta}_{(i)}) = 0$, indicating $\mathbf{z}_0'\hat{\boldsymbol{\beta}}_{(i)}$ is an *unbiased predictor* of Y_{0i}. The forecast errors have covariances

$$\begin{aligned} E\left(Y_{0i} - \mathbf{z}_0'\hat{\boldsymbol{\beta}}_{(i)}\right)\left(Y_{0k} - \mathbf{z}_0'\hat{\boldsymbol{\beta}}_{(k)}\right) &= E\left(\varepsilon_{0i} - \mathbf{z}_0'\left(\hat{\boldsymbol{\beta}}_{(i)} - \boldsymbol{\beta}_{(i)}\right)\right)\left(\varepsilon_{0k} - \mathbf{z}_0'\left(\hat{\boldsymbol{\beta}}_{(k)} - \boldsymbol{\beta}_{(k)}\right)\right) \\ &= E(\varepsilon_{0i}\varepsilon_{0k}) + \mathbf{z}_0'E\left(\hat{\boldsymbol{\beta}}_{(i)} - \boldsymbol{\beta}_{(i)}\right)\left(\hat{\boldsymbol{\beta}}_{(k)} - \boldsymbol{\beta}_{(k)}\right)'\mathbf{z}_0 \\ &\quad - \mathbf{z}_0'E\left(\left(\hat{\boldsymbol{\beta}}_{(i)} - \boldsymbol{\beta}_{(i)}\right)\varepsilon_{0k}\right) - E\left(\varepsilon_{0i}\left(\hat{\boldsymbol{\beta}}_{(k)} - \boldsymbol{\beta}_{(k)}\right)'\right)\mathbf{z}_0 \\ &= \sigma_{ik}\left(1 + \mathbf{z}_0'(\mathbf{Z}'\mathbf{Z})^{-1}\mathbf{z}_0\right) \end{aligned} \tag{7-37}$$

Note that $E((\hat{\boldsymbol{\beta}}_{(i)} - \boldsymbol{\beta}_{(i)})\varepsilon_{0k}) = \mathbf{0}$ since $\hat{\boldsymbol{\beta}}_{(i)} = (\mathbf{Z}'\mathbf{Z})^{-1}\mathbf{Z}'\boldsymbol{\varepsilon}_{(i)} + \boldsymbol{\beta}_{(i)}$ is independent of $\boldsymbol{\varepsilon}_0$. A similar result holds for $E(\varepsilon_{0i}(\boldsymbol{\beta}_{(k)} - \boldsymbol{\beta}_{(k)})')$.

Maximum likelihood estimators and their distributions can be obtained when the errors, $\boldsymbol{\varepsilon}$, have a normal distribution.

Result 7.10. Let the multivariate multiple regression model in (7-24) hold with full rank$(\mathbf{Z}) = r + 1 < n$ and let the errors, $\boldsymbol{\varepsilon}$, have a normal distribution. Then

$$\hat{\boldsymbol{\beta}} = (\mathbf{Z}'\mathbf{Z})^{-1}\mathbf{Z}'\mathbf{Y}$$

is the maximum likelihood estimator of $\boldsymbol{\beta}$ and $\hat{\boldsymbol{\beta}}$ has a normal distribution with $E(\hat{\boldsymbol{\beta}}) = \boldsymbol{\beta}$ and $\text{Cov}(\hat{\boldsymbol{\beta}}_{(i)}, \hat{\boldsymbol{\beta}}_{(k)}) = \sigma_{ik}(\mathbf{Z}'\mathbf{Z})^{-1}$. Also, $\hat{\boldsymbol{\beta}}$ is independent of the maximum likelihood estimator of $\boldsymbol{\Sigma}$ given by

$$\hat{\boldsymbol{\Sigma}} = \frac{\hat{\boldsymbol{\varepsilon}}'\hat{\boldsymbol{\varepsilon}}}{n} = \frac{(\mathbf{Y} - \mathbf{Z}\hat{\boldsymbol{\beta}})'(\mathbf{Y} - \mathbf{Z}\hat{\boldsymbol{\beta}})}{n}$$

and

$$n\hat{\boldsymbol{\Sigma}} \quad \text{is distributed as} \quad W_{n-r-1}(\cdot \mid \boldsymbol{\Sigma})$$

Proof. According to the regression model, the likelihood is determined from the data $\mathbf{Y} = [\mathbf{Y}_1, \mathbf{Y}_2, \ldots, \mathbf{Y}_n]'$ whose rows are independent with $\mathbf{Y}_j$ distributed as $N_m(\boldsymbol{\beta}'\mathbf{z}_j, \boldsymbol{\Sigma})$. We first note that $\mathbf{Y} - \mathbf{Z}\boldsymbol{\beta} = [\mathbf{Y}_1 - \boldsymbol{\beta}'\mathbf{z}_1, \mathbf{Y}_2 - \boldsymbol{\beta}'\mathbf{z}_2, \ldots, \mathbf{Y}_n - \boldsymbol{\beta}'\mathbf{z}_n]'$ so

$$(\mathbf{Y} - \mathbf{Z}\boldsymbol{\beta})'(\mathbf{Y} - \mathbf{Z}\boldsymbol{\beta}) = \sum_{j=1}^{n} (\mathbf{Y}_j - \boldsymbol{\beta}'\mathbf{z}_j)(\mathbf{Y}_j - \boldsymbol{\beta}'\mathbf{z}_j)'$$

and

$$\begin{aligned}\sum_{j=1}^{n} (\mathbf{Y}_j - \boldsymbol{\beta}'\mathbf{z}_j)'\boldsymbol{\Sigma}^{-1}(\mathbf{Y}_j - \boldsymbol{\beta}'\mathbf{z}_j) &= \sum_{j=1}^{n} \operatorname{tr}\left[(\mathbf{Y}_j - \boldsymbol{\beta}'\mathbf{z}_j)'\boldsymbol{\Sigma}^{-1}(\mathbf{Y}_j - \boldsymbol{\beta}'\mathbf{z}_j)\right] \\ &= \sum_{j=1}^{n} \operatorname{tr}\left[\boldsymbol{\Sigma}^{-1}(\mathbf{Y}_j - \boldsymbol{\beta}'\mathbf{z}_j)(\mathbf{Y}_j - \boldsymbol{\beta}'\mathbf{z}_j)'\right] \\ &= \operatorname{tr}\left[\boldsymbol{\Sigma}^{-1}(\mathbf{Y} - \mathbf{Z}\boldsymbol{\beta})'(\mathbf{Y} - \mathbf{Z}\boldsymbol{\beta})\right]\end{aligned} \tag{7-38}$$

Another preliminary calculation will enable us to express the likelihood in a simple form. Since $\hat{\boldsymbol{\varepsilon}} = \mathbf{Y} - \mathbf{Z}\hat{\boldsymbol{\beta}}$ satisfies $\mathbf{Z}'\hat{\boldsymbol{\varepsilon}} = \mathbf{0}$ [see (7-30)],

$$\begin{aligned}(\mathbf{Y} - \mathbf{Z}\boldsymbol{\beta})'(\mathbf{Y} - \mathbf{Z}\boldsymbol{\beta}) &= \left[\mathbf{Y} - \mathbf{Z}\hat{\boldsymbol{\beta}} + \mathbf{Z}(\hat{\boldsymbol{\beta}} - \boldsymbol{\beta})\right]'\left[\mathbf{Y} - \mathbf{Z}\hat{\boldsymbol{\beta}} + \mathbf{Z}(\hat{\boldsymbol{\beta}} - \boldsymbol{\beta})\right] \\ &= (\mathbf{Y} - \mathbf{Z}\hat{\boldsymbol{\beta}})'(\mathbf{Y} - \mathbf{Z}\hat{\boldsymbol{\beta}}) + (\hat{\boldsymbol{\beta}} - \boldsymbol{\beta})'\mathbf{Z}'\mathbf{Z}(\hat{\boldsymbol{\beta}} - \boldsymbol{\beta}) \\ &= \hat{\boldsymbol{\varepsilon}}'\hat{\boldsymbol{\varepsilon}} + (\hat{\boldsymbol{\beta}} - \boldsymbol{\beta})'\mathbf{Z}'\mathbf{Z}(\hat{\boldsymbol{\beta}} - \boldsymbol{\beta})\end{aligned} \tag{7-39}$$

Using (7-38) and (7-39), we obtain the likelihood

$$\begin{aligned}L(\boldsymbol{\beta}, \boldsymbol{\Sigma}) &= \prod_{j=1}^{n} \frac{1}{(2\pi)^{m/2}} \frac{1}{|\boldsymbol{\Sigma}|^{1/2}} e^{-\frac{1}{2}(\mathbf{y}_j - \boldsymbol{\beta}'\mathbf{z}_j)'\boldsymbol{\Sigma}^{-1}(\mathbf{y}_j - \boldsymbol{\beta}'\mathbf{z}_j)} \\ &= \frac{1}{(2\pi)^{mn/2}} \frac{1}{|\boldsymbol{\Sigma}|^{n/2}} e^{-\frac{1}{2}\operatorname{tr}[\boldsymbol{\Sigma}^{-1}(\hat{\boldsymbol{\varepsilon}}'\hat{\boldsymbol{\varepsilon}} + (\hat{\boldsymbol{\beta}} - \boldsymbol{\beta})'\mathbf{Z}'\mathbf{Z}(\hat{\boldsymbol{\beta}} - \boldsymbol{\beta}))]} \\ &= \frac{1}{(2\pi)^{mn/2}} \frac{1}{|\boldsymbol{\Sigma}|^{n/2}} e^{-\frac{1}{2}\operatorname{tr}[\boldsymbol{\Sigma}^{-1}\hat{\boldsymbol{\varepsilon}}'\hat{\boldsymbol{\varepsilon}}] - \frac{1}{2}\operatorname{tr}[\mathbf{Z}(\hat{\boldsymbol{\beta}} - \boldsymbol{\beta})\boldsymbol{\Sigma}^{-1}(\hat{\boldsymbol{\beta}} - \boldsymbol{\beta})'\mathbf{Z}']}\end{aligned}$$

The matrix $\mathbf{Z}(\hat{\boldsymbol{\beta}} - \boldsymbol{\beta})\boldsymbol{\Sigma}^{-1}(\hat{\boldsymbol{\beta}} - \boldsymbol{\beta})'\mathbf{Z}'$ is of the form $\mathbf{A}'\mathbf{A}$ with $\mathbf{A} = \boldsymbol{\Sigma}^{-1/2}(\hat{\boldsymbol{\beta}} - \boldsymbol{\beta})'\mathbf{Z}'$, and, by Exercise 2.16, it is nonnegative definite. Therefore its eigenvalues are nonnegative. Since, by Result 4.9, $\operatorname{tr}[\mathbf{Z}(\hat{\boldsymbol{\beta}} - \boldsymbol{\beta})\boldsymbol{\Sigma}^{-1}(\hat{\boldsymbol{\beta}} - \boldsymbol{\beta})'\mathbf{Z}']$ is the sum of its eigenvalues, this trace will equal its minimum value, zero, if $\boldsymbol{\beta} = \hat{\boldsymbol{\beta}}$. This choice is unique because $\mathbf{Z}$ is of full rank and $\hat{\boldsymbol{\beta}}_{(i)} - \boldsymbol{\beta}_{(i)} \neq \mathbf{0}$ implies $\mathbf{Z}(\hat{\boldsymbol{\beta}}_{(i)} - \boldsymbol{\beta}_{(i)}) \neq \mathbf{0}$, in which case $\operatorname{tr}[\mathbf{Z}(\hat{\boldsymbol{\beta}} - \boldsymbol{\beta})\boldsymbol{\Sigma}^{-1}(\hat{\boldsymbol{\beta}} - \boldsymbol{\beta})'\mathbf{Z}'] \geq \mathbf{c}'\boldsymbol{\Sigma}^{-1}\mathbf{c} > 0$, where $\mathbf{c}'$ is any nonzero row of $\mathbf{Z}(\hat{\boldsymbol{\beta}} - \boldsymbol{\beta})$. Applying Result 4.10 with $\mathbf{B} = \hat{\boldsymbol{\varepsilon}}'\hat{\boldsymbol{\varepsilon}}$, $b = n/2$, and $p = m$, we find that $\hat{\boldsymbol{\beta}}$ and $\hat{\boldsymbol{\Sigma}} = \hat{\boldsymbol{\varepsilon}}'\hat{\boldsymbol{\varepsilon}}/n$ are the maximum likelihood estimators of $\boldsymbol{\beta}$ and $\boldsymbol{\Sigma}$, respectively,

and

$$L(\hat{\boldsymbol\beta}, \hat{\boldsymbol\Sigma}) = \frac{1}{(2\pi)^{mn/2}} \frac{(n)^{mn/2}}{|\hat{\boldsymbol\varepsilon}'\hat{\boldsymbol\varepsilon}|^{n/2}} e^{-nm/2} = \frac{e^{-nm/2}}{(2\pi)^{mn/2}|\hat{\boldsymbol\Sigma}|^{n/2}} \tag{7-40}$$

It remains to establish the distributional results. From (7-34) we know that $\hat{\boldsymbol\beta}_{(i)}$ and $\hat{\boldsymbol\varepsilon}_{(i)}$ are linear combinations of the elements of $\boldsymbol\varepsilon$. Specifically,

$$\hat{\boldsymbol\beta}_{(i)} = (\mathbf{Z}'\mathbf{Z})^{-1}\mathbf{Z}'\boldsymbol\varepsilon_{(i)} + \boldsymbol\beta_{(i)}$$
$$\hat{\boldsymbol\varepsilon}_{(i)} = \left[\mathbf{I} - \mathbf{Z}(\mathbf{Z}'\mathbf{Z})^{-1}\mathbf{Z}'\right]\boldsymbol\varepsilon_{(i)}, \qquad i = 1, 2, \ldots, m$$

Therefore, by Result 4.3, $\hat{\boldsymbol\beta}_{(1)}, \hat{\boldsymbol\beta}_{(2)}, \ldots, \hat{\boldsymbol\beta}_{(m)}, \hat{\boldsymbol\varepsilon}_{(1)}, \hat{\boldsymbol\varepsilon}_{(2)}, \ldots, \hat{\boldsymbol\varepsilon}_{(m)}$ are jointly normal. Their mean vectors and covariance matrices are given in Result 7.9. Since $\hat{\boldsymbol\varepsilon}$ and $\hat{\boldsymbol\beta}$ have a zero covariance matrix, by Result 4.5 they are independent. Further, as in (7-13), $[\mathbf{I} - \mathbf{Z}(\mathbf{Z}'\mathbf{Z})^{-1}\mathbf{Z}'] = \sum_{\ell=1}^{n-r-1} \mathbf{e}_\ell\mathbf{e}_\ell'$, where $\mathbf{e}_\ell'\mathbf{e}_k = 0$, $\ell \neq k$, and $\mathbf{e}_\ell'\mathbf{e}_\ell = 1$. Set $\mathbf{V}_\ell = \boldsymbol\varepsilon'\mathbf{e}_\ell = [\boldsymbol\varepsilon_{(1)}'\mathbf{e}_\ell, \boldsymbol\varepsilon_{(2)}'\mathbf{e}_\ell, \ldots, \boldsymbol\varepsilon_{(m)}'\mathbf{e}_\ell]' = e_{\ell 1}\boldsymbol\varepsilon_1 + e_{\ell 2}\boldsymbol\varepsilon_2 + \cdots + e_{\ell n}\boldsymbol\varepsilon_n$. Because $\mathbf{V}_\ell$, $\ell = 1, 2, \ldots, n-r-1$, are linear combinations of the elements of $\boldsymbol\varepsilon$, they have a joint normal distribution with $E(\mathbf{V}_\ell) = E(\boldsymbol\varepsilon')\mathbf{e}_\ell = \mathbf{0}$. Also, by Result 4.8, $\mathbf{V}_\ell$ and $\mathbf{V}_k$ have covariance matrix $(\mathbf{e}_\ell'\mathbf{e}_k)\boldsymbol\Sigma = (0)\boldsymbol\Sigma = \mathbf{0}$ if $\ell \neq k$. Consequently, the $\mathbf{V}_\ell$ are independently distributed as $N_m(\mathbf{0}, \boldsymbol\Sigma)$. Finally

$$\hat{\boldsymbol\varepsilon}'\hat{\boldsymbol\varepsilon} = \boldsymbol\varepsilon'\left[\mathbf{I} - \mathbf{Z}(\mathbf{Z}'\mathbf{Z})^{-1}\mathbf{Z}'\right]\boldsymbol\varepsilon = \sum_{\ell=1}^{n-r-1} \boldsymbol\varepsilon'\mathbf{e}_\ell\mathbf{e}_\ell'\boldsymbol\varepsilon = \sum_{\ell=1}^{n-r-1} \mathbf{V}_\ell\mathbf{V}_\ell'$$

which has the $W_{n-r-1}(\cdot\,|\,\boldsymbol\Sigma)$ distribution by (4-22). ■

Result 7.10 provides additional support for using least squares estimates. When the errors are normally distributed, $\hat{\boldsymbol\beta}$ and $\hat{\boldsymbol\varepsilon}'\hat{\boldsymbol\varepsilon}/n$ are the maximum likelihood estimators of $\boldsymbol\beta$ and $\boldsymbol\Sigma$, respectively. Therefore, for large samples, they have the smallest possible variances.

Comment. The multivariate multiple regression model poses no new estimation problems. Least squares (maximum likelihood) estimates, $\hat{\boldsymbol\beta}_{(i)} = (\mathbf{Z}'\mathbf{Z})^{-1}\mathbf{Z}'\mathbf{y}_{(i)}$, are computed individually for each response variable. Note, however, the model requires that the *same* predictor variables be used for all responses.

Once a multivariate multiple regression model has been fit to the data, it should be subjected to the diagnostic checks described in Section 7.6 for the single-response model. The residual vector $[\hat{\varepsilon}_{j1}, \hat{\varepsilon}_{j2}, \ldots, \hat{\varepsilon}_{jm}]$ can be examined for normality or outliers using the techniques in Section 4.6.

The remainder of this section is devoted to brief discussions of inference for the normal theory multivariate multiple regression model. Extended accounts of these procedures appear in [1] and [20].

The multiresponse analog of (7-15), the hypothesis that the responses do not depend on $z_{q+1}, z_{q+2}, \ldots, z_r$, becomes

$$H_0: \boldsymbol{\beta}_{(2)} = \mathbf{0} \quad \text{where} \quad \boldsymbol{\beta} = \begin{bmatrix} \underset{((q+1)\times m)}{\boldsymbol{\beta}_{(1)}} \\ \hline \underset{((r-q)\times m)}{\boldsymbol{\beta}_{(2)}} \end{bmatrix} \tag{7-41}$$

Setting $\mathbf{Z} = \left[\underset{(n\times(q+1))}{\mathbf{Z}_1} \;\vdots\; \underset{(n\times(r-q))}{\mathbf{Z}_2} \right]$, the general model can be written as

$$E(\mathbf{Y}) = \mathbf{Z}\boldsymbol{\beta} = [\mathbf{Z}_1 \;\vdots\; \mathbf{Z}_2] \begin{bmatrix} \boldsymbol{\beta}_{(1)} \\ \hline \boldsymbol{\beta}_{(2)} \end{bmatrix} = \mathbf{Z}_1\boldsymbol{\beta}_{(1)} + \mathbf{Z}_2\boldsymbol{\beta}_{(2)}$$

Under H_0: $\boldsymbol{\beta}_{(2)} = \mathbf{0}$, $\mathbf{Y} = \mathbf{Z}_1\boldsymbol{\beta}_{(1)} + \boldsymbol{\varepsilon}$ and the likelihood ratio test of H_0 is based on the quantities involved in the

$$\begin{aligned} \text{extra sum of squares and crossproducts} \\ &= (\mathbf{Y} - \mathbf{Z}_1\hat{\boldsymbol{\beta}}_{(1)})'(\mathbf{Y} - \mathbf{Z}_1\hat{\boldsymbol{\beta}}_{(1)}) - (\mathbf{Y} - \mathbf{Z}\hat{\boldsymbol{\beta}})'(\mathbf{Y} - \mathbf{Z}\hat{\boldsymbol{\beta}}) \\ &= n(\hat{\boldsymbol{\Sigma}}_1 - \hat{\boldsymbol{\Sigma}}) \end{aligned}$$

where $\hat{\boldsymbol{\beta}}_{(1)} = (\mathbf{Z}_1'\mathbf{Z}_1)^{-1}\mathbf{Z}_1'\mathbf{Y}$ and $\hat{\boldsymbol{\Sigma}}_1 = (\mathbf{Y} - \mathbf{Z}_1\hat{\boldsymbol{\beta}}_{(1)})'(\mathbf{Y} - \mathbf{Z}_1\hat{\boldsymbol{\beta}}_{(1)})/n$.

From (7-40) the likelihood ratio, Λ, can be expressed in terms of generalized variances, so

$$\Lambda = \frac{\max_{\boldsymbol{\beta}_{(1)}, \boldsymbol{\Sigma}} L(\boldsymbol{\beta}_{(1)}, \boldsymbol{\Sigma})}{\max_{\boldsymbol{\beta}, \boldsymbol{\Sigma}} L(\boldsymbol{\beta}, \boldsymbol{\Sigma})} = \frac{L(\hat{\boldsymbol{\beta}}_{(1)}, \hat{\boldsymbol{\Sigma}}_1)}{L(\hat{\boldsymbol{\beta}}, \hat{\boldsymbol{\Sigma}})} = \left(\frac{|\hat{\boldsymbol{\Sigma}}|}{|\hat{\boldsymbol{\Sigma}}_1|} \right)^{n/2} \tag{7-42}$$

Equivalently, the Wilk's lambda statistic

$$\Lambda^{2/n} = \frac{|\hat{\boldsymbol{\Sigma}}|}{|\hat{\boldsymbol{\Sigma}}_1|}$$

can be used.

Result 7.11. Let the multivariate multiple regression model of (7-24) hold with $\mathbf{Z}$ of full rank $r + 1 < n$. Let the errors $\boldsymbol{\varepsilon}$ be normally distributed. Under H_0: $\boldsymbol{\beta}_{(2)} = \mathbf{0}$, $n\hat{\boldsymbol{\Sigma}}$ is distributed as $W_{n-r-1}(\cdot \mid \boldsymbol{\Sigma})$ independently of $n(\hat{\boldsymbol{\Sigma}}_1 - \hat{\boldsymbol{\Sigma}})$ which, in turn, is distributed as $W_{r-q}(\cdot \mid \boldsymbol{\Sigma})$. The likelihood ratio test of H_0 is equivalent to rejecting H_0 for large values of

$$-2\ln\Lambda = -n\ln\left(\frac{|\hat{\boldsymbol{\Sigma}}|}{|\hat{\boldsymbol{\Sigma}}_1|} \right) = -n\ln\frac{|n\hat{\boldsymbol{\Sigma}}|}{|n\hat{\boldsymbol{\Sigma}} + n(\hat{\boldsymbol{\Sigma}}_1 - \hat{\boldsymbol{\Sigma}})|}$$

For n large,[5] the modified statistic

$$-\left(n-r-1-\frac{1}{2}(m-q)\right)\ln\left(\frac{|\hat{\boldsymbol{\Sigma}}|}{|\hat{\boldsymbol{\Sigma}}_1|}\right)$$

has, to a close approximation, a chi-square distribution with $m(q+1)$ d.f.

Proof. We know that $\hat{\boldsymbol{\Sigma}} = \mathbf{Y}'(\mathbf{I}-\mathbf{Z}(\mathbf{Z}'\mathbf{Z})^{-1}\mathbf{Z}')\mathbf{Y}$ and, under H_0, $n\hat{\boldsymbol{\Sigma}}_1 = \mathbf{Y}'[\mathbf{I}-\mathbf{Z}_1(\mathbf{Z}_1'\mathbf{Z}_1)^{-1}\mathbf{Z}_1']\mathbf{Y}$ with $\mathbf{Y} = \mathbf{Z}_1\boldsymbol{\beta}_{(1)} + \boldsymbol{\varepsilon}$. Set $\mathbf{P} = [\mathbf{I}-\mathbf{Z}(\mathbf{Z}'\mathbf{Z})^{-1}\mathbf{Z}']$. Since $\mathbf{0} = [\mathbf{I}-\mathbf{Z}(\mathbf{Z}'\mathbf{Z})^{-1}\mathbf{Z}']\mathbf{Z} = [\mathbf{I}-\mathbf{Z}(\mathbf{Z}'\mathbf{Z})^{-1}\mathbf{Z}'][\mathbf{Z}_1 \mid \mathbf{Z}_2] = [\mathbf{P}\mathbf{Z}_1 \mid \mathbf{P}\mathbf{Z}_2]$, the columns of $\mathbf{Z}_1$ are perpendicular to $\mathbf{P}$. Thus we can write

$$n\hat{\boldsymbol{\Sigma}} = (\mathbf{Z}\boldsymbol{\beta}+\boldsymbol{\varepsilon})'\mathbf{P}(\mathbf{Z}\boldsymbol{\beta}+\boldsymbol{\varepsilon}) = \boldsymbol{\varepsilon}'\mathbf{P}\boldsymbol{\varepsilon}$$

$$n\hat{\boldsymbol{\Sigma}}_1 = \left(\mathbf{Z}_1\boldsymbol{\beta}_{(1)}+\boldsymbol{\varepsilon}\right)'\mathbf{P}_1\left(\mathbf{Z}_1\boldsymbol{\beta}_{(1)}+\boldsymbol{\varepsilon}\right) = \boldsymbol{\varepsilon}'\mathbf{P}_1\boldsymbol{\varepsilon}$$

where $\mathbf{P}_1 = \mathbf{I}-\mathbf{Z}_1(\mathbf{Z}_1'\mathbf{Z}_1)^{-1}\mathbf{Z}_1'$. Use the Gram–Schmidt process (see Result 2A.3) to construct the orthonormal vectors $[\mathbf{g}_1, \mathbf{g}_2, \ldots, \mathbf{g}_{q+1}] = \mathbf{G}$ from the columns of $\mathbf{Z}_1$. Continue, obtaining the orthonormal set from $[\mathbf{G}, \mathbf{Z}_2]$, and finally complete the set to n dimensions by constructing an arbitrary orthonormal set of $n-r-1$ vectors orthogonal to the previous vectors. Consequently we have

$$\underbrace{\mathbf{g}_1, \mathbf{g}_2, \ldots, \mathbf{g}_{q+1}}_{\substack{\text{from columns}\\ \text{of } \mathbf{Z}_1}},\ \underbrace{\mathbf{g}_{q+2}, \mathbf{g}_{q+3}, \ldots, \mathbf{g}_{r+1}}_{\substack{\text{from columns of } \mathbf{Z}_2\\ \text{but perpendicular}\\ \text{to columns of } \mathbf{Z}_1}},\ \underbrace{\mathbf{g}_{r+2}, \mathbf{g}_{r+3}, \ldots, \mathbf{g}_n}_{\substack{\text{arbitrary set of}\\ \text{orthonormal}\\ \text{vectors orthogonal}\\ \text{to columns of } \mathbf{Z}}}$$

Let $(\lambda, \mathbf{e})$ be an eigenvalue-eigenvector pair of $\mathbf{Z}_1(\mathbf{Z}_1'\mathbf{Z}_1)^{-1}\mathbf{Z}_1'$. Then, since $[\mathbf{Z}_1(\mathbf{Z}_1'\mathbf{Z}_1)^{-1}\mathbf{Z}_1'][\mathbf{Z}_1(\mathbf{Z}_1'\mathbf{Z}_1)^{-1}\mathbf{Z}_1'] = \mathbf{Z}_1(\mathbf{Z}_1'\mathbf{Z}_1)^{-1}\mathbf{Z}_1$, it follows that

$$\lambda\mathbf{e} = \mathbf{Z}_1(\mathbf{Z}_1'\mathbf{Z}_1)^{-1}\mathbf{Z}_1'\mathbf{e} = \left(\mathbf{Z}_1(\mathbf{Z}_1'\mathbf{Z}_1)^{-1}\mathbf{Z}_1'\right)^2\mathbf{e} = \lambda\left(\mathbf{Z}_1(\mathbf{Z}_1'\mathbf{Z}_1)^{-1}\mathbf{Z}_1'\right)\mathbf{e} = \lambda^2\mathbf{e}$$

and the eigenvalues of $\mathbf{Z}_1(\mathbf{Z}_1'\mathbf{Z}_1)^{-1}\mathbf{Z}_1'$ are 0 or 1. Moreover, $\text{tr}(\mathbf{Z}_1(\mathbf{Z}_1'\mathbf{Z}_1)^{-1}\mathbf{Z}_1') = \text{tr}((\mathbf{Z}_1'\mathbf{Z}_1)^{-1}\mathbf{Z}_1'\mathbf{Z}_1) = \text{tr}\left(\underset{(q+1)\times(q+1)}{\mathbf{I}}\right) = q+1 = \lambda_1 + \lambda_2 + \cdots + \lambda_{q+1}$, where $\lambda_1 \geq \lambda_2 \geq \cdots \geq \lambda_{q+1} > 0$ are the eigenvalues of $\mathbf{Z}_1(\mathbf{Z}_1'\mathbf{Z}_1)^{-1}\mathbf{Z}_1'$. This shows $\mathbf{Z}_1(\mathbf{Z}_1'\mathbf{Z}_1)^{-1}\mathbf{Z}_1'$ has $q+1$ eigenvalues equal to 1. Now $(\mathbf{Z}_1(\mathbf{Z}_1'\mathbf{Z}_1)^{-1}\mathbf{Z}_1')\mathbf{Z}_1 = \mathbf{Z}_1$, so any linear combination $\mathbf{Z}_1\mathbf{b}_\ell$, of unit length, is an eigenvector corresponding to the eigenvalue 1. The orthonormal vectors $\mathbf{g}_\ell$, $\ell = 1, 2, \ldots, q+1$, are therefore eigenvectors of $\mathbf{Z}_1(\mathbf{Z}_1'\mathbf{Z}_1)^{-1}\mathbf{Z}_1'$ since they are formed by taking particular linear combinations of the columns of $\mathbf{Z}_1$. By the spectral decomposition (2-16), we have $\mathbf{Z}_1(\mathbf{Z}_1'\mathbf{Z}_1)^{-1}\mathbf{Z}_1' = \sum_{\ell=1}^{q+1}\mathbf{g}_\ell\mathbf{g}_\ell'$. Similarly, by writing $(\mathbf{Z}(\mathbf{Z}'\mathbf{Z})^{-1}\mathbf{Z}')\mathbf{Z} = \mathbf{Z}$, it follows that the linear combination $\mathbf{Z}\mathbf{b}_\ell = \mathbf{g}_\ell$, for example, is an eigenvector of $\mathbf{Z}(\mathbf{Z}'\mathbf{Z})^{-1}\mathbf{Z}'$ with eigenvalue $\lambda = 1$ so $\mathbf{Z}(\mathbf{Z}'\mathbf{Z})^{-1}\mathbf{Z}' = \sum_{\ell=1}^{r+1}\mathbf{g}_\ell\mathbf{g}_\ell'$.

[5] Technically, both $n-r$ and $n-m$ should also be large to obtain a good chi-square approximation.

Continuing, $\mathbf{PZ} = [\mathbf{I} - \mathbf{Z}(\mathbf{Z}'\mathbf{Z})^{-1}\mathbf{Z}']\mathbf{Z} = \mathbf{Z} - \mathbf{Z} = \mathbf{0}$ so $\mathbf{g}_\ell = \mathbf{Zb}_\ell$, $\ell \le r + 1$, are eigenvectors of $\mathbf{P}$ with eigenvalues $\lambda = 0$. Also from the way the $\mathbf{g}_\ell$, $\ell > r + 1$, were constructed, $\mathbf{Z}'\mathbf{g}_\ell = \mathbf{0}$ so that $\mathbf{Pg}_\ell = \mathbf{g}_\ell$. Consequently, these $\mathbf{g}_\ell$'s are eigenvectors of $\mathbf{P}$ corresponding to the $n - r - 1$ unit eigenvalues. By the spectral decomposition (2-16), $\mathbf{P} = \sum_{\ell=r+2}^{n} \mathbf{g}_\ell\mathbf{g}'_\ell$ and

$$n\hat{\mathbf{\Sigma}} = \boldsymbol{\varepsilon}'\mathbf{P}\boldsymbol{\varepsilon} = \sum_{\ell=r+2}^{n} (\boldsymbol{\varepsilon}'\mathbf{g}_\ell)(\boldsymbol{\varepsilon}'\mathbf{g}_\ell)' = \sum_{\ell=r+2}^{n} \mathbf{V}_\ell\mathbf{V}'_\ell$$

where, because $\text{Cov}(V_{i\ell}, V_{kj}) = E(\mathbf{g}'_\ell\boldsymbol{\varepsilon}_{(i)}\boldsymbol{\varepsilon}'_{(k)}\mathbf{g}_j) = \sigma_{ik}\mathbf{g}'_\ell\mathbf{g}_j = 0$, $\ell \ne j$, the $\boldsymbol{\varepsilon}'\mathbf{g}_\ell = \mathbf{V}_\ell = [V_{1\ell}, \ldots, V_{i\ell}, \ldots, V_{m\ell}]'$ are independently distributed as $N_m(\mathbf{0}, \mathbf{\Sigma})$. Consequently, by (4-22), $n\hat{\mathbf{\Sigma}}$ is distributed as $W_{n-r-1}(\cdot \mid \mathbf{\Sigma})$. In the same manner

$$\mathbf{P}_1\mathbf{g}_\ell = \begin{cases} \mathbf{g}_\ell & \ell > q + 1 \\ \mathbf{0} & \ell \le q + 1 \end{cases}$$

so $\mathbf{P}_1 = \sum_{\ell=q+2}^{n} \mathbf{g}_\ell\mathbf{g}'_\ell$. We can write the extra sum of squares and crossproducts

$$n(\hat{\mathbf{\Sigma}}_1 - \hat{\mathbf{\Sigma}}) = \boldsymbol{\varepsilon}'(\mathbf{P}_1 - \mathbf{P})\boldsymbol{\varepsilon} = \sum_{\ell=q+2}^{r+1} (\boldsymbol{\varepsilon}'\mathbf{g}_\ell)(\boldsymbol{\varepsilon}'\mathbf{g}_\ell)' = \sum_{\ell=q+2}^{r+1} \mathbf{V}_\ell\mathbf{V}'_\ell$$

where the $\mathbf{V}_\ell$ are independently distributed as $N_m(\mathbf{0}, \mathbf{\Sigma})$. By (4-22), $n(\hat{\mathbf{\Sigma}}_1 - \hat{\mathbf{\Sigma}})$ is distributed as $W_{r-q}(\cdot \mid \mathbf{\Sigma})$ independently of $n\hat{\mathbf{\Sigma}}$, since $n(\hat{\mathbf{\Sigma}}_1 - \hat{\mathbf{\Sigma}})$ involves a different set of independent $\mathbf{V}_\ell$'s.

The large sample distribution for $-(n - r - 1 - \frac{1}{2}(m - q))\ln(|\hat{\mathbf{\Sigma}}|/|\hat{\mathbf{\Sigma}}_1|)$ follows from Result 5.2, with $\nu - \nu_0 = (r + 1)(r + 2)/2 + m(r + 1) - (r + 1)(r + 2)/2 - m(r + 1 - q - 1) = m(q + 1)$ d.f. The use of $(n - r - 1 - \frac{1}{2}(m - q))$ instead of n in the test statistic is due to Bartlett [3] following Box [4], and it improves the chi-square approximation. ■

If $\mathbf{Z}$ is not of full rank but has rank $r_1 + 1$, then $\hat{\boldsymbol{\beta}} = (\mathbf{Z}'\mathbf{Z})^{-}\mathbf{Z}'\mathbf{Y}$, where $(\mathbf{Z}'\mathbf{Z})^{-}$ is the *generalized inverse* discussed in [18] (see Exercise 7.6 also). The distributional conclusions stated in Result 7.11 remain the same, provided r is replaced by r_1 and $q_1 + 1$ by rank$(\mathbf{Z}_1)$. However, not all hypotheses concerning $\boldsymbol{\beta}$ can be tested due to the lack of uniqueness in the identification of $\boldsymbol{\beta}$ caused by the linear dependencies among the columns of $\mathbf{Z}$. Nevertheless, the generalized inverse allows all of the important MANOVA models to be analyzed as special cases of the multivariate multiple regression model.

Example 7.9

The service in three locations of a large restaurant chain was rated according to two measures of quality by male and female patrons. The first service-quality index was introduced in Example 7.5. Suppose we consider a regression model that allows for the effects of location, sex, and the location-sex interaction on both service quality indices. The design matrix (see Example 7.5) remains the same for the two-response situation. We shall illustrate the test of no location-sex interaction in either response using Result 7.11. A computer program provides

$$\begin{pmatrix}\text{Residual sum of squares}\\ \text{and crossproducts}\end{pmatrix} = n\hat{\mathbf{\Sigma}} = \begin{bmatrix} 2977.39 & 1021.72 \\ 1021.72 & 2050.95 \end{bmatrix}$$

$$\begin{pmatrix}\text{Extra sum of squares}\\ \text{and crossproducts}\end{pmatrix} = n(\hat{\mathbf{\Sigma}}_1 - \hat{\mathbf{\Sigma}}) = \begin{bmatrix} 441.76 & 246.16 \\ 246.16 & 366.12 \end{bmatrix}$$

Let $\boldsymbol{\beta}_{(2)}$ be the matrix of interaction parameters for the two responses. Although the sample size $n = 18$ is not large, we shall illustrate the calculations involved in the test of H_0: $\boldsymbol{\beta}_{(2)} = \mathbf{0}$ given in Result 7.11. Setting $\alpha = .05$, we test H_0 by referring

$$-\left(n - r_1 - 1 - \frac{1}{2}(m - q_1)\right)\ln\left(\frac{|n\hat{\mathbf{\Sigma}}|}{|n\hat{\mathbf{\Sigma}} + n(\hat{\mathbf{\Sigma}}_1 - \hat{\mathbf{\Sigma}})|}\right)$$

$$= -\left(18 - 5 - 1 - \frac{1}{2}(2 - 3)\right)\ln(.7605) = 3.42$$

to a chi-square percentage point with $m(q_1 + 1) = 2(4) = 8$ d.f. Since $3.42 < \chi^2_8(.05) = 15.51$, we do not reject H_0 at the 5% level. The interaction terms are not needed. ■

More generally, we could consider a null hypothesis of the form H_0: $\mathbf{C}\boldsymbol{\beta} = \mathbf{\Gamma}_0$ where $\mathbf{C}$ is $(q + 1) \times (r + 1)$, with $q + 1 \le r + 1$, and is of full rank $q + 1$. For the choices $\mathbf{C} = \left[\mathbf{0} \;\middle|\; \underset{(r-q)\times(r-q)}{\mathbf{I}}\right]$ and $\mathbf{\Gamma}_0 = \mathbf{0}$, this null hypothesis becomes H_0: $\mathbf{C}\boldsymbol{\beta} = \boldsymbol{\beta}_{(2)} = \mathbf{0}$, the case considered earlier. It can be shown that the extra sum of squares and crossproducts generated by the hypothesis H_0 is

$$n(\hat{\mathbf{\Sigma}}_1 - \hat{\mathbf{\Sigma}}) = (\mathbf{C}\hat{\boldsymbol{\beta}} - \mathbf{\Gamma}_0)'(\mathbf{C}(\mathbf{Z}'\mathbf{Z})^{-1}\mathbf{C}')^{-1}(\mathbf{C}\hat{\boldsymbol{\beta}} - \mathbf{\Gamma}_0)$$

Under the null hypothesis, the statistic $n(\hat{\mathbf{\Sigma}}_1 - \hat{\mathbf{\Sigma}})$ is distributed as $W_{r-q}(\cdot\,|\,\mathbf{\Sigma})$ independently of $\hat{\mathbf{\Sigma}}$. This distribution theory can be employed to develop a test of H_0: $\mathbf{C}\boldsymbol{\beta} = \mathbf{\Gamma}_0$ similar to the test discussed in Result 7.11 (see, for example, [20]).

Tests other than the likelihood ratio test have been proposed for the multivariate multiple regression model. The most prominent alternative is Roy's test, which rejects H_0: $\boldsymbol{\beta}_{(2)} = \mathbf{0}$ for large values of η, where η is the largest root of

$$|(\hat{\mathbf{\Sigma}}_1 - \hat{\mathbf{\Sigma}}) - \eta\hat{\mathbf{\Sigma}}|$$

Charts and tables of critical values are available for the related quantity $\theta = \eta/(1 + \eta)$ (see [16] and [12]) so, in particular, the upper $(100\alpha)\%$ critical value for η is given by $\eta_\alpha = \theta_\alpha/(1 - \theta_\alpha)$. We encountered a special case of the largest root statistic when constructing simultaneous confidence intervals in the context of one-way MANOVA (see Result 6.5).

Predictions from Multivariate Regressions

Suppose the model $\mathbf{Y} = \mathbf{Z}\boldsymbol{\beta} + \boldsymbol{\varepsilon}$, with normal errors $\boldsymbol{\varepsilon}$, has been fit and checked for any inadequacies. If the model is adequate, it can be employed for predictive purposes.

One problem is to predict the mean responses corresponding to fixed values $\mathbf{z}_0$ of the predictor variables. Inferences about the mean responses can be made using the distribution theory in Result 7.10. From this result we determine

$$\hat{\boldsymbol{\beta}}'\mathbf{z}_0 \quad \text{is distributed as} \quad N_m\left(\boldsymbol{\beta}'\mathbf{z}_0, \mathbf{z}_0'(\mathbf{Z}'\mathbf{Z})^{-1}\mathbf{z}_0\boldsymbol{\Sigma}\right)$$

and

$$n\hat{\boldsymbol{\Sigma}} \quad \text{is independently distributed as} \quad W_{n-r-1}(\cdot \mid \boldsymbol{\Sigma})$$

The unknown value of the regression function at $\mathbf{z}_0$ is $\boldsymbol{\beta}'\mathbf{z}_0$. So, from the discussion of the T^2-statistic in Section 5.2, we can write

$$T^2 = \left(\frac{\hat{\boldsymbol{\beta}}'\mathbf{z}_0 - \boldsymbol{\beta}'\mathbf{z}_0}{\sqrt{\mathbf{z}_0'(\mathbf{Z}'\mathbf{Z})^{-1}\mathbf{z}_0}}\right)'\left(\frac{n\hat{\boldsymbol{\Sigma}}}{n-r-1}\right)^{-1}\left(\frac{\hat{\boldsymbol{\beta}}'\mathbf{z}_0 - \boldsymbol{\beta}'\mathbf{z}_0}{\sqrt{\mathbf{z}_0'(\mathbf{Z}'\mathbf{Z})^{-1}\mathbf{z}_0}}\right) \tag{7-43}$$

and the $100(1-\alpha)\%$ confidence ellipsoid for $\boldsymbol{\beta}'\mathbf{z}_0$ is provided by the inequality

$$\left(\boldsymbol{\beta}'\mathbf{z}_0 - \hat{\boldsymbol{\beta}}'\mathbf{z}_0\right)'\left(\frac{n\hat{\boldsymbol{\Sigma}}}{n-r-1}\right)^{-1}\left(\boldsymbol{\beta}'\mathbf{z}_0 - \hat{\boldsymbol{\beta}}'\mathbf{z}_0\right)$$
$$\leq \mathbf{z}_0'(\mathbf{Z}'\mathbf{Z})^{-1}\mathbf{z}_0\left[\left(\frac{m(n-r-1)}{n-r-m}\right)F_{m,\,n-r-m}(\alpha)\right] \tag{7-44}$$

where $F_{m,\,n-r-m}(\alpha)$ is the upper (100α)th percentile of an F-distribution with m and $n-r-m$ d.f.

The $100(1-\alpha)\%$ *simultaneous* confidence intervals for $E(Y_i) = \mathbf{z}_0'\boldsymbol{\beta}_{(i)}$ are

$$\mathbf{z}_0'\hat{\boldsymbol{\beta}}_{(i)} \pm \sqrt{\left(\frac{m(n-r-1)}{n-r-m}\right)F_{m,\,n-r-m}(\alpha)}\sqrt{\mathbf{z}_0'(\mathbf{Z}'\mathbf{Z})^{-1}\mathbf{z}_0\left(\frac{n}{n-r-1}\hat{\sigma}_{ii}\right)},$$
$$i = 1, 2, \ldots, m \tag{7-45}$$

where $\hat{\boldsymbol{\beta}}_{(i)}$ is the ith column of $\hat{\boldsymbol{\beta}}$ and $\hat{\sigma}_{ii}$ is the ith diagonal element of $\hat{\boldsymbol{\Sigma}}$.

The second prediction problem is concerned with forecasting new responses $\mathbf{Y}_0 = \boldsymbol{\beta}'\mathbf{z}_0 + \boldsymbol{\varepsilon}_0$ at $\mathbf{z}_0$. Here $\boldsymbol{\varepsilon}_0$ is independent of $\boldsymbol{\varepsilon}$. We have

$$\mathbf{Y}_0 - \hat{\boldsymbol{\beta}}'\mathbf{z}_0 = (\boldsymbol{\beta} - \hat{\boldsymbol{\beta}})'\mathbf{z}_0 + \boldsymbol{\varepsilon}_0 \quad \text{is distributed as} \quad N_m\left(\mathbf{0}, \left(1 + \mathbf{z}_0'(\mathbf{Z}'\mathbf{Z})^{-1}\mathbf{z}_0\right)\boldsymbol{\Sigma}\right)$$

independently of $n\hat{\boldsymbol{\Sigma}}$, so the $100(1-\alpha)\%$ *prediction ellipsoid* for $\mathbf{Y}_0$ becomes

$$\left(\mathbf{Y}_0 - \hat{\boldsymbol{\beta}}'\mathbf{z}_0\right)'\left(\frac{n\hat{\boldsymbol{\Sigma}}}{n-r-1}\right)^{-1}\left(\mathbf{Y}_0 - \hat{\boldsymbol{\beta}}'\mathbf{z}_0\right)$$
$$\leq \left(1 + \mathbf{z}_0'(\mathbf{Z}'\mathbf{Z})^{-1}\mathbf{z}_0\right)\left[\left(\frac{m(n-r-1)}{n-r-m}\right)F_{m,\,n-r-m}(\alpha)\right] \tag{7-46}$$

The $100(1-\alpha)\%$ *simultaneous prediction intervals* for the individual responses Y_{0i}

are

$$\mathbf{z}_0'\hat{\boldsymbol{\beta}}_{(i)} \pm \sqrt{\left(\frac{m(n-r-1)}{n-r-m}\right)F_{m,n-r-m}(\alpha)}\sqrt{\left(1+\mathbf{z}_0'(\mathbf{Z}'\mathbf{Z})^{-1}\mathbf{z}_0\right)\left(\frac{n}{n-r-1}\hat{\sigma}_{ii}\right)},$$

$$i = 1, 2, \ldots, m \qquad (7\text{-}47)$$

where $\hat{\boldsymbol{\beta}}_{(i)}$, $\hat{\sigma}_{ii}$ and $F_{m,n-r-m}(\alpha)$ are the same quantities appearing in (7-45). Comparing (7-45) and (7-47), we see that the prediction intervals for the *actual* values of the response variables are wider than the corresponding intervals for the *expected* values. The extra width comes from the additional factor $[n/(n-r-1)]\hat{\sigma}_{ii}$ and reflects the presence of the random error ε_{0i}.

Example 7.10

A second response variable was measured for the computer-requirement problem discussed in Example 7.6. Measurements on the response Y_2, disk input/output capacity, corresponding to the z_1 and z_2 values in Example 7.6, were

$$\mathbf{y}_2 = [301.8, 396.1, 328.2, 307.4, 362.4, 369.5, 229.1]'$$

Obtain the 95% confidence ellipse for $\boldsymbol{\beta}'\mathbf{z}_0$ and the 95% prediction ellipse for $\mathbf{Y}_0 = [Y_{01}, Y_{02}]'$ for a site with the configuration $\mathbf{z}_0 = [1, 130, 7.5]'$.

Computer calculations provide the fitted equation

$$\hat{y}_2 = 14.14 + 2.25z_1 + 5.67z_2$$

with $s = 1.812$. Thus $\hat{\boldsymbol{\beta}}_{(2)} = [14.14, 2.25, 5.67]'$. From Example 7.6

$$\hat{\boldsymbol{\beta}}_{(1)} = [8.42, 1.08, .42]', \quad \mathbf{z}_0'\hat{\boldsymbol{\beta}}_{(1)} = 151.97, \quad \text{and} \quad \mathbf{z}_0'(\mathbf{Z}'\mathbf{Z})^{-1}\mathbf{z}_0 = .34725$$

We find

$$\mathbf{z}_0'\hat{\boldsymbol{\beta}}_{(2)} = 14.14 + 2.25(130) + 5.67(7.5) = 349.17$$

and

$$n\hat{\boldsymbol{\Sigma}} = \begin{bmatrix} (\mathbf{y}_{(1)} - \mathbf{Z}\hat{\boldsymbol{\beta}}_{(1)})'(\mathbf{y}_{(1)} - \mathbf{Z}\hat{\boldsymbol{\beta}}_{(1)}) & (\mathbf{y}_{(1)} - \mathbf{Z}\hat{\boldsymbol{\beta}}_{(1)})'(\mathbf{y}_{(2)} - \mathbf{Z}\hat{\boldsymbol{\beta}}_{(2)}) \\ (\mathbf{y}_{(2)} - \mathbf{Z}\hat{\boldsymbol{\beta}}_{(2)})'(\mathbf{y}_{(1)} - \mathbf{Z}\hat{\boldsymbol{\beta}}_{(1)}) & (\mathbf{y}_{(2)} - \mathbf{Z}\hat{\boldsymbol{\beta}}_{(2)})'(\mathbf{y}_{(2)} - \mathbf{Z}\hat{\boldsymbol{\beta}}_{(2)}) \end{bmatrix}$$

$$= \begin{bmatrix} 5.80 & 5.30 \\ 5.30 & 13.13 \end{bmatrix}$$

Since

$$\hat{\boldsymbol{\beta}}'\mathbf{z}_0 = \begin{bmatrix} \hat{\boldsymbol{\beta}}_{(1)}' \\ \hline \hat{\boldsymbol{\beta}}_{(2)}' \end{bmatrix}\mathbf{z}_0 = \begin{bmatrix} \mathbf{z}_0'\hat{\boldsymbol{\beta}}_{(1)} \\ \hline \mathbf{z}_0'\hat{\boldsymbol{\beta}}_{(2)} \end{bmatrix} = \begin{bmatrix} 151.97 \\ 349.17 \end{bmatrix}$$

$n = 7$, $r = 2$, and $m = 2$, a 95% confidence ellipse for $\boldsymbol{\beta}'\mathbf{z}_0 = \begin{bmatrix} \mathbf{z}_0'\boldsymbol{\beta}_{(1)} \\ \hline \mathbf{z}_0'\boldsymbol{\beta}_{(2)} \end{bmatrix}$ is, from

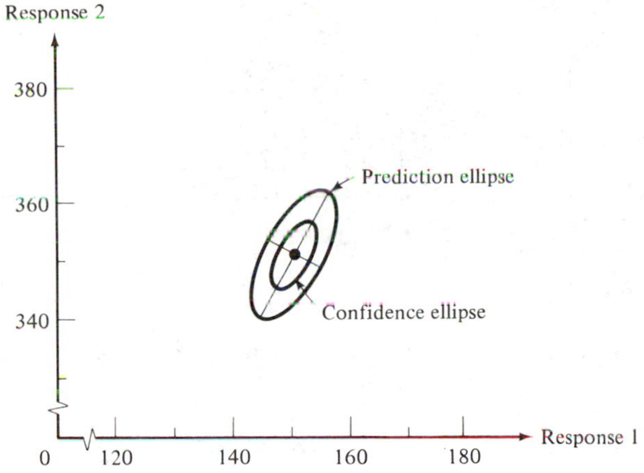

Figure 7.5 95% confidence and prediction ellipses for the computer data with two responses.

(7-44), the set

$$[\mathbf{z}_0'\boldsymbol{\beta}_{(1)} - 151.97, \mathbf{z}_0'\boldsymbol{\beta}_{(2)} - 349.17](4)\begin{bmatrix} 5.80 & 5.30 \\ 5.30 & 13.13 \end{bmatrix}^{-1}\begin{bmatrix} \mathbf{z}_0'\boldsymbol{\beta}_{(1)} - 151.97 \\ \mathbf{z}_0'\boldsymbol{\beta}_{(2)} - 349.17 \end{bmatrix}$$

$$\le (.34725)\left[\left(\frac{2(4)}{3}\right)F_{2,3}(.05)\right]$$

with $F_{2,3}(.05) = 9.55$. This ellipse is centered at (151.97, 349.17). Its orientation and the lengths of the major and minor axes can be determined from the eigenvalues and eigenvectors of $n\hat{\boldsymbol{\Sigma}}$.

Comparing (7-44) and (7-46), the only change required for the calculation of the 95% prediction ellipse is to replace $\mathbf{z}_0'(\mathbf{Z}'\mathbf{Z})^{-1}\mathbf{z}_0 = .34725$ with $1 + \mathbf{z}_0'(\mathbf{Z}'\mathbf{Z})^{-1}\mathbf{z}_0 = 1.34725$. Thus the 95% prediction ellipse for $\mathbf{Y}_0 = [Y_{01}, Y_{02}]'$ is also centered at (151.97, 349.17) but is larger than the confidence ellipse. Both ellipses are sketched in Figure 7.5.

It is the *prediction* ellipse that is relevant to the determination of computer requirements for a particular site with the given $\mathbf{z}_0$. ■

7.8 THE CONCEPT OF LINEAR REGRESSION

The classical linear regression model is concerned with the association between a single dependent variable Y and a collection of predictor variables $z_1, z_2, \ldots, z_r$. The regression model that we have considered treats Y as a random variable whose mean depends upon *fixed* values of the z_i's. This mean is assumed to be a linear function of the regression *coefficients* $\beta_0, \beta_1, \ldots, \beta_r$.

The linear regression model also arises in a different setting. Suppose all the variables $Y, Z_1, Z_2, \ldots, Z_r$ are random and have a joint distribution, not necessarily normal, with mean vector $\underset{(r+1)\times 1}{\boldsymbol{\mu}}$ and covariance matrix $\underset{(r+1)\times(r+1)}{\boldsymbol{\Sigma}}$. Partitioning

$\boldsymbol{\mu}$ and $\boldsymbol{\Sigma}$ in an obvious fashion, we write

$$\boldsymbol{\mu} = \begin{bmatrix} \underset{(1\times 1)}{\mu_Y} \\ \hdashline \underset{(r\times 1)}{\boldsymbol{\mu}_Z} \end{bmatrix} \quad \text{and} \quad \boldsymbol{\Sigma} = \left[\begin{array}{c:c} \underset{(1\times 1)}{\sigma_{YY}} & \underset{(1\times r)}{\boldsymbol{\sigma}'_{ZY}} \\ \hdashline \underset{(r\times 1)}{\boldsymbol{\sigma}_{ZY}} & \underset{(r\times r)}{\boldsymbol{\Sigma}_{ZZ}} \end{array}\right]$$

with

$$\boldsymbol{\sigma}_{ZY} = \left[\sigma_{YZ_1}, \sigma_{YZ_2}, \ldots, \sigma_{YZ_r}\right]' \tag{7-48}$$

and $\boldsymbol{\Sigma}_{ZZ}$ can be taken to have full rank.[6] Consider the problem of predicting Y using the

$$\textit{linear predictor} = b_0 + b_1 Z_1 + \cdots + b_r Z_r = b_0 + \mathbf{b}'\mathbf{Z} \tag{7-49}$$

For a given predictor of the form of (7-49), the error in the prediction of Y is

$$\textit{prediction error} = Y - b_0 - b_1 Z_1 - \cdots - b_r Z_r = Y - b_0 - \mathbf{b}'\mathbf{Z} \tag{7-50}$$

Because this error is random, it is customary to select b_0 and $\mathbf{b}$ to minimize

$$\textit{mean square error} = E(Y - b_0 - \mathbf{b}'\mathbf{Z})^2 \tag{7-51}$$

Now the mean square error depends on the joint distribution of Y and $\mathbf{Z}$ only through the parameters $\boldsymbol{\mu}$ and $\boldsymbol{\Sigma}$. It is possible to express the "optimal" linear predictor in terms of these latter quantities.

Result 7.12. The linear predictor $\beta_0 + \boldsymbol{\beta}'\mathbf{Z}$ with coefficients

$$\boldsymbol{\beta} = \boldsymbol{\Sigma}_{ZZ}^{-1}\boldsymbol{\sigma}_{ZY}, \qquad \beta_0 = \mu_Y - \boldsymbol{\beta}'\boldsymbol{\mu}_Z$$

has minimum mean square among all *linear* predictors of the response Y. Its mean square error is

$$E(Y - \beta_0 - \boldsymbol{\beta}'\mathbf{Z})^2 = E\left(Y - \mu_Y - \boldsymbol{\sigma}'_{ZY}\boldsymbol{\Sigma}_{ZZ}^{-1}(\mathbf{Z} - \boldsymbol{\mu}_Z)\right)^2 = \sigma_{YY} - \boldsymbol{\sigma}'_{ZY}\boldsymbol{\Sigma}_{ZZ}^{-1}\boldsymbol{\sigma}_{ZY}$$

Also, $\beta_0 + \boldsymbol{\beta}'\mathbf{Z} = \mu_Y + \boldsymbol{\sigma}'_{ZY}\boldsymbol{\Sigma}_{ZZ}^{-1}(\mathbf{Z} - \boldsymbol{\mu}_Z)$ is the linear predictor having maximum correlation with Y,

$$\begin{aligned} \text{Corr}(Y, \beta_0 + \boldsymbol{\beta}'\mathbf{Z}) &= \max_{b_0, \mathbf{b}} \text{Corr}(Y, b_0 + \mathbf{b}'\mathbf{Z}) \\ &= \sqrt{\frac{\boldsymbol{\beta}'\boldsymbol{\Sigma}_{ZZ}\boldsymbol{\beta}}{\sigma_{YY}}} = \sqrt{\frac{\boldsymbol{\sigma}'_{ZY}\boldsymbol{\Sigma}_{ZZ}^{-1}\boldsymbol{\sigma}_{ZY}}{\sigma_{YY}}} \end{aligned}$$

[6] If $\boldsymbol{\Sigma}_{ZZ}$ is not of full rank, one variable—for example, Z_k, can be written as a linear combination of the other Z_i's and thus is redundant in forming the linear regression function $\mathbf{Z}'\boldsymbol{\beta}$. That is, $\mathbf{Z}$ may be replaced by any subset of components whose nonsingular covariance matrix has the same rank as $\boldsymbol{\Sigma}_{ZZ}$.

Proof. Writing $b_0 + \mathbf{b}'\mathbf{Z} = b_0 + \mathbf{b}'\mathbf{Z} - (\mu_Y - \mathbf{b}'\boldsymbol{\mu}_Z) + (\mu_Y - \mathbf{b}'\boldsymbol{\mu}_Z)$ we get

$$\begin{aligned} E(Y - b_0 - \mathbf{b}'\mathbf{Z})^2 &= E\left[Y - \mu_Y - (\mathbf{b}'\mathbf{Z} - \mathbf{b}'\boldsymbol{\mu}_Z) + (\mu_Y - b_0 - \mathbf{b}'\boldsymbol{\mu}_Z)\right]^2 \\ &= E(Y - \mu_Y)^2 + E(\mathbf{b}'(\mathbf{Z} - \boldsymbol{\mu}_Z))^2 + (\mu_Y - b_0 - \mathbf{b}'\boldsymbol{\mu}_Z)^2 \\ &\quad -2E\left[\mathbf{b}'(\mathbf{Z} - \boldsymbol{\mu}_Z)(Y - \mu_Y)\right] \\ &= \sigma_{YY} + \mathbf{b}'\boldsymbol{\Sigma}_{ZZ}\mathbf{b} + (\mu_Y - b_0 - \mathbf{b}'\boldsymbol{\mu}_Z)^2 - 2\mathbf{b}'\boldsymbol{\sigma}_{ZY} \end{aligned}$$

Adding and subtracting $\boldsymbol{\sigma}'_{ZY}\boldsymbol{\Sigma}_{ZZ}^{-1}\boldsymbol{\sigma}_{ZY}$, we obtain

$$\begin{aligned} E(Y - b_0 - \mathbf{b}'\mathbf{Z})^2 &= \sigma_{YY} - \boldsymbol{\sigma}'_{ZY}\boldsymbol{\Sigma}_{ZZ}^{-1}\boldsymbol{\sigma}_{ZY} + (\mu_Y - b_0 - \mathbf{b}'\boldsymbol{\mu}_Z)^2 \\ &\quad + \left(\mathbf{b} - \boldsymbol{\Sigma}_{ZZ}^{-1}\boldsymbol{\sigma}_{ZY}\right)'\boldsymbol{\Sigma}_{ZZ}\left(\mathbf{b} - \boldsymbol{\Sigma}_{ZZ}^{-1}\boldsymbol{\sigma}_{ZY}\right) \end{aligned}$$

The mean square error is minimized by taking $\mathbf{b} = \boldsymbol{\Sigma}_{ZZ}^{-1}\boldsymbol{\sigma}_{ZY} = \boldsymbol{\beta}$, making the last term zero, and then choosing $b_0 = \mu_Y - (\boldsymbol{\Sigma}_{ZZ}^{-1}\boldsymbol{\sigma}_{ZY})'\boldsymbol{\mu}_Z = \beta_0$ to make the third term zero. The minimum mean square error is thus $\sigma_{YY} - \boldsymbol{\sigma}'_{ZY}\boldsymbol{\Sigma}_{ZZ}^{-1}\boldsymbol{\sigma}_{ZY}$.

Next we note that $\text{Cov}(b_0 + \mathbf{b}'\mathbf{Z}, Y) = \text{Cov}(\mathbf{b}'\mathbf{Z}, Y) = \mathbf{b}'\boldsymbol{\sigma}_{ZY}$ so

$$\left[\text{Corr}(b_0 + \mathbf{b}'\mathbf{Z}, Y)\right]^2 = \frac{[\mathbf{b}'\boldsymbol{\sigma}_{ZY}]^2}{\sigma_{YY}(\mathbf{b}'\boldsymbol{\Sigma}_{ZZ}\mathbf{b})}, \qquad \text{for all } b_0, \mathbf{b}$$

Employing the extended Cauchy-Schwarz inequality of (2-49) with $\mathbf{B} = \boldsymbol{\Sigma}_{ZZ}$,

$$(\mathbf{b}'\boldsymbol{\sigma}_{ZY})^2 \leq \mathbf{b}'\boldsymbol{\Sigma}_{ZZ}\mathbf{b}\boldsymbol{\sigma}'_{ZY}\boldsymbol{\Sigma}_{ZZ}^{-1}\boldsymbol{\sigma}_{ZY}$$

or

$$\left[\text{Corr}(b_0 + \mathbf{b}'\mathbf{Z}, Y)\right]^2 \leq \frac{\boldsymbol{\sigma}'_{ZY}\boldsymbol{\Sigma}_{ZZ}^{-1}\boldsymbol{\sigma}_{ZY}}{\sigma_{YY}}$$

with equality for $\mathbf{b} = \boldsymbol{\Sigma}_{ZZ}^{-1}\boldsymbol{\sigma}_{ZY} = \boldsymbol{\beta}$. The alternative expression for the maximum correlation follows from $\boldsymbol{\sigma}'_{ZY}\boldsymbol{\Sigma}_{ZZ}^{-1}\boldsymbol{\sigma}_{ZY} = \boldsymbol{\sigma}'_{ZY}\boldsymbol{\beta} = \boldsymbol{\sigma}'_{ZY}\boldsymbol{\Sigma}_{ZZ}^{-1}\boldsymbol{\Sigma}_{ZZ}\boldsymbol{\beta} = \boldsymbol{\beta}'\boldsymbol{\Sigma}_{ZZ}\boldsymbol{\beta}$. ■

The correlation between Y and its best linear predictor is called the *population multiple correlation coefficient*:

$$\rho_{Y(\mathbf{Z})} = +\sqrt{\frac{\boldsymbol{\sigma}'_{ZY}\boldsymbol{\Sigma}_{ZZ}^{-1}\boldsymbol{\sigma}_{ZY}}{\sigma_{YY}}} \tag{7-52}$$

Its square $\rho^2_{Y(\mathbf{Z})}$, is called the *population coefficient of determination*. Note that, unlike other correlation coefficients, the multiple correlation coefficient is a *positive* square root so $0 \leq \rho_{Y(\mathbf{Z})} \leq 1$.

The population coefficient of determination has an important interpretation. From Result 7.12, the mean square error in using $\beta_0 + \boldsymbol{\beta}'\mathbf{Z}$ to forecast Y is

$$\sigma_{YY} - \boldsymbol{\sigma}'_{ZY}\boldsymbol{\Sigma}_{ZZ}^{-1}\boldsymbol{\sigma}_{ZY} = \sigma_{YY} - \sigma_{YY}\left(\frac{\boldsymbol{\sigma}'_{ZY}\boldsymbol{\Sigma}_{ZZ}^{-1}\boldsymbol{\sigma}_{ZY}}{\sigma_{YY}}\right) = \sigma_{YY}\left(1 - \rho^2_{Y(\mathbf{Z})}\right) \tag{7-53}$$

If $\rho^2_{Y(\mathbf{Z})} = 0$, there is no predictive power in $\mathbf{Z}$. At the other extreme, $\rho^2_{Y(\mathbf{Z})} = 1$ implies Y can be predicted with no error.

Example 7.11

Given the mean vector and covariance matrix of Y, Z_1, Z_2,

$$\boldsymbol{\mu} = \begin{bmatrix} \mu_Y \\ \hline \boldsymbol{\mu}_Z \end{bmatrix} = \begin{bmatrix} 5 \\ \hline 2 \\ 0 \end{bmatrix} \quad \text{and} \quad \boldsymbol{\Sigma} = \left[\begin{array}{c|c} \sigma_{YY} & \boldsymbol{\sigma}'_{ZY} \\ \hline \boldsymbol{\sigma}_{ZY} & \boldsymbol{\Sigma}_{ZZ} \end{array}\right] = \left[\begin{array}{c|cc} 10 & 1 & -1 \\ \hline 1 & 7 & 3 \\ -1 & 3 & 2 \end{array}\right]$$

determine (a) the best linear predictor $\beta_0 + \beta_1 Z_1 + \beta_2 Z_2$, (b) its mean square error, and (c) the multiple correlation coefficient. Also verify the relationship: mean square error $= \sigma_{YY}(1 - \rho^2_{Y(\mathbf{Z})})$.

First

$$\boldsymbol{\beta} = \boldsymbol{\Sigma}_{ZZ}^{-1}\boldsymbol{\sigma}_{ZY} = \begin{bmatrix} 7 & 3 \\ 3 & 2 \end{bmatrix}^{-1}\begin{bmatrix} 1 \\ -1 \end{bmatrix} = \begin{bmatrix} .4 & -.6 \\ -.6 & 1.4 \end{bmatrix}\begin{bmatrix} 1 \\ -1 \end{bmatrix} = \begin{bmatrix} 1 \\ -2 \end{bmatrix}$$

$$\beta_0 = \mu_Y - \boldsymbol{\beta}'\boldsymbol{\mu}_Z = 5 - [1 \quad -2]\begin{bmatrix} 2 \\ 0 \end{bmatrix} = 3$$

so the best linear predictor is $\beta_0 + \boldsymbol{\beta}'\mathbf{Z} = 3 + Z_1 - 2Z_2$. The mean square error is

$$\sigma_{YY} - \boldsymbol{\sigma}'_{ZY}\boldsymbol{\Sigma}_{ZZ}^{-1}\boldsymbol{\sigma}_{ZY} = 10 - [1, -1]\begin{bmatrix} .4 & -.6 \\ -.6 & 1.4 \end{bmatrix}\begin{bmatrix} 1 \\ -1 \end{bmatrix} = 10 - 3 = 7$$

and the multiple correlation coefficient is

$$\rho_{Y(\mathbf{Z})} = \sqrt{\frac{\boldsymbol{\sigma}'_{ZY}\boldsymbol{\Sigma}_{ZZ}^{-1}\boldsymbol{\sigma}_{ZY}}{\sigma_{YY}}} = \sqrt{\frac{3}{10}} = .548$$

Note $\sigma_{YY}(1 - \rho^2_{Y(\mathbf{Z})}) = 10(1 - \frac{3}{10}) = 7 =$ mean square error. ■

It is possible to show (see Exercise 7.5) that

$$\rho^2_{Y(\mathbf{Z})} = \frac{1}{\rho^{YY}} \tag{7-54}$$

where ρ^{YY} is the upper left-hand corner of the inverse of the correlation matrix determined from $\boldsymbol{\Sigma}$.

The restriction to linear predictors is closely connected to the assumption of normality. Specifically, if we take

$$\begin{bmatrix} Y \\ Z_1 \\ Z_2 \\ \vdots \\ Z_r \end{bmatrix} \quad \text{to be distributed as} \quad N_{r+1}(\boldsymbol{\mu}, \boldsymbol{\Sigma})$$

then the conditional distribution of Y with $z_1, z_2, \ldots, z_r$ fixed (see Result 4.6) is

$$N\left(\mu_Y + \boldsymbol{\sigma}'_{ZY}\boldsymbol{\Sigma}_{ZZ}^{-1}(\mathbf{z} - \boldsymbol{\mu}_Z), \sigma_{YY} - \boldsymbol{\sigma}'_{ZY}\boldsymbol{\Sigma}_{ZZ}^{-1}\boldsymbol{\sigma}_{ZY}\right)$$

The mean of this conditional distribution is the linear predictor in Result 7.12. That is,

$$E(Y \mid z_1, z_2, \ldots, z_r) = \mu_Y + \boldsymbol{\sigma}'_{ZY}\boldsymbol{\Sigma}_{ZZ}^{-1}(\mathbf{z} - \boldsymbol{\mu}_Z)$$
$$= \beta_0 + \boldsymbol{\beta}'\mathbf{z} \tag{7-55}$$

and we conclude that $E(Y \mid z_1, z_2, \ldots, z_r)$ is the best linear predictor of Y when the population is $N_{r+1}(\boldsymbol{\mu}, \boldsymbol{\Sigma})$. The conditional expectation of Y in (7-55) is called the *linear regression function*.

When the population is *not* normal, the regression function $E(Y \mid z_1, z_2, \ldots, z_r)$ need not be of the form $\beta_0 + \boldsymbol{\beta}'\mathbf{z}$. Yet it can be shown (see [17]) that $E(Y \mid z_1, z_2, \ldots, z_r)$, whatever its form, predicts Y with the smallest mean square error. Fortunately, this wider optimality among all estimators is possessed by the *linear* predictor when the population is normal.

Result 7.13. Suppose the joint distribution of Y and $\mathbf{Z}$ is $N_{r+1}(\boldsymbol{\mu}, \boldsymbol{\Sigma})$. Let

$$\hat{\boldsymbol{\mu}} = \begin{bmatrix} \bar{Y} \\ \bar{\mathbf{Z}} \end{bmatrix} \quad \text{and} \quad \mathbf{S} = \begin{bmatrix} s_{YY} & \mathbf{s}'_{ZY} \\ \mathbf{s}_{ZY} & \mathbf{S}_{ZZ} \end{bmatrix}$$

be the sample mean vector and sample covariance matrix for a random sample of size n from this population. The maximum likelihood estimators of the coefficients in the linear predictor are

$$\hat{\boldsymbol{\beta}} = \mathbf{S}_{ZZ}^{-1}\mathbf{s}_{ZY}, \qquad \hat{\beta}_0 = \bar{Y} - \mathbf{s}'_{ZY}\mathbf{S}_{ZZ}^{-1}\bar{\mathbf{Z}} = \bar{Y} - \hat{\boldsymbol{\beta}}'\bar{\mathbf{Z}}$$

Consequently, the maximum likelihood estimator of the regression function is

$$\hat{\beta}_0 + \hat{\boldsymbol{\beta}}'\mathbf{z} = \bar{Y} + \mathbf{s}'_{ZY}\mathbf{S}_{ZZ}^{-1}(\mathbf{z} - \bar{\mathbf{Z}})$$

The maximum likelihood estimator of the mean square error, $E[Y - \beta_0 - \boldsymbol{\beta}'\mathbf{Z}]^2$, is

$$\hat{\sigma}_{YY \cdot Z} = \frac{n-1}{n}\left(s_{YY} - \mathbf{s}'_{ZY}\mathbf{S}_{ZZ}^{-1}\mathbf{s}_{ZY}\right)$$

Proof. We use Result 4.11 and the invariance property of maximum likelihood estimators [see (4-20)]. Since, from Result 7.12,

$$\beta_0 = \mu_Y - (\boldsymbol{\Sigma}_{ZZ}^{-1}\boldsymbol{\sigma}_{ZY})'\boldsymbol{\mu}_Z,$$

$$\boldsymbol{\beta} = \boldsymbol{\Sigma}_{ZZ}^{-1}\boldsymbol{\sigma}_{ZY}, \qquad \beta_0 + \boldsymbol{\beta}'\mathbf{z} = \mu_Y + \boldsymbol{\sigma}'_{ZY}\boldsymbol{\Sigma}_{ZZ}^{-1}(\mathbf{z} - \boldsymbol{\mu}_Z)$$

and

$$\text{mean square error} = \sigma_{YY \cdot Z} = \sigma_{YY} - \boldsymbol{\sigma}'_{ZY}\boldsymbol{\Sigma}_{ZZ}^{-1}\boldsymbol{\sigma}_{ZY}$$

the conclusions in Result 7.13 follow upon substitution of the maximum likelihood estimators

$$\hat{\boldsymbol{\mu}} = \begin{bmatrix} \bar{Y} \\ \bar{\mathbf{Z}} \end{bmatrix} \quad \text{and} \quad \hat{\boldsymbol{\Sigma}} = \begin{bmatrix} \hat{\sigma}_{YY} & \hat{\boldsymbol{\sigma}}'_{ZY} \\ \hat{\boldsymbol{\sigma}}_{ZY} & \hat{\boldsymbol{\Sigma}}_{ZZ} \end{bmatrix} = \left(\frac{n-1}{n}\right)\mathbf{S}$$

for

$$\boldsymbol{\mu} = \begin{bmatrix} \mu_Y \\ \boldsymbol{\mu}_Z \end{bmatrix} \quad \text{and} \quad \boldsymbol{\Sigma} = \begin{bmatrix} \sigma_{YY} & \boldsymbol{\sigma}'_{ZY} \\ \boldsymbol{\sigma}_{ZY} & \boldsymbol{\Sigma}_{ZZ} \end{bmatrix} \qquad \blacksquare$$

It is customary to change the divisor from n to $n - (r + 1)$ in the estimator of the mean square error, $\sigma_{YY\cdot Z} = E(Y - \beta_0 - \boldsymbol{\beta}'\mathbf{Z})^2$, in order to obtain the *unbiased* estimator

$$\left(\frac{n-1}{n-r-1}\right)\left(s_{YY} - \mathbf{s}'_{ZY}\mathbf{S}_{ZZ}^{-1}\mathbf{s}_{ZY}\right) = \frac{\sum_{j=1}^{n}\left(Y_j - \hat{\beta}_0 - \hat{\boldsymbol{\beta}}'\mathbf{Z}_j\right)^2}{n-r-1} \tag{7-56}$$

Example 7.12

For the computer data of Example 7.6, the $n = 7$ observations on Y (CPU time), Z_1 (orders), and Z_2 (add-delete items) give the sample mean vector and sample covariance matrix

$$\hat{\boldsymbol{\mu}} = \begin{bmatrix} \bar{y} \\ \hline \bar{\mathbf{z}} \end{bmatrix} = \begin{bmatrix} 150.44 \\ \hline 130.24 \\ 3.547 \end{bmatrix}$$

$$\mathbf{S} = \left[\begin{array}{c|c} s_{YY} & \mathbf{s}'_{ZY} \\ \hline \mathbf{s}_{ZY} & \mathbf{S}_{ZZ} \end{array}\right] = \left[\begin{array}{c|cc} 467.913 & 418.763 & 35.983 \\ \hline 418.763 & 377.200 & 28.034 \\ 35.983 & 28.034 & 13.657 \end{array}\right]$$

Assuming Y, Z_1 and Z_2 are jointly normal, obtain the estimated regression function and the estimated mean square error.

Result 7.13 gives the maximum likelihood estimates

$$\hat{\boldsymbol{\beta}} = \mathbf{S}_{ZZ}^{-1}\mathbf{s}_{ZY} = \begin{bmatrix} .003128 & -.006422 \\ -.006422 & .086404 \end{bmatrix}\begin{bmatrix} 418.763 \\ 35.983 \end{bmatrix} = \begin{bmatrix} 1.079 \\ .420 \end{bmatrix}$$

$$\hat{\beta}_0 = \bar{y} - \hat{\boldsymbol{\beta}}'\bar{\mathbf{z}} = 150.44 - [1.079, .420]\begin{bmatrix} 130.24 \\ 3.547 \end{bmatrix} = 150.44 - 142.019 = 8.421$$

and the estimated regression function

$$\hat{\beta}_0 + \hat{\boldsymbol{\beta}}'\mathbf{z} = 8.42 - 1.08z_1 + .42z_2$$

The maximum likelihood estimate of the mean square error arising from the prediction of Y with this regression function is

$$\left(\frac{n-1}{n}\right)\left(s_{YY} - \mathbf{s}'_{ZY}\mathbf{S}_{ZZ}^{-1}\mathbf{s}_{ZY}\right)$$

$$= \left(\tfrac{6}{7}\right)\left(467.913 - [418.763, 35.983]\begin{bmatrix} .003128 & -.006422 \\ -.006422 & .086404 \end{bmatrix}\begin{bmatrix} 418.763 \\ 35.983 \end{bmatrix}\right)$$

$$= .894$$

■

Prediction of Several Variables

The extension of the previous results to the prediction of several responses $Y_1, Y_2, \ldots, Y_m$ is almost immediate. We present this extension for normal populations.

Suppose

$$\begin{bmatrix} \underset{(m\times 1)}{\mathbf{Y}} \\ \hdashline \underset{(r\times 1)}{\mathbf{Z}} \end{bmatrix} \quad \text{is distributed as} \quad N_{m+r}(\boldsymbol{\mu}, \boldsymbol{\Sigma})$$

with

$$\boldsymbol{\mu} = \begin{bmatrix} \underset{(m\times 1)}{\boldsymbol{\mu}_Y} \\ \hdashline \underset{(r\times 1)}{\boldsymbol{\mu}_Z} \end{bmatrix} \quad \text{and} \quad \boldsymbol{\Sigma} = \begin{bmatrix} \underset{(m\times m)}{\boldsymbol{\Sigma}_{YY}} & \underset{(m\times r)}{\boldsymbol{\Sigma}_{YZ}} \\ \hdashline \underset{(r\times m)}{\boldsymbol{\Sigma}_{ZY}} & \underset{(r\times r)}{\boldsymbol{\Sigma}_{ZZ}} \end{bmatrix}$$

By Result 4.6, the conditional expectation of $[Y_1, Y_2, \ldots, Y_m]'$, given the fixed values $z_1, z_2, \ldots, z_r$ of the predictor variables, is

$$E[\mathbf{Y} \mid z_1, z_2, \ldots, z_r] = \boldsymbol{\mu}_Y + \boldsymbol{\Sigma}_{YZ}\boldsymbol{\Sigma}_{ZZ}^{-1}(\mathbf{z} - \boldsymbol{\mu}_Z) \tag{7-57}$$

This conditional expected value, considered as a function of $z_1, z_2, \ldots, z_r$, is called the *multivariate regression* of the vector $\mathbf{Y}$ on $\mathbf{Z}$. It is composed of m univariate regressions. For instance, the first component of the conditional mean vector is $\mu_{Y_1} + \boldsymbol{\Sigma}_{Y_1Z}\boldsymbol{\Sigma}_{ZZ}^{-1}(\mathbf{z} - \boldsymbol{\mu}_Z) = E(Y_1 \mid z_1, z_2, \ldots, z_r)$, which minimizes the mean square error for the prediction of Y_1. The $m \times r$ matrix $\boldsymbol{\beta} = \boldsymbol{\Sigma}_{YZ}\boldsymbol{\Sigma}_{ZZ}^{-1}$ is called the matrix of *regression coefficients*.

The error of prediction vector

$$\mathbf{Y} - \boldsymbol{\mu}_Y - \boldsymbol{\Sigma}_{YZ}\boldsymbol{\Sigma}_{ZZ}^{-1}(\mathbf{Z} - \boldsymbol{\mu}_Z)$$

has the expected squares and crossproducts matrix

$$\begin{aligned}
\boldsymbol{\Sigma}_{YY\cdot Z} &= E\big[\mathbf{Y} - \boldsymbol{\mu}_Y - \boldsymbol{\Sigma}_{YZ}\boldsymbol{\Sigma}_{ZZ}^{-1}(\mathbf{Z} - \boldsymbol{\mu}_Z)\big]\big[\mathbf{Y} - \boldsymbol{\mu}_Y - \boldsymbol{\Sigma}_{YZ}\boldsymbol{\Sigma}_{ZZ}^{-1}(\mathbf{Z} - \boldsymbol{\mu}_Z)\big]' \\
&= \boldsymbol{\Sigma}_{YY} - \boldsymbol{\Sigma}_{YZ}\boldsymbol{\Sigma}_{ZZ}^{-1}(\boldsymbol{\Sigma}_{YZ})' - \boldsymbol{\Sigma}_{YZ}\boldsymbol{\Sigma}_{ZZ}^{-1}\boldsymbol{\Sigma}_{ZY} + \boldsymbol{\Sigma}_{YZ}\boldsymbol{\Sigma}_{ZZ}^{-1}\boldsymbol{\Sigma}_{ZZ}\boldsymbol{\Sigma}_{ZZ}^{-1}(\boldsymbol{\Sigma}_{YZ})' \\
&= \boldsymbol{\Sigma}_{YY} - \boldsymbol{\Sigma}_{YZ}\boldsymbol{\Sigma}_{ZZ}^{-1}\boldsymbol{\Sigma}_{ZY}
\end{aligned} \tag{7-58}$$

Because $\boldsymbol{\mu}$ and $\boldsymbol{\Sigma}$ are typically unknown, they must be estimated from a random sample in order to construct the multivariate linear predictor and determine expected prediction errors.

Result 7.14. Suppose $\mathbf{Y}$ and $\mathbf{Z}$ are distributed as $N_{m+r}(\boldsymbol{\mu}, \boldsymbol{\Sigma})$. The regression of the vector $\mathbf{Y}$ on $\mathbf{Z}$ is

$$\boldsymbol{\beta}_0 + \boldsymbol{\beta}\mathbf{z} = \boldsymbol{\mu}_Y - \boldsymbol{\Sigma}_{YZ}\boldsymbol{\Sigma}_{ZZ}^{-1}\boldsymbol{\mu}_Z + \boldsymbol{\Sigma}_{YZ}\boldsymbol{\Sigma}_{ZZ}^{-1}\mathbf{z} = \boldsymbol{\mu}_Y + \boldsymbol{\Sigma}_{YZ}\boldsymbol{\Sigma}_{ZZ}^{-1}(\mathbf{z} - \boldsymbol{\mu}_Z)$$

The expected squares and crossproducts matrix for the errors is

$$E(\mathbf{Y} - \boldsymbol{\beta}_0 - \boldsymbol{\beta}\mathbf{Z})(\mathbf{Y} - \boldsymbol{\beta}_0 - \boldsymbol{\beta}\mathbf{Z})' = \boldsymbol{\Sigma}_{YY\cdot Z} = \boldsymbol{\Sigma}_{YY} - \boldsymbol{\Sigma}_{YZ}\boldsymbol{\Sigma}_{ZZ}^{-1}\boldsymbol{\Sigma}_{ZY}$$

Based on a random sample of size n, the maximum likelihood estimator of the regression function is

$$\hat{\boldsymbol{\beta}}_0 + \hat{\boldsymbol{\beta}}\mathbf{z} = \overline{\mathbf{Y}} + \mathbf{S}_{YZ}\mathbf{S}_{ZZ}^{-1}(\mathbf{z} - \overline{\mathbf{Z}})$$

and the maximum likelihood estimator of $\boldsymbol{\Sigma}_{YY\cdot Z}$ is

$$\hat{\boldsymbol{\Sigma}}_{YY\cdot Z} = \left(\frac{n-1}{n}\right)(\mathbf{S}_{YY} - \mathbf{S}_{YZ}\mathbf{S}_{ZZ}^{-1}\mathbf{S}_{ZY})$$

Proof. The regression function and the covariance matrix for the prediction errors follow from Result 4.6. Using the relationships

$$\boldsymbol{\beta}_0 = \boldsymbol{\mu}_Y - \boldsymbol{\Sigma}_{YZ}\boldsymbol{\Sigma}_{ZZ}^{-1}\boldsymbol{\mu}_Z, \qquad \boldsymbol{\beta} = \boldsymbol{\Sigma}_{YZ}\boldsymbol{\Sigma}_{ZZ}^{-1}$$

$$\boldsymbol{\beta}_0 + \boldsymbol{\beta}\mathbf{z} = \boldsymbol{\mu}_Y + \boldsymbol{\Sigma}_{YZ}\boldsymbol{\Sigma}_{ZZ}^{-1}(\mathbf{z} - \boldsymbol{\mu}_Z)$$

$$\boldsymbol{\Sigma}_{YY\cdot Z} = \boldsymbol{\Sigma}_{YY} - \boldsymbol{\Sigma}_{YZ}\boldsymbol{\Sigma}_{ZZ}^{-1}\boldsymbol{\Sigma}_{ZY} = \boldsymbol{\Sigma}_{YY} - \boldsymbol{\beta}\boldsymbol{\Sigma}_{ZZ}\boldsymbol{\beta}'$$

the maximum likelihood statements follow from the invariance property [see (4-20)] of maximum likelihood estimators upon substitution of

$$\hat{\boldsymbol{\mu}} = \begin{bmatrix} \bar{\mathbf{Y}} \\ \hline \bar{\mathbf{Z}} \end{bmatrix}; \qquad \hat{\boldsymbol{\Sigma}} = \left[\begin{array}{c|c} \hat{\boldsymbol{\Sigma}}_{YY} & \hat{\boldsymbol{\Sigma}}_{YZ} \\ \hline \hat{\boldsymbol{\Sigma}}_{ZY} & \hat{\boldsymbol{\Sigma}}_{ZZ} \end{array}\right] = \left(\frac{n-1}{n}\right)\mathbf{S} = \left(\frac{n-1}{n}\right)\left[\begin{array}{c|c} \mathbf{S}_{YY} & \mathbf{S}_{YZ} \\ \hline \mathbf{S}_{ZY} & \mathbf{S}_{ZZ} \end{array}\right] \qquad \blacksquare$$

It can be shown that the unbiased estimator of $\boldsymbol{\Sigma}_{YY\cdot Z}$ is

$$\left(\frac{n-1}{n-r-1}\right)(\mathbf{S}_{YY} - \mathbf{S}_{YZ}\mathbf{S}_{ZZ}^{-1}\mathbf{S}_{ZY}) = \frac{\sum_{j=1}^{n}(\mathbf{Y}_j - \hat{\boldsymbol{\beta}}_0 - \hat{\boldsymbol{\beta}}\mathbf{Z}_j)(\mathbf{Y}_j - \hat{\boldsymbol{\beta}}_0 - \hat{\boldsymbol{\beta}}\mathbf{Z}_j)'}{n-r-1} \tag{7-59}$$

Example 7.13

We return to the computer data given in Examples 7.6 and 7.10. For Y_1 = CPU time, Y_2 = disc I/O capacity, Z_1 = orders, and Z_2 = add-delete items, we have

$$\hat{\boldsymbol{\mu}} = \begin{bmatrix} \bar{\mathbf{y}} \\ \hline \bar{\mathbf{z}} \end{bmatrix} = \begin{bmatrix} 150.44 \\ 327.79 \\ \hline 130.24 \\ 3.547 \end{bmatrix}$$

and

$$\mathbf{S} = \left[\begin{array}{c|c} \mathbf{S}_{YY} & \mathbf{S}_{YZ} \\ \hline \mathbf{S}_{ZY} & \mathbf{S}_{ZZ} \end{array}\right] = \left[\begin{array}{cc|cc} 467.913 & 1148.556 & 418.763 & 35.983 \\ 1148.556 & 3072.491 & 1008.976 & 140.558 \\ \hline 418.763 & 1008.976 & 377.200 & 28.034 \\ 35.983 & 140.558 & 28.034 & 13.657 \end{array}\right]$$

Assuming normality, the estimated regression function is

$$\hat{\boldsymbol{\beta}}_0 + \hat{\boldsymbol{\beta}}\mathbf{z} = \bar{\mathbf{y}} + \mathbf{S}_{YZ}\mathbf{S}_{ZZ}^{-1}(\mathbf{z} - \bar{\mathbf{z}})$$

$$= \begin{bmatrix} 150.44 \\ 327.79 \end{bmatrix} + \begin{bmatrix} 418.763 & 35.983 \\ 1008.976 & 140.558 \end{bmatrix}$$

$$\times \begin{bmatrix} .003128 & -.006422 \\ -.006422 & .086404 \end{bmatrix} \begin{bmatrix} z_1 - 130.24 \\ z_2 - 3.547 \end{bmatrix}$$

$$= \begin{bmatrix} 150.44 \\ 327.79 \end{bmatrix} + \begin{bmatrix} 1.079(z_1 - 130.24) + .420(z_2 - 3.547) \\ 2.254(z_1 - 130.24) + 5.665(z_2 - 3.547) \end{bmatrix}$$

Thus the minimum mean square error predictor of Y_1 is

$$150.44 + 1.079(z_1 - 130.24) + .420(z_2 - 3.547) = 8.42 + 1.08z_1 + .42z_2$$

Similarly, the best predictor of Y_2 is

$$14.14 + 2.25z_1 + 5.67z_2$$

The maximum likelihood estimate of the expected squared errors and cross-products matrix $\mathbf{\Sigma}_{YY \cdot Z}$ is given by

$$\begin{aligned}
\left(\frac{n-1}{n}\right)&(\mathbf{S}_{YY} - \mathbf{S}_{YZ}\mathbf{S}_{ZZ}^{-1}\mathbf{S}_{ZY}) \\
&= \left(\frac{6}{7}\right)\left(\begin{bmatrix} 467.913 & 1148.536 \\ 1148.536 & 3072.491 \end{bmatrix}\right. \\
&\qquad \left. - \begin{bmatrix} 418.763 & 35.983 \\ 1008.976 & 140.558 \end{bmatrix}\begin{bmatrix} .003128 & -.006422 \\ -.006422 & .086404 \end{bmatrix}\begin{bmatrix} 418.763 & 1008.976 \\ 35.983 & 140.558 \end{bmatrix}\right) \\
&= \left(\frac{6}{7}\right)\begin{bmatrix} 1.043 & 1.042 \\ 1.042 & 2.572 \end{bmatrix} = \begin{bmatrix} .894 & .893 \\ .893 & 2.205 \end{bmatrix}
\end{aligned}$$

The first estimated regression function, $8.42 + 1.08z_1 + .42z_2$, and the associated mean square error, .894, are the same as those in Example 7.12 for the single response case. Similarly, the second estimated regression function, $14.14 + 2.25z_1 + 5.67z_2$, is the same as that given in Example 7.10.

We see that the data enable us to predict the first response, Y_1, with smaller error than the second response, Y_2. The positive covariance .893 indicates that overprediction (underprediction) of CPU time tends to be accompanied by overprediction (underprediction) of disc capacity. ■

Comment. Result 7.14 states that the assumption of a joint normal distribution for the whole collection $Y_1, Y_2, \ldots, Y_m, Z_1, Z_2, \ldots, Z_r$ leads to the prediction equations

$$\begin{aligned}
\hat{y}_1 &= \hat{\beta}_{01} + \hat{\beta}_{11}z_1 + \cdots + \hat{\beta}_{r1}z_r \\
\hat{y}_2 &= \hat{\beta}_{02} + \hat{\beta}_{12}z_1 + \cdots + \hat{\beta}_{r2}z_r \\
&\ \ \vdots \qquad\qquad \vdots \\
\hat{y}_m &= \hat{\beta}_{0m} + \hat{\beta}_{1m}z_1 + \cdots + \hat{\beta}_{rm}z_r
\end{aligned}$$

We note the following:

1. The same values $z_1, z_2, \ldots, z_r$ are used to predict each Y_i.
2. The $\hat{\beta}_{ik}$ are estimates of the (i, k) entry of the regression coefficient matrix $\boldsymbol{\beta} = \mathbf{\Sigma}_{YZ}\mathbf{\Sigma}_{ZZ}^{-1}$ for $i, k \geq 1$.

We conclude this discussion of the regression problem by introducing one further correlation coefficient.

Partial Correlation Coefficient

Consider the pair of errors

$$Y_1 - \mu_{Y_1} - \boldsymbol{\Sigma}_{Y_1Z}\boldsymbol{\Sigma}_{ZZ}^{-1}(\mathbf{Z} - \boldsymbol{\mu}_Z)$$

$$Y_2 - \mu_{Y_2} - \boldsymbol{\Sigma}_{Y_2Z}\boldsymbol{\Sigma}_{ZZ}^{-1}(\mathbf{Z} - \boldsymbol{\mu}_Z)$$

obtained from using the best linear predictors to predict Y_1 and Y_2. Their correlation, determined from the error covariance matrix $\boldsymbol{\Sigma}_{YY\cdot Z} = \boldsymbol{\Sigma}_{YY} - \boldsymbol{\Sigma}_{YZ}\boldsymbol{\Sigma}_{ZZ}^{-1}\boldsymbol{\Sigma}_{ZY}$, measures the association between Y_1 and Y_2 after eliminating the effects of $Z_1, Z_2, \ldots, Z_r$.

We define the *partial correlation coefficient* between Y_1 and Y_2, eliminating $Z_1, Z_2, \ldots, Z_r$, by

$$\rho_{Y_1Y_2\cdot Z} = \frac{\sigma_{Y_1Y_2\cdot Z}}{\sqrt{\sigma_{Y_1Y_1\cdot Z}}\sqrt{\sigma_{Y_2Y_2\cdot Z}}} \tag{7-60}$$

where $\sigma_{Y_iY_k\cdot Z}$ is the (i, k) entry in the matrix $\boldsymbol{\Sigma}_{YY\cdot Z} = \boldsymbol{\Sigma}_{YY} - \boldsymbol{\Sigma}_{YZ}\boldsymbol{\Sigma}_{ZZ}^{-1}\boldsymbol{\Sigma}_{ZY}$. The corresponding *sample partial correlation coefficient* is

$$r_{Y_1Y_2\cdot Z} = \frac{s_{Y_1Y_2\cdot Z}}{\sqrt{s_{Y_1Y_1\cdot Z}}\sqrt{s_{Y_2Y_2\cdot Z}}} \tag{7-61}$$

with $s_{Y_iY_k\cdot Z}$ the (i, k) element of $\mathbf{S}_{YY} - \mathbf{S}_{YZ}\mathbf{S}_{ZZ}^{-1}\mathbf{S}_{ZY}$. Assuming $\mathbf{Y}$ and $\mathbf{Z}$ have a joint multivariate normal distribution, the sample partial correlation coefficient in (7-61) is the maximum likelihood estimator of the partial correlation coefficient in (7-60).

Example 7.14

From the computer data in Example 7.13

$$\mathbf{S}_{YY} - \mathbf{S}_{YZ}\mathbf{S}_{ZZ}^{-1}\mathbf{S}_{ZY} = \begin{bmatrix} 1.043 & 1.042 \\ 1.042 & 2.572 \end{bmatrix}$$

Therefore

$$r_{Y_1Y_2\cdot Z} = \frac{s_{Y_1Y_2\cdot Z}}{\sqrt{s_{Y_1Y_1\cdot Z}}\sqrt{s_{Y_2Y_2\cdot Z}}} = \frac{1.042}{\sqrt{1.043}\sqrt{2.572}} = .64$$

Calculating the ordinary correlation coefficient, we obtain $r_{Y_1Y_2} = .96$. Comparing the two correlation coefficients, we see that the association between Y_1 and Y_2 has been sharply reduced after eliminating the effects of the variables $\mathbf{Z}$ on both responses. ■

7.9 COMPARING THE TWO FORMULATIONS OF THE REGRESSION MODEL

In Sections 7.2 and 7.7 we presented the multiple regression models for one and several response variables, respectively. In these treatments the predictor variables have *fixed* values $\mathbf{z}_j$ at the jth trial. Alternatively, we can start—as in Section

7.8—with a set of variables that have a joint normal distribution. The process of conditioning on one subset of variables in order to predict values of the other set leads to a conditional expectation that is a multiple regression model. The two approaches to multiple regression are related. To show this relationship explicitly, we introduce two minor variants of the regression model formulation.

Mean Corrected Form of the Regression Model

For any response variable Y, the multiple regression model asserts that

$$Y_j = \beta_0 + \beta_1 z_{1j} + \cdots + \beta_r z_{rj} + \varepsilon_j$$

The predictor variables can be "centered" by subtracting their means. For instance $\beta_1 z_{1j} = \beta_1(z_{1j} - \bar{z}_1) + \beta_1 \bar{z}_1$ and we can write

$$\begin{aligned} Y_j &= (\beta_0 + \beta_1\bar{z}_1 + \cdots + \beta_r\bar{z}_r) + \beta_1(z_{1j} - \bar{z}_1) + \cdots + \beta_r(z_{rj} - \bar{z}_r) + \varepsilon_j \\ &= \beta_* + \beta_1(z_{1j} - \bar{z}_1) + \cdots + \beta_r(z_{rj} - \bar{z}_r) + \varepsilon_j \qquad (7\text{-}62) \end{aligned}$$

with $\beta_* = \beta_0 + \beta_1\bar{z}_1 + \cdots + \beta_r\bar{z}_r$. The *mean corrected* design matrix corresponding to the reparameterization in (7-62) is

$$\mathbf{Z}_c = \begin{bmatrix} 1 & z_{11} - \bar{z}_1 & \cdots & z_{1r} - \bar{z}_r \\ 1 & z_{21} - \bar{z}_1 & \cdots & z_{2r} - \bar{z}_r \\ \vdots & \vdots & & \vdots \\ 1 & z_{n1} - \bar{z}_1 & \cdots & z_{nr} - \bar{z}_r \end{bmatrix}$$

where the last r columns are each perpendicular to the first column since

$$\sum_{j=1}^{n} 1(z_{ji} - \bar{z}_i) = 0, \; i = 1, 2, \ldots, r$$

Further, setting $\mathbf{Z}_c = [\mathbf{1} \mid \mathbf{Z}_{c2}]$ with $\mathbf{Z}_{c2}'\mathbf{1} = \mathbf{0}$,

$$\mathbf{Z}_c'\mathbf{Z}_c = \begin{bmatrix} \mathbf{1}'\mathbf{1} & \mathbf{1}'\mathbf{Z}_{c2} \\ \mathbf{Z}_{c2}'\mathbf{1} & \mathbf{Z}_{c2}'\mathbf{Z}_{c2} \end{bmatrix} = \begin{bmatrix} n & \mathbf{0}' \\ \mathbf{0} & \mathbf{Z}_{c2}'\mathbf{Z}_{c2} \end{bmatrix}$$

so

$$\begin{bmatrix} \hat{\beta}_* \\ \hline \hat{\beta}_1 \\ \vdots \\ \hat{\beta}_r \end{bmatrix} = (\mathbf{Z}_c'\mathbf{Z}_c)^{-1}\mathbf{Z}_c'\mathbf{y} = \begin{bmatrix} \frac{1}{n} & \mathbf{0}' \\ \mathbf{0} & (\mathbf{Z}_{c2}'\mathbf{Z}_{c2})^{-1} \end{bmatrix} \begin{bmatrix} \mathbf{1}'\mathbf{y} \\ \mathbf{Z}_{c2}'\mathbf{y} \end{bmatrix} = \begin{bmatrix} \bar{y} \\ \hline (\mathbf{Z}_{c2}'\mathbf{Z}_{c2})^{-1}\mathbf{Z}_{c2}'\mathbf{y} \end{bmatrix} \qquad (7\text{-}63)$$

That is, the regression coefficients $[\beta_1, \beta_2, \ldots, \beta_r]'$ are unbiasedly estimated by $(\mathbf{Z}_{c2}'\mathbf{Z}_{c2})^{-1}\mathbf{Z}_{c2}'\mathbf{y}$ and β_* is estimated by $\bar{y}$. Because the definitions of $\beta_1, \beta_2, \ldots, \beta_r$ remain unchanged by the reparameterization in (7-62), their best estimates computed from the design matrix $\mathbf{Z}_c$ are exactly the same as the best estimates computed from the design matrix $\mathbf{Z}$. Thus, setting $\hat{\boldsymbol{\beta}}_c = [\hat{\beta}_1, \hat{\beta}_2, \ldots, \hat{\beta}_r]'$, the linear predictor of Y can be written as

$$\hat{y} = \hat{\beta}_* + \hat{\boldsymbol{\beta}}_c'(\mathbf{z} - \bar{\mathbf{z}}) = \bar{y} + \mathbf{y}'\mathbf{Z}_{c2}(\mathbf{Z}_{c2}'\mathbf{Z}_{c2})^{-1}(\mathbf{z} - \bar{\mathbf{z}}) \tag{7-64}$$

with $(\mathbf{z} - \bar{\mathbf{z}}) = [z_1 - \bar{z}_1, z_2 - \bar{z}_2, \ldots, z_r - \bar{z}_r]'$. Finally,

$$\begin{bmatrix} \operatorname{Var}(\hat{\beta}_*) & \operatorname{Cov}(\hat{\beta}_*, \hat{\boldsymbol{\beta}}_c) \\ \operatorname{Cov}(\hat{\boldsymbol{\beta}}_c, \hat{\beta}_*) & \operatorname{Cov}(\hat{\boldsymbol{\beta}}_c) \end{bmatrix} = (\mathbf{Z}_c'\mathbf{Z}_c)^{-1}\sigma^2 = \begin{bmatrix} \dfrac{\sigma^2}{n} & \mathbf{0}' \\ \mathbf{0} & (\mathbf{Z}_{c2}'\mathbf{Z}_{c2})^{-1}\sigma^2 \end{bmatrix} \tag{7-65}$$

Comment. The *multivariate* multiple regression model yields the same mean corrected design matrix for each response. The least squares estimates of the coefficient vectors, $\hat{\boldsymbol{\beta}}_{(i)}$, for the ith response are given by

$$\hat{\boldsymbol{\beta}}_{(i)} = \begin{bmatrix} \bar{y}_{(i)} \\ \hline (\mathbf{Z}_{c2}'\mathbf{Z}_{c2})^{-1}\mathbf{Z}_{c2}'\mathbf{y}_{(i)} \end{bmatrix}, \qquad i = 1, 2, \ldots, m \tag{7-66}$$

Sometimes, for even further numerical stability, "standardized" input variables $(z_{ji} - \bar{z}_i) \Big/ \sqrt{\sum_{j=1}^{n} (z_{ji} - \bar{z}_i)^2} = (z_{ji} - \bar{z}_i) \Big/ \sqrt{(n-1)s_{z_i z_i}}$ are used. In this case, the slope coefficients β_i in the regression model are replaced by $\tilde{\beta}_i = \beta_i\sqrt{(n-1)s_{z_i z_i}}$. The least squares estimates of the *beta coefficients* $\tilde{\beta}_i$ become $\hat{\tilde{\beta}}_i = \hat{\beta}_i\sqrt{(n-1)s_{z_i z_i}}$, $i = 1, 2, \ldots, r$. These relationships hold for each response in the multivariate multiple regression situation as well.

Relating the Formulations

When the variables $Y, Z_1, Z_2, \ldots, Z_r$ are jointly normal, we determined that the estimated predictor of Y (see Result 7.13) is

$$\hat{\beta}_0 + \hat{\boldsymbol{\beta}}'\mathbf{z} = \bar{y} + \mathbf{s}_{ZY}'\mathbf{S}_{ZZ}^{-1}(\mathbf{z} - \bar{\mathbf{z}}) = \hat{\mu}_Y + \hat{\boldsymbol{\sigma}}_{ZY}'\hat{\boldsymbol{\Sigma}}_{ZZ}^{-1}(\mathbf{z} - \hat{\boldsymbol{\mu}}_Z) \tag{7-67}$$

where the estimation procedure leads naturally to the introduction of centered z_i's.

Recall from the mean correlated form of the regression model that the best linear predictor of Y [see (7-64)] is

$$\hat{y} = \hat{\beta}_* + \hat{\boldsymbol{\beta}}_c'(\mathbf{z} - \bar{\mathbf{z}})$$

with $\hat{\beta}_* = \bar{y}$ and $\hat{\boldsymbol{\beta}}_c' = \mathbf{y}'\mathbf{Z}_{c2}(\mathbf{Z}_{c2}'\mathbf{Z}_{c2})^{-1}$. Comparing (7-64) and (7-67), $\hat{\beta}_* = \bar{y} = \hat{\beta}_0$ and $\hat{\boldsymbol{\beta}}_c = \hat{\boldsymbol{\beta}}$ since[7]

$$\mathbf{s}_{ZY}'\mathbf{S}_{ZZ}^{-1} = \mathbf{y}'\mathbf{Z}_{c2}(\mathbf{Z}_{c2}'\mathbf{Z}_{c2})^{-1} \tag{7-68}$$

[7]The identity in (7-68) is established by writing $\mathbf{y} = (\mathbf{y} - \bar{y}\mathbf{1}) + \bar{y}\mathbf{1}$ so that

$$\mathbf{y}'\mathbf{Z}_{c2} = (\mathbf{y} - \bar{y}\mathbf{1})'\mathbf{Z}_{c2} + \bar{y}\mathbf{1}'\mathbf{Z}_{c2} = (\mathbf{y} - \bar{y}\mathbf{1})'\mathbf{Z}_{c2} + \mathbf{0}' = (\mathbf{y} - \bar{y}\mathbf{1})'\mathbf{Z}_{c2}$$

Consequently,

$$\mathbf{y}'\mathbf{Z}_{c2}(\mathbf{Z}_{c2}'\mathbf{Z}_{c2})^{-1} = (\mathbf{y} - \bar{y}\mathbf{1})'\mathbf{Z}_{c2}(\mathbf{Z}_{c2}'\mathbf{Z}_{c2})^{-1} = (n-1)\mathbf{s}_{ZY}'[(n-1)\mathbf{S}_{ZZ}]^{-1} = \mathbf{s}_{ZY}'\mathbf{S}_{ZZ}^{-1}$$

Therefore, both the normal theory conditional mean and the classical regression model approaches yield exactly the *same* linear predictors.

A similar argument indicates the best linear predictors of the responses in the two multivariate multiple regression setups are also exactly the same.

Example 7.15

The computer data with the single response Y_1 = CPU time was analyzed in Example 7.6 using the classical linear regression model. The same data were analyzed again in Example 7.12, assuming the variables Y_1, Z_1, and Z_2 were jointly normal so that the best predictor of Y_1 is the conditional mean of Y_1 given z_1 and z_2. Both approaches yielded the same predictor,

$$\hat{y}_1 = 8.42 + 1.08z_1 + .42z_2 \qquad \blacksquare$$

Although the two formulations of the linear prediction problem yield the same predictor equations, conceptually they are quite different. For the models in (7-3) or (7-24), the values of the input variables are assumed to be set by the experimenter. In the conditional mean model of (7-55) or (7-57), the values of the predictor variables are random variables that are observed along with the values of the response variable(s). The assumptions underlying the second approach are more stringent, but they yield an *optimal* predictor among *all* choices rather than merely among linear predictors.

We close by noting that the multivariate regression calculations in either case can be couched in terms of the sample mean vectors $\bar{\mathbf{y}}$ and $\bar{\mathbf{z}}$ and the sample sums of squares and crossproducts

$$\left[\begin{array}{c|c} \sum_{j=1}^{n} (\mathbf{y}_j - \bar{\mathbf{y}})(\mathbf{y}_j - \bar{\mathbf{y}})' & \sum_{j=1}^{n} (\mathbf{y}_j - \bar{\mathbf{y}})(\mathbf{z}_j - \bar{\mathbf{z}})' \\ \hline \sum_{j=1}^{n} (\mathbf{z}_j - \bar{\mathbf{z}})(\mathbf{y}_j - \bar{\mathbf{y}})' & \sum_{j=1}^{n} (\mathbf{z}_j - \bar{\mathbf{z}})(\mathbf{z}_j - \bar{\mathbf{z}})' \end{array}\right] = \left[\begin{array}{c|c} \mathbf{Y}_c'\mathbf{Y}_c & \mathbf{Y}_c'\mathbf{Z}_{c2} \\ \hline \mathbf{Z}_{c2}'\mathbf{Y}_c & \mathbf{Z}_{c2}'\mathbf{Z}_{c2} \end{array}\right]$$

$$= n\left[\begin{array}{c|c} \hat{\boldsymbol{\Sigma}}_{YY} & \hat{\boldsymbol{\Sigma}}_{YZ} \\ \hline \hat{\boldsymbol{\Sigma}}_{ZY} & \hat{\boldsymbol{\Sigma}}_{ZZ} \end{array}\right]$$

This is the only information necessary to compute the estimated regression coefficients and their estimated covariances. Of course, an important part of regression analysis is model checking. This requires the residuals (errors), which must be calculated using all the original data.

7.10 PATH ANALYSIS—AN ANALYSIS OF CAUSAL RELATIONS

The method of path analysis was developed by the geneticist Sewell Wright in 1918–1921 to explain causal relations in population genetics. His 1925 application of path analysis to corn and hog prices also pioneers the use of structural equations in economics. The goal of path analysis (or structural equation analysis) is to provide plausible explanations of observed correlations by constructing models of cause-and-effect relations among variables.

The fact that a "significant" correlation coefficient does not imply a causal relationship is repeatedly emphasized in discussions of correlation, often with humorous examples such as the positive association between bubble gum sales and crime rates. Indeed, an observed correlation can never be used as proof of a causal relationship. Yet very convincing arguments for causality can be constructed from statistical inference, together with postulated relationships developed from knowledge of subject matter and common sense. For instance, the classical theory of pricing attributes an increase in the price of corn to an increase in demand or a decrease in supply. The two variables, demand and supply, are treated as the causes of changes in corn prices.

When one variable X_1 preceeds another variable X_2 in time, it may be postulated that X_1 causes X_2. Diagrammatically, we could write $X_1 \rightarrow X_2$. Allowing for error, ε_2, in the relationship, the *path diagram* is

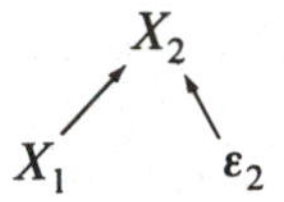

In terms of a linear model, $X_2 = \beta_0 + \beta_1 X_1 + \varepsilon_2$, where X_1 is now considered to be a causal (or exogenous) variable that is not influenced by other variables. The notion of a causal relation between X_1 and X_2 requires that all other possible causal factors be ruled out. Statistically, we specify that X_1 and ε_2 be uncorrelated, where ε_2 represents the collective effect of all unmeasured variables that could conceivably influence X_1 and X_2.

Typically, the regression $X_2 = \beta_0 + \beta_1 X_1 + \varepsilon_2$ is written in the standardized form, with obvious notation,

$$\frac{X_2 - \mu_2}{\sqrt{\sigma_{22}}} = \beta_1 \sqrt{\frac{\sigma_{11}}{\sigma_{22}}} \left(\frac{X_1 - \mu_1}{\sqrt{\sigma_{11}}} \right) + \sqrt{\frac{\sigma_{\varepsilon\varepsilon}}{\sigma_{22}}} \frac{\varepsilon_2}{\sqrt{\sigma_{\varepsilon\varepsilon}}}$$

or

$$Z_2 = p_{21} Z_1 + p_{2\varepsilon}\varepsilon \tag{7-69}$$

where even the standardized error, ε, has a coefficient. It is customary to call the parameters, p, in the standardized model *path coefficients*. The causal model in (7-69) implies

$$\rho_{12} = \text{Corr}(X_2, X_1) = \text{Corr}(Z_2, Z_1) = p_{21}$$

$$1 = \text{Var}(Z_2) = p_{21}^2 \text{Var}(Z_1) + p_{2\varepsilon}^2 \text{Var}(\varepsilon) = p_{21}^2 + p_{2\varepsilon}^2$$

The second equation states that the postulated path diagram is self-contained, or completely determined by the variables shown, since the contributions to the variance of Z_2 have a sum of one.

Mathematically, it is equally logical to postulate that X_2 causes X_1 or to postulate a third model that includes a common factor—for example, F_3—that is responsible for the observed correlation between X_1 and X_2. In the latter case the correlation between X_1 and X_2 is spurious and not a cause-effect correlation. The path diagram is

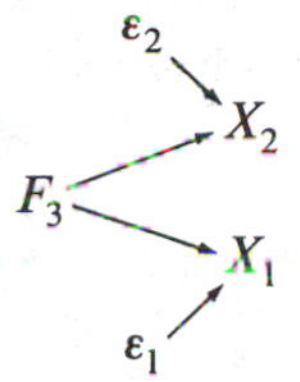

where we again allow for errors in the relationship. In terms of standardized variables, the linear model implied by the path diagram above becomes

$$\begin{aligned} Z_1 &= p_{13}F_3 + p_{1\varepsilon_1}\varepsilon_1 \\ Z_2 &= p_{23}F_3 + p_{2\varepsilon_2}\varepsilon_2 \end{aligned} \tag{7-70}$$

where the standardized errors ε_1 and ε_2 are uncorrelated with each other and with F_3. As a consequence, the correlations are related to the path coefficients by

$$\rho_{12} = \text{Corr}(X_1, X_2) = \text{Corr}(Z_1, Z_2) = p_{13}p_{23}$$
$$1 = \text{Var}(Z_1) = p_{13}^2 + p_{1\varepsilon_1}^2$$
$$1 = \text{Var}(Z_2) = p_{23}^2 + p_{2\varepsilon_2}^2$$

and

$$\rho_{13} = \text{Corr}(Z_1, F_3) = p_{13}, \qquad \rho_{23} = \text{Corr}(Z_2, F_3) = p_{23}$$

The postulated causal model in (7-70) is different from the model in (7-69), so it is not surprising that the relations between the correlations and path coefficients differ.

Path analysis contains two major components: (1) the path diagram, and (2) the decomposition of observed correlations into a sum of path coefficient terms representing simple and compound paths. These features enable us to measure both the direct and indirect effect that one variable has upon another.

Construction of a Path Diagram

A distinction is made between variables that are not influenced by other variables in the system (*exogenous* variables) and those variables that are affected by others (*endogenous* variables). With each of the latter dependent variables is associated a residual. Certain conventions govern the drawing of a path diagram. Directed arrows represent a path. The path diagram is constructed as follows.

1. A straight arrow is drawn to each dependent (endogenous) variable from each of its sources.
2. A straight arrow is also drawn to each dependent variable from its residual.
3. A curved, double-headed arrow is drawn between each pair of independent (exogenous) variables thought to have nonzero correlation.

The curved arrow for correlation is indicative of the symmetrical nature of a correlation coefficient. The other connections are directional, as indicated by the single-headed arrow.

When constructing path diagrams, it is customary to use variables that have been standardized to have mean 0 and variance 1. In the context of multiple regression, the model is

$$\frac{Y-\mu_Y}{\sqrt{\sigma_{YY}}} = \beta_1\frac{\sqrt{\sigma_{11}}}{\sqrt{\sigma_{YY}}}\left(\frac{X_1-\mu_1}{\sqrt{\sigma_{11}}}\right) + \beta_2\frac{\sqrt{\sigma_{22}}}{\sqrt{\sigma_{YY}}}\left(\frac{X_2-\mu_2}{\sqrt{\sigma_{22}}}\right) + \cdots$$

$$+\beta_r\frac{\sqrt{\sigma_{rr}}}{\sqrt{\sigma_{YY}}}\left(\frac{X_r-\mu_r}{\sqrt{\sigma_{rr}}}\right) + \frac{\sqrt{\sigma_{\varepsilon\varepsilon}}}{\sqrt{\sigma_{YY}}}\left(\frac{\varepsilon}{\sqrt{\sigma_{\varepsilon\varepsilon}}}\right)$$

or

$$Y_s = p_{Y1}Z_1 + p_{Y2}Z_2 + \cdots + p_{Yr}Z_r + p_{Y\varepsilon}\varepsilon_s \tag{7-71}$$

where the *path coefficients*, $p_{Yk} = \beta_k\sqrt{\sigma_{kk}}/\sqrt{\sigma_{YY}}$, are the regression coefficients for the standardized predictors and $p_{Y\varepsilon} = \sqrt{\sigma_{\varepsilon\varepsilon}}/\sqrt{\sigma_{YY}}$.

To illustrate the construction of path diagrams, we first draw the diagram that describes the multiple regression situation with $r = 3$ predictor variables.

When each Z_k is treated as a causal variable, the correlations between pairs of these exogenous variables are represented by curved double-headed arrows. Straight arrows go from each causal variable to Y. The error ε and each Z_k are uncorrelated (by assumption), so no arrows link these variables. The path diagram for $r = 3$ predictor variables is given in Figure 7.6.

Another simple, yet interesting, situation is the factor analysis model with one unobserved common factor (see Chapter 9). According to this model, a single unobserved factor, F, is responsible for the correlations between response variables. With three response variables, the model can be written in terms of the standardized variables F, ε_1, ε_2, ε_3 and Z_1, Z_2, Z_3 as

$$\begin{aligned} Z_1 &= p_{1F}F + p_{1\varepsilon_1}\varepsilon_1 \\ Z_2 &= p_{2F}F + p_{2\varepsilon_2}\varepsilon_2 \\ Z_3 &= p_{3F}F + p_{3\varepsilon_3}\varepsilon_3 \end{aligned} \tag{7-72}$$

where F, ε_1, ε_2, and ε_3 are all uncorrelated. The path diagram is shown in Figure 7.7.

Construction of the path diagrams can aid the investigator to think substantively about a problem and to picture the important components of observed correlations.

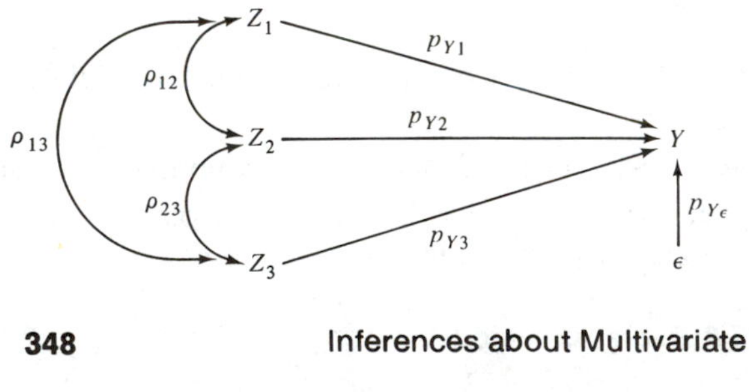

Figure 7.6 The path diagram for multiple regression with $r = 3$ predictor variables.

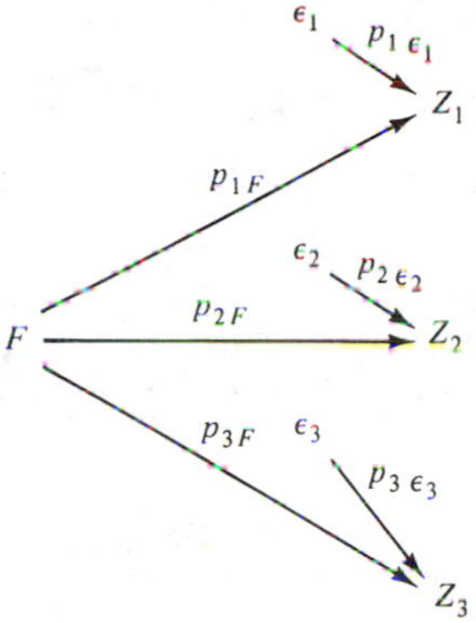

Figure 7.7 Path diagram for the factor analysis model with one common factor.

Decomposition of Observed Correlations

Estimation of the path coefficients will enable us to assess both the direct and indirect effect that one variable has on another. From the linear models that express the causal relations, we can obtain expressions that relate the path coefficients and correlations.

Example 7.16 (Path Analysis of Regression Model)

From the standardized form of the multiple regression model [see (7-71)], the correlations between Y and each Z_k can be decomposed as

$$\rho_{Yk} = \text{Corr}(Y, Z_k) = \text{Cov}\left(\sum_{i=1}^{r} p_{Yi} Z_i, Z_k\right) = \sum_{i=1}^{r} p_{Yi}\rho_{ik}, \qquad k = 1, 2, \ldots, r \tag{7-73}$$

Also, when the path diagram is self-contained so Y is completely determined by the variables in the diagram, we obtain the equation of *complete determination*

$$1 = \text{Var}(Y) = \text{Var}\left(\sum_{i=1}^{r} p_{Yi} Z_i + p_{Y\varepsilon}\varepsilon\right) = \sum_{i=1}^{r}\sum_{k=1}^{r} p_{Yi}\rho_{ik}p_{Yk} + p_{Y\varepsilon}^2$$

$$= \sum_{i=1}^{r} p_{Yi}^2 + 2\sum_{i=1}^{r}\sum_{k=i+1}^{r} p_{Yi}\rho_{ik}p_{Yk} + p_{Y\varepsilon}^2 \tag{7-74}$$

$$\begin{pmatrix}\text{total}\\ \text{variance}\\ \text{of } Y\end{pmatrix} = \begin{pmatrix}\text{variance proportion}\\ \text{given directly}\\ \text{by path coefficients}\end{pmatrix} + \begin{pmatrix}\text{variance proportion due}\\ \text{to intercorrelations}\\ \text{among independent variables}\end{pmatrix} + \begin{pmatrix}\text{variance}\\ \text{proportion}\\ \text{due to}\\ \text{error}\end{pmatrix}$$

Setting $\boldsymbol{\rho}_{ZY} = [\rho_{Y1}, \rho_{Y2}, \ldots, \rho_{Yr}]'$, the $r \times r$ matrix $\boldsymbol{\rho}_{ZZ} = \{\rho_{ik}\}$ and $\mathbf{p}_Y = [p_{Y1}, p_{Y2}, \ldots, p_{Yr}]'$, Equation (7-73) can be written in matrix notation as

$\boldsymbol{\rho}_{ZY} = \boldsymbol{\rho}_{ZZ}\mathbf{p}_Y$, so

$$\mathbf{p}_Y = \boldsymbol{\rho}_{ZZ}^{-1}\boldsymbol{\rho}_{ZY}$$

Moreover, the error term $p_{Y\varepsilon}\varepsilon$ in (7-71) has variance $p_{Y\varepsilon}^2 \text{ Var}(\varepsilon) = p_{Y\varepsilon}^2$, which, from (7-74), becomes

$$p_{Y\varepsilon}^2 = 1 - \boldsymbol{\rho}_{ZY}'\boldsymbol{\rho}_{ZZ}^{-1}\boldsymbol{\rho}_{ZY} = 1 - \boldsymbol{\rho}_{ZY}'\mathbf{p}_Y$$

That is, the squared path coefficient $p_{Y\varepsilon}^2$ is related to the multiple correlation coefficient since

$$p_{Y\varepsilon}^2 = \frac{\left(1 - \boldsymbol{\rho}_{ZY}'\boldsymbol{\rho}_{ZZ}^{-1}\boldsymbol{\rho}_{ZY}\right)}{1} = 1 - \rho_{Y(\mathbf{Z})}^2$$

For the computer data of Example 7.6, we pose the following path diagram based on presumed causal relations between Z_1, Z_2, and Y:

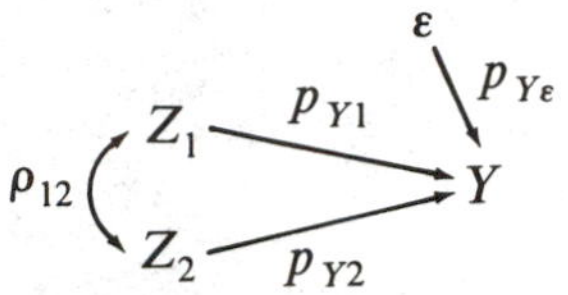

This diagram leads to the linear model (in terms of standardized variables)

$$Y = p_{Y1}Z_1 + p_{Y2}Z_2 + p_{Y\varepsilon}\varepsilon$$

Consequently, Equations (7-73) and (7-74) become

$$\rho_{Y1} = p_{Y1}(1) + p_{Y2}\rho_{12}$$
$$\rho_{Y2} = p_{Y1}\rho_{12} + p_{Y2}(1)$$

and $$1 = \text{Var}(Y) = p_{Y1}^2 + p_{Y2}^2 + p_{Y\varepsilon}^2 + 2p_{Y1}\rho_{12}p_{Y2}$$

Substituting the *sample* correlations (see Example 7.12 for **S**) $r_{Y1} = r_{YZ_1} = .997$, $r_{Y2} = r_{YZ_2} = .450$ and $r_{12} = r_{Z_1Z_2} = .391$ for the corresponding population quantities above, we can estimate the path coefficients p_{Y1} and p_{Y2} by solving

$$.997 = p_{Y1} + .391p_{Y2}$$
$$.450 = .391p_{Y1} + p_{Y2}$$

Equivalently, we can use

$$\hat{\mathbf{p}}_Y = \begin{bmatrix} \hat{p}_{Y1} \\ \hat{p}_{Y2} \end{bmatrix} = \hat{\boldsymbol{\rho}}_{ZZ}^{-1}\hat{\boldsymbol{\rho}}_{ZY} = \begin{bmatrix} 1 & .391 \\ .391 & 1 \end{bmatrix}^{-1}\begin{bmatrix} .997 \\ .450 \end{bmatrix} = \begin{bmatrix} .969 \\ .071 \end{bmatrix}$$

Finally

$$\hat{p}_{Y\varepsilon}^2 = 1 - \hat{\boldsymbol{\rho}}_{ZY}'\hat{\mathbf{p}}_Y = 1 - [.997, .450]\begin{bmatrix} .969 \\ .071 \end{bmatrix} = .002$$

and

$$\hat{p}_{Y\varepsilon} = \sqrt{.002} = .044$$

Thus the observed correlations between the response Y = CPU time and the predictor variables Z_1 = orders and Z_2 = add-delete items can be decomposed into pieces representing direct and indirect effects. For example, Z_1 affects Y directly (represented by the path coefficient $\hat{p}_{Y1}$) and also affects Y indirectly through Z_2 (represented by the product term $\hat{\rho}_{12}\hat{p}_{Y2}$). Substituting the numbers on the path diagram, we have

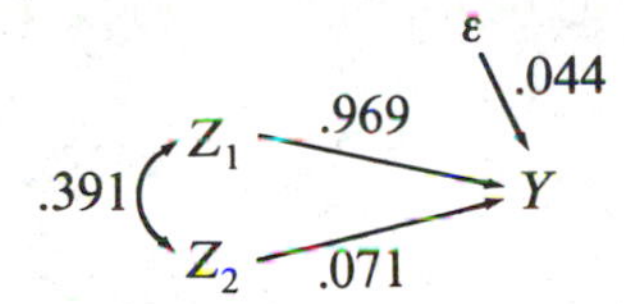

It is convenient to use a table to show the decomposition of the effects of the predictor variables on the response.

	Indirect effect	Direct effect	Total effect
Z_1 (orders)	.028	.969	.997
Z_2 (add-delete items)	.379	.071	.450

Note that the path coefficients measuring the direct effect of the Z_k's on Y are the regression coefficients for the standardized variables. ■

Example 7.17 (Path Analysis of Factor Analysis Model with One Common Factor)

The single factor model in (7-72) for three response variables yields the relations for the decomposition of observed correlations

$$\rho_{ik} = \text{Corr}(Z_i, Z_k) = \text{Cov}(p_{iF}F + p_{i\varepsilon_i}\varepsilon_i, p_{kF}F + p_{k\varepsilon_k}\varepsilon_k) = p_{iF}p_{kF},$$
$$i \neq k = 1, 2, 3$$

and the complete determination equations

$$1 = \text{Var}(Z_k) = \text{Var}(p_{kF}F + p_{k\varepsilon_k}\varepsilon_k) = p_{kF}^2 + p_{k\varepsilon_k}^2, \qquad k = 1, 2, 3$$

These six equations are easily solved for the path coefficients in terms of the estimated correlations.

Example 8.4 gives the sample covariance matrix **S** for three dimensions of turtle shells, from which we determine $r_{12} = .951$, $r_{13} = .942$, and $r_{14} = .911$. Assuming a single (growth) factor causes the shell dimensions, we can write

$$\left.\begin{aligned} .951 &= \hat{p}_{1F}\hat{p}_{2F} \\ .942 &= \hat{p}_{1F}\hat{p}_{3F} \\ .911 &= \hat{p}_{2F}\hat{p}_{3F} \end{aligned}\right\} \qquad \text{so} \qquad \frac{(.951)(.942)}{.911} = \frac{\hat{p}_{1F}\hat{p}_{2F}\hat{p}_{1F}\hat{p}_{3F}}{\hat{p}_{2F}\hat{p}_{3F}} = \hat{p}_{1F}^2$$

and $\hat{p}_{1F} = .992$. Also $\hat{p}_{1\varepsilon_1}^2 = 1 - \hat{p}_{1F}^2 = .017$ and $\hat{p}_{1\varepsilon_1} = .129$. Similarly, $\hat{p}_{2F} = \sqrt{(.951)(.911)/.942} = .959$, $\hat{p}_{2\varepsilon_2} = \sqrt{1 - (.959)^2} = .283$, $\hat{p}_{3F} = .950$, and

$\hat{p}_{3\varepsilon_3} = .312$. All of the path coefficients for the common factor are large compared to the error path coefficients. This suggests a strong causal mechanism if this causal model is correct. In addition, the path coefficients $\hat{p}_{kF}$ are nearly equal, although $Z_1 = \ln(\text{length})$ is influenced slightly more by F. The path diagram with the estimated path coefficients is displayed below.

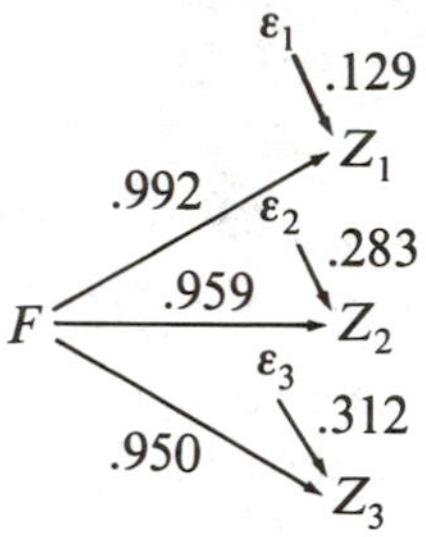

■

To summarize, path analysis takes substantive theories for causal orderings and uses path diagrams to obtain decompositions of observed correlations into direct and indirect influences. The path coefficients help determine the relative importance of the direct and indirect influences. The conclusions of a path analysis will depend on the assumed causal relationships.

The literature on path analysis is growing rapidly. Good references for additional reading are [19], [8], [2], [13], and [23].

EXERCISES

7.1. Given the data

z_1	10	5	7	19	11	8
y	15	9	3	25	7	13

fit the linear regression model $Y_j = \beta_0 + \beta_1 z_{j1} + \varepsilon_j$, $j = 1, 2, \ldots, 6$. Specifically, calculate the least squares estimates $\hat{\boldsymbol{\beta}}$, the fitted values $\hat{\mathbf{y}}$, the residuals $\hat{\boldsymbol{\varepsilon}}$, and the residual sum of squares, $\hat{\boldsymbol{\varepsilon}}'\hat{\boldsymbol{\varepsilon}}$.

7.2. Given the data

z_1	10	5	7	19	11	18
z_2	2	3	3	6	7	9
y	15	9	3	25	7	13

fit the regression model

$$Y_j = \beta_1 z_{j1} + \beta_2 z_{j2} + \varepsilon_j, \qquad j = 1, 2, \ldots, 6$$

to the *standardized* form of the variables y, z_1, and z_2. From this fit, deduce the corresponding fitted regression equation for the original (not standardized) variables.

7.3. (Weighted least squares estimators.) Let

$$\underset{(n\times 1)}{\mathbf{Y}} = \underset{(n\times(r+1))}{\mathbf{Z}} \; \underset{((r+1)\times 1)}{\boldsymbol{\beta}} + \underset{(n\times 1)}{\boldsymbol{\varepsilon}}$$

where $E(\boldsymbol{\varepsilon}) = \mathbf{0}$ but $E(\boldsymbol{\varepsilon}\boldsymbol{\varepsilon}') = \sigma^2\mathbf{V}$, with $\mathbf{V}$ $(n \times n)$ known. Show that the *weighted least squares* estimator is

$$\hat{\boldsymbol{\beta}}_W = (\mathbf{Z}'\mathbf{V}^{-1}\mathbf{Z})^{-1}\mathbf{Z}'\mathbf{V}^{-1}\mathbf{Y}$$

If σ^2 is unknown, it may be estimated, unbiasedly, by $(\mathbf{Y} - \mathbf{Z}\hat{\boldsymbol{\beta}}_W)'\mathbf{V}^{-1}(\mathbf{Y} - \mathbf{Z}\hat{\boldsymbol{\beta}}_W)/(n - r - 1)$.
(*Hint*: $\mathbf{V}^{-1/2}\mathbf{Y} = (\mathbf{V}^{-1/2}\mathbf{Z})\boldsymbol{\beta} + \mathbf{V}^{-1/2}\boldsymbol{\varepsilon}$ is of the classical linear regression form $\mathbf{Y}^* = \mathbf{Z}^*\boldsymbol{\beta} + \boldsymbol{\varepsilon}^*$, with $E(\boldsymbol{\varepsilon}^*) = \mathbf{0}$ and $E(\boldsymbol{\varepsilon}^*\boldsymbol{\varepsilon}^{*\prime}) = \sigma^2\mathbf{I}$. Thus $\hat{\boldsymbol{\beta}}_W = \hat{\boldsymbol{\beta}}^* = (\mathbf{Z}^{*\prime}\mathbf{Z}^*)^{-1}\mathbf{Z}^{*\prime}\mathbf{Y}^*$.)

7.4. Use the weighted least squares estimator in Exercise 7.3 to derive an expression for the estimate of the slope β in the model $Y_j = \beta z_j + \varepsilon_j, j = 1, 2, \ldots, n$, when (a) $\text{Var}(\varepsilon_j) = \sigma^2$, (b) $\text{Var}(\varepsilon_j) = \sigma^2 z_j$, and (c) $\text{Var}(\varepsilon_j) = \sigma^2 z_j^2$. Comment on the manner in which the unequal variances for the errors influence the optimal choice of $\hat{\beta}_W$.

7.5. Establish (7-54): $\rho^2_{Y(\mathbf{Z})} = 1 - 1/\rho^{YY}$.
(*Hint*: From (7-53) and Exercise 4.8,

$$1 - \rho^2_{Y(\mathbf{Z})} = \frac{\sigma_{YY} - \boldsymbol{\sigma}'_{ZY}\boldsymbol{\Sigma}^{-1}_{ZZ}\boldsymbol{\sigma}_{ZY}}{\sigma_{YY}} = \frac{|\boldsymbol{\Sigma}_{ZZ}|\left(\sigma_{YY} - \boldsymbol{\sigma}'_{ZY}\boldsymbol{\Sigma}^{-1}_{ZZ}\boldsymbol{\sigma}_{ZY}\right)}{|\boldsymbol{\Sigma}_{ZZ}|\,\sigma_{YY}} = \frac{|\boldsymbol{\Sigma}|}{|\boldsymbol{\Sigma}_{ZZ}|\,\sigma_{YY}}$$

From Result 2A.8(c), $\sigma^{YY} = |\boldsymbol{\Sigma}_{ZZ}|/|\boldsymbol{\Sigma}|$, where σ^{YY} is the entry of $\boldsymbol{\Sigma}^{-1}$ in the first row and first column. Since (see Exercise 2.21) $\boldsymbol{\rho} = \mathbf{V}^{-1/2}\boldsymbol{\Sigma}\mathbf{V}^{-1/2}$ and $\boldsymbol{\rho}^{-1} = (\mathbf{V}^{-1/2}\boldsymbol{\Sigma}\mathbf{V}^{-1/2})^{-1} = \mathbf{V}^{1/2}\boldsymbol{\Sigma}^{-1}\mathbf{V}^{1/2}$, the entry in the $(1, 1)$ position of $\boldsymbol{\rho}^{-1}$ is $\rho^{YY} = \sigma^{YY}\sigma_{YY}$.)

***7.6.** (Generalized inverse of $\mathbf{Z}'\mathbf{Z}$.) A matrix $(\mathbf{Z}'\mathbf{Z})^-$ is called a *generalized inverse* of $\mathbf{Z}'\mathbf{Z}$ if $\mathbf{Z}'\mathbf{Z}(\mathbf{Z}'\mathbf{Z})^-\mathbf{Z}'\mathbf{Z} = \mathbf{Z}'\mathbf{Z}$. Suppose $\lambda_1 \geq \lambda_2 \geq \cdots \geq \lambda_{r_1+1} > 0$ are the nonzero eigenvalues of $\mathbf{Z}'\mathbf{Z}$ with corresponding eigenvectors $\mathbf{e}_1, \mathbf{e}_2, \ldots, \mathbf{e}_{r_1+1}$.

(a) Show that

$$(\mathbf{Z}'\mathbf{Z})^- = \sum_{i=1}^{r_1+1} \lambda_i^{-1}\mathbf{e}_i\mathbf{e}_i'$$

is a generalized inverse of $\mathbf{Z}'\mathbf{Z}$.

(b) The coefficients $\hat{\boldsymbol{\beta}}$ that minimize the sum of squared errors $(\mathbf{y} - \mathbf{Z}\boldsymbol{\beta})'(\mathbf{y} - \mathbf{Z}\boldsymbol{\beta})$ satisfy the *normal equations* $(\mathbf{Z}'\mathbf{Z})\hat{\boldsymbol{\beta}} = \mathbf{Z}'\mathbf{y}$. Show that these equations are satisfied for any $\hat{\boldsymbol{\beta}}$ such that $\mathbf{Z}\hat{\boldsymbol{\beta}}$ is the projection of $\mathbf{y}$ on the column of $\mathbf{Z}$.

(c) Show that $\mathbf{Z}\hat{\boldsymbol{\beta}} = \mathbf{Z}(\mathbf{Z}'\mathbf{Z})^-\mathbf{Z}'\mathbf{y}$ is the projection of $\mathbf{y}$ on the columns of $\mathbf{Z}$ (see Footnote 2 in Chapter 7).

(d) Show directly that $\hat{\boldsymbol{\beta}} = (\mathbf{Z}'\mathbf{Z})^-\mathbf{Z}'\mathbf{y}$ is a solution to the normal equations $(\mathbf{Z}'\mathbf{Z})[(\mathbf{Z}'\mathbf{Z})^-\mathbf{Z}'\mathbf{y}] = \mathbf{Z}'\mathbf{y}$.

(*Hint*: **(b)** If $\mathbf{Z}\hat{\boldsymbol{\beta}}$ is the projection, then $\mathbf{y} - \mathbf{Z}\hat{\boldsymbol{\beta}}$ is perpendicular to the columns of $\mathbf{Z}$.
(d) The eigenvalue-eigenvector requirement implies $(\mathbf{Z}'\mathbf{Z})(\lambda_i^{-1}\mathbf{e}_i) = \mathbf{e}_i$ for $i \leq r_1 + 1$ or $0 = \mathbf{e}_i'(\mathbf{Z}'\mathbf{Z})\mathbf{e}_i$ for $i > r_1 + 1$. Therefore $(\mathbf{Z}'\mathbf{Z})(\lambda_i^{-1}\mathbf{e}_i)\mathbf{e}_i'\mathbf{Z}' = \mathbf{e}_i\mathbf{e}_i'\mathbf{Z}'$. Summing over i gives

$$(\mathbf{Z}'\mathbf{Z})(\mathbf{Z}'\mathbf{Z})^-\mathbf{Z}' = \mathbf{Z}'\mathbf{Z}\left(\sum_{i=1}^{r_1+1}\lambda_i^{-1}\mathbf{e}_i\mathbf{e}_i'\right)\mathbf{Z}' = \left(\sum_{i=1}^{r_1+1}\mathbf{e}_i\mathbf{e}_i'\right)\mathbf{Z}' = \left(\sum_{i=1}^{r+1}\mathbf{e}_i\mathbf{e}_i'\right)\mathbf{Z}' = \mathbf{I}\mathbf{Z}' = \mathbf{Z}',$$

since $\mathbf{e}_i'\mathbf{Z}' = \mathbf{0}$ for $i > r_1 + 1$.)

7.7. Suppose the classical regression model is written as

$$\underset{(n\times 1)}{\mathbf{Y}} = \underset{(n\times(q+1))}{\mathbf{Z}_1}\ \underset{((q+1)\times 1)}{\boldsymbol{\beta}_{(1)}} + \underset{(n\times(r-q))}{\mathbf{Z}_2}\ \underset{((r-q)\times 1)}{\boldsymbol{\beta}_{(2)}} + \underset{(n\times 1)}{\boldsymbol{\varepsilon}}$$

If the parameters $\boldsymbol{\beta}_{(2)}$ are identified beforehand as being of primary interest, show that a

$100(1-\alpha)\%$ confidence region for $\boldsymbol{\beta}_{(2)}$ is given by

$$(\hat{\boldsymbol{\beta}}_{(2)} - \boldsymbol{\beta}_{(2)})'[\mathbf{Z}_2'\mathbf{Z}_2 - \mathbf{Z}_2'\mathbf{Z}_1(\mathbf{Z}_1'\mathbf{Z}_1)^{-1}\mathbf{Z}_1'\mathbf{Z}_2](\hat{\boldsymbol{\beta}}_{(2)} - \boldsymbol{\beta}_{(2)})$$

$$\leq s^2(r-q)F_{r-q,\,n-r-1}(\alpha)$$

[*Hint:* By Exercise 4.9, with 1's and 2's interchanged,

$$\mathbf{C}^{22} = [\mathbf{Z}_2'\mathbf{Z}_2 - \mathbf{Z}_2'\mathbf{Z}_1(\mathbf{Z}_1'\mathbf{Z}_1)^{-1}\mathbf{Z}_1'\mathbf{Z}_2]^{-1}, \quad \text{where } (\mathbf{Z}'\mathbf{Z})^{-1} = \begin{bmatrix} \mathbf{C}^{11} & \mathbf{C}^{12} \\ \mathbf{C}^{21} & \mathbf{C}^{22} \end{bmatrix}$$

Multiplying by the square root matrix $(\mathbf{C}^{22})^{-1/2}$, conclude that $(\mathbf{C}^{22})^{-1/2}(\hat{\boldsymbol{\beta}}_{(2)} - \boldsymbol{\beta}_{(2)})/\sigma^2$ is $N(\mathbf{0}, \mathbf{I})$ so $(\hat{\boldsymbol{\beta}}_{(2)} - \boldsymbol{\beta}_{(2)})'\mathbf{C}^{22}(\hat{\boldsymbol{\beta}}_{(2)} - \boldsymbol{\beta}_{(2)})$ is χ^2_{r-q}.]

7.8. Given the following data on one predictor variable z_1 and two responses Y_1 and Y_2,

z_1	-2	-1	0	1	2
y_1	5	3	4	2	1
y_2	-3	-1	-1	2	3

determine the least squares estimates of the parameters in the straight-line regression model

$$Y_{j1} = \beta_{01} + \beta_{11}z_{j1} + \varepsilon_{j1}$$

$$Y_{j2} = \beta_{02} + \beta_{12}z_{j1} + \varepsilon_{j2}, \qquad j = 1,2,3,4,5$$

Also calculate the matrices of fitted values $\hat{\mathbf{Y}}$, and residuals, $\hat{\boldsymbol{\varepsilon}}$; with $\mathbf{Y} = [\mathbf{y}_1 \mid \mathbf{y}_2]$. Verify the sum of squares and crossproducts breakup

$$\mathbf{Y}'\mathbf{Y} = \hat{\mathbf{Y}}'\hat{\mathbf{Y}} + \hat{\boldsymbol{\varepsilon}}'\hat{\boldsymbol{\varepsilon}}$$

7.9. Using the results from Exercise 7.8, calculate each of the following.

(a) A 95% confidence interval for the mean response, $E(Y_{01}) = \beta_{01} + \beta_{11}z_{01}$, corresponding to $z_{01} = 0.5$.

(b) A 95% prediction interval for the actual response Y_{01} corresponding to $z_{01} = 0.5$.

(c) A 95% prediction region for the actual responses Y_{01} and Y_{02} corresponding to $z_{01} = 0.5$.

7.10. (Generalized least squares for multivariate multiple regression.) Let $\mathbf{A}$ be a positive definite matrix so $d_j^2(\mathbf{B}) = (\mathbf{y}_j - \mathbf{B}'\mathbf{z}_j)'\mathbf{A}(\mathbf{y}_j - \mathbf{B}'\mathbf{z}_j)$ is a squared statistical distance from the jth observation $\mathbf{y}_j$ to its regression $\mathbf{B}'\mathbf{z}_j$. Show that the choice $\mathbf{B} = \hat{\boldsymbol{\beta}} = (\mathbf{Z}'\mathbf{Z})^{-1}\mathbf{Z}'\mathbf{Y}$ minimizes the sum of squared statistical distances, $\sum_{j=1}^{n} d_j^2(\mathbf{B})$, for any choice of positive definite $\mathbf{A}$. Choices for $\mathbf{A}$ include $\boldsymbol{\Sigma}^{-1}$ and $\mathbf{I}$.

[*Hint:* Repeat the steps in (7-38) and (7-39) with $\boldsymbol{\Sigma}^{-1}$ replaced by $\mathbf{A}$.]

7.11. Given the mean vector and covariance matrix of Y, Z_1, and Z_2,

$$\boldsymbol{\mu} = \begin{bmatrix} \mu_Y \\ \hline \boldsymbol{\mu}_Z \end{bmatrix} = \begin{bmatrix} 4 \\ \hline 3 \\ -2 \end{bmatrix} \quad \text{and} \quad \boldsymbol{\Sigma} = \begin{bmatrix} \sigma_{YY} & \boldsymbol{\sigma}_{ZY}' \\ \boldsymbol{\sigma}_{ZY} & \boldsymbol{\Sigma}_{ZZ} \end{bmatrix} = \begin{bmatrix} 9 & 3 & 1 \\ 3 & 2 & 1 \\ 1 & 1 & 1 \end{bmatrix}$$

determine each of the following.

(a) The best linear predictor $\beta_0 + \beta_1 Z_1 + \beta_2 Z_2$ of Y.

(b) The mean square error of the best linear predictor.

(c) The population multiple correlation coefficient.

(d) The partial correlation coefficient $\rho_{YZ_1 \cdot Z_2}$.

7.12. The test scores for college students described in Example 5.4 have

$$\bar{\mathbf{z}} = \begin{bmatrix} \bar{z}_1 \\ \bar{z}_2 \\ \bar{z}_3 \end{bmatrix} = \begin{bmatrix} 527.74 \\ 54.69 \\ 25.13 \end{bmatrix}, \qquad \mathbf{S} = \begin{bmatrix} 5691.34 & & \\ 600.51 & 126.05 & \\ 217.25 & 23.37 & 23.11 \end{bmatrix}$$

Assume joint normality.

(a) Obtain the maximum likelihood estimates of the parameters for predicting Z_1 from Z_2 and Z_3.

(b) Evaluate the estimated multiple correlation coefficient $R_{Z_1(Z_2, Z_3)}$.

(c) Determine the estimated partial correlation coefficient, $r_{Z_1, Z_2 \cdot Z_3}$.

7.13. Twenty-five portfolio managers were evaluated in terms of their performance. Suppose Y represents the rate of return achieved over a period of time; Z_1 is the manager's attitude toward risk measured on a five-point scale from "very conservative" to "very risky"; and Z_2 is years of experience in the investment business. The observed correlation coefficients between pairs of variables are:

$$\begin{array}{cc} & \begin{array}{ccc} Y & Z_1 & Z_2 \end{array} \\ \mathbf{R} = & \begin{bmatrix} 1.0 & -.35 & .82 \\ -.35 & 1.0 & -.60 \\ .82 & -.60 & 1.0 \end{bmatrix} \end{array}$$

(a) Interpret the sample correlation coefficients $r_{YZ_1} = -.35$ and $r_{YZ_2} = .82$.

(b) Calculate the partial correlation coefficient $r_{YZ_1 \cdot Z_2}$ and interpret this quantity with respect to the interpretation provided for r_{YZ_1} in Part a.

7.14. Sociologists are often interested in the relationship between the final occupational level of offspring and the amounts of education of the offspring and parents. Let X_1 = final occupational level of son, X_2 = education level of son, and X_3 = education level of father. Using an index of occupational level, suppose the sample correlations for the pairs of variables are

$$\begin{array}{cc} & \begin{array}{ccc} X_1 & X_2 & X_3 \end{array} \\ \mathbf{R} = & \begin{bmatrix} 1 & .62 & .29 \\ .62 & 1 & .51 \\ .29 & .51 & 1 \end{bmatrix} \end{array}$$

(a) Construct a path diagram, assuming X_2 and X_3 "cause" X_1. (Allow for the influence of error in X_1 and for the association between X_2 and X_3.)

(b) Evaluate the path coefficients and display them on the path diagram. Discuss the results.

The following exercises may require the use of a computer.

7.15. Use the real-estate data in Table 7.1 and the linear regression model in Example 7.4.

(a) Verify the results in Example 7.4.

(b) Analyze the residuals to check the adequacy of the model (see Section 7.6).

(c) Generate a 95% prediction interval for the selling price (Y_0) corresponding to total dwelling size $z_1 = 17$ and assessed value, $z_2 = 46$.

(d) Carry out a likelihood ratio test of H_0: $\beta_2 = 0$ with a significance level of $\alpha = .05$. Should the original model be modified? Discuss.

7.16. Calculate a C_p plot corresponding to the possible linear regressions involving the real-estate data in Table 7.1.

7.17. Satellite applications motivated the development of a silver-zinc battery. Table 7.4 contains failure data collected to characterize the cycle life performance. Use these data.

TABLE 7.4 BATTERY-FAILURE DATA

Z_1 Charge rate (amps)	Z_2 Discharge rate (amps)	Z_3 Depth of discharge (% of rated ampere-hours)	Z_4 Temperature (°C)	Z_5 End of charge voltage (volts)	Y Cycles to failure
0.375	3.13	60.0	40.	2.00	101.
1.000	3.13	76.8	30.	1.99	141.
1.000	3.13	60.0	20.	2.00	96.
1.000	3.13	60.0	20.	1.98	125.
1.625	3.13	43.2	10.	2.01	43.
1.625	3.13	60.0	20.	2.00	16.
1.625	3.13	60.0	20.	2.02	188.
0.375	5.00	76.8	10.	2.01	10.
1.000	5.00	43.2	10.	1.99	3.
1.000	5.00	43.2	30.	2.01	386.
1.000	5.00	100.0	20.	2.00	45.
1.625	5.00	76.8	10.	1.99	2.
0.375	1.25	76.8	10.	2.01	76.
1.000	1.25	43.2	10.	1.99	78.
1.000	1.25	76.8	30.	2.00	160.
1.000	1.25	60.0	0.	2.00	3.
1.625	1.25	43.2	30.	1.99	216.
1.625	1.25	60.0	20.	2.00	73.
0.375	3.13	76.8	30.	1.99	314.
0.375	3.13	60.0	20.	2.00	170.

SOURCE: Selected from S. Sidik, H. Leibecki, and J. Bozek, "Failure of Silver-Zinc Cells with Competing Failure Modes—Preliminary Data Analysis," NASA Technical Memorandum 81556 (Cleveland, Ohio: Lewis Research Center, 1980).

(a) Find the estimated linear regression of $\ln(Y)$ on an appropriate ("best") subset of predictor variables.

(b) Plot the residuals from the fitted model chosen in Part a to check the normal assumption.

7.18. Using the battery-failure data in Table 7.4, regress $\ln(Y)$ on the first principal component of the predictor variables $z_1, z_2, \ldots, z_5$ (see Section 8.3). Compare the result with the fitted model obtained in Exercise 7.17(a).

7.19. Consider the air-pollution data in Table 1.2. Let $Y_1 = NO_2$ and $Y_2 = O_3$ be the two responses (pollutants) corresponding to the predictor variables Z_1 = wind and Z_2 = solar radiation.

(a) Perform a regression analysis using only the first response Y_1.

(i) Suggest and fit appropriate linear regression models.

(ii) Analyze the residuals.

(iii) Construct a 95% prediction interval for NO_2 corresponding to $z_1 = 10$ and $z_2 = 80$.

(b) Perform a multivariate multiple regression analysis using both responses Y_1 and Y_2.

(i) Suggest and fit appropriate linear regression models.

(ii) Analyze the residuals.

(iii) Construct a 95% prediction ellipse for both NO_2 and O_3 for $z_1 = 10$ and $z_2 = 80$. Compare this ellipse with the prediction interval in Part a(iii). Comment.

7.20. (Canonical correlations and canonical variables.) A canonical correlation analysis seeks to identify and quantify the associations between *two sets* of variables $\underset{(m\times 1)}{\mathbf{Y}}$ and $\underset{(s\times 1)}{\mathbf{Z}}$. Canonical correlation includes and extends the multiple correlation coefficient. The analysis focuses on the correlation between a *linear combination*, $U = \mathbf{a}'\mathbf{Y}$, of the variables in one set and a *linear combination*, $V = \mathbf{b}'\mathbf{Z}$, of the variables in the other set. In the notation of Section 7.8, the pair of linear combinations has sample correlation

$$r_{U,V} = \frac{\mathbf{a}'\mathbf{S}_{YZ}\mathbf{b}}{\sqrt{\mathbf{a}'\mathbf{S}_{YY}\mathbf{a}}\,\sqrt{\mathbf{b}'\mathbf{S}_{ZZ}\mathbf{b}}}$$

The *first pair of sample canonical variates* is the pair of linear combinations $U_1 = \mathbf{a}_1'\mathbf{Y}$; $V_1 = \mathbf{b}_1'\mathbf{Z}$, having unit sample variances, that *maximize* the sample correlation $r_{U,V}$. The *first sample canonical correlation* is $\hat{\rho}_1^* = r_{U_1,V_1}$.

The *kth pair of sample canonical variates* are the linear combinations $U_k = \mathbf{a}_k'\mathbf{Y}$; $V_k = \mathbf{b}_k'\mathbf{Z}$, with unit sample variances, which maximize the sample correlation $r_{U,V}$ among those linear combinations uncorrelated with the previous $k-1$ sample canonical variates. The *kth sample canonical correlation* is $\hat{\rho}_k^* = r_{U_k,V_k}$.

The $k \le \min(m, s)$ sample canonical variates and canonical correlations can be obtained from the sample covariance matrix. In particular, the $\hat{\rho}_k^*$ can be obtained as solutions to the equation

$$0 = |\mathbf{S}_{YZ}\mathbf{S}_{ZZ}^{-1}\mathbf{S}_{ZY} - (\hat{\rho}^*)^2\mathbf{S}_{YY}|$$

where $\hat{\rho}_1^* \ge \hat{\rho}_2^* \ge \cdots \ge \hat{\rho}_k^* \ge 0$. The coefficient vectors $\mathbf{a}_1, \mathbf{b}_1; \mathbf{a}_2, \mathbf{b}_2; \cdots; \mathbf{a}_k, \mathbf{b}_k$ are then determined by

$$\begin{bmatrix} -\hat{\rho}_k^*\mathbf{S}_{YY} & \mathbf{S}_{YZ} \\ \mathbf{S}_{ZY} & -\hat{\rho}_k^*\mathbf{S}_{ZZ} \end{bmatrix} \begin{bmatrix} \mathbf{a}_k \\ \mathbf{b}_k \end{bmatrix} = \mathbf{0}$$

This algebra is most conveniently performed by a computer program.

From the covariance matrix in Example 7.13:

(a) Find the sample canonical correlations.

(b) Determine the first sample canonical variate pair U_1, V_1 and interpret these quantities.

REFERENCES

[1] Anderson, T. W., *An Introduction to Multivariate Statistical Analysis*, New York: John Wiley, 1958.

[2] Asher, H. E., *Causal Modeling*. Sage University Paper Series on Quantitative Applications in the Social Sciences, 07–003. Beverly Hills and London: Sage Publications, 1976.

[3] Bartlett, M. S., "A Note on Multiplying Factors for Various Chi-Squared Approximations," *Journal of the Royal Statistical Society (B)*, **16** (1954), 296–298.

[4] Box, G. E. P., "A General Distribution Theory for a Class of Likelihood Criteria," *Biometrika*, **36** (1949), 317–346.

[5] Chatterjee, S., and B. Price, *Regression Analysis by Example*, New York: John Wiley, 1977.

[6] Daniel, C., and F. S. Wood, *Fitting Equations to Data* (2nd ed.), New York: John Wiley, 1980.

[7] Draper, N. R., and H. Smith, *Applied Regression Analysis* (2nd ed.), New York: John Wiley, 1981.

[8] Duncan, O. D., *Introduction to Structural Equation Models*, New York: Academic Press, 1975.

[9] Durbin, J., and G. S. Watson, "Testing for Serial Correlation in Least Squares Regression, II," *Biometrika*, **38** (1951), 159–178.

[10] Galton, F., "Regression Toward Mediocrity in Heredity Stature," *Journal of the Anthropological Institute*, **15** (1885), 246–263.

[11] Goldberger, A. S., *Econometric Theory*, New York: John Wiley, 1964.

[12] Heck, D. L., "Charts of Some Upper Percentage Points of the Distribution of the Largest Characteristic Root," *Annals of Mathematical Statistics*, **31** (1960), 625–642.

[13] Li, C. C., *Path Analysis*, Pacific Grove, Calif.: Boxwood Press, 1975.

[14] Miller, R. B., and D. W. Wichern, *Intermediate Business Statistics: Analysis of Variance, Regression and Time Series*, New York: Holt, Rinehart and Winston, 1977.

[15] Neter, J., and W. Wasserman, *Applied Linear Statistical Models*, Homewood, Ill.: Richard D. Irwin, 1974.

[16] Pillai, K. C. S., "Upper Percentage Points of the Largest Root of a Matrix in Multivariate Analysis," *Biometrika*, **54** (1967), 189–193.

[17] Rao, C. R., *Linear Statistical Inference and Its Applications* (2nd ed.), New York: John Wiley, 1973.

[18] Seber, G. A. F., *Linear Regression Analysis*, New York: John Wiley, 1977.

[19] Simon, H. A., "Spurious Correlations: A Causal Interpretation," *Journal of the American Statistical Association*, **49** (1954), 467–479.

[20] Timm, N. H., *Multivariate Analysis with Applications in Education and Psychology*, Monterey, Calif.: Brooks/Cole, 1975.

[21] Wright, S., "Statistical Methods in Biology," *Journal of the American Statistical Association Supplement; Papers and Proceedings of the 92nd Annual Meeting*, **26** (1931), 155–163.

[22] Wright, S., "The Method of Path Coefficients," *Annals of Mathematical Statistics*, **5** (1934), 161–215.

[23] Wright, S., *Evolution and the Genetics of Population*, vol. 1, Chicago, Ill.: University of Chicago Press, 1968.

Part III

Analysis of Covariance Structure

8

Principal Components

8.1 INTRODUCTION

A principal component analysis is concerned with explaining the variance-covariance structure through a few *linear* combinations of the original variables. Its general objectives are (1) data reduction, and (2) interpretation.

Although p components are required to reproduce the total system variability, often much of this variability can be accounted for by a small number, k, of the principal components. If so, there is (almost) as much information in the k components as there is in the original p variables. The k principal components can then replace the initial p variables, and the original data set, consisting of n measurements on p variables, is reduced to one consisting of n measurements on k principal components.

An analysis of principal components often reveals relationships that were not previously suspected and thereby allows interpretations that would not ordinarily result. A good example of this is provided by the stock market data discussed in Example 8.5. Other real-world situations that are clearly amenable to a principal component analysis are given by Scenarios 1E.1, 1E.9, 1E.11, and 1E.12 in Section 1.2.

Analyses of principal components are more of a means to an end rather than an end in themselves because they frequently serve as intermediate steps in much larger investigations. For example, principal components may be inputs to a multiple regression (see Chapter 7) or cluster analysis (see Chapter 11). Moreover, (scaled) principal components are one "factoring" of the covariance matrix for the factor analysis model considered in Chapter 9.

8.2 POPULATION PRINCIPAL COMPONENTS

Algebraically, principal components are particular linear combinations of the p random variables $X_1, X_2, \ldots, X_p$. Geometrically, these linear combinations represent the selection of a new coordinate system obtained by rotating the original system with $X_1, X_2, \ldots, X_p$ as the coordinate axes. The new axes represent the directions with maximum variability and provide a simpler and more parsimonious description of the covariance structure.

As we shall see, principal components depend solely on the covariance matrix $\mathbf{\Sigma}$ (or the correlation matrix $\boldsymbol{\rho}$) of $X_1, X_2, \ldots, X_p$. Their development does not require a multivariate normal assumption. On the other hand, principal components derived for multivariate normal populations have useful interpretations in terms of the constant density ellipsoids. Further, inferences can be made from the sample components when the population is multivariate normal (see Section 8.5).

Let the random vector $\mathbf{X}' = [X_1, X_2, \ldots, X_p]$ have the covariance matrix $\mathbf{\Sigma}$ with eigenvalues $\lambda_1 \geq \lambda_2 \geq \cdots \geq \lambda_p \geq 0$.

Consider the linear combinations

$$\begin{aligned} Y_1 &= \boldsymbol{\ell}_1'\mathbf{X} = \ell_{11}X_1 + \ell_{21}X_2 + \cdots + \ell_{p1}X_p \\ Y_2 &= \boldsymbol{\ell}_2'\mathbf{X} = \ell_{12}X_1 + \ell_{22}X_2 + \cdots + \ell_{p2}X_p \\ &\vdots \qquad\qquad \vdots \\ Y_p &= \boldsymbol{\ell}_p'\mathbf{X} = \ell_{1p}X_1 + \ell_{2p}X_2 + \cdots + \ell_{pp}X_p \end{aligned} \tag{8-1}$$

Then, using (2-45),

$$\mathrm{Var}(Y_i) = \boldsymbol{\ell}_i'\mathbf{\Sigma}\boldsymbol{\ell}_i \qquad i = 1, 2, \ldots, p \tag{8-2}$$

$$\mathrm{Cov}(Y_i, Y_k) = \boldsymbol{\ell}_i'\mathbf{\Sigma}\boldsymbol{\ell}_k \qquad i, k = 1, 2, \ldots, p \tag{8-3}$$

The principal components are those *uncorrelated* linear combinations $Y_1, Y_2, \ldots, Y_p$ whose variances in (8-2) are as large as possible.

The first principal component is the linear combination with maximum variance. That is, it maximizes $\mathrm{Var}(Y_1) = \boldsymbol{\ell}_1'\mathbf{\Sigma}\boldsymbol{\ell}_1$. It is clear that $\mathrm{Var}(Y_1) = \boldsymbol{\ell}_1'\mathbf{\Sigma}\boldsymbol{\ell}_1$ can be increased by multiplying any $\boldsymbol{\ell}_1$ by some constant. To eliminate this indeterminacy, it is convenient to restrict attention to coefficient vectors of unit length. We therefore define

First principal component = linear combination $\boldsymbol{\ell}_1'\mathbf{X}$ that maximizes $\mathrm{Var}(\boldsymbol{\ell}_1'\mathbf{X})$ subject to $\boldsymbol{\ell}_1'\boldsymbol{\ell}_1 = 1$

Second principal component = linear combination $\boldsymbol{\ell}_2'\mathbf{X}$ that maximizes $\mathrm{Var}(\boldsymbol{\ell}_2'\mathbf{X})$ subject to $\boldsymbol{\ell}_2'\boldsymbol{\ell}_2 = 1$ and $\mathrm{Cov}(\boldsymbol{\ell}_1'\mathbf{X}, \boldsymbol{\ell}_2'\mathbf{X}) = 0$

At the ith step

$$i\text{th principal component} = \text{linear combination } \boldsymbol{\ell}_i'\mathbf{X} \text{ that maximizes}$$
$$\text{Var}(\boldsymbol{\ell}_i'\mathbf{X}) \text{ subject to } \boldsymbol{\ell}_i'\boldsymbol{\ell}_i = 1 \text{ and}$$
$$\text{Cov}(\boldsymbol{\ell}_i'\mathbf{X}, \boldsymbol{\ell}_k'\mathbf{X}) = 0 \quad \text{for } k < i$$

Result 8.1. Let $\boldsymbol{\Sigma}$ be the covariance matrix associated with the random vector $\mathbf{X}' = [X_1, X_2, \ldots, X_p]$. Let $\boldsymbol{\Sigma}$ have the eigenvalue-eigenvector pairs $(\lambda_1, \mathbf{e}_1)$, $(\lambda_2, \mathbf{e}_2), \ldots, (\lambda_p, \mathbf{e}_p)$ where $\lambda_1 \geq \lambda_2 \geq \cdots \geq \lambda_p \geq 0$. The *ith principal component* is given by

$$Y_i = \mathbf{e}_i'\mathbf{X} = e_{1i}X_1 + e_{2i}X_2 + \cdots + e_{pi}X_p, \qquad i = 1, 2, \ldots, p \tag{8-4}$$

With these choices,

$$\begin{aligned} \text{Var}(Y_i) &= \mathbf{e}_i'\boldsymbol{\Sigma}\mathbf{e}_i = \lambda_i \qquad & i = 1, 2, \ldots, p \\ \text{Cov}(Y_i, Y_k) &= \mathbf{e}_i'\boldsymbol{\Sigma}\mathbf{e}_k = 0 \qquad & i \neq k \end{aligned} \tag{8-5}$$

If some λ_i are equal, the choice of the corresponding coefficient vectors $\mathbf{e}_i$, and hence Y_i, are not unique.

Proof. We know from (2-51) with $\mathbf{B} = \boldsymbol{\Sigma}$,

$$\max_{\boldsymbol{\ell} \neq \mathbf{0}} \frac{\boldsymbol{\ell}'\boldsymbol{\Sigma}\boldsymbol{\ell}}{\boldsymbol{\ell}'\boldsymbol{\ell}} = \lambda_1 \qquad (\text{attained when } \boldsymbol{\ell} = \mathbf{e}_1)$$

But $\mathbf{e}_1'\mathbf{e}_1 = 1$ since the eigenvectors are normalized. Thus

$$\max_{\boldsymbol{\ell} \neq \mathbf{0}} \frac{\boldsymbol{\ell}'\boldsymbol{\Sigma}\boldsymbol{\ell}}{\boldsymbol{\ell}'\boldsymbol{\ell}} = \lambda_1 = \frac{\mathbf{e}_1'\boldsymbol{\Sigma}\mathbf{e}_1}{\mathbf{e}_1'\mathbf{e}_1} = \mathbf{e}_1'\boldsymbol{\Sigma}\mathbf{e}_1 = \text{Var}(Y_1)$$

Similarly, using (2-52),

$$\max_{\boldsymbol{\ell} \perp \mathbf{e}_1, \mathbf{e}_2, \ldots, \mathbf{e}_k} \frac{\boldsymbol{\ell}'\boldsymbol{\Sigma}\boldsymbol{\ell}}{\boldsymbol{\ell}'\boldsymbol{\ell}} = \lambda_{k+1} \qquad k = 1, 2, \ldots, p-1$$

For the choice $\boldsymbol{\ell} = \mathbf{e}_{k+1}$, where $\mathbf{e}_{k+1}'\mathbf{e}_k = 0$, $k = 1, 2, \ldots, p-1$,

$$\mathbf{e}_{k+1}'\boldsymbol{\Sigma}\mathbf{e}_{k+1}/\mathbf{e}_{k+1}'\mathbf{e}_{k+1} = \mathbf{e}_{k+1}'\boldsymbol{\Sigma}\mathbf{e}_{k+1} = \text{Var}(Y_{k+1})$$

But $\mathbf{e}_{k+1}'(\boldsymbol{\Sigma}\mathbf{e}_{k+1}) = \lambda_{k+1}\mathbf{e}_{k+1}'\mathbf{e}_{k+1} = \lambda_{k+1}$ so $\text{Var}(Y_{k+1}) = \lambda_{k+1}$. It remains to show that $\mathbf{e}_i$ perpendicular to $\mathbf{e}_k$ (that is, $\mathbf{e}_i'\mathbf{e}_k = 0$, $i \neq k$) gives $\text{Cov}(Y_i, Y_k) = 0$. Now the eigenvectors of $\boldsymbol{\Sigma}$ are orthogonal if all the eigenvalues $\lambda_1, \lambda_2, \ldots, \lambda_p$ are distinct. If the eigenvalues are not all distinct, the eigenvectors corresponding to common eigenvalues may be chosen to be orthogonal. Therefore, for any two eigenvectors $\mathbf{e}_i$ and $\mathbf{e}_k$, $\mathbf{e}_i'\mathbf{e}_k = 0$, $i \neq k$. Since $\boldsymbol{\Sigma}\mathbf{e}_k = \lambda_k\mathbf{e}_k$, premultiplication by $\mathbf{e}_i'$ gives

$$\text{Cov}(Y_i, Y_k) = \mathbf{e}_i'\boldsymbol{\Sigma}\mathbf{e}_k = \mathbf{e}_i'\lambda_k\mathbf{e}_k = \lambda_k\mathbf{e}_i'\mathbf{e}_k = 0$$

for any $i \neq k$ and the proof is complete. ■

From Result 8.1, the principal components are uncorrelated and have variances equal to the eigenvalues of $\boldsymbol{\Sigma}$.

Result 8.2. Let $\mathbf{X}' = [X_1, X_2, \ldots, X_p]$ have covariance matrix $\boldsymbol{\Sigma}$, with eigenvalue-eigenvector pairs $(\lambda_1, \mathbf{e}_1), (\lambda_2, \mathbf{e}_2), \ldots, (\lambda_p, \mathbf{e}_p)$ where $\lambda_1 \geq \lambda_2 \geq \cdots \geq \lambda_p \geq 0$. Let $Y_1 = \mathbf{e}_1'\mathbf{X}$, $Y_2 = \mathbf{e}_2'\mathbf{X}, \ldots, Y_p = \mathbf{e}_p'\mathbf{X}$ be the principal components. Then

$$\sigma_{11} + \sigma_{22} + \cdots + \sigma_{pp} = \sum_{i=1}^{p} \text{Var}(X_i) = \lambda_1 + \lambda_2 + \cdots + \lambda_p = \sum_{i=1}^{p} \text{Var}(Y_i)$$

Proof. From Definition 2A.28, $\sigma_{11} + \sigma_{22} + \cdots + \sigma_{pp} = \text{tr}(\boldsymbol{\Sigma})$. From (2-20) with $\mathbf{A} = \boldsymbol{\Sigma}$, we can write $\boldsymbol{\Sigma} = \mathbf{P}\boldsymbol{\Lambda}\mathbf{P}'$ where $\boldsymbol{\Lambda}$ is the diagonal matrix of eigenvalues and $\mathbf{P} = [\mathbf{e}_1, \mathbf{e}_2, \ldots, \mathbf{e}_p]$ so that $\mathbf{P}\mathbf{P}' = \mathbf{P}'\mathbf{P} = \mathbf{I}$. Using Result 2A.12(c), we have

$$\text{tr}(\boldsymbol{\Sigma}) = \text{tr}(\mathbf{P}\boldsymbol{\Lambda}\mathbf{P}') = \text{tr}(\boldsymbol{\Lambda}\mathbf{P}'\mathbf{P}) = \text{tr}(\boldsymbol{\Lambda}) = \lambda_1 + \lambda_2 + \cdots + \lambda_p$$

Thus

$$\sum_{i=1}^{p} \text{Var}(X_i) = \text{tr}(\boldsymbol{\Sigma}) = \text{tr}(\boldsymbol{\Lambda}) = \sum_{i=1}^{p} \text{Var}(Y_i)$$ ■

Result 8.2 says

$$\text{Total population variance} = \sigma_{11} + \sigma_{22} + \cdots + \sigma_{pp} = \lambda_1 + \lambda_2 + \cdots + \lambda_p \tag{8-6}$$

and consequently, the proportion of total variance due to (explained by) the kth principal component is

$$\begin{pmatrix} \text{Proportion of total} \\ \text{population variance} \\ \text{due to } k\text{th principal} \\ \text{component} \end{pmatrix} = \frac{\lambda_k}{\lambda_1 + \lambda_2 + \cdots + \lambda_p} \qquad k = 1, 2, \ldots, p \tag{8-7}$$

If most (for instance, 80 to 90%) of the total population variance, for large p, can be attributed to the first one, two, or three components, then these components can "replace" the original p variables without much loss of information.

Each component of the coefficient vector $\mathbf{e}_i' = [e_{1i}, \ldots, e_{ki}, \ldots, e_{pi}]$ also merits inspection. The magnitude of e_{ki} measures the importance of the kth variable to the ith principal component. In particular, e_{ki} is proportional to the correlation coefficient between Y_i and X_k.

Result 8.3. If $Y_1 = \mathbf{e}_1'\mathbf{X}$, $Y_2 = \mathbf{e}_2'\mathbf{X}, \ldots, Y_p = \mathbf{e}_p'\mathbf{X}$ are the principal components obtained from the covariance matrix $\boldsymbol{\Sigma}$, then

$$\rho_{Y_i, X_k} = \frac{e_{ki}\sqrt{\lambda_i}}{\sqrt{\sigma_{kk}}} \qquad i, k = 1, 2, \ldots, p \tag{8-8}$$

are the correlation coefficients between the components Y_i and the variables X_k. Here $(\lambda_1, \mathbf{e}_1), (\lambda_2, \mathbf{e}_2), \ldots, (\lambda_p, \mathbf{e}_p)$ are the eigenvalue-eigenvector pairs for $\boldsymbol{\Sigma}$.

Proof. Set $\boldsymbol{\ell}_k' = [0, \ldots, 0, 1, 0, \ldots, 0]$ so that $X_k = \boldsymbol{\ell}_k'\mathbf{X}$ and $\text{Cov}(X_k, Y_i) = \text{Cov}(\boldsymbol{\ell}_k'\mathbf{X}, \mathbf{e}_i'\mathbf{X}) = \boldsymbol{\ell}_k'\boldsymbol{\Sigma}\mathbf{e}_i$ according to (2-45). Since $\boldsymbol{\Sigma}\mathbf{e}_i = \lambda_i\mathbf{e}_i$, $\text{Cov}(X_k, Y_i) = \boldsymbol{\ell}_k'\lambda_i\mathbf{e}_i$

$= \lambda_i e_{ki}$. Then $\text{Var}(Y_i) = \lambda_i$ [see (8-5)] and $\text{Var}(X_k) = \sigma_{kk}$ yield

$$\rho_{Y_i, X_k} = \frac{\text{Cov}(Y_i, X_k)}{\sqrt{\text{Var}(Y_i)}\sqrt{\text{Var}(X_k)}} = \frac{\lambda_i e_{ki}}{\sqrt{\lambda_i}\sqrt{\sigma_{kk}}} = \frac{e_{ki}\sqrt{\lambda_i}}{\sqrt{\sigma_{kk}}} \qquad i, k = 1, 2, \ldots, p \quad \blacksquare$$

The following hypothetical example will illustrate the contents of Results 8.1, 8.2, and 8.3.

Example 8.1

Suppose the random variables X_1, X_2, and X_3 have the covariance matrix

$$\mathbf{\Sigma} = \begin{bmatrix} 1 & -2 & 0 \\ -2 & 5 & 0 \\ 0 & 0 & 2 \end{bmatrix}$$

It may be verified that the eigenvalue-eigenvector pairs are

$$\begin{aligned} \lambda_1 &= 5.83, & \mathbf{e}_1' &= [.383, -.924, 0] \\ \lambda_2 &= 2.00, & \mathbf{e}_2' &= [0, 0, 1] \\ \lambda_3 &= 0.17, & \mathbf{e}_3' &= [.924, .383, 0] \end{aligned}$$

Therefore the principal components become

$$\begin{aligned} Y_1 &= \mathbf{e}_1'\mathbf{X} = .383X_1 - .924X_2 \\ Y_2 &= \mathbf{e}_2'\mathbf{X} = X_3 \\ Y_3 &= \mathbf{e}_3'\mathbf{X} = .924X_1 + .383X_2 \end{aligned}$$

The variable X_3 is one of the principal components because it is uncorrelated with the other two variables.

Equation (8-5) can be demonstrated from first principles. For example,

$$\begin{aligned} \text{Var}(Y_1) &= \text{Var}(.383X_1 - .924X_2) = (.383)^2\,\text{Var}(X_1) + (-.924)^2\,\text{Var}(X_2) \\ &\quad + 2(.383)(-.924)\,\text{Cov}(X_1, X_2) \\ &= .147(1) + .854(5) - .708(-2) \\ &= 5.83 = \lambda_1 \\ \text{Cov}(Y_1, Y_2) &= \text{Cov}(.383X_1 - .924X_2, X_3) \\ &= .383\,\text{Cov}(X_1, X_3) - .924\,\text{Cov}(X_2, X_3) \\ &= .383(0) - .924(0) = 0 \end{aligned}$$

It is also readily apparent that

$$\sigma_{11} + \sigma_{22} + \sigma_{33} = 1 + 5 + 2 = \lambda_1 + \lambda_2 + \lambda_3 = 5.83 + 2.00 + .17$$

validating Equation (8-6) for this example. The proportion of total variance accounted for by the first principal component is $\lambda_1/(\lambda_1 + \lambda_2 + \lambda_3) = 5.83/8 = .73$. The first two components account for a proportion $(5.83 + 2)/8 = .98$

of the population variance. In this case the components Y_1 and Y_2 could replace the three original variables with little loss of information.

Finally, using (8-8),

$$\rho_{Y_1, X_1} = \frac{e_{11}\sqrt{\lambda_1}}{\sqrt{\sigma_{11}}} = \frac{.383\sqrt{5.83}}{\sqrt{1}} = .925$$

$$\rho_{Y_1, X_2} = \frac{e_{21}\sqrt{\lambda_1}}{\sqrt{\sigma_{22}}} = \frac{-.924\sqrt{5.83}}{\sqrt{5}} = -.998$$

We conclude that X_1 and X_2 are about equally important to the first principal component. Also

$$\rho_{Y_2, X_1} = \rho_{Y_2, X_2} = 0 \quad \text{and} \quad \rho_{Y_2, X_3} = \frac{\sqrt{\lambda_2}}{\sqrt{\sigma_{33}}} = \frac{\sqrt{2}}{\sqrt{2}} = 1 \qquad \text{(as it should)}$$

The remaining correlations can be neglected since the third component is unimportant. ■

It is informative to consider principal components derived from multivariate normal random variables. Suppose $\mathbf{X}$ is distributed as $N_p(\boldsymbol{\mu}, \boldsymbol{\Sigma})$. We know from (4-7) that the $\boldsymbol{\mu}$ centered ellipsoids

$$(\mathbf{x} - \boldsymbol{\mu})'\boldsymbol{\Sigma}^{-1}(\mathbf{x} - \boldsymbol{\mu}) = c^2$$

of constant density have axes $\pm c\sqrt{\lambda_i}\,\mathbf{e}_i$, $i = 1, 2, \ldots, p$, where the $(\lambda_i, \mathbf{e}_i)$ are the eigenvalue-eigenvector pairs of $\boldsymbol{\Sigma}$. A point lying on the ith axis of the ellipsoid will have coordinates proportional to $\mathbf{e}_i' = [e_{1i}, e_{2i}, \ldots, e_{pi}]$ in the coordinate system with origin $\boldsymbol{\mu}$ and axes $x_1, x_2, \ldots, x_p$. It will be convenient to set $\boldsymbol{\mu} = \mathbf{0}$ in the argument that follows.[1]

From our discussion in Section 2.3 with $\mathbf{A} = \boldsymbol{\Sigma}^{-1}$, we can write

$$c^2 = \mathbf{x}'\boldsymbol{\Sigma}^{-1}\mathbf{x} = \frac{1}{\lambda_1}(\mathbf{e}_1'\mathbf{x})^2 + \frac{1}{\lambda_2}(\mathbf{e}_2'\mathbf{x})^2 + \cdots + \frac{1}{\lambda_p}(\mathbf{e}_p'\mathbf{x})^2$$

where $\mathbf{e}_1'\mathbf{x}, \mathbf{e}_2'\mathbf{x}, \ldots, \mathbf{e}_p'\mathbf{x}$ are recognized as the principal components of $\mathbf{x}$. Setting $y_1 = \mathbf{e}_1'\mathbf{x}$, $y_2 = \mathbf{e}_2'\mathbf{x}, \ldots, y_p = \mathbf{e}_p'\mathbf{x}$, we have

$$c^2 = \frac{1}{\lambda_1}y_1^2 + \frac{1}{\lambda_2}y_2^2 + \cdots + \frac{1}{\lambda_p}y_p^2$$

and this equation defines an ellipsoid (since $\lambda_1, \lambda_2, \ldots, \lambda_p$ are positive) in a coordinate system with axes $y_1, y_2, \ldots, y_p$ lying in the directions of $\mathbf{e}_1, \mathbf{e}_2, \ldots, \mathbf{e}_p$ respectively. If λ_1 is the largest eigenvalue, then the major axis lies in the direction $\mathbf{e}_1$. The remaining minor axes lie in the directions defined by $\mathbf{e}_2, \ldots, \mathbf{e}_p$.

To summarize, the principal components $y_1 = \mathbf{e}_1'\mathbf{x}$, $y_2 = \mathbf{e}_2'\mathbf{x}, \ldots, y_p = \mathbf{e}_p'\mathbf{x}$ lie in the directions of the axes of the constant density ellipsoid. Thus any point on the ith

[1]This can be done without loss of generality because the normal random vector $\mathbf{X}$ can always be translated to the normal random vector $\mathbf{W} = \mathbf{X} - \boldsymbol{\mu}$ and $E(\mathbf{W}) = \mathbf{0}$. However $\text{Cov}(\mathbf{X}) = \text{Cov}(\mathbf{W})$.

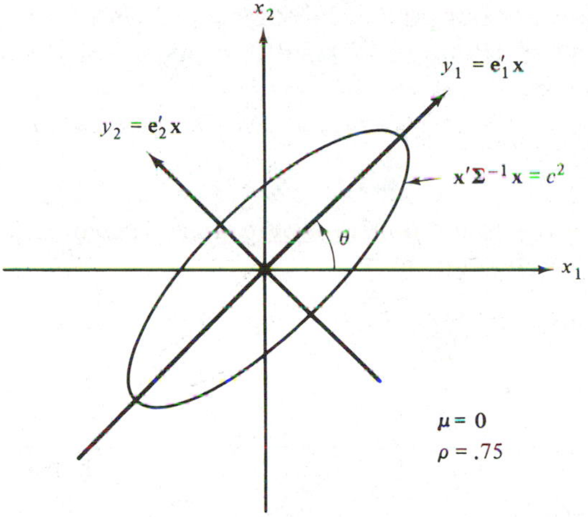

Figure 8.1 The constant density ellipse, $\mathbf{x}'\mathbf{\Sigma}^{-1}\mathbf{x} = c^2$, and the principal components y_1, y_2 for a bivariate normal random vector $\mathbf{X}$.

ellipsoid axis has $\mathbf{x}$ coordinates proportional to $\mathbf{e}_i' = [e_{1i}, e_{2i}, \ldots, e_{pi}]$ and, necessarily, principal component coordinates of the form $[0, \ldots, 0, y_i, 0, \ldots, 0]$.

A constant density ellipse and the principal components for a bivariate normal random vector with $\boldsymbol{\mu} = \mathbf{0}$ and $\rho = .75$ are shown in Figure 8.1. We see that the principal components are obtained by rotating the original coordinate axes through an angle θ until they coincide with the axes of the constant density ellipse. This result holds for $p > 2$ dimensions as well.

Principal Components Obtained from Standardized Variables

Principal components may also be obtained for the standardized variables

$$\begin{aligned} Z_1 &= \frac{(X_1 - \mu_1)}{\sqrt{\sigma_{11}}} \\ Z_2 &= \frac{(X_2 - \mu_2)}{\sqrt{\sigma_{22}}} \\ &\vdots \\ Z_p &= \frac{(X_p - \mu_p)}{\sqrt{\sigma_{pp}}} \end{aligned} \tag{8-9}$$

In matrix notation,

$$\mathbf{Z} = (\mathbf{V}^{1/2})^{-1}(\mathbf{X} - \boldsymbol{\mu}) \tag{8-10}$$

where the diagonal standard deviation matrix $\mathbf{V}^{1/2}$ is defined in (2-35). Clearly $E(\mathbf{Z}) = \mathbf{0}$ and

$$\text{Cov}(\mathbf{Z}) = (\mathbf{V}^{1/2})^{-1}\mathbf{\Sigma}(\mathbf{V}^{1/2})^{-1} = \boldsymbol{\rho}$$

by (2-37). The principal components of $\mathbf{Z}$ may be obtained from the eigenvectors of the *correlation* matrix $\boldsymbol{\rho}$ of $\mathbf{X}$. All of our previous results apply, with some simplifications since the variance of each Z_i is unity. We shall continue to use the notation Y_i to refer to the ith principal component and $(\lambda_i, \mathbf{e}_i)$ for the eigenvalue-eigenvector pair. *However, these quantities derived from $\boldsymbol{\Sigma}$ are, in general, not the same as the ones derived from $\boldsymbol{\rho}$.*

Result 8.4. The ith principal component of the standardized variables $\mathbf{Z}' = [Z_1, Z_2, \ldots, Z_p]$, with $\text{Cov}(\mathbf{Z}) = \boldsymbol{\rho}$, is given by

$$Y_i = \mathbf{e}_i'\mathbf{Z} = \mathbf{e}_i'(\mathbf{V}^{1/2})^{-1}(\mathbf{X} - \boldsymbol{\mu}), \qquad i = 1, 2, \ldots, p$$

Moreover,

$$\sum_{i=1}^{p} \text{Var}(Y_i) = \sum_{i=1}^{p} \text{Var}(Z_i) = p \tag{8-11}$$

and

$$\rho_{Y_i, Z_k} = e_{ki}\sqrt{\lambda_i}, \qquad i, k = 1, 2, \ldots, p$$

In this case, $(\lambda_1, \mathbf{e}_1), (\lambda_2, \mathbf{e}_2), \ldots, (\lambda_p, \mathbf{e}_p)$ are the eigenvalue-eigenvector pairs for $\boldsymbol{\rho}$ with $\lambda_1 \geq \lambda_2 \geq \cdots \geq \lambda_p \geq 0$.

Proof. Result 8.4 follows from Results 8.1, 8.2, and 8.3, with $Z_1, Z_2, \ldots, Z_p$ in place of $X_1, X_2, \ldots, X_p$ and $\boldsymbol{\rho}$ in place of $\boldsymbol{\Sigma}$. ■

We see from (8-11) that the total (standardized variables) population variance is simply p, the sum of the diagonal elements of the matrix $\boldsymbol{\rho}$. Using (8-7) with $\mathbf{Z}$ in place of $\mathbf{X}$, the proportion of total variance explained by the kth principal component of $\mathbf{Z}$ is

$$\begin{pmatrix} \text{Proportion of (standardized)} \\ \text{population variance due} \\ \text{to } k\text{th principal component} \end{pmatrix} = \frac{\lambda_k}{p}, \qquad k = 1, 2, \ldots, p \tag{8-12}$$

where the λ_k's are the eigenvalues of $\boldsymbol{\rho}$.

Example 8.2 (Principal Components Obtained from Covariance and Correlation Matrices)

Consider the covariance matrix

$$\boldsymbol{\Sigma} = \begin{bmatrix} 1 & 4 \\ 4 & 100 \end{bmatrix}$$

and the derived correlation matrix

$$\boldsymbol{\rho} = \begin{bmatrix} 1 & .4 \\ .4 & 1 \end{bmatrix}$$

The eigenvalue-eigenvector pairs from $\boldsymbol{\Sigma}$ are

$$\lambda_1 = 100.16, \qquad \mathbf{e}_1' = [.040, .999]$$
$$\lambda_2 = \quad .84, \qquad \mathbf{e}_2' = [.999, -.040]$$

Similarly, the eigenvalue-eigenvector pairs from $\boldsymbol{\rho}$ are

$$\lambda_1 = 1 + \rho = 1.4, \qquad \mathbf{e}_1' = [.707, .707]$$
$$\lambda_2 = 1 - \rho = .6, \qquad \mathbf{e}_2' = [.707, -.707]$$

The respective principal components become

$$\boldsymbol{\Sigma}: \quad \begin{aligned} Y_1 &= .040X_1 + .999X_2 \\ Y_2 &= .999X_1 - .040X_2 \end{aligned}$$

$$\boldsymbol{\rho}: \quad \begin{aligned} Y_1 = .707Z_1 + .707Z_2 &= .707\left(\frac{X_1 - \mu_1}{1}\right) + .707\left(\frac{X_2 - \mu_2}{10}\right) \\ &= .707(X_1 - \mu_1) + .0707(X_2 - \mu_2) \\ Y_2 = .707Z_1 - .707Z_2 &= .707\left(\frac{X_1 - \mu_1}{1}\right) - .707\left(\frac{X_2 - \mu_2}{10}\right) \\ &= .707(X_1 - \mu_1) - .0707(X_2 - \mu_2) \end{aligned}$$

Because of its large variance, X_2 completely dominates the first principal component determined from $\boldsymbol{\Sigma}$. Moreover, this first principal component explains a proportion

$$\frac{\lambda_1}{\lambda_1 + \lambda_2} = \frac{100.16}{101} = .992$$

of the total population variance.

When the variables X_1 and X_2 are standardized however, the resulting variables contribute equally to the principal components determined from $\boldsymbol{\rho}$. Using Result 8.4

$$\rho_{Y_1, Z_1} = e_{11}\sqrt{\lambda_1} = .707\sqrt{1.4} = .837$$

and

$$\rho_{Y_1, Z_2} = e_{21}\sqrt{\lambda_1} = .707\sqrt{1.4} = .837$$

In this case, the first principal component explains a proportion

$$\frac{\lambda_1}{p} = \frac{1.4}{2} = .7$$

of the total (standardized) population variance.

Most strikingly we see that the relative importance of the variables to, for instance, the first principal component is greatly affected by the standardization. When the principal components obtained from $\boldsymbol{\rho}$ are expressed in terms of X_1 and X_2, the relative magnitudes of the weights .707 and .0707 are in direct opposition to those of the weights .040 and .999 attached to these variables in the principal components obtained from $\boldsymbol{\Sigma}$. ■

The preceding example demonstrates that the principal components derived from $\boldsymbol{\Sigma}$ are different from those derived from $\boldsymbol{\rho}$. Furthermore, one set of principal components is not a simple function of the other. This suggests that the standardization is not inconsequential.

Variables should probably be standardized if they are measured on scales with widely differing ranges or if the measurement units are not commensurate. For example, if X_1 represents annual sales in the \$10,000 to \$350,000 range and X_2 is the ratio (net annual income)/(total assets) that falls in the .01 to .60 range, then the total variation will be due almost exclusively to dollar sales. In this case, we would expect a single (important) principal component with a heavy weighting on X_1. Alternatively, if both variables are standardized, their subsequent magnitudes will be of the same order and X_2 (or Z_2) will play a larger role in the construction of the components. This behavior was observed in Example 8.2.

Principal Components for Covariance Matrices with Special Structures

There are certain patterned covariance and correlation matrices whose principal components can be expressed in simple forms. Suppose $\mathbf{\Sigma}$ is the diagonal matrix

$$\mathbf{\Sigma} = \begin{bmatrix} \sigma_{11} & 0 & \cdots & 0 \\ 0 & \sigma_{22} & \cdots & 0 \\ \vdots & \vdots & \ddots & \vdots \\ 0 & 0 & \cdots & \sigma_{pp} \end{bmatrix} \tag{8-13}$$

Setting $\mathbf{e}_i' = [0,\ldots,0,1,0,\ldots,0]$, with 1 in the ith position, we observe that

$$\begin{bmatrix} \sigma_{11} & 0 & \cdots & 0 \\ 0 & \sigma_{22} & \cdots & 0 \\ \vdots & \vdots & \ddots & \vdots \\ 0 & 0 & \cdots & \sigma_{pp} \end{bmatrix} \begin{bmatrix} 0 \\ \vdots \\ 0 \\ 1 \\ 0 \\ \vdots \\ 0 \end{bmatrix} = \begin{bmatrix} 0 \\ \vdots \\ 0 \\ 1\sigma_{ii} \\ 0 \\ \vdots \\ 0 \end{bmatrix} \quad \text{or} \quad \mathbf{\Sigma e}_i = \sigma_{ii}\mathbf{e}_i$$

and we conclude that $(\sigma_{ii}, \mathbf{e}_i)$ is the ith eigenvalue-eigenvector pair. Since the linear combination $\mathbf{e}_i'\mathbf{X} = X_i$, the set of principal components is just the original set of uncorrelated random variables.

For a covariance matrix with the pattern of (8-13), nothing is gained by extracting the principal components. From another point of view, if $\mathbf{X}$ is distributed as $N_p(\boldsymbol{\mu}, \mathbf{\Sigma})$, the contours of constant density are ellipsoids whose axes already lie in the directions of maximum variation. Consequently, there is no need to rotate the coordinate system.

Standardization does not substantially alter the situation for the $\mathbf{\Sigma}$ in (8-13). In this case, $\boldsymbol{\rho} = \mathbf{I}$, the $p \times p$ identity matrix. Clearly, $\boldsymbol{\rho}\mathbf{e}_i = 1\mathbf{e}_i$, so the eigenvalue 1 has multiplicity p and $\mathbf{e}_i' = [0,\ldots,0,1,0,\ldots,0]$, $i = 1,2,\ldots,p$, are convenient choices for the eigenvectors. Consequently, the principal components determined for $\boldsymbol{\rho}$ are also the original variables $Z_1, \ldots, Z_p$. Moreover, in this case of equal eigenvalues, the multivariate normal ellipsoids of constant density are spheroids.

Another patterned covariance matrix, which often describes the correspondence among certain biological variables such as the sizes of living things, has the general form

$$\boldsymbol{\Sigma} = \begin{bmatrix} \sigma^2 & \rho\sigma^2 & \cdots & \rho\sigma^2 \\ \rho\sigma^2 & \sigma^2 & \cdots & \rho\sigma^2 \\ \vdots & \vdots & \ddots & \vdots \\ \rho\sigma^2 & \rho\sigma^2 & \cdots & \sigma^2 \end{bmatrix} \tag{8-14}$$

The resulting correlation matrix,

$$\boldsymbol{\rho} = \begin{bmatrix} 1 & \rho & \cdots & \rho \\ \rho & 1 & \cdots & \rho \\ \vdots & \vdots & \ddots & \vdots \\ \rho & \rho & \cdots & 1 \end{bmatrix} \tag{8-15}$$

is also the covariance matrix of the standardized variables. The matrix in (8-15) implies that the variables $X_1, X_2, \ldots, X_p$ are equally correlated.

It is not difficult to show (see Exercise 8.5) that the p eigenvalues of the correlation matrix (8-15) can be divided into two groups. When ρ is positive, the largest is

$$\lambda_1 = 1 + (p - 1)\rho \tag{8-16}$$

with associated eigenvector

$$\mathbf{e}_1' = \left[\frac{1}{\sqrt{p}}, \frac{1}{\sqrt{p}}, \ldots, \frac{1}{\sqrt{p}}\right] \tag{8-17}$$

The remaining $p - 1$ eigenvalues are

$$\lambda_2 = \lambda_3 = \cdots = \lambda_p = 1 - \rho$$

and one choice for their eigenvectors is

$$\mathbf{e}_2' = \left[\frac{1}{\sqrt{1 \times 2}}, \frac{-1}{\sqrt{1 \times 2}}, 0, \ldots, 0\right]$$

$$\mathbf{e}_3' = \left[\frac{1}{\sqrt{2 \times 3}}, \frac{1}{\sqrt{2 \times 3}}, \frac{-2}{\sqrt{2 \times 3}}, 0, \ldots, 0\right]$$

$$\vdots$$

$$\mathbf{e}_i' = \left[\frac{1}{\sqrt{(i-1)i}}, \ldots, \frac{1}{\sqrt{(i-1)i}}, \frac{-(i-1)}{\sqrt{(i-1)i}}, 0, \ldots, 0\right]$$

$$\vdots$$

$$\mathbf{e}_p' = \left[\frac{1}{\sqrt{(p-1)p}}, \ldots, \frac{1}{\sqrt{(p-1)p}}, \frac{-(p-1)}{\sqrt{(p-1)p}}\right]$$

The first principal component

$$Y_1 = \mathbf{e}_1'\mathbf{X} = \frac{1}{\sqrt{p}} \sum_{i=1}^{p} X_i$$

is proportional to the sum of the p original variables. It might be regarded as an "index" with equal weights. This principal component explains a proportion

$$\frac{\lambda_1}{p} = \frac{1 + (p-1)\rho}{p} = \rho + \frac{1-\rho}{p} \tag{8-18}$$

of the total population variation. We see that $\lambda_1/p \doteq \rho$ for ρ close to 1 or p large. For example, if $\rho = .80$ and $p = 5$, the first component explains 84% of the total variance. When ρ is near 1, the last $p - 1$ components, collectively, contribute very little to the total variance and can often be neglected.

If the standardized variables $Z_1, Z_2, \ldots, Z_p$ have a multivariate normal distribution with a covariance matrix given by (8-15), then the ellipsoids of constant density are "cigar-shaped" with the major axis along the first principal component $Y_1 = (1/\sqrt{p})[1, 1, \ldots, 1]\mathbf{X}$. This principal component is the projection of $\mathbf{X}$ on the equiangular line $\mathbf{1}' = [1, 1, \ldots, 1]$. The minor axes (and remaining principal components) occur in spherically symmetric directions perpendicular to the major axis (and first principal component).

8.3 SUMMARIZING SAMPLE VARIATION BY PRINCIPAL COMPONENTS

We now have the framework necessary to study the problem of summarizing the variation in n measurements on p variables with a few judiciously chosen linear combinations.

Assume the data $\mathbf{x}_1, \mathbf{x}_2, \ldots, \mathbf{x}_n$ represent n independent drawings from some p-dimensional population with mean vector $\boldsymbol{\mu}$ and covariance matrix $\boldsymbol{\Sigma}$. These data yield the sample mean vector $\bar{\mathbf{x}}$, the sample covariance matrix $\mathbf{S}$ and the sample correlation matrix $\mathbf{R}$.

Our objective in this section will be to construct uncorrelated linear combinations of the measured characteristics that account for much of the variation in the sample. The uncorrelated combinations with the largest variances will be called the *sample principal components*.

Recall the n values of any linear combination

$$\boldsymbol{\ell}_1'\mathbf{x}_j = \ell_{11}x_{1j} + \ell_{21}x_{2j} + \cdots + \ell_{p1}x_{pj}, \qquad j = 1, 2, \ldots, n$$

have sample mean $\boldsymbol{\ell}_1'\bar{\mathbf{x}}$ and sample variance $\boldsymbol{\ell}_1'\mathbf{S}\boldsymbol{\ell}_1$. Also, the pairs of values $(\boldsymbol{\ell}_1'\mathbf{x}_j, \boldsymbol{\ell}_2'\mathbf{x}_j)$, for two linear combinations, have sample covariance $\boldsymbol{\ell}_1'\mathbf{S}\boldsymbol{\ell}_2$ [see (3-36)].

The sample principal components are defined as those linear combinations which have maximum sample variance. As with the population quantities, we restrict the coefficient vectors ℓ_i to satisfy $\ell_i'\ell_i = 1$. Specifically,

First *sample* principal component	=	linear combination $\ell_1'\mathbf{x}_j$ which maximizes the sample variance of $\ell_1'\mathbf{x}_j$ subject to $\ell_1'\ell_1 = 1$
Second *sample* principal component	=	linear combination $\ell_2'\mathbf{x}_j$ which maximizes the sample variance of $\ell_2'\mathbf{x}_j$ subject to $\ell_2'\ell_2 = 1$ and zero sample covariance for the pairs $(\ell_1'\mathbf{x}_j, \ell_2'\mathbf{x}_j)$

At the ith step

ith *sample* principal component	=	linear combination $\ell_i'\mathbf{x}_j$ which maximizes the sample variance of $\ell_i'\mathbf{x}_j$ subject to $\ell_i'\ell_i = 1$ and zero sample covariance for all pairs $(\ell_i'\mathbf{x}_j, \ell_k'\mathbf{x}_j)$, $k < i$

The first principal component maximizes $\ell_1'\mathbf{S}\ell_1$ or, equivalently,

$$\frac{\ell_1'\mathbf{S}\ell_1}{\ell_1'\ell_1} \tag{8-19}$$

By (2-51), the maximum is the largest eigenvalue, $\hat{\lambda}_1$, attained for the choice $\ell_1 =$ eigenvector, $\hat{\mathbf{e}}_1$, of $\mathbf{S}$. Successive choices of ℓ_i maximize (8-19) subject to $0 = \ell_i'\mathbf{S}\hat{\mathbf{e}}_k = \ell_i'\hat{\lambda}_k\hat{\mathbf{e}}_k$, or ℓ_i perpendicular to $\hat{\mathbf{e}}_k$. Thus, as in the proof of Results 8.1–8.3, we obtain the following results concerning sample principal components.

> If $\mathbf{S} = \{s_{ik}\}$ is the $p \times p$ sample covariance matrix with eigenvalue-eigenvector pairs $(\hat{\lambda}_1, \hat{\mathbf{e}}_1), (\hat{\lambda}_2, \hat{\mathbf{e}}_2), \ldots, (\hat{\lambda}_p, \hat{\mathbf{e}}_p)$, the ith sample principal component is given by
>
> $$\hat{y}_i = \hat{\mathbf{e}}_i'\mathbf{x} = \hat{e}_{1i}x_1 + \hat{e}_{2i}x_2 + \cdots + \hat{e}_{pi}x_p, \qquad i = 1, 2, \ldots, p$$
>
> where $\hat{\lambda}_1 \geq \hat{\lambda}_2 \geq \cdots \geq \hat{\lambda}_p \geq 0$ and $\mathbf{x}$ is any observation on the variables $X_1, X_2, \ldots, X_p$.
>
> Also
>
> $$\begin{aligned} \text{Sample variance}(\hat{y}_k) &= \hat{\lambda}_k \qquad k = 1, 2, \ldots, p \\ \text{Sample covariance}(\hat{y}_i, \hat{y}_k) &= 0 \qquad i \neq k \end{aligned} \tag{8-20}$$
>
> In addition,
>
> $$\text{Total sample variance} = \sum_{i=1}^{p} s_{ii} = \hat{\lambda}_1 + \hat{\lambda}_2 + \cdots + \hat{\lambda}_p$$
>
> and
>
> $$r_{\hat{y}_i, x_k} = \frac{\hat{e}_{ki}\sqrt{\hat{\lambda}_i}}{\sqrt{s_{kk}}} \qquad i, k = 1, 2, \ldots, p$$

We shall denote the sample principal components by $\hat{y}_1, \hat{y}_2, \ldots, \hat{y}_p$, irrespective of whether they are obtained from **S** or **R**.[2] The components constructed from **S** and **R** are *not* the same, in general, but it will be clear from the context which matrix is being used and the single notation $\hat{y}_i$ is convenient. It is also convenient to label the component coefficient vectors $\hat{\mathbf{e}}_i$ and the component variances $\hat{\lambda}_i$ for both situations.

The observations $\mathbf{x}_j$ are often "centered" by subtracting $\bar{\mathbf{x}}$. This has no effect on the sample covariance matrix **S** and gives the ith principal component

$$\hat{y}_i = \hat{\mathbf{e}}_i'(\mathbf{x} - \bar{\mathbf{x}}), \qquad i = 1, 2, \ldots, p \tag{8-21}$$

for any observation vector **x**. If we consider the *values* of the ith component

$$\hat{y}_{ij} = \hat{\mathbf{e}}_i'(\mathbf{x}_j - \bar{\mathbf{x}}) \qquad i = 1, 2, \ldots, p \tag{8-22}$$

generated by substituting each observation $\mathbf{x}_j$ for the arbitrary **x** in (8-21), then

$$\begin{aligned}
\bar{\hat{y}}_i &= \frac{1}{n} \sum_{j=1}^{n} \hat{\mathbf{e}}_i'(\mathbf{x}_j - \bar{\mathbf{x}}) \\
&= \frac{1}{n}\left\{\hat{e}_{1i} \sum_{j=1}^{n} (x_{1j} - \bar{x}_1) + \hat{e}_{2i} \sum_{j=1}^{n} (x_{2j} - \bar{x}_2) + \cdots + \hat{e}_{pi} \sum_{j=1}^{n} (x_{pj} - \bar{x}_p)\right\} \\
&= 0
\end{aligned} \tag{8-23}$$

That is, the sample means of the p components are all zero. The sample variances are still given by the $\hat{\lambda}_i$'s, as in (8-20).

Example 8.3

The 1970 census provided tract information on 5 socioeconomic variables for the Madison, Wisconsin, area. The data from 14 tracts are listed in Table 8.2 in the exercises at the end of this chapter. These data produced the following summary statistics.

$\bar{\mathbf{x}}' =$ [4.32,	14.01,	1.95,	2.17,	2.45]
total population (thousands)	median school years	total employment (thousands)	health services employment (hundreds)	median home value ($10,000s)

and

$$\mathbf{S} = \begin{bmatrix} 4.308 & 1.683 & 1.803 & 2.155 & -.253 \\ 1.683 & 1.768 & .588 & .177 & .176 \\ 1.803 & .588 & .801 & 1.065 & -.158 \\ 2.155 & .177 & 1.065 & 1.970 & -.357 \\ -.253 & .176 & -.158 & -.357 & .504 \end{bmatrix}$$

[2]Sample principal components can also be obtained from $\hat{\boldsymbol{\Sigma}} = \mathbf{S}_n$, the maximum likelihood estimate of the covariance matrix $\boldsymbol{\Sigma}$, if the $\mathbf{X}_j$ are normally distributed (see Result 4.11). In this case, provided the eigenvalues of $\boldsymbol{\Sigma}$ are distinct, the sample principal components can be viewed as the maximum likelihood estimates of the corresponding population counterparts (see [1]). We shall not consider $\hat{\boldsymbol{\Sigma}}$ because the assumption of normality is not required in this section. Also, $\hat{\boldsymbol{\Sigma}}$ has eigenvalues $[(n-1)/n]\hat{\lambda}_i$ and corresponding eigenvectors $\hat{\mathbf{e}}_i$, where $(\hat{\lambda}_i, \hat{\mathbf{e}}_i)$ are the eigenvalue-eigenvector pairs for **S**. Thus both **S** and $\hat{\boldsymbol{\Sigma}}$ give the same sample principal components $\hat{\mathbf{e}}_i'\mathbf{x}$ [see (8-20)] and the same proportion of explained variance $\hat{\lambda}_i/(\hat{\lambda}_1 + \hat{\lambda}_2 + \cdots + \hat{\lambda}_p)$. Finally, both **S** and $\hat{\boldsymbol{\Sigma}}$ give the same sample correlation matrix **R**, so if the variables are standardized, the choice of **S** or $\hat{\boldsymbol{\Sigma}}$ is irrelevant.

Can the sample variation be summarized by one or two principal components?

We find the following.

COEFFICIENTS FOR THE PRINCIPAL COMPONENTS
(Correlation Coefficients in Parentheses)

Variable	$\hat{\mathbf{e}}_1(r_{\hat{y}_1, x_k})$	$\hat{\mathbf{e}}_2(r_{\hat{y}_2, x_k})$	$\hat{\mathbf{e}}_3$	$\hat{\mathbf{e}}_4$	$\hat{\mathbf{e}}_5$
Total population	.781 (.99)	−.071 (−.04)	.004	.542	−.302
Median school years	.306 (.61)	−.764 (−.76)	−.162	−.545	−.010
Total employment	.334 (.98)	.083 (.12)	.015	.050	.937
Health services employment	.426 (.80)	.579 (.55)	.220	−.636	−.173
Median home value	−.054 (−.20)	−.262 (−.49)	.962	−.051	.024
Variance ($\hat{\lambda}_i$):	6.931	1.786	.390	.230	.014
Cumulative percentage of total variance:	74.1	93.2	97.4	99.9	100

The first principal component explains 74.1% of the total sample variance. The first two principal components, collectively, explain 93.2% of the total sample variance. Consequently, sample variation is summmarized very well by two principal components and a reduction in the data from 14 observations on 5 variables to 14 observations on 2 principal components is reasonable.

Given the component coefficients above, the first principal component appears to be essentially a weighted average of the first four variables. The second principal component appears to contrast health services employment with a weighted average of median school years and median home value. ■

When attempting a subject matter interpretation of the principal components, the correlations $r_{\hat{y}_i, x_k}$ are ordinarily more reliable guides than the component coefficients $\hat{e}_{ki}$. The correlations allow for differences in the variances of the original variables and therefore avoid the interpretive problem caused by different measurement scales. In Example 8.3, the correlation coefficients displayed in the table confirm the interpretation provided by the component coefficients.

Example 8.4

In a study of size and shape relationships for painted turtles, Jolicoeur and Mosimann [10] measured carapace length, width, and height. Their data, reproduced in Exercise 6.13, Table 6.5, suggests an analysis in terms of logarithms. (Jolicoeur [9] generally suggests a logarithmic transformation in studies of size-and-shape relationships.) Perform a principal component analysis.

The natural logarithms of the dimensions of 24 male turtles have sample mean vector $\bar{\mathbf{x}}' = [4.725, 4.478, 3.703]$ and covariance matrix

$$\mathbf{S} = 10^{-3}\begin{bmatrix} 11.555 & 8.367 & 8.508 \\ 8.367 & 6.697 & 6.264 \\ 8.508 & 6.264 & 7.061 \end{bmatrix}$$

A principal component analysis yields the following summary.

COEFFICIENTS FOR PRINCIPAL COMPONENTS
(Correlation Coefficients in Parentheses)

Variable	$\hat{\mathbf{e}}_1(r_{\hat{y}_1, x_k})$	$\hat{\mathbf{e}}_2$	$\hat{\mathbf{e}}_3$
ln (length)	.683 (.99)	.162	.712
ln (width)	.510 (.97)	.591	−.624
ln (height)	.522 (.97)	−.790	−.321
Variance ($\hat{\lambda}_i$):	24.31×10^{-3}	$.63 \times 10^{-3}$	$.38 \times 10^{-3}$
Cumulative percentage of total variance:	96.0	98.5	100

The first principal component, which explains 96% of the total variance, has an interesting subject-matter interpretation. Since

$$\begin{aligned} \hat{y}_1 &= .683 \ln(\text{length}) + .51 \ln(\text{width}) + .522 \ln(\text{height}) \\ &= \ln\left[(\text{length})^{.683}(\text{width})^{.510}(\text{height})^{.522}\right] \end{aligned}$$

the first principal component may be viewed as the ln(volume) of a box with adjusted dimensions. For instance, the adjusted height is $(\text{height})^{.522}$, which accounts, in some sense, for the rounded shape of the carapace. ■

Geometrically the data can be plotted as n points in p-dimensional space. If $\mathbf{S}$ is positive definite, all $p \times 1$ vectors $\mathbf{x}$ satisfying

$$(\mathbf{x} - \bar{\mathbf{x}})'\mathbf{S}^{-1}(\mathbf{x} - \bar{\mathbf{x}}) = c^2 \tag{8-24}$$

define a hyperellipsoid centered at $\bar{\mathbf{x}}$, whose axes are given by the eigenvectors of $\mathbf{S}^{-1}$ or, equivalently, of $\mathbf{S}$ (see Section 2.3 and Result 4.1 with $\mathbf{S}$ in place of $\mathbf{\Sigma}$). The lengths of these axes are proportional to $\sqrt{\hat{\lambda}_i}$, $i = 1, 2, \ldots, p$, where $\hat{\lambda}_1 \geq \hat{\lambda}_2 \geq \cdots \geq \hat{\lambda}_p > 0$ are the eigenvalues of $\mathbf{S}$.

The sample principal components bear the same relationship to the axes of the constant distance ellipsoids in (8-24) as the population components do to the constant density ellipsoids for $N_p(\boldsymbol{\mu}, \mathbf{\Sigma})$ variables. That is, the sample components lie along the axes of the constant distance ellipsoids. These components can be viewed as the result of rotating the original coordinate system until the coordinate axes pass through the scatter in the directions of maximum variance.

The absolute value of the ith principal component, $|\hat{y}_i| = |\hat{\mathbf{e}}_i'(\mathbf{x} - \bar{\mathbf{x}})|$, gives the length of the projection of the vector $(\mathbf{x} - \bar{\mathbf{x}})$ on the unit vector $\hat{\mathbf{e}}_i$ [see (2-8) and (2-9)]. This geometrical interpretation of the sample principal components is illustrated in Figure 8.2 for $p = 2$.

Figure 8.2(a) shows an ellipse of constant distance, centered at $\bar{\mathbf{x}}$, with $\hat{\lambda}_1 > \hat{\lambda}_2$. The sample principal components are well determined. They lie along the axes of the ellipse in the perpendicular directions of maximum sample variance. Figure 8.2(b) shows a constant distance ellipse, centered at $\bar{\mathbf{x}}$, with $\hat{\lambda}_1 \doteq \hat{\lambda}_2$. In this case, the axes of the ellipse (circle) of constant distance are not uniquely determined and can lie in any two perpendicular directions, including the directions of the original coordinate axes. Similarly, the sample principal components can lie in any two perpendicular directions, including those of the original coordinate axes. When the contours of constant distance are nearly circular or, equivalently, when the eigenvalues of $\mathbf{S}$ are nearly equal, the sample variation is homogeneous in all directions. It is then not possible to represent the data in fewer than p dimensions.

If $\mathbf{x}_1, \mathbf{x}_2, \ldots, \mathbf{x}_n$ can be regarded as a sample from a normal population, the sample principal components, $\hat{y}_i = \hat{\mathbf{e}}_i'(\mathbf{x} - \bar{\mathbf{x}})$ are realizations of population principal components $Y_i = \mathbf{e}_i'(\mathbf{X} - \boldsymbol{\mu})$, which have an $N_p(\mathbf{0}, \boldsymbol{\Lambda})$ distribution. The diagonal matrix $\boldsymbol{\Lambda}$ has entries $\lambda_1, \lambda_2, \ldots, \lambda_p$ and $(\lambda_i, \mathbf{e}_i)$ are the eigenvalue-eigenvector pairs of $\boldsymbol{\Sigma}$. Also the constant distance ellipsoids (8-24) are estimates of the constant density ellipsoids $(\mathbf{x} - \boldsymbol{\mu})'\boldsymbol{\Sigma}^{-1}(\mathbf{x} - \boldsymbol{\mu}) = c^2$. The normality assumption is useful for the inference procedures discussed in Section 8.5, but it is not required for the development of the properties of the sample principal components summarized in (8-20).

Sample principal components are, in general, not invariant with respect to changes in scale (see Exercise 8.2). As we mentioned in the treatment of population components, variables measured on different scales or a common scale with widely

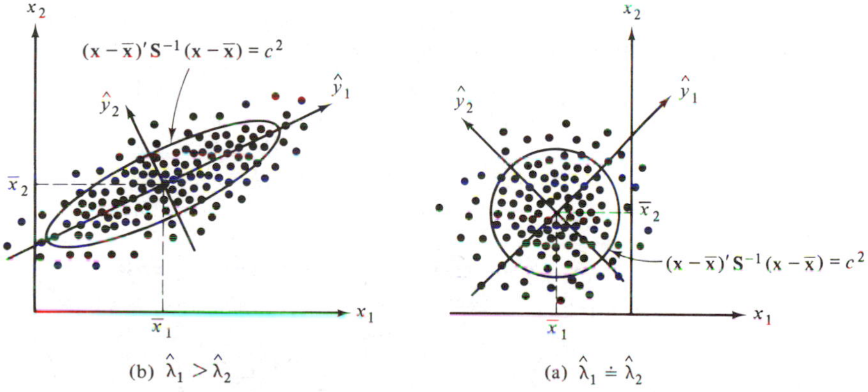

Figure 8.2 Sample principal components and ellipses of constant distance.

differing ranges are often standardized. For the sample, standardization is accomplished by constructing

$$\mathbf{z}_j = \mathbf{D}^{-1/2}(\mathbf{x}_j - \bar{\mathbf{x}}) = \begin{bmatrix} \dfrac{x_{1j} - \bar{x}_1}{\sqrt{s_{11}}} \\ \dfrac{x_{2j} - \bar{x}_2}{\sqrt{s_{22}}} \\ \vdots \\ \dfrac{x_{pj} - \bar{x}_p}{\sqrt{s_{pp}}} \end{bmatrix} \qquad j = 1, 2, \ldots, n \tag{8-25}$$

The $p \times n$ data matrix of standardized observations

$$\mathbf{Z} = [\mathbf{z}_1, \mathbf{z}_2, \ldots, \mathbf{z}_n] = \begin{bmatrix} z_{11} & z_{12} & \cdots & z_{1n} \\ z_{21} & z_{22} & \cdots & z_{2n} \\ \vdots & \vdots & & \vdots \\ z_{p1} & z_{p2} & \cdots & z_{pn} \end{bmatrix}$$

$$= \begin{bmatrix} \dfrac{x_{11} - \bar{x}_1}{\sqrt{s_{11}}} & \dfrac{x_{12} - \bar{x}_1}{\sqrt{s_{11}}} & \cdots & \dfrac{x_{1n} - \bar{x}_1}{\sqrt{s_{11}}} \\ \dfrac{x_{21} - \bar{x}_2}{\sqrt{s_{22}}} & \dfrac{x_{22} - \bar{x}_2}{\sqrt{s_{22}}} & \cdots & \dfrac{x_{2n} - \bar{x}_2}{\sqrt{s_{22}}} \\ \vdots & \vdots & & \vdots \\ \dfrac{x_{p1} - \bar{x}_p}{\sqrt{s_{pp}}} & \dfrac{x_{p2} - \bar{x}_p}{\sqrt{s_{pp}}} & \cdots & \dfrac{x_{pn} - \bar{x}_p}{\sqrt{s_{pp}}} \end{bmatrix} \tag{8-26}$$

yields the sample mean vector [see (3-24)]

$$\bar{\mathbf{z}} = \frac{1}{n}\mathbf{Z}\mathbf{1} = \frac{1}{n} \begin{bmatrix} \displaystyle\sum_{j=1}^{n} \frac{x_{1j} - \bar{x}_1}{\sqrt{s_{11}}} \\ \displaystyle\sum_{j=1}^{n} \frac{x_{2j} - \bar{x}_2}{\sqrt{s_{22}}} \\ \vdots \\ \displaystyle\sum_{j=1}^{n} \frac{x_{pj} - \bar{x}_p}{\sqrt{s_{pp}}} \end{bmatrix} = \mathbf{0} \tag{8-27}$$

and sample covariance matrix [see (3-27)]

$$\mathbf{S}_z = \frac{1}{n-1}\left(\mathbf{Z} - \frac{1}{n}\mathbf{Z}\mathbf{1}\mathbf{1}'\right)\left(\mathbf{Z} - \frac{1}{n}\mathbf{Z}\mathbf{1}\mathbf{1}'\right)' = \frac{1}{n-1}(\mathbf{Z} - \bar{\mathbf{z}}\mathbf{1}')(\mathbf{Z} - \bar{\mathbf{z}}\mathbf{1}')'$$

$$= \frac{1}{n-1}\mathbf{Z}\mathbf{Z}'$$

$$= \frac{1}{n-1}\begin{bmatrix} \frac{(n-1)s_{11}}{s_{11}} & \frac{(n-1)s_{12}}{\sqrt{s_{11}}\sqrt{s_{22}}} & \cdots & \frac{(n-1)s_{1p}}{\sqrt{s_{11}}\sqrt{s_{pp}}} \\ \frac{(n-1)s_{12}}{\sqrt{s_{11}}\sqrt{s_{22}}} & \frac{(n-1)s_{22}}{s_{22}} & \cdots & \frac{(n-1)s_{2p}}{\sqrt{s_{22}}\sqrt{s_{pp}}} \\ \vdots & \vdots & & \vdots \\ \frac{(n-1)s_{1p}}{\sqrt{s_{11}}\sqrt{s_{pp}}} & \frac{(n-1)s_{2p}}{\sqrt{s_{22}}\sqrt{s_{pp}}} & \cdots & \frac{(n-1)s_{pp}}{s_{pp}} \end{bmatrix} = \mathbf{R} \qquad (8\text{-}28)$$

The sample principal components of the standardized observations are given by (8-20), with the matrix **R** in place of **S**. Since the observations are already "centered" by construction, there is no need to write the components in the form of (8-21).

If $\mathbf{z}_1, \mathbf{z}_2, \ldots, \mathbf{z}_n$ are standardized observations with covariance matrix **R**, the ith sample principal components is

$$\hat{y}_i = \hat{\mathbf{e}}_i'\mathbf{z} = \hat{e}_{1i}z_1 + \hat{e}_{2i}z_2 + \cdots + \hat{e}_{pi}z_p, \qquad i = 1, 2, \ldots, p$$

where $(\hat{\lambda}_i, \hat{\mathbf{e}}_i)$ is the ith eigenvalue-eigenvector pair of **R** with $\hat{\lambda}_1 \geq \hat{\lambda}_2 \geq \ldots \geq \hat{\lambda}_p \geq 0$.

Also, (8-29)

$$\text{Sample variance } (\hat{y}_i) = \hat{\lambda}_i \qquad i = 1, 2, \ldots, p$$

$$\text{Sample covariance } (\hat{y}_i, \hat{y}_k) = 0 \qquad i \neq k$$

In addition,

$$\begin{aligned}&\text{Total (standardized)}\\ &\quad \text{sample variance} = \text{tr}(\mathbf{R}) = p = \hat{\lambda}_1 + \hat{\lambda}_2 + \cdots + \hat{\lambda}_p\end{aligned}$$

and

$$r_{\hat{y}_i, z_k} = \hat{e}_{ki}\sqrt{\hat{\lambda}_i} \qquad i, k = 1, 2, \ldots, p$$

Using (8-29), the proportion of total sample variance explained by the ith sample principal component is

$$\begin{pmatrix}\text{Proportion of (standardized)}\\ \text{sample variance due to } i\text{th}\\ \text{sample principal component}\end{pmatrix} = \frac{\hat{\lambda}_i}{p} \qquad i = 1, 2, \ldots, p \qquad (8\text{-}30)$$

A rule of thumb suggests retaining only those components whose variances, $\hat{\lambda}_i$, are greater than unity or, equivalently, only those components which, individually, explain at least a proportion $1/p$ of the total variance. This rule doesn't have a great deal of theoretical support, however, and it should not be applied blindly.

Example 8.5

The weekly rates of return for 5 stocks (Allied Chemical, DuPont, Union Carbide, Exxon, and Texaco) listed on the New York Stock Exchange were determined for the period January 1975 through December 1976. The weekly rates of return are defined as (current Friday closing price − previous Friday closing price)/(previous Friday closing price) adjusted for stock splits and dividends. The data are listed in Table 8.1 in the exercises. The observations in 100 successive weeks appear to be independently distributed, but the rates of return *across* stocks are correlated, since, as one expects, stocks tend to move together in response to general economic conditions.

Let $x_1, x_2, \ldots, x_5$ denote observed weekly rates of return for Allied Chemical, DuPont, Union Carbide, Exxon, and Texaco, respectively. Then

$$\bar{\mathbf{x}}' = [.0054, .0048, .0057, .0063, .0037]$$

and

$$\mathbf{R} = \begin{bmatrix} 1.000 & .577 & .509 & .387 & .462 \\ .577 & 1.000 & .599 & .389 & .322 \\ .509 & .599 & 1.000 & .436 & .426 \\ .387 & .389 & .436 & 1.000 & .523 \\ .462 & .322 & .426 & .523 & 1.000 \end{bmatrix}$$

We note that $\mathbf{R}$ is the covariance matrix of the standardized observations

$$z_1 = \frac{x_1 - \bar{x}_1}{\sqrt{s_{11}}},\ z_2 = \frac{x_2 - \bar{x}_2}{\sqrt{s_{22}}}, \ldots, z_5 = \frac{x_5 - \bar{x}_5}{\sqrt{s_{55}}}$$

The eigenvalues and corresponding normalized eigenvectors of $\mathbf{R}$ were determined by a computer and are given below.

$$\hat{\lambda}_1 = 2.857, \qquad \hat{\mathbf{e}}_1' = [.464, .457, .470, .421, .421]$$

$$\hat{\lambda}_2 = .809, \qquad \hat{\mathbf{e}}_2' = [.240, .509, .260, -.526, -.582]$$

$$\hat{\lambda}_3 = .540, \qquad \hat{\mathbf{e}}_3' = [-.612, .178, .335, .541, -.435]$$

$$\hat{\lambda}_4 = .452, \qquad \hat{\mathbf{e}}_4' = [.387, .206, -.662, .472, -.382]$$

$$\hat{\lambda}_5 = .343, \qquad \hat{\mathbf{e}}_5' = [-.451, .676, -.400, -.176, .385]$$

Using the standardized variables, we obtain the first two sample principal components

$$\hat{y}_1 = \hat{\mathbf{e}}_1'\mathbf{z} = .464z_1 + .457z_2 + .470z_3 + .421z_4 + .421z_5$$
$$\hat{y}_2 = \hat{\mathbf{e}}_2'\mathbf{z} = .240z_1 + .509z_2 + .260z_3 - .526z_4 - .582z_5$$

These components, which account for

$$\left(\frac{\hat{\lambda}_1 + \hat{\lambda}_2}{p}\right)100\% = \left(\frac{2.857 + .809}{5}\right)100\% = 73\%$$

of the total (standardized) sample variance, have interesting interpretations. The first component is a (roughly) equally weighted sum, or "index," of the five stocks. This component might be called a *general stock-market component*, or simply a *market component*. (In fact, these five stocks are included in the Dow Jones Industrial Average).

The second component represents a contrast between the chemical stocks (Allied Chemical, DuPont, and Union Carbide) and the oil stocks (Exxon and Texaco). It might be called an *industry component*. Thus we see that most of the variation in these stock returns is due to market activity and uncorrelated industry activity. This interpretation of stock price behavior has also been suggested by King [11].

The remaining components are not easy to interpret and, collectively, represent variation that is probably specific to each stock. In any event, they do not explain much of the total sample variance.

This example provides a case where it seems sensible to retain a component ($\hat{y}_2$) associated with an eigenvalue less than 1. ■

Example 8.6

Geneticists are often concerned with the inheritance of characteristics that can be measured several times during an animal's lifetime. Body weight (in grams) for $n = 150$ female mice were obtained immediately after birth of their first 4 litters.[3] The sample mean vector and sample correlation matrix were

$$\bar{\mathbf{x}}' = [39.88, 45.08, 48.11, 49.95]$$

$$\mathbf{R} = \begin{bmatrix} 1.000 & .7501 & .6329 & .6363 \\ .7501 & 1.000 & .6925 & .7386 \\ .6329 & .6925 & 1.000 & .6625 \\ .6363 & .7386 & .6625 & 1.000 \end{bmatrix}$$

The eigenvalues of this matrix are

$$\hat{\lambda}_1 = 3.058, \quad \hat{\lambda}_2 = .382, \quad \hat{\lambda}_3 = .342, \quad \text{and} \quad \hat{\lambda}_4 = .217$$

We note that the first eigenvalue is nearly equal to $1 + (p - 1)\bar{r} = 1 + (4 - 1)(.6854) = 3.056$, where $\bar{r}$ is the arithmetic average of the off-diagonal elements of $\mathbf{R}$. The remaining eigenvalues are small and about equal, although $\hat{\lambda}_4$ is somewhat smaller than $\hat{\lambda}_2$ and $\hat{\lambda}_3$. Thus there is some evidence that the corresponding population correlation matrix $\boldsymbol{\rho}$ may be of the "equal-correlation" form of (8-15). This notion is explored further in Example 8.9.

The first principal component

$$\hat{y}_1 = \hat{\mathbf{e}}_1'\mathbf{z} = .49z_1 + .52z_2 + .49z_3 + .50z_4$$

[3]Data courtesy of J. J. Rutledge.

accounts for $100(\hat{\lambda}_1/p)\% = 100(3.058/4)\% = 76\%$ of the total variance. Although the average post-birth weights increase over time, the *variation* in weights is fairly well explained by the first principal component with (nearly) equal coefficients. ■

Comment. An unusually small value for the *last* eigenvalue from either the sample covariance or correlation matrix can indicate an unnoticed linear dependency in the data set. If this occurs, one (or more) of the variables is redundant and should be deleted. Consider a situation where x_1, x_2, and x_3 are subtest scores and the total score x_4 is the sum $x_1 + x_2 + x_3$. Then, although the linear combination $\mathbf{e}'\mathbf{x} = [1, 1, 1, -1]\mathbf{x} = x_1 + x_2 + x_3 - x_4$ is always zero, rounding error in the computation of eigenvalues may lead to a small nonzero value. If the linear expression relating x_4 to (x_1, x_2, x_3) was initially overlooked, the smallest eigenvalue-eigenvector pair should provide a clue to its existence.

Thus, although "large" eigenvalues and the corresponding eigenvectors are important in a principal component analysis, eigenvalues very close to zero should not be routinely ignored. The eigenvectors associated with these latter eigenvalues may point out linear dependencies in the data set that can cause interpretive and computational problems in a subsequent analysis.

8.4 GRAPHING THE PRINCIPAL COMPONENTS

Plots of the principal components can reveal suspect observations, as well as provide checks on the assumption of normality. Since the principal components are linear combinations of the original variables, it is not unreasonable to expect them to be nearly normal. It is often necessary to verify that the first few principal components are approximately normally distributed when they are to be used as the input data for additional analyses.

The last principal components can help pinpoint suspect observations. Each observation $\mathbf{x}_j$ can be expressed as a linear combination

$$\begin{aligned}\mathbf{x}_j &= (\mathbf{x}_j'\hat{\mathbf{e}}_1)\hat{\mathbf{e}}_1 + (\mathbf{x}_j'\hat{\mathbf{e}}_2)\hat{\mathbf{e}}_2 + \cdots + (\mathbf{x}_j'\hat{\mathbf{e}}_p)\hat{\mathbf{e}}_p \\ &= \hat{y}_{1j}\hat{\mathbf{e}}_1 + \hat{y}_{2j}\hat{\mathbf{e}}_2 + \cdots + \hat{y}_{pj}\hat{\mathbf{e}}_p\end{aligned}$$

of the complete set of eigenvectors $\hat{\mathbf{e}}_1, \hat{\mathbf{e}}_2, \ldots, \hat{\mathbf{e}}_p$ of $\mathbf{S}$. Thus the magnitudes of the last principal components determine how well the first few fit the observations. That is, $\hat{y}_{1j}\hat{\mathbf{e}}_1 + \hat{y}_{2j}\hat{\mathbf{e}}_2 + \cdots + \hat{y}_{q-1,j}\hat{\mathbf{e}}_{q-1}$ differs from $\mathbf{x}_j$ by $\hat{y}_{qj}\hat{\mathbf{e}}_q + \cdots + \hat{y}_{pj}\hat{\mathbf{e}}_p$, whose squared length is $\hat{y}_{qj}^2 + \cdots + \hat{y}_{pj}^2$. Suspect observations will often be such that at least one of the coordinates $\hat{y}_{qj}, \ldots, \hat{y}_{pj}$ contributing to this squared length will be large (see Supplement 8A).

The following statements summarize these ideas.

1. To help check the normal assumption, construct scatter diagrams for pairs of the first few principal components. Also make *Q-Q* plots from the sample values generated by *each* principal component.

2. Construct scatter diagrams and Q-Q plots for the last few principal components. These help identify suspect observations.

Example 8.7

We illustrate the plotting of principal components for the male-turtle data discussed in Example 8.4. The three sample principal components are

$$\hat{y}_1 = .683(x_1 - 4.725) + .510(x_2 - 4.478) + .522(x_3 - 3.703)$$
$$\hat{y}_2 = .162(x_1 - 4.725) + .591(x_2 - 4.478) - .790(x_3 - 3.703)$$
$$\hat{y}_3 = .712(x_1 - 4.725) - .624(x_2 - 4.478) - .321(x_3 - 3.703)$$

where $x_1 = \ln(\text{length})$, $x_2 = \ln(\text{width})$, and $x_3 = \ln(\text{height})$, respectively.

Figure 8.3 shows the Q-Q plot for $\hat{y}_2$ and Figure 8.4 shows the scatterplot of $(\hat{y}_1, \hat{y}_2)$. The observation for the first turtle is circled and lies to the lower left corner of the scatterplot and in the lower left corner of the Q-Q plot; it may be suspect. This point should have been checked for recording errors or the turtle examined for any structural anomalies. Apart from the first turtle, the scatterplot appears to be reasonably elliptical. The plots for other sets of principal components do not indicate any substantial departures from normality. ■

The diagnostics involving principal components apply equally well to the checking of assumptions for a multivariate multiple regression model. In fact, having

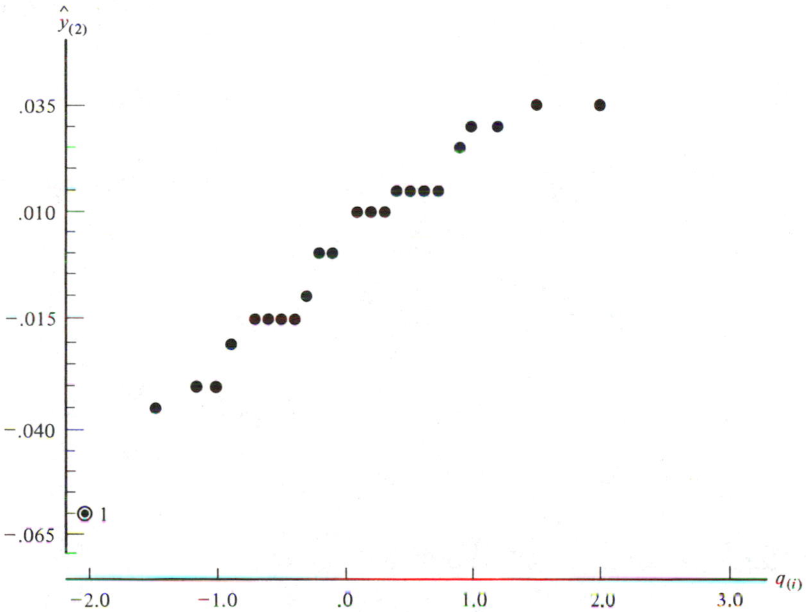

Figure 8.3 A Q-Q plot for the second principal component, $\hat{y}_2$, from the male-turtle data.

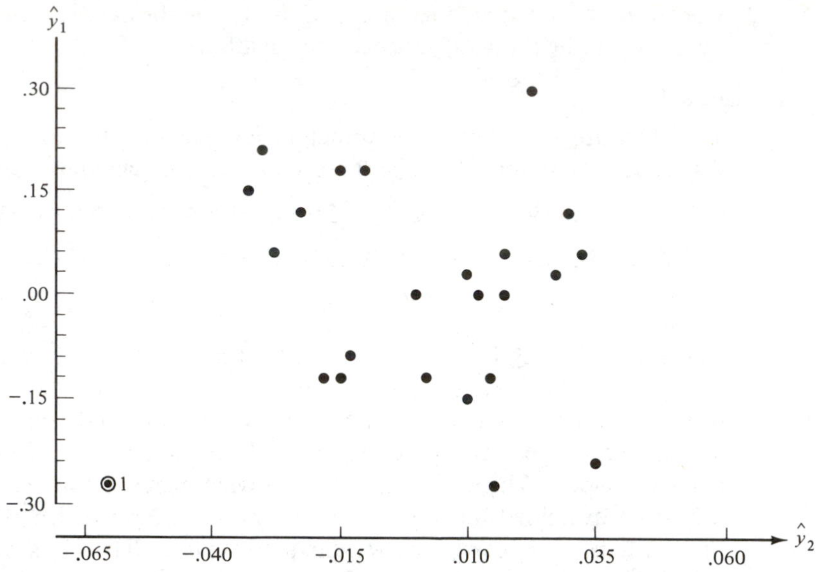

Figure 8.4 Scatterplot of the principal components $\hat{y}_1$ and $\hat{y}_2$, male-turtle data.

fit any model by any method of estimation, it is prudent to consider the

$$\text{residual vector} = (\text{observation vector}) - \begin{pmatrix} \text{vector of predicted} \\ \text{(estimated) values} \end{pmatrix}$$

or

$$\underset{(p\times 1)}{\hat{\boldsymbol{\varepsilon}}_j} = \underset{(p\times 1)}{\mathbf{y}_j} - \underset{(p\times 1)}{\mathbf{z}_j'\hat{\boldsymbol{\beta}}}, \qquad j = 1, 2, \ldots, n \tag{8-31}$$

for the multivariate linear model. Principal components, derived from the residual matrix,

$$\frac{\sum_{j=1}^{n} (\hat{\boldsymbol{\varepsilon}}_j - \bar{\hat{\boldsymbol{\varepsilon}}}_j)(\hat{\boldsymbol{\varepsilon}}_j - \bar{\hat{\boldsymbol{\varepsilon}}}_j)'}{n-p} \tag{8-32}$$

can be scrutinized in the same manner as those determined from a random sample. You should be aware that there *are* linear dependencies among the residuals from a linear regression analysis, so the last eigenvalues will be zero within rounding error.

8.5 LARGE SAMPLE INFERENCES

We have seen that the eigenvalues and eigenvectors of the covariance (correlation) matrix are the essence of a principal components analysis. The eigenvectors determine the directions of maximum variability and the eigenvalues specify the variances. When the first few eigenvalues are much larger than the rest, most of the total variance can be "explained" in fewer than p dimensions.

In practice, decisions regarding quality of the principal component approximation must be made on the basis of the eigenvalue-eigenvector pairs $(\hat{\lambda}_i, \hat{\mathbf{e}}_i)$ extracted from $\mathbf{S}$ or $\mathbf{R}$. Because of sampling variation, these eigenvalues and eigenvectors will differ from their underlying population counterparts. The sampling distributions of $\hat{\lambda}_i$ and $\hat{\mathbf{e}}_i$ are difficult to derive and beyond the scope of this book. If you are interested, you can find some of these derivations for multivariate normal populations in [1], [2], and [4]. We shall simply summarize the pertinent large-sample results.

Large-Sample Properties of $\hat{\lambda}_i$ and $\hat{\mathbf{e}}_i$

Currently available results concerning large-sample confidence intervals for $\hat{\lambda}_i$ and $\hat{\mathbf{e}}_i$ assume the observations $\mathbf{X}_1, \mathbf{X}_2, \ldots, \mathbf{X}_n$ are a random sample from a normal population. It must also be assumed that the (unknown) eigenvalues from $\mathbf{\Sigma}$ are distinct and positive, so that $\lambda_1 > \lambda_2 > \cdots > \lambda_p > 0$. The one exception is the case where the number of equal eigenvalues is known. Usually the conclusions for distinct eigenvalues are applied unless there is a strong reason to believe $\mathbf{\Sigma}$ has a special structure that yields equal eigenvalues. Even when the normal assumption is violated, the confidence intervals obtained in this manner still provide some indication of the uncertainty in $\hat{\lambda}_i$ and $\hat{\mathbf{e}}_i$.

Anderson [2] and Girshick [4] have established the following large-sample distribution theory for the eigenvalues $\hat{\boldsymbol{\lambda}}' = [\hat{\lambda}_1, \ldots, \hat{\lambda}_p]$ and eigenvectors $\hat{\mathbf{e}}_1, \ldots, \hat{\mathbf{e}}_p$ of $\mathbf{S}$.

1. Let $\mathbf{\Lambda}$ be the diagonal matrix of eigenvalues $\lambda_1, \ldots, \lambda_p$ of $\mathbf{\Sigma}$, then $\sqrt{n}\,(\hat{\boldsymbol{\lambda}} - \boldsymbol{\lambda})$ is approximately $N_p(\mathbf{0}, 2\mathbf{\Lambda}^2)$.
2. Let
$$\mathbf{E}_i = \lambda_i \sum_{\substack{k=1 \\ k \neq i}}^{p} \frac{\lambda_k}{(\lambda_k - \lambda_i)^2} \mathbf{e}_k \mathbf{e}_k'$$
then $\sqrt{n}\,(\hat{\mathbf{e}}_i - \mathbf{e}_i)$ is approximately $N_p(\mathbf{0}, \mathbf{E}_i)$.
3. Each $\hat{\lambda}_i$ is distributed independently of the elements of the associated $\hat{\mathbf{e}}_i$.

Result 1 implies that, for n large, the $\hat{\lambda}_i$ are independently distributed. Moreover, $\hat{\lambda}_i$ has an approximate $N(\lambda_i, 2\lambda_i^2/n)$ distribution. Using this normal distribution $P[|\hat{\lambda}_i - \lambda_i| \leq z(\alpha/2)\lambda_i\sqrt{2/n}\,] = 1 - \alpha$. A large sample $100(1 - \alpha)\%$ confidence interval for λ_i is thus provided by

$$\frac{\hat{\lambda}_i}{\left(1 + z(\alpha/2)\sqrt{2/n}\,\right)} \leq \lambda_i \leq \frac{\hat{\lambda}_i}{\left(1 - z(\alpha/2)\sqrt{2/n}\,\right)} \qquad (8\text{-}33)$$

where $z(\alpha/2)$ is the upper $(100\alpha/2)$th percentile of a standard normal distribution. Bonferroni-type simultaneous $100(1 - \alpha)\%$ intervals for m λ_i's are obtained by replacing $z(\alpha/2)$ with $z(\alpha/2m)$ (see Section 5.4).

Result 2 implies that the $\hat{\mathbf{e}}_i$'s are normally distributed about the corresponding $\mathbf{e}_i$'s for large samples. The elements of each $\hat{\mathbf{e}}_i$ are correlated and the correlation depends to a large extent on the separation of the eigenvalues $\lambda_1, \lambda_2, \ldots, \lambda_p$ (which is unknown) and the sample size n. Approximate standard errors for the coefficients $\hat{e}_{ki}$ are given by the diagonal elements of $(1/n)\hat{\mathbf{E}}_i$ where $\hat{\mathbf{E}}_i$ is derived from $\mathbf{E}_i$ by substituting $\hat{\lambda}_i$'s for the λ_i's.

Example 8.8

We shall obtain a 95% confidence interval for λ_1, the variance of the first population principal component, using the stock price data listed in Table 8.1.

Assume the stock rates of return represent independent drawings from an $N_5(\boldsymbol{\mu}, \boldsymbol{\Sigma})$ population, where $\boldsymbol{\Sigma}$ is positive definite with distinct eigenvalues $\lambda_1 > \lambda_2 > \cdots > \lambda_5 > 0$. Since $n = 100$ is large, we can use (8-33) with $i = 1$ to construct a 95% confidence interval for λ_1. From Exercise 8.10, $\hat{\lambda}_1 = .0036$ and, in addition, $z(.025) = 1.96$. Therefore, with 95% confidence,

$$\frac{.0036}{\left(1 + 1.96\sqrt{\frac{2}{100}}\right)} \le \lambda_1 \le \frac{.0036}{\left(1 - 1.96\sqrt{\frac{2}{100}}\right)} \quad \text{or} \quad .0028 \le \lambda_1 \le .0050 \qquad \blacksquare$$

Whenever an eigenvalue is large, such as 100 or even 1000, the intervals generated by (8-33) can be quite wide, for reasonable confidence levels, even though n is fairly large. In general, the confidence interval gets wider at the same rate that $\hat{\lambda}_i$ gets larger. Consequently, some care must be exercised in dropping or retaining principal components based on an examination of the $\hat{\lambda}_i$'s.

Testing for the Equal Correlation Structure

The special correlation structure $\text{Cov}(X_i, X_k) = \sqrt{\sigma_{ii}\sigma_{kk}}\,\rho$, or $\text{Corr}(X_i, X_k) = \rho$, all $i \ne k$, is one important structure where the eigenvalues of $\boldsymbol{\Sigma}$ are not distinct and the previous results do not apply.

To test for this structure, let

$$H_0: \boldsymbol{\rho} = \underset{(p \times p)}{\boldsymbol{\rho}_0} = \begin{bmatrix} 1 & \rho & \cdots & \rho \\ \rho & 1 & \cdots & \rho \\ \vdots & \vdots & \ddots & \vdots \\ \rho & \rho & \cdots & 1 \end{bmatrix}$$

and

$$H_1: \quad \boldsymbol{\rho} \ne \boldsymbol{\rho}_0$$

A test of H_0 versus H_1 may be based on a likelihood ratio statistic, but Lawley [12] has demonstrated that an equivalent test procedure can be constructed from the off-diagonal elements of $\mathbf{R}$.

Lawley's procedure requires the quantities

$$\bar{r}_k = \frac{1}{p-1} \sum_{\substack{i=1 \\ i \neq k}}^{p} r_{ik} \qquad k = 1, 2, \ldots, p; \qquad \bar{r} = \frac{2}{p(p-1)} \sum\sum_{i<k} r_{ik} \tag{8-34}$$

$$\hat{\gamma} = \frac{(p-1)^2[1-(1-\bar{r})^2]}{p-(p-2)(1-\bar{r})^2}$$

It is evident that $\bar{r}_k$ is the average of the off-diagonal elements in the kth column (or row) of $\mathbf{R}$ and $\bar{r}$ is the overall average of the off-diagonal elements.

The large sample approximate α-level test has the form: Reject H_0 in favor of H_1 if

$$T = \frac{(n-1)}{(1-\bar{r})^2}\left[\sum\sum_{i<k}(r_{ik}-\bar{r})^2 - \hat{\gamma}\sum_{k=1}^{p}(\bar{r}_k-\bar{r})^2\right] > \chi^2_{(p+1)(p-2)/2}(\alpha) \tag{8-35}$$

where $\chi^2_{(p+1)(p-2)/2}(\alpha)$ is the upper (100α)th percentile of a chi-square distribution with $(p+1)(p-2)/2$ d.f.

Example 8.9

The sample correlation matrix constructed from the post-birth weights of female mice discussed in Example 8.6 is reproduced below.

$$\mathbf{R} = \begin{bmatrix} 1.0 & .7501 & .6329 & .6363 \\ .7501 & 1.0 & .6925 & .7386 \\ .6329 & .6925 & 1.0 & .6625 \\ .6363 & .7386 & .6625 & 1.0 \end{bmatrix}$$

We shall use this correlation matrix to illustrate the large sample test in (8-35).

Here $p = 4$ and we set

$$H_0: \ \boldsymbol{\rho} = \boldsymbol{\rho}_0 = \begin{bmatrix} 1 & \rho & \rho & \rho \\ \rho & 1 & \rho & \rho \\ \rho & \rho & 1 & \rho \\ \rho & \rho & \rho & 1 \end{bmatrix}$$

$$H_1: \ \boldsymbol{\rho} \neq \boldsymbol{\rho}_0$$

Using (8-34) and (8-35),

$$\bar{r}_1 = \tfrac{1}{3}(.7501 + .6329 + .6363) = .6731, \qquad \bar{r}_2 = .7271, \qquad \bar{r}_3 = .6626,$$

$$\bar{r}_4 = .6791$$

$$\bar{r} = \frac{2}{4(3)}(.7501 + .6329 + .6363 + .6925 + .7386 + .6625) = .6855$$

$$\sum\sum_{i<k}(r_{ik}-\bar{r})^2 = (.7501 - .6855)^2 + (.6329 - .6855)^2$$

$$+ \cdots + (.6625 - .6855)^2 = .01277$$

$$\sum_{k=1}^{4} (\bar{r}_k - \bar{r})^2 = (.6731 - .6855)^2 + \cdots + (.6791 - .6855)^2 = .00245$$

$$\hat{\gamma} = \frac{(4-1)^2[1-(1-.6855)^2]}{4-(4-2)(1-.6855)^2} = 2.1329$$

and

$$T = \frac{(150-1)}{(1-.6855)^2}[.01277 - (2.1329)(.00245)] = 11.4$$

Since $(p+1)(p-2)/2 = 5(2)/2 = 5$, the 5% critical value for the test in (8-35) is $\chi_5^2(.05) = 11.07$. The value of our test statistic is approximately equal to the large-sample 5% critical point, so the evidence against H_0 (equal correlations) is strong but not overwhelming.

As we saw in Example 8.6, the smallest eigenvalues $\hat{\lambda}_2$, $\hat{\lambda}_3$, and $\hat{\lambda}_4$ are slightly different, with $\hat{\lambda}_4$ being somewhat smaller than the other two. Consequently, with the large sample size in this problem, small differences from the equal correlation structure show up as statistically significant. ■

Supplement 8A: The Geometry of the Sample Principal Component Approximation

We shall present interpretations for approximations to the data based on the first r sample principal components. The interpretations of both the p-dimensional scatterplot and the n-dimensional representation rely on the algebraic result below. We consider approximations of the form $\underset{(p\times n)}{\mathbf{A}} = [\mathbf{a}_1, \mathbf{a}_2, \ldots, \mathbf{a}_n]$ to the mean corrected data matrix

$$[\mathbf{x}_1 - \bar{\mathbf{x}}, \mathbf{x}_2 - \bar{\mathbf{x}}, \ldots, \mathbf{x}_n - \bar{\mathbf{x}}]$$

The error of approximation is quantified as the sum of the np squared errors

$$\sum_{j=1}^{n} (\mathbf{x}_j - \bar{\mathbf{x}} - \mathbf{a}_j)'(\mathbf{x}_j - \bar{\mathbf{x}} - \mathbf{a}_j) = \sum_{i=1}^{p} \sum_{j=1}^{n} (x_{ij} - \bar{x}_i - a_{ij})^2 \tag{8A-1}$$

Result 8A.1. Let $\underset{(p\times n)}{\mathbf{A}}$ be any matrix with rank(A) $\leq r < \min(p, n)$. The error of approximation sum of squares in (8A-1) is minimized by the choice

$$\hat{\mathbf{A}} = \hat{\mathbf{E}}\hat{\mathbf{E}}'[\mathbf{x}_1 - \bar{\mathbf{x}}, \ldots, \mathbf{x}_n - \bar{\mathbf{x}}] = \hat{\mathbf{E}} \begin{bmatrix} \hat{\mathbf{y}}_1' \\ \hat{\mathbf{y}}_2' \\ \vdots \\ \hat{\mathbf{y}}_r' \end{bmatrix}$$

so the jth column of $\hat{\mathbf{A}}$ is

$$\hat{\mathbf{a}}_j = \hat{y}_{1j}\hat{\mathbf{e}}_1 + \hat{y}_{2j}\hat{\mathbf{e}}_2 + \cdots + \hat{y}_{rj}\hat{\mathbf{e}}_r \quad \text{where}$$

$$\left[\hat{y}_{1j}, \hat{y}_{2j}, \ldots, \hat{y}_{rj}\right] = \left[\hat{\mathbf{e}}_1'(\mathbf{x}_j - \bar{\mathbf{x}}), \hat{\mathbf{e}}_2'(\mathbf{x}_j - \bar{\mathbf{x}}), \ldots, \hat{\mathbf{e}}_r'(\mathbf{x}_j - \bar{\mathbf{x}})\right]$$

are the values of the first r sample principal components for the jth unit. Moreover,

$$\sum_{j=1}^{n} (\mathbf{x}_j - \bar{\mathbf{x}} - \hat{\mathbf{a}}_j)'(\mathbf{x}_j - \bar{\mathbf{x}} - \hat{\mathbf{a}}_j) = (n-1)\left(\hat{\lambda}_{r+1} + \cdots + \hat{\lambda}_p\right)$$

where $\hat{\lambda}_{r+1} \geq \cdots \geq \hat{\lambda}_p$ are the smallest eigenvalues of $\mathbf{S}$.

Proof. Consider first any $\mathbf{A}$ whose columns are a linear combination of a *fixed* set of r perpendicular vectors $\boldsymbol{\ell}_1, \boldsymbol{\ell}_2, \ldots, \boldsymbol{\ell}_r$, so $\mathbf{L} = [\boldsymbol{\ell}_1, \boldsymbol{\ell}_2, \ldots, \boldsymbol{\ell}_r]$ satisfies $\mathbf{L}'\mathbf{L} = \mathbf{I}$. For fixed $\mathbf{L}$, $\mathbf{x}_j - \bar{\mathbf{x}}$ is best approximated by its projection on the space spanned by

$\boldsymbol{\ell}_1, \boldsymbol{\ell}_2, \ldots, \boldsymbol{\ell}_r$ (see Result 2A.3), or

$$(\mathbf{x}_j - \bar{\mathbf{x}})'\boldsymbol{\ell}_1\boldsymbol{\ell}_1 + (\mathbf{x}_j - \bar{\mathbf{x}})'\boldsymbol{\ell}_2\boldsymbol{\ell}_2 + \cdots + (\mathbf{x}_j - \bar{\mathbf{x}})'\boldsymbol{\ell}_r\boldsymbol{\ell}_r$$

$$= [\boldsymbol{\ell}_1, \boldsymbol{\ell}_2, \ldots, \boldsymbol{\ell}_r] \begin{bmatrix} \boldsymbol{\ell}_1'(\mathbf{x}_j - \bar{\mathbf{x}}) \\ \boldsymbol{\ell}_2'(\mathbf{x}_j - \bar{\mathbf{x}}) \\ \vdots \\ \boldsymbol{\ell}_r'(\mathbf{x}_j - \mathbf{x}) \end{bmatrix} = \mathbf{LL}'(\mathbf{x}_j - \bar{\mathbf{x}}) \tag{8A-2}$$

This follows because, for an arbitrary vector $\mathbf{b}_j$,

$$\begin{aligned} \mathbf{x}_j - \bar{\mathbf{x}} - \mathbf{Lb}_j &= \mathbf{x}_j - \bar{\mathbf{x}} - \mathbf{LL}'(\mathbf{x}_j - \bar{\mathbf{x}}) + \mathbf{LL}'(\mathbf{x}_j - \bar{\mathbf{x}}) - \mathbf{Lb}_j \\ &= (\mathbf{I} - \mathbf{LL}')(\mathbf{x}_j - \bar{\mathbf{x}}) + \mathbf{L}(\mathbf{L}'(\mathbf{x}_j - \bar{\mathbf{x}}) - \mathbf{b}_j) \end{aligned}$$

so the error sum of squares is

$$\begin{aligned} (\mathbf{x}_j - \bar{\mathbf{x}} - \mathbf{Lb}_j)'(\mathbf{x}_j - \bar{\mathbf{x}} - \mathbf{Lb}_j) &= (\mathbf{x}_j - \bar{\mathbf{x}})'(\mathbf{I} - \mathbf{LL}')(\mathbf{x}_j - \bar{\mathbf{x}}) + 0 \\ &\quad + (\mathbf{LL}'(\mathbf{x}_j - \bar{\mathbf{x}}) - \mathbf{Lb}_j)'(\mathbf{LL}'(\mathbf{x}_j - \bar{\mathbf{x}}) - \mathbf{Lb}_j) \end{aligned}$$

where the cross product vanishes because $(\mathbf{I} - \mathbf{LL}')\mathbf{L} = \mathbf{L} - \mathbf{LL}'\mathbf{L} = \mathbf{L} - \mathbf{L} = \mathbf{0}$. The last term is positive unless $\mathbf{b}_j$ is chosen so $\mathbf{Lb}_j = \mathbf{LL}'(\mathbf{x}_j - \bar{\mathbf{x}})$ is the projection.

Further, with the choice $\mathbf{a}_j = \mathbf{Lb}_j = \mathbf{LL}'(\mathbf{x}_j - \bar{\mathbf{x}})$, (8A-1) becomes

$$\begin{aligned} &\sum_{j=1}^{n} (\mathbf{x}_j - \bar{\mathbf{x}} - \mathbf{LL}'(\mathbf{x}_j - \bar{\mathbf{x}}))'(\mathbf{x}_j - \bar{\mathbf{x}} - \mathbf{LL}'(\mathbf{x}_j - \bar{\mathbf{x}})) \\ &\qquad = \sum_{j=1}^{n} (\mathbf{x}_j - \bar{\mathbf{x}})'(\mathbf{I} - \mathbf{LL}')(\mathbf{x}_j - \bar{\mathbf{x}}) \\ &\qquad = \sum_{j=1}^{n} (\mathbf{x}_j - \bar{\mathbf{x}})'(\mathbf{x}_j - \bar{\mathbf{x}}) - \sum_{j=1}^{n} (\mathbf{x}_j - \bar{\mathbf{x}})'\mathbf{LL}'(\mathbf{x}_j - \bar{\mathbf{x}}) \end{aligned} \tag{8A-3}$$

We are now in a position to minimize the error over choices of $\mathbf{L}$ by maximizing the last term in (8A-3). By the properties of trace (see Result 2A.12)

$$\begin{aligned} \sum_{j=1}^{n} (\mathbf{x}_j - \bar{\mathbf{x}})'\mathbf{LL}'(\mathbf{x}_j - \bar{\mathbf{x}}) &= \sum_{j=1}^{n} \operatorname{tr}\left[(\mathbf{x}_j - \bar{\mathbf{x}})'\mathbf{LL}'(\mathbf{x}_j - \bar{\mathbf{x}})\right] \\ &= \sum_{j=1}^{n} \operatorname{tr}\left[\mathbf{LL}'(\mathbf{x}_j - \bar{\mathbf{x}})(\mathbf{x}_j - \bar{\mathbf{x}})'\right] \\ &= (n-1)\operatorname{tr}[\mathbf{LL}'\mathbf{S}] = (n-1)\operatorname{tr}[\mathbf{L}'\mathbf{SL}] \end{aligned} \tag{8A-4}$$

That is, the best choice for $\mathbf{L}$ maximizes the sum of the diagonal elements of $\mathbf{L}'\mathbf{SL}$. From (8-19), selecting $\boldsymbol{\ell}_1$ to maximize $\boldsymbol{\ell}_1'\mathbf{S}\boldsymbol{\ell}_1$, the first diagonal element of $\mathbf{L}'\mathbf{SL}$, gives $\boldsymbol{\ell}_1 = \hat{\mathbf{e}}_1$. For $\boldsymbol{\ell}_2$ perpendicular to $\hat{\mathbf{e}}_1$, $\boldsymbol{\ell}_2'\mathbf{S}\boldsymbol{\ell}_2$ is maximized by $\hat{\mathbf{e}}_2$ [see (2-52)].

Continuing, we find $\hat{\mathbf{L}} = [\hat{\mathbf{e}}_1, \hat{\mathbf{e}}_2, \ldots, \hat{\mathbf{e}}_r] = \hat{\mathbf{E}}$ and $\hat{\mathbf{A}} = \hat{\mathbf{E}}\hat{\mathbf{E}}'[\mathbf{x}_1 - \bar{\mathbf{x}}, \mathbf{x}_2 - \bar{\mathbf{x}}, \ldots, \mathbf{x}_n - \bar{\mathbf{x}}]$, as asserted.

We have that $\text{tr}[\hat{\mathbf{L}}'\mathbf{S}\hat{\mathbf{L}}] = \hat{\lambda}_1 + \hat{\lambda}_2 + \cdots + \hat{\lambda}_r$. Also, $\sum_{j=1}^{n} (\mathbf{x}_j - \bar{\mathbf{x}})'(\mathbf{x}_j - \bar{\mathbf{x}}) = \text{tr}[\sum_{j=1}^{n} (\mathbf{x}_j - \bar{\mathbf{x}})(\mathbf{x}_j - \bar{\mathbf{x}})'] = (n-1)\,\text{tr}(\mathbf{S}) = (n-1)(\hat{\lambda}_1 + \hat{\lambda}_2 + \cdots + \hat{\lambda}_p)$. Let $\mathbf{L} = \hat{\mathbf{L}}$ in (8A-3) and the error bound follows. ∎

The p-dimensional Geometrical Interpretation

The geometrical interpretations involve the determination of best approximating planes to the p-dimensional scatter plot. The plane through the origin, determined by $\boldsymbol{\ell}_1, \boldsymbol{\ell}_2, \ldots, \boldsymbol{\ell}_r$, consists of all points $\mathbf{x}$ with

$$\mathbf{x} = b_1\boldsymbol{\ell}_1 + b_2\boldsymbol{\ell}_2 + \cdots + b_r\boldsymbol{\ell}_r = \mathbf{L}\mathbf{b}, \qquad \text{for some } \mathbf{b}$$

This plane, translated to pass though $\mathbf{a}$, becomes $\mathbf{a} + \mathbf{L}\mathbf{b}$ for some $\mathbf{b}$.

We want to select the r-dimensional plane $\mathbf{a} + \mathbf{L}\mathbf{b}$ that *minimizes the sum of squared distances between the observations* $\mathbf{x}_j$ *and the plane.* If $\mathbf{x}_j$ is approximated by $\mathbf{a} + \mathbf{L}\mathbf{b}_j$ with $\sum_{j=1}^{n} \mathbf{b}_j = \mathbf{0}$,[4]

$$\begin{aligned}\sum_{j=1}^{n} (\mathbf{x}_j - \mathbf{a} - \mathbf{L}\mathbf{b}_j)'(\mathbf{x}_j - \mathbf{a} - \mathbf{L}\mathbf{b}_j) &= \sum_{j=1}^{n} (\mathbf{x}_j - \bar{\mathbf{x}} - \mathbf{L}\mathbf{b}_j + \bar{\mathbf{x}} - \mathbf{a})'(\mathbf{x}_j - \bar{\mathbf{x}} - \mathbf{L}\mathbf{b}_j + \bar{\mathbf{x}} - \mathbf{a}) \\ &= \sum_{j=1}^{n} (\mathbf{x}_j - \bar{\mathbf{x}} - \mathbf{L}\mathbf{b}_j)'(\mathbf{x}_j - \bar{\mathbf{x}} - \mathbf{L}\mathbf{b}_j) + n(\bar{\mathbf{x}} - \mathbf{a})'(\bar{\mathbf{x}} - \mathbf{a}) \\ &\geq \sum_{j=1}^{n} \left(\mathbf{x}_j - \bar{\mathbf{x}} - \hat{\mathbf{E}}\hat{\mathbf{E}}'(\mathbf{x}_j - \bar{\mathbf{x}})\right)'\left(\mathbf{x}_j - \bar{\mathbf{x}} - \hat{\mathbf{E}}\hat{\mathbf{E}}'(\mathbf{x}_j - \bar{\mathbf{x}})\right)\end{aligned}$$

by Result 8A.1, since $[\mathbf{L}\mathbf{b}_1, \ldots, \mathbf{L}\mathbf{b}_r] = \mathbf{A}$ has $\text{rank}(\mathbf{A}) \leq r$. The lower bound is reached by taking $\mathbf{a} = \bar{\mathbf{x}}$, so the plane passes through the sample mean. This plane is determined by $\hat{\mathbf{e}}_1, \hat{\mathbf{e}}_2, \ldots, \hat{\mathbf{e}}_r$. The coefficients of $\hat{\mathbf{e}}_k$ are $\hat{\mathbf{e}}_k'(\mathbf{x}_j - \bar{\mathbf{x}}) = \hat{y}_{kj}$, the kth sample principal component evaluated at the jth observation.

The approximating plane interpretation of sample principal components first suggested by K. Pearson is illustrated in Figure 8.5.

An alternative interpretation, can be given. The investigator places a plane, through $\bar{\mathbf{x}}$, and moves it about to obtain the *largest spread* among the shadows of the observations. From (8A-2), the projection of the deviation $\mathbf{x}_j - \bar{\mathbf{x}}$ on the plane $\mathbf{L}\mathbf{b}$ is $\mathbf{v}_j = \mathbf{L}\mathbf{L}'(\mathbf{x}_j - \bar{\mathbf{x}})$. Now $\bar{\mathbf{v}} = \mathbf{0}$ and the *sum of the squared lengths of the projection*

[4] If $\sum_{j=1}^{n} \mathbf{b}_j = n\bar{\mathbf{b}} \neq \mathbf{0}$, use $\mathbf{a} + \mathbf{L}\mathbf{b}_j = (\mathbf{a} + \mathbf{L}\bar{\mathbf{b}}) + \mathbf{L}(\mathbf{b}_j - \bar{\mathbf{b}}) = \mathbf{a}^* + \mathbf{L}\mathbf{b}_j^*$.

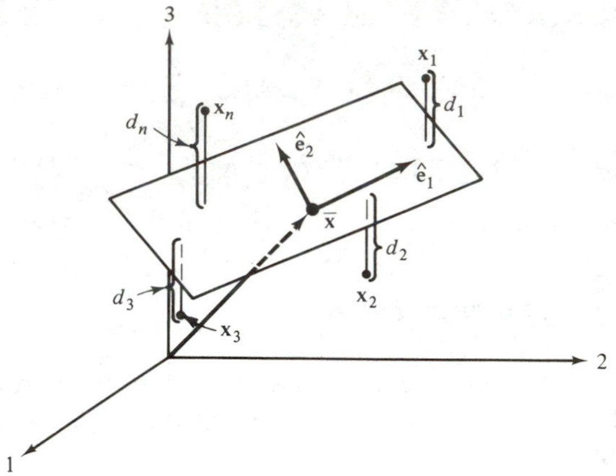

Figure 8.5 The $r = 2$-dimensional plane that approximates the scatterplot by minimizing $\sum_{j=1}^{n} d_j^2$.

deviations

$$\sum_{j=1}^{n} \mathbf{v}_j'\mathbf{v}_j = \sum_{j=1}^{n} (\mathbf{x}_j - \bar{\mathbf{x}})'\mathbf{L}\mathbf{L}'(\mathbf{x}_j - \bar{\mathbf{x}}) = (n-1)\,\text{tr}[\mathbf{L}'\mathbf{S}\mathbf{L}]$$

is maximized by $\mathbf{L} = \hat{\mathbf{E}}$. Also, since $\bar{\mathbf{v}} = \mathbf{0}$,

$$(n-1)\mathbf{S}_v = \sum_{j=1}^{n} (\mathbf{v}_j - \bar{\mathbf{v}})(\mathbf{v}_j - \bar{\mathbf{v}})' = \sum_{j=1}^{n} \mathbf{v}_j\mathbf{v}_j'$$

and this plane also maximizes the total variance

$$\text{tr}(\mathbf{S}_v) = \frac{1}{(n-1)}\,\text{tr}\left[\sum_{j=1}^{n} \mathbf{v}_j\mathbf{v}_j'\right] = \frac{1}{n-1}\,\text{tr}\left[\sum_{j=1}^{n} \mathbf{v}_j'\mathbf{v}_j\right].$$

The n-dimensional Geometrical Interpretation

Let us now consider the approximation in Result 8A.1 row by row. For $r = 1$, the ith row $[x_{i1} - \bar{x}_i, x_{i2} - \bar{x}_i, \ldots, x_{in} - \bar{x}_i]'$ is approximated by a multiple $c_i\mathbf{b}'$ of a fixed vector $\mathbf{b} = [b_1, b_2, \ldots, b_n]'$. The squared length of the error of approximation is the squared length

$$L_i^2 = \sum_{j=1}^{n} (x_{ij} - \bar{x}_i - c_i b_j)^2$$

Considering $\underset{(p\times n)}{\mathbf{A}} = \{a_{ij}\}$, with $a_{ij} = c_i b_j$, we conclude that

$$\begin{aligned}\hat{\mathbf{A}} &= [\hat{\mathbf{e}}_1\hat{\mathbf{e}}_1'(\mathbf{x}_1 - \bar{\mathbf{x}}), \hat{\mathbf{e}}_1\hat{\mathbf{e}}_1'(\mathbf{x}_2 - \bar{\mathbf{x}}), \ldots, \hat{\mathbf{e}}_1\hat{\mathbf{e}}_1'(\mathbf{x}_n - \bar{\mathbf{x}})] \\ &= \hat{\mathbf{e}}_1[\hat{y}_{11}, \hat{y}_{12}, \ldots, \hat{y}_{1n}]\end{aligned}$$

minimizes the sum of squared lengths $\sum_{i=1}^{p} L_i^2$. That is, the best direction is de-

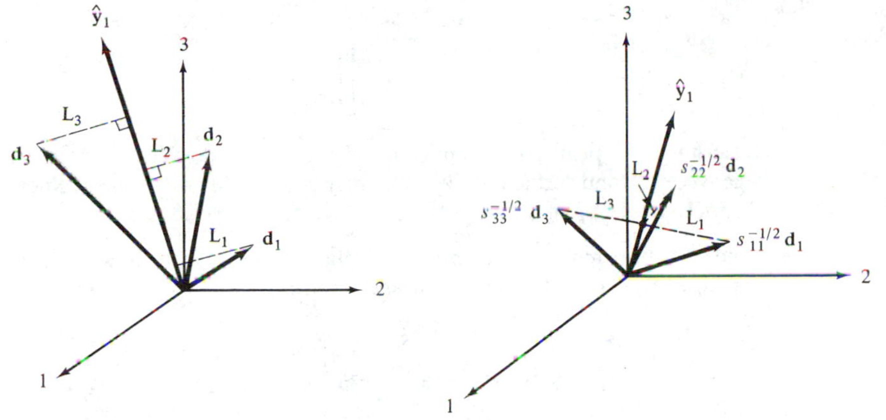

(a) Principal component of **S**

(b) Principal component of **R**

Figure 8.6 The first sample principal component, $\hat{\mathbf{y}}_1$, minimizes the sum of squared distances, L_i^2, from the deviation vectors, $[x_{i1} - \bar{x}_i, x_{i2} - \bar{x}_i, \ldots, x_{in} - \bar{x}_i]'$ to a line.

termined by the vector of values of the first principal component. This is illustrated in Figure 8.6(a). Note that the longer deviation vectors (largest s_{ii}) have the most influence on the minimization of $\sum_{i=1}^{p} L_i^2$.

If the variables are first standardized, the vector $[(x_{i1} - \bar{x}_i)/\sqrt{s_{ii}}, (x_{i2} - \bar{x}_i)/\sqrt{s_{ii}}, \ldots, (x_{in} - \bar{x}_i)/\sqrt{s_{ii}}]$ has length 1 for all variables, and each exerts equal influence on the choice of direction [see Figure 8.6(b)].

In either case, the vector **b** is moved around in n-space to minimize the sum of squared distances between $[x_{i1} - \bar{x}_i, x_{i2} - \bar{x}_i, \ldots, x_{in} - \bar{x}_i]'$ and its projection on the line determined by **b**. The second principal component minimizes the same quantity among all vectors perpendicular to the first choice.

EXERCISES

8.1. Determine the population principal components Y_1 and Y_2 for the covariance matrix

$$\boldsymbol{\Sigma} = \begin{bmatrix} 5 & 2 \\ 2 & 2 \end{bmatrix}$$

Also calculate the proportion of total population variance explained by the first principal component.

8.2. Convert the covariance matrix in Exercise 8.1 to a correlation matrix $\boldsymbol{\rho}$.

(a) Determine the principal components Y_1 and Y_2 from $\boldsymbol{\rho}$ and compute the proportion of total population variance explained by Y_1.

(b) Compare the components with those obtained in Exercise 8.1. Are they the same? Should they be?

(c) Compute the correlations ρ_{Y_1, Z_1}, ρ_{Y_1, Z_2}, and ρ_{Y_2, Z_1}.

8.3. Let

$$\mathbf{\Sigma} = \begin{bmatrix} 2 & 0 & 0 \\ 0 & 4 & 0 \\ 0 & 0 & 4 \end{bmatrix}$$

Determine the principal components Y_1, Y_2, and Y_3. What can you say about the eigenvectors (and principal components) associated with eigenvalues that are not distinct?

8.4. Find the principal components, and the proportion of total population variance explained by each, when the covariance matrix is

$$\mathbf{\Sigma} = \begin{bmatrix} \sigma^2 & \sigma^2\rho & 0 \\ \sigma^2\rho & \sigma^2 & \sigma^2\rho \\ 0 & \sigma^2\rho & \sigma^2 \end{bmatrix}, \qquad -\frac{1}{\sqrt{2}} < \rho < \frac{1}{\sqrt{2}}$$

8.5. **(a)** Find the eigenvalues of the correlation matrix

$$\boldsymbol{\rho} = \begin{bmatrix} 1 & \rho & \rho \\ \rho & 1 & \rho \\ \rho & \rho & 1 \end{bmatrix}$$

Are your results consistent with (8-16) and (8-17)?

(b) Verify the eigenvalue-eigenvector pairs for the $p \times p$ matrix $\boldsymbol{\rho}$ given in (8-15).

8.6. Data on x_1 = assets and x_2 = net income for the 10 largest U.S. industrial corporations were listed in Exercise 1.4 of Chapter 1.

From Example 4.11

$$\bar{\mathbf{x}} = \begin{bmatrix} 19.32 \\ 1.51 \end{bmatrix}, \qquad \mathbf{S} = \begin{bmatrix} 70.41 & 5.87 \\ 5.87 & .97 \end{bmatrix}$$

(a) Determine the sample principal components and their variances for these data. (You may need the quadratic formula to solve for the eigenvalues of $\mathbf{S}$.)

(b) Find the proportion of total sample variance explained by $\hat{y}_1$.

(c) Sketch the constant density ellipse $(\mathbf{x} - \bar{\mathbf{x}})'\mathbf{S}^{-1}(\mathbf{x} - \bar{\mathbf{x}}) = 1.4$ and indicate the principal components $\hat{y}_1$ and $\hat{y}_2$ on your graph.

(d) Compute the correlation coefficients $r_{\hat{y}_1, x_k}$, $k = 1, 2$.

What interpretation, if any, can you give to the first principal component?

8.7. Convert the covariance matrix $\mathbf{S}$ in Exercise 8.6 to a sample correlation matrix $\mathbf{R}$.

(a) Find the sample principal components $\hat{y}_1$, $\hat{y}_2$ and their variances.

(b) Compute the proportion of total sample variance explained by $\hat{y}_1$.

(c) Compute the correlation coefficients $r_{\hat{y}_1, z_k}$, $k = 1, 2$. Interpret $\hat{y}_1$.

(d) Compare the components obtained in Part a with those obtained in Exercise 8.6(a). Given the original data displayed in Exercise 1.4, do you feel it is better to determine principal components from the sample covariance matrix or sample correlation matrix? Explain.

8.8. Use the results in Example 8.5.

(a) Compute the correlations $r_{\hat{y}_i, z_k}$ for $i = 1, 2$ and $k = 1, 2, \ldots, 5$. Do these correlations reinforce the interpretations given to the first two components? Explain.

(b) Test the hypothesis

$$H_0: \boldsymbol{\rho} = \boldsymbol{\rho}_0 = \begin{bmatrix} 1 & \rho & \rho & \rho & \rho \\ \rho & 1 & \rho & \rho & \rho \\ \rho & \rho & 1 & \rho & \rho \\ \rho & \rho & \rho & 1 & \rho \\ \rho & \rho & \rho & \rho & 1 \end{bmatrix}$$

versus

$$H_1: \boldsymbol{\rho} \neq \boldsymbol{\rho}_0$$

at the 5% level of significance. List any assumptions required to be justified in carrying out this test.

8.9. (A test that all variables are independent.)

(a) Consider the normal theory likelihood ratio test of H_0: $\boldsymbol{\Sigma}$ is the diagonal matrix

$$\boldsymbol{\Sigma}_0 = \begin{bmatrix} \sigma_{11} & 0 & \cdots & 0 \\ 0 & \sigma_{22} & \cdots & 0 \\ \vdots & \vdots & \ddots & \vdots \\ 0 & 0 & \cdots & \sigma_{pp} \end{bmatrix}$$

Show that the test is: Reject H_0 if

$$\Lambda = \frac{|\mathbf{S}|^{n/2}}{\prod_{i=1}^{p} s_{ii}^{n/2}} = |\mathbf{R}|^{n/2} < c$$

For a large sample size, $-2 \ln \Lambda$ is approximately $\chi^2_{p(p-1)/2}$. Bartlett [3] suggests that the test statistic $-2[1 - (2p + 11)/6n] \ln \Lambda$ be used in place of $-2 \ln \Lambda$. This results in an improved chi-square approximation. The large sample α critical point is $\chi^2_{p(p-1)/2}(\alpha)$. Note that testing $\boldsymbol{\Sigma} = \boldsymbol{\Sigma}_0$ is the same as testing $\boldsymbol{\rho} = \mathbf{I}$.

(b) Show that the likelihood ratio test of H_0: $\boldsymbol{\Sigma} = \sigma^2 \mathbf{I}$ rejects H_0 if

$$\Lambda = \frac{|\mathbf{S}|^{n/2}}{(\text{tr}(\mathbf{S})/p)^{np/2}} = \left[\frac{\prod_{i=1}^{p} \hat{\lambda}_i}{\left(\frac{1}{p} \sum_{i=1}^{p} \hat{\lambda}_i \right)^p} \right]^{n/2} = \left[\frac{\text{geometric mean } \hat{\lambda}_i}{\text{arithmetic mean } \hat{\lambda}_i} \right]^{np/2} < c$$

For a large sample size, Bartlett [3] suggests that $-2[1 - (2p^2 + p + 2)/6pn] \ln \Lambda$ is approximately $\chi^2_{(p+2)(p-1)/2}$. Thus the large sample α critical point is $\chi^2_{(p+2)(p-1)/2}(\alpha)$. This test is called a *sphericity test* because the constant density contours are spheres when $\boldsymbol{\Sigma} = \sigma^2 \mathbf{I}$.

(*Hint*:

(a) $\max_{\boldsymbol{\mu}, \boldsymbol{\Sigma}} L(\boldsymbol{\mu}, \boldsymbol{\Sigma})$ is given by (5-9) and $\max L(\boldsymbol{\mu}, \boldsymbol{\Sigma}_0)$ is the product of the univariate likelihoods $\max_{\mu_i} (2\pi)^{-n/2} \sigma_{ii}^{-n/2} \exp\left[-\sum_{j=1}^{n} (x_{ij} - \mu_i)^2 / \sigma_{ii} \right]$. Hence $\hat{\mu}_i = (1/n) \sum_{j=1}^{n} x_{ij}$ and $\hat{\sigma}_{ii} = (1/n) \sum_{j=1}^{n} (x_{ij} - \bar{x}_i)^2$. The divisor n cancels, so $\mathbf{S}$ may be used.

(b) Verify $\hat{\sigma}^2 = \left[\sum_{j=1}^{n} (x_{1j} - \bar{x}_1)^2 + \cdots + \sum_{j=1}^{n} (x_{pj} - \bar{x}_p)^2\right]/np$ under H_0. Again the divisors n cancel in the statistic, so $\mathbf{S}$ may be used. Use Result 5.2 to calculate the chi-square degrees of freedom.)

The following exercises require the use of a computer.

8.10. The weekly rates of return for five stocks listed on the New York Stock Exchange are given in Table 8.1.

(a) Construct the sample covariance matrix $\mathbf{S}$ and find the sample principal components in (8-20). (Note the sample mean vector $\bar{\mathbf{x}}$ is displayed in Example 8.5.)

(b) Determine the proportion of total sample variance explained by the first three principal components. Interpret these components.

(c) Construct Bonferroni simultaneous 90% confidence intervals for the variances λ_1, λ_2, and λ_3 of the first three population components Y_1, Y_2, and Y_3.

(d) Given the results in Parts a–c, do you feel the stock rates of return data can be summarized in fewer than five dimensions? Explain.

8.11. Consider the census-tract data listed in Table 8.2. Suppose the observations on X_5 = median value home were recorded in thousands, rather than ten thousands of dollars; that is, multiply all the numbers listed in the sixth column of Table 8.2 by 10.

(a) Construct the sample covariance matrix, $\mathbf{S}$, for the census-tract data when X_5 = median value home is recorded in thousands of dollars. (Note that this covariance matrix can be obtained from the covariance matrix given in Example 8.3 by multiplying the off-diagonal elements in the fifth column and row by 10 and the diagonal element s_{55} by 100. Why?)

(b) Obtain the eigenvalue-eigenvector pairs and the first two sample principal components for the covariance matrix in Part a.

(c) Compute the proportion of total variance explained by the first two principal components obtained in Part b. Calculate the correlation coefficients, $r_{\hat{y}_i, x_k}$ and interpret these components if possible. Compare your results with the results in Example 8.3. What can you say about the effects of this change in scale on the principal components?

8.12. Consider the air-pollution data listed in Table 1.2. Your job is to summarize this data in fewer than $p = 7$ dimensions if possible. Conduct a principal component analysis of this data using both the covariance matrix $\mathbf{S}$ and the correlation matrix $\mathbf{R}$. What have you learned? Does it make any difference which matrix is chosen for analysis? Can the data be summarized in three or fewer dimensions? Can you interpret the principal components?

8.13. The radiotherapy data is listed in Table 1.4. The $n = 98$ observations on $p = 6$ variables represent patients' reactions to radiotherapy.

(a) Obtain the covariance and correlation matrices $\mathbf{S}$ and $\mathbf{R}$ for this data.

(b) Pick one of the matrices $\mathbf{S}$ or $\mathbf{R}$ (justify your choice) and determine the eigenvalues and eigenvectors. Prepare a table showing, in decreasing order of size, the percent that each eigenvalue contributes to the total sample variance.

(c) Given the results in (b), decide on the number of important sample principal components. Is it possible to summarize the radiotherapy data with a single reaction-index component? Explain.

(d) Prepare a table of the correlation coefficients between each principal component you decide to retain and the original variables. If possible, interpret the components.

TABLE 8.1 STOCK-PRICE DATA
(WEEKLY RATE OF RETURN JANUARY 1975 THROUGH DECEMBER 1976)

Week	Allied Chemical	DuPont	Union Carbide	Exxon	Texaco
1	.000000	.000000	.000000	.039473	−.000000
2	.027027	−.044855	−.003030	−.014466	.043478
3	.122807	.060773	.088146	.086238	.078124
4	.057031	.029948	.066808	.013513	.019512
5	.063670	−.003793	−.039788	−.018644	−.024154
6	.003521	.050761	.082873	.074265	.049504
7	−.045614	−.033007	.002551	−.009646	−.028301
8	.058823	.041719	.081425	−.014610	.014563
9	.000000	−.019417	.002353	.001647	−.028708
10	.006944	−.025990	.007042	−.041118	−.024630
11	.010345	.006353	.083916	.010291	−.000000
12	−.030717	.020202	−.040860	−.039049	−.050505
13	−.003521	.118812	.089686	.060070	.021276
14	.060071	.079646	.028807	.036666	.026041
15	−.003333	−.001025	.028000	.028938	−.010152
16	.055596	.091282	.042759	.059375	−.015812
17	.051282	−.007519	−.041431	−.016269	.058510
18	−.060976	−.043561	.023576	.004566	−.015075
19	−.035714	.018170	−.021113	−.007575	−.010204
20	.000000	−.021569	−.007843	.088549	.082474
21	−.006734	−.015030	−.086956	−.021037	−.019047
22	.000000	−.017294	.017316	.054441	.033980
23	.030508	.047619	.055319	−.008152	.032863
24	.023026	.012846	−.002016	.013698	−.031518
25	−.061093	−.043902	−.042424	−.029729	−.014084
26	.041096	.016326	.048523	.018105	.071428
27	−.013158	−.004016	−.038229	−.042407	−.048888
28	.003333	−.008065	−.014992	.000000	−.028037
29	−.056478	−.014228	−.038627	−.005714	−.019607
30	.051899	.018557	.066964	.020302	−.015000
31	−.013559	−.029352	.012552	−.008571	−.010152
32	−.037801	.003252	−.012397	−.020172	−.025641
33	−.021429	.031446	.039749	.016176	.005263
34	−.014599	−.024390	−.010060	.004341	−.005235
35	−.014815	−.020833	−.091463	−.007204	−.015789
36	.011278	−.017021	.064877	.065312	.026737
37	−.096654	−.075758	−.073529	−.053133	−.026041
38	.020576	.058548	.018141	.063309	.016042
39	.088710	.046460	.022272	.004059	−.000000
40	.007407	.019027	.045752	−.008086	.052631
41	−.022059	.002075	−.017272	−.021739	−.045000
42	−.031579	.010352	.012848	−.013888	.010695
43	.039370	.054303	−.014799	.011428	−.005291
44	.015151	.029154	−.021459	−.009887	−.021276
45	.000000	−.010466	.035088	−.014265	.038043
46	−.037313	−.024038	−.019068	−.024602	−.010471
47	.015504	−.027586	.006479	.022255	−.026455
48	.034351	.024316	.034335	.020319	.005434
49	−.036900	.011869	.014523	.007112	.016216
50	.068965	.014663	.016360	.038135	.063829

TABLE 8.1 (*continued*)

Week	Allied Chemical	DuPont	Union Carbide	Exxon	Texaco
51	.089606	.079961	.102616	.002721	.020000
52	.000000	.016949	.029197	.002713	.004901
53	.059210	.077193	.019504	−.012178	.039024
54	.027950	.009772	.000000	−.000000	−.000265
55	−.004196	.014516	−.031696	−.004445	−.014354
56	.018405	−.046900	.061594	−.043235	−.029126
57	.069277	.056888	.040956	.040816	.020000
58	−.016901	−.018268	−.008197	−.005602	−.019607
59	−.017192	−.001618	−.001653	−.016901	.005000
60	−.040816	−.035656	.000000	.014326	.004975
61	−.018237	−.003361	−.028146	.035310	.014851
62	−.003096	−.021922	−.027257	.005457	.039024
63	.018634	.025862	−.017513	.018995	−.004694
64	−.057927	−.018487	.000000	−.023968	−.037735
65	.087379	.049657	.033868	.047748	.039215
66	.000000	−.011419	−.010345	−.005208	.028301
67	−.019367	−.011551	−.022817	.007853	.013761
68	−.046012	.035893	.044964	.040612	.004608
69	−.077170	−.004029	−.003442	.003797	−.027522
70	.034843	−.008157	−.018998	.008827	−.014151
71	−.006734	−.019737	−.026408	.023749	.014354
72	−.023729	−.019295	−.032550	−.001221	.023584
73	.065972	.024807	.057944	.020782	.004608
74	.000000	−.036728	−.014134	−.007185	.004587
75	−.052117	−.058925	−.069892	.009650	.009132
76	.054983	−.003683	.026975	−.002389	.009049
77	−.003257	−.009242	−.022514	.005988	−.013452
78	.022876	.033582	.001919	.026190	.004545
79	−.003195	−.005415	−.003831	−.013921	−.000165
80	.043590	−.014519	−.013385	.021176	.013824
81	−.009317	.013812	.021654	−.014927	−.009090
82	−.056426	−.005557	−.003854	−.023696	−.018348
83	.003322	−.041475	−.029014	−.002427	−.004672
84	.016556	.017308	.033864	.034063	.009389
85	−.009772	−.016068	−.003854	.014117	.013953
86	.026316	−.016330	−.009671	.032482	.027522
87	.009615	.009766	.017578	.016247	.017857
88	−.047619	−.027079	−.051823	−.045468	−.021929
89	−.026667	−.061630	−.056680	−.013452	−.040358
90	.010274	.023305	.034335	−.018181	−.004672
91	−.044068	.020704	−.006224	−.018518	.004694
92	.039007	.038540	.024988	−.028301	.032710
93	−.039457	−.029297	−.065844	−.015837	−.045758
94	.039568	.024145	−.006608	.028423	−.009661
95	−.031142	−.007941	.011080	.007537	.014634
96	.000000	−.020080	−.006579	.029925	−.004807
97	.021429	.049180	.006622	−.002421	.028985
98	.045454	.046375	.074561	.014563	.018779
99	.050167	.036380	.004082	−.011961	.009216
100	.019108	−.033303	.008362	.033898	.004566

TABLE 8.2 CENSUS-TRACT DATA

Tract	Total population (thousands)	Median school years	Total employment (thousands)	Health services employment (hundreds)	Median value home ($10,000s)
1	5.935	14.2	2.265	2.27	2.91
2	1.523	13.1	.597	.75	2.62
3	2.599	12.7	1.237	1.11	1.72
4	4.009	15.2	1.649	.81	3.02
5	4.687	14.7	2.312	2.50	2.22
6	8.044	15.6	3.641	4.51	2.36
7	2.766	13.3	1.244	1.03	1.97
8	6.538	17.0	2.618	2.39	1.85
9	6.451	12.9	3.147	5.52	2.01
10	3.314	12.2	1.606	2.18	1.82
11	3.777	13.0	2.119	2.83	1.80
12	1.530	13.8	.798	.84	4.25
13	2.768	13.6	1.336	1.75	2.64
14	6.585	14.9	2.763	1.91	3.17

NOTE: Observations from adjacent census tracts are likely to be correlated. That is, these 14 observations may not constitute a random sample.

8.14. Perform a principal component analysis using the sample covariance matrix of the sweat data given in Example 5.2. Construct a *Q-Q* plot for each of the important principal components. Are there any suspect observations? Explain.

8.15. The four sample standard deviations for the post-birth weights discussed in Example 8.6 are

$$\sqrt{s_{11}} = 32.9909, \quad \sqrt{s_{22}} = 33.5918, \quad \sqrt{s_{33}} = 36.5534, \quad \text{and} \quad \sqrt{s_{44}} = 37.3517$$

Use these and the correlations given in Example 8.6 to construct the sample covariance matrix **S**. Perform a principal component analysis using **S**.

REFERENCES

[1] Anderson, T. W., *An Introduction to Multivariate Statistical Analysis*, New York: John Wiley, 1958.

[2] Anderson, T. W., "Asymptotic Theory for Principal Components Analysis," *Annals of Mathematical Statistics*, **34** (1963), 122–148.

[3] Bartlett, M. S., "A Note on Multiplying Factors for Various Chi-Squared Approximations," *Journal of the Royal Statistical Society (B)*, **16** (1954), 296–298.

[4] Girschick, M. A., "On the Sampling Theory of Roots of Determinantal Equations," *Annals of Mathematical Statistics*, **10** (1939), 203–224.

[5] Hotelling, H., "Analysis of a Complex of Statistical Variables into Principal Components," *Journal of Educational Psychology*, **24** (1933), 417–441, 498–520.

[6] Hotelling, H., "The Most Predictable Criterion," *Journal of Educational Psychology*, **26** (1935), 139–142.

[7] Hotelling, H., "Simplified Calculation of Principal Components," *Psychometrika*, **1** (1936), 27–35.

[8] Hotelling, H., "Relations Between Two Sets of Variates," *Biometrika*, **28** (1936), 321–377.

[9] Jolicoeur, P., "The Multivariate Generalization of the Allometry Equation," *Biometrics*, **19** (1963), 497–499.

[10] Jolicoeur, P., and J. E. Mosimann, "Size and Shape Variation in the Painted Turtle: A Principal Component Analysis," *Growth*, **24** (1960), 339–354.

[11] King, B., "Market and Industry Factors in Stock Price Behavior," *Journal of Business*, **39** (1966), 139–190.

[12] Lawley, D. N., "On Testing a Set of Correlation Coefficients for Equality," *Annals of Mathematical Statistics*, **34** (1963), 149–151.

[13] Maxwell, A. E., *Multivariate Analysis in Behavioural Research*, London: Chapman and Hall, 1977.

[14] Rao, C. R., *Linear Statistical Inference and Its Applications* (2nd ed.), New York: John Wiley, 1973.

9

Factor Analysis

9.1 INTRODUCTION

Factor analysis has provoked rather turbulent controversy throughout its history. Its modern beginnings lie in the early twentieth-century attempts of Karl Pearson, Charles Spearman, and others to define and measure "intelligence." Because of this early association with constructs such as intelligence, factor analysis was nurtured and developed primarily by scientists interested in psychometric measurement. Arguments over the psychological interpretations of several early studies and the lack of powerful computing facilities impeded its initial development as a statistical method. The advent of high-speed computers has generated a renewed interest in the theoretical and computational aspects of factor analysis. Most of the original techniques have been abandoned and early controversies resolved in the wake of recent developments. It is still true that each application of the technique must be examined on its own merits to determine its success.

The essential purpose of factor analysis is to describe, if possible, the covariance relationships among many variables in terms of a few underlying, but unobservable, random quantities called *factors*. Basically, the factor model is motivated by the following argument. Suppose variables can be grouped by their correlations. That is, all variables within a particular group are highly correlated among themselves but have relatively small correlations with variables in a different group. It is conceivable that each group of variables represents a single underlying construct, or factor, that is responsible for the observed correlations. For example, correlations from the group of test scores in classics, French, English, mathematics, and music collected by Spearman suggested an underlying "intelligence" factor. A second group of variables, representing physical-fitness scores, if available, might correspond to another factor. It is this type of structure that factor analysis seeks to confirm.

Factor analysis can be considered as an extension of principal component analysis. Both can be viewed as attempts to approximate the covariance matrix $\boldsymbol{\Sigma}$. However, the approximation based on the factor analysis model is more elaborate. The primary question in factor analysis is whether the data are consistent with a prescribed structure.

9.2 THE ORTHOGONAL FACTOR MODEL

The observable random vector $\mathbf{X}$, with p components, has mean $\boldsymbol{\mu}$ and covariance matrix $\boldsymbol{\Sigma}$. The factor model postulates that $\mathbf{X}$ is linearly dependent upon a few unobservable random variables $F_1, F_2, \ldots, F_m$, called *common factors*, and p additional sources of variation $\varepsilon_1, \varepsilon_2, \ldots, \varepsilon_p$, called *errors* or, sometimes, *specific factors.*[1] In particular, the factor analysis model is

$$\begin{aligned} X_1 - \mu_1 &= \ell_{11}F_1 + \ell_{12}F_2 + \cdots + \ell_{1m}F_m + \varepsilon_1 \\ X_2 - \mu_2 &= \ell_{21}F_1 + \ell_{22}F_2 + \cdots + \ell_{2m}F_m + \varepsilon_2 \\ &\vdots \qquad\qquad \vdots \\ X_p - \mu_p &= \ell_{p1}F_1 + \ell_{p2}F_2 + \cdots + \ell_{pm}F_m + \varepsilon_p \end{aligned} \tag{9-1}$$

or, in matrix notation,

$$\underset{(p\times 1)}{\mathbf{X} - \boldsymbol{\mu}} = \underset{(p\times m)}{\mathbf{L}} \; \underset{(m\times 1)}{\mathbf{F}} + \underset{(p\times 1)}{\boldsymbol{\varepsilon}} \tag{9-2}$$

The coefficient ℓ_{ij} is called the *loading* of the ith variable on the jth factor, so the matrix $\mathbf{L}$ is the *matrix of factor loadings*. Note that the ith specific factor ε_i is associated only with the ith response X_i. The p deviations $X_1 - \mu_1, X_2 - \mu_2, \ldots, X_p - \mu_p$ are expressed in terms of $p + m$ random variables $F_1, F_2, \ldots, F_m, \varepsilon_1, \varepsilon_2, \ldots, \varepsilon_p$ which are *unobservable*. This distinguishes the factor model of (9-2) from the multivariate regression model in (7-24), in which the independent variables [whose position is occupied by $\mathbf{F}$ in (9-2)] can be observed.

With so many unobservable quantities, a direct verification of the factor model from observations on $X_1, X_2, \ldots, X_p$ is hopeless, However, with some additional assumptions about the random vectors $\mathbf{F}$ and $\boldsymbol{\varepsilon}$, the model in (9-2) implies certain covariance relationships, which can be checked.

[1]As Maxwell [11] points out, in many investigations the ε_i tend to be combinations of measurement error and factors that are uniquely associated with the individual variables.

We assume that

$$E(\mathbf{F}) = \underset{(m\times 1)}{\mathbf{0}}, \qquad \text{Cov}(\mathbf{F}) = E[\mathbf{FF}'] = \underset{(m\times m)}{\mathbf{I}}$$

$$E(\boldsymbol{\varepsilon}) = \underset{(p\times 1)}{\mathbf{0}}, \qquad \text{Cov}(\boldsymbol{\varepsilon}) = E[\boldsymbol{\varepsilon\varepsilon}'] = \underset{(p\times p)}{\boldsymbol{\Psi}} = \begin{bmatrix} \psi_1 & 0 & \cdots & 0 \\ 0 & \psi_2 & \cdots & 0 \\ \vdots & \vdots & \ddots & \vdots \\ 0 & 0 & \cdots & \psi_p \end{bmatrix} \tag{9-3}$$

and that $\mathbf{F}$ and $\boldsymbol{\varepsilon}$ are independent so

$$\text{Cov}(\boldsymbol{\varepsilon}, \mathbf{F}) = E(\boldsymbol{\varepsilon}\mathbf{F}') = \underset{(p\times m)}{\mathbf{0}}$$

These assumptions and the relation in (9-2) constitute the *orthogonal factor model*.[2]

Orthogonal Factor Model with m Common Factors

$$\underset{(p\times 1)}{\mathbf{X}} = \underset{(p\times 1)}{\boldsymbol{\mu}} + \underset{(p\times m)}{\mathbf{L}}\ \underset{(m\times 1)}{\mathbf{F}} + \underset{(p\times 1)}{\boldsymbol{\varepsilon}}$$

μ_i = *mean* of variable i

ε_i = ith *specific factor*

F_j = jth *common factor*

ℓ_{ij} = *loading* of the ith variable on the jth factor (9-4)

The unobservable random vectors $\mathbf{F}$ and $\boldsymbol{\varepsilon}$ satisfy

$\mathbf{F}$ and $\boldsymbol{\varepsilon}$ are independent

$E(\mathbf{F}) = \mathbf{0}$, $\text{Cov}(\mathbf{F}) = \mathbf{I}$

$E(\boldsymbol{\varepsilon}) = \mathbf{0}$, $\text{Cov}(\boldsymbol{\varepsilon}) = \boldsymbol{\Psi}$, where $\boldsymbol{\Psi}$ is a diagonal matrix

The orthogonal factor model implies a covariance structure for $\mathbf{X}$. From the model in (9-4),

$$\begin{aligned}(\mathbf{X} - \boldsymbol{\mu})(\mathbf{X} - \boldsymbol{\mu})' &= (\mathbf{LF} + \boldsymbol{\varepsilon})(\mathbf{LF} + \boldsymbol{\varepsilon})' \\ &= (\mathbf{LF} + \boldsymbol{\varepsilon})((\mathbf{LF})' + \boldsymbol{\varepsilon}') \\ &= \mathbf{LF}(\mathbf{LF})' + \boldsymbol{\varepsilon}(\mathbf{LF})' + \mathbf{LF}\boldsymbol{\varepsilon}' + \boldsymbol{\varepsilon\varepsilon}'\end{aligned}$$

so that

$$\begin{aligned}\boldsymbol{\Sigma} = \text{Cov}(\mathbf{X}) &= E(\mathbf{X} - \boldsymbol{\mu})(\mathbf{X} - \boldsymbol{\mu})' \\ &= \mathbf{L}E(\mathbf{FF}')\mathbf{L}' + E(\boldsymbol{\varepsilon}\mathbf{F}')\mathbf{L}' + \mathbf{L}E(\mathbf{F}\boldsymbol{\varepsilon}') + E(\boldsymbol{\varepsilon\varepsilon}') \\ &= \mathbf{LL}' + \boldsymbol{\Psi}\end{aligned}$$

according to (9-3).

[2]Allowing the factors $\mathbf{F}$ to be correlated so that $\text{Cov}(\mathbf{F})$ is *not* diagonal gives the oblique factor model. The oblique model presents some additional estimation difficulties and will not be discussed in the book. See [9].

Also by the model in (9-4), $(\mathbf{X} - \boldsymbol{\mu})\mathbf{F}' = (\mathbf{LF} + \boldsymbol{\varepsilon})\mathbf{F}' = \mathbf{LFF}' + \boldsymbol{\varepsilon}\mathbf{F}'$, so $\text{Cov}(\mathbf{X}, \mathbf{F}) = E(\mathbf{X} - \boldsymbol{\mu})\mathbf{F}' = \mathbf{L}E(\mathbf{FF}') + E(\boldsymbol{\varepsilon}\mathbf{F}') = \mathbf{L}$.

Covariance Structure for the Orthogonal Factor Model

1. $\text{Cov}(\mathbf{X}) = \mathbf{LL}' + \boldsymbol{\Psi}$

or

$$\begin{aligned}\text{Var}(X_i) &= \ell_{i1}^2 + \cdots + \ell_{im}^2 + \psi_i \\ \text{Cov}(X_i, X_k) &= \ell_{i1}\ell_{k1} + \cdots + \ell_{im}\ell_{km}\end{aligned} \tag{9-5}$$

2. $\text{Cov}(\mathbf{X}, \mathbf{F}) = \mathbf{L}$

or

$$\text{Cov}(X_i, F_j) = \ell_{ij}$$

The model $\mathbf{X} - \boldsymbol{\mu} = \mathbf{LF} + \boldsymbol{\varepsilon}$ is *linear* in the common factors. If the p responses $\mathbf{X}$ are, in fact, related to underlying factors but the relationship is nonlinear such as in $X_1 - \mu_1 = \ell_{11}F_1F_3 + \varepsilon_1$, $X_2 - \mu_2 = \ell_{21}F_2F_3 + \varepsilon_2$, and so forth, then the covariance structure $\mathbf{LL}' + \boldsymbol{\Psi}$ given by (9-5) may not be adequate. The very important assumption of linearity is inherent in the formulation of the traditional factor model.

That portion of the variance of the ith variable contributed by the m common factors is called the ith *communality*. That portion of $\text{Var}(X_i) = \sigma_{ii}$ due to the specific factor is often called the *uniqueness*, or *specific variance*. Denoting the ith communality by h_i^2, we see from (9-5) that

$$\underbrace{\sigma_{ii}}_{\text{Var}(X_i)} = \underbrace{\ell_{i1}^2 + \ell_{i2}^2 + \cdots + \ell_{im}^2}_{\text{communality}} + \underbrace{\psi_i}_{\text{specific variance}}$$

or

$$h_i^2 = \ell_{i1}^2 + \ell_{i2}^2 + \cdots + \ell_{im}^2 \tag{9-6}$$

and

$$\sigma_{ii} = h_i^2 + \psi_i \qquad i = 1, 2, \ldots, p$$

The ith communality is the sum of squares of the loadings of the ith variable on the m common factors.

Example 9.1

Consider the covariance matrix

$$\boldsymbol{\Sigma} = \begin{bmatrix} 19 & 30 & 2 & 12 \\ 30 & 57 & 5 & 23 \\ 2 & 5 & 38 & 47 \\ 12 & 23 & 47 & 68 \end{bmatrix}$$

The equality

$$\begin{bmatrix} 19 & 30 & 2 & 12 \\ 30 & 57 & 5 & 23 \\ 2 & 5 & 38 & 47 \\ 12 & 23 & 47 & 68 \end{bmatrix} = \begin{bmatrix} 4 & 1 \\ 7 & 2 \\ -1 & 6 \\ 1 & 8 \end{bmatrix} \begin{bmatrix} 4 & 7 & -1 & 1 \\ 1 & 2 & 6 & 8 \end{bmatrix} + \begin{bmatrix} 2 & 0 & 0 & 0 \\ 0 & 4 & 0 & 0 \\ 0 & 0 & 1 & 0 \\ 0 & 0 & 0 & 3 \end{bmatrix}$$

or

$$\boldsymbol{\Sigma} = \mathbf{LL}' + \boldsymbol{\Psi}$$

may be verified by matrix algebra. Therefore $\boldsymbol{\Sigma}$ has the structure produced by an $m = 2$ orthogonal factor model. Since

$$\mathbf{L} = \begin{bmatrix} \ell_{11} & \ell_{12} \\ \ell_{21} & \ell_{22} \\ \ell_{31} & \ell_{32} \\ \ell_{41} & \ell_{42} \end{bmatrix} = \begin{bmatrix} 4 & 1 \\ 7 & 2 \\ -1 & 6 \\ 1 & 8 \end{bmatrix}, \qquad \boldsymbol{\Psi} = \begin{bmatrix} \psi_1 & 0 & 0 & 0 \\ 0 & \psi_2 & 0 & 0 \\ 0 & 0 & \psi_3 & 0 \\ 0 & 0 & 0 & \psi_4 \end{bmatrix} = \begin{bmatrix} 2 & 0 & 0 & 0 \\ 0 & 4 & 0 & 0 \\ 0 & 0 & 1 & 0 \\ 0 & 0 & 0 & 3 \end{bmatrix}$$

the communality of $\mathbf{X}_1$ is, from (9-6),

$$h_1^2 = \ell_{11}^2 + \ell_{12}^2 = 4^2 + 1^2 = 17$$

and the variance of $\mathbf{X}_1$ can be decomposed as

$$\sigma_{11} = \left(\ell_{11}^2 + \ell_{12}^2\right) + \psi_1 = h_1^2 + \psi_1$$

or

$$\underbrace{19}_{\text{Variance}} = \underbrace{4^2 + 1^2}_{\text{communality}} + \underbrace{2}_{\substack{\text{specific}\\ \text{variance}}} = 17 + 2$$

A similar breakdown occurs for the other variables. ■

The factor model assumes that the $p + p(p-1)/2 = p(p+1)/2$ variances and covariances for $\mathbf{X}$ can be reproduced from the pm factor loadings ℓ_{ij} and the p specific variances ψ_i. When $m = p$, any covariance matrix $\boldsymbol{\Sigma}$ can be reproduced exactly as $\mathbf{LL}'$ [see (9-11)], so $\boldsymbol{\Psi}$ can be the zero matrix. However, it is when m is small relative to p that factor analysis is most useful. In this case the factor model provides a "simple" explanation of the covariation in $\mathbf{X}$ with fewer parameters than the $p(p+1)/2$ parameters in $\boldsymbol{\Sigma}$. For example, if $\mathbf{X}$ contains $p = 12$ variables and the factor model in (9-4) with $m = 2$ is appropriate, the $p(p+1)/2 = 12(13)/2 = 78$ elements of $\boldsymbol{\Sigma}$ are described in terms of the $mp + p = 12(2) + 12 = 36$ parameters ℓ_{ij} and ψ_i of the factor model.

Unfortunately for the factor analyst, most covariance matrices cannot be factored as $\mathbf{LL}' + \boldsymbol{\Psi}$, where the number of factors m is much less than p. The following example demonstrates one of the problems that can arise when attempting to determine the parameters ℓ_{ij} and ψ_i from the variances and covariances of the observable variables.

Example 9.2 (Nonexistence of a Proper Solution)

Let $p = 3$ and $m = 1$ and suppose the random variables X_1, X_2 and X_3 have the positive definite covariance matrix

$$\boldsymbol{\Sigma} = \begin{bmatrix} 1 & .9 & .7 \\ .9 & 1 & .4 \\ .7 & .4 & 1 \end{bmatrix}$$

Using the factor model in (9-4),

$$X_1 - \mu_1 = \ell_{11}F_1 + \varepsilon_1$$
$$X_2 - \mu_2 = \ell_{21}F_1 + \varepsilon_2$$
$$X_3 - \mu_2 = \ell_{31}F_1 + \varepsilon_3$$

The covariance structure in (9-5) implies.

$$\mathbf{\Sigma} = \mathbf{LL}' + \mathbf{\Psi}$$

or

$$1 = \ell_{11}^2 + \psi_1 \qquad .90 = \ell_{11}\ell_{21} \qquad .70 = \ell_{11}\ell_{31}$$
$$1 = \ell_{21}^2 + \psi_2 \qquad .40 = \ell_{21}\ell_{31}$$
$$1 = \ell_{31}^2 + \psi_3$$

The pair of equations

$$.70 = \ell_{11}\ell_{31}$$
$$.40 = \ell_{21}\ell_{31}$$

imply

$$\ell_{21} = \left(\frac{.40}{.70}\right)\ell_{11}$$

Substituting this result for ℓ_{21} in the equation

$$.90 = \ell_{11}\ell_{21}$$

yields $\ell_{11}^2 = 1.575$ or $\ell_{11} = \pm 1.255$. Since $\text{Var}(F_1) = 1$ (by assumption) and $\text{Var}(X_1) = 1$, $\ell_{11} = \text{Cov}(X_1, F_1) = \text{Corr}(X_1, F_1)$. A correlation coefficient cannot be greater than unity (in absolute value) so, from this point of view, $|\ell_{11}| = 1.255$ is "too large." Also the equation

$$1 = \ell_{11}^2 + \psi_1 \quad \text{or} \quad \psi_1 = 1 - \ell_{11}^2$$

gives

$$\psi_1 = 1 - 1.575 = -.575$$

which is unsatisfactory since it gives a negative value for $\text{Var}(\varepsilon_1) = \psi_1$.

Thus, for this example with $m = 1$, it is possible to get unique numerical solutions to the equations $\mathbf{\Sigma} = \mathbf{LL}' + \mathbf{\Psi}$. However, the solution is not consistent with the statistical interpretation of the coefficients, so it is not a proper solution. ■

When $m > 1$, there is always some inherent ambiguity associated with the factor model. To see this, let $\mathbf{T}$ be any $m \times m$ orthogonal matrix so that $\mathbf{TT}' = \mathbf{T}'\mathbf{T} = \mathbf{I}$. The expression in (9-2) can be written

$$\mathbf{X} - \boldsymbol{\mu} = \mathbf{LF} + \boldsymbol{\varepsilon} = \mathbf{LTT}'\mathbf{F} + \boldsymbol{\varepsilon} = \mathbf{L}^*\mathbf{F}^* + \boldsymbol{\varepsilon} \tag{9-7}$$

where

$$\mathbf{L}^* = \mathbf{LT} \quad \text{and} \quad \mathbf{F}^* = \mathbf{T}'\mathbf{F}$$

Since

$$E(\mathbf{F}^*) = \mathbf{T}'E(\mathbf{F}) = \mathbf{0}$$

and

$$\text{Cov}(\mathbf{F}^*) = \mathbf{T}'\text{Cov}(\mathbf{F})\mathbf{T} = \mathbf{T}'\mathbf{T} = \underset{(m\times m)}{\mathbf{I}}$$

it is impossible, on the basis of observations on **X**, to distinguish the loadings **L** from the loadings **L***. That is, the factors **F** and $\mathbf{F}^* = \mathbf{T}'\mathbf{F}$ have the same statistical properties, and even though the loadings **L*** are, in general, different from the loadings **L**, they both generate the same covariance matrix $\mathbf{\Sigma}$. That is,

$$\mathbf{\Sigma} = \mathbf{LL}' + \mathbf{\Psi} = \mathbf{LTT}'\mathbf{L}' + \mathbf{\Psi} = (\mathbf{L}^*)(\mathbf{L}^*)' + \mathbf{\Psi} \qquad (9\text{-}8)$$

This ambiguity provides the rationale for "factor rotation," since orthogonal matrices correspond to rotations (and reflections) of the coordinate system for **X**.

> Factor loadings **L** are determined only up to an orthogonal matrix **T**. Thus, loadings
>
> $$\mathbf{L}^* = \mathbf{LT} \quad \text{and} \quad \mathbf{L} \qquad (9\text{-}9)$$
>
> both give the same representation. The communalities, given by the diagonal elements of $\mathbf{LL}' = (\mathbf{L}^*)(\mathbf{L}^*)'$, are also unaffected by the choice of **T**.

The analysis of the factor model proceeds by imposing conditions that allow one to uniquely estimate **L** and $\mathbf{\Psi}$. The loading matrix is then rotated (multiplied by an orthogonal matrix), where the rotation is determined by some "ease-of-interpretation" criterion. Once the loadings and specific variances are obtained, factors are identified and estimated values for the factors themselves (called *factor scores*) are frequently constructed.

9.3 METHODS OF ESTIMATION

Given observations $\mathbf{x}_1, \mathbf{x}_2, \ldots, \mathbf{x}_n$ on p generally correlated variables, factor analysis seeks to answer the question, Does the factor model of (9-4), with a small number of factors, adequately represent the data? In essence, we tackle this statistical model-building problem by trying to verify the covariance relationship in (9-5).

The sample covariance matrix **S** is an estimator of the unknown population covariance matrix $\mathbf{\Sigma}$. If the off-diagonal elements of **S** are small or those of the sample correlation matrix **R** essentially zero, the variables are not related and a factor analysis will not prove useful. In these circumstances, the *specific* factors play the dominant role, whereas the major aim of the factor analysis is to determine a few important *common* factors.

If $\mathbf{\Sigma}$ appears to deviate significantly from a diagonal matrix, then a factor model can be entertained and the initial problem is one of estimating the factor loadings ℓ_{ij} and specific variances ψ_i. We shall consider two of the most popular methods of parameter estimation, the *principal component* (and the related *principal factor*) *method* and the *maximum likelihood method*. The solution from either method can be rotated in order to simplify the interpretation of factors, as described in Section 9.4. It is always prudent to try more than one method of solution. If the factor model is appropriate for the problem at hand, the solutions should be consistent with one another.

Current estimation and rotation methods require iterative calculations that must be done on a computer. Several computer programs are now available for this purpose.

The Principal Component (and Principal Factor) Method

The spectral decomposition of (2-20) provides us with one factoring of the covariance matrix $\mathbf{\Sigma}$. Let $\mathbf{\Sigma}$ have eigenvalue-eigenvector pairs $(\lambda_i, \mathbf{e}_i)$ with $\lambda_1 \geq \lambda_2 \geq \cdots \geq \lambda_p \geq 0$. Then

$$\mathbf{\Sigma} = \lambda_1 \mathbf{e}_1 \mathbf{e}_1' + \lambda_2 \mathbf{e}_2 \mathbf{e}_2' + \cdots + \lambda_p \mathbf{e}_p \mathbf{e}_p'$$

$$= \left[\sqrt{\lambda_1}\, \mathbf{e}_1 \;\vdots\; \sqrt{\lambda_2}\, \mathbf{e}_2 \;\vdots\; \cdots \;\vdots\; \sqrt{\lambda_p}\, \mathbf{e}_p \right] \begin{bmatrix} \sqrt{\lambda_1}\, \mathbf{e}_1' \\ \hdashline \sqrt{\lambda_2}\, \mathbf{e}_2' \\ \hdashline \vdots \\ \hdashline \sqrt{\lambda_p}\, \mathbf{e}_p' \end{bmatrix} \tag{9-10}$$

This fits the prescribed covariance structure for the factor analysis model having as many factors as variables $(m = p)$ and specific variances $\psi_i = 0$ for all i. The loading matrix has jth column given by $\sqrt{\lambda_j}\, \mathbf{e}_j$. That is, we can write

$$\underset{(p\times p)}{\mathbf{\Sigma}} = \underset{(p\times p)}{\mathbf{L}}\;\underset{(p\times p)}{\mathbf{L}'} + \underset{(p\times p)}{\mathbf{0}} = \mathbf{L}\mathbf{L}' \tag{9-11}$$

Apart from the scale factor $\sqrt{\lambda_j}$, the factor loadings on the jth factor are the coefficients for the population jth principal component.

Although the factor analysis representation of $\mathbf{\Sigma}$ in (9-11) is exact, it is not particularly useful. It employs as many common factors as there are variables and does not allow for any variation in the specific factors $\boldsymbol{\varepsilon}$ in (9-4). We prefer models that explain the covariance structure in terms of just a few common factors. One approach, when the last $p - m$ eigenvalues are small, is to neglect the contribution of $\lambda_{m+1} \mathbf{e}_{m+1} \mathbf{e}_{m+1}' + \cdots + \lambda_p \mathbf{e}_p \mathbf{e}_p'$ to $\mathbf{\Sigma}$ in (9-10). Neglecting this contribution, we obtain the approximation

$$\mathbf{\Sigma} \doteq \left[\sqrt{\lambda_1}\, \mathbf{e}_1 \;\vdots\; \sqrt{\lambda_2}\, \mathbf{e}_2 \;\vdots\; \cdots \;\vdots\; \sqrt{\lambda_m}\, \mathbf{e}_m \right] \begin{bmatrix} \sqrt{\lambda_1}\, \mathbf{e}_1' \\ \hdashline \sqrt{\lambda_2}\, \mathbf{e}_2' \\ \hdashline \vdots \\ \hdashline \sqrt{\lambda_m}\, \mathbf{e}_m' \end{bmatrix} = \underset{(p\times m)}{\mathbf{L}}\;\underset{(m\times p)}{\mathbf{L}'} \tag{9-12}$$

The approximate representation in (9-12) assumes that the specific factors $\boldsymbol{\varepsilon}$ in (9-4) are of minor importance and can also be ignored in the factoring of $\boldsymbol{\Sigma}$. If specific factors are included in the model, their variances may be taken to be the diagonal elements of $\boldsymbol{\Sigma} - \mathbf{LL}'$, where $\mathbf{LL}'$ is defined in (9-12).

Allowing for specific factors, the approximation becomes

$$\boldsymbol{\Sigma} \doteq \mathbf{LL}' + \boldsymbol{\Psi}$$

$$= \left[\sqrt{\lambda_1}\,\mathbf{e}_1 \;\vdots\; \sqrt{\lambda_2}\,\mathbf{e}_2 \;\vdots\; \cdots \;\vdots\; \sqrt{\lambda_m}\,\mathbf{e}_m \right] \begin{bmatrix} \sqrt{\lambda_1}\,\mathbf{e}_1' \\ \hline \sqrt{\lambda_2}\,\mathbf{e}_2' \\ \hline \vdots \\ \hline \sqrt{\lambda_m}\,\mathbf{e}_m' \end{bmatrix} + \begin{bmatrix} \psi_1 & 0 & \cdots & 0 \\ 0 & \psi_2 & \cdots & 0 \\ \vdots & \vdots & & \vdots \\ 0 & 0 & \cdots & \psi_p \end{bmatrix} \tag{9-13}$$

where $\psi_i = \sigma_{ii} - \sum_{j=1}^{m} \ell_{ij}^2$ for $i = 1, 2, \ldots, p$.

To apply this approach to a data set $\mathbf{x}_1, \mathbf{x}_2, \ldots, \mathbf{x}_n$, it is customary first to center the observations by subtracting the sample mean $\bar{\mathbf{x}}$. The centered observations

$$\mathbf{x}_j - \bar{\mathbf{x}} = \begin{bmatrix} x_{1j} \\ x_{2j} \\ \vdots \\ x_{pj} \end{bmatrix} - \begin{bmatrix} \bar{x}_1 \\ \bar{x}_2 \\ \vdots \\ \bar{x}_p \end{bmatrix} = \begin{bmatrix} x_{1j} - \bar{x}_1 \\ x_{2j} - \bar{x}_2 \\ \vdots \\ x_{pj} - \bar{x}_p \end{bmatrix}, \qquad j = 1, 2, \ldots, n \tag{9-14}$$

have the same sample covariance matrix, $\mathbf{S}$, as the original observations.

In cases where the units of the variables are not commensurate, it is usually desirable to work with the standardized variables

$$\mathbf{z}_j = \begin{bmatrix} \dfrac{(x_{1j} - \bar{x}_1)}{\sqrt{s_{11}}} \\ \dfrac{(x_{2j} - \bar{x}_2)}{\sqrt{s_{22}}} \\ \vdots \\ \dfrac{(x_{pj} - \bar{x}_p)}{\sqrt{s_{pp}}} \end{bmatrix}, \qquad j = 1, 2, \ldots, n$$

whose sample covariance matrix is the sample correlation matrix, $\mathbf{R}$, of the observations $\mathbf{x}_1, \mathbf{x}_2, \ldots, \mathbf{x}_n$. Standardization avoids the problems of having one variable with large variance unduly influencing the determination of factor loadings.

The representation in (9-13), when applied to the sample covariance matrix $\mathbf{S}$ or the sample correlation matrix $\mathbf{R}$, is known as the *principal component solution*. The name follows from the fact that the factor loadings are the scaled coefficients of the first sample principal components (see Chapter 8).

Principal Component Solution of the Factor Model

The principal component factor analysis of the sample covariance matrix $\mathbf{S}$ is specified in terms of its eigenvalue-eigenvector pairs $(\hat{\lambda}_1, \hat{\mathbf{e}}_1)$, $(\hat{\lambda}_2, \hat{\mathbf{e}}_2), \ldots, (\hat{\lambda}_p, \hat{\mathbf{e}}_p)$ where $\hat{\lambda}_1 \geq \hat{\lambda}_2 \geq \cdots \geq \hat{\lambda}_p$. Let $m < p$ be the number of common factors. The matrix of estimated factor loadings $\{\tilde{\ell}_{ij}\}$ is given by

$$\tilde{\mathbf{L}} = \left[\sqrt{\hat{\lambda}_1}\,\hat{\mathbf{e}}_1 \;\vdots\; \sqrt{\hat{\lambda}_2}\,\hat{\mathbf{e}}_2 \;\vdots\; \cdots \;\vdots\; \sqrt{\hat{\lambda}_m}\,\hat{\mathbf{e}}_m \right] \tag{9-15}$$

The estimated specific variances are provided by the diagonal elements of the matrix $\mathbf{S} - \tilde{\mathbf{L}}\tilde{\mathbf{L}}'$, so

$$\tilde{\mathbf{\Psi}} = \begin{bmatrix} \tilde{\psi}_1 & 0 & \cdots & 0 \\ 0 & \tilde{\psi}_2 & \cdots & 0 \\ \vdots & \vdots & & \vdots \\ 0 & 0 & \cdots & \tilde{\psi}_p \end{bmatrix} \quad \text{with} \quad \tilde{\psi}_i = s_{ii} - \sum_{j=1}^{m} \tilde{\ell}_{ij}^2 \tag{9-16}$$

Communalities are estimated as

$$\tilde{h}_i^2 = \tilde{\ell}_{i1}^2 + \tilde{\ell}_{i2}^2 + \cdots + \tilde{\ell}_{im}^2 \tag{9-17}$$

The principal component factor analysis of the sample correlation matrix is obtained by starting with $\mathbf{R}$ in place of $\mathbf{S}$.

For the principal component solution, the estimated factor loadings for a given factor do not change as the number of factors is increased. For example, if $m = 1$, $\tilde{\mathbf{L}} = \left[\sqrt{\hat{\lambda}_1}\,\hat{\mathbf{e}}_1\right]$ and if $m = 2$, $\tilde{\mathbf{L}} = \left[\sqrt{\hat{\lambda}_1}\,\hat{\mathbf{e}}_1 \;\vdots\; \sqrt{\hat{\lambda}_2}\,\hat{\mathbf{e}}_2\right]$, where $(\hat{\lambda}_1, \hat{\mathbf{e}}_1)$ and $(\hat{\lambda}_2, \hat{\mathbf{e}}_2)$ are the first two eigenvalue-eigenvector pairs for $\mathbf{S}$ (or $\mathbf{R}$).

By the definition of $\tilde{\psi}_i$, the diagonal elements of $\mathbf{S}$ are equal to the diagonal elements of $\tilde{\mathbf{L}}\tilde{\mathbf{L}}' + \tilde{\mathbf{\Psi}}$. However, the off-diagonal elements of $\mathbf{S}$ are not usually reproduced by $\tilde{\mathbf{L}}\tilde{\mathbf{L}}' + \tilde{\mathbf{\Psi}}$. How then, do we select the number of factors m?

If the number of common factors is not determined by a priori considerations, such as by theory or the work of other researchers, the choice of m can be based on the estimated eigenvalues in much the same manner as with principal components. Consider the *residual matrix*

$$\mathbf{S} - (\tilde{\mathbf{L}}\tilde{\mathbf{L}}' + \tilde{\mathbf{\Psi}}) \tag{9-18}$$

resulting from the approximation of $\mathbf{S}$ by the principal component solution. The diagonal elements are zero and if the other elements are also small, we may subjectively take the m factor model to be appropriate. Analytically, we have (see Exercise 9.5)

$$\text{Sum of squared entries of } \left(\mathbf{S} - (\tilde{\mathbf{L}}\tilde{\mathbf{L}}' + \tilde{\mathbf{\Psi}})\right) \leq \hat{\lambda}_{m+1}^2 + \cdots + \hat{\lambda}_p^2 \tag{9-19}$$

Consequently, a small value for the sum of squares of the neglected eigenvalues implies a small value for the sum of squared errors of approximation.

Ideally, the contributions of the first few factors to the sample variances of the variables should be large. The contribution to the sample variance s_{ii} from the first common factor is $\tilde{\ell}_{i1}^2$. The contribution to the *total* sample variance, $s_{11} + s_{22} + \cdots + s_{pp} = \text{tr}(\mathbf{S})$, from the first common factor is then

$$\tilde{\ell}_{11}^2 + \tilde{\ell}_{21}^2 + \cdots + \tilde{\ell}_{p1}^2 = \left(\sqrt{\hat{\lambda}_1}\, \hat{\mathbf{e}}_1\right)'\left(\sqrt{\hat{\lambda}_1}\, \hat{\mathbf{e}}_1\right) = \hat{\lambda}_1$$

since the eigenvector $\hat{\mathbf{e}}_1$ has unit length. In general

$$\begin{pmatrix} \text{Proportion of total} \\ \text{sample variance due} \\ \text{to } j\text{th factor} \end{pmatrix} = \begin{cases} \dfrac{\hat{\lambda}_j}{s_{11} + s_{22} + \cdots + s_{pp}} & \text{for a factor analysis of } \mathbf{S} \\[2ex] \dfrac{\hat{\lambda}_j}{p} & \text{for a factor analysis of } \mathbf{R} \end{cases} \tag{9-20}$$

Criterion (9-20) is frequently used as a heuristic device for determining the appropriate number of common factors. The number of common factors retained in the model is increased until a "suitable proportion" of the total sample variance has been explained.

Another convention, frequently encountered in packaged computer programs, is to set m equal to the number of eigenvalues of $\mathbf{R}$ greater than one if the sample correlation matrix is factored, or equal to the number of positive eigenvalues of $\mathbf{S}$ if the sample covariance matrix is factored. These rules of thumb should not be applied indiscriminantly. For example, $m = p$ if the rule for $\mathbf{S}$ is obeyed, since all the eigenvalues are expected to be positive for large sample sizes. The best approach is to retain few rather than many factors, assuming they provide a satisfactory interpretation of the data and yield a satisfactory fit to $\mathbf{S}$ or $\mathbf{R}$.

Example 9.3

In a consumer-preference study, a random sample of customers were asked to rate several attributes of a new product. The responses on a 7-point semantic differential scale were tabulated and the attribute correlation matrix constructed. The correlation matrix is presented below.

Attribute (Variable)		1	2	3	4	5
Taste	1	1.00	.02	(.96)	.42	.01
Good buy for money	2	.02	1.00	.13	.71	(.85)
Flavor	3	.96	.13	1.00	.50	.11
Suitable for snack	4	.42	.71	.50	1.00	(.79)
Provides lots of energy	5	.01	.85	.11	.79	1.00

It is clear from the circled entries in the correlation matrix that variables 1 and 3 and variables 2 and 5 form groups. Variable 4 is "closer" to the (2, 5) group than the (1, 3) group. Given these results and the small number of variables, we might expect that the apparent linear relationships between the variables can be explained in terms of, at most, two or three common factors.

The first two eigenvalues $\hat{\lambda}_1 = 2.85$ and $\hat{\lambda}_2 = 1.81$ of $\mathbf{R}$ are the only eigenvalues greater than unity. Moreover, $m = 2$ common factors will account for a cumulative proportion

$$\frac{\hat{\lambda}_1 + \hat{\lambda}_2}{p} = \frac{2.85 + 1.81}{5} = .93$$

of the total (standardized) sample variance. The estimated factor loadings, communalities, and specific variances, obtained using (9-15), (9-16), and (9-17), are given in Table 9.1.

Now

$$\tilde{\mathbf{L}}\tilde{\mathbf{L}}' + \tilde{\mathbf{\Psi}} = \begin{bmatrix} .56 & .82 \\ .78 & -.53 \\ .65 & .75 \\ .94 & -.11 \\ .80 & -.54 \end{bmatrix} \begin{bmatrix} .56 & .78 & .65 & .94 & .80 \\ .82 & -.53 & .75 & -.11 & -.54 \end{bmatrix} +$$

$$\begin{bmatrix} .02 & 0 & 0 & 0 & 0 \\ 0 & .12 & 0 & 0 & 0 \\ 0 & 0 & .02 & 0 & 0 \\ 0 & 0 & 0 & .11 & 0 \\ 0 & 0 & 0 & 0 & .07 \end{bmatrix} = \begin{bmatrix} 1.00 & .01 & .97 & .44 & .00 \\ & 1.00 & .11 & .79 & .91 \\ & & 1.00 & .53 & .11 \\ & & & 1.00 & .81 \\ & & & & 1.00 \end{bmatrix}$$

nearly reproduces the correlation matrix $\mathbf{R}$. Thus on a purely descriptive basis,

TABLE 9.1

Variable	Estimated factor loadings $\tilde{\ell}_{ij} = \sqrt{\hat{\lambda}_j}\hat{e}_{ij}$ F_1	F_2	Communalities $\tilde{h}_i^2$	Specific variances $\tilde{\psi}_i = 1 - \tilde{h}_i^2$
1. Taste	.56	.82	.98	.02
2. Good buy for money	.78	−.53	.88	.12
3. Flavor	.65	.75	.98	.02
4. Suitable for snack	.94	−.11	.89	.11
5. Provides lots of energy	.80	−.54	.93	.07
Eigenvalues	2.85	1.81		
Cumulative proportion of total (standardized) sample variance	.571	.932		

we would judge a two-factor model with the factor loadings displayed above as providing a good fit to the data. The communalities (.98, .88, .98, .89, .93) indicate that the two factors account for a large percentage of the sample variance of each variable.

We shall not interpret the factors at this point. As we noted in Section 9.2, the factors (and loadings) are unique up to an orthogonal rotation. Rotating factors often reveals a simple structure and aids interpretation. We shall consider this example again (Example 9.9) after factor rotation has been discussed. ■

Example 9.4

Stock-price data consisting of $n = 100$ weekly rates of return on $p = 5$ stocks were introduced in Example 8.5. In that example the first two sample principal components were obtained from $\mathbf{R}$. Taking $m = 1$ and $m = 2$, principal component solutions to the orthogonal factor model can be easily obtained. Specifically, the estimated factor loadings are the sample principal component coefficients (eigenvectors of $\mathbf{R}$) scaled by the square root of the corresponding eigenvalues. The estimated factor loadings, communalities, specific variances, and proportion of total (standardized) sample variance explained by each factor for the $m = 1$ and $m = 2$ factor solutions are displayed in Table 9.2. The communalities are given by (9-17). So, for example, with $m = 2$, $\tilde{h}_1^2 = \tilde{\ell}_{11}^2 + \tilde{\ell}_{12}^2 = (.783)^2 + (-.217)^2 = .66$.

The residual matrix corresponding to the solution for $m = 2$ factors is

$$\mathbf{R} - \tilde{\mathbf{L}}\tilde{\mathbf{L}}' - \tilde{\mathbf{\Psi}} = \begin{bmatrix} 0 & -.127 & -.164 & -.069 & .017 \\ -.127 & 0 & -.122 & .055 & .012 \\ -.164 & -.122 & 0 & -.019 & -.017 \\ -.069 & .055 & -.019 & 0 & -.232 \\ .017 & .012 & -.017 & -.232 & 0 \end{bmatrix}$$

The proportion of total variance explained by the two-factor solution is

TABLE 9.2

	One-factor solution		Two-factor solution		
Variable	Estimated factor loadings F_1	Specific variances $\tilde{\psi}_i = 1 - \tilde{h}_i^2$	Estimated factor loadings F_1	F_2	Specific variances $\tilde{\psi}_i = 1 - \tilde{h}_i^2$
1. Allied Chemical	.783	.39	.783	−.217	.34
2. DuPont	.773	.40	.773	−.458	.19
3. Union Carbide	.794	.37	.794	−.234	.31
4. Exxon	.713	.49	.713	.472	.27
5. Texaco	.712	.49	.712	.524	.22
Cumulative proportion of total (standardized) sample variance explained	.571		.571	.733	

appreciably larger than that for the one-factor solution. However, for $m = 2$, $\tilde{\mathbf{L}}\tilde{\mathbf{L}}'$ produces numbers that are, in general, larger than the sample correlations. This is particularly true for r_{45}.

It seems fairly clear that the first factor, F_1, represents general economic conditions and might be called a *market factor*. All of the stocks load highly on this factor and the loadings are about equal. The second factor contrasts the chemical stocks with the oil stocks (the chemicals have relatively large negative loadings and the oils have large positive loadings on the factor). Thus F_2 seems to differentiate stocks in different industries and might be called an *industry factor*. To summarize, rates of return appear to be determined by general market conditions and activities that are unique to the different industries, as well as a residual or firm specific factor. This is essentially the conclusion reached by an examination of the sample principal components in Example 8.5. ■

A Modified Approach—the Principal Factor Solution

A modification of the principal component approach is sometimes considered. We describe the reasoning in terms of a factor analysis of **R**, although the procedure is also appropriate for **S**. If the factor model $\boldsymbol{\rho} = \mathbf{L}\mathbf{L}' + \boldsymbol{\Psi}$ is correctly specified, the m *common* factors should account for the *off-diagonal* elements of $\boldsymbol{\rho}$, as well as the *communality portions* of the diagonal elements

$$\rho_{ii} = 1 = h_i^2 + \psi_i$$

If the specific factor contribution ψ_i is removed from the diagonal or, equivalently, the 1 replaced by h_i^2, the resulting matrix is $\boldsymbol{\rho} - \boldsymbol{\Psi} = \mathbf{L}\mathbf{L}'$.

Suppose initial estimates, ψ_i^*, of the specific variances are available. Then replacing the ith diagonal element of **R** by $h_i^{*2} = 1 - \psi_i^*$, we obtain a "reduced" sample correlation matrix

$$\mathbf{R}_r = \begin{bmatrix} h_1^{*2} & r_{12} & \cdots & r_{1p} \\ r_{12} & h_2^{*2} & \cdots & r_{2p} \\ \vdots & \vdots & & \vdots \\ r_{1p} & r_{2p} & \cdots & h_p^{*2} \end{bmatrix}$$

Now, apart from sampling variation, all of the elements of the reduced sample correlation matrix $\mathbf{R}_r$ should be accounted for by the m common factors. In particular, $\mathbf{R}_r$ is factored as

$$\mathbf{R}_r \doteq \mathbf{L}_r^* \mathbf{L}_r^{*\prime} \tag{9-21}$$

where $\mathbf{L}_r^* = \{\ell_{ij}^*\}$ are the estimated loadings.

The *principal factor method* of factor analysis employs the estimates

$$\mathbf{L}_r^* = \left[\sqrt{\hat{\lambda}_1^*}\,\hat{\mathbf{e}}_1^* \;\vdots\; \sqrt{\hat{\lambda}_2^*}\,\hat{\mathbf{e}}_2^* \;\vdots\; \cdots \;\vdots\; \sqrt{\hat{\lambda}_m^*}\,\hat{\mathbf{e}}_m^* \right]$$

$$\psi_i^* = 1 - \sum_{j=1}^{m} \ell_{ij}^{*2} \tag{9-22}$$

where $(\hat{\lambda}_i^*, \hat{\mathbf{e}}_i^*)$, $i = 1, 2, \ldots, m$ are the (largest) eigenvalue-eigenvector pairs determined from $\mathbf{R}_r$. In turn, the communalities would then be (re)estimated by

$$\tilde{h}_i^{*2} = \sum_{j=1}^{m} \ell_{ij}^{*2} \tag{9-23}$$

The principal factor procedure can be used iteratively, with the communality estimates of (9-23) becoming the initial estimates for the next stage.

In the spirit of the principal component solution, consideration of the estimated eigenvalues $\hat{\lambda}_1^*, \hat{\lambda}_2^*, \ldots, \hat{\lambda}_p^*$ helps determine the number of common factors to retain. An added complication is that now some of the eigenvalues may be negative due to the use of initial communality estimates. Ideally, we should take the number of common factors equal to the rank of the reduced *population* matrix. Unfortunately, this rank is not always well determined from $\mathbf{R}_r$ and some judgment is necessary.

Although there are many choices for initial specific variances estimates, the most popular choice, when working with a correlation matrix, is $\psi_i^* = 1/r^{ii}$, where r^{ii} is the ith diagonal element of $\mathbf{R}^{-1}$. The initial communality estimates become

$$h_i^{*2} = 1 - \psi_i^* = 1 - \frac{1}{r^{ii}} \tag{9-24}$$

which is equal to the squared multiple correlation coefficient between X_i and the other $p - 1$ variables. The relation to the multiple correlation coefficient means that h_i^{*2} can be calculated even when $\mathbf{R}$ is not of full rank. For factoring $\mathbf{S}$, the initial specific variance estimates are s^{ii}, where s^{ii} are the ith diagonal elements of $\mathbf{S}^{-1}$. Further discussion of these and other initial estimates is contained in [6].

Although the principal components method for $\mathbf{R}$ can be regarded as a principal factor method with *initial* communality estimates of unity, or specific variances equal to zero, the two are philosophically and geometrically different (see [6]). In practice, however, the two frequently produce comparable factor loadings if the number of variables is large and the number of common factors is small.

We do not pursue the principal factor solution since, to our minds, the solution methods that have the most to recommend them are the principal component method and the maximum likelihood method, which we discuss next.

The Maximum Likelihood Method

If the common factors $\mathbf{F}$ and the specific factors $\boldsymbol{\varepsilon}$ can be assumed to be normally distributed, then maximum likelihood estimates of the factor loadings and specific variances may be obtained. When $\mathbf{F}_j$ and $\boldsymbol{\varepsilon}_j$ are jointly normal, the observations

$\mathbf{X}_j - \boldsymbol{\mu} = \mathbf{LF}_j + \boldsymbol{\varepsilon}_j$ are then normal, and from (4-16), the likelihood is

$$L(\boldsymbol{\mu}, \boldsymbol{\Sigma}) = (2\pi)^{-np/2} |\boldsymbol{\Sigma}|^{-n/2} e^{-\frac{1}{2}\operatorname{tr}\left[\boldsymbol{\Sigma}^{-1}\left(\sum_{j=1}^{n} (\mathbf{x}_j - \bar{\mathbf{x}})(\mathbf{x}_j - \bar{\mathbf{x}})' + n(\bar{\mathbf{x}} - \boldsymbol{\mu})(\bar{\mathbf{x}} - \boldsymbol{\mu})'\right)\right]}$$

$$= (2\pi)^{-(n-1)p/2} |\boldsymbol{\Sigma}|^{-(n-1)/2} e^{-\frac{1}{2}\operatorname{tr}\left[\boldsymbol{\Sigma}^{-1}\left(\sum_{j=1}^{n} (\mathbf{x}_j - \bar{\mathbf{x}})(\mathbf{x}_j - \bar{\mathbf{x}})'\right)\right]}$$

$$\times (2\pi)^{-p/2} |\boldsymbol{\Sigma}|^{-1/2} e^{-(n/2)(\bar{\mathbf{x}} - \boldsymbol{\mu})'\boldsymbol{\Sigma}^{-1}(\bar{\mathbf{x}} - \boldsymbol{\mu})} \qquad (9\text{-}25)$$

which depends on $\mathbf{L}$ and $\boldsymbol{\Psi}$ through $\boldsymbol{\Sigma} = \mathbf{LL}' + \boldsymbol{\Psi}$. This model is still not well defined because of the multiplicity of choices for $\mathbf{L}$ made possible by orthogonal transformations. It is desirable to make $\mathbf{L}$ well defined by imposing the computationally convenient *uniqueness condition*

$$\mathbf{L}'\boldsymbol{\Psi}^{-1}\mathbf{L} = \boldsymbol{\Delta}, \quad \text{a diagonal matrix} \qquad (9\text{-}26)$$

The maximum likelihood estimates $\hat{\mathbf{L}}$ and $\hat{\boldsymbol{\Psi}}$ must be obtained by numerical maximization of (9-25). Fortunately efficient computer programs now exist (see [7]) that enable one to get the maximum likelihood estimates rather easily.

We summarize some facts about maximum likelihood estimators and, for now, rely on a computer to perform the numerical detail.

Result 9.1. Let $\mathbf{X}_1, \mathbf{X}_2, \ldots, \mathbf{X}_n$ be a random sample from $N_p(\boldsymbol{\mu}, \boldsymbol{\Sigma})$, where $\boldsymbol{\Sigma} = \mathbf{LL}' + \boldsymbol{\Psi}$ is the covariance matrix for the m common factor model of (9-4). The maximum likelihood estimators $\hat{\mathbf{L}}$, $\hat{\boldsymbol{\Psi}}$, and $\hat{\boldsymbol{\mu}} = \bar{\mathbf{x}}$ maximize (9-25) subject to $\hat{\mathbf{L}}'\hat{\boldsymbol{\Psi}}^{-1}\hat{\mathbf{L}}$ being diagonal.

The maximum likelihood estimates of the communalities are

$$\hat{h}_i^2 = \hat{\ell}_{i1}^2 + \hat{\ell}_{i2}^2 + \cdots + \hat{\ell}_{im}^2 \qquad \text{for } i = 1, 2, \ldots, p \qquad (9\text{-}27)$$

so

$$\begin{pmatrix}\text{Proportion of total sample} \\ \text{variance due to } j\text{th factor}\end{pmatrix} = \frac{\hat{\ell}_{1j}^2 + \hat{\ell}_{2j}^2 + \cdots + \hat{\ell}_{pj}^2}{s_{11} + s_{22} + \cdots + s_{pp}} \qquad (9\text{-}28)$$

Proof. By the invariance property of maximum likelihood estimates (see Section 4.3), functions of $\mathbf{L}$ and $\boldsymbol{\Psi}$ are estimated by the same functions of $\hat{\mathbf{L}}$ and $\hat{\boldsymbol{\Psi}}$. In particular, the communalities $h_i^2 = \ell_{i1}^2 + \cdots + \ell_{im}^2$ have maximum likelihood estimates $\hat{h}_i^2 = \hat{\ell}_{i1}^2 + \cdots + \hat{\ell}_{im}^2$. ■

If, as in (8-10), the variables are standardized so that $\mathbf{Z} = \mathbf{V}^{-1/2}(\mathbf{X} - \boldsymbol{\mu})$, the covariance matrix, $\boldsymbol{\rho}$, of $\mathbf{Z}$ has the representation

$$\boldsymbol{\rho} = \mathbf{V}^{-1/2}\boldsymbol{\Sigma}\mathbf{V}^{-1/2} = (\mathbf{V}^{-1/2}\mathbf{L})(\mathbf{V}^{-1/2}\mathbf{L})' + \mathbf{V}^{-1/2}\boldsymbol{\Psi}\mathbf{V}^{-1/2} \qquad (9\text{-}29)$$

Thus $\boldsymbol{\rho}$ has a factorization analogous to (9-5) with loading matrix $\mathbf{L}_z = \mathbf{V}^{-1/2}\mathbf{L}$ and specific variance matrix $\boldsymbol{\Psi}_z = \mathbf{V}^{-1/2}\boldsymbol{\Psi}\mathbf{V}^{-1/2}$. By the invariance property of maximum likelihood estimators, the maximum likelihood estimator of $\boldsymbol{\rho}$ is

$$\hat{\boldsymbol{\rho}} = (\hat{\mathbf{V}}^{-1/2}\hat{\mathbf{L}})(\hat{\mathbf{V}}^{-1/2}\hat{\mathbf{L}})' + \hat{\mathbf{V}}^{-1/2}\hat{\boldsymbol{\Psi}}\hat{\mathbf{V}}^{-1/2}$$

$$= \hat{\mathbf{L}}_z\hat{\mathbf{L}}_z' + \hat{\boldsymbol{\Psi}}_z \qquad (9\text{-}30)$$

where $\hat{\mathbf{V}}^{-1/2}$ and $\hat{\mathbf{L}}$ are the maximum likelihood estimators of $\mathbf{V}^{-1/2}$ and $\mathbf{L}$, respectively (see Supplement 9A).

As a consequence of factorization of (9-30), whenever the maximum likelihood analysis pertains to the correlation matrix, we call

$$\hat{h}_i^2 = \hat{\ell}_{i1}^2 + \hat{\ell}_{i2}^2 + \cdots + \hat{\ell}_{im}^2, \qquad i = 1, 2, \ldots, p \tag{9-31}$$

the maximum likelihood estimates of the communalities, and we evaluate the importance of factors on the basis of

$$\begin{pmatrix} \text{Proportion of total (standardized)} \\ \text{sample variance due to the } j\text{th factor} \end{pmatrix} = \frac{\hat{\ell}_{1j}^2 + \hat{\ell}_{2j}^2 + \cdots + \hat{\ell}_{pj}^2}{p} \tag{9-32}$$

To avoid more tedious notations, the $\hat{\ell}_{ij}$'s above denote the elements of $\hat{\mathbf{L}}_z$.

Comment. Ordinarily the observations are standardized and a sample correlation matrix is factor analyzed. The sample correlation matrix, $\mathbf{R}$, is inserted for $[(n-1)/n]\mathbf{S}$ in the likelihood function of (9-25) and the maximum likelihood estimates $\hat{\mathbf{L}}_z$ and $\hat{\mathbf{\Psi}}_z$ are obtained by computer using the procedures outlined in Supplement 9A. Although the likelihood in (9-25) is appropriate for $\mathbf{S}$, not $\mathbf{R}$, surprisingly, this practice is equivalent to obtaining the maximum likelihood estimates $\hat{\mathbf{L}}$ and $\hat{\mathbf{\Psi}}$ based on the sample covariance matrix $\mathbf{S}$, setting $\hat{\mathbf{L}}_z = \hat{\mathbf{V}}^{-1/2}\hat{\mathbf{L}}$ and $\hat{\mathbf{\Psi}}_z = \hat{\mathbf{V}}^{-1/2}\hat{\mathbf{\Psi}}\hat{\mathbf{V}}^{-1/2}$. Here $\hat{\mathbf{V}}^{-1/2}$ is the diagonal matrix with the reciprocal of the sample standard deviations (computed with the divisor $\sqrt{n}$) on the main diagonal. This equivalency has apparently been confused in many published discussions of factor analysis. (The mathematical details are contained in Supplement 9A.)

Example 9.5

The stock-price data of Examples 8.5 and 9.4 were reanalyzed, assuming a $m = 2$ factor model and using the *maximum likelihood method*. The estimated factor loadings, communalities, specific variances, and proportion of total (standardized) sample variance explained by each factor are in Table 9.3. The

TABLE 9.3

	Maximum likelihood			Principal components		
	Estimated factor loadings		Specific variances	Estimated factor loadings		Specific variances
Variable	F_1	F_2	$\hat{\psi}_i = 1 - \hat{h}_i^2$	F_1	F_2	$\tilde{\psi}_i = 1 - \tilde{h}_i^2$
1. Allied Chemical	.684	.189	.50	.783	−.217	.34
2. DuPont	.694	.517	.25	.773	−.458	.19
3. Union Carbide	.681	.248	.47	.794	−.234	.31
4. Exxon	.621	−.073	.61	.713	.412	.27
5. Texaco	.792	−.442	.18	.712	.524	.22
Cumulative proportion of total (standardized) sample variance explained	.485	.598		.571	.733	

corresponding figures for the $m = 2$ factor solution obtained by the *principal components method* (see Example 9.4) are also provided. The communalities corresponding to the maximum likelihood factoring of $\mathbf{R}$ are of the form [see (9-31)] $\hat{h}_i^2 = \hat{\ell}_{i1}^2 + \hat{\ell}_{i2}^2$.

So, for example,

$$\hat{h}_1^2 = (.684)^2 + (.189)^2 = .50$$

The residual matrix is

$$\mathbf{R} - \hat{\mathbf{L}}\hat{\mathbf{L}}' - \hat{\mathbf{\Psi}} = \begin{bmatrix} 0 & .005 & -.004 & -.024 & -.004 \\ .005 & 0 & -.003 & -.004 & .000 \\ -.004 & -.003 & 0 & .031 & -.004 \\ -.024 & -.004 & .031 & 0 & -.000 \\ -.004 & .000 & -.004 & -.000 & 0 \end{bmatrix}$$

The elements of $\mathbf{R} - \hat{\mathbf{L}}\hat{\mathbf{L}}' - \hat{\mathbf{\Psi}}$ are much smaller than those of the residual matrix corresponding to the principal component factoring of $\mathbf{R}$ presented in Example 9.4. On this basis, we prefer the maximum likelihood approach.

The cumulative proportion of total sample variance explained by the factors is larger for principal component factoring than for maximum likelihood factoring. It is not surprising that this criterion typically favors principal component factoring. Loadings obtained by a principal component factor analysis are related to the principal components which have, by design, a variance optimizing property [see the discussion above (8-19)].

Focusing attention on the maximum likelihood solution, we see that all variables have large positive loadings on F_1. We call this factor the *market factor*, as we did for the principal component solution. The interpretation of the second factor, however, is not as clear as it appeared to be for the principal component solution. The signs of the factor loadings are consistent with a contrast, or *industry factor*, but the magnitudes are small in some cases and one might identify this factor as a comparison between DuPont and Texaco. The pattern of the initial factor loadings for the maximum likelihood solution are constrained by the uniqueness condition, $\hat{\mathbf{L}}'\hat{\mathbf{\Psi}}^{-1}\hat{\mathbf{L}}$ is a diagonal matrix. Therefore useful factor patterns are often not revealed until the factors are rotated (see Section 9.4). ■

Example 9.6

Linden [10] conducted a factor-analytic study of Olympic decathlon scores since World War II. Altogether, 160 complete starts were made by 139 athletes.[3] The scores for each of the 10 decathlon events were standardized and a sample correlation matrix was factor-analyzed by the methods of principal components and maximum likelihood. Linden reports that the "distributions of standard scores were normal or approximately normal for each of the ten decathlon events." The sample correlation matrix, based on $n = 160$

[3]Because of the potential correlation between successive scores by athletes who competed in more than one Olympic games, an analysis was also done using 139 scores representing *different* athletes. The score for an athlete who participated more than once was selected at random. The results were virtually identical to those based on all 160 scores.

starts, is

$$\mathbf{R} = \begin{bmatrix} 1.0 & .59 & .35 & .34 & .63 & .40 & .28 & .20 & .11 & -.07 \\ & 1.0 & .42 & .51 & .49 & .52 & .31 & .36 & .21 & .09 \\ & & 1.0 & .38 & .19 & .36 & .73 & .24 & .44 & -.08 \\ & & & 1.0 & .29 & .46 & .27 & .39 & .17 & .18 \\ & & & & 1.0 & .34 & .17 & .23 & .13 & .39 \\ & & & & & 1.0 & .32 & .33 & .18 & .00 \\ & & & & & & 1.0 & .24 & .34 & -.02 \\ & & & & & & & 1.0 & .24 & .17 \\ & & & & & & & & 1.0 & -.00 \\ & & & & & & & & & 1.0 \end{bmatrix}$$

(Columns: 100-m run, Long jump, Shot put, High jump, 400-m run, 110-m hurdles, Dis-cus, Pole vault, Jave-lin, 1500-m run)

From a principal component factor analysis perspective, the first four eigenvalues, 3.78, 1.52, 1.11, .91, of $\mathbf{R}$ suggest a factor solution with $m = 3$ or $m = 4$. A subsequent interpretation of the factor loadings reinforces the choice $m = 4$.

The principal component and maximum likelihood solution methods were applied to Linden's correlation matrix and yielded the estimated factor loadings, communalities, and specific variance contributions in Table 9.4.

In this case, the two solution methods produced very different results. For the principal component factorization, all events except the 1500-meter run have large positive loadings on the first factor. This factor might be labeled *general athletic ability*. The remaining factors cannot, be easily interpreted to our minds. Factor 2 appears to contrast running ability with throwing ability,

TABLE 9.4

	Principle component					Maximum likelihood				
	Estimated factor loadings				Specific variances	Estimated factor loadings				Specific variances
Variable	F_1	F_2	F_3	F_4	$\tilde{\psi}_i = 1 - \tilde{h}_i^2$	F_1	F_2	F_3	F_4	$\hat{\psi}_i = 1 - \hat{h}_i^2$
1. 100-m run	.691	.217	−.520	−.206	.16	−.090	.341	.830	−.169	.16
2. Long jump	.789	.184	−.193	.092	.30	.065	.433	.595	.275	.38
3. Shot put	.702	−.535	.047	−.175	.19	−.139	.990	.000	.000	.00
4. High jump	.674	.134	.139	.396	.35	.156	.406	.336	.445	.50
5. 400-m run	.620	.551	−.084	−.419	.13	.376	.245	.671	−.137	.33
6. 100-m hurdles	.687	.042	−.161	.345	.38	−.021	.361	.425	.388	.54
7. Discus	.621	−.521	.109	−.234	.28	−.063	.728	.030	.019	.46
8. Pole vault	.538	.087	.411	.440	.34	.155	.264	.229	.394	.70
9. Javelin	.434	−.439	.372	−.235	.43	−.026	.441	−.010	.098	.80
10. 1500-m run	.147	.596	.658	−.279	.11	.998	.059	.000	.000	.00
Cumulative proportion of total variance explained	.38	.53	.64	.73		.12	.37	.55	.61	

or "arm strength." Factor 3 appears to contrast running endurance (1500-meter run) with running speed (100-meter run) although there is a relatively high pole vault loading on this factor. Factor 4 is a mystery at this point.

For the maximum likelihood method, the 1500-meter run is the only variable with a large loading on the first factor. This factor might be called a *running endurance* factor. The second factor appears to be primarily a *strength* factor (discus and shot put load highly on this factor) and the third factor might be *running speed*, since the 100-meter and 400-meter runs load highly on this factor. Again, the fourth factor is not easily identified, although it may have something to do with jumping ability or *leg strength*. We shall return to an interpretation of the factors in Example 9.11 after a discussion of factor rotation.

The four-factor principal component solution accounts for much of the total (standardized) sample variance, although the estimated specific variances are large in some cases (for example, the javelin and hurdles). This suggests that some events might require *unique* or specific attributes not required for the other events. The four-factor maximum likelihood solution accounts for less of the total sample variance but, as the residual matrices below indicate, the maximum likelihood estimates $\hat{\mathbf{L}}$ and $\hat{\boldsymbol{\Psi}}$ do a better job of reproducing $\mathbf{R}$ than the principal component estimates $\tilde{\mathbf{L}}$ and $\tilde{\boldsymbol{\Psi}}$.

The following are the residual matrices.

Principal component:

$$\mathbf{R} - \tilde{\mathbf{L}}\tilde{\mathbf{L}}' - \tilde{\boldsymbol{\Psi}} =$$

$$\begin{bmatrix}
0 \\
-.075 & 0 \\
-.030 & -.010 & 0 \\
-.001 & -.056 & .042 & 0 \\
-.047 & -.077 & -.020 & -.024 & 0 \\
-.096 & -.092 & -.032 & -.122 & .022 & 0 \\
-.027 & -.041 & -.031 & -.001 & -.017 & .014 & 0 \\
.114 & -.042 & -.034 & -.215 & .067 & -.129 & .009 & 0 \\
.051 & .042 & -.158 & -.022 & .036 & .041 & -.254 & -.005 & 0 \\
-.016 & .017 & .056 & .020 & -.091 & .076 & .062 & -.109 & -.112 & 0
\end{bmatrix}$$

Maximum likelihood:

$$\mathbf{R} - \hat{\mathbf{L}}\hat{\mathbf{L}}' - \hat{\boldsymbol{\Psi}} =$$

$$\begin{bmatrix}
0 \\
.000 & 0 \\
.000 & .000 & 0 \\
.012 & .002 & .000 & 0 \\
.000 & -.002 & .000 & -.033 & 0 \\
-.012 & .006 & -.000 & .001 & .028 & 0 \\
.004 & -.025 & -.000 & -.034 & -.002 & .036 & 0 \\
.000 & -.009 & -.000 & .006 & .008 & -.012 & .043 & 0 \\
-.018 & -.000 & -.000 & -.045 & .052 & -.013 & .016 & .091 & 0 \\
.000 & .000 & .000 & .000 & .000 & .000 & .000 & .000 & .000 & 0
\end{bmatrix}$$

■

A Large Sample Test for the Number of Common Factors

The assumption of a normal population leads directly to a test of model adequacy. Suppose the m common factor model holds. In this case $\boldsymbol{\Sigma} = \mathbf{LL}' + \boldsymbol{\Psi}$, and testing the adequacy of the m common factor model is equivalent to testing

$$H_0: \underset{(p\times p)}{\boldsymbol{\Sigma}} = \underset{(p\times m)}{\mathbf{L}} \; \underset{(m\times p)}{\mathbf{L}'} + \underset{(p\times p)}{\boldsymbol{\Psi}} \tag{9-33}$$

versus H_1: $\boldsymbol{\Sigma}$ any other positive definite matrix. When $\boldsymbol{\Sigma}$ does not have any special form, the maximum of the likelihood function [see (4-18) and Result 4.11 with $\hat{\boldsymbol{\Sigma}} = [(n-1)/n]\mathbf{S} = \mathbf{S}_n$] is proportional to

$$|\mathbf{S}_n|^{-n/2} e^{-np/2} \tag{9-34}$$

Under H_0, $\boldsymbol{\Sigma}$ is restricted to have the form of (9-33). In this case the maximum of the likelihood function [see (9-25) with $\hat{\boldsymbol{\mu}} = \bar{\mathbf{x}}$ and $\hat{\boldsymbol{\Sigma}} = \hat{\mathbf{L}}\hat{\mathbf{L}}' + \hat{\boldsymbol{\Psi}}$, where $\hat{\mathbf{L}}$ and $\hat{\boldsymbol{\Psi}}$ are the maximum likelihood estimates of $\mathbf{L}$ and $\boldsymbol{\Psi}$, respectively] is proportional to

$$|\hat{\boldsymbol{\Sigma}}|^{-n/2} e^{-\frac{1}{2}\operatorname{tr}\left[\hat{\boldsymbol{\Sigma}}^{-1}\left(\sum_{j=1}^{n}(\mathbf{x}_j-\bar{\mathbf{x}})(\mathbf{x}_j-\bar{\mathbf{x}})'\right)\right]}$$

$$= |\hat{\mathbf{L}}\hat{\mathbf{L}}' + \hat{\boldsymbol{\Psi}}|^{-n/2} e^{-\frac{1}{2}n\operatorname{tr}[(\hat{\mathbf{L}}\hat{\mathbf{L}}'+\hat{\boldsymbol{\Psi}})^{-1}\mathbf{S}_n]} \tag{9-35}$$

Using Result 5.2, (9-34), and (9-35), the likelihood ratio statistic for testing H_0 is

$$-2\ln\Lambda = -2\ln\left[\frac{\text{maximized likelihood under } H_0}{\text{maximized likelihood}}\right]$$

$$= -2\ln\left(\frac{|\hat{\boldsymbol{\Sigma}}|}{|\mathbf{S}_n|}\right)^{-n/2} + n\left[\operatorname{tr}(\hat{\boldsymbol{\Sigma}}^{-1}\mathbf{S}_n) - p\right] \tag{9-36}$$

with

$$\nu - \nu_0 = \tfrac{1}{2}p(p+1) - \left[p(m+1) - \tfrac{1}{2}m(m-1)\right] = \tfrac{1}{2}\left[(p-m)^2 - p - m\right] \tag{9-37}$$

d.f. It is shown in Supplement 9A that $\operatorname{tr}(\hat{\boldsymbol{\Sigma}}^{-1}\mathbf{S}_n) - p = 0$ provided $\hat{\boldsymbol{\Sigma}} = \hat{\mathbf{L}}\hat{\mathbf{L}}' + \hat{\boldsymbol{\Psi}}$ is the maximum likelihood estimate of $\boldsymbol{\Sigma} = \mathbf{LL}' + \boldsymbol{\Psi}$. Thus we have

$$-2\ln\Lambda = n\ln\left(\frac{|\hat{\boldsymbol{\Sigma}}|}{|\mathbf{S}_n|}\right) \tag{9-38}$$

Bartlett [3] has shown that the chi-square approximation to the sampling distribution of $-2\ln\Lambda$ can be improved by replacing n in (9-38) with the multiplicative factor $(n - 1 - (2p + 4m + 5)/6)$.

Using Bartlett's correction,[4] we reject H_0 at the α level of significance if

$$(n - 1 - (2p + 4m + 5)/6)\ln\frac{|\hat{\mathbf{L}}\hat{\mathbf{L}}' + \hat{\boldsymbol{\Psi}}|}{|\mathbf{S}_n|} > \chi^2_{[(p-m)^2-p-m]/2}(\alpha) \tag{9-39}$$

[4]Many factor analysts obtain an approximate maximum likelihood estimate by replacing $\mathbf{S}_n$ with the unbiased estimate $\mathbf{S} = [n/(n-1)]\mathbf{S}_n$ and then minimizing $\ln|\boldsymbol{\Sigma}| + \operatorname{tr}[\boldsymbol{\Sigma}^{-1}\mathbf{S}]$. The dual substitution of $\mathbf{S}$ and the approximate maximum likelihood estimator into the test statistic of (9-39) does not affect its large sample properties.

provided n and $n - p$ are large. Since the number of degrees of freedom, $\frac{1}{2}[(p-m)^2 - p - m]$, must be positive, it follows that

$$m < \tfrac{1}{2}\left(2p + 1 - \sqrt{8p + 1}\right) \tag{9-40}$$

in order to apply the test (9-39).

Comment. In implementing the test in (9-39), we are testing for the adequacy of the m common factor model by comparing the generalized variances $|\hat{\mathbf{L}}\hat{\mathbf{L}}' + \hat{\boldsymbol{\Psi}}|$ and $|\mathbf{S}_n|$. If n is large and m is small relative to p, the hypothesis H_0 will usually be rejected, leading to a retention of more common factors. However, $\boldsymbol{\Sigma} = \hat{\mathbf{L}}\hat{\mathbf{L}}' + \hat{\boldsymbol{\Psi}}$ may be close enough to $\mathbf{S}_n$ so that adding more factors does not provide additional insights, even though they are "significant." Some judgment must be exercised in the choice of m.

Example 9.7

The two-factor maximum likelihood analysis of the stock-price data was presented in Example 9.5. The residual matrix there suggests that a two-factor solution may be adequate. Test the hypotheses H_0: $\boldsymbol{\Sigma} = \mathbf{L}\mathbf{L}' + \boldsymbol{\Psi}$, with $m = 2$, at level $\alpha = .05$.

The test statistic in (9-39) is based on the ratio of generalized variances

$$\frac{|\hat{\boldsymbol{\Sigma}}|}{|\mathbf{S}_n|} = \frac{|\hat{\mathbf{L}}\hat{\mathbf{L}}' + \hat{\boldsymbol{\Psi}}|}{|\mathbf{S}_n|}$$

Let $\hat{\mathbf{V}}^{-1/2}$ be the diagonal matrix such that $\hat{\mathbf{V}}^{-1/2}\mathbf{S}_n\hat{\mathbf{V}}^{-1/2} = \mathbf{R}$. By the properties of determinants (see Result 2A.11),

$$|\hat{\mathbf{V}}^{-1/2}||\hat{\mathbf{L}}\hat{\mathbf{L}}' + \hat{\boldsymbol{\Psi}}||\hat{\mathbf{V}}^{-1/2}| = |\hat{\mathbf{V}}^{-1/2}\hat{\mathbf{L}}\hat{\mathbf{L}}'\hat{\mathbf{V}}^{-1/2} + \hat{\mathbf{V}}^{-1/2}\hat{\boldsymbol{\Psi}}\hat{\mathbf{V}}^{-1/2}|$$

and

$$|\hat{\mathbf{V}}^{-1/2}||\mathbf{S}_n||\hat{\mathbf{V}}^{-1/2}| = |\hat{\mathbf{V}}^{-1/2}\mathbf{S}_n\hat{\mathbf{V}}^{-1/2}|$$

Consequently,

$$\frac{|\hat{\boldsymbol{\Sigma}}|}{|\mathbf{S}_n|} = \frac{|\hat{\mathbf{V}}^{-1/2}|}{|\hat{\mathbf{V}}^{-1/2}|}\,\frac{|\hat{\mathbf{L}}\hat{\mathbf{L}}' + \hat{\boldsymbol{\Psi}}|}{|\mathbf{S}_n|}\,\frac{|\hat{\mathbf{V}}^{-1/2}|}{|\hat{\mathbf{V}}^{-1/2}|} = \frac{|\hat{\mathbf{V}}^{-1/2}\hat{\mathbf{L}}\hat{\mathbf{L}}'\hat{\mathbf{V}}^{-1/2} + \hat{\mathbf{V}}^{-1/2}\hat{\boldsymbol{\Psi}}\hat{\mathbf{V}}^{-1/2}|}{|\hat{\mathbf{V}}^{-1/2}\mathbf{S}_n\hat{\mathbf{V}}^{-1/2}|}$$

$$= \frac{|\hat{\mathbf{L}}_z\hat{\mathbf{L}}_z' + \hat{\boldsymbol{\Psi}}_z|}{|\mathbf{R}|} \tag{9-41}$$

by (9-30). From Example 9.5 we determine

$$\frac{|\hat{\mathbf{L}}_z\hat{\mathbf{L}}_z' + \hat{\boldsymbol{\Psi}}_z|}{|\mathbf{R}|} = \frac{\begin{vmatrix} 1.000 & & & & \\ .572 & 1.000 & & & \\ .513 & .602 & 1.000 & & \\ .411 & .393 & .405 & 1.000 & \\ .458 & .322 & .430 & .523 & 1.000 \end{vmatrix}}{\begin{vmatrix} 1.000 & & & & \\ .577 & 1.000 & & & \\ .509 & .599 & 1.000 & & \\ .387 & .389 & .436 & 1.000 & \\ .462 & .322 & .426 & .523 & 1.000 \end{vmatrix}} = \frac{.194414}{.193163} = 1.0065$$

Using Barlett's correction, we evaluate the test statistic in (9-39)

$$(n - 1 - (2p + 4m + 5)/6) \ln \frac{|\hat{\mathbf{L}}\hat{\mathbf{L}}' + \hat{\mathbf{\Psi}}|}{|\mathbf{S}_n|}$$

$$= \left(100 - 1 - \frac{(10 + 8 + 5)}{6}\right) \ln(1.0065) = .62$$

Since $\frac{1}{2}[(p - m)^2 - p - m] = \frac{1}{2}[(5 - 2)^2 - 5 - 2] = 1$, the 5% critical value $\chi_1^2(.05) = 3.84$ is not exceeded and we fail to reject H_0. We conclude the data do not contradict a two-factor model. In fact, the observed significance level $P[\chi_1^2 > .62] \doteq .43$ implies H_0 would not be rejected at *any* reasonable level. ■

Large sample variances and covariances for the maximum likelihood estimates $\hat{\ell}_{ij}$, $\hat{\psi}_i$, have been derived when these estimates have been determined from the sample covariance matrix **S** (see [9]). These expressions are, in general, quite complicated.

9.4 FACTOR ROTATION

As we indicated in Section 9.2, all factor loadings obtained from the initial loadings by an orthogonal transformation have the same ability to reproduce the covariance (or correlation) matrix [see Equation (9-8)]. From matrix algebra, we know that an orthogonal transformation corresponds to a rigid rotation (or reflection) of the coordinate axes. For this reason an orthogonal transformation of the factor loadings, and the implied orthogonal transformation of the factors, is called *factor rotation*.

If $\hat{\mathbf{L}}$ is the $p \times m$ matrix of estimated factor loadings obtained by any method (principal component, maximum likelihood, and so forth) then

$$\hat{\mathbf{L}}^* = \hat{\mathbf{L}}\mathbf{T}, \qquad \text{where } \mathbf{T}\mathbf{T}' = \mathbf{T}'\mathbf{T} = \mathbf{I} \tag{9-42}$$

is a $p \times m$ matrix of "rotated" loadings. Moreover, the estimated covariance (or correlation) matrix remains unchanged, since

$$\hat{\mathbf{L}}\hat{\mathbf{L}}' + \hat{\mathbf{\Psi}} = \hat{\mathbf{L}}\mathbf{T}\mathbf{T}'\hat{\mathbf{L}} + \hat{\mathbf{\Psi}} = \hat{\mathbf{L}}^*\hat{\mathbf{L}}^{*\prime} + \hat{\mathbf{\Psi}} \tag{9-43}$$

Equation (9-43) indicates that the residual matrix, $\mathbf{S}_n - \hat{\mathbf{L}}\hat{\mathbf{L}}' - \hat{\mathbf{\Psi}} = \mathbf{S}_n - \hat{\mathbf{L}}^*\hat{\mathbf{L}}^{*\prime} - \hat{\mathbf{\Psi}}$, remains unchanged. Moreover, the specific variances $\hat{\psi}_i$, and hence the communalities $\hat{h}_i^2$, are unaltered. Thus from a mathematical viewpoint, it is immaterial whether $\hat{\mathbf{L}}$ or $\hat{\mathbf{L}}^*$ is obtained.

Since the original loadings may not be readily interpretable, it is usual practice to rotate them until a "simple structure" is achieved. The rationale is very much akin to sharpening the focus of microscope in order to see the detail more clearly.

Ideally we should like to see a pattern of loadings such that each variable loads highly on a single factor and has small-to-moderate loadings on the remaining factors. It is not always possible to get this simple structure, although the rotated loadings for the decathlon data discussed in Example 9.11 provide a nearly ideal pattern.

We shall concentrate on graphical and analytical methods for determining an orthogonal rotation to simple structure. When $m = 2$, or the common factors are considered two at a time, the transformation to simple structure can frequently be determined graphically. The uncorrelated common factors are regarded as unit vectors along perpendicular coordinate axes. A plot of the pairs of factor loadings, $(\hat{\ell}_{i1}, \hat{\ell}_{i2})$, yields p points, each point corresponding to a variable. The coordinate axes can then be visually rotated through an angle, call it, ϕ, and the new rotated loadings $\hat{\ell}_{ij}^*$ are determined from the relationships

$$\underset{(p\times2)}{\hat{\mathbf{L}}^*} = \underset{(p\times2)}{\hat{\mathbf{L}}}\ \underset{(2\times2)}{\mathbf{T}} \tag{9-44}$$

where
$$\begin{cases} \mathbf{T} = \begin{bmatrix} \cos\phi & \sin\phi \\ -\sin\phi & \cos\phi \end{bmatrix} & \begin{array}{l}\text{clockwise}\\ \text{rotation}\end{array} \\[20pt] \mathbf{T} = \begin{bmatrix} \cos\phi & -\sin\phi \\ \sin\phi & \cos\phi \end{bmatrix} & \begin{array}{l}\text{counterclockwise}\\ \text{rotation}\end{array} \end{cases}$$

The relationship in (9-44) is rarely implemented in a two-dimensional graphical analysis. In this situation, clusters of variables are often apparent by eye and these clusters enable one to identify the common factors without having to inspect the magnitudes of the rotated loadings. On the other hand, for $m > 2$, orientations are not easily visualized and the magnitudes of the *rotated* loadings must be inspected to find a meaningful interpretation of the original data. The choice of an orthogonal matrix $\mathbf{T}$ that satisfies an *analytical* measure of simple structure will be considered shortly.

Example 9.8

Lawley and Maxwell [9] present the sample correlation matrix of examination scores in $p = 6$ subject areas for $n = 220$ male students. They give

$$\mathbf{R} = \begin{array}{cccccc} \text{Gaelic} & \text{English} & \text{history} & \text{arithmetic} & \text{algebra} & \text{geometry} \\ \begin{bmatrix} 1.0 & .439 & .410 & .288 & .329 & .248 \\ & 1.0 & .351 & .354 & .320 & .329 \\ & & 1.0 & .164 & .190 & .181 \\ & & & 1.0 & .595 & .470 \\ & & & & 1.0 & .464 \\ & & & & & 1.0 \end{bmatrix} \end{array}$$

and a maximum likelihood solution for $m = 2$ common factors yields Table 9.5.

TABLE 9.5

Variable	Estimated factor loadings		Communalities
	F_1	F_2	$\hat{h}_i^2$
1. Gaelic	.553	.429	.490
2. English	.568	.288	.406
3. History	.392	.450	.356
4. Arithmetic	.740	−.273	.623
5. Algebra	.724	−.211	.569
6. Geometry	.595	−.132	.372

All of the variables have positive loadings on the first factor. Lawley and Maxwell suggest this factor reflects the overall response of the students to instruction and might be labeled a *general intelligence* factor. Half the loadings are positive and half are negative on the second factor. A factor with this pattern of loadings is called a bipolar factor. (The assignment of negative and positive poles is arbitrary because the signs of the loadings on a factor can be reversed without affecting the analysis.) This factor is not easily identified, but is such that individuals who get above-average scores on the verbal tests get above-average scores on the factor. Individuals with above-average scores on the mathematical tests get below-average scores on the factor. Perhaps this can be classified as a "math, nonmath" factor.

The factor loading pairs $(\hat{\ell}_{i1}, \hat{\ell}_{i2})$ are plotted as points in Figure 9.1. The points are labeled with the numbers of the corresponding variables. Also shown is a clockwise orthogonal rotation of the coordinate axes through an angle of $\phi \doteq 20°$. This angle was chosen so that one of the new axes passes through $(\hat{\ell}_{41}, \hat{\ell}_{42})$. When this is done, all the points fall in the first quadrant (the factor loadings are all positive) and the two distinct clusters of variables are more clearly revealed.

The mathematical test variables load highly on F_1^* and have negligible loadings on F_2^*. The first factor might be called a *mathematical-ability* factor. Similarly the three verbal test variables have high loadings on F_2^* and moderate-to-small loadings on F_1^*. The second factor might be labeled a *verbal-ability* factor. The *general-intelligence* factor identified initially is submerged in the factors F_1^* and F_2^*.

The rotated factor loadings obtained from (9-44) with $\phi \doteq 20°$ and the corresponding communality estimates are shown in Table 9.6. The magnitudes of the rotated factor loadings reinforce the interpretation of the factors suggested by Figure 9.1.

The communality estimates are unchanged by the orthogonal rotation, since $\hat{\mathbf{L}}\hat{\mathbf{L}}' = \hat{\mathbf{L}}\mathbf{T}\mathbf{T}'\hat{\mathbf{L}}' = \hat{\mathbf{L}}^*\hat{\mathbf{L}}^{*\prime}$, and the communalities are the diagonal elements of these matrices.

We point out that Figure 9.1 suggests an *oblique rotation* of the coordinates. One new axis would pass through the cluster $\{1, 2, 3\}$ and the other

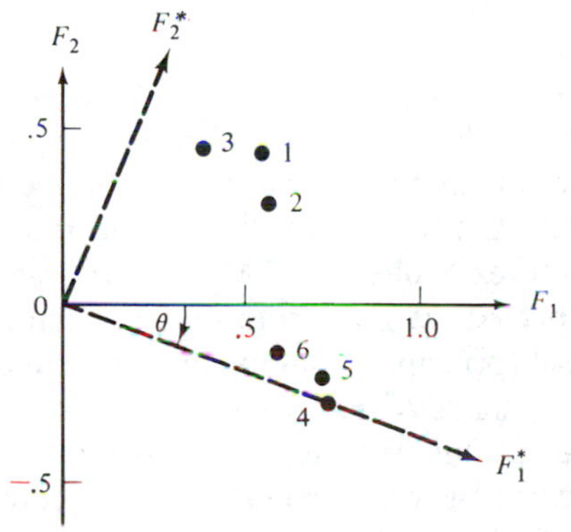

Figure 9.1 Factor rotation for test scores.

TABLE 9.6

Variable	Estimated rotated factor loadings F_1^*	F_2^*	Communalities $\hat{h}_i^{*2} = \hat{h}_i^2$
1. Gaelic	.369	.594	.490
2. English	.433	.467	.406
3. History	.211	.558	.356
4. Arithmetic	.789	.001	.623
5. Algebra	.752	.054	.568
6. Geometry	.604	.083	.372

through the $\{4, 5, 6\}$ group. Oblique rotations are so named because they correspond to a *nonrigid* rotation of coordinate axes leading to new axes which are not perpendicular. It is apparent, however, that the interpretation of the oblique factors for this example would be much the same as that given above for an orthogonal rotation. ■

Kaiser [8] has suggested an analytical measure of simple structure known as the *varimax* (or normal varimax) *criterion*. Define $\tilde{\ell}_{ij}^* = \hat{\ell}_{ij}^* / \hat{h}_i$ to be the final rotated coefficients scaled by the square root of the communalities. The (normal) varimax procedure selects the orthogonal transformation **T** that makes

$$V = \frac{1}{p} \sum_{j=1}^{m} \left[\sum_{i=1}^{p} \tilde{\ell}_{ij}^{*4} - \left(\sum_{i=1}^{p} \tilde{\ell}_{ij}^{*2} \right)^2 \Big/ p \right] \tag{9-45}$$

as large as possible.

Scaling the rotated coefficients $\hat{\ell}_{ij}^*$ has the effect of giving variables with small communalities relatively more weight in the determination of simple structure. After the transformation **T** is determined, the loadings $\tilde{\ell}_{ij}^*$ are multiplied by $\hat{h}_i$ so that the original communalities are preserved.

Although (9-45) looks rather forbidding, it has a simple interpretation. In words

$$V \propto \sum_{j=1}^{m} \left(\begin{array}{c} \text{variance of squares of (scaled) loadings for} \\ j\text{th factor} \end{array} \right) \tag{9-46}$$

Effectively, maximizing V corresponds to "spreading out" the squares of the loadings on each factor as much as possible. Therefore we hope to find groups of large and negligible coefficients in any *column* of the rotated loadings matrix $\hat{\mathbf{L}}^*$.

Computing algorithms exist for maximizing V and most popular factor analysis computer programs (for example, the BMDP series in [4]) provide varimax rotations. As might be expected, varimax rotations of factor loadings obtained by different solution methods (principal components, maximum likelihood and so forth) will not, in general, coincide. Also, the pattern of rotated loadings may change considerably if additional common factors are included in the rotation. If a dominant single factor exists, it will generally be obscured by any orthogonal

TABLE 9.7

Variable	Estimated factor loadings F_1	F_2	Rotated estimated factor loadings F_1^*	F_2^*	Communalities $\hat{h}_i^2$
1. Taste	.56	.82	.02	.99	.98
2. Good buy for money	.78	−.52	.94	−.01	.88
3. Flavor	.65	.75	.13	.98	.98
4. Suitable for snack	.94	−.11	.84	.43	.89
5. Provides lots of energy	.80	−.54	.97	−.02	.93
Cumulative proportion of total (standardized) sample variance explained	.571	.934	.507	.934	

rotation. On the other hand, it can always be held fixed and the remaining factors rotated.

Example 9.9

Let us return to the marketing data discussed in Example 9.3. The original factor loadings (obtained by the principal component method), the communalities, and the (varimax) rotated factor loadings are shown in Table 9.7.

It is clear that variables 2, 4, and 5 define factor 1 (high loadings on factor 1, small or negligible loadings on factor 2) while variables 1 and 3 define factor 2 (high loadings on factor 2, small or negligible loadings on factor 1). Variable 4 is most closely aligned with factor 1, although it has aspects of the trait represented by factor 2. We might call factor 1 a *nutritional* factor and factor 2 a *taste* factor.

The factor loadings for the variables are pictured with respect to the original and (varimax) rotated factor axes in Figure 9.2. ■

Rotation of factor loadings is particularly recommended for loadings obtained by maximum likelihood since the initial values are constrained to satisfy the uniqueness condition, $\hat{\mathbf{L}}'\hat{\mathbf{\Psi}}^{-1}\hat{\mathbf{L}}$ is a diagonal matrix. This condition is convenient for computational purposes but may not lead to factors that can easily be interpreted.

Example 9.10

Table 9.8 shows the initial and rotated maximum likelihood estimates of the factor loadings for the stock-price data (see Examples 8.5 and 9.5). An $m = 2$ factor model is assumed. The estimated specific variances and cumulative proportions of total (standardized) sample variance explained by each factor are also given.

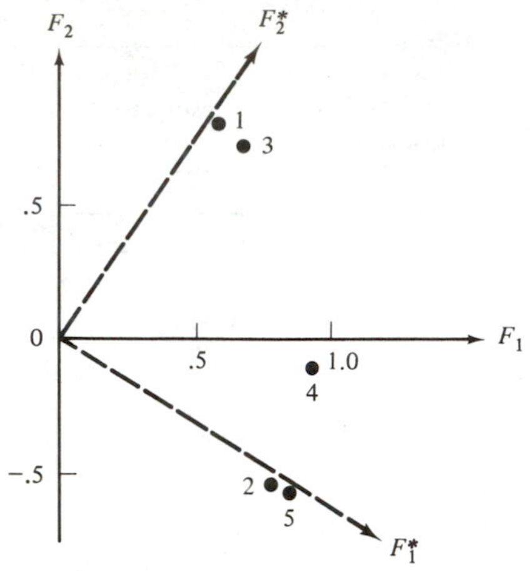

Figure 9.2 Factor rotation for hypothetical marketing data.

An interpretation of the factors suggested by the unrotated loadings was presented in Example 9.5. We identified *market* and *industry* factors.

The rotated loadings indicate that the chemical stocks (Allied Chemical, DuPont and Union Carbide) load highly on the first factor, while the oil stocks (Exxon and Texaco) load highly on the second factor. (Although they are not displayed, the same phenomenon is observed for the rotated loadings obtained from the principal component solution.) The two rotated factors, together, differentiate the industries. It is difficult for us to label these factors intelligently. Factor 1 represents those unique economic forces that cause chemical stocks to move together. Factor 2 appears to represent economic conditions affecting oil stocks.

We note that a general factor (that is, one on which *all* the variables load highly) tends to be "destroyed after rotation." For this reason, in cases where a

TABLE 9.8

Variable	Maximum likelihood estimates of factor loadings F_1	F_2	Rotated estimated factor loadings F_1^*	F_2^*	Specific variances $\hat{\psi}_i = 1 - \hat{h}_i^2$
Allied Chemical	.684	.189	.601	.377	.50
DuPont	.694	.517	.850	.164	.25
Union Carbide	.681	.248	.643	.335	.47
Exxon	.621	−.073	.365	.507	.61
Texaco	.792	−.442	.208	.883	.18
Cumulative proportion of total sample variance explained	.485	.598	.335	.598	

general factor is evident, an orthogonal rotation is sometimes performed with the general factor loadings fixed.[5] ■

Example 9.11

The estimated factor loadings and specific variances for the Olympic-decathlon data were presented in Example 9.6. These quantities were derived for an $m = 4$ factor model, using both principal component and maximum likelihood solution methods. The interpretation of all the underlying factors was not immediately evident. A varimax rotation [see (9-45)] was performed to see if the rotated factor loadings provided additional insights. The varimax rotated loadings for the $m = 4$ factor solutions are displayed in Table 9.9 along with the specific variances. Apart from the estimated loadings, rotation will only affect the *distribution* of the proportions of total sample variance explained by each factor. The cumulative proportion of total sample variance explained for *all* factors does not change.

The rotated factor loadings for both methods of solution point to the same underlying attributes, although factors 1 and 2 are not in the same order. We see that shot put, discus, and javelin load highly on a factor and, following Linden [10], this factor might be called *explosive arm strength*. Similarly, high jump, 110-meter hurdles, pole vault, and—to some extent—long jump, load highly on another factor. Linden labeled this factor *explosive leg strength*. The 100-meter run, 400-meter run, and—again to some extent—the long jump, load highly on a third factor. This factor could be called *running speed*. Finally, the 1500-meter run loads highly and the 400-meter run loads moderately on the fourth factor. Linden called this factor *running endurance*. As Linden notes, "The basic functions indicated in this study are mainly consistent with the traditional classification of track and field athletics."

Plots of rotated maximum likelihood loadings for factors pairs $(1, 2)$ and $(1, 3)$ are displayed in Figure 9.3. The points are generally grouped along the factor axes. Plots of rotated principal component loadings are very similar. ■

Oblique Rotations

Orthogonal rotations are appropriate for a factor model in which the common factors are assumed to be independent. Many investigators in social sciences consider *oblique* (nonorthogonal) rotations, as well as orthogonal rotations. The former are often suggested after viewing the estimated factor loadings and do not follow from our postulated model. Nevertheless, an oblique rotation is frequently a useful aid in factor analysis.

[5]Some general-purpose factor analysis programs allow one to fix loadings associated with certain factors and to rotate the remaining factors.

TABLE 9.9

Variable	Principal component					Maximum likelihood				
	Estimated rotated factor loadings, $\tilde{\ell}_{ij}^*$				Specific variances	Estimated rotated factor loadings, $\hat{\ell}_{ij}^*$				Specific variances
	F_1^*	F_2^*	F_3^*	F_4^*	$\tilde{\psi}_i = 1 - \tilde{h}_i^2$	F_1^*	F_2^*	F_3^*	F_4^*	$\hat{\psi}_i = 1 - \hat{h}_i^2$
100-m run	.884	.136	.156	−.113	.16	.167	.857	.246	−.138	.16
Long jump	.631	.194	.515	−.006	.30	.240	.477	.580	.011	.38
Shot put	.245	.825	.223	−.148	.19	.966	.154	.200	−.058	.00
High jump	.239	.150	.750	.076	.35	.242	.173	.632	.113	.50
400-m run	.797	.075	.102	.468	.13	.055	.709	.236	.330	.33
110-m hurdles	.404	.153	.635	−.170	.38	.205	.261	.589	−.071	.54
Discus	.186	.814	.147	−.079	.28	.697	.133	.180	−.009	.46
Pole vault	−.036	.176	.762	.217	.34	.137	.078	.51[illegible]	.116	.70
Javelin	−.048	.735	.110	.141	.43	.416	.019	.175	.002	.80
1500-m run	.045	−.041	.112	.934	.11	−.055	.056	.113	.990	.00
Cumulative proportion of total sample variance explained	.21	.42	.61	.73		.18	.34	.50	.61	

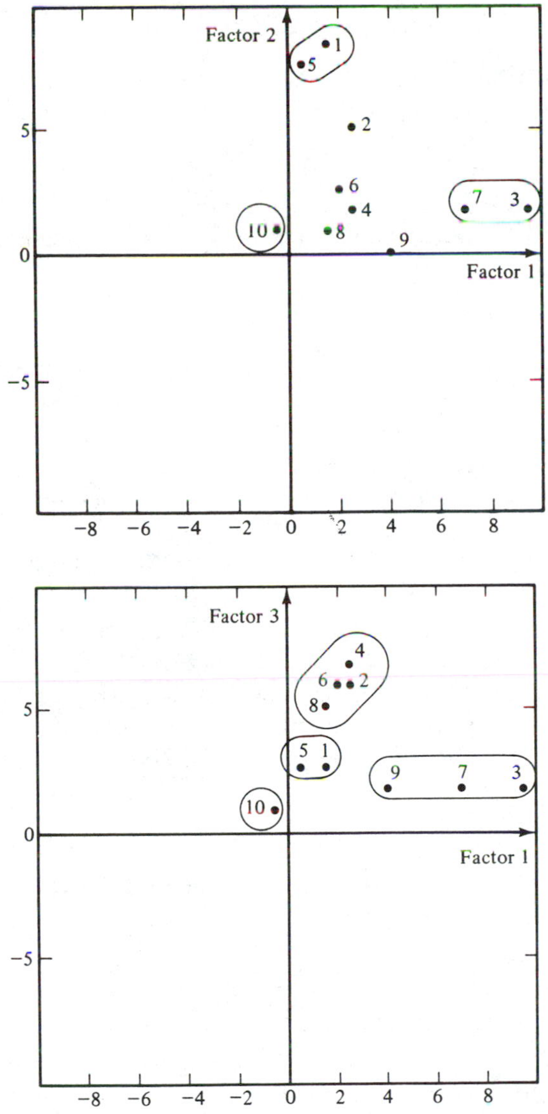

Figure 9.3 Rotated maximum likelihood loadings for factor pairs (1, 2) and (1, 3)—decathlon data. (The numbers in the figures correspond to variables.)

If we regard the m common factors as coordinate axes, the point with the m coordinates $(\hat{\ell}_{i1}, \hat{\ell}_{i2}, \ldots, \hat{\ell}_{im})$ represents the position of the ith variable in the *factor space*. Assuming the variables are grouped into nonoverlapping clusters, an orthogonal rotation to a simple structure corresponds to a *rigid* rotation of the coordinate axes such that the axes, after rotation, pass as closely to the clusters as possible. An oblique rotation to simple structure corresponds to a *nonrigid* rotation of the coordinate system such that the rotated axes (no longer perpendicular) pass (nearly) through the clusters. An oblique rotation seeks to express each variable in terms of a minimum number of factors, preferably a single factor. Oblique rotations are discussed in several sources (see, for example [6] or [9]) and will not be pursued in this book.

In factor analysis, interest is usually centered on the parameters in the factor model. However, the estimated values of the common factors, called *factor scores*, may also be required. These quantities are often used for diagnostic purposes as well as inputs to a subsequent analysis.

Factor scores are not estimates of unknown parameters in the usual sense. Rather, they are estimates of values for the unobserved random factor vectors $\mathbf{F}_j$, $j = 1, 2, \ldots, n$. That is, factor scores

$$\hat{\mathbf{f}}_j = \text{estimate of the value, } \mathbf{f}_j \text{, attained by } \mathbf{F}_j \text{ (} j\text{th case)}$$

The estimation situation is complicated by the fact that the unobserved quantities $\mathbf{f}_j$ and $\boldsymbol{\varepsilon}_j$ outnumber the observed $\mathbf{x}_j$. To overcome this difficulty, some rather heuristic, but reasoned, approaches to the problem of estimating factor values have been advanced. We describe two of these approaches.

Both of the factor score approaches have two elements in common.

1. They treat the estimated factor loadings, $\hat{\ell}_{ij}$, and specific variances, $\hat{\psi}_i$, as if they were the true values.
2. They involve linear transformations of the original data, perhaps centered or standardized. Typically, the estimated *rotated* loadings rather than the original estimated loadings are used to compute factor scores. The computational formulas, as given in this section, do not change when rotated loadings are substituted for unrotated loadings and we will not differentiate between them.

The Weighted Least Squares Method

Suppose first that the mean vector, $\boldsymbol{\mu}$, the factor loadings, $\mathbf{L}$, and the specific variances, $\boldsymbol{\Psi}$, are known for the factor model

$$\underset{(p\times 1)}{\mathbf{X}} - \underset{(p\times 1)}{\boldsymbol{\mu}} = \underset{(p\times m)}{\mathbf{L}}\ \underset{(m\times 1)}{\mathbf{F}} + \underset{(p\times 1)}{\boldsymbol{\varepsilon}}$$

Further, regard the specific factors $\boldsymbol{\varepsilon}' = [\varepsilon_1, \varepsilon_2, \ldots, \varepsilon_p]$ as errors. Since $\text{Var}(\varepsilon_i) = \psi_i$, $i = 1, 2, \ldots, p$, need not be equal, Bartlett [2] has suggested that weighted least squares be used to estimate the common factor values.

The sum of squared errors, weighted by the reciprocal of their variances, is

$$\sum_{i=1}^{p} \frac{\varepsilon_i^2}{\psi_i} = \boldsymbol{\varepsilon}'\boldsymbol{\Psi}^{-1}\boldsymbol{\varepsilon} = (\mathbf{x} - \boldsymbol{\mu} - \mathbf{L}\mathbf{f})'\boldsymbol{\Psi}^{-1}(\mathbf{x} - \boldsymbol{\mu} - \mathbf{L}\mathbf{f}) \qquad (9\text{-}47)$$

Bartlett proposed choosing the estimates $\hat{\mathbf{f}}$ of $\mathbf{f}$ to minimize (9-47). The solution (see Exercise 7.3) is

$$\hat{\mathbf{f}} = (\mathbf{L}'\boldsymbol{\Psi}^{-1}\mathbf{L})^{-1}\mathbf{L}'\boldsymbol{\Psi}^{-1}(\mathbf{x} - \boldsymbol{\mu}) \qquad (9\text{-}48)$$

Motivated by (9-48), we take the estimates $\hat{\mathbf{L}}$, $\hat{\boldsymbol{\Psi}}$, and $\hat{\boldsymbol{\mu}} = \bar{\mathbf{x}}$ as the true values and obtain the factor scores for the jth case as

$$\hat{\mathbf{f}}_j = (\hat{\mathbf{L}}'\hat{\boldsymbol{\Psi}}^{-1}\hat{\mathbf{L}})^{-1}\hat{\mathbf{L}}'\hat{\boldsymbol{\Psi}}^{-1}(\mathbf{x}_j - \bar{\mathbf{x}}) \qquad (9\text{-}49)$$

When $\hat{\mathbf{L}}$ and $\hat{\mathbf{\Psi}}$ are determined by the maximum likelihood method, these estimates must satisfy the uniqueness condition, $\hat{\mathbf{L}}'\hat{\mathbf{\Psi}}^{-1}\hat{\mathbf{L}} = \hat{\mathbf{\Delta}}$, a diagonal matrix. We have the following.

Factor Scores Obtained by Weighted Least Squares from the Maximum Likelihood Estimates

$$\hat{\mathbf{f}}_j = (\hat{\mathbf{L}}'\hat{\mathbf{\Psi}}^{-1}\hat{\mathbf{L}})^{-1}\hat{\mathbf{L}}'\hat{\mathbf{\Psi}}^{-1}(\mathbf{x}_j - \hat{\boldsymbol{\mu}}) = \hat{\mathbf{\Delta}}^{-1}\hat{\mathbf{L}}'\hat{\mathbf{\Psi}}^{-1}(\mathbf{x}_j - \bar{\mathbf{x}}), \qquad j = 1, 2, \ldots, n$$

or, if the correlation matrix is factored, (9-50)

$$\hat{\mathbf{f}}_j = (\hat{\mathbf{L}}_z'\hat{\mathbf{\Psi}}_z^{-1}\hat{\mathbf{L}}_z)^{-1}\hat{\mathbf{L}}_z'\hat{\mathbf{\Psi}}_z^{-1}\mathbf{z}_j = \hat{\mathbf{\Delta}}_z^{-1}\hat{\mathbf{L}}_z'\hat{\mathbf{\Psi}}_z^{-1}\mathbf{z}_j, \qquad j = 1, 2, \ldots, n$$

where $\mathbf{z}_j = \mathbf{D}^{-1/2}(\mathbf{x}_j - \bar{\mathbf{x}})$ and $\hat{\boldsymbol{\rho}} = \hat{\mathbf{L}}_z\hat{\mathbf{L}}_z' + \hat{\mathbf{\Psi}}_z$.

The factor scores generated by (9-50) have sample mean vector $\mathbf{0}$ and zero sample covariances (see Exercise 9.15).

If rotated loadings $\hat{\mathbf{L}}^* = \hat{\mathbf{L}}\mathbf{T}$ are used in place of the original loadings in (9-50), the subsequent factor scores, $\hat{\mathbf{f}}_j^*$, are related to $\hat{\mathbf{f}}_j$ by $\hat{\mathbf{f}}_j^* = \mathbf{T}'\hat{\mathbf{f}}_j$, $j = 1, 2, \ldots, n$.

Comment. If the factor loadings are estimated by the principal component method, it is customary to generate factor scores using an unweighted (ordinary) least squares procedure. Implicitly this amounts to assuming the ψ_i are equal or nearly equal. The factor scores are then

$$\hat{\mathbf{f}}_j = (\hat{\mathbf{L}}'\hat{\mathbf{L}})^{-1}\hat{\mathbf{L}}'(\mathbf{x}_j - \bar{\mathbf{x}})$$

$$\text{or} \quad \hat{\mathbf{f}}_j = \left(\hat{\mathbf{L}}_z'\hat{\mathbf{L}}_z\right)^{-1}\hat{\mathbf{L}}_z'\mathbf{z}_j$$

for standardized data. Since $\mathbf{L} = \left[\sqrt{\hat{\lambda}_1}\,\hat{\mathbf{e}}_1 \,\vdots\, \sqrt{\hat{\lambda}_2}\,\hat{\mathbf{e}}_2 \,\vdots\, \cdots \,\vdots\, \sqrt{\hat{\lambda}_m}\,\hat{\mathbf{e}}_m\right]$ [see (9-15)], we have

$$\hat{\mathbf{f}}_j = \begin{bmatrix} \dfrac{1}{\sqrt{\hat{\lambda}_1}}\hat{\mathbf{e}}_1'(\mathbf{x}_j - \bar{\mathbf{x}}) \\ \dfrac{1}{\sqrt{\hat{\lambda}_2}}\hat{\mathbf{e}}_2'(\mathbf{x}_j - \bar{\mathbf{x}}) \\ \vdots \\ \dfrac{1}{\sqrt{\hat{\lambda}_m}}\hat{\mathbf{e}}_m'(\mathbf{x}_j - \bar{\mathbf{x}}) \end{bmatrix} \tag{9-51}$$

For these factor scores,

$$\frac{1}{n}\sum_{j=1}^{n}\hat{\mathbf{f}}_j = \mathbf{0} \qquad \text{(sample mean)}$$

and

$$\frac{1}{n-1}\sum_{j=1}^{n}\hat{\mathbf{f}}_j\hat{\mathbf{f}}_j' = \mathbf{I} \qquad \text{(sample covariance)}$$

Comparing (9-51) with (8-21), we see that the $\hat{\mathbf{f}}_j$ is nothing more than the first m (scaled) principal components evaluated at $\mathbf{x}_j$.

The Regression Method

Starting again with the original factor model $\mathbf{X} - \boldsymbol{\mu} = \mathbf{LF} + \boldsymbol{\varepsilon}$, we initially treat the loadings matrix $\mathbf{L}$ and specific variance matrix $\boldsymbol{\Psi}$ as known. When the common factors $\mathbf{F}$ and the specific factors (or errors) $\boldsymbol{\varepsilon}$ are jointly normally distributed with means and covariances given by (9-3), the linear combination $\mathbf{X} - \boldsymbol{\mu} = \mathbf{LF} + \boldsymbol{\varepsilon}$ has a $N_p(\mathbf{0}, \mathbf{LL}' + \boldsymbol{\Psi})$ distribution (see Result 4.3). Moreover, the joint distribution of $(\mathbf{X} - \boldsymbol{\mu})$ and $\mathbf{F}$ is $N_{(m+p)\times p}(\mathbf{0}, \boldsymbol{\Sigma}^*)$, where

$$\underset{(m+p)\times(m+p)}{\boldsymbol{\Sigma}^*} = \left[\begin{array}{c|c} \boldsymbol{\Sigma} = \underset{(p\times p)}{\mathbf{LL}' + \boldsymbol{\Psi}} & \underset{(p\times m)}{\mathbf{L}} \\ \hline \underset{(m\times p)}{\mathbf{L}'} & \underset{(m\times m)}{\mathbf{I}} \end{array}\right] \tag{9-52}$$

and $\mathbf{0}$ is a $(m + p) \times p$ matrix of zeros. Using Result 4.6, the conditional distribution of $\mathbf{F}\,|\,\mathbf{x}$ is multivariate normal with

$$\text{mean} = E(\mathbf{F}\,|\,\mathbf{x}) = \mathbf{L}'\boldsymbol{\Sigma}^{-1}(\mathbf{x} - \boldsymbol{\mu}) = \mathbf{L}'(\mathbf{LL}' + \boldsymbol{\Psi})^{-1}(\mathbf{x} - \boldsymbol{\mu}) \tag{9-53}$$

and

$$\text{covariance} = \text{Cov}(\mathbf{F}\,|\,\mathbf{x}) = \mathbf{I} - \mathbf{L}'\boldsymbol{\Sigma}^{-1}\mathbf{L} = \mathbf{I} - \mathbf{L}'(\mathbf{LL}' + \boldsymbol{\Psi})^{-1}\mathbf{L} \tag{9-54}$$

The quantities $\mathbf{L}'(\mathbf{LL}' + \boldsymbol{\Psi})^{-1}$ in (9-53) are the coefficients in a (multivariate) regression of the factors on the variables. Estimates of these coefficients produce factor scores that are analogous to the estimates of the conditional mean values in multivariate regression analysis (see Chapter 7). Consequently, given any vector of observations $\mathbf{x}_j$ and taking the maximum likelihood estimates $\hat{\mathbf{L}}$ and $\hat{\boldsymbol{\Psi}}$ as the true values, the jth factor score vector is given by

$$\hat{\mathbf{f}}_j = \hat{\mathbf{L}}'\boldsymbol{\Sigma}^{-1}(\mathbf{x}_j - \bar{\mathbf{x}}) = \hat{\mathbf{L}}'(\hat{\mathbf{L}}\hat{\mathbf{L}}' + \hat{\boldsymbol{\Psi}})^{-1}(\mathbf{x}_j - \bar{\mathbf{x}}), \qquad j = 1, 2, \ldots, n \tag{9-55}$$

The calculation of $\hat{\mathbf{f}}_j$ in (9-55) can be simplified by using the matrix identity (see Exercise 9.6)

$$\underset{(m\times p)}{\hat{\mathbf{L}}'}\left(\underset{(p\times p)}{\hat{\mathbf{L}}\hat{\mathbf{L}}' + \hat{\boldsymbol{\Psi}}}\right)^{-1} = \left(\underset{(m\times m)}{\mathbf{I} + \hat{\mathbf{L}}'\hat{\boldsymbol{\Psi}}^{-1}\hat{\mathbf{L}}}\right)^{-1} \underset{(m\times p)}{\hat{\mathbf{L}}'} \underset{(p\times p)}{\hat{\boldsymbol{\Psi}}^{-1}} \tag{9-56}$$

This identity allows us to compare the factor scores in (9-55), generated by the regression argument, with those generated by the weighted least squares procedure [see (9-50)]. Temporarily, we denote the former by $\hat{\mathbf{f}}_j^R$ and the latter by $\hat{\mathbf{f}}_j^{LS}$. Then, using (9-56),

$$\hat{\mathbf{f}}_j^{LS} = (\hat{\mathbf{L}}'\hat{\boldsymbol{\Psi}}^{-1}\hat{\mathbf{L}})^{-1}(\mathbf{I} + \hat{\mathbf{L}}'\hat{\boldsymbol{\Psi}}^{-1}\hat{\mathbf{L}})\mathbf{f}_j^R = \left(\mathbf{I} + (\hat{\mathbf{L}}'\hat{\boldsymbol{\Psi}}^{-1}\hat{\mathbf{L}})^{-1}\right)\mathbf{f}_j^R \tag{9-57}$$

For maximum likelihood estimates $(\hat{\mathbf{L}}'\hat{\mathbf{\Psi}}^{-1}\hat{\mathbf{L}})^{-1} = \hat{\mathbf{\Delta}}^{-1}$ and if the elements of this diagonal matrix are close to zero, the regression and generalized least squares methods will give nearly the same factor scores.

In an attempt to reduce the effects of a (possibly) incorrect determination of the number of factors, practitioners tend to calculate the factor scores in (9-55) by using $\mathbf{S}$ (the original sample covariance matrix) instead of $\hat{\mathbf{\Sigma}} = \hat{\mathbf{L}}\hat{\mathbf{L}}' + \hat{\mathbf{\Psi}}$. We have the following.

Factor Scores Obtained by Regression

$$\hat{\mathbf{f}}_j = \hat{\mathbf{L}}'\mathbf{S}^{-1}(\mathbf{x}_j - \bar{\mathbf{x}}) \qquad j = 1, 2, \ldots, n$$

or, if a correlation matrix is factored, (9-58)

$$\hat{\mathbf{f}}_j = \hat{\mathbf{L}}_z'\mathbf{R}^{-1}\mathbf{z}_j \qquad j = 1, 2, \ldots, n$$

where

$$\mathbf{z}_j = \mathbf{D}^{-1/2}(\mathbf{x}_j - \bar{\mathbf{x}}) \quad \text{and} \quad \hat{\boldsymbol{\rho}} = \hat{\mathbf{L}}_z\hat{\mathbf{L}}_z' + \hat{\mathbf{\Psi}}_z$$

Again, if rotated loadings $\hat{\mathbf{L}}^* = \hat{\mathbf{L}}\mathbf{T}$ are used in place of the original loadings in (9-58), the subsequent factor scores $\hat{\mathbf{f}}_j^*$ are related to $\hat{\mathbf{f}}_j$ by

$$\hat{\mathbf{f}}_j^* = \mathbf{T}'\hat{\mathbf{f}}_j, \qquad j = 1, 2, \ldots, n$$

A numerical measure of agreement between the factor scores generated from two *different* calculation methods is provided by the sample correlation coefficient between scores on the same factor. Of the methods presented, none is recommended as uniformly superior.

Example 9.12

We shall illustrate the computation of factor scores by the least squares and regression methods using the stock-price data discussed in Example 9.10. A maximum likelihood solution from $\mathbf{R}$ gave the estimated rotated loadings and specific variances

$$\hat{\mathbf{L}}^* = \begin{bmatrix} .601 & .377 \\ .850 & .164 \\ .643 & .335 \\ .365 & .507 \\ .208 & .883 \end{bmatrix} \quad \text{and} \quad \hat{\mathbf{\Psi}} = \begin{bmatrix} .50 & 0 & 0 & 0 & 0 \\ 0 & .25 & 0 & 0 & 0 \\ 0 & 0 & .47 & 0 & 0 \\ 0 & 0 & 0 & .61 & 0 \\ 0 & 0 & 0 & 0 & .18 \end{bmatrix}$$

The vector of standardized observations,

$$\mathbf{z}' = [.50, -1.40, -.20, -.70, 1.40]$$

yields the following scores on factors 1 and 2.

Weighted least squares (9-50):

$$\hat{\mathbf{f}} = (\hat{\mathbf{L}}_z^{*\prime}\hat{\mathbf{\Psi}}_z^{-1}\hat{\mathbf{L}}_z^*)^{-1}\hat{\mathbf{L}}_z^{*\prime}\hat{\mathbf{\Psi}}_z^{-1}\mathbf{z} = \begin{bmatrix} -1.8 \\ 2.0 \end{bmatrix}$$

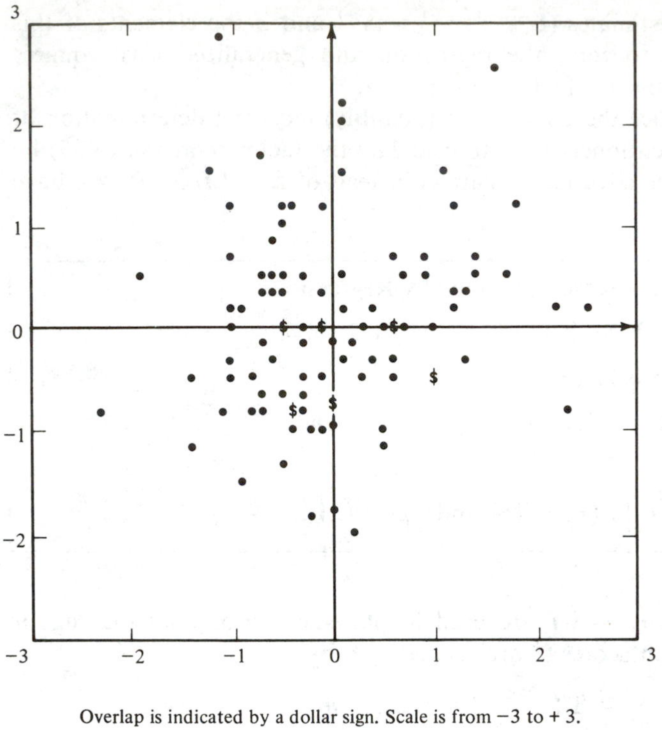

Overlap is indicated by a dollar sign. Scale is from −3 to + 3.

Figure 9.4 Factor scores using (9-58) for factors 1 and 2 of the stock-price data. (Maximum likelihood estimates of the factor loadings.)

Regression (9-58):

$$\hat{\mathbf{f}} = \hat{\mathbf{L}}_z^{*\prime}\mathbf{R}^{-1}\mathbf{z} = \begin{bmatrix} .187 & .657 & .222 & .050 & -.210 \\ .037 & -.185 & .013 & .107 & .864 \end{bmatrix} \begin{bmatrix} .50 \\ -1.40 \\ -.20 \\ -.70 \\ 1.40 \end{bmatrix} = \begin{bmatrix} -1.2 \\ 1.4 \end{bmatrix}$$

In this case, the two methods produce somewhat different results. All of the regression factor scores, obtained using (9-58), are plotted in Figure 9.4. ■

Comment. Factor scores with a rather pleasing intuitive property can be constructed very simply. Group the variables with high (say, greater than .40) loadings on a factor. The scores for factor 1 are then formed by summing the (standardized) observed values of the variables in the group combined according to the sign of the loadings. The factor scores for factor 2 are the sums of the standardized observations corresponding to variables with high loadings on factor 2, and so forth. Data reduction is accomplished by replacing the standardized data by these simple factor scores. The simple factor scores are frequently highly correlated with the factor scores obtained by the more complex least squares and regression methods.

Example 9.13 (Creating Simple Summary Scores from Factor Analysis Groupings)

The principal component factor analysis of the stock price data in Example 9.4 produced the estimated loadings

$$\tilde{\mathbf{L}} = \begin{bmatrix} .784 & -.216 \\ .773 & -.458 \\ .795 & -.234 \\ .712 & .473 \\ .712 & .524 \end{bmatrix} \quad \text{and} \quad \tilde{\mathbf{L}}^* = \tilde{\mathbf{L}}\mathbf{T} = \begin{bmatrix} .746 & .323 \\ .889 & .128 \\ .766 & .316 \\ .258 & .815 \\ .226 & .854 \end{bmatrix}$$

For each factor, take the largest loadings in $\tilde{\mathbf{L}}$ as equal in magnitude and neglect the smaller loadings. Thus we create the linear combinations.

$$\hat{f}_1 = x_1 + x_2 + x_3 + x_4 + x_5$$
$$\hat{f}_2 = x_4 + x_5 - x_2$$

as a summary. In practice, we would standardize these new variables.

If instead of $\tilde{\mathbf{L}}$, we start with the varimax rotated loadings, $\tilde{\mathbf{L}}^*$, the simple factor scores would be

$$\hat{f}_1 = x_1 + x_2 + x_3$$
$$\hat{f}_2 = x_4 + x_5$$

The identification of high loadings and negligible loadings is really quite subjective. Linear compounds that make subject-matter sense are preferable. ■

Although multivariate normality is often assumed for the variables in a factor analysis, it is very difficult to justify the assumption for a large number of variables. As we pointed out in Chapter 4, marginal transformations may help. Similarly, the factor scores may or may not be normally distributed. Bivariate scatterplots of factor scores can produce all sorts of nonelliptical shapes. Plots of factor scores should be examined prior to using these scores in other analyses. They can reveal outlying values and the extent of the (possible) nonnormality.

9.6 PERSPECTIVES AND A STRATEGY FOR FACTOR ANALYSIS

There are many decisions that must be made in any factor analytic study. Probably the most important decision is the choice of m, the number of common factors. Although a large sample test of model adequacy is available for a given m, it is suitable only for data that are approximately normally distributed. Moreover, the test will most assuredly reject model adequacy for small m if the number of variables and observations is large. Yet this is the situation when factor analysis provides a useful approximation. Most often, the final choice of m is based on some combination of (1) the proportion of sample variance explained, (2) subject matter knowledge, and (3) the "reasonableness" of the results.

The choice of solution method and type of rotation are less crucial decisions. In fact, the most satisfactory factor analyses are those where rotations are tried with more than one method and all the results substantially confirm the same factor structure.

At the present time, factor analysis still maintains the flavor of an art and no single strategy should yet be chiseled into stone. We suggest and illustrate one reasonable option.

1. *Perform a principal component factor analysis.* This method is particularly appropriate for a first pass through the data. (It is not required that **R** or **S** be nonsingular.)
   (a) Look for suspicious observations by plotting the factor scores. Also calculate standardized scores for each observation and squared distances as described in Section 4.6.
   (b) Try a varimax rotation.
2. *Perform a maximum likelihood factor analysis including a varimax rotation.*
3. *Compare the factor analyses solutions.*
   (a) Do the loadings group in the same manner?
   (b) Plot factor scores obtained for principal components against scores from the maximum likelihood analysis.
4. *Repeat the first three steps for other numbers of common factors m.* Do extra factors necessarily contribute to the understanding and interpretation of the data?
5. *For large data sets, split them in half and perform a factor analysis on each part.* Compare the two solutions with each other and that obtained from the complete data set to check solution stability. (The data might be divided at random or by placing odd-numbered cases in one group and even-numbered cases in the other group.)

Example 9.14

We present the results of several factor analyses on bone and skull measurements of White Leghorn fowl. The original data were taken from Dunn [5]. Factor analysis of Dunn's data was originally considered by Wright [14], who started his analysis from a different correlation matrix than the one we use.

The full data set consists of $n = 276$ measurements on bone dimensions:

$$\text{Head: } \begin{cases} X_1 = \text{skull length} \\ X_2 = \text{skull breadth} \end{cases}$$

$$\text{Leg: } \begin{cases} X_3 = \text{femur length} \\ X_4 = \text{tibia length} \end{cases}$$

$$\text{Wing: } \begin{cases} X_5 = \text{humerus length} \\ X_6 = \text{ulna length} \end{cases}$$

The sample correlation matrix

$$\mathbf{R} = \begin{bmatrix} 1.000 & .505 & .569 & .602 & .621 & .603 \\ .505 & 1.000 & .422 & .467 & .482 & .450 \\ .569 & .422 & 1.000 & .926 & .877 & .878 \\ .602 & .467 & .926 & 1.000 & .874 & .894 \\ .621 & .482 & .877 & .874 & 1.000 & .937 \\ .603 & .450 & .878 & .894 & .937 & 1.000 \end{bmatrix}$$

was factor analyzed by the principal component and maximum likelihood methods for an $m = 3$ factor model. The results are given in Table 9.10.

TABLE 9.10 FACTOR ANALYSIS OF FOWL DATA

Principal Component

Variable	Estimated factor loadings			Rotated estimated loadings			
	F_1	F_2	F_3	F_1^*	F_2^*	F_3^*	$\tilde{\psi}_i$
1. Skull length	.741	.350	.573	.355	.244	.902	.00
2. Skull breadth	.604	.720	−.340	.235	.949	.211	.00
3. Femur length	.929	−.233	−.075	.921	.164	.218	.08
4. Tibia length	.943	−.175	−.067	.904	.212	.252	.08
5. Humerus length	.948	−.143	−.045	.888	.228	.283	.08
6. Ulna length	.945	−.189	−.047	.908	.192	.264	.07
Cumulative proportion of total (standardized) sample variance explained	.743	.873	.950	.576	.763	.950	

Maximum Likelihood

Variable	Estimated factor loadings			Rotated estimated loadings			
	F_1	F_2	F_3	F_1^*	F_2^*	F_2^*	$\hat{\psi}_i$
1. Skull length	.602	.214	.286	.467	.506	.128	.51
2. Skull breadth	.467	.177	.652	.211	.792	.050	.33
3. Femur length	.926	.145	−.057	.890	.289	.084	.12
4. Tibia length	1.000	.000	−.000	.936	.345	−.073	.00
5. Humerus length	.874	.463	−.012	.831	.362	.396	.02
6. Ulna length	.894	.336	−.039	.857	.325	.272	.09
Cumulative proportion of total (standardized) sample variance explained	.667	.738	.823	.559	.779	.823	

After rotation, the two methods of solution appear to give somewhat different results. Focusing attention on the principal component method and the cumulative proportion of total sample variance explained, a three-factor solution appears to be warranted. The third factor explains a "significant" amount of additional sample variation. The first factor appears to be a *body-size* factor dominated by wing and leg dimensions. The second and third factors, collectively, represent skull dimension and might be given the same names as the variables, *skull breadth* and *skull length*, respectively.

The rotated maximum likelihood factor loadings are consistent with those generated by the principal component method for the first factor, but not for factors two and three. For the maximum likelihood method, the second factor appears to represent head size. The meaning of the third factor is unclear and it is probably not needed.

Further support for retaining three or fewer factors is provided by the residual matrix obtained from the maximum likelihood estimates:

$$\mathbf{R} - \hat{\mathbf{L}}_z\hat{\mathbf{L}}_z' - \hat{\mathbf{\Psi}}_z = \begin{bmatrix} .000 & & & & & \\ -.000 & .000 & & & & \\ -.003 & .001 & .000 & & & \\ .000 & .000 & .000 & .000 & & \\ -.001 & .000 & .000 & .000 & .000 & \\ .004 & -.001 & -.001 & .000 & -.000 & .000 \end{bmatrix}$$

All the entries in this matrix are very small. We shall pursue the $m = 3$ factor model in this example. A $m = 2$ factor model is considered in Exercise 9.10.

Factor scores for factors 1 and 2 produced from (9-58) with the rotated maximum likelihood estimates are plotted in Figure 9.5. Plots of this kind allow us to identify observations that, for one reason or another, are not consistent with the remaining observations. Potential outliers are circled in Figure 9.5.

It is also of interest to plot pairs of factor scores obtained using the principal component and maximum likelihood estimates of factor loadings. For the chicken-bones data, plots of pairs of factor scores are given in Figure 9.6. If the factor loadings on a particular factor agree, the pairs of scores should cluster tightly about the 45° line through the origin. Sets of loadings that do not agree will produce factor scores that deviate from this pattern. If the latter occurs, it is usually associated with the last factors and may suggest that the number of factors is too large. That is, the last factors are not meaningful. This seems to be the case with the third factor in the chicken-bones data, as indicated by Plot (c) in Figure 9.6.

Plots of pairs of factor scores using estimated loadings from two solution methods are also good tools for detecting outliers. If the sets of loadings for a factor tend to agree, outliers will appear as points in the neighborhood of the 45° line, but far from the origin and the cluster of the remaining points. It is clear from Plot (b) in Figure 9.6 that one of the 276 observations is not consistent with the others. It has an unusually large F_2 score. When this point,

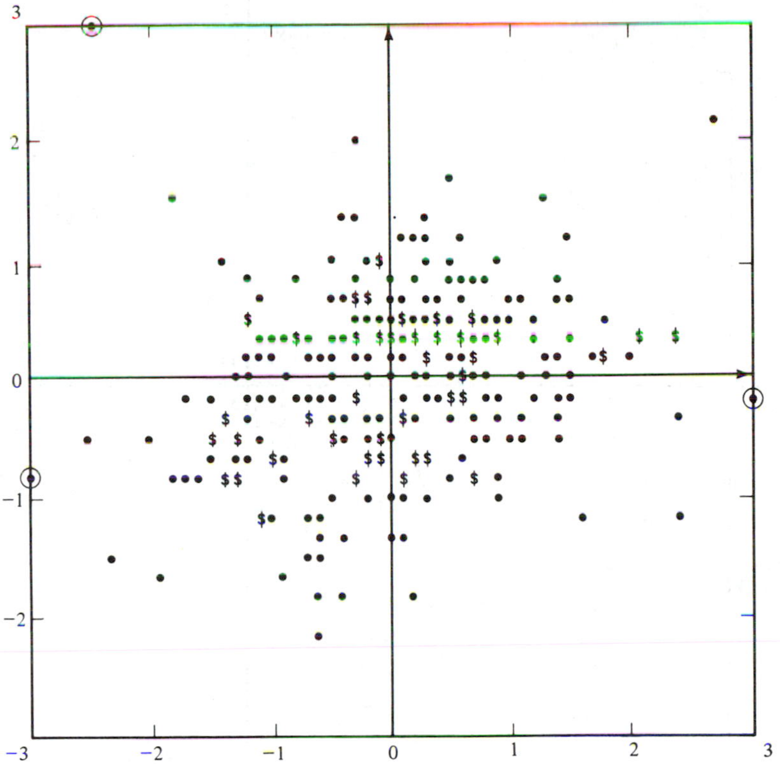

Overlap is indicated by a dollar sign. Scale is from −3 to +3.
Factor scores grater than 3 are plotted as 3. Factor scores
less than −3 are plotted as −3.

Figure 9.5 Factor scores for first two factors of chicken-bones data.

[39.1, 39.3, 75.7, 115, 73.4, 69.1], was removed and the analysis repeated, the loadings were not altered appreciably.

When the data set is large, it should be divided into two (roughly) equal sets, and a factor analysis should be performed on each half. The results of these analyses can be compared with each other and with the analysis for the full data set to test the stability of the solution. If the results are consistent with one another, confidence in the solution is increased.

The chicken-bones data were divided into two sets of $n_1 = 137$ and $n_2 = 139$ observations, respectively. The resulting sample correlation matrices were

$$\mathbf{R}_1 = \begin{bmatrix} 1.000 & & & & & \\ .696 & 1.000 & & & & \\ .588 & .540 & 1.000 & & & \\ .639 & .575 & .901 & 1.000 & & \\ .694 & .606 & .844 & .835 & 1.000 & \\ .660 & .584 & .866 & .863 & .931 & 1.000 \end{bmatrix}$$

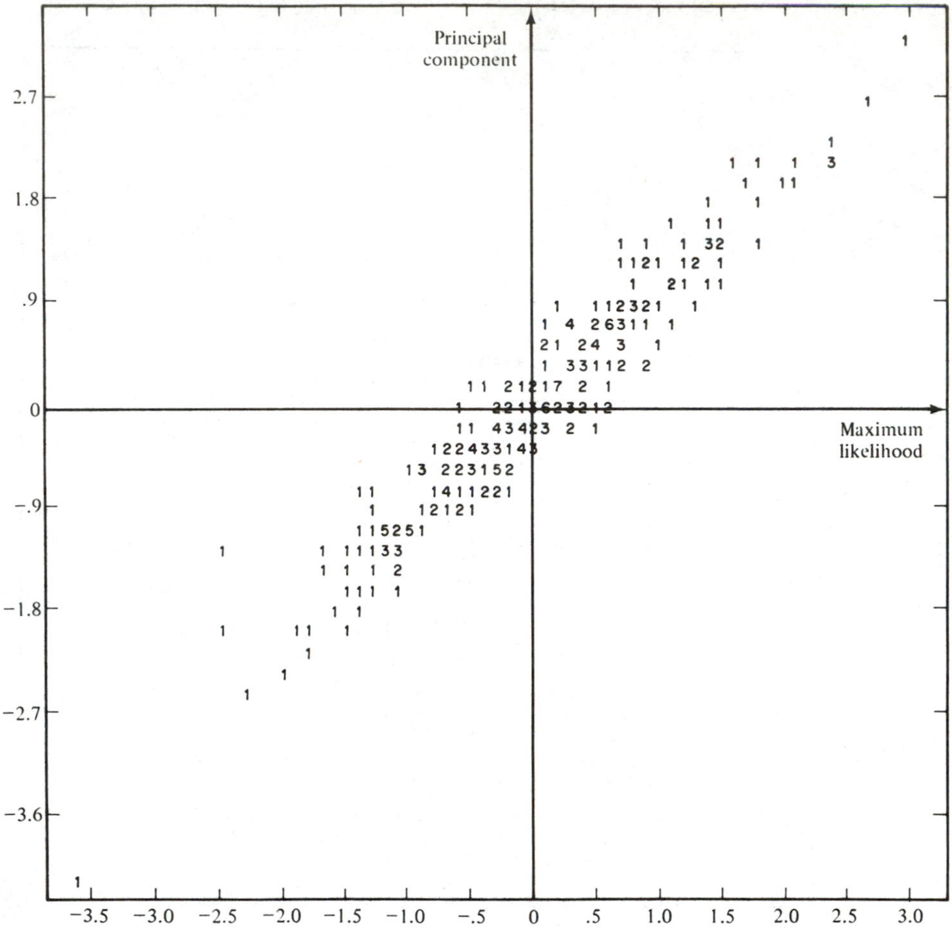

(a) First factor

Figure 9.6 Pairs of factor scores for the chicken-bones data. (Loadings are estimated by principal component and maximum likelihood methods.)

$$\mathbf{R}_2 = \begin{bmatrix} 1.000 & & & & & \\ .366 & 1.000 & & & & \\ .572 & .352 & 1.000 & & & \\ .587 & .406 & .950 & 1.000 & & \\ .587 & .420 & .909 & .911 & 1.000 & \\ .598 & .386 & .894 & .927 & .940 & 1.000 \end{bmatrix}$$

The rotated estimated loadings, specific variances and proportion of total (standardized) sample variance explained for a principal component solution of an $m = 3$ factor model are given in Table 9.11 on page 445.

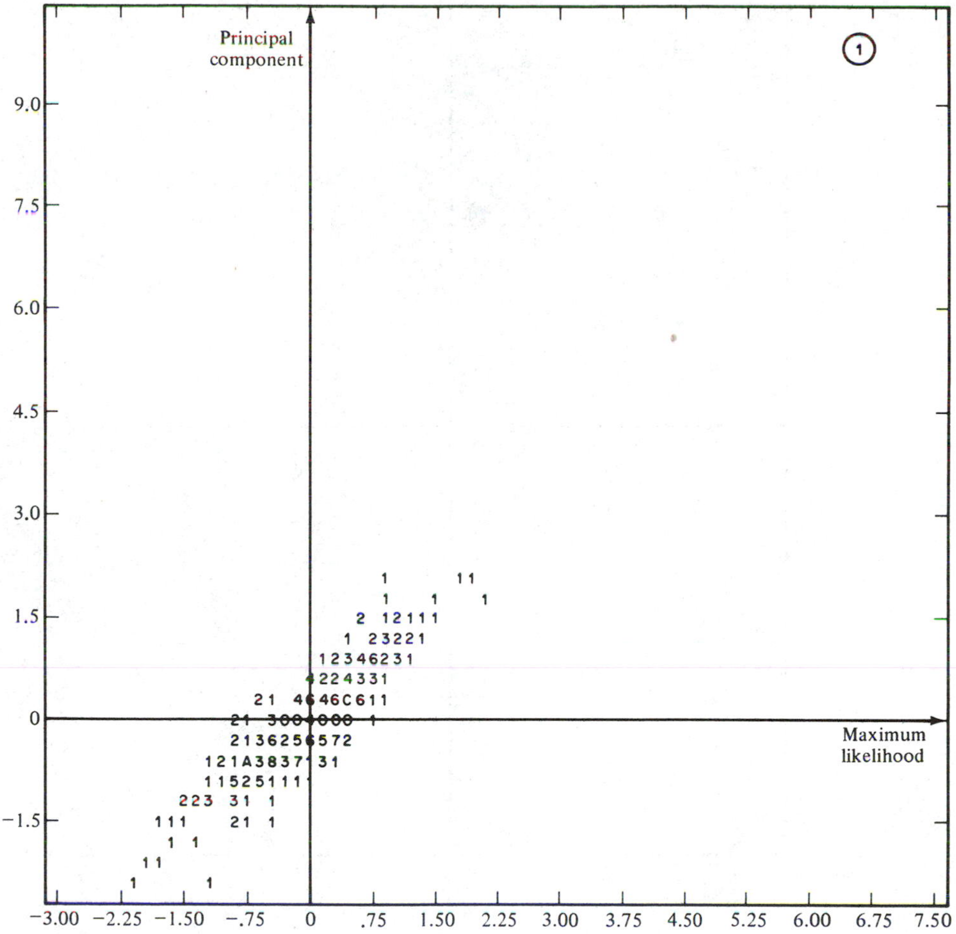

(b) Second factor

Figure 9.6 (*continued*)

The results for the two halves of the chicken-bone measurements are very similar. Factors F_2^* and F_3^* interchange with respect to their labels, skull length and skull breadth, but they collectively seem to represent *head size*. The first factor, F_1^*, again appears to be a *body-size* factor dominated by leg and wing dimensions. These are the same interpretations we gave to the results from a principal component factor analysis of the entire set of data. The solution is remarkably stable and we can be fairly confident the large loadings are "real." As we have pointed out however, three factors are probably too many. A one- or two-factor model is surely sufficient for the chicken-bones data and you are encouraged to repeat the analyses here with fewer factors and alternative solution methods (see Exercise 9.10). ■

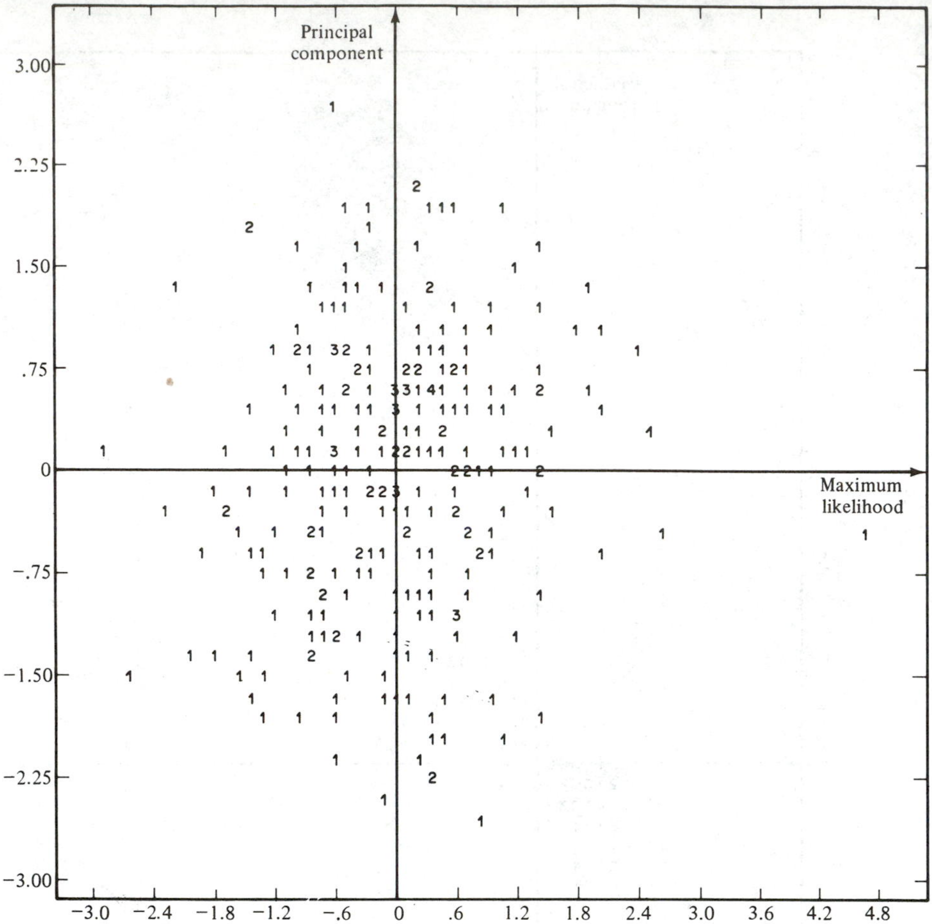

(c) Third factor

Figure 9.6 *(continued)*

Factor analysis has a tremendous intuitive appeal for the behavioral and social sciences. In these areas, it is natural to regard multivariate observations on animal and human processes as manifestations of underlying unobservable "traits." Factor analysis provides a way of explaining the observed variability in behavior in terms of these traits.

When all is said and done, factor analysis remains very subjective. Our examples, in common with most published sources, consist of situations where the factor analysis model provides reasonable explanations in terms of a few interpretable factors. In practice, the vast majority of attempted factor analyses do not yield such clear-cut results. Unfortunately, the criterion for judging the quality of any factor analysis has not been well quantified. Rather, it seems to depend on a

WOW criterion

If, while scrutinizing the factor analysis, the investigator can shout "Wow, I understand these factors" the application is deemed successful.

TABLE 9.11

	First set ($n_1 = 137$ observations) Rotated estimated factor loadings				Second set ($n_2 = 139$ observations) Rotated estimated factor loadings			
Variable	F_1^*	F_2^*	F_3^*	$\tilde{\psi}_i$	F_1^*	F_2^*	F_3^*	$\tilde{\psi}_i$
1. Skull length	.360	.361	.853	.01	.352	.921	.167	.00
2. Skull breadth	.303	.899	.312	.00	.203	.145	.968	.00
3. Femur length	.914	.238	.175	.08	.930	.239	.130	.06
4. Tibia length	.877	.270	.242	.10	.925	.248	.187	.05
5. Humerus length	.830	.247	.395	.11	.912	.252	.208	.06
6. Ulna length	.871	.231	.332	.08	.914	.272	.168	.06
Cumulative proportion of total (standardized) sample variance explained	.546	.743	.940		.593	.780	.962	

Supplement 9A: Some Computational Details for Maximum Likelihood Estimation

We now present some rather detailed computations that enable us to perform a systematic two-step iterative maximization of the likelihood. Although a simple analytical expression cannot be obtained for the maximum likelihood estimators, $\hat{\mathbf{L}}$ and $\hat{\boldsymbol{\Psi}}$, they can be shown to satisfy certain equations. Not surprisingly, the conditions are stated in terms of the maximum likelihood estimator, $\mathbf{S}_n = (1/n) \sum_{j=1}^{n} (\mathbf{X}_j - \overline{\mathbf{X}})(\mathbf{X}_j - \overline{\mathbf{X}})'$, of an unstructured covariance matrix. Some factor analysts employ the usual sample covariance $\mathbf{S}$, but still use the title *maximum likelihood* to refer to resulting estimates. This modification, referenced in Footnote 3 of Chapter 9, amounts to employing the likelihood obtained from the Wishart distribution of $\sum_{j=1}^{n} (\mathbf{X}_j - \overline{\mathbf{X}})(\mathbf{X}_j - \overline{\mathbf{X}})'$ and ignoring the minor contribution due to the normal density for $\overline{\mathbf{X}}$. The factor analysis of $\mathbf{R}$ is, of course, unaffected by the choice of $\mathbf{S}_n$ or $\mathbf{S}$, since they both produce the same correlation matrix.

Result 9A.1. Let $\mathbf{x}_1, \mathbf{x}_2, \ldots, \mathbf{x}_n$ be a random sample from a normal population. The maximum likelihood estimates $\hat{\mathbf{L}}$ and $\hat{\boldsymbol{\Psi}}$ are obtained by maximizing (9-25) subject to the uniqueness condition in (9-26). They satisfy

$$\left(\hat{\boldsymbol{\Psi}}^{-1/2}\mathbf{S}_n\hat{\boldsymbol{\Psi}}^{-1/2}\right)\left(\hat{\boldsymbol{\Psi}}^{-1/2}\hat{\mathbf{L}}\right) = \left(\hat{\boldsymbol{\Psi}}^{-1/2}\hat{\mathbf{L}}\right)\left(\mathbf{I} + \hat{\boldsymbol{\Delta}}\right) \qquad \text{(9A-1)}$$

so the jth column of $\hat{\boldsymbol{\Psi}}^{-1/2}\hat{\mathbf{L}}$ is the (nonnormalized) eigenvector of $\hat{\boldsymbol{\Psi}}^{-1/2}\mathbf{S}_n\hat{\boldsymbol{\Psi}}^{-1/2}$ corresponding to eigenvalue $1 + \Delta_i$. Here

$$\mathbf{S}_n = n^{-1} \sum_{j=1}^{n} (\mathbf{x}_j - \overline{\mathbf{x}})(\mathbf{x}_j - \overline{\mathbf{x}})' = n^{-1}(n-1)\mathbf{S} \quad \text{and} \quad \hat{\Delta}_1 \geq \hat{\Delta}_2 \geq \cdots \geq \hat{\Delta}_m.$$

Also, at convergence

$$\hat{\psi}_i = i\text{th diagonal element of } \mathbf{S}_n - \hat{\mathbf{L}}\hat{\mathbf{L}}' \qquad \text{(9A-2)}$$

and

$$\operatorname{tr}\left(\hat{\boldsymbol{\Sigma}}\mathbf{S}_n^{-1}\right) = p$$

Proof. It is evident that $\hat{\boldsymbol{\mu}} = \overline{\mathbf{x}}$ and consideration of the log-likelihood leads to the maximization of $-(n/2)[\ln|\boldsymbol{\Sigma}| + \operatorname{tr}(\boldsymbol{\Sigma}^{-1}\mathbf{S}_n)]$ over $\mathbf{L}$ and $\boldsymbol{\Psi}$. Equivalently, since $\mathbf{S}_n$ and p are constant with respect to the maximization, we minimize

$$h(\hat{\boldsymbol{\mu}}, \boldsymbol{\Psi}, \mathbf{L}) = \ln|\boldsymbol{\Sigma}| - \ln|\mathbf{S}_n| + \operatorname{tr}\left(\boldsymbol{\Sigma}^{-1}\mathbf{S}_n\right) - p \qquad \text{(9A-3)}$$

subject to $\mathbf{L}'\boldsymbol{\Psi}^{-1}\mathbf{L} = \boldsymbol{\Delta}$, a diagonal matrix.

We take a two-stage approach to the minimization of (9A-3). First consider $\boldsymbol{\Psi}$ as fixed and introduce $\boldsymbol{\Psi}^{-1/2}\boldsymbol{\Psi}^{1/2} = \mathbf{I}$, so $|\boldsymbol{\Psi}|^{-1/2}|\boldsymbol{\Psi}|^{1/2} = 1$. Then the objective function may be expressed as

$$h(\hat{\boldsymbol{\mu}},\boldsymbol{\Psi},\mathbf{L}) = \ln|\boldsymbol{\Psi}^{-1/2}\boldsymbol{\Sigma}\boldsymbol{\Psi}^{-1/2}| - \ln|\boldsymbol{\Psi}^{-1/2}\mathbf{S}_n\boldsymbol{\Psi}^{-1/2}| + \text{tr}\left[\mathbf{S}_n\boldsymbol{\Psi}^{-1/2}\boldsymbol{\Psi}^{1/2}\boldsymbol{\Sigma}^{-1}\boldsymbol{\Psi}^{1/2}\boldsymbol{\Psi}^{-1/2}\right] - p$$

$$= \ln\left(\frac{|\boldsymbol{\Psi}^{-1/2}\boldsymbol{\Sigma}\boldsymbol{\Psi}^{-1/2}|}{|\boldsymbol{\Psi}^{-1/2}\mathbf{S}_n\boldsymbol{\Psi}^{-1/2}|}\right) + \text{tr}\left[\boldsymbol{\Psi}^{-1/2}\mathbf{S}_n\boldsymbol{\Psi}^{-1/2}\boldsymbol{\Psi}^{1/2}\boldsymbol{\Sigma}^{-1}\boldsymbol{\Psi}^{1/2}\right] - p. \tag{9A-4}$$

In order to simplify $h(\hat{\boldsymbol{\mu}}, \boldsymbol{\Psi}, \mathbf{L})$, we note that

$$\boldsymbol{\Psi}^{-1/2}\boldsymbol{\Sigma}^{-1/2} = \boldsymbol{\Psi}^{-1/2}(\mathbf{L}\mathbf{L}' + \boldsymbol{\Psi})\boldsymbol{\Psi}^{-1/2} = (\boldsymbol{\Psi}^{-1/2}\mathbf{L})(\boldsymbol{\Psi}^{-1/2}\mathbf{L})' + \mathbf{I}$$

By Exercise 4.8 with $\underset{(p\times m)}{\mathbf{A}_{12}} = \boldsymbol{\Psi}^{-1/2}\mathbf{L} = -\mathbf{A}'_{21}$, $\underset{(p\times p)}{\mathbf{A}_{11}} = \mathbf{I}$, and $\underset{(m\times m)}{\mathbf{A}_{22}} = \mathbf{I}$,

$$|\boldsymbol{\Psi}^{-1/2}\boldsymbol{\Sigma}\boldsymbol{\Psi}^{-1/2}| = |\mathbf{I} + (\boldsymbol{\Psi}^{-1/2}\mathbf{L})(\boldsymbol{\Psi}^{-1/2}\mathbf{L})'|$$
$$= |\mathbf{I} + (\boldsymbol{\Psi}^{-1/2}\mathbf{L})'(\boldsymbol{\Psi}^{-1/2}\mathbf{L})|$$
$$= |\mathbf{I} + \boldsymbol{\Delta}| = \prod_{i=1}^{m}(1 + \Delta_i) \tag{9A-5}$$

where $\mathbf{L}'\boldsymbol{\Psi}^{-1}\mathbf{L} = \boldsymbol{\Delta}$ by the uniqueness condition. From Exercise 9.6,

$$\boldsymbol{\Psi}^{1/2}\boldsymbol{\Sigma}^{-1}\boldsymbol{\Psi}^{1/2} = (\boldsymbol{\Psi}^{-1/2}\boldsymbol{\Sigma}\boldsymbol{\Psi}^{-1/2})^{-1}$$
$$= (\boldsymbol{\Psi}^{-1/2}\mathbf{L}\mathbf{L}'\boldsymbol{\Psi}^{-1/2} + \mathbf{I})^{-1} = \boldsymbol{\Psi}^{1/2}(\mathbf{L}\mathbf{L}' + \boldsymbol{\Psi})^{-1}\boldsymbol{\Psi}^{1/2}$$
$$= \mathbf{I} - \boldsymbol{\Psi}^{-1/2}\mathbf{L}(\mathbf{I} + \mathbf{L}'\boldsymbol{\Psi}^{-1/2}\boldsymbol{\Psi}^{-1/2}\mathbf{L})^{-1}\mathbf{L}'\boldsymbol{\Psi}^{-1/2}$$
$$= \mathbf{I} - \boldsymbol{\Psi}^{-1/2}\mathbf{L}(\mathbf{I} + \boldsymbol{\Delta})^{-1}\mathbf{L}'\boldsymbol{\Psi}^{-1/2} \tag{9A-6}$$

Next, setting $\mathbf{S}^* = \boldsymbol{\Psi}^{-1/2}\mathbf{S}_n\boldsymbol{\Psi}^{-1/2}$, (9A-6) allows us to write

$$\text{tr}\left[\mathbf{S}^*(\boldsymbol{\Psi}^{1/2}\boldsymbol{\Sigma}^{-1}\boldsymbol{\Psi}^{1/2})\right] = \text{tr}(\mathbf{S}^*) - \text{tr}\left[\mathbf{S}^*\boldsymbol{\Psi}^{-1/2}\mathbf{L}(\mathbf{I} + \boldsymbol{\Delta})^{-1}\mathbf{L}'\boldsymbol{\Psi}^{-1/2}\right]$$
$$= \text{tr}(\mathbf{S}^*) - \text{tr}\left[(\mathbf{I} + \boldsymbol{\Delta})^{-1}(\mathbf{L}'\boldsymbol{\Psi}^{-1/2})\mathbf{S}^*(\boldsymbol{\Psi}^{-1/2}\mathbf{L})\right]$$

The ith diagonal element of $(\boldsymbol{\Psi}^{-1/2}\mathbf{L})'\mathbf{S}^*(\boldsymbol{\Psi}^{-1/2}\mathbf{L})$ is of the form $\boldsymbol{\eta}_i'\mathbf{S}^*\boldsymbol{\eta}_i$, where $\boldsymbol{\eta}_i$ is the ith column of $\boldsymbol{\Psi}^{-1/2}\mathbf{L}$. Because $(\mathbf{I} + \boldsymbol{\Delta})$ is diagonal, the contribution of the ith diagonal element of the product $(\mathbf{I} + \boldsymbol{\Delta})^{-1}(\boldsymbol{\Psi}^{-1/2}\mathbf{L})'\mathbf{S}^*(\boldsymbol{\Psi}^{-1/2}\mathbf{L})$ to the trace is

$$\frac{1}{1 + \Delta_i}\boldsymbol{\eta}_i'\mathbf{S}^*\boldsymbol{\eta}_i, \qquad i = 1, 2, \ldots, m \tag{9A-7}$$

Let the components be labeled so that $\boldsymbol{\Delta}$ has the diagonal elements $\Delta_1 \geq \Delta_2 \geq \cdots \geq \Delta_m$ and denote the eigenvalue-eigenvector pairs of $\mathbf{S}^*$ by $(\lambda_i^*, \mathbf{e}_i^*)$, where $\lambda_1^* \geq \lambda_2^* \geq \cdots \geq \lambda_p^*$. By the uniqueness condition, $\boldsymbol{\eta}_i'\boldsymbol{\eta}_i = \Delta_i$, so $\Delta_i^{-1/2}\boldsymbol{\eta}_i$ has unit length, and the terms (9A-7) can be written as

$$\frac{\Delta_i}{1 + \Delta_i}(\Delta_i^{-1/2}\boldsymbol{\eta}_i)'\mathbf{S}^*(\Delta_i^{-1/2}\boldsymbol{\eta}_i)$$

The maximization lemma of (2-51) asserts that $\Delta_1^{-1/2}\boldsymbol{\eta}_1 = \mathbf{e}_1^*$ maximizes the quadratic form $(\Delta_1^{-1/2}\boldsymbol{\eta}_1)'\mathbf{S}^*(\Delta_1^{-1/2}\boldsymbol{\eta}_1)$, so the contribution to the trace from the maximized first term is

$$\frac{\Delta_1}{1+\Delta_1}\mathbf{e}_1^{*\prime}\mathbf{S}^*\mathbf{e}_1^* = \left(\frac{\Delta_1}{1+\Delta_1}\right)\lambda_1^*$$

For the second term, the uniqueness condition requires that $\boldsymbol{\eta}_2$ be perpendicular to $\boldsymbol{\eta}_1$, so the maximization result (2-52) then yields

$$\lambda_2^*$$

as the largest possible value for $(\Delta_2^{-1/2}\boldsymbol{\eta}_2)'\mathbf{S}^*(\Delta_2^{-1/2}\boldsymbol{\eta}_2)$. Continuing in the same manner, for each of the m terms in (9A-7), we determine an attained lower bound for the objective function $h(\hat{\boldsymbol{\mu}}, \boldsymbol{\Psi}, \mathbf{L})$. Since $|\mathbf{S}^*| = \prod_{i=1}^{p}\lambda_i^*$ and $\text{tr}[\mathbf{S}^*] = \sum_{i=1}^{p}\lambda_i^*$,

$$\begin{aligned}
h(\hat{\boldsymbol{\mu}}, \boldsymbol{\Psi}, \mathbf{L}) &= \ln\left[\frac{\prod_{i=1}^{m}(1+\Delta_i)}{|\mathbf{S}^*|}\right] + \text{tr}[\mathbf{S}^*] - \frac{\Delta_1}{1+\Delta_1}\lambda_1^* \\
&\qquad - \frac{\Delta_2}{1+\Delta_2}\lambda_2^* - \cdots - \frac{\Delta_m}{1+\Delta_m}\lambda_m^* - p \\
&= \sum_{i=1}^{m}\ln\left(\frac{1+\Delta_i}{\lambda_i^*}\right) - \sum_{i=m+1}^{p}\ln(\lambda_i^*) + \sum_{i=1}^{m}\lambda_i^* + \sum_{i=m+1}^{p}\lambda_i^* \\
&\qquad - \sum_{i=1}^{m}\frac{\Delta_i}{1+\Delta_i}\lambda_i^* - p \\
&= \sum_{i=1}^{m}\left[\frac{\lambda_i^*}{1+\Delta_i} - \ln\left(\frac{\lambda_i^*}{1+\Delta_i}\right) - 1\right] \\
&\qquad + \sum_{i=m+1}^{p}\left[\lambda_i^* - \ln(\lambda_i^*) - 1\right]
\end{aligned} \tag{9A-8}$$

Now $x - \ln(x) - 1$ is never negative for $x > 0$ and has a minimum at $x = 1$, so the function in (9A-8) is minimized for fixed $\boldsymbol{\Psi}$ by the choice $\lambda_i^*/(1+\Delta_i) = 1$, or $\Delta_i = \lambda_i^* - 1$, $i = 1, 2, \ldots, m$. With these choices, $\mathbf{S}^*\mathbf{e}_i^* = (\Delta_i + 1)\mathbf{e}_i^*$, $i = 1, 2, \ldots, m$, and the ith column of $\boldsymbol{\Psi}^{-1/2}\mathbf{L}$ is proportional to $\mathbf{e}_i^*$, or

$$\boldsymbol{\Psi}^{-1/2}\mathbf{S}_n\boldsymbol{\Psi}^{-1/2}(\boldsymbol{\Psi}^{-1/2}\mathbf{L}) = (\boldsymbol{\Psi}^{-1/2}\mathbf{L})(\mathbf{I} + \boldsymbol{\Delta}) \tag{9A-9}$$

It remains to vary the p diagonal elements of $\boldsymbol{\Psi}$ to minimize the remaining summation,

$$\sum_{i=m+1}^{p}(\lambda_i^* - \ln(\lambda_i^*) - 1) \tag{9A-10}$$

in (9A-8). This is accomplished on a computer using iterative search routines designed to locate the minimum of a function of several variables.

The calculation process then returns to (9A-9) with the new $\boldsymbol{\Psi}$ and continues until convergence. The condition

$$\text{Diagonal elements of } \hat{\boldsymbol{\Psi}} = \text{diagonal elements of } \mathbf{S}_n - \hat{\mathbf{L}}\hat{\mathbf{L}}' \qquad \text{(9A-11)}$$

that, at convergence, the optimal values $\hat{\boldsymbol{\Psi}}$ and $\hat{\mathbf{L}}$ must satisfy is established by calculus (see [9]).

To obtain $\text{tr}[\mathbf{S}_n\hat{\boldsymbol{\Sigma}}^{-1}] = p$, we note from the uniqueness condition that

$$\hat{\boldsymbol{\Sigma}}\hat{\boldsymbol{\Psi}}^{-1}\hat{\mathbf{L}} = (\hat{\mathbf{L}}\hat{\mathbf{L}}' + \hat{\boldsymbol{\Psi}})\hat{\boldsymbol{\Psi}}^{-1}\hat{\mathbf{L}} = \hat{\mathbf{L}}(\hat{\mathbf{L}}'\hat{\boldsymbol{\Psi}}^{-1}\hat{\mathbf{L}}) + \hat{\mathbf{L}} = \hat{\mathbf{L}}(\hat{\boldsymbol{\Delta}} + \mathbf{I})$$

and by (9A-9),

$$\mathbf{S}_n\hat{\boldsymbol{\Psi}}^{-1}\hat{\mathbf{L}} = \hat{\boldsymbol{\Psi}}^{1/2}\left(\hat{\boldsymbol{\Psi}}^{-1/2}\mathbf{S}_n\hat{\boldsymbol{\Psi}}^{-1/2}\right)\hat{\boldsymbol{\Psi}}^{-1/2}\hat{\mathbf{L}} = \hat{\boldsymbol{\Psi}}^{1/2}\hat{\boldsymbol{\Psi}}^{-1/2}\hat{\mathbf{L}}(\hat{\boldsymbol{\Delta}} + \mathbf{I}) = \hat{\mathbf{L}}(\hat{\boldsymbol{\Delta}} + \mathbf{I})$$

Subtraction yields

$$\left(\hat{\boldsymbol{\Sigma}} - \mathbf{S}_n\right)\hat{\boldsymbol{\Psi}}^{-1}\hat{\mathbf{L}} = \mathbf{0} \qquad \text{(9A-12)}$$

Finally, with $\hat{\boldsymbol{\Sigma}}^{-1} = (\hat{\mathbf{L}}\hat{\mathbf{L}}' + \hat{\boldsymbol{\Psi}})^{-1} = \hat{\boldsymbol{\Psi}}^{-1} - \hat{\boldsymbol{\Psi}}^{-1}\hat{\mathbf{L}}(\mathbf{I} + \hat{\mathbf{L}}'\hat{\boldsymbol{\Psi}}^{-1}\hat{\mathbf{L}})\hat{\mathbf{L}}'\hat{\boldsymbol{\Psi}}^{-1}$ (see Exercise 9.6) and the uniqueness condition $\hat{\mathbf{L}}'\hat{\boldsymbol{\Psi}}^{-1}\hat{\mathbf{L}} = \hat{\boldsymbol{\Delta}}$, we have

$$\begin{aligned}
\mathbf{S}_n\hat{\boldsymbol{\Sigma}}^{-1} &= \mathbf{I} - \left(\hat{\boldsymbol{\Sigma}} - \mathbf{S}_n\right)\hat{\boldsymbol{\Sigma}}^{-1} \\
&= \mathbf{I} - \left(\hat{\boldsymbol{\Sigma}} - \mathbf{S}_n\right)\left[\hat{\boldsymbol{\Psi}}^{-1} - \hat{\boldsymbol{\Psi}}^{-1}\hat{\mathbf{L}}(\mathbf{I} + \hat{\Delta})\hat{\mathbf{L}}'\hat{\boldsymbol{\Psi}}^{-1}\right] \\
&= \mathbf{I} - \left(\hat{\boldsymbol{\Sigma}} - \mathbf{S}_n\right)\hat{\boldsymbol{\Psi}}^{-1}
\end{aligned}$$

The last equality follows from (9A-12). Since, according to (9A-11), $\hat{\boldsymbol{\Sigma}} - \mathbf{S}_n$ has zeros as its diagonal elements, the diagonal elements of $(\hat{\boldsymbol{\Sigma}} - \mathbf{S}_n)\hat{\boldsymbol{\Psi}}^{-1}$ are zeros, and consequently

$$\text{tr}\left[\mathbf{S}_n\hat{\boldsymbol{\Sigma}}^{-1}\right] = \text{tr}[\mathbf{I}] - \text{tr}\left[\left(\hat{\boldsymbol{\Sigma}} - \mathbf{S}_n\right)\hat{\boldsymbol{\Psi}}^{-1}\right] = \text{tr}[\mathbf{I}] = p \qquad \blacksquare$$

Comment. Lawley and Maxwell [9], along with many others who do factor analysis, use the unbiased estimate $\mathbf{S}$ of the covariance matrix instead of the maximum likelihood estimate $\mathbf{S}_n$. Now $(n - 1)\mathbf{S}$ has, for normal data, a Wishart distribution [see (4-21) and (4-23)]. If we ignore the contribution to the likelihood in (9-25) from the second term involving $(\boldsymbol{\mu} - \bar{\mathbf{x}})$, then maximizing the reduced likelihood over $\mathbf{L}$ and $\boldsymbol{\Psi}$ is equivalent to maximizing the Wishart likelihood

$$\text{Likelihood} \propto |\boldsymbol{\Sigma}|^{-(n-1)/2}e^{-[(n-1)/2]\,\text{tr}[\boldsymbol{\Sigma}^{-1}\mathbf{S}]}$$

over $\mathbf{L}$ and $\boldsymbol{\Psi}$. Equivalently, we can minimize

$$\ln|\boldsymbol{\Sigma}| + \text{tr}(\boldsymbol{\Sigma}^{-1}\mathbf{S})$$

or, as in (9A-3),

$$\ln|\boldsymbol{\Sigma}| + \text{tr}[\boldsymbol{\Sigma}^{-1}\mathbf{S}] - \ln|\mathbf{S}| - p$$

Under these conditions, Result 9A-1 holds with $\mathbf{S}$ in place of $\mathbf{S}_n$. Also, for large n, $\mathbf{S}$ and $\mathbf{S}_n$ are almost identical, and the corresponding maximum likelihood estimates, $\hat{\mathbf{L}}$ and $\hat{\boldsymbol{\Psi}}$, would be similar. For testing the factor model [see (9-39)], $|\hat{\mathbf{L}}\hat{\mathbf{L}}' + \hat{\boldsymbol{\Psi}}|$

should be compared with $|\mathbf{S}_n|$ if the actual likelihood of (9-25) is employed, and $|\hat{\mathbf{L}}\hat{\mathbf{L}}' + \hat{\mathbf{\Psi}}|$ should be compared with $|\mathbf{S}|$ if the Wishart likelihood above is used to derive $\hat{\mathbf{L}}$ and $\hat{\mathbf{\Psi}}$.

Recommended Computational Scheme

For $m > 1$, the condition $\mathbf{L}'\mathbf{\Psi}^{-1}\mathbf{L} = \mathbf{\Delta}$ effectively imposes $m(m-1)/2$ constraints on the elements of $\mathbf{L}$ and $\mathbf{\Psi}$, and the likelihood equations are solved, subject to these constraints, in an iterative fashion. One procedure is the following.

1. Compute initial estimates of the specific variances $\psi_1, \psi_2, \ldots, \psi_p$. Jöreskog [7] suggests setting

$$\hat{\psi}_i = \left(1 - \frac{1}{2} \cdot \frac{m}{p}\right)\left(\frac{1}{s^{ii}}\right) \tag{9A-13}$$

   where s^{ii} is the ith diagonal element of $\mathbf{S}^{-1}$.

2. Given $\hat{\mathbf{\Psi}}$, compute the first m distinct eigenvalues, $\hat{\lambda}_1 > \hat{\lambda}_2 > \cdots > \hat{\lambda}_m > 1$, and corresponding eigenvectors, $\hat{\mathbf{e}}_1, \hat{\mathbf{e}}_2, \ldots, \hat{\mathbf{e}}_m$, of the "uniqueness-rescaled" covariance matrix

$$\mathbf{S}^* = \hat{\mathbf{\Psi}}^{-1/2}\mathbf{S}_n\hat{\mathbf{\Psi}}^{-1/2} \tag{9A-14}$$

   Let $\hat{\mathbf{E}} = [\hat{\mathbf{e}}_1 \mid \hat{\mathbf{e}}_2 \mid \cdots \mid \hat{\mathbf{e}}_m]$ be the $p \times m$ matrix of *normalized* eigenvectors and $\hat{\mathbf{\Lambda}} = \text{diag}[\hat{\lambda}_1, \hat{\lambda}_2, \ldots, \hat{\lambda}_m]$ be the $m \times m$ diagonal matrix of eigenvalues. From (9A-1), $\hat{\mathbf{\Lambda}} = \mathbf{I} + \hat{\mathbf{\Delta}}$ and $\hat{\mathbf{E}} = \hat{\mathbf{\Psi}}^{-1/2}\hat{\mathbf{L}}\hat{\mathbf{\Delta}}^{-1/2}$. Thus we obtain the estimates

$$\hat{\mathbf{L}} = \hat{\mathbf{\Psi}}^{1/2}\hat{\mathbf{E}}\hat{\mathbf{\Delta}}^{1/2} = \hat{\mathbf{\Psi}}^{1/2}\hat{\mathbf{E}}(\hat{\mathbf{\Lambda}} - \mathbf{I})^{1/2} \tag{9A-15}$$

3. Substitute $\hat{\mathbf{L}}$ obtained in (9A-15) into the likelihood function (9A-3) and minimize the result with respect to $\hat{\psi}_1, \hat{\psi}_2, \ldots, \hat{\psi}_p$. A numerical search routine must be used. The values $\hat{\psi}_1, \hat{\psi}_2, \ldots, \hat{\psi}_p$ obtained from this minimization are employed at Step (2) to create a new $\hat{\mathbf{L}}$. Steps (2) and (3) are repeated until convergence; that is, until the differences between successive values of $\hat{\ell}_{ij}$ and $\hat{\psi}_i$ are negligible.

Comment. It often happens that the objective function in (9A-3) has a relative minimum corresponding to *negative* values for some $\hat{\psi}_i$. This solution is clearly inadmissible and is said to be improper, or a *Heywood case*. For most packaged computer programs, negative $\hat{\psi}_i$, if they occur on a particular iteration, are changed to small positive numbers before proceeding with the next step.

Maximum Likelihood Estimators of $\boldsymbol{\rho} = \mathbf{L}_z\mathbf{L}_z' + \mathbf{\Psi}_z$

When $\mathbf{\Sigma}$ has the factor analysis structure $\mathbf{\Sigma} = \mathbf{L}\mathbf{L}' + \mathbf{\Psi}$, then $\boldsymbol{\rho}$ can be factored as $\boldsymbol{\rho} = \mathbf{V}^{-1/2}\mathbf{\Sigma}\mathbf{V}^{-1/2} = (\mathbf{V}^{-1/2}\mathbf{L})(\mathbf{V}^{-1/2}\mathbf{L})' + \mathbf{V}^{-1/2}\mathbf{\Psi}\mathbf{V}^{-1/2} = \mathbf{L}_z\mathbf{L}_z' + \mathbf{\Psi}_z$. The loading matrix for the standardized variables is $\mathbf{L}_z = \mathbf{V}^{-1/2}\mathbf{L}$ and the corresponding specific variance matrix is $\mathbf{\Psi}_z = \mathbf{V}^{-1/2}\mathbf{\Psi}\mathbf{V}^{-1/2}$, where $\mathbf{V}^{-1/2}$ is the diagonal matrix with ith diagonal element $\sigma_{ii}^{-1/2}$. If $\mathbf{R}$ is substituted for $\mathbf{S}_n$ in the objective function of (9A-3), the investigator minimizes

$$\ln\left(\frac{|\mathbf{L}_z\mathbf{L}_z' + \mathbf{\Psi}_z|}{|\mathbf{R}|}\right) + \text{tr}\left[(\mathbf{L}_z\mathbf{L}_z' + \mathbf{\Psi}_z)^{-1}\mathbf{R}\right] - p \qquad \text{(9A-16)}$$

Introducing the diagonal matrix $\hat{\mathbf{V}}^{1/2}$, whose ith diagonal element is the square root of the ith diagonal element of $\mathbf{S}_n$, we can write the objective function in (9A-16) as

$$\begin{aligned}
&\ln\left(\frac{|\hat{\mathbf{V}}^{1/2}||\mathbf{L}_z\mathbf{L}_z' + \mathbf{\Psi}_z||\hat{\mathbf{V}}^{1/2}|}{|\hat{\mathbf{V}}^{1/2}||\mathbf{R}||\hat{\mathbf{V}}^{1/2}|}\right) + \text{tr}\left[(\mathbf{L}_z\mathbf{L}_z' + \mathbf{\Psi}_z)^{-1}\hat{\mathbf{V}}^{-1/2}\hat{\mathbf{V}}^{1/2}\mathbf{R}\hat{\mathbf{V}}^{1/2}\hat{\mathbf{V}}^{-1/2}\right] - p \\
&\qquad = \ln\left(\frac{|(\hat{\mathbf{V}}^{1/2}\mathbf{L}_z)(\hat{\mathbf{V}}^{1/2}\mathbf{L}_z)' + \hat{\mathbf{V}}^{1/2}\mathbf{\Psi}_z\hat{\mathbf{V}}^{1/2}|}{|\mathbf{S}_n|}\right) \\
&\qquad\quad + \text{tr}\left[\left((\hat{\mathbf{V}}^{1/2}\mathbf{L}_z)(\hat{\mathbf{V}}^{1/2}\mathbf{L}_z)' + \hat{\mathbf{V}}^{1/2}\mathbf{\Psi}_z\hat{\mathbf{V}}^{1/2}\right)^{-1}\mathbf{S}_n\right] - p \\
&\qquad \geq \ln\left(\frac{|\hat{\mathbf{L}}\hat{\mathbf{L}}' + \hat{\mathbf{\Psi}}|}{|\mathbf{S}_n|}\right) + \text{tr}\left[(\hat{\mathbf{L}}\hat{\mathbf{L}}' + \hat{\mathbf{\Psi}})^{-1}\mathbf{S}_n\right] - p
\end{aligned} \qquad \text{(9A-17)}$$

The last inequality follows because the maximum likelihood estimates $\hat{\mathbf{L}}$ and $\hat{\mathbf{\Psi}}$ minimize the objective function (9A-3). [Equality holds in (9A-17) for $\hat{\mathbf{L}}_z = \hat{\mathbf{V}}^{-1/2}\hat{\mathbf{L}}$ and $\hat{\mathbf{\Psi}}_z = \hat{\mathbf{V}}^{-1/2}\hat{\mathbf{\Psi}}\hat{\mathbf{V}}^{-1/2}$.] Therefore minimizing (9A-16) over $\mathbf{L}_z$ and $\mathbf{\Psi}_z$ is equivalent to obtaining $\hat{\mathbf{L}}$ and $\hat{\mathbf{\Psi}}$ from $\mathbf{S}_n$ and estimating $\mathbf{L}_z = \mathbf{V}^{-1/2}\mathbf{L}$ by $\hat{\mathbf{L}}_z = \hat{\mathbf{V}}^{-1/2}\hat{\mathbf{L}}$ and $\mathbf{\Psi}_z = \mathbf{V}^{-1/2}\mathbf{\Psi}\mathbf{V}^{-1/2}$ by $\hat{\mathbf{\Psi}}_z = \hat{\mathbf{V}}^{-1/2}\hat{\mathbf{\Psi}}\hat{\mathbf{V}}^{-1/2}$. The rationale for the latter procedure comes from the invariance property of maximum likelihood estimators [see (4-20)].

EXERCISES

9.1. Show that the covariance matrix

$$\boldsymbol{\rho} = \begin{bmatrix} 1.0 & .63 & .45 \\ .63 & 1.0 & .35 \\ .45 & .35 & 1.0 \end{bmatrix}$$

for the $p = 3$ standardized random variables Z_1, Z_2, and Z_3 can be generated by the $m = 1$ factor model

$$Z_1 = .9F_1 + \varepsilon_1$$
$$Z_2 = .7F_1 + \varepsilon_2$$
$$Z_3 = .5F_1 + \varepsilon_3$$

where $\text{Var}(F_1) = 1$, $\text{Cov}(\boldsymbol{\varepsilon}, F_1) = \mathbf{0}$, and

$$\mathbf{\Psi} = \text{Cov}(\boldsymbol{\varepsilon}) = \begin{bmatrix} .19 & 0 & 0 \\ 0 & .51 & 0 \\ 0 & 0 & .75 \end{bmatrix}$$

That is, write $\boldsymbol{\rho}$ in the form $\boldsymbol{\rho} = \mathbf{LL}' + \mathbf{\Psi}$.

9.2. Use the information in Exercise 9.1.

(a) Calculate communalities h_i^2, $i = 1, 2, 3$, and interpret these quantities.

(b) Calculate $\text{Corr}(Z_i, F_1)$ for $i = 1, 2, 3$. Which variable might carry the greatest weight in "naming" the common factor? Why?

9.3. The eigenvalues and eigenvectors of the correlation matrix $\boldsymbol{\rho}$ in Exercise 9.1 are

$$\lambda_1 = 1.96, \quad \mathbf{e}_1' = [.625, .593, .507], \quad \lambda_2 = .68, \quad \mathbf{e}_2' = [-.219, -.491, .843]$$
$$\lambda_3 = .36, \quad \mathbf{e}_3' = [.749, -.638, -.177]$$

(a) Assuming an $m = 1$ factor model, calculate the loading matrix $\mathbf{L}$ and matrix of specific variances $\boldsymbol{\Psi}$ using the principal component solution method. Compare the results with those in Exercise 9.1.

(b) What proportion of the total population variance is explained by the first common factor.

9.4. Given $\boldsymbol{\rho}$ and $\boldsymbol{\Psi}$ in Exercise 9.1 and an $m = 1$ factor model, calculate the reduced correlation matrix $\tilde{\boldsymbol{\rho}} = \boldsymbol{\rho} - \boldsymbol{\Psi}$ and the principal factor solution for the loading matrix $\mathbf{L}$. Is the result consistent with the information in Exercise 9.1? Should it be?

9.5. Establish the inequality (9-19). (*Hint*: Since $\mathbf{S} - \tilde{\mathbf{L}}\tilde{\mathbf{L}}' - \tilde{\boldsymbol{\Psi}}$ has zeros on the diagonal, (sum of squared entries of $\mathbf{S} - \tilde{\mathbf{L}}\tilde{\mathbf{L}}' - \tilde{\boldsymbol{\Psi}}$) $\leq$ (sum of squared entries of $\mathbf{S} - \tilde{\mathbf{L}}\tilde{\mathbf{L}}'$). Now, $\mathbf{S} - \tilde{\mathbf{L}}\tilde{\mathbf{L}}' = \hat{\lambda}_{m+1}\hat{\mathbf{e}}_{m+1}\hat{\mathbf{e}}_{m+1}' + \cdots + \hat{\lambda}_p\hat{\mathbf{e}}_p\hat{\mathbf{e}}_p' = \hat{\mathbf{P}}_{(2)}\hat{\boldsymbol{\Lambda}}_{(2)}\hat{\mathbf{P}}_{(2)}'$, where $\hat{\mathbf{P}}_{(2)} = [\hat{\mathbf{e}}_{m+1} \mid \cdots \mid \hat{\mathbf{e}}_p]$ and $\hat{\boldsymbol{\Lambda}}_{(2)}$ is the diagonal matrix with elements $\hat{\lambda}_{m+1}, \ldots, \hat{\lambda}_p$. Use (sum of squared entries of $\mathbf{A}$) $= \text{tr}\,\mathbf{A}\mathbf{A}'$ and $\text{tr}[\hat{\mathbf{P}}_{(2)}\hat{\boldsymbol{\Lambda}}_{(2)}\hat{\boldsymbol{\Lambda}}_{(2)}\hat{\mathbf{P}}_{(2)}'] = \text{tr}[\hat{\boldsymbol{\Lambda}}_{(2)}\hat{\boldsymbol{\Lambda}}_{(2)}]$.)

9.6. Verify the following matrix identities.

(a) $(\mathbf{I} + \mathbf{L}'\boldsymbol{\Psi}^{-1}\mathbf{L})^{-1}\mathbf{L}'\boldsymbol{\Psi}^{-1}\mathbf{L} = \mathbf{I} - (\mathbf{I} + \mathbf{L}'\boldsymbol{\Psi}^{-1}\mathbf{L})^{-1}$

[*Hint*: Premultiply both sides by $(\mathbf{I} + \mathbf{L}'\boldsymbol{\Psi}^{-1}\mathbf{L})$.]

(b) $(\mathbf{L}\mathbf{L}' + \boldsymbol{\Psi})^{-1} = \boldsymbol{\Psi}^{-1} - \boldsymbol{\Psi}^{-1}\mathbf{L}(\mathbf{I} + \mathbf{L}'\boldsymbol{\Psi}^{-1}\mathbf{L})^{-1}\mathbf{L}'\boldsymbol{\Psi}^{-1}$

[*Hint*: Postmultiply both sides by $(\mathbf{L}\mathbf{L}' + \boldsymbol{\Psi})$ and use (a).]

(c) $\mathbf{L}'(\mathbf{L}\mathbf{L}' + \boldsymbol{\Psi})^{-1} = (\mathbf{I} + \mathbf{L}'\boldsymbol{\Psi}^{-1}\mathbf{L})^{-1}\mathbf{L}'\boldsymbol{\Psi}^{-1}$

[*Hint*: Postmultiply the result in (b) by $\mathbf{L}$, use (a), and take the transpose, noting that $(\mathbf{L}\mathbf{L}' + \boldsymbol{\Psi})^{-1}$, $\boldsymbol{\Psi}^{-1}$, and $(\mathbf{I} + \mathbf{L}'\boldsymbol{\Psi}^{-1}\mathbf{L})^{-1}$ are symmetric matrices.]

9.7. (The factor model parameterization need not be unique.) Let the factor model with $p = 2$ and $m = 1$ prevail. Show that

$$\sigma_{11} = \ell_{11}^2 + \psi_1, \qquad \sigma_{12} = \sigma_{21} = \ell_{11}\ell_{21}$$
$$\sigma_{22} = \ell_{21}^2 + \psi_2$$

and, for given σ_{11}, σ_{22}, and σ_{12}, there is an infinity of choices for $\mathbf{L}$ and $\boldsymbol{\Psi}$.

9.8. (Unique but improper solution: Heywood case.)
Consider an $m = 1$ factor model for the population with covariance matrix

$$\boldsymbol{\Sigma} = \begin{bmatrix} 1 & .9 & .7 \\ .9 & 1 & .4 \\ .7 & .4 & 1 \end{bmatrix}$$

Show that there is a unique choice of $\mathbf{L}$ and $\boldsymbol{\Psi}$ with $\boldsymbol{\Sigma} = \mathbf{L}\mathbf{L}' + \boldsymbol{\Psi}$, but that $\psi_i < 0$, so the choice is not admissible.

9.9. In a study of liquor preference in France, Stoetzel [13] collected preference rankings of $p = 9$ liquor types from $n = 1442$ individuals. A factor analysis of the 9×9 sample correlation matrix of rank orderings gave the following estimated loadings.

Variable (X_i)	Estimated factor loadings F_1	F_2	F_3
Liquors	.64	.02	.16
Kirsch	.50	−.06	−.10
Mirabelle	.46	−.24	−.19
Rum	.17	.74	.97*
Marc	−.29	.66	−.39
Whiskey	−.29	−.08	.09
Calvados	−.49	.20	−.04
Cognac	−.52	−.03	.42
Armagnac	−.60	−.17	.14

*This figure is too high. It exceeds the maximum value of .64, as a result of an approximate method for obtaining the estimated factor loadings used by Stoetzel.

Given the results above, Stoetzel concluded: The major principle of liquor preference in France is the distinction between sweet and strong liquors. The second motivating element is price, which can be understood by remembering that liquor is both an expensive commodity and an item of conspicuous consumption. Except in the case of the two most popular and least expensive items (Rum and Marc), this second factor plays a much smaller role in producing preference judgments. The third factor concerns the sociological and primarily the regional variability of the judgments [13], p. 11.

(a) Given what you know about the various liquors involved, does Stoetzel's interpretation seem reasonable?

(b) Plot the loading pairs for the first two factors. Conduct a graphical orthogonal rotation of the factor axes. Generate approximate rotated loadings. Interpret the rotated loadings for the first two factors. Does your interpretation agree with Stoetzel's interpretation of these factors from the unrotated loadings? Explain.

9.10. The correlation matrix for chicken-bone measurements (see Example 9.14) is

$$\begin{bmatrix} 1.000 & & & & & \\ .505 & 1.000 & & & & \\ .569 & .422 & 1.000 & & & \\ .602 & .467 & .926 & 1.000 & & \\ .621 & .482 & .877 & .874 & 1.000 & \\ .603 & .450 & .878 & .894 & .937 & 1.000 \end{bmatrix}$$

The following estimated factor loadings were extracted by the maximum likelihood procedure.

Variable	Estimated factor loadings F_1	F_2	Varimax rotated estimated factor loadings F_1^*	F_2^*
1. Skull length	.602	.200	.484	.411
2. Skull breadth	.467	.154	.375	.319
3. Femur length	.926	.143	.603	.717
4. Tibia length	1.000	.000	.519	.855
5. Humerus length	.874	.476	.861	.499
6. Ulna length	.894	.327	.744	.594

Using the *unrotated* estimated factor loadings, obtain the maximum likelihood estimates of the following.

(a) The specific variances.

(b) The communalities.

(c) The proportion of variance explained by each factor.

(d) The residual matrix $\mathbf{R} - \hat{\mathbf{L}}_z\hat{\mathbf{L}}_z' - \hat{\mathbf{\Psi}}_z$.

9.11. Refer to Exercise 9.10. Compute the value of the varimax criterion using both unrotated and rotated estimated factor loadings. Comment on the results.

9.12. The *covariance* matrix for the logarithms of turtle measurements (see Example 8.4) is:

$$\mathbf{S} = 10^{-3}\begin{bmatrix} 11.555 & & \\ 8.367 & 6.697 & \\ 8.508 & 6.264 & 7.061 \end{bmatrix}$$

The following maximum likelihood estimates of the factor loadings for an $m = 1$ model were obtained.

Variable	Estimated factor loadings F_1
1. ln(length)	.107
2. ln(width)	.078
3. ln(height)	.080

Using the estimated factor loadings, obtain the maximum likelihood estimates of each.

(a) Specific variances.

(b) Communalities.

(c) Proportion of variance explained by the factor.

(d) The residual matrix $\mathbf{S}_n - \hat{\mathbf{L}}\hat{\mathbf{L}}' - \hat{\mathbf{\Psi}}$. (*Hint:* Convert $\mathbf{S}$ to $\mathbf{S}_n$.)

9.13. Refer to Exercise 9.12. Compute the test statistic in (9-39). Indicate why a test of H_0: $\mathbf{\Sigma} = \mathbf{LL}' + \mathbf{\Psi}$ (with $m = 1$) versus H_1 : $\mathbf{\Sigma}$ unrestricted cannot be carried out for this example [see (9-40)].

9.14. The maximum likelihood factor loading estimates are given in (9A-15) by

$$\hat{\mathbf{L}} = \hat{\mathbf{\Psi}}^{1/2}\hat{\mathbf{E}}\hat{\mathbf{\Delta}}^{1/2}$$

Verify, for this choice, that

$$\hat{\mathbf{L}}'\hat{\mathbf{\Psi}}^{-1}\hat{\mathbf{L}} = \hat{\mathbf{\Delta}}$$

where $\hat{\mathbf{\Delta}} = \hat{\mathbf{\Lambda}} - \mathbf{I}$ is a diagonal matrix.

9.15. Verify that factor scores constructed according to (9-50) have sample mean vector $\mathbf{0}$ and zero sample covariances.

The following exercises require the use of a computer.

9.16. A firm is attempting to evaluate the quality of its sales staff and is trying to find an examination, or series of tests, that may reveal the potential for good performance in sales. The firm has selected a random sample of 50 salespeople and has evaluated each on 3 measures of performance: growth of sales, profitability of sales, and new account sales. These measures have been converted to a scale, on which 100 indicates "average" performance. Each of the 50 individuals took each of 4 tests, which purported to measure creativity, mechanical reasoning, abstract reasoning, and mathematical ability, respectively. The $n = 50$ observations on $p = 7$ variables are listed in Table 9.12.

TABLE 9.12 SALESPEOPLE DATA

Salesperson	Index of: Sales growth (x_1)	Sales profitability (x_2)	New account sales (x_3)	Score on: Creativity test (x_4)	Mechanical reasoning test (x_5)	Abstract reasoning test (x_6)	Mathematics test (x_7)
1	93.0	96.0	97.8	09	12	09	20
2	88.8	91.8	96.8	07	10	10	15
3	95.0	100.3	99.0	08	12	09	26
4	101.3	103.8	106.8	13	14	12	29
5	102.0	107.8	103.0	10	15	12	32
6	95.8	97.5	99.3	10	14	11	21
7	95.5	99.5	99.0	09	12	09	25
8	110.8	122.0	115.3	18	20	15	51
9	102.8	108.3	103.8	10	17	13	31
10	106.8	120.5	102.0	14	18	11	39
11	103.3	109.8	104.0	12	17	12	32
12	99.5	111.8	100.3	10	18	08	31
13	103.5	112.5	107.0	16	17	11	34
14	99.5	105.5	102.3	08	10	11	34
15	100.0	107.0	102.8	13	10	08	34
16	81.5	93.5	95.0	07	09	05	16
17	101.3	105.3	102.8	11	12	11	32
18	103.3	110.8	103.5	11	14	11	35
19	95.3	104.3	103.0	05	14	13	30
20	99.5	105.3	106.3	17	17	11	27
21	88.5	95.3	95.8	10	12	07	15
22	99.3	115.0	104.3	05	11	11	42
23	87.5	92.5	95.8	09	09	07	16
24	105.3	114.0	105.3	12	15	12	37
25	107.0	121.0	109.0	16	19	12	39
26	93.3	102.0	97.8	10	15	07	23
27	106.8	118.0	107.3	14	16	12	39
28	106.8	120.0	104.8	10	16	11	49
29	92.3	90.8	99.8	08	10	13	17
30	106.3	121.0	104.5	09	17	11	44
31	106.0	119.5	110.5	18	15	10	43
32	88.3	92.8	96.8	13	11	08	10
33	96.0	103.3	100.5	07	15	11	27
34	94.3	94.5	99.0	10	12	11	19
35	106.5	121.5	110.5	18	17	10	42
36	106.5	115.5	107.0	08	13	14	47
37	92.0	99.5	103.5	18	16	08	18
38	102.0	99.8	103.3	13	12	14	28
39	108.3	122.3	108.5	15	19	12	41
40	106.8	119.0	106.8	14	20	12	37
41	102.5	109.3	103.8	09	17	13	32
42	92.5	102.5	99.3	13	15	06	23
43	102.8	113.8	106.8	17	20	10	32
44	83.3	87.3	96.3	01	05	09	15
45	94.8	101.8	99.8	07	16	11	24
46	103.5	112.0	110.8	18	13	12	37
47	89.5	96.0	97.3	07	15	11	14
48	84.3	89.8	94.3	08	08	08	09
49	104.3	109.5	106.5	14	12	12	36
50	106.0	118.5	105.0	12	16	11	39

(a) Assume an orthogonal factor model for the *standardized* variables $Z_i = (X_i - \mu_i)/\sqrt{\sigma_{ii}}$, $i = 1, 2, \ldots, 7$. Obtain either the principal component solution or the maximum likelihood solution for $m = 2$ and $m = 3$ common factors.

(b) Given your solution in (a), obtain the rotated loadings for $m = 2$ and $m = 3$. Compare the two sets of rotated loadings. Interpret the $m = 2$ and $m = 3$ factor solutions.

(c) List the estimated communalities, specific variances, and $\hat{\mathbf{L}}\hat{\mathbf{L}}' + \hat{\mathbf{\Psi}}$ for the $m = 2$ and $m = 3$ solutions. Compare the results. Which choice of m do you prefer at this point? Why?

(d) Conduct a test of $H_0: \mathbf{\Sigma} = \mathbf{LL}' + \mathbf{\Psi}$ versus $H_1: \mathbf{\Sigma} \neq \mathbf{LL}' + \mathbf{\Psi}$ for both $m = 2$ and $m = 3$ at the $\alpha = .01$ level. With these results and those in Parts b and c, which choice of m appears to be the best?

(e) Suppose a new salesperson, selected at random, obtains the test scores $\mathbf{x}' = [x_1, x_2, \ldots, x_7] = [110, 98, 105, 15, 18, 12, 35]$. Calculate the salesperson's factor score using the weighted least squares method and the regression method. (*Note*: The components of $\mathbf{x}$ must be standardized using the sample means and variances calculated from the original data.)

9.17. Using the air-pollution variables X_1, X_2, X_5, and X_6 given in Table 1.2, generate the sample *covariance* matrix and

(a) Obtain the principal component solution to a factor model with $m = 1$ and $m = 2$.

(b) Find the maximum likelihood estimates of $\mathbf{L}$ and $\mathbf{\Psi}$ for $m = 1$ and $m = 2$.

(c) Compare the factorization obtained by the principal component and maximum likelihood methods.

9.18. Perform a varimax rotation of both $m = 2$ solutions in Exercise 9.17. Interpret the results. Are the principal component and maximum likelihood solutions consistent with each other?

9.19. Refer to Exercise 9.17.

(a) Calculate the factor scores from the $m = 2$ maximum likelihood estimates by (i) weighted least squares in (9-50) and by (ii) the regression approach of (9-58).

(b) Find the factor scores from the principal component solution using (9-51).

(c) Compare the three sets of factor scores.

9.20. Repeat Exercise 9.17 starting from the sample *correlation* matrix. Interpret the factors for the $m = 1$ and $m = 2$ solutions. Does it make a difference if $\mathbf{R}$, rather than $\mathbf{S}$, is factored? Explain.

9.21. Perform a factor analysis of the census-tract data in Table 8.2. Start with $\mathbf{R}$ and obtain both the maximum likelihood and principal component solutions. Comment on your choice of m. Your analysis should include factor rotation and the computation of factor scores.

9.22. Perform a factor analysis of the "stiffness" measurements given in Table 4.3 and discussed in Example 4.13. Compute factor scores and check for outliers in the data. Use the sample covariance matrix $\mathbf{S}$.

9.23. Consider the mice-weight data in Example 8.6. Start with the sample *covariance* matrix (see Exercise 8.15 for $\sqrt{s_{ii}}$).

(a) Obtain the principal component solution to the factor model with $m = 1$ and $m = 2$.

(b) Find the maximum likelihood estimates of the loadings and specific variances for $m = 1$ and $m = 2$.

(c) Perform a varimax rotation of the solutions in Parts a and b.

9.24. Repeat Exercise 9.23 by factoring **R** instead of the sample covariance matrix **S**. Also, for the mouse with standardized weights $[.8, -.2, -.6, 1.5]$, obtain the factor scores using the maximum likelihood estimates of the loadings and Equation (9-58).

REFERENCES

[1] Anderson, T. W., *An Introduction to Multivariate Statistical Methods*, New York: John Wiley, 1958.

[2] Bartlett, M. S., "The Statistical Conception of Mental Factors," *British Journal of Psychology*, **28** (1937), 97–104.

[3] Bartlett, M. S., "A Note on Multiplying Factors for Various Chi-Squared Approximations," *Journal of the Royal Statistical Society (B)*, **16** (1954), 296–298.

[4] Dixon, W. J., ed., *BMDP Biomedical Computer Programs*, Berkeley, Calif.: University of California Press, 1979.

[5] Dunn, L. C., "The Effect of Inbreeding on the Bones of the Fowl," *Storrs Agricultural Experimental Station Bulletin*, **52** (1928), 1–112.

[6] Harmon, H. H., *Modern Factor Analysis*, Chicago, Ill.: The University of Chicago Press, 1967.

[7] Jöreskog, K. G., "Factor Analysis by Least Squares and Maximum Likelihood" in *Statistical Methods for Digital Computers*, K. Enslein, A. Ralston, and H. S. Wilf (eds.), New York: John Wiley, 1975.

[8] Kaiser, H. F., "The Varimax Criterion for Analytic Rotation in Factor Analysis," *Psychometrika*, **23** (1958), 187–200.

[9] Lawley, D. N., and A. E. Maxwell, *Factor Analysis as a Statistical Method* (2nd ed.), New York: American Elsevier Publishing Co., 1971.

[10] Linden, M., "A Factor Analytic Study of Olympic Decathlon Data," *Research Quarterly*, **48**, no. 3 (1977), 562–568.

[11] Maxwell, A. E., *Multivariate Analysis in Behavioral Research*, London: Chapman and Hall, 1977.

[12] Morrison, D. F., *Multivariate Statistical Methods* (2nd ed.), New York: McGraw-Hill, 1976.

[13] Stoetzel, J., "A Factor Analysis of Liquor Preference," *Journal of Advertising Research*, **1** (1960), 7–11.

[14] Wright, S., "The Interpretation of Multivariate Systems" in *Statistics and Mathematics in Biology*, O. Kempthorne and others (eds.), Ames, Iowa: Iowa State University Press, 1954, 11–33.

Part IV

Classification and Grouping Techniques

10

Discrimination and Classification

10.1 INTRODUCTION

Discriminant analysis and classification are multivariate techniques concerned with *separating* distinct sets of objects (or observations) and with *allocating* new objects (observations) to previously defined groups. Discriminant analysis is rather exploratory in nature. As a separatory procedure, it is often employed on a one-time basis in order to investigate observed differences when causal relationships are not well understood. Classification procedures are less exploratory in the sense that they lead to well-defined rules, which can be used for assigning new objects. Classification ordinarily requires more problem structure than discrimination.

Thus, the immediate goals of discrimination and classification, respectively, are as follows.

Goal 1. To describe either graphically (in three or fewer dimensions) or algebraically, the differential features of objects (observations) from several known collections (populations). We try to find "discriminants" whose numerical values are such that the collections are separated as much as possible.

Goal 2. To sort objects (observations) into two or more labeled classes. The emphasis is on deriving a rule that can be used to optimally assign a *new* object to the labeled classes.

We shall follow convention and use the term *discrimination* to refer to Goal 1. This terminology was introduced by R. A. Fisher [6] in the first modern treatment of separatory problems. A more descriptive term for this goal, however, is *separation*. We shall refer to the second goal as *classification*, or *allocation*.

A function that separates may sometimes serve as an allocator, and, conversely, an allocatory rule may suggest a discriminatory procedure. In practice, Goals 1 and 2 frequently overlap and the distinction between separation and allocation becomes blurred. In the next section, we treat discrimination and classification together, using methods originally proposed by Fisher. Later, we shall attempt to treat the goals separately, dealing first with classification and then with discrimination.

10.2 SEPARATION AND CLASSIFICATION FOR TWO POPULATIONS: FISHER'S METHOD

To fix ideas, we list below situations where one may be interested in (1) separating two classes of objects, or (2) assigning a new object to one of the two classes (or both). It is convenient to label the classes π_1 and π_2. The objects are ordinarily separated or classified on the basis of measurements on, for instance, p associated random variables $\mathbf{X}' = [X_1, X_2, \ldots, X_p]$. The observed values of $\mathbf{X}$ differ to some extent from one class to the other.[1] We can think of the totality of values from the first class as being the population of $\mathbf{x}$ values for π_1 and those from the second class as the population of $\mathbf{x}$ values for π_2. These two populations can then be described by probability density functions $f_1(\mathbf{x})$ and $f_2(\mathbf{x})$, and, consequently, we can talk of assigning observations to populations or objects to classes interchangeably.

You may recall that some of the examples of the following separation-classification situations were introduced in Chapter 1.

Populations π_1 and π_2	Measured variables $\mathbf{X}$
1. Solvent and distressed property-liability insurance companies.	Total assets, cost of stocks and bonds, market value of stocks and bonds, loss expenses, surplus, amount of premiums written.
2. Nonulcer dyspeptics (those with upset stomach problems) and controls ("normal").	Measures of anxiety, dependence, guilt, perfectionism.
3. Federalist papers written by James Madison and those written by Alexander Hamilton.	Frequencies of different words and length of sentences.
4. Two species of chickweed.	Sepal and petal length, petal cleft depth, bract length, scarious tip length, pollen diameter.
5. Purchasers of a new product and laggards (those "slow" to purchase).	Education, income, family size, amount of previous brand switching.

[1]If the values of $\mathbf{X}$ were not very different for objects in π_1 and π_2, there would be no problem; that is, the classes would be indistinguishable and new objects could be assigned to either class indiscriminately.

Populations π_1 and π_2	Measured variables **X**
6. Successful or unsuccessful (fail to graduate) college students.	Entrance examination scores, high-school grade-point average, number of high school activities.
7. Males and females.	Anthropological measurements like circumference and volume on ancient skulls.
8. Good and poor credit risks.	Income, age, number of credit cards, family size.

We see from 5, for example, that objects (consumers) are to be separated into two labeled classes ("purchasers" and "laggards") on the basis of observed values of presumably relevant variables (education, income, and so forth). In the terminology of *observation* and *population*, we want to identify an observation of the form $\mathbf{x}' = [x_1(\text{education}), x_2(\text{income}), x_3(\text{family size}), x_4(\text{amount of brand switching})]$ as population π_1: purchasers, or population π_2: laggards.

Fisher's idea was to transform the multivariate observations **x** to univariate observations y such that the y's derived from populations π_1 and π_2 were separated as much as possible. Fisher suggested taking linear combinations of **x** to create the y's because they are simple functions of **x** and are easily handled mathematically. If we let μ_{1Y} be the mean of the Y's obtained from **X**'s belonging to π_1 and μ_{2Y} be the mean of the Y's obtained from **X**'s belonging to π_2, then Fisher selected the linear combinations that maximized the (squared) distance between μ_{1Y} and μ_{2Y} relative to the variability of the Y's.

We begin by defining

$$\boldsymbol{\mu}_1 = E(\mathbf{X} \mid \pi_1) = \text{expected value of a multivariate observation from } \pi_1$$
$$\boldsymbol{\mu}_2 = E(\mathbf{X} \mid \pi_2) = \text{expected value of a multivariate observation from } \pi_2 \tag{10-1}$$

and supposing the covariance matrix

$$\boldsymbol{\Sigma} = E(\mathbf{X} - \boldsymbol{\mu}_i)(\mathbf{X} - \boldsymbol{\mu}_i)', \qquad i = 1, 2 \tag{10-2}$$

is the same for both populations.[2] We then consider the linear combination

$$\underset{(1\times 1)}{Y} = \underset{(1\times p)}{\boldsymbol{\ell}'}\ \underset{(p\times 1)}{\mathbf{X}} \tag{10-3}$$

Using results from Section 2.6, Y has a mean

$$\mu_{1Y} = E(Y \mid \pi_1) = E(\boldsymbol{\ell}'\mathbf{X} \mid \pi_1) = \boldsymbol{\ell}'\boldsymbol{\mu}_1$$

or (10-4)

$$\mu_{2Y} = E(Y \mid \pi_2) = E(\boldsymbol{\ell}'\mathbf{X} \mid \pi_2) = \boldsymbol{\ell}'\boldsymbol{\mu}_2$$

[2] The assumption of a common covariance matrix is somewhat critical. This assumption is often violated in practice.

depending on the underlying population, but its variance

$$\sigma_Y^2 = \text{Var}(\boldsymbol{\ell}'\mathbf{X}) = \boldsymbol{\ell}' \,\text{Cov}(\mathbf{X})\boldsymbol{\ell} = \boldsymbol{\ell}'\boldsymbol{\Sigma}\boldsymbol{\ell} \tag{10-5}$$

is the same for both populations.

The best linear combination is derived from the ratio

$$\frac{\begin{pmatrix}\text{Squared distance}\\ \text{between means of } Y\end{pmatrix}}{(\text{Variance of } Y)} = \frac{(\mu_{1Y} - \mu_{2Y})^2}{\sigma_Y^2} = \frac{(\boldsymbol{\ell}'\boldsymbol{\mu}_1 - \boldsymbol{\ell}'\boldsymbol{\mu}_2)^2}{\boldsymbol{\ell}\boldsymbol{\Sigma}\boldsymbol{\ell}}$$

$$= \frac{\boldsymbol{\ell}'(\boldsymbol{\mu}_1 - \boldsymbol{\mu}_2)(\boldsymbol{\mu}_1 - \boldsymbol{\mu}_2)'\boldsymbol{\ell}}{\boldsymbol{\ell}'\boldsymbol{\Sigma}\boldsymbol{\ell}} = \frac{(\boldsymbol{\ell}'\boldsymbol{\delta})^2}{\boldsymbol{\ell}'\boldsymbol{\Sigma}\boldsymbol{\ell}} \tag{10-6}$$

where $\boldsymbol{\delta} = (\boldsymbol{\mu}_1 - \boldsymbol{\mu}_2)$ is the difference in mean vectors. Note the $p \times p$ matrix $\boldsymbol{\delta}\boldsymbol{\delta}' = (\boldsymbol{\mu}_1 - \boldsymbol{\mu}_2)(\boldsymbol{\mu}_1 - \boldsymbol{\mu}_2)'$ contains the squares and crossproducts of the component differences between the means of populations π_1 and π_2. Fisher's linear combination coefficients $\boldsymbol{\ell}' = [\ell_1, \ell_2, \ldots, \ell_p]$ are those that maximize the ratio in (10-6).

Result 10.1. Let $\boldsymbol{\delta} = \boldsymbol{\mu}_1 - \boldsymbol{\mu}_2$ and $Y = \boldsymbol{\ell}'\mathbf{X}$, then

$$\frac{\begin{pmatrix}\text{Squared distance}\\ \text{between means of } Y\end{pmatrix}}{(\text{Variance of } Y)} = \frac{(\boldsymbol{\ell}'\boldsymbol{\delta})^2}{\boldsymbol{\ell}'\boldsymbol{\Sigma}\boldsymbol{\ell}}$$

is maximized by the choice

$$\boldsymbol{\ell} = c\boldsymbol{\Sigma}^{-1}\boldsymbol{\delta} = c\boldsymbol{\Sigma}^{-1}(\boldsymbol{\mu}_1 - \boldsymbol{\mu}_2)$$

for any $c \neq 0$. Choosing $c = 1$ produces the linear combination

$$Y = \boldsymbol{\ell}'\mathbf{X} = (\boldsymbol{\mu}_1 - \boldsymbol{\mu}_2)'\boldsymbol{\Sigma}^{-1}\mathbf{X} \tag{10-7}$$

which is known as *Fisher's linear discriminant function.*[3]

The maximum of the ratio is given by

$$\max_{\boldsymbol{\ell}} \frac{(\boldsymbol{\ell}'\boldsymbol{\delta})^2}{\boldsymbol{\ell}'\boldsymbol{\Sigma}\boldsymbol{\ell}} = \boldsymbol{\delta}'\boldsymbol{\Sigma}^{-1}\boldsymbol{\delta} \tag{10-8}$$

Proof. Since $\boldsymbol{\Sigma}$ is a positive definite matrix, the maximum of the ratio can be determined directly by applying (2-50). ■

The linear discriminant function converts the π_1 and π_2 multivariate populations into univariate populations such that the corresponding univariate population means are separated as much as possible relative to the population variance.

We can also employ (10-7) as a classification device. Let $y_0 = (\boldsymbol{\mu}_1 - \boldsymbol{\mu}_2)'\boldsymbol{\Sigma}^{-1}\mathbf{x}_0$ be the value of the discriminant function for a new observation $\mathbf{x}_0$ and let

$$m = \tfrac{1}{2}(\mu_{1Y} + \mu_{2Y}) = \tfrac{1}{2}(\boldsymbol{\ell}'\boldsymbol{\mu}_1 + \boldsymbol{\ell}'\boldsymbol{\mu}_2) = \tfrac{1}{2}(\boldsymbol{\mu}_1 - \boldsymbol{\mu}_2)'\boldsymbol{\Sigma}^{-1}(\boldsymbol{\mu}_1 + \boldsymbol{\mu}_2) \tag{10-9}$$

[3]Fisher actually worked with the sample analog of (10-7) that appears in Equation (10-16).

be the midpoint between the two univariate population means. It can be shown that

$$E(Y_0 \mid \pi_1) - m \geq 0$$

and (10-10)

$$E(Y_0 \mid \pi_2) - m < 0$$

That is, if $\mathbf{X}_0$ is from π_1, Y_0 is expected to be larger than the midpoint. If $\mathbf{X}_0$ is from π_2, Y_0 is expected to be smaller than the midpoint. Thus the classification rule is:

$$\begin{aligned} &\text{Allocate } \mathbf{x}_0 \text{ to } \pi_1 \text{ if } y_0 = (\boldsymbol{\mu}_1 - \boldsymbol{\mu}_2)'\boldsymbol{\Sigma}^{-1}\mathbf{x}_0 \geq m \\ &\text{Allocate } \mathbf{x}_0 \text{ to } \pi_2 \text{ if } y_0 = (\boldsymbol{\mu}_1 - \boldsymbol{\mu}_2)'\boldsymbol{\Sigma}^{-1}\mathbf{x}_0 < m \end{aligned} \qquad (10\text{-}11)$$

Alternatively, we can subtract m from y_0 and compare the result with zero. In this case, the rule becomes:

$$\begin{aligned} &\text{Allocate } \mathbf{x}_0 \text{ to } \pi_1 \text{ if } y_0 - m \geq 0 \\ &\text{Allocate } \mathbf{x}_0 \text{ to } \pi_2 \text{ if } y_0 - m < 0 \end{aligned} \qquad (10\text{-}12)$$

Of course, the population quantities $\boldsymbol{\mu}_1$, $\boldsymbol{\mu}_2$, and $\boldsymbol{\Sigma}$ are rarely known. Therefore the rules in (10-11) and (10-12) cannot be implemented unless ℓ and m can be estimated from observations that have already been correctly classified.

Suppose, then, that we have n_1 observations of the multivariate random variable $\mathbf{X}' = [X_1, X_2, \ldots, X_p]$ from π_1 and n_2 measurements of this quantity from π_2. The respective data matrices are

$$\begin{aligned} \underset{(p \times n_1)}{\mathbf{X}_1} &= [\mathbf{x}_{11}, \mathbf{x}_{12}, \ldots, \mathbf{x}_{1n_1}] \\ \underset{(p \times n_2)}{\mathbf{X}_2} &= [\mathbf{x}_{21}, \mathbf{x}_{22}, \ldots, \mathbf{x}_{2n_2}] \end{aligned} \qquad (10\text{-}13)$$

From these data matrices, the sample mean vectors and covariance matrices are determined by

$$\begin{aligned} \underset{(p \times 1)}{\bar{\mathbf{x}}_1} &= \frac{1}{n_1} \sum_{j=1}^{n_1} \mathbf{x}_{1j}; \qquad \underset{(p \times p)}{\mathbf{S}_1} = \frac{1}{n_1 - 1} \sum_{j=1}^{n_1} (\mathbf{x}_{1j} - \bar{\mathbf{x}}_1)(\mathbf{x}_{1j} - \bar{\mathbf{x}}_1)' \\ \underset{(p \times 1)}{\bar{\mathbf{x}}_2} &= \frac{1}{n_2} \sum_{j=1}^{n_2} \mathbf{x}_{2j}; \qquad \underset{(p \times p)}{\mathbf{S}_2} = \frac{1}{n_2 - 1} \sum_{j=1}^{n_2} (\mathbf{x}_{2j} - \bar{\mathbf{x}}_2)(\mathbf{x}_{2j} - \bar{\mathbf{x}}_2)' \end{aligned}$$

(10-14)

Since it is assumed that the parent populations have the same covariance matrix $\boldsymbol{\Sigma}$, the sample covariance matrices $\mathbf{S}_1$ and $\mathbf{S}_2$ are combined (pooled) to derive a single, unbiased estimate of $\boldsymbol{\Sigma}$ as in (6-21). In particular, the weighted average

$$\begin{aligned} \mathbf{S}_{\text{pooled}} &= \left[\frac{n_1 - 1}{(n_1 - 1) + (n_2 - 1)}\right]\mathbf{S}_1 + \left[\frac{n_2 - 1}{(n_1 - 1) + (n_2 - 1)}\right]\mathbf{S}_2 \\ &= \frac{(n_1 - 1)\mathbf{S}_1 + (n_2 - 1)\mathbf{S}_2}{(n_1 + n_2 - 2)} \end{aligned} \qquad (10\text{-}15)$$

is an unbiased estimate of $\mathbf{\Sigma}$ if the data matrices $\mathbf{X}_1$ and $\mathbf{X}_2$ contain *random* samples from the populations π_1 and π_2, respectively.

The sample quantities $\bar{\mathbf{x}}_1$, $\bar{\mathbf{x}}_2$, and $\mathbf{S}_{\text{pooled}}$ are substituted for $\boldsymbol{\mu}_1$, $\boldsymbol{\mu}_2$ and $\mathbf{\Sigma}$ in (10-7) to give the following.

Fisher's Sample Linear Discriminant Function[4]

$$y = \hat{\boldsymbol{\ell}}'\mathbf{x} = (\bar{\mathbf{x}}_1 - \bar{\mathbf{x}}_2)'\mathbf{S}_{\text{pooled}}^{-1}\mathbf{x} \qquad (10\text{-}16)$$

The midpoint, $\hat{m}$, between the two univariate sample means, $\bar{y}_1 = \hat{\boldsymbol{\ell}}'\bar{\mathbf{x}}_1$ and $\bar{y}_2 = \hat{\boldsymbol{\ell}}'\bar{\mathbf{x}}_2$ is given by

$$\hat{m} = \tfrac{1}{2}(\bar{y}_1 + \bar{y}_2) = \tfrac{1}{2}(\bar{\mathbf{x}}_1 - \bar{\mathbf{x}}_2)'\mathbf{S}_{\text{pooled}}^{-1}(\bar{\mathbf{x}}_1 + \bar{\mathbf{x}}_2) \qquad (10\text{-}17)$$

and the classification rule based on the samples becomes the following.

An Allocation Rule Based on Fisher's Discriminant Function

Allocate $\mathbf{x}_0$ to π_1 if

$$\begin{aligned} y_0 &= (\bar{\mathbf{x}}_1 - \bar{\mathbf{x}}_2)'\mathbf{S}_{\text{pooled}}^{-1}\mathbf{x}_0 \\ &\geq \hat{m} = \tfrac{1}{2}(\bar{\mathbf{x}}_1 - \bar{\mathbf{x}}_2)'\mathbf{S}_{\text{pooled}}^{-1}(\bar{\mathbf{x}}_1 + \bar{\mathbf{x}}_2) \end{aligned}$$

or

$$y_0 - \hat{m} \geq 0$$

Allocate $\mathbf{x}_0$ to π_2 if (10-18)

$$y_0 < \hat{m}$$

or

$$y_0 - \hat{m} < 0$$

Fisher's solution to the separation and classification problems is illustrated schematically for $p = 2$ in Figure 10.1.

The sample linear discriminant function in (10-16) has the following "optimal" property.

Result 10.2. The particular linear combination $y = \hat{\boldsymbol{\ell}}'\mathbf{x} = (\bar{\mathbf{x}}_1 - \bar{\mathbf{x}}_2)'\mathbf{S}_{\text{pooled}}^{-1}\mathbf{x}$ maximizes the ratio

$$\frac{\left[\begin{array}{c}\text{Squared distance}\\ \text{between sample means of } y\end{array}\right]}{[\text{Sample variance of } y]} = \frac{(\bar{y}_1 - \bar{y}_2)^2}{s_y^2}$$
$$= \frac{(\hat{\boldsymbol{\ell}}'\bar{\mathbf{x}}_1 - \hat{\boldsymbol{\ell}}'\bar{\mathbf{x}}_2)^2}{\hat{\boldsymbol{\ell}}'\mathbf{S}_{\text{pooled}}\hat{\boldsymbol{\ell}}} = \frac{(\hat{\boldsymbol{\ell}}'\mathbf{d})^2}{\hat{\boldsymbol{\ell}}'\mathbf{S}_{\text{pooled}}\hat{\boldsymbol{\ell}}} \qquad (10\text{-}19)$$

where $\mathbf{d} = (\bar{\mathbf{x}}_1 - \bar{\mathbf{x}}_2)$.

[4]We must have $(n_1 + n_2 - 2) > p$, or otherwise $\mathbf{S}_{\text{pooled}}$ is singular and the usual inverse, $\mathbf{S}_{\text{pooled}}^{-1}$, does not exist.

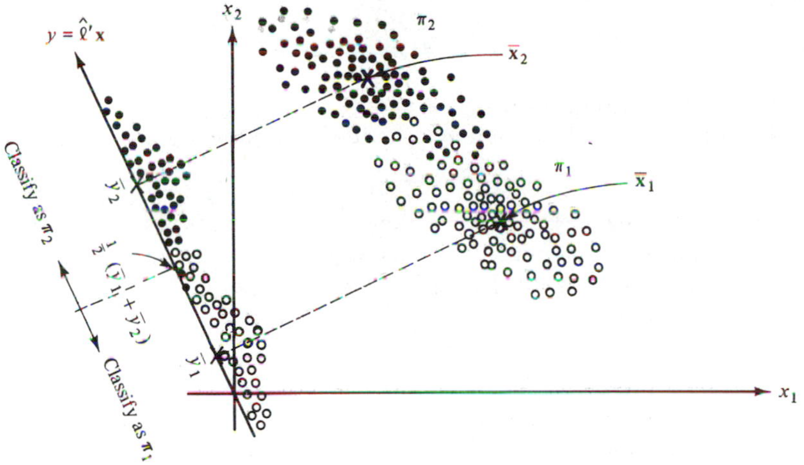

Figure 10.1 A pictorial representation of Fisher's procedure for two populations with $p = 2$.

Proof. This result follows directly from (2-50) as in Result 10.1. ■

The ratio in (10-19) is the sample analog of the ratio in (10-6). Also, s_y^2 may be calculated as

$$s_y^2 = \left(\frac{\sum_{j=1}^{n_1} (y_{1j} - \bar{y}_1)^2 + \sum_{j=1}^{n_2} (y_{2j} - \bar{y}_2)^2}{n_1 + n_2 - 2} \right) \tag{10-20}$$

with $y_{1j} = \hat{\ell}'\mathbf{x}_{1j}$ and $y_{2j} = \hat{\ell}'\mathbf{x}_{2j}$.

The maximum value of the population ratio in (10-6) is, from (10-8), $\boldsymbol{\delta}'\boldsymbol{\Sigma}^{-1}\boldsymbol{\delta} = (\boldsymbol{\mu}_1 - \boldsymbol{\mu}_2)'\boldsymbol{\Sigma}^{-1}(\boldsymbol{\mu}_1 - \boldsymbol{\mu}_2)$. This is the squared distance, Δ^2, between two populations. The maximum of the sample ratio in (10-19) is given by setting $\hat{\ell} = \mathbf{S}_{\text{pooled}}^{-1}(\bar{\mathbf{x}}_1 - \bar{\mathbf{x}}_2)$. Thus

$$\max_{\hat{\ell}} \frac{(\hat{\ell}'\mathbf{d})^2}{\hat{\ell}'\mathbf{S}_{\text{pooled}}\hat{\ell}} = \mathbf{d}'\mathbf{S}_{\text{pooled}}^{-1}\mathbf{d} = (\bar{\mathbf{x}}_1 - \bar{\mathbf{x}}_2)'\mathbf{S}_{\text{pooled}}^{-1}(\bar{\mathbf{x}}_1 - \bar{\mathbf{x}}_2) = D^2 \tag{10-21}$$

where D^2 is the sample squared distance (see Section 6.3 and Exercise 10.5).

For two populations the maximum relative separation that can be obtained by considering linear combinations of the multivariate observations is equal to the distance D. This is convenient because D^2 can be used, in certain situations, to test whether the population means $\boldsymbol{\mu}_1$ and $\boldsymbol{\mu}_2$ differ significantly. Consequently, a test for differences in mean vectors can be viewed as a test for the "significance" of the separation that can be achieved.

Suppose the populations π_1 and π_2 are multivariate normal *with a common covariance matrix* $\boldsymbol{\Sigma}$. Then, as in Section 6.3, a test of $H_0: \boldsymbol{\mu}_1 = \boldsymbol{\mu}_2$ versus $H_1: \boldsymbol{\mu}_1 \neq \boldsymbol{\mu}_2$ is accomplished by referring

$$\left(\frac{n_1 + n_2 - p - 1}{(n_1 + n_2 - 2)p}\right)\left(\frac{n_1 n_2}{n_1 + n_2}\right) D^2$$

to an F-distribution with $\nu_1 = p$ and $\nu_2 = n_1 + n_2 - p - 1$ d.f. If H_0 is rejected, we can conclude the separation between the two populations π_1 and π_2 is significant.

Comment. Significant separation does not necessarily imply good classification. As we shall see in later sections, the efficacy of a classification procedure can be evaluated independently of any test of separation. On the other hand, if the separation is not significant, the search for a useful classification rule will probably prove fruitless.

Example 10.1

This example is adapted from a study [3] concerned with the detection of hemophilia A carriers.

In order to construct a procedure for detecting potential hemophilia A carriers, blood samples were assayed for two groups of women and measurements on the two variables,

$$X_1 = \log_{10}(\text{AHF activity})$$
$$X_2 = \log_{10}(\text{AHF-like antigen})$$

recorded. The first group of $n_1 = 30$ women were selected from a population of women who did not carry the hemophilia gene. This group was called the *normal* group. The second group of $n_2 = 22$ women was selected from known hemophilia A carriers (daughters of hemophiliacs, mothers with more than one hemophilic son, and mothers with one hemophilic son and other hemophilic relatives.) This group was called the *obligatory carriers*. The pairs of observations (x_1, x_2) for the two groups are plotted in Figure 10.2. Also shown are estimated contours containing 50% and 95% of the probability for bivariate normal distributions centered at $\bar{\mathbf{x}}_1$ and $\bar{\mathbf{x}}_2$, respectively. Their common covariance matrix was taken as the pooled sample covariance matrix, $\mathbf{S}_{\text{pooled}}$. In this example, bivariate normal distributions seem to fit the data fairly well.

The investigators (see [3]) provide the information

$$\bar{\mathbf{x}}_1 = \begin{bmatrix} -.0065 \\ -.0390 \end{bmatrix}, \qquad \bar{\mathbf{x}}_2 = \begin{bmatrix} -.2483 \\ .0262 \end{bmatrix}$$

and

$$\mathbf{S}_{\text{pooled}}^{-1} = \begin{bmatrix} 131.158 & -90.423 \\ -90.423 & 108.147 \end{bmatrix}$$

Therefore Fisher's discriminant function is

$$\begin{aligned} y = \hat{\boldsymbol{\ell}}'\mathbf{x} &= [\bar{\mathbf{x}}_1 - \bar{\mathbf{x}}_2]'\mathbf{S}_{\text{pooled}}^{-1}\mathbf{x} \\ &= [.2418 \quad -.0652]\begin{bmatrix} 131.158 & -90.423 \\ -90.423 & 108.147 \end{bmatrix}\begin{bmatrix} x_1 \\ x_2 \end{bmatrix} \\ &= 37.61x_1 - 28.92x_2 \end{aligned}$$

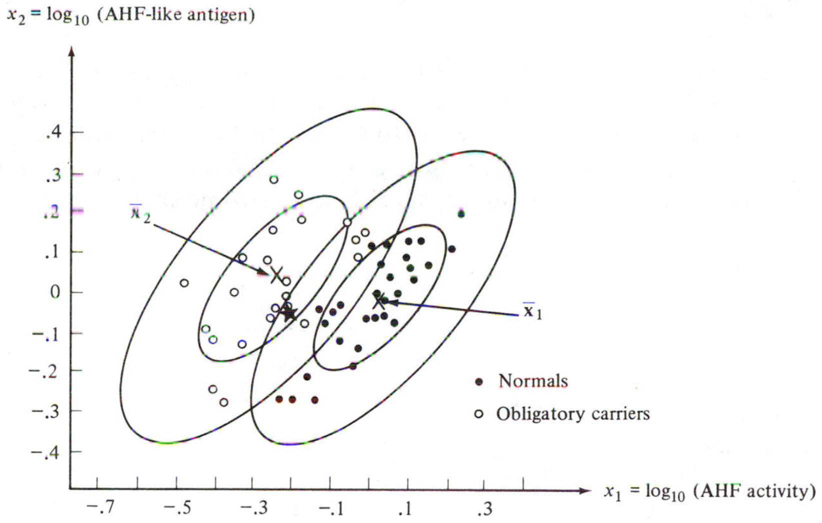

Figure 10.2 Scatterplots of [$\log_{10}$ (AHF activity), $\log_{10}$ (AHF-like antigen)] for the normal group and obligatory hemophilia A carriers.

Moreover,

$$\bar{y}_1 = \hat{\boldsymbol{\ell}}'\bar{\mathbf{x}}_1 = [37.61 \quad -28.92]\begin{bmatrix} -.0065 \\ -.0390 \end{bmatrix} = .88$$

$$\bar{y}_2 = \hat{\boldsymbol{\ell}}'\bar{\mathbf{x}}_2 = [37.61 \quad -28.92]\begin{bmatrix} -.2483 \\ .0262 \end{bmatrix} = -10.10$$

and the midpoint between these means is

$$\hat{m} = \tfrac{1}{2}(\bar{y}_1 + \bar{y}_2) = \tfrac{1}{2}(.88 - 10.10) = -4.61$$

Measurements of AHF activity and AHF-like antigen on a woman who may be a hemophilia A carrier give $x_1 = -.210$ and $x_2 = -.044$. Should this woman be classified as π_1: normal or π_2: obligatory carrier?

Using (10-18), we obtain:

$$\text{Allocate } \mathbf{x}_0 \text{ to } \pi_1 \text{ if } y_0 = \hat{\boldsymbol{\ell}}'\mathbf{x}_0 \geq \hat{m} = -4.61$$

$$\text{Allocate } \mathbf{x}_0 \text{ to } \pi_2 \text{ if } y_0 = \hat{\boldsymbol{\ell}}'\mathbf{x}_0 < \hat{m} = -4.61$$

where $\mathbf{x}_0' = [-.210, -.044]$. Since

$$y_0 = \hat{\boldsymbol{\ell}}'\mathbf{x}_0 = [37.61 \quad -28.92]\begin{bmatrix} -.210 \\ -.044 \end{bmatrix} = -6.62 < -4.61$$

we classify the woman as π_2: obligatory carrier. The new observation is indicated by a star in Figure 10.2. We see that it falls within the .50 probability contour of population π_2 and about on the .95 probability contour of population π_1. Thus the classification is not clear-cut. ■

Scaling

The coefficient vectors $\ell = \boldsymbol{\Sigma}^{-1}(\boldsymbol{\mu}_1 - \boldsymbol{\mu}_2)$ and $\hat{\ell} = \mathbf{S}_{\text{pooled}}^{-1}(\overline{\mathbf{x}}_1 - \overline{\mathbf{x}}_2)$ are not unique. They are only unique up to a multiplicative constant so, for $c \neq 0$, any vectors $c\ell$ and $c\hat{\ell}$ will also maximize the ratios in (10-6) and (10-19), respectively.

The vector $\hat{\ell}$ is frequently "scaled" or "normalized" to ease the interpretation of its elements. Two of the most commonly employed normalizations are:

1. Set

$$\hat{\ell}^* = \frac{\hat{\ell}}{\sqrt{\hat{\ell}'\hat{\ell}}} \qquad \text{(10-22)}$$

so $\hat{\ell}^*$ has unit length.

2. Set

$$\hat{\ell}^* = \frac{\hat{\ell}}{\hat{\ell}_1} \qquad \text{(10-23)}$$

so that the first element of the new coefficient vector $\hat{\ell}^*$ is 1.

In both cases, $\hat{\ell}^*$ is of the form $c\hat{\ell}$. For normalization (1), $c = (\hat{\ell}'\hat{\ell})^{-1/2}$ and for (2), $c = \hat{\ell}_1^{-1}$.

The magnitudes of $\hat{\ell}_1^*, \hat{\ell}_2^*, \ldots, \hat{\ell}_p^*$ in (10-22) all lie in the interval $[-1, 1]$. In (10-23), $\hat{\ell}_1^* = 1$ and $\hat{\ell}_2^*, \ldots, \hat{\ell}_p^*$ are expressed as multiples of $\hat{\ell}_1^*$. Constraining the $\hat{\ell}_i^*$ to the interval $[-1, 1]$ usually facilitates a visual comparison of the coefficients. Similarly, expressing the coefficients as multiples of $\hat{\ell}_1^*$ allows one to readily assess the relative importance (vis-à-vis X_1) of variables $X_2, \ldots, X_p$ as discriminators.

Normalizing the $\hat{\ell}_i$'s is recommended only if the X variables have been standardized. If this is not the case, a great deal of care must be exercised in interpreting the results.

Example 10.2

Returning to the discriminant functions derived in Example 10.1, suppose we normalize $\hat{\ell}$ using (10-22) and (10-23). The results are, respectively,

$$\hat{\ell}^* = \frac{\hat{\ell}}{\sqrt{\hat{\ell}'\hat{\ell}}} = \frac{1}{47.44}\begin{bmatrix} 37.61 \\ -28.92 \end{bmatrix} = \begin{bmatrix} .79 \\ -.61 \end{bmatrix}$$

and

$$\hat{\ell}^* = \frac{\hat{\ell}}{\hat{\ell}_1} = \frac{1}{37.61}\begin{bmatrix} 37.61 \\ -28.92 \end{bmatrix} = \begin{bmatrix} 1.00 \\ -.77 \end{bmatrix}$$

Variable X_1 receives more weight than X_2 in the discriminant function although, apart from sign, the difference is reasonably small. Both variables are measured in units per milliliter.

Corresponding to the two choices of the normalized vectors $\hat{\ell}^*$, the midpoints are

$$\hat{m}^* = \frac{1}{2}(\bar{y}_1^* + \bar{y}_2^*) = \frac{1}{2}(\hat{\ell}^{*\prime}\bar{\mathbf{x}}_1 + \hat{\ell}^{*\prime}\bar{\mathbf{x}}_2)$$

$$= \frac{1}{2}\left[[.79, -.61]\begin{bmatrix} -.0065 \\ -.0390 \end{bmatrix} + [.79, -.61]\begin{bmatrix} -.2483 \\ .0262 \end{bmatrix}\right] = -.10$$

$$\hat{m}^* = \frac{1}{2}(\bar{y}_1^* + \bar{y}_2^*) = \frac{1}{2}(\hat{\ell}^{*\prime}\bar{\mathbf{x}}_1 + \hat{\ell}^{*\prime}\bar{\mathbf{x}}_2)$$

$$= \frac{1}{2}\left[[1.00, -.77]\begin{bmatrix} -.0065 \\ -.0390 \end{bmatrix} + [1.00, -.77]\begin{bmatrix} -.2483 \\ .0262 \end{bmatrix}\right] = -.12$$

respectively. If we classify $\mathbf{x}_0' = [-.210, -.044]$ with the function $y_0^* = \hat{\ell}^{*\prime}\mathbf{x}_0$ we find, in the two cases,

$$y_0^* = \hat{\ell}^{*\prime}\mathbf{x}_0 = [.79, -.61]\begin{bmatrix} -.210 \\ -.044 \end{bmatrix} = -.14$$

and

$$y_0^* = \hat{\ell}^{*\prime}\mathbf{x}_0 = [1.00, -.77]\begin{bmatrix} -.210 \\ -.044 \end{bmatrix} = -.18$$

Since both y_0^*'s are less than their respective midpoints $\hat{m}^*$ we would classify $\mathbf{x}_0$ as π_2: obligatory carrier, using either normalized classification function. This is to be expected since the normalizations will not change the "optimality" property of the procedure. (If you are observant, you will note that the values of the y_0^*'s and $\hat{m}^*$'s for the new observation $\mathbf{x}_0$ could have been obtained from the original y_0 and $\hat{m}$ by multiplying these latter quantities by the normalization constants $c = 1/\sqrt{\hat{\ell}'\hat{\ell}} = 1/47.44$ and $c = 1/\hat{\ell}_1 = 1/37.61$, respectively.) ■

10.3. THE GENERAL CLASSIFICATION PROBLEM

At this point, we shall concentrate on classification, returning to separation in Section 10.8. Moreover, we shall concentrate on classification for two populations. Most of the ideas presented can immediately be generalized to more than two populations.

Allocation or classification rules are usually developed from "learning" samples. Measured characteristics of randomly selected objects *known* to come from each of the two populations are examined for differences. Essentially, the set of all possible sample outcomes is divided into two regions, R_1 and R_2, such that if a *new* observation falls in R_1, it is allocated to population π_1 and if it falls in R_2, we allocate it to population π_2. Thus one set of observed values favors π_1, the other set of values favors π_2.

You may wonder at this point how it is we *know* some observations belong to a particular population but we are unsure about others (This, of course, is what makes classification a problem!). There are several conditions that can give rise to this apparent anomaly (see [12]).

1. *Incomplete knowledge of future performance.*
   Examples: In the past, extreme values of certain financial variables were observed 2 years prior to subsequent bankruptcy. Classifying another firm as *sound* or *distressed* on the basis of observed values of these leading indicators may allow the officers to take corrective action, if necessary, before it is too late.
   A medical school applications office might want to classify an applicant as *likely to become M.D.* and *unlikely to become M.D.* on the basis of test scores and other college records. Here the actual determination can be made only at the end of several years of training.
2. *"Perfect" information requires destroying object.*
   Example: The life length of a calculator battery is determined by using it until it fails and the strength of a piece of lumber is obtained by loading it until it breaks. Failed products cannot be sold. One would like to classify products as *good* or *bad* (not meeting specifications) on the basis of certain preliminary measurements.
3. *Unavailable or expensive information.*
   Examples: It is assumed that certain of the Federalist papers were written by James Madison or Alexander Hamilton because they signed them. Other papers, however, were unsigned and it is of interest to determine which of the two men wrote the unsigned papers. Clearly, we cannot ask them. Word frequencies and sentence lengths may help classify the disputed papers.
   Many medical problems can be identified conclusively only by conducting an expensive operation. Usually, one would like to diagnose an illness from easily observed, yet potentially fallible, external symptoms. This approach helps avoid needless, and expensive, operations.

It should be clear from the examples presented above that classification rules cannot usually provide an error-free method of assignment. This is because there may not be a clear distinction between the measured characteristics of the populations; that is, the groups may overlap. It is then possible, for example, to incorrectly classify a π_2 object as belonging to π_1 or a π_1 object as belonging to π_2.

Example 10.3

Consider two groups in a city—π_1: riding-mower owners, and π_2: those without riding mowers; that is, nonowners. In order to identify the best sales prospects for an intensive sales campaign, a riding-mower manufacturer is interested in classifying families as prospective owners or nonowners on the basis of x_1 = income and x_2 = lot-size data. Random samples of $n_1 = 12$ current owners and $n_2 = 12$ current nonowners yield the values in Table 10.1.

These data are plotted in Figure 10.3. We see that riding-mower owners tend to have larger incomes and bigger lots than nonowners, although income seems to be a better "discriminator" than lot size. On the other hand, there is some overlap between the two groups. If, for example, we were to allocate

TABLE 10.1

π_1: Riding-mower owners		π_2: Nonowners	
x_1 (Income in \$1000s)	x_2 (Lot size in 1000 sq ft)	x_1 (Income in \$1000s)	x_2 (Lot size in 1000 sq ft)
20.0	9.2	25.0	9.8
28.5	8.4	17.6	10.4
21.6	10.8	21.6	8.6
20.5	10.4	14.4	10.2
29.0	11.8	28.0	8.8
36.7	9.6	16.4	8.8
36.0	8.8	19.8	8.0
27.6	11.2	22.0	9.2
23.0	10.0	15.8	8.2
31.0	10.4	11.0	9.4
17.0	11.0	17.0	7.0
27.0	10.0	21.0	7.4

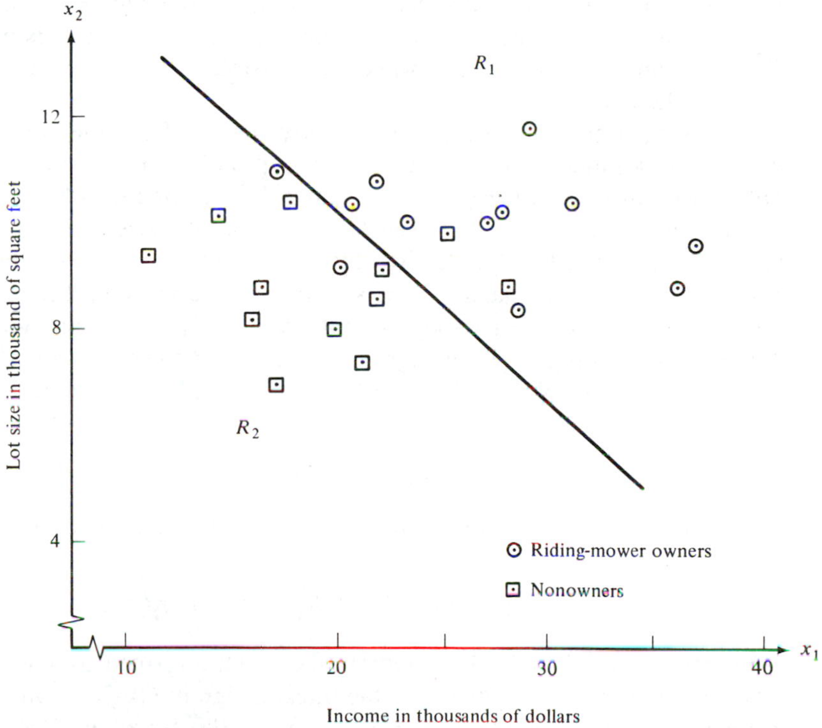

Figure 10.3 Income and lot size for riding-mower owners and nonowners.

those values of (x_1, x_2) that fall into region R_1 (as determined by the dashed line in the figure) to π_1: mower owners, and those (x_1, x_2) values which fall into R_2 to π_2: nonowners, we would make some mistakes. Some riding-mower owners will be incorrectly classified as nonowners and, conversely, some nonowners as owners. The idea is to create a rule (regions R_1 and R_2) which minimizes the chances of making these mistakes (see Exercise 10.2). ■

A good classification procedure should result in few misclassifications. In other words, the chances, or probabilities, of misclassification should be small. As we shall see, there are additional features that an "optimal" classification rule should possess.

It may be that one class or population has a greater likclihood of occurrence than another because one of the two populations is relatively much larger than the other. For example, there tend to be more financially sound firms than bankrupt firms. As another example, one species of chickweed may be more prevalent than another. An optimal classification rule should take these "prior probabilities of occurrence" into account. If we really believe that the (prior) probability of a financially distressed and ultimately bankrupted firm is very small, then one should classify a randomly selected firm as nonbankrupt unless the data overwhelmingly favors bankruptcy.

Another aspect of classification is cost. Suppose that classifying a π_1 object as belonging to π_2 represents a more serious error than classifying a π_2 object as belonging to π_1. Then one should be cautious about making the former assignment. As an example, failing to diagnose a potentially fatal illness is substantially more "costly" than concluding the disease is present when, in fact, it is not. An optimal classification procedure should, whenever possible, account for the costs associated with misclassification.

Let $f_1(\mathbf{x})$ and $f_2(\mathbf{x})$ be the probability density functions associated with the $p \times 1$ vector random variable $\mathbf{X}$ for the populations π_1 and π_2, respectively. An object, with associated measurements $\mathbf{x}$, *must* be assigned to either π_1 or π_2. Let Ω be the sample space; that is, the collection of all possible observations $\mathbf{x}$. Let R_1 be that set of $\mathbf{x}$ values for which we classify objects as π_1 and $R_2 = \Omega - R_1$ be the remaining $\mathbf{x}$ values for which we classify objects as π_2. Since every object must be assigned to one and only one of the two populations, the sets R_1 and R_2 are mutually exclusive and exhaustive. For $p = 2$, we might have a case like the one pictured in Figure 10.4.

The conditional probability, $P(2\,|\,1)$, of classifying an object as π_2 when, in fact, it is from π_1 is

$$P(2\,|\,1) = P(\mathbf{X} \in R_2\,|\,\pi_1) = \int_{R_2=\Omega-R_1} f_1(\mathbf{x})\,d\mathbf{x} \tag{10-24}$$

Similarly, the conditional probability, $P(1\,|\,2)$, of classifying an object as π_1 when it is really from π_2 is

$$P(1\,|\,2) = P(\mathbf{X} \in R_1\,|\,\pi_2) = \int_{R_1} f_2(\mathbf{x})\,d\mathbf{x} \tag{10-25}$$

The integral sign in (10-24) represents the volume formed by the density function $f_1(\mathbf{x})$ over the region R_2. Similarly, the integral sign in (10-25) represents the volume formed by $f_2(\mathbf{x})$ over the region R_1. This is illustrated in Figure 10.5 for the univariate case, $p = 1$.

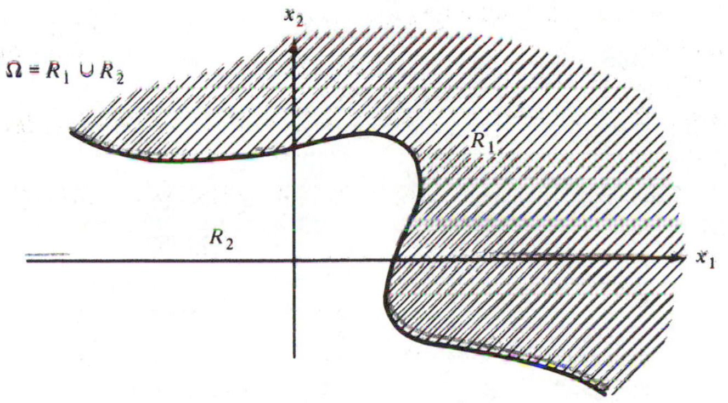

Figure 10.4 Classification regions for two populations.

Let p_1 be the *prior* probability of π_1 and p_2 be the *prior* probability of π_2, where $p_1 + p_2 = 1$. The overall probabilities of correctly or incorrectly classifying objects can be derived as the product of the prior and conditional classification probabilities:

$$
\begin{aligned}
P(\text{correctly classified as } \pi_1) &= P(\text{observation comes from } \pi_1 \text{ and is correctly} \\
&\qquad \text{classified as } \pi_1) \\
&= P(\mathbf{X} \in R_1 \mid \pi_1)P(\pi_1) = P(1 \mid 1)p_1 \\
P(\text{misclassified as } \pi_1) &= P(\text{observation comes from } \pi_2 \text{ and is} \\
&\qquad \text{misclassified as } \pi_1) \\
&= P(\mathbf{X} \in R_1 \mid \pi_2)P(\pi_2) = P(1 \mid 2)p_2 \qquad (10\text{-}26) \\
P(\text{correctly classified as } \pi_2) &= P(\text{observation comes from } \pi_2 \text{ and is correctly} \\
&\qquad \text{classified as } \pi_2) \\
&= P(\mathbf{X} \in R_2 \mid \pi_2)P(\pi_2) = P(2 \mid 2)p_2 \\
P(\text{misclassified as } \pi_2) &= P(\text{observation comes from } \pi_1 \text{ and is} \\
&\qquad \text{misclassified as } \pi_2) \\
&= P(\mathbf{X} \in R_2 \mid \pi_1)P(\pi_1) = P(2 \mid 1)p_1
\end{aligned}
$$

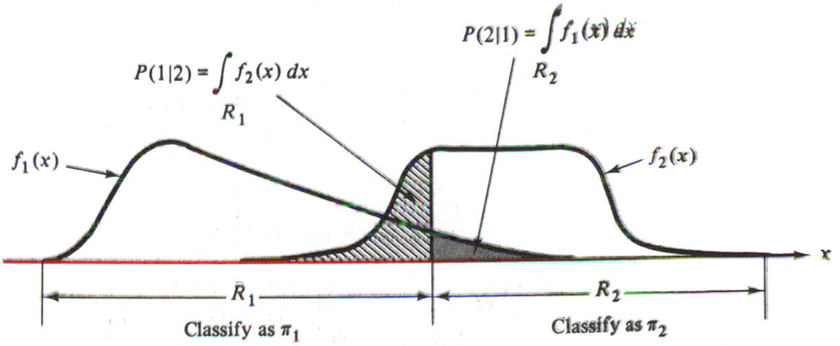

Figure 10.5 Misclassification probabilities for hypothetical classification regions when $p = 1$.

Classification schemes are often evaluated in terms of their misclassification probabilities (see Section 10.6), but this ignores misclassification cost. If, for example, the probability of misclassifying an observation as π_1 is small but the cost of making this incorrect assignment is high, a rule that ignores costs may cause problems.

The costs of misclassification can be defined by a cost matrix.

		Classify as:	
		π_1	π_2
True population:	π_1	0	$c(2\mid 1)$
	π_2	$c(1\mid 2)$	0

(10-27)

The costs are: (1) zero for correct classification, (2) $c(1\mid 2)$ when an observation from π_2 is incorrectly classified as π_1, and (3) $c(2\mid 1)$ when a π_1 observation is incorrectly classified as π_2.

For any rule, the average, or *expected cost of misclassification* (ECM) is provided by multiplying the off-diagonal entries in (10-27) by their probabilities of occurrence, obtained from (10-26). Consequently,

$$\text{ECM} = c(2\mid 1)P(2\mid 1)p_1 + c(1\mid 2)P(1\mid 2)p_2 \tag{10-28}$$

A reasonable classification rule should have an ECM as small, or nearly as small, as possible. We shall present rules of this kind in the next section.

10.4 OPTIMAL CLASSIFICATION RULES FOR TWO POPULATIONS

We have suggested that a sensible classification rule could be determined by minimizing the ECM. In other words, the assignment regions R_1 and R_2 must be chosen so that the ECM is as small as possible.

Result 10.3. The regions R_1 and R_2 that minimize the ECM are defined by the values $\mathbf{x}$ for which the following inequalities hold.

$$R_1: \quad \frac{f_1(\mathbf{x})}{f_2(\mathbf{x})} \geq \left[\frac{c(1\mid 2)}{c(2\mid 1)}\right]\left[\frac{p_2}{p_1}\right]$$

$$\begin{bmatrix}\text{Density}\\\text{ratio}\end{bmatrix} \geq \begin{bmatrix}\text{cost}\\\text{ratio}\end{bmatrix}\begin{bmatrix}\text{prior}\\\text{probability}\\\text{ratio}\end{bmatrix} \tag{10-29}$$

$$R_2: \quad \frac{f_1(\mathbf{x})}{f_2(\mathbf{x})} < \left[\frac{c(1\mid 2)}{c(2\mid 1)}\right]\left[\frac{p_2}{p_1}\right]$$

$$\begin{bmatrix}\text{Density}\\\text{ratio}\end{bmatrix} < \begin{bmatrix}\text{cost}\\\text{ratio}\end{bmatrix}\begin{bmatrix}\text{prior}\\\text{probability}\\\text{ratio}\end{bmatrix}$$

Proof. Substituting the integral expressions for $P(2 \mid 1)$ and $P(1 \mid 2)$ given by (10-24) and (10-25) into (10-28) gives

$$\text{ECM} = c(2 \mid 1)p_1 \int_{R_2} f_1(\mathbf{x})\, d\mathbf{x} + c(1 \mid 2)p_2 \int_{R_1} f_2(\mathbf{x})\, d\mathbf{x}.$$

Noting that $\Omega = R_1 \cup R_2$ so that the total probability

$$1 = \int_{\Omega} f_1(\mathbf{x})\, d\mathbf{x} = \int_{R_1} f_1(\mathbf{x})\, d\mathbf{x} + \int_{R_2} f_1(\mathbf{x})\, d\mathbf{x}$$

we can write

$$\text{ECM} = c(2 \mid 1)p_1 \left[1 - \int_{R_1} f_1(\mathbf{x})\, d\mathbf{x} \right] + c(1 \mid 2)p_2 \int_{R_1} f_2(\mathbf{x})\, d\mathbf{x}$$

By the additive property of integrals (volumes),

$$\text{ECM} = \int_{R_1} \left[c(1 \mid 2)p_2 f_2(\mathbf{x}) - c(2 \mid 1)p_1 f_1(\mathbf{x}) \right] d\mathbf{x} + c(2 \mid 1)p_1$$

Now p_1, p_2, $c(1 \mid 2)$, and $c(2 \mid 1)$ are nonnegative. In addition, $f_1(\mathbf{x})$ and $f_2(\mathbf{x})$ are nonnegative for all $\mathbf{x}$ and are the only quantities in ECM that depend on $\mathbf{x}$. Thus ECM is minimized if R_1 includes those values $\mathbf{x}$ for which the integrand

$$\left[c(1 \mid 2)p_2 f_2(\mathbf{x}) - c(2 \mid 1)p_1 f_1(\mathbf{x}) \right] \le 0$$

and excludes those $\mathbf{x}$ for which this quantity is positive. That is, R_1 must be the set of points $\mathbf{x}$ such that

$$c(1 \mid 2)p_2 f_2(\mathbf{x}) \le c(2 \mid 1)p_1 f_1(\mathbf{x})$$

or

$$\frac{f_1(\mathbf{x})}{f_2(\mathbf{x})} \ge \left[\frac{c(1 \mid 2)}{c(2 \mid 1)} \right] \left[\frac{p_2}{p_1} \right]$$

Since R_2 is the complement of R_1 in Ω, R_2 must be the set of points $\mathbf{x}$ for which

$$\frac{f_1(\mathbf{x})}{f_2(\mathbf{x})} < \left[\frac{c(1 \mid 2)}{c(2 \mid 1)} \right] \left[\frac{p_2}{p_1} \right] \qquad \blacksquare$$

It is clear from (10-29) that the implementation of the minimum ECM rule requires (1) the density function ratio evaluated at a new observation $\mathbf{x}_0$, (2) the cost ratio, and (3) the prior probability ratio. The appearance of ratios in the definition of the optimal classification regions is significant. Often it is much easier to specify the ratios than their component parts.

For example, it may be difficult to specify the costs (in appropriate units) of classifying a student as college material when, in fact, he or she is not and classifying a student as not-college material, when, in fact, he or she is. The cost to taxpayers of educating a college dropout for 2 years, for instance, can be roughly assessed. The cost to the university and society of not educating a capable student is more difficult to determine. However, it may be that a realistic number for the ratio of these misclassification costs can be obtained. Whatever the units of measurement, not admitting a prospective college graduate may be five times more costly, over a suitable time horizon, than admitting an eventual dropout. In this case, the cost ratio is five.

It is interesting to consider the classification regions defined in (10-29) for some special cases.

Special Cases of Minimum Expected Cost Regions

(a) $(p_2/p_1) = 1$ (equal prior probabilities)

$$R_1: \frac{f_1(\mathbf{x})}{f_2(\mathbf{x})} \geq \frac{c(1\mid 2)}{c(2\mid 1)}; \qquad R_2: \frac{f_1(\mathbf{x})}{f_2(\mathbf{x})} < \frac{c(1\mid 2)}{c(2\mid 1)}$$

(b) $[c(1\mid 2)/c(2\mid 1)] = 1$ (equal misclassification costs)

$$R_1: \frac{f_1(\mathbf{x})}{f_2(\mathbf{x})} \geq \frac{p_2}{p_1}; \qquad R_2: \frac{f_1(\mathbf{x})}{f_2(\mathbf{x})} < \frac{p_2}{p_1} \tag{10-30}$$

(c) $[p_2/p_1] = [c(1\mid 2)/c(2\mid 1)] = 1$ or $[p_2/p_1] = 1/[c(1\mid 2)/c(2\mid 1)]$
(equal prior probabilities and equal misclassification costs)

$$R_1: \frac{f_1(\mathbf{x})}{f_2(\mathbf{x})} \geq 1; \qquad R_2: \frac{f_1(\mathbf{x})}{f_2(\mathbf{x})} < 1$$

When the prior probabilities are unknown, they are often taken to be equal and the minimum ECM rule involves comparing the ratio of the population densities to the ratio of the appropriate misclassification costs. If the misclassification cost ratio is indeterminate, it is usually taken to be unity and the population density ratio is compared with the ratio of the prior probabilities. (Note the prior probabilities are in the reverse order of the densities.) Finally, when both the prior probability and misclassification cost ratios are unity or one ratio is the reciprocal of the other, the optimal classification regions are determined simply by comparing the values of the density functions. In this case, if $\mathbf{x}_0$ is a new observation and $f_1(\mathbf{x}_0)/f_2(\mathbf{x}_0) \geq 1$ [that is, $f_1(\mathbf{x}_0) \geq f_2(\mathbf{x}_0)$], we assign $\mathbf{x}_0$ to π_1. On the other hand, if $f_1(\mathbf{x}_0)/f_2(\mathbf{x}_0) < 1$ [$f_1(\mathbf{x}_0) < f_2(\mathbf{x}_0)$], we assign $\mathbf{x}_0$ to π_2.

It is common practice to arbitrarily use case (c) in (10-30) for classification. This is tantamount to assuming equal prior probabilities and equal misclassification costs for the minimum ECM rule.[5] As we shall see, allocation rules appropriate for the case involving equal prior probabilities and equal misclassification costs correspond to functions designed to maximally separate populations. It is in this situation that we begin to lose the distinction between classification and separation.

[5]This is the justification generally provided. It is also equivalent to assuming the prior probability ratio is the reciprocal of the misclassification cost ratio.

Example 10.4

A researcher has enough data available to estimate the density functions $f_1(\mathbf{x})$ and $f_2(\mathbf{x})$ associated with populations π_1 and π_2, respectively. Suppose $c(2\,|\,1) = 5$ units and $c(1\,|\,2) = 10$ units. In addition, it is known that about 20% of *all* objects (for which the measurements $\mathbf{x}$ can be recorded) belong to π_2. Thus the prior probabilities are $p_1 = .8$ and $p_2 = .2$.

Given the prior probabilities and costs of misclassification, we can use (10-29) to derive the classification regions R_1 and R_2. Specifically,

$$R_1: \quad \frac{f_1(\mathbf{x})}{f_2(\mathbf{x})} \geq \left[\frac{10}{5}\right]\left[\frac{.2}{.8}\right] = .5$$

$$R_2: \quad \frac{f_1(\mathbf{x})}{f_2(\mathbf{x})} < \left[\frac{10}{5}\right]\left[\frac{.2}{.8}\right] = .5$$

Suppose the density functions evaluated at a new observation $\mathbf{x}_0$ give $f_1(\mathbf{x}_0) = .3$ and $f_2(\mathbf{x}_0) = .4$. Do we classify the new observation as π_1 or π_2? To answer the question, we form the ratio

$$\frac{f_1(\mathbf{x}_0)}{f_2(\mathbf{x}_0)} = \frac{.3}{.4} = .75$$

and compare it with .5 obtained above. Since

$$\frac{f_1(\mathbf{x}_0)}{f_2(\mathbf{x}_0)} = .75 > \left[\frac{c(1\,|\,2)}{c(2\,|\,1)}\right]\left[\frac{p_2}{p_1}\right] = .5$$

we find that $\mathbf{x}_0 \in R_1$ and classify it as belonging to π_1. ■

Criteria other than the expected cost of misclassification can be used to derive "optimal" classification procedures. For example, one might ignore the costs of misclassification and choose R_1 and R_2 to minimize the *total probability of misclassification* (TPM),

$$\begin{aligned}
\text{TPM} &= P(\text{misclassifying a } \pi_1 \text{ observation } \textit{or} \text{ misclassifying a } \pi_2 \text{ observation}) \\
&= P(\text{observation comes from } \pi_1 \text{ and is misclassified}) \\
&\qquad + P(\text{observation comes from } \pi_2 \text{ and is misclassified}) \\
&= p_1 \int_{R_2} f_1(\mathbf{x})\, d\mathbf{x} + p_2 \int_{R_1} f_2(\mathbf{x})\, d\mathbf{x} \qquad (10\text{-}31)
\end{aligned}$$

Mathematically, this problem is equivalent to minimizing the expected cost of misclassification when the costs of misclassification are equal. Consequently, the optimal regions in this case are given by (b) in (10-30).

We could also allocate a new observation $\mathbf{x}_0$ to the population with the largest "posterior" probability $P(\pi_i \mid \mathbf{x}_0)$, where

$$P(\pi_1 \mid \mathbf{x}_0) = \frac{P(\pi_1 \text{ occurs and observe } \mathbf{x}_0)}{P(\text{observe } \mathbf{x}_0)}$$

$$= \frac{P(\text{observe } \mathbf{x}_0 \mid \pi_1)P(\pi_1)}{P(\text{observe } \mathbf{x}_0 \mid \pi_1)P(\pi_1) + P(\text{observe } \mathbf{x}_0 \mid \pi_2)P(\pi_2)}$$

$$= \frac{p_1 f_1(\mathbf{x}_0)}{p_1 f_1(\mathbf{x}_0) + p_2 f_2(\mathbf{x}_0)}$$

$$P(\pi_2 \mid \mathbf{x}_0) = 1 - P(\pi_1 \mid \mathbf{x}_0) = \frac{p_2 f_2(\mathbf{x}_0)}{p_1 f_1(\mathbf{x}_0) + p_2 f_2(\mathbf{x}_0)} \tag{10-32}$$

Classifying an observation $\mathbf{x}_0$ as π_1 when $P(\pi_1 \mid \mathbf{x}_0) > P(\pi_2 \mid \mathbf{x}_0)$ is equivalent to using the (b) rule for total probability of misclassification in (10-30) because the denominators in (10-32) are the same. However, computing the probabilities of the populations π_1 and π_2 after observing $\mathbf{x}_0$ (hence the name *posterior* probabilities) is frequently useful for purposes of identifying the less clear-cut assignments.

10.5. CLASSIFICATION WITH TWO MULTIVARIATE NORMAL POPULATIONS

We now assume $f_1(\mathbf{x})$ and $f_2(\mathbf{x})$ are multivariate normal densities; the first with mean vector $\boldsymbol{\mu}_1$ and covariance matrix $\boldsymbol{\Sigma}_1$ and the second with mean vector $\boldsymbol{\mu}_2$ and covariance matrix $\boldsymbol{\Sigma}_2$.

Let $\boldsymbol{\Sigma}_1 = \boldsymbol{\Sigma}_2 = \boldsymbol{\Sigma}$

Fisher's linear discriminant function can be used for classification in this case since it was developed under the assumption that the two populations, whatever their form, have a common covariance matrix. Consequently, it may not be surprising that Fisher's method corresponds to a particular case of the minimum expected cost of misclassification rule which we now develop.

Let the joint densities of $\mathbf{X}' = [X_1, X_2, \ldots, X_p]$ for populations π_1 and π_2 be given by

$$f_i(\mathbf{x}) = \frac{1}{(2\pi)^{p/2} |\boldsymbol{\Sigma}|^{1/2}} \exp\left[-\frac{1}{2}(\mathbf{x} - \boldsymbol{\mu}_i)'\boldsymbol{\Sigma}^{-1}(\mathbf{x} - \boldsymbol{\mu}_i)\right], \quad \text{for} \quad i = 1, 2 \tag{10-33}$$

Suppose the population parameters $\boldsymbol{\mu}_1$, $\boldsymbol{\mu}_2$, and $\boldsymbol{\Sigma}$ are known.

After cancellation of the terms $(2\pi)^{p/2}|\boldsymbol{\Sigma}|^{1/2}$ and a rearrangement of the exponents in the multivariate normal densities in (10-33), the minimum ECM regions in (10-29) become

$$R_1:\quad \exp\left[-\tfrac{1}{2}(\mathbf{x}-\boldsymbol{\mu}_1)'\boldsymbol{\Sigma}^{-1}(\mathbf{x}-\boldsymbol{\mu}_1)+\tfrac{1}{2}(\mathbf{x}-\boldsymbol{\mu}_2)'\boldsymbol{\Sigma}^{-1}(\mathbf{x}-\boldsymbol{\mu}_2)\right] \geq \left[\frac{c(1\,|\,2)}{c(2\,|\,1)}\right]\left[\frac{p_2}{p_1}\right]$$

$$R_2:\quad \exp\left[-\tfrac{1}{2}(\mathbf{x}-\boldsymbol{\mu}_1)'\boldsymbol{\Sigma}^{-1}(\mathbf{x}-\boldsymbol{\mu}_1)+\tfrac{1}{2}(\mathbf{x}-\boldsymbol{\mu}_2)'\boldsymbol{\Sigma}^{-1}(\mathbf{x}-\boldsymbol{\mu}_2)\right] < \left[\frac{c(1\,|\,2)}{c(2\,|\,1)}\right]\left[\frac{p_2}{p_1}\right] \tag{10-34}$$

Given the regions R_1 and R_2 above, we can construct the following classification rule.

Result 10.4. Let the populations π_1 and π_2 be described by multivariate normal densities of the form (10-33). The allocation rule that minimizes the ECM is given by:

Allocate $\mathbf{x}_0$ to π_1 if

$$(\boldsymbol{\mu}_1-\boldsymbol{\mu}_2)'\boldsymbol{\Sigma}^{-1}\mathbf{x}_0-\frac{1}{2}(\boldsymbol{\mu}_1-\boldsymbol{\mu}_2)'\boldsymbol{\Sigma}^{-1}(\boldsymbol{\mu}_1+\boldsymbol{\mu}_2) \geq \ln\left[\left(\frac{c(1\,|\,2)}{c(2\,|\,1)}\right)\left(\frac{p_2}{p_1}\right)\right] \tag{10-35}$$

Allocate $\mathbf{x}_0$ to π_2 otherwise.

Proof. Since the quantities in (10-34) are nonnegative for all $\mathbf{x}$, we can take their natural logarithms and preserve the order of the inequalities. Moreover, (see Exercise 10.6),

$$-\frac{1}{2}(\mathbf{x}-\boldsymbol{\mu}_1)'\boldsymbol{\Sigma}^{-1}(\mathbf{x}-\boldsymbol{\mu}_1)+\frac{1}{2}(\mathbf{x}-\boldsymbol{\mu}_2)'\boldsymbol{\Sigma}^{-1}(\mathbf{x}-\boldsymbol{\mu}_2) = (\boldsymbol{\mu}_1-\boldsymbol{\mu}_2)'\boldsymbol{\Sigma}^{-1}\mathbf{x}-\frac{1}{2}(\boldsymbol{\mu}_1-\boldsymbol{\mu}_2)'\boldsymbol{\Sigma}^{-1}(\boldsymbol{\mu}_1+\boldsymbol{\mu}_2) \tag{10-36}$$

and, consequently,

$$R_1:\quad (\boldsymbol{\mu}_1-\boldsymbol{\mu}_2)'\boldsymbol{\Sigma}^{-1}\mathbf{x}-\frac{1}{2}(\boldsymbol{\mu}_1-\boldsymbol{\mu}_2)'\boldsymbol{\Sigma}^{-1}(\boldsymbol{\mu}_1+\boldsymbol{\mu}_2) \geq \ln\left[\left(\frac{c(1\,|\,2)}{c(2\,|\,1)}\right)\left(\frac{p_2}{p_1}\right)\right]$$

$$R_2:\quad (\boldsymbol{\mu}_1-\boldsymbol{\mu}_2)'\boldsymbol{\Sigma}^{-1}\mathbf{x}-\frac{1}{2}(\boldsymbol{\mu}_1-\boldsymbol{\mu}_2)'\boldsymbol{\Sigma}^{-1}(\boldsymbol{\mu}_1+\boldsymbol{\mu}_2) < \ln\left[\left(\frac{c(1\,|\,2)}{c(2\,|\,1)}\right)\left(\frac{p_2}{p_1}\right)\right] \tag{10-37}$$

The minimum ECM classification rule follows. ■

Comparing the minimum ECM rule of (10-35) with the population analog of Fisher's method summarized by (10-9), (10-11), and (10-12), it is clear the two procedures are identical when

$$\left[\left(\frac{c(1|2)}{c(2|1)}\right)\left(\frac{p_2}{p_1}\right)\right] = 1$$

since $\ln(1) = 0$.

In most practical situations, the population quantities $\boldsymbol{\mu}_1$, $\boldsymbol{\mu}_2$, and $\boldsymbol{\Sigma}$ are unknown, so the rule (10-35) must be modified. Wald [18] and Anderson [1] have suggested replacing the population parameters by their sample counterparts. Substituting $\bar{\mathbf{x}}_1$ for $\boldsymbol{\mu}_1$, $\bar{\mathbf{x}}_2$ for $\boldsymbol{\mu}_2$, and $\mathbf{S}_{\text{pooled}}$ for $\boldsymbol{\Sigma}$ in (10-35) gives the "sample" classification rule.

The Estimated Minimum ECM Rule for Two Normal Populations

Allocate $\mathbf{x}_0$ to π_1 if

$$(\bar{\mathbf{x}}_1 - \bar{\mathbf{x}}_2)'\mathbf{S}_{\text{pooled}}^{-1}\mathbf{x}_0 - \frac{1}{2}(\bar{\mathbf{x}}_1 - \bar{\mathbf{x}}_2)'\mathbf{S}_{\text{pooled}}^{-1}(\bar{\mathbf{x}}_1 + \bar{\mathbf{x}}_2) \geq \ln\left[\left(\frac{c(1|2)}{c(2|1)}\right)\left(\frac{p_2}{p_1}\right)\right]$$

Allocate $\mathbf{x}_0$ to π_2 otherwise. (10-38)

The first term, $y = (\bar{\mathbf{x}}_1 - \bar{\mathbf{x}}_2)'\mathbf{S}_{\text{pooled}}^{-1}\mathbf{x}$ in (10-38) is the linear function obtained by Fisher that maximizes the univariate "between" samples variability relative to the "within" samples variability [see (10-19)]. The entire expression

$$\begin{aligned} w &= (\bar{\mathbf{x}}_1 - \bar{\mathbf{x}}_2)'\mathbf{S}_{\text{pooled}}^{-1}\mathbf{x} - \tfrac{1}{2}(\bar{\mathbf{x}}_1 - \bar{\mathbf{x}}_2)'\mathbf{S}_{\text{pooled}}^{-1}(\bar{\mathbf{x}}_1 + \bar{\mathbf{x}}_2) \\ &= (\bar{\mathbf{x}}_1 - \bar{\mathbf{x}}_2)'\mathbf{S}_{\text{pooled}}^{-1}\left[\mathbf{x} - \tfrac{1}{2}(\bar{\mathbf{x}}_1 + \bar{\mathbf{x}}_2)\right] \end{aligned} \qquad (10\text{-}39)$$

is often called *Anderson's classification function (statistic)*. Once again, if $[(c(1|2)/c(2|1))(p_2/p_1)] = 1$ so that $\ln[(c(1|2)/c(2|1))(p_2/p_1)] = 0$, Rule (10-38) is comparable to Rule (10-18) based on Fisher's linear discriminant function. Thus, provided the two normal populations have the same covariance matrix, Fisher's classification rule is equivalent to the minimum ECM rule with equal prior probabilities and equal costs of misclassification.

Once parameter estimates are inserted for the corresponding unknown population quantities, there is no assurance the resulting rule will minimize the expected cost of misclassification in a particular application. This is because the optimal rule in (10-35) was derived, assuming the multivariate normal densities $f_1(\mathbf{x})$ and $f_2(\mathbf{x})$ were known completely. Expression (10-38) is simply an estimate of the optimal rule. However, it seems reasonable to expect that it should perform well if the sample sizes are large.[6]

[6]As the sample size increase, $\bar{\mathbf{x}}_1$, $\bar{\mathbf{x}}_2$ and $\mathbf{S}_{\text{pooled}}$ become, with probability approaching 1, indistinguishable from $\boldsymbol{\mu}_1$, $\boldsymbol{\mu}_2$, and $\boldsymbol{\Sigma}$, respectively [see (4-26) and (4-27)].

To summarize, if the data appear to be multivariate normal,[7] the classification statistic w in (10-39) can be calculated for each new observation $\mathbf{x}_0$. These observations are classified by comparing the values of w with $\ln[(c(1|2)/c(2|1))(p_2/p_1)]$ as in (10-38).

Example 10.5

The scatterplots of the data for hemophilia A carriers and noncarriers in Figure 10.2 suggest that these data are multivariate normal. [Note that the original data were transformed by taking logarithms (see Example 10.1)]. Let us use the results obtained in Example 10.1 to classify a woman as π_1: normal, or π_2: obligatory carrier, when the prior probabilities of group membership are known. We shall assume, somewhat unrealistically, that the costs of misclassification are equal so that $c(1|2) = c(2|1)$.

Suppose blood is drawn from a maternal first cousin of a hemophiliac and the results of $x_1 = \log_{10}(\text{AHF activity})$ and $x_2 = \log_{10}(\text{AHF-like antigen})$ assays give $x_1 = -.210$ and $x_2 = -.044$. (These are the same data used to classify a woman in Example 10.1). However, the genetic chance of being a hemophilia A carrier for a maternal first cousin of a hemophiliac is .25. Thus we set $p_1 = .75$ and $p_2 = .25$. The classification statistic w is

$$w = (\bar{\mathbf{x}}_1 - \bar{\mathbf{x}}_2)'\mathbf{S}_{\text{pooled}}^{-1}\mathbf{x}_0 - \tfrac{1}{2}(\bar{\mathbf{x}}_1 - \bar{\mathbf{x}}_2)'\mathbf{S}_{\text{pooled}}^{-1}(\bar{\mathbf{x}}_1 + \bar{\mathbf{x}}_2)$$

or $w = \hat{\boldsymbol{\ell}}'\mathbf{x}_0 - \hat{m}$, where $y_0 = \hat{\boldsymbol{\ell}}'\mathbf{x}_0$ is Fisher's linear discriminant function and $\hat{m}$ is the midpoint between the sample means of the y's. From Example 10.1, with $\mathbf{x}_0' = [-.210, -.044]$, we found $\hat{m} = -4.61$ and $\hat{\boldsymbol{\ell}}'\mathbf{x}_0 = -6.62$. Consequently,

$$w = -6.62 - (-4.61) = -2.01$$

Applying (10-38) we see that

$$w = -2.01 < \ln\left[\frac{p_2}{p_1}\right] = \ln\left[\frac{.25}{.75}\right] = -1.10$$

and we classify the woman as π_2: obligatory carrier. ■

Let $\Sigma_1 \neq \Sigma_2$

As might be expected, the classification rules are more complicated when the population covariance matrices are unequal.

Consider the multivariate normal densities in (10-33) with $\boldsymbol{\Sigma}_i$, $i = 1, 2$, replacing $\boldsymbol{\Sigma}$. Thus the covariance matrices, as well as the mean vectors, are different from one another for the two populations. As we have seen, the minimum ECM and minimum total probability of misclassification (TPM) regions depend on the ratio of

[7] At the very least the marginal frequency distributions of the observations on each variable can be checked for normality. This must be done for the samples from both populations. Often some variables must be transformed in order to make them more "normal looking" (see Sections 4.6 and 4.7).

the densities, $f_1(\mathbf{x})/f_2(\mathbf{x})$, or, equivalently, the natural logarithms of the density ratio, $\ln[f_1(\mathbf{x})/f_2(\mathbf{x})] = \ln[f_1(\mathbf{x})] - \ln[f_2(\mathbf{x})]$. When the multivariate normal densities have different covariance structures, the terms in the density ratio involving $|\mathbf{\Sigma}_i|^{1/2}$ do not cancel as they do when $\mathbf{\Sigma}_1 = \mathbf{\Sigma}_2$. Moreover, the quadratic forms in the exponents of $f_1(\mathbf{x})$ and $f_2(\mathbf{x})$ do not combine to give the rather simple result in (10-36).

Substituting multivariate normal densities with different covariance matrices into (10-29) gives, after taking natural logarithms and simplifying (see Exercise 10.7), the classification regions

$$\begin{aligned} R_1:\quad & -\tfrac{1}{2}\mathbf{x}'(\mathbf{\Sigma}_1^{-1} - \mathbf{\Sigma}_2^{-1})\mathbf{x} + (\boldsymbol{\mu}_1'\mathbf{\Sigma}_1^{-1} - \boldsymbol{\mu}_2'\mathbf{\Sigma}_2^{-1})\mathbf{x} - k \\ & \qquad \geq \ln\left[\left(\frac{c(1|2)}{c(2|1)}\right)\left(\frac{p_2}{p_1}\right)\right] \\ R_2:\quad & -\tfrac{1}{2}\mathbf{x}'(\mathbf{\Sigma}_1^{-1} - \mathbf{\Sigma}_2^{-1})\mathbf{x} + (\boldsymbol{\mu}_1'\mathbf{\Sigma}_1^{-1} - \boldsymbol{\mu}_2'\mathbf{\Sigma}_2^{-1})\mathbf{x} - k \\ & \qquad < \ln\left[\left(\frac{c(1|2)}{c(2|1)}\right)\left(\frac{p_2}{p_1}\right)\right] \end{aligned} \tag{10-40}$$

where

$$k = \frac{1}{2}\ln\left(\frac{|\mathbf{\Sigma}_1|}{|\mathbf{\Sigma}_2|}\right) + \frac{1}{2}(\boldsymbol{\mu}_1'\mathbf{\Sigma}_1^{-1}\boldsymbol{\mu}_1 - \boldsymbol{\mu}_2'\mathbf{\Sigma}_2^{-1}\boldsymbol{\mu}_2) \tag{10-41}$$

The classification regions are defined by *quadratic* functions of $\mathbf{x}$. When $\mathbf{\Sigma}_1 = \mathbf{\Sigma}_2$, the quadratic term, $-\frac{1}{2}\mathbf{x}'(\mathbf{\Sigma}_1^{-1} - \mathbf{\Sigma}_2^{-1})\mathbf{x}$, disappears and the regions defined by (10-40) reduce to those defined by (10-37).

The classification rule for general multivariate normal populations follows directly from (10-40).

Result 10.5. Let the populations π_1 and π_2 be described by multivariate normal densities with mean vectors and covariance matrices $\boldsymbol{\mu}_1$, $\mathbf{\Sigma}_1$ and $\boldsymbol{\mu}_2$, $\mathbf{\Sigma}_2$, respectively. The allocation rule that minimizes the expected cost of misclassification is given by:

Allocate $\mathbf{x}_0$ to π_1 if

$$-\tfrac{1}{2}\mathbf{x}_0'(\mathbf{\Sigma}_1^{-1} - \mathbf{\Sigma}_2^{-1})\mathbf{x}_0 + (\boldsymbol{\mu}_1'\mathbf{\Sigma}_1^{-1} - \boldsymbol{\mu}_2'\mathbf{\Sigma}_2^{-1})\mathbf{x}_0 - k \geq \ln\left[\left(\frac{c(1|2)}{c(2|1)}\right)\left(\frac{p_2}{p_1}\right)\right]$$

Allocate $\mathbf{x}_0$ to π_2 otherwise.

Here k is set out in (10-41). ■

In practice, the classification rule in Result 10.5 is implemented by substituting the sample quantities $\bar{\mathbf{x}}_1$, $\bar{\mathbf{x}}_2$, $\mathbf{S}_1$, and $\mathbf{S}_2$ (see (10-14)) for $\boldsymbol{\mu}_1$, $\boldsymbol{\mu}_2$, $\mathbf{\Sigma}_1$, and $\mathbf{\Sigma}_2$, respectively.[8]

[8] The inequalities $n_1 > p$ and $n_2 > p$ must both hold for $\mathbf{S}_1^{-1}$ and $\mathbf{S}_2^{-1}$ to exist. These quantities are used in place of $\mathbf{\Sigma}_1^{-1}$ and $\mathbf{\Sigma}_2^{-1}$, respectively, in the sample analog (10-42).

Quadratic Classification Rule
(Normal populations with unequal covariance matrices)

Allocate $\mathbf{x}_0$ to π_1 if

$$-\tfrac{1}{2}\mathbf{x}_0'(\mathbf{S}_1^{-1} - \mathbf{S}_2^{-1})\mathbf{x}_0 + (\bar{\mathbf{x}}_1'\mathbf{S}_1^{-1} - \bar{\mathbf{x}}_2'\mathbf{S}_2^{-1})\mathbf{x}_0 - k \geq \ln\left[\left(\frac{c(1\,|\,2)}{c(2\,|\,1)}\right)\left(\frac{p_2}{p_1}\right)\right]$$

Allocate $\mathbf{x}_0$ to π_2 otherwise. (10-42)

Classification with quadratic functions is rather awkward in more than two dimensions and can lead to some strange results. This is particularly true when the data are not (essentially) multivariate normal.

If the data are not multivariate normal, two options are available. First, the nonnormal data can be transformed to data more nearly normal and a test for the equality of covariance matrices can be conducted to see if the linear rule (10-38) or the quadratic rule in (10-42) is appropriate. Transformations are discussed in Chapter 4. (The usual tests for covariance homogeneity are greatly affected by nonnormality. The conversion of nonnormal data to normal data must be done before this testing is carried out.)

Secondly, we can use a linear (or quadratic) rule without worrying about the form of the parent populations and hope that it will work reasonably well. Fisher's procedure, for example, did not depend on the form of the parent populations, apart from the requirement of identical covariance structures. Studies (see [13] and [14]) have shown, however, that there are nonnormal cases where Fisher's linear classification function performs poorly even though the population covariance matrices are the same. The moral is to always check the performance of any classification procedure. At the very least, this should be done with the data sets used to build the classifier. Ideally, there will be enough data available to provide for "training" samples and "validation" samples. The training samples can be used to develop the classification function and the validation samples can be used to evaluate its performance.

10.6 EVALUATING CLASSIFICATION FUNCTIONS

One important way of judging the performance of any classification procedure is to calculate its "error rates," or misclassification probabilities. When the forms of the parent populations are known completely, misclassification probabilities can be calculated with relative ease, as we show below in Example 10.6. Because parent populations are rarely known, we shall concentrate on the error rates associated with the sample classification function. Once this classification function is constructed, a measure of its performance in *future* samples is of interest.

From (10-31) the TPM is

$$\text{TPM} = p_1 \int_{R_2} f_1(\mathbf{x})\, d\mathbf{x} + p_2 \int_{R_1} f_2(\mathbf{x})\, d\mathbf{x}$$

The smallest value of this quantity, obtained by a judicious choice of R_1 and R_2, is called the optimum error rate (OER).

$$\text{Optimum error rate (OER)} = p_1 \int_{R_2} f_1(\mathbf{x})\, d\mathbf{x} + p_2 \int_{R_1} f_2(\mathbf{x})\, d\mathbf{x}$$

where R_1 and R_2 are determined by case (b) in (10-30). (10-43)

Thus the OER is the error rate for the minimum TPM classification rule.

Example 10.6

Let us derive an expression for the optimum error rate when $p_1 = p_2 = \frac{1}{2}$ and $f_1(\mathbf{x})$ and $f_2(\mathbf{x})$ are the multivariate normal densities in (10-33).

Now, the minimum ECM and minimum TPM classification rules coincide when $c(1\,|\,2) = c(2\,|\,1)$. Because the prior probabilities are also equal, the minimum TPM classification regions are defined for normal populations by (10-35), with $\ln\left[\left(\dfrac{c(1\,|\,2)}{c(2\,|\,1)}\right)\left(\dfrac{p_2}{p_1}\right)\right] = 0$. We find

$$R_1: \quad (\boldsymbol{\mu}_1 - \boldsymbol{\mu}_2)'\boldsymbol{\Sigma}^{-1}\mathbf{x} - \tfrac{1}{2}(\boldsymbol{\mu}_1 - \boldsymbol{\mu}_2)'\boldsymbol{\Sigma}^{-1}(\boldsymbol{\mu}_1 + \boldsymbol{\mu}_2) \geq 0$$

$$R_2: \quad (\boldsymbol{\mu}_1 - \boldsymbol{\mu}_2)'\boldsymbol{\Sigma}^{-1}\mathbf{x} - \tfrac{1}{2}(\boldsymbol{\mu}_1 - \boldsymbol{\mu}_2)'\boldsymbol{\Sigma}^{-1}(\boldsymbol{\mu}_1 + \boldsymbol{\mu}_2) < 0$$

These sets can be expressed in terms of $y = (\boldsymbol{\mu}_1 - \boldsymbol{\mu}_2)'\boldsymbol{\Sigma}^{-1}\mathbf{x} = \boldsymbol{\ell}'\mathbf{x}$ as

$$R_1(y): \quad y \geq \tfrac{1}{2}(\boldsymbol{\mu}_1 - \boldsymbol{\mu}_2)'\boldsymbol{\Sigma}^{-1}(\boldsymbol{\mu}_1 + \boldsymbol{\mu}_2)$$

$$R_2(y): \quad y < \tfrac{1}{2}(\boldsymbol{\mu}_1 - \boldsymbol{\mu}_2)'\boldsymbol{\Sigma}^{-1}(\boldsymbol{\mu}_1 + \boldsymbol{\mu}_2)$$

But Y is a linear combination of normal random variables, so the probability densities of Y, $f_1(y)$ and $f_2(y)$, are univariate normal (see Result 4.2) with means and a variance given by

$$\mu_{1Y} = \boldsymbol{\ell}'\boldsymbol{\mu}_1 = (\boldsymbol{\mu}_1 - \boldsymbol{\mu}_2)'\boldsymbol{\Sigma}^{-1}\boldsymbol{\mu}_1$$

$$\mu_{2Y} = \boldsymbol{\ell}'\boldsymbol{\mu}_2 = (\boldsymbol{\mu}_1 - \boldsymbol{\mu}_2)'\boldsymbol{\Sigma}^{-1}\boldsymbol{\mu}_2$$

$$\sigma_Y^2 = \boldsymbol{\ell}'\boldsymbol{\Sigma}\boldsymbol{\ell} = (\boldsymbol{\mu}_1 - \boldsymbol{\mu}_2)'\boldsymbol{\Sigma}^{-1}(\boldsymbol{\mu}_1 - \boldsymbol{\mu}_2) = \Delta^2$$

Now,

$$\begin{aligned}\text{TPM} &= \tfrac{1}{2}P[\text{misclassify a } \pi_1 \text{ observation as } \pi_2] \\ &\quad + \tfrac{1}{2}P[\text{misclassify a } \pi_2 \text{ observation as } \pi_1]\end{aligned}$$

But,

$$\begin{aligned}P[\text{misclassify a } \pi_1 \text{ observation as } \pi_2] &= P(2|1)\\ &= P\left[Y < \tfrac{1}{2}(\boldsymbol{\mu}_1 - \boldsymbol{\mu}_2)'\boldsymbol{\Sigma}^{-1}(\boldsymbol{\mu}_1 + \boldsymbol{\mu}_2)\right]\\ &= P\left(\frac{Y - \mu_{1Y}}{\sigma_Y} < \frac{\frac{1}{2}(\boldsymbol{\mu}_1 - \boldsymbol{\mu}_2)'\boldsymbol{\Sigma}^{-1}(\boldsymbol{\mu}_1 + \boldsymbol{\mu}_2) - (\boldsymbol{\mu}_1 - \boldsymbol{\mu}_2)'\boldsymbol{\Sigma}^{-1}\boldsymbol{\mu}_1}{\Delta}\right)\\ &= P\left(Z < \frac{-\frac{1}{2}\Delta^2}{\Delta}\right) = \Phi\left(\frac{-\Delta}{2}\right)\end{aligned}$$

where $\Phi(\cdot)$ is the cumulative distribution function of a standard normal random variable. Similarly,

$$\begin{aligned}&P[\text{misclassify a } \pi_2 \text{ observation as } \pi_1]\\ &\qquad = P(1\,|\,2) = P\left[Y \geq \tfrac{1}{2}(\boldsymbol{\mu}_1 - \boldsymbol{\mu}_2)'\boldsymbol{\Sigma}^{-1}(\boldsymbol{\mu}_1 + \boldsymbol{\mu}_2)\right]\\ &\qquad = P\left(Z \geq \frac{\Delta}{2}\right) = 1 - \Phi\left(\frac{\Delta}{2}\right) = \Phi\left(\frac{-\Delta}{2}\right)\end{aligned}$$

These misclassification probabilities are shown in Figure 10.6. Therefore the optimum error rate is

$$\text{OER} = \text{minimum TPM} = \frac{1}{2}\Phi\left(\frac{-\Delta}{2}\right) + \frac{1}{2}\Phi\left(\frac{-\Delta}{2}\right) = \Phi\left(\frac{-\Delta}{2}\right) \tag{10-44}$$

If, for example, $\Delta^2 = (\boldsymbol{\mu}_1 - \boldsymbol{\mu}_2)'\boldsymbol{\Sigma}^{-1}(\boldsymbol{\mu}_1 - \boldsymbol{\mu}_2) = 2.56$, then $\Delta = \sqrt{2.56} = 1.6$ and using Table 1 in Appendix 1,

$$\text{minimum TPM} = \Phi\left(\frac{-1.6}{2}\right) = \Phi(-.8) = .2119$$

The optimal classification rule here will incorrectly allocate, to one population or the other, about 21% of the items. ■

Example 10.6 illustrates how the optimum error rate can be calculated when the population density functions are known. If, as is usually the case, certain

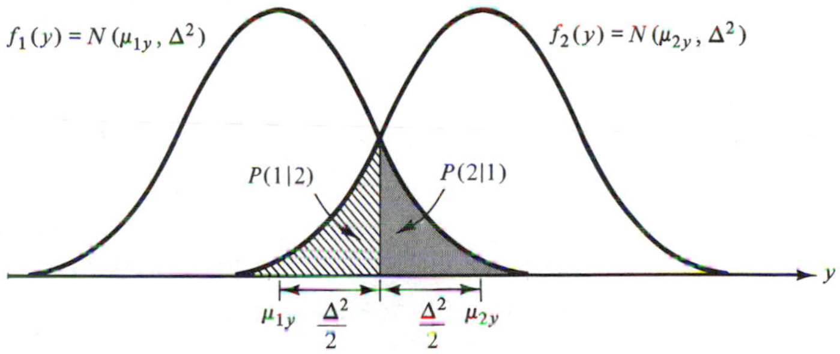

Figure 10.6 The misclassification probabilities based on Y.

$\Delta^2 = (\boldsymbol{\mu}_1 - \boldsymbol{\mu}_2)'\boldsymbol{\Sigma}^{-1}(\boldsymbol{\mu}_1 - \boldsymbol{\mu}_2)$

population parameters appearing in allocation rules must be estimated from the sample, then the evaluation of error rates is not straightforward.

The performance of *sample* classification functions can, in principle, be evaluated by calculating the actual error rate (AER),

$$\text{AER} = p_1 \int_{\hat{R}_2} f_1(\mathbf{x})\, d\mathbf{x} + p_2 \int_{\hat{R}_1} f_2(\mathbf{x})\, d\mathbf{x} \tag{10-45}$$

where $\hat{R}_1$ and $\hat{R}_2$ represent the classification regions determined by samples of size n_1 and n_2, respectively. For example, if the classification function in (10-39) is employed, the regions $\hat{R}_1$ and $\hat{R}_2$ are defined by the set of $\mathbf{x}$'s for which the following inequalities are satisfied.

$$\hat{R}_1: \quad (\bar{\mathbf{x}}_1 - \bar{\mathbf{x}}_2)'\mathbf{S}_{\text{pooled}}^{-1}\mathbf{x} - \frac{1}{2}(\bar{\mathbf{x}}_1 - \bar{\mathbf{x}}_2)'\mathbf{S}_{\text{pooled}}^{-1}(\bar{\mathbf{x}}_1 + \bar{\mathbf{x}}_2) \geq \ln\left[\left(\frac{c(1\,|\,2)}{c(2\,|\,1)}\right)\left(\frac{p_2}{p_1}\right)\right]$$

$$\hat{R}_2: \quad (\bar{\mathbf{x}}_1 - \bar{\mathbf{x}}_2)'\mathbf{S}_{\text{pooled}}^{-1}\mathbf{x} - \frac{1}{2}(\bar{\mathbf{x}}_1 - \bar{\mathbf{x}}_2)'\mathbf{S}_{\text{pooled}}^{-1}(\bar{\mathbf{x}}_1 + \bar{\mathbf{x}}_2) < \ln\left[\left(\frac{c(1\,|\,2)}{c(2\,|\,1)}\right)\left(\frac{p_2}{p_1}\right)\right]$$

The AER indicates how the sample classification function will perform in future samples. Like the optimal error rate, it cannot, in general, be calculated because it depends on the unknown density functions $f_1(\mathbf{x})$ and $f_2(\mathbf{x})$. However, an estimate of a quantity related to the actual error rate can be calculated, and this estimate will be discussed shortly.

There is a measure of performance that does not depend on the form of the parent populations and that can be calculated for *any* classification procedure. This measure, called the *apparent error rate* (APER), is defined as the fraction of observations in the *training* sample that are misclassified by the sample classification function.

The apparent error rate can be easily calculated from the *confusion matrix*, which shows actual versus predicted group membership. For n_1 observations from π_1 and n_2 observations from π_2, the confusion matrix has the form

		Predicted membership π_1	Predicted membership π_2	
Actual membership	π_1	n_{1C}	$n_{1M} = n_1 - n_{1C}$	n_1
	π_2	$n_{2M} = n_2 - n_{2C}$	n_{2C}	n_2

(10-46)

where

n_{1C} = number of π_1 items correctly classified as π_1 items
n_{1M} = number of π_1 items misclassified as π_2 items
n_{2C} = number of π_2 items correctly classified
n_{2M} = number of π_2 items misclassified

The apparent error rate is then

$$\text{APER} = \frac{n_{1M} + n_{2M}}{n_1 + n_2} \tag{10-47}$$

which is recognized as the *proportion* of items in the training set that are misclassified.

Example 10.7

Consider the classification regions R_1 and R_2 shown in Figure 10.3 for the riding-mower data. In this case, observations northeast of the dashed line are classified as π_1: mower owners; observations southwest of the dashed line are classified as π_2: nonowners. Notice that some observations are misclassified. The confusion matrix is

		Predicted membership		
		π_1: riding-mower owners	π_2: nonowners	
Actual membership	π_1: riding-mower owners	$n_{1C} = 10$	$n_{1M} = 2$	$n_1 = 12$
	π_2: nonowners	$n_{2M} = 2$	$n_{2C} = 10$	$n_2 = 12$

The apparent error rate, expressed as a percentage, is

$$\text{APER} = \left(\frac{2+2}{12+12}\right)100\% = \left(\frac{4}{24}\right)100\% = 16.7\%$$ ■

The APER is intuitively appealing and easy to calculate. Unfortunately, it tends to underestimate the AER, and this problem does not disappear unless the sample sizes n_1 and n_2 are very large. Essentially, this optimistic estimate occurs because the data used to build the classification function are also used to evaluate it.

Error-rate estimates can be constructed that are better than the apparent error rate, remain relatively easy to calculate, and do not require distributional assumptions. One procedure is to split the total sample into a training sample and a validation sample. The training sample is used to construct the classification function and the validation sample is used to evaluate it. The error rate is determined by the proportion misclassified in the validation sample. Although this method overcomes the bias problem by not using the same data to both build and judge the classification function, it suffers from two main defects.

(i) It requires large samples.

(ii) The function evaluated is not the function of interest. Ultimately, almost *all* of the data must be used to construct the classification function. If not, valuable information may be lost.

A second approach that seems to work well is called Lachenbruch's "holdout" procedure[9] (see also Lachenbruch and Mickey [15]).

1. Start with the π_1 group of observations. Omit one observation from this group and develop a classification function based on the remaining $n_1 - 1$, n_2 observations.
2. Classify the "holdout" observation using the function constructed in Step 1.
3. Repeat Steps 1 and 2 until all of the π_1 observations are classified. Let $n_{1M}^{(H)}$ be the number of holdout (H) observations misclassified in this group.
4. Repeat Steps 1 through 3 for the π_2 observations. Let $n_{2M}^{(H)}$ be the number of holdout observations misclassified in this group.

Estimates $\hat{P}(2 \mid 1)$ and $\hat{P}(1 \mid 2)$ of the conditional misclassification probabilities in (10-24) and (10-25) are then given by

$$\hat{P}(2 \mid 1) = \frac{n_{1M}^{(H)}}{n_1}$$

$$\hat{P}(1 \mid 2) = \frac{n_{2M}^{(H)}}{n_2} \tag{10-48}$$

and the total proportion misclassified, $(n_{1M}^{(H)} + n_{2M}^{(H)})/(n_1 + n_2)$ is, for moderate samples, a nearly unbiased estimate of the *expected* actual error rate, $E(\text{AER})$.

$$\hat{E}(\text{AER}) = \frac{n_{1M}^{(H)} + n_{2M}^{(H)}}{n_1 + n_2} \tag{10-49}$$

Lachenbruch's holdout method is computationally feasible when used in conjunction with the linear classification statistics in (10-16) and (10-39). It is offered as an option in some "canned" discriminant analysis computer programs.

Example 10.8

We shall illustrate Lachenbruch's holdout procedure and the calculation of error rate estimates for the equal prior and equal cost version of (10-38) or, equivalently, (10-18). Consider the data matrices and descriptive statistics given below. We shall assume the $n_1 = n_2 = 3$ bivariate observations were selected randomly from two populations π_1 and π_2 with a common covariance matrix.

$$\mathbf{X}_1 = \begin{bmatrix} 2 & 4 & 3 \\ 12 & 10 & 8 \end{bmatrix}; \quad \bar{\mathbf{x}}_1 = \begin{bmatrix} 3 \\ 10 \end{bmatrix}, \quad 2\mathbf{S}_1 = \begin{bmatrix} 2 & -2 \\ -2 & 8 \end{bmatrix}$$

$$\mathbf{X}_2 = \begin{bmatrix} 5 & 3 & 4 \\ 7 & 9 & 5 \end{bmatrix}; \quad \bar{\mathbf{x}}_2 = \begin{bmatrix} 4 \\ 7 \end{bmatrix}, \quad 2\mathbf{S}_2 = \begin{bmatrix} 2 & -2 \\ -2 & 8 \end{bmatrix}$$

[9] Lachenbruch's holdout procedure is sometimes referred to as *jackknifing*.

The pooled covariance matrix is

$$\mathbf{S}_{\text{pooled}} = \frac{1}{4}(2\mathbf{S}_1 + 2\mathbf{S}_2) = \begin{bmatrix} 1 & -1 \\ -1 & 4 \end{bmatrix}$$

Using $\mathbf{S}_{\text{pooled}}$, the rest of the data, and Rule (10-18), the sample observations may be classified. You may verify (see Exercise 10.17) that the confusion matrix is

		Classify as: π_1	π_2
True population:	π_1	2	1
	π_2	1	2

and consequently,

$$\text{APER (apparent error rate)} = \tfrac{2}{6} = .33$$

Holding out the first observation $\mathbf{x}'_H = [2, 12]$ from $\mathbf{X}_1$, we calculate

$$\mathbf{X}_{1H} = \begin{bmatrix} 4 & 3 \\ 10 & 8 \end{bmatrix}; \quad \bar{\mathbf{x}}_{1H} = \begin{bmatrix} 3.5 \\ 9 \end{bmatrix} \quad \text{and} \quad 1\mathbf{S}_{1H} = \begin{bmatrix} .5 & 1 \\ 1 & 2 \end{bmatrix}$$

The new pooled covariance matrix, $\mathbf{S}_{H,\text{pooled}}$, is

$$\mathbf{S}_{H,\text{pooled}} = \frac{1}{3}[1\mathbf{S}_{1H} + 2\mathbf{S}_2] = \frac{1}{3}\begin{bmatrix} 2.5 & -1 \\ -1 & 10 \end{bmatrix}$$

with inverse[10]

$$\mathbf{S}_{H,\text{pooled}}^{-1} = \frac{1}{8}\begin{bmatrix} 10 & 1 \\ 1 & 2.5 \end{bmatrix}$$

It is computationally quicker to classify the holdout observation, $\mathbf{x}_{1H}$, on the basis of its sample distances from the group means $\bar{\mathbf{x}}_{1H}$ and $\bar{\mathbf{x}}_2$. This procedure is equivalent to computing the value of the linear function $\hat{y} = \hat{\boldsymbol{\ell}}'_H\mathbf{x}_H = (\bar{\mathbf{x}}_{1H} - \bar{\mathbf{x}}_2)'\mathbf{S}_{H,\text{pooled}}^{-1}\mathbf{x}_H$ and comparing it to the midpoint $\hat{m}_H = \frac{1}{2}(\bar{\mathbf{x}}_{1H} - \bar{\mathbf{x}}_2)'\mathbf{S}_{H,\text{pooled}}^{-1}(\bar{\mathbf{x}}_{1H} + \bar{\mathbf{x}}_2)$ [see (10-18)].

Thus with $\mathbf{x}'_H = [2, 12]$, we have

$$\text{Squared distance from } \bar{\mathbf{x}}_{1H} = (\mathbf{x}_H - \bar{\mathbf{x}}_{1H})'\mathbf{S}_{H,\text{pooled}}^{-1}(\mathbf{x}_H - \bar{\mathbf{x}}_{1H})$$

$$= [2 - 3.5 \quad 12 - 9]\frac{1}{8}\begin{bmatrix} 10 & 1 \\ 1 & 2.5 \end{bmatrix}\begin{bmatrix} 2 - 3.5 \\ 12 - 9 \end{bmatrix} = 4.5$$

$$\text{Squared distance from } \bar{\mathbf{x}}_2 = (\mathbf{x}_H - \bar{\mathbf{x}}_2)'\mathbf{S}_{H,\text{pooled}}^{-1}(\mathbf{x}_H - \bar{\mathbf{x}}_2)$$

$$= [2 - 4 \quad 12 - 7]\frac{1}{8}\begin{bmatrix} 10 & 1 \\ 1 & 2.5 \end{bmatrix}\begin{bmatrix} 2 - 4 \\ 12 - 7 \end{bmatrix} = 10.3$$

Since the distance from $\mathbf{x}_H$ to $\bar{\mathbf{x}}_{1H}$ is smaller than the distance from $\mathbf{x}_H$ to $\bar{\mathbf{x}}_2$, we classify $\mathbf{x}_H$ as a π_1 observation. In this case the classification is correct.

[10]A matrix identity due to Bartlett [2] allows for the quick calculation of $\mathbf{S}_{H,\text{pooled}}^{-1}$ directly from $\mathbf{S}_{\text{pooled}}^{-1}$. Thus one does not have to recompute the inverse after withholding each observation.

If $\mathbf{x}'_H = [4, 10]$ is withheld, $\bar{\mathbf{x}}_{1H}$ and $\mathbf{S}^{-1}_{H,\text{pooled}}$ become

$$\bar{\mathbf{x}}_{1H} = \begin{bmatrix} 2.5 \\ 10 \end{bmatrix} \quad \text{and} \quad \mathbf{S}^{-1}_{H,\text{pooled}} = \frac{1}{8}\begin{bmatrix} 16 & 4 \\ 4 & 2.5 \end{bmatrix}$$

We find

$$(\mathbf{x}_H - \bar{\mathbf{x}}_{1H})'\mathbf{S}^{-1}_{H,\text{pooled}}(\mathbf{x}_H - \bar{\mathbf{x}}_{1H}) = [4 - 2.5 \quad 10 - 10]\frac{1}{8}\begin{bmatrix} 16 & 4 \\ 4 & 2.5 \end{bmatrix}\begin{bmatrix} 4 - 2.5 \\ 10 - 10 \end{bmatrix} = 4.5$$

$$(\mathbf{x}_H - \bar{\mathbf{x}}_2)'\mathbf{S}^{-1}_{H,\text{pooled}}(\mathbf{x}_H - \bar{\mathbf{x}}_2) = [4 - 4 \quad 10 - 7]\frac{1}{8}\begin{bmatrix} 16 & 4 \\ 4 & 2.5 \end{bmatrix}\begin{bmatrix} 4 - 4 \\ 10 - 7 \end{bmatrix} = 2.8$$

and consequently we would incorrectly assign $\mathbf{x}'_H = [4, 10]$ to π_2. Holding out $\mathbf{x}'_H = [3, 8]$ leads to incorrectly assigning this observation to π_2 as well. Thus $n^{(H)}_{1M} = 2$.

Turning to the second group, suppose $\mathbf{x}'_H = [5, 7]$ is withheld. Then

$$\mathbf{X}_{2H} = \begin{bmatrix} 3 & 4 \\ 9 & 5 \end{bmatrix}; \quad \bar{\mathbf{x}}_{2H} = \begin{bmatrix} 3.5 \\ 7 \end{bmatrix} \quad \text{and} \quad 1\mathbf{S}_{2H} = \begin{bmatrix} .5 & -2 \\ -2 & 8 \end{bmatrix}$$

The new pooled covariance matrix is

$$\mathbf{S}_{H,\text{pooled}} = \frac{1}{3}[2\mathbf{S}_1 + 1\mathbf{S}_{2H}] = \frac{1}{3}\begin{bmatrix} 2.5 & -4 \\ -4 & 16 \end{bmatrix}$$

with inverse

$$\mathbf{S}^{-1}_{H,\text{pooled}} = \frac{3}{24}\begin{bmatrix} 16 & 4 \\ 4 & 2.5 \end{bmatrix}$$

We find

$$(\mathbf{x}_H - \bar{\mathbf{x}}_1)'\mathbf{S}^{-1}_{H,\text{pooled}}(\mathbf{x}_H - \bar{\mathbf{x}}_1) = [5 - 3 \quad 7 - 10]\frac{3}{24}\begin{bmatrix} 16 & 4 \\ 4 & 2.5 \end{bmatrix}\begin{bmatrix} 5 - 3 \\ 7 - 10 \end{bmatrix} = 4.8$$

$$(\mathbf{x}_H - \bar{\mathbf{x}}_{2H})'\mathbf{S}^{-1}_{H,\text{pooled}}(\mathbf{x}_H - \bar{\mathbf{x}}_{2H}) = [5 - 3.5 \quad 7 - 7]\frac{3}{24}\begin{bmatrix} 16 & 4 \\ 4 & 2.5 \end{bmatrix}\begin{bmatrix} 5 - 3.5 \\ 7 - 7 \end{bmatrix} = 4.5$$

and $\mathbf{x}'_H = [5, 7]$ is correctly assigned to π_2.

When $\mathbf{x}'_H = [3, 9]$ is withheld,

$$(\mathbf{x}_H - \bar{\mathbf{x}}_1)'\mathbf{S}^{-1}_{H,\text{pooled}}(\mathbf{x}_H - \bar{\mathbf{x}}_1) = [3 - 3 \quad 9 - 10]\frac{3}{24}\begin{bmatrix} 10 & 1 \\ 1 & 2.5 \end{bmatrix}\begin{bmatrix} 3 - 3 \\ 9 - 10 \end{bmatrix} = .3$$

$$(\mathbf{x}_H - \bar{\mathbf{x}}_{2H})'\mathbf{S}^{-1}_{H,\text{pooled}}(\mathbf{x}_H - \bar{\mathbf{x}}_{2H}) = [3 - 4.5 \quad 9 - 6]\frac{3}{24}\begin{bmatrix} 10 & 1 \\ 1 & 2.5 \end{bmatrix}\begin{bmatrix} 3 - 4.5 \\ 9 - 6 \end{bmatrix} = 4.5$$

and $\mathbf{x}'_H = [3, 9]$ is incorrectly assigned to π_1. Finally, withholding $\mathbf{x}'_H = [4, 5]$ leads to correctly classifying this observation as π_2. Thus $n_{2M}^{(H)} = 1$.

An estimate of the expected actual error rate is provided by

$$\hat{E}(\text{AER}) = \frac{n_{1M}^{(H)} + n_{2M}^{(H)}}{n_1 + n_2} = \frac{2 + 1}{3 + 3} = .5$$

Thus we see that the apparent error rate, APER = .33, is an optimistic measure of performance. Of course, in practice sample sizes are larger than those we have considered here and the difference between APER and $\hat{E}(\text{AER})$ may not be as large. ■

If you are interested in pursuing the approaches to estimating classification error rates, see [14].

We point out that for two populations, observations can be classified by flipping a coin: If heads, allocate the observation to π_1 and if tails, allocate it to π_2. The expected error rate here is 50% and we should be suspicious of any classification scheme that has an estimated error rate near this figure.

Finally, it should be intuitively clear that good classification (low error rates) will depend upon the separation of the populations. The farther apart the groups, the more likely it is that a *useful* classification rule can be developed. This separatory goal, alluded to in Section 10.1, is explored further in Section 10.8.

10.7 CLASSIFICATION WITH SEVERAL POPULATIONS

In theory, the generalization of classification procedures from 2 to $g \geq 2$ groups is straightforward. However, not much is known about the properties of the corresponding *sample* classification functions and, in particular, their error rates have not been fully investigated.

The "robustness" of the *two* group linear classification statistics to, for instance, unequal covariances or nonnormal distributions can be studied with computer generated sampling experiments.[11] For more than two populations, this approach does not lead to general conclusions because the properties depend on where the populations are located and there are far too many configurations to study conveniently.

As before, our approach in this section will be to develop the theoretically optimal rules and then indicate the modifications required for real-world applications.

[11]Here *robustness* refers to the deterioration in error rates caused by using a classification procedure with data that do not conform to the assumptions on which the procedure was based.

It is very difficult to study the robustness of classification procedures analytically. However, data from a wide variety of distributions with different covariance structures can be easily generated on a computer. The performance of various classification rules can then be evaluated using computer generated "samples" from these distributions.

The Minimum Expected Cost of Misclassification Method

Let $f_i(\mathbf{x})$ be the density associated with population π_i, $i = 1, 2, \ldots, g$. [For the most part, we shall take $f_i(\mathbf{x})$ to be a multivariate normal density, but this is unnecessary for the development of the general theory.] Let

$$\begin{aligned} p_i &= \text{the prior probability of population } \pi_i, \qquad i = 1, 2, \ldots, g \\ c(k \mid i) &= \text{the cost of allocating an item to } \pi_k \text{ when, in fact,} \\ &\qquad \text{it belongs to } \pi_i, \qquad \text{for } k, i = 1, 2, \ldots, g \end{aligned}$$

For $k = i$, $c(i \mid i) = 0$. Finally, let R_k be the set of $\mathbf{x}$'s classified as π_k and

$$P(k \mid i) = P(\text{classify item as } \pi_k \mid \pi_i) = \int_{R_k} f_i(\mathbf{x})\, d\mathbf{x}$$

for $k, i = 1, 2, \ldots, g$ with $P(i \mid i) = 1 - \sum_{\substack{\ell=1 \\ \ell \neq i}}^{g} P(\ell \mid i)$.

The conditional expected cost of misclassifying an $\mathbf{x}$ from π_1 into π_2, or $\pi_3, \ldots,$ or π_g is

$$\begin{aligned} \text{ECM}(1) &= P(2 \mid 1)c(2 \mid 1) + P(3 \mid 1)c(3 \mid 1) + \cdots + P(g \mid 1)c(g \mid 1) \\ &= \sum_{\ell=2}^{g} P(\ell \mid 1)c(\ell \mid 1) \end{aligned}$$

This conditional expected cost occurs with prior probability p_1, the probability of π_1.

In a similar manner, we can obtain the conditional expected costs of misclassification, $\text{ECM}(2), \ldots, \text{ECM}(g)$. Multiplying each conditional ECM by its prior probability and summing gives the overall ECM.

$$\begin{aligned} \text{ECM} &= p_1 \text{ECM}(1) + p_2 \text{ECM}(2) + \cdots + p_g \text{ECM}(g) \\ &= p_1 \left(\sum_{\ell=2}^{g} P(\ell \mid 1)c(\ell \mid 1) \right) + p_2 \left(\sum_{\substack{\ell=1 \\ \ell \neq 2}}^{g} P(\ell \mid 2)c(\ell \mid 2) \right) \\ &\qquad + \cdots + p_g \left(\sum_{\ell=1}^{g-1} P(\ell \mid g)c(\ell \mid g) \right) \\ &= \sum_{i=1}^{g} p_i \left(\sum_{\substack{\ell=1 \\ \ell \neq i}}^{g} P(\ell \mid i)c(\ell \mid i) \right) \end{aligned} \tag{10-50}$$

Determining an optimal classification procedure amounts to choosing the mutually exclusive and exhaustive classification regions $R_1, R_2, \ldots, R_g$ such that (10-50) is a minimum.

Result 10.6. The classification regions that minimize the ECM (10-50) are defined by allocating $\mathbf{x}$ to that population π_k, $k = 1, 2, \ldots, g$ for which

$$\sum_{\substack{i=1 \\ i \neq k}}^{g} p_i f_i(\mathbf{x}) c(k \mid i) \tag{10-51}$$

is smallest. If a tie occurs, $\mathbf{x}$ can be assigned to any of the tied populations.

Proof. See Anderson [1]. ■

Suppose all the misclassification costs are equal. (Without loss of generality we can set them equal to 1.) Using the argument leading to (10-51), we would allocate $\mathbf{x}$ to that population π_k, $k = 1, 2, \ldots, g$, for which

$$\sum_{\substack{i=1 \\ i \neq k}}^{g} p_i f_i(\mathbf{x}) \tag{10-52}$$

is smallest. Now (10-52) will be smallest when the omitted term, $p_k f_k(\mathbf{x})$, is *largest*. Consequently, when the misclassification costs are the same, the minimum expected cost of misclassification rule has the following rather simple form

Minimum ECM Classification Rule with Equal Misclassification Costs

Allocate $\mathbf{x}$ to π_k if

$$p_k f_k(\mathbf{x}) > p_i f_i(\mathbf{x}) \quad \text{for all } i \neq k \tag{10-53}$$

or, equivalently,

Allocate $\mathbf{x}$ to π_k if

$$\ln p_k f_k(\mathbf{x}) > \ln p_i f_i(\mathbf{x}) \quad \text{for all } i \neq k \tag{10-54}$$

It is interesting to note that the classification rule in (10-53) is identical to the one that maximizes the "posterior" probability, $P(\pi_k \mid \mathbf{x}) = P(\mathbf{x}$ comes from π_k given that $\mathbf{x}$ was observed), where

$$P(\pi_k \mid \mathbf{x}) = \frac{p_k f_k(\mathbf{x})}{\sum_{\ell=1}^{g} p_\ell f_\ell(\mathbf{x})} = \frac{(\text{prior}) \times (\text{likelihood})}{\Sigma[(\text{prior}) \times (\text{likelihood})]} \quad \text{for } k = 1, 2, \ldots, g \tag{10-55}$$

Equation (10-55) is the generalization of Equation (10-32) to $g \geq 2$ groups.

You should keep in mind that, in general, the minimum ECM rules have three components: prior probabilities, misclassification costs, and density functions. These components must be specified (or estimated) before the rules can be implemented.

Example 10.9

Let us assign an observation $\mathbf{x}_0$ to one of the $g = 3$ populations π_1, π_2, or π_3, given the hypothetical prior probabilities, misclassification costs, and density values below. We shall use the minimum ECM procedures.

		True population		
		π_1	π_2	π_3
Classify as:	π_1	$c(1\mid 1) = 0$	$c(1\mid 2) = 500$	$c(1\mid 3) = 100$
	π_2	$c(2\mid 1) = 10$	$c(2\mid 2) = 0$	$c(2\mid 3) = 50$
	π_3	$c(3\mid 1) = 50$	$c(3\mid 2) = 200$	$c(3\mid 3) = 0$
Prior probabilities:		$p_1 = .05$	$p_2 = .60$	$p_3 = .35$
Densities at $\mathbf{x}_0$:		$f_1(\mathbf{x}_0) = .01$	$f_2(\mathbf{x}_0) = .85$	$f_3(\mathbf{x}_0) = 2$

The values of $\sum_{\substack{i=1 \\ i \neq k}}^{3} p_i f_i(\mathbf{x}_0) c(k \mid i)$ [see (10-51)] are

$$\begin{aligned}
k = 1:&\quad p_2 f_2(\mathbf{x}_0) c(1 \mid 2) + p_3 f_3(\mathbf{x}_0) c(1 \mid 3) \\
&\quad = (.60)(.85)(500) + (.35)(2)(100) = 325 \\
k = 2:&\quad p_1 f_1(\mathbf{x}_0) c(2 \mid 1) + p_3 f_3(\mathbf{x}_0) c(2 \mid 3) \\
&\quad = (.05)(.01)(10) + (.35)(2)(50) = 35.055 \\
k = 3:&\quad p_1 f_1(\mathbf{x}_0) c(3 \mid 1) + p_2 f_2(\mathbf{x}_0) c(3 \mid 2) \\
&\quad = (.05)(.01)(50) + (.60)(.85)(200) = 102.025
\end{aligned}$$

Since $\sum_{\substack{i=1 \\ i \neq k}}^{3} p_i f_i(\mathbf{x}_0) c(k \mid i)$ is smallest for $k = 2$, we would allocate $\mathbf{x}_0$ to π_2.

If all of the costs of misclassification were equal, we would assign $\mathbf{x}_0$ according to (10-53), which requires only the products

$$\begin{aligned}
p_1 f_1(\mathbf{x}_0) &= (.05)(.01) = .0005 \\
p_2 f_2(\mathbf{x}_0) &= (.60)(.85) = .510 \\
p_3 f_3(\mathbf{x}_0) &= (.35)(2) = .700
\end{aligned}$$

Since

$$p_3 f_3(\mathbf{x}_0) = .700 \geq p_i f_i(\mathbf{x}_0) \qquad i = 1, 2$$

we should allocate $\mathbf{x}_0$ to π_3. Equivalently, calculating the posterior probabilities, [see (10-55)]

$$\begin{aligned}
P(\pi_1 \mid \mathbf{x}_0) &= \frac{p_1 f_1(\mathbf{x}_0)}{\sum_{\ell=1}^{3} p_\ell f_\ell(\mathbf{x}_0)} \\
&= \frac{(.05)(.01)}{(.05)(.01) + (.60)(.85) + (.35)(2)} = \frac{.0005}{1.2105} = .0004
\end{aligned}$$

$$P(\pi_2 \mid \mathbf{x}_0) = \frac{p_2 f_2(\mathbf{x}_0)}{\sum_{\ell=1}^{3} p_\ell f_\ell(\mathbf{x}_0)} = \frac{(.60)(.85)}{1.2105} = \frac{.510}{1.2105} = .421$$

$$P(\pi_3 \mid \mathbf{x}_0) = \frac{p_3 f_3(\mathbf{x}_0)}{\sum_{\ell=1}^{3} p_\ell f_\ell(\mathbf{x}_0)} = \frac{(.35)(2)}{1.2105} = \frac{.700}{1.2105} = .578$$

we see that $\mathbf{x}_0$ is allocated to π_3, the population with the largest posterior probability. ■

Classification with Normal Populations

An important special case occurs when the

$$f_i(\mathbf{x}) = \frac{1}{(2\pi)^{p/2} |\boldsymbol{\Sigma}_i|^{1/2}} \exp\left[-\frac{1}{2}(\mathbf{x} - \boldsymbol{\mu}_i)'\boldsymbol{\Sigma}_i^{-1}(\mathbf{x} - \boldsymbol{\mu}_i)\right], \qquad i = 1, 2, \ldots, g \tag{10-56}$$

are multivariate normal densities with mean vectors $\boldsymbol{\mu}_i$ and covariance matrices $\boldsymbol{\Sigma}_i$. If further $c(i \mid i) = 0$, $c(k \mid i) = 1$, $k \neq i$ (or equivalently, the misclassification costs are all equal), (10-54) becomes:

Allocate $\mathbf{x}$ to π_k if

$$\begin{aligned} \ln p_k f_k(\mathbf{x}) &= \ln p_k - \left(\frac{p}{2}\right)\ln(2\pi) - \frac{1}{2}\ln|\boldsymbol{\Sigma}_k| - \frac{1}{2}(\mathbf{x} - \boldsymbol{\mu}_k)'\boldsymbol{\Sigma}_k^{-1}(\mathbf{x} - \boldsymbol{\mu}_k) \\ &= \max_i \ln p_i f_i(\mathbf{x}) \end{aligned} \tag{10-57}$$

The constant $(p/2)\ln(2\pi)$ can be ignored in (10-57) since it is the same for all populations. We therefore define the *quadratic discrimination score* for the ith population to be

$$d_i^Q(\mathbf{x}) = -\tfrac{1}{2}\ln|\boldsymbol{\Sigma}_i| - \tfrac{1}{2}(\mathbf{x} - \boldsymbol{\mu}_i)'\boldsymbol{\Sigma}_i^{-1}(\mathbf{x} - \boldsymbol{\mu}_i) + \ln p_i, \qquad i = 1, 2, \ldots, g \tag{10-58}$$

The quadratic score, $d_i^Q(\mathbf{x})$, is composed of contributions from the generalized variance $|\boldsymbol{\Sigma}_i|$, the prior probability p_i, and the squared distance from $\mathbf{x}$ to the population mean $\boldsymbol{\mu}_i$. Using discriminant scores the classification rule of (10-57) becomes the following.

> **Minimum Total Probability of Misclassification Rule for Normal Populations**
>
> Allocate $\mathbf{x}$ to π_k if
>
> the quadratic score $d_k^Q(\mathbf{x})$ = largest of $d_1^Q(\mathbf{x}), d_2^Q(\mathbf{x}), \ldots, d_g^Q(\mathbf{x})$ (10-59)
>
> where $d_i^Q(\mathbf{x})$ is given by (10-58), $i = 1, 2, \ldots, g$.

In practice, the $\boldsymbol{\mu}_i$ and $\boldsymbol{\Sigma}_i$ are unknown, but a training set of correctly classified observations is often available for the construction of estimates. The relevant sample quantities for population π_i are

$$\bar{\mathbf{x}}_i = \text{sample mean vector}$$
$$\mathbf{S}_i = \text{sample covariance matrix}$$

and

$$n_i = \text{sample size}$$

The estimate of the quadratic discrimination score $\hat{d}_i^Q(\mathbf{x})$ is then

$$\hat{d}_i^Q(\mathbf{x}) = -\tfrac{1}{2}\ln|\mathbf{S}_i| - \tfrac{1}{2}(\mathbf{x} - \bar{\mathbf{x}}_i)'\mathbf{S}_i^{-1}(\mathbf{x} - \bar{\mathbf{x}}_i) + \ln p_i \qquad \text{(10-60)}$$

and the classification rule based on the sample is as follows.

> **Estimated Minimum TPM Rule for Several Normal Populations**
>
> Allocate $\mathbf{x}$ to π_k if
>
> the quadratic score $\hat{d}_k^Q(\mathbf{x}) = $ the largest of $\hat{d}_1^Q(\mathbf{x}), \hat{d}_2^Q(\mathbf{x}), \ldots, \hat{d}_g^Q(\mathbf{x})$
>
> where $\hat{d}_i^Q(\mathbf{x})$ is given by (10-60), $i = 1, 2, \ldots, g$. (10-61)

A simplification is possible if the population covariance matrices, $\boldsymbol{\Sigma}_i$, are equal. When $\boldsymbol{\Sigma}_i = \boldsymbol{\Sigma}$, for $i = 1, 2, \ldots, g$, the discriminant score in (10-58) becomes

$$d_i^Q(\mathbf{x}) = -\tfrac{1}{2}\ln|\boldsymbol{\Sigma}| - \tfrac{1}{2}\mathbf{x}'\boldsymbol{\Sigma}^{-1}\mathbf{x} + \boldsymbol{\mu}_i'\boldsymbol{\Sigma}^{-1}\mathbf{x} - \tfrac{1}{2}\boldsymbol{\mu}_i'\boldsymbol{\Sigma}^{-1}\boldsymbol{\mu}_i + \ln p_i$$

The first two terms are the same for $d_1^Q(\mathbf{x}), d_2^Q(\mathbf{x}), \ldots, d_g^Q(\mathbf{x})$, and, consequently, they can be ignored for allocatory purposes. The remaining terms consist of a constant $c_i = \ln p_i - \frac{1}{2}\boldsymbol{\mu}_i'\boldsymbol{\Sigma}^{-1}\boldsymbol{\mu}_i$ and a *linear* combination of the components of $\mathbf{x}$. Defining the *linear discriminant score*

$$d_i(\mathbf{x}) = \boldsymbol{\mu}_i'\boldsymbol{\Sigma}^{-1}\mathbf{x} - \tfrac{1}{2}\boldsymbol{\mu}_i'\boldsymbol{\Sigma}^{-1}\boldsymbol{\mu}_i + \ln p_i \qquad \text{(10-62)}$$

we obtain the following form of the allocation rule.

> **Minimum Total Probability of Misclassification Rule for Equal Covariance Normal Populations**
>
> Allocate $\mathbf{x}$ to π_k if
>
> the linear discriminant score $d_k(\mathbf{x}) = $ largest of $d_1(\mathbf{x}), d_2(\mathbf{x}), \ldots, d_g(\mathbf{x})$ (10-63)
>
> where $d_i(\mathbf{x})$ is given by (10-62), $i = 1, 2, \ldots, g$.

The estimate $\hat{d}_i(\mathbf{x})$, of the linear discriminant score $d_i(\mathbf{x})$ is based on the pooled estimate of $\boldsymbol{\Sigma}$,

$$\mathbf{S}_{\text{pooled}} = \frac{(n_1 - 1)\mathbf{S}_1 + (n_2 - 1)\mathbf{S}_2 + \cdots + (n_g - 1)\mathbf{S}_g}{n_1 + n_2 + \cdots + n_g - g} \qquad \text{(10-64)}$$

and is given by

$$\hat{d}_i(\mathbf{x}) = \bar{\mathbf{x}}_i' \mathbf{S}_{\text{pooled}}^{-1} \mathbf{x} - \tfrac{1}{2} \bar{\mathbf{x}}_i' \mathbf{S}_{\text{pooled}}^{-1} \bar{\mathbf{x}}_i + \ln p_i. \qquad (10\text{-}65)$$

Consequently, we have the following.

Estimated Minimum TPM Rule for Equal Covariance Normal Populations

Allocate $\mathbf{x}$ to π_k if

the linear discriminant score $\hat{d}_k(\mathbf{x})$ = the largest of $\hat{d}_1(\mathbf{x}), \hat{d}_2(\mathbf{x}), \ldots, \hat{d}_g(\mathbf{x})$

with $\hat{d}_i(\mathbf{x})$ given by (10-65), $i = 1, 2, \ldots, g$. (10-66)

Comment. Expressions (10-62) and (10-65) are convenient linear functions of $\mathbf{x}$. An equivalent classifier for the equal covariance case can be obtained from (10-58) by ignoring the constant term, $-\frac{1}{2}\ln|\mathbf{\Sigma}|$. The result can then be interpreted in terms of the squared distances

$$D_i^2(\mathbf{x}) = (\mathbf{x} - \bar{\mathbf{x}}_i)' \mathbf{S}_{\text{pooled}}^{-1} (\mathbf{x} - \bar{\mathbf{x}}_i) \qquad (10\text{-}67)$$

from $\mathbf{x}$ to the sample mean vector $\bar{\mathbf{x}}_i$. The allocatory rule is then:

Assign $\mathbf{x}$ to the population π_i for which $-\frac{1}{2}D_i^2(\mathbf{x}) + \ln p_i$ is largest

(10-68)

We see that this rule—or, equivalently, (10-66)—assigns $\mathbf{x}$ to the "closest" population. (The distance measure is penalized by $\ln p_i$.)

If the prior probabilities are unknown, the usual procedure is to set $p_1 = p_2 = \cdots = p_g = 1/g$. An observation is then assigned to the closest population.

Example 10.10

Let us calculate the linear discriminant scores based on data from $g = 3$ populations assumed to be bivariate normal with a common covariance matrix.

Random samples from the populations π_1, π_2, and π_3 are displayed below, along with the sample mean vectors and covariance matrices.

$$\pi_1: \quad \mathbf{X}_1 = \begin{bmatrix} -2 & 0 & -1 \\ 5 & 3 & 1 \end{bmatrix} \text{ so } n_1 = 3, \quad \bar{\mathbf{x}}_1 = \begin{bmatrix} -1 \\ 3 \end{bmatrix}, \text{ and } \mathbf{S}_1 = \begin{bmatrix} 1 & -1 \\ -1 & 4 \end{bmatrix}$$

$$\pi_2: \quad \mathbf{X}_2 = \begin{bmatrix} 0 & 2 & 1 \\ 6 & 4 & 2 \end{bmatrix} \text{ so } n_2 = 3, \quad \bar{\mathbf{x}}_2 = \begin{bmatrix} 1 \\ 4 \end{bmatrix}, \text{ and } \mathbf{S}_2 = \begin{bmatrix} 1 & -1 \\ -1 & 4 \end{bmatrix}$$

$$\pi_3: \quad \mathbf{X}_3 = \begin{bmatrix} 1 & 0 & -1 \\ -2 & 0 & -4 \end{bmatrix} \text{ so } n_3 = 3, \quad \bar{\mathbf{x}}_3 = \begin{bmatrix} 0 \\ -2 \end{bmatrix}, \text{ and } \mathbf{S}_3 = \begin{bmatrix} 1 & 1 \\ 1 & 4 \end{bmatrix}$$

Given $p_1 = p_2 = .25$, and $p_3 = .50$, let us classify the observation $\mathbf{x}_0' = [x_{01}, x_{02}] = [-2 \quad -1]$ according to (10-66). From (10-64)

$$\mathbf{S}_{\text{pooled}} = \frac{(3-1)\begin{bmatrix} 1 & -1 \\ -1 & 4 \end{bmatrix} + (3-1)\begin{bmatrix} 1 & -1 \\ -1 & 4 \end{bmatrix} + (3-1)\begin{bmatrix} 1 & 1 \\ 1 & 4 \end{bmatrix}}{9-3}$$

$$= \frac{2}{6}\begin{bmatrix} 1+1+1 & -1-1+1 \\ -1-1+1 & 4+4+4 \end{bmatrix} = \begin{bmatrix} 1 & -\frac{1}{3} \\ -\frac{1}{3} & 4 \end{bmatrix}$$

so

$$\mathbf{S}_{\text{pooled}}^{-1} = \frac{9}{35}\begin{bmatrix} 4 & \frac{1}{3} \\ \frac{1}{3} & 1 \end{bmatrix} = \frac{1}{35}\begin{bmatrix} 36 & 3 \\ 3 & 9 \end{bmatrix}$$

Next,

$$\bar{\mathbf{x}}_1'\mathbf{S}_{\text{pooled}}^{-1} = [-1 \quad 3]\frac{1}{35}\begin{bmatrix} 36 & 3 \\ 3 & 9 \end{bmatrix} = \frac{1}{35}[-27 \quad 24]$$

and

$$\bar{\mathbf{x}}_1'\mathbf{S}_{\text{pooled}}^{-1}\bar{\mathbf{x}}_1 = \frac{1}{35}[-27 \quad 24]\begin{bmatrix} -1 \\ 3 \end{bmatrix} = \frac{99}{35}$$

so

$$\hat{d}_1(\mathbf{x}_0) = \ln p_1 + \bar{\mathbf{x}}_1'\mathbf{S}_{\text{pooled}}^{-1}\mathbf{x}_0 - \frac{1}{2}\bar{\mathbf{x}}_1'\mathbf{S}_{\text{pooled}}^{-1}\bar{\mathbf{x}}_1$$
$$= \ln(.25) + \left(\frac{-27}{35}\right)x_{01} + \left(\frac{24}{35}\right)x_{02} - \frac{1}{2}\left(\frac{99}{35}\right)$$

Notice the linear form of $\hat{d}_1(\mathbf{x}_0)$ = constant + (constant)x_{01} + (constant)x_{02}. In a similar manner

$$\bar{\mathbf{x}}_2'\mathbf{S}_{\text{pooled}}^{-1} = [1 \quad 4]\frac{1}{35}\begin{bmatrix} 36 & 3 \\ 3 & 9 \end{bmatrix} = \frac{1}{35}[48 \quad 39]$$
$$\bar{\mathbf{x}}_2'\mathbf{S}_{\text{pooled}}^{-1}\bar{\mathbf{x}}_2 = \frac{1}{35}[48 \quad 39]\begin{bmatrix} 1 \\ 4 \end{bmatrix} = \frac{204}{35}$$

and

$$\hat{d}_2(\mathbf{x}_0) = \ln(.25) + \left(\frac{48}{35}\right)x_{01} + \left(\frac{39}{35}\right)x_{02} - \frac{1}{2}\left(\frac{204}{35}\right)$$

Finally

$$\bar{\mathbf{x}}_3'\mathbf{S}_{\text{pooled}}^{-1} = [0 \quad -2]\frac{1}{35}\begin{bmatrix} 36 & 3 \\ 3 & 9 \end{bmatrix} = \frac{1}{35}[-6 \quad -18]$$
$$\bar{\mathbf{x}}_3'\mathbf{S}_{\text{pooled}}^{-1}\bar{\mathbf{x}}_3 = \frac{1}{35}[-6 \quad -18]\begin{bmatrix} 0 \\ -2 \end{bmatrix} = \frac{36}{35}$$

and

$$\hat{d}_3(\mathbf{x}_0) = \ln(.50) + \left(\frac{-6}{35}\right)x_{01} + \left(\frac{-18}{35}\right)x_{02} - \frac{1}{2}\left(\frac{36}{35}\right)$$

Substituting the numerical values $x_{01} = -2$ and $x_{02} = -1$ gives

$$\hat{d}_1(\mathbf{x}_0) = -1.386 + \left(\frac{-27}{35}\right)(-2) + \left(\frac{24}{35}\right)(-1) - \frac{99}{70} = -1.943$$

$$\hat{d}_2(\mathbf{x}_0) = -1.386 + \left(\frac{48}{35}\right)(-2) + \left(\frac{39}{35}\right)(-1) - \frac{204}{70} = -8.158$$

$$\hat{d}_3(\mathbf{x}_0) = -.693 + \left(\frac{-6}{35}\right)(-2) + \left(\frac{-18}{35}\right)(-1) - \frac{36}{70} = -.350$$

Since $\hat{d}_3(\mathbf{x}_0) = -.350$ is the largest discriminant score, we allocate $\mathbf{x}_0$ to π_3. ■

Example 10.11

The admissions officer of a business school has used an "index" of undergraduate grade point average (GPA) and graduate management aptitude test (GMAT) scores to help decide which applicants should be admitted to the school's graduate programs. Figure 10.7 shows pairs of x_1 = GPA, x_2 = GMAT values for groups of recent applicants who have been categorized as π_1: admit; π_2: not admit; and π_3: borderline.[12] The data pictured are listed in Table 10.5 (see Exercise 10.22). These data yield

$$n_1 = 31 \qquad n_2 = 28 \qquad n_3 = 26$$

$$\bar{\mathbf{x}}_1 = \begin{bmatrix} 3.40 \\ 561.23 \end{bmatrix} \qquad \bar{\mathbf{x}}_2 = \begin{bmatrix} 2.48 \\ 447.07 \end{bmatrix} \qquad \bar{\mathbf{x}}_3 = \begin{bmatrix} 2.99 \\ 446.23 \end{bmatrix}$$

$$\bar{\mathbf{x}} = \begin{bmatrix} 2.97 \\ 488.45 \end{bmatrix} \qquad \mathbf{S}_{\text{pooled}} = \begin{bmatrix} .0361 & -2.0188 \\ -2.0188 & 3655.9024 \end{bmatrix}$$

Suppose a new applicant has an undergraduate GPA of $x_1 = 3.21$ and a GMAT score of $x_2 = 497$. Let us classify this applicant using the rule in (10-68) with equal prior probabilities.

With $\mathbf{x}_0' = [3.21, 497]$ the sample distances are

$$D_1^2(\mathbf{x}_0) = (\mathbf{x}_0 - \bar{\mathbf{x}}_1)'\mathbf{S}_{\text{pooled}}^{-1}(\mathbf{x}_0 - \bar{\mathbf{x}}_1)$$

$$= [3.21 - 3.40, 497 - 561.23]\begin{bmatrix} 28.5835 & .0158 \\ .0158 & .0003 \end{bmatrix}\begin{bmatrix} 3.21 - 3.40 \\ 497 - 561.23 \end{bmatrix}$$

$$= 2.65$$

$$D_2^2(\mathbf{x}_0) = (\mathbf{x}_0 - \bar{\mathbf{x}}_2)'\mathbf{S}_{\text{pooled}}^{-1}(\mathbf{x}_0 - \bar{\mathbf{x}}_2) = 17.13$$

$$D_3^2(\mathbf{x}_0) = (\mathbf{x}_0 - \bar{\mathbf{x}}_3)'\mathbf{S}_{\text{pooled}}^{-1}(\mathbf{x}_0 - \bar{\mathbf{x}}_3) = 2.51$$

[12] In this case, the populations are artificial in the sense that they have been created by the admissions officer. On the other hand, experience has shown that applicants with high GPA and high GMAT scores generally do well in a graduate program; those with low readings on these variables generally experience difficulty.

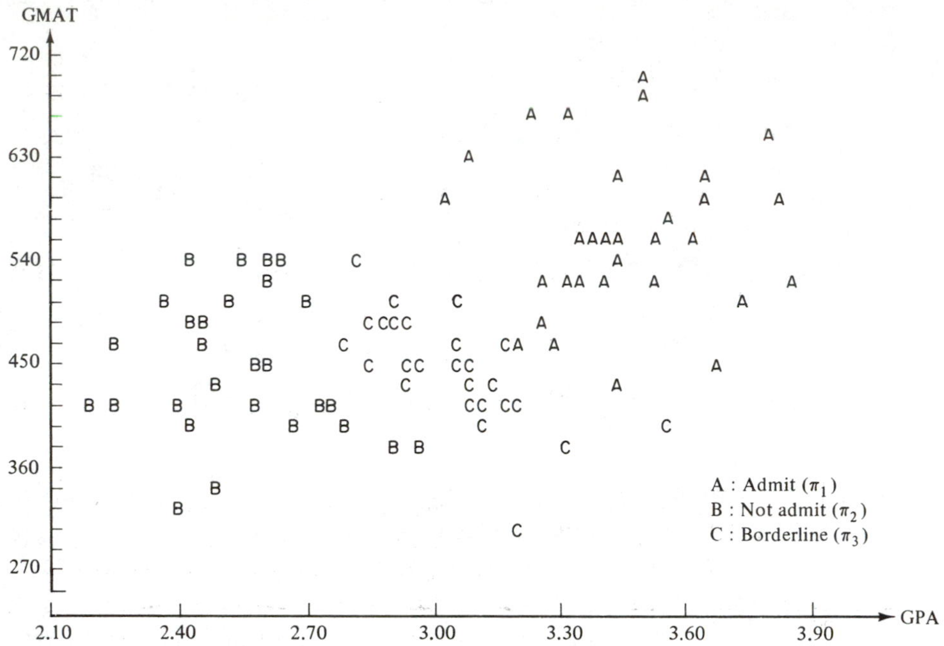

Figure 10.7 Scatterplot of (x_1 = GPA, x_2 = GMAT) for applicants to a graduate school of business who have been classified as admit, not admit, or borderline.

Since the distance from $\mathbf{x}_0' = [3.21, 497]$ to the group mean $\bar{\mathbf{x}}_3$ is smallest, we assign this applicant to π_3: borderline. ■

A second form of the classification rule in (10-63), obtained by comparing the $d_i(\mathbf{x})$ two at a time, merits consideration. The condition $d_k(\mathbf{x})$ is the largest linear discriminant score among $d_1(\mathbf{x}), d_2(\mathbf{x}), \ldots, d_g(\mathbf{x})$ is equivalent to

$$0 \le d_k(\mathbf{x}) - d_i(\mathbf{x})$$
$$= (\boldsymbol{\mu}_k - \boldsymbol{\mu}_i)'\boldsymbol{\Sigma}^{-1}\mathbf{x} - \frac{1}{2}(\boldsymbol{\mu}_k - \boldsymbol{\mu}_i)'\boldsymbol{\Sigma}^{-1}(\boldsymbol{\mu}_k + \boldsymbol{\mu}_i) + \ln\left(\frac{p_k}{p_i}\right)$$

for all $i = 1, 2, \ldots, g$.

Adding $-\ln(p_k/p_i) = \ln(p_i/p_k)$ to both sides of the inequality above gives the alternate form of the classification rule which minimizes the total probability of misclassification. Thus we:

Allocate $\mathbf{x}$ to π_k if

$$(\boldsymbol{\mu}_k - \boldsymbol{\mu}_i)'\boldsymbol{\Sigma}^{-1}\mathbf{x} - \frac{1}{2}(\boldsymbol{\mu}_k - \boldsymbol{\mu}_i)'\boldsymbol{\Sigma}^{-1}(\boldsymbol{\mu}_k + \boldsymbol{\mu}_i) \ge \ln\left(\frac{p_i}{p_k}\right) \tag{10-69}$$

for all $i = 1, 2, \ldots, g$.

Denote the left-hand side of (10-69) by $d_{ki}(\mathbf{x})$. The conditions in (10-69) define classification regions $R_1, R_2, \ldots, R_g$, which are separated by (hyper) planes. This follows because $d_{ki}(\mathbf{x})$ is a linear combination of the components of $\mathbf{x}$. For example, when $g = 3$, the classification region R_1 consists of all $\mathbf{x}$ satisfying

$$R_1: d_{1i}(\mathbf{x}) \geq \ln\left(\frac{p_i}{p_1}\right) \qquad \text{for } i = 2, 3$$

That is, R_1 consists of those $\mathbf{x}$ for which

$$d_{12}(\mathbf{x}) = (\boldsymbol{\mu}_1 - \boldsymbol{\mu}_2)'\boldsymbol{\Sigma}^{-1}\mathbf{x} - \frac{1}{2}(\boldsymbol{\mu}_1 - \boldsymbol{\mu}_2)'\boldsymbol{\Sigma}^{-1}(\boldsymbol{\mu}_1 + \boldsymbol{\mu}_2) \geq \ln\left(\frac{p_2}{p_1}\right)$$

and, *simultaneously*,

$$d_{13}(\mathbf{x}) = (\boldsymbol{\mu}_1 - \boldsymbol{\mu}_3)'\boldsymbol{\Sigma}^{-1}\mathbf{x} - \frac{1}{2}(\boldsymbol{\mu}_1 - \boldsymbol{\mu}_3)'\boldsymbol{\Sigma}^{-1}(\boldsymbol{\mu}_1 + \boldsymbol{\mu}_3) \geq \ln\left(\frac{p_3}{p_1}\right)$$

Assuming $\boldsymbol{\mu}_1$, $\boldsymbol{\mu}_2$, and $\boldsymbol{\mu}_3$ do not lie along a straight line, the equations $d_{12}(\mathbf{x}) = \ln(p_2/p_1)$ and $d_{13}(\mathbf{x}) = \ln(p_3/p_1)$ define two intersecting hyperplanes that delineate R_1 in the p-dimensional variable space. The term $\ln(p_2/p_1)$ places the plane closer to $\boldsymbol{\mu}_1$ than $\boldsymbol{\mu}_2$ if p_2 is greater than p_1. The regions R_1, R_2, and R_3 are shown in Figure 10.8 for the case of two variables. The picture is the same for more variables if we graph the plane that contains the three mean vectors.

Since (10-63) and (10-69) are just two equivalent forms of the minimum TPM rule, the classification regions are the same for both cases.

The sample version of the alternative form in (10-69) is obtained by substituting $\bar{\mathbf{x}}_i$ for $\boldsymbol{\mu}_i$ and the pooled sample covariance matrix $\mathbf{S}_{\text{pooled}}$ for $\boldsymbol{\Sigma}$. When $\sum_{i=1}^{g}(n_i - 1) > p$, so that $\mathbf{S}_{\text{pooled}}^{-1}$ exists, this sample analog becomes:

Allocate $\mathbf{x}$ to π_k if

$$\hat{d}_{ki}(\mathbf{x}) = (\bar{\mathbf{x}}_k - \bar{\mathbf{x}}_i)'\mathbf{S}_{\text{pooled}}^{-1}\mathbf{x} - \frac{1}{2}(\bar{\mathbf{x}}_k - \bar{\mathbf{x}}_i)'\mathbf{S}_{\text{pooled}}^{-1}(\bar{\mathbf{x}}_k + \bar{\mathbf{x}}_i) \geq \ln\left(\frac{p_i}{p_k}\right) \qquad \text{for all } i \neq k \tag{10-70}$$

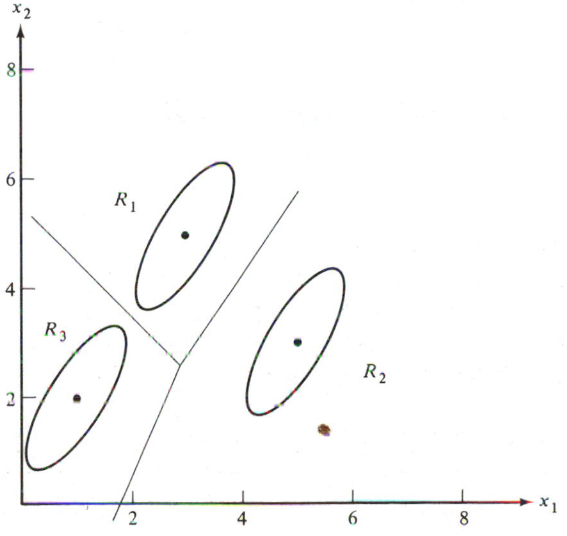

Figure 10.8 The classification regions R_1, R_2, and R_3 for the linear minimum TPM rule ($p_1 = \frac{1}{4}$, $p_2 = \frac{1}{2}$, $p_3 = \frac{1}{4}$).

Given the fixed training set values $\bar{\mathbf{x}}_i$ and $\mathbf{S}_{\text{pooled}}$, $\hat{d}_{ki}(\mathbf{x})$ is a linear function of the components of $\mathbf{x}$. Therefore, the classification regions defined by (10-70)—or, equivalently, by (10-66)—are also bounded by hyperplanes as in Figure 10.8.

As with the sample linear discriminant rule of (10-66), if the prior probabilities are difficult to assess, they are frequently all taken to be equal. In this case, $\ln(p_i/p_k) = 0$ for all pairs.

Because they employ estimates of population parameters, the sample classification rules of (10-61) and (10-66) may no longer be optimal. Their performance, however, can be evaluated using Lachenbruch's holdout procedure. If $n_{iM}^{(H)}$ is the number of misclassified holdout observations in the ith group, $i = 1, 2, \ldots, g$, then an estimate of the expected actual error rate, $E(\text{AER})$, is provided by

$$\hat{E}(\text{AER}) = \frac{\sum_{i=1}^{g} n_{iM}^{(H)}}{\sum_{i=1}^{g} n_i} \tag{10-71}$$

Our discussion has tended to emphasize the linear discriminant rule of (10-66) or (10-70), and many commercial computer programs are based upon it. Although the linear discriminant rule has a simple structure, you must remember that it was derived under the rather strong assumptions of multivariate normality and equal covariances. Before implementing a linear classification rule, these tentative assumptions should be checked in the order: multivariate normality, then equality of covariances. If one or both of these assumptions is violated, improved classification is probably possible if the data are first suitably transformed.

The quadratic rules are an alternative to classification with linear discriminant functions. They are appropriate if normality appears to hold but the assumption of equal covariance matrices is seriously violated. However, the assumption of normality seems to be more critical for quadratic rules than linear rules. If doubt exists as to the appropriateness of a linear or quadratic rule, both rules can be constructed and their error rates examined using Lachenbruch's holdout procedure.

10.8 FISHER'S METHOD FOR DISCRIMINATING AMONG SEVERAL POPULATIONS

Fisher also proposed a several population extension of his discriminant method discussed in Section 10.2. The motivation behind the Fisher discriminant analysis is the need to obtain a reasonable representation of the population that involves only a *few* linear combinations of the observations, such as $\boldsymbol{\ell}_1'\mathbf{x}$, $\boldsymbol{\ell}_2'\mathbf{x}$, and $\boldsymbol{\ell}_3'\mathbf{x}$. His approach has several advantages when one is interested in *separating* several populations for (1) visual inspection or (2) graphical descriptive purposes. It allows for the following.

1. Convenient representations of the g populations that reduce the dimension from a very large number of characteristics to a relatively few linear combinations. Of course, some information—needed for optimal classification—may

be lost unless the population means lie completely in the lower-dimensional space selected.

2. Plotting of the means of the first two or three linear combinations (discriminants). This helps display the relationships and possible groupings of the populations.
3. Scatterplots of the sample values of the first two discriminants, which can indicate outliers or other abnormalities in the data.

The primary purpose of Fisher's discriminant analysis is to *separate* populations. It can, however, also be used to classify, and we shall indicate this use. It is not necessary to assume that the g populations are multivariate normal. However, we do assume the $p \times p$ population covariance matrices are equal and of full rank.[13] That is, $\mathbf{\Sigma}_1 = \mathbf{\Sigma}_2 = \cdots = \mathbf{\Sigma}_g = \mathbf{\Sigma}$.

Let $\bar{\boldsymbol{\mu}}$ denote the mean vector of the combined groups and $\mathbf{B}_0$ the between groups sums of crossproducts so that

$$\mathbf{B}_0 = \sum_{i=1}^{g} (\boldsymbol{\mu}_i - \bar{\boldsymbol{\mu}})(\boldsymbol{\mu}_i - \bar{\boldsymbol{\mu}})', \qquad \text{where } \bar{\boldsymbol{\mu}} = \frac{1}{g}\sum_{i=1}^{g} \boldsymbol{\mu}_i \tag{10-72}$$

We consider the linear combination

$$Y = \boldsymbol{\ell}'\mathbf{X}$$

which has expected value

$$E(Y) = \boldsymbol{\ell}'E(\mathbf{X} \mid \pi_i) = \boldsymbol{\ell}'\boldsymbol{\mu}_i, \qquad \text{for population } \pi_i$$

and variance

$$\text{Var}(Y) = \boldsymbol{\ell}'\,\text{Cov}(\mathbf{X})\boldsymbol{\ell} = \boldsymbol{\ell}'\mathbf{\Sigma}\boldsymbol{\ell} \qquad \text{for all populations}$$

Consequently, the expected value $\mu_{iY} = \boldsymbol{\ell}'\boldsymbol{\mu}_i$ changes as the population from which $\mathbf{X}$ is selected changes. We first define the overall mean

$$\begin{aligned}\bar{\mu}_Y &= \frac{1}{g}\sum_{i=1}^{g} \mu_{iY} = \frac{1}{g}\sum_{i=1}^{g} \boldsymbol{\ell}'\boldsymbol{\mu}_i = \boldsymbol{\ell}'\left(\frac{1}{g}\sum_{i=1}^{g} \boldsymbol{\mu}_i\right) \\ &= \boldsymbol{\ell}'\bar{\boldsymbol{\mu}}\end{aligned}$$

and form the ratio

$$\frac{\begin{pmatrix}\text{Sum of squared distances from} \\ \text{populations to overall mean of } Y\end{pmatrix}}{(\text{Variance of } Y)} = \frac{\sum_{i=1}^{g} (\mu_{iY} - \bar{\mu}_Y)^2}{\sigma_Y^2} = \frac{\sum_{i=1}^{g} (\boldsymbol{\ell}'\boldsymbol{\mu}_i - \boldsymbol{\ell}'\bar{\boldsymbol{\mu}})^2}{\boldsymbol{\ell}'\mathbf{\Sigma}\boldsymbol{\ell}}$$

$$= \frac{\boldsymbol{\ell}'\left(\sum_{i=1}^{g} (\boldsymbol{\mu}_i - \bar{\boldsymbol{\mu}})(\boldsymbol{\mu}_i - \bar{\boldsymbol{\mu}})'\right)\boldsymbol{\ell}}{\boldsymbol{\ell}'\mathbf{\Sigma}\boldsymbol{\ell}}$$

[13] If not, we let $\mathbf{P} = [\mathbf{e}_1, \ldots, \mathbf{e}_q]$ be the eigenvectors of $\mathbf{\Sigma}$ corresponding to nonzero eigenvalues $[\lambda_1, \ldots, \lambda_q]$. Then we replace $\mathbf{X}$ by $\mathbf{P}'\mathbf{X}$, which has a full rank covariance matrix $\mathbf{P}'\mathbf{\Sigma}\mathbf{P}$.

or

$$\frac{\sum_{i=1}^{g} (\mu_{iY} - \bar{\mu}_Y)^2}{\sigma_Y^2} = \frac{\ell'\mathbf{B}_0\ell}{\ell'\mathbf{\Sigma}\ell} \tag{10-73}$$

The ratio in (10-73), a multigroup generalization of (10-6), measures the variability *between* the groups of Y-values relative to the common variability *within* groups (see Exercise 10.8). Analogous to the two-population case, we can select ℓ to maximize the ratio of (10-73). It is convenient to scale ℓ so that $\ell'\mathbf{\Sigma}\ell = 1$.

Result 10.7. Let $\lambda_1 \geq \lambda_2 \geq \cdots \geq \lambda_s > 0$ denote the $s \leq \min(g-1, p)$ nonzero eigenvalues of $\mathbf{\Sigma}^{-1}\mathbf{B}_0$ and $\mathbf{e}_1, \mathbf{e}_2, \ldots, \mathbf{e}_s$ the corresponding eigenvectors (scaled so that $\mathbf{e}'\mathbf{\Sigma}\mathbf{e} = 1$). Then the vector of coefficients ℓ that maximizes the ratio

$$\frac{\ell'\mathbf{B}_0\ell}{\ell'\mathbf{\Sigma}\ell} = \frac{\ell'\left[\sum_{i=1}^{g} (\boldsymbol{\mu}_i - \bar{\boldsymbol{\mu}})(\boldsymbol{\mu}_i - \bar{\boldsymbol{\mu}})'\right]\ell}{\ell'\mathbf{\Sigma}\ell}$$

is given by $\ell_1 = \mathbf{e}_1$. The linear combination $\ell_1'\mathbf{X}$ is called the *first discriminant*.

The value $\ell_2 = \mathbf{e}_2$ maximizes the ratio subject to $\text{Cov}(\ell_1'\mathbf{X}, \ell_2'\mathbf{X}) = 0$. The linear combination $\ell_2'\mathbf{X}$ is called the *second discriminant*. Continuing, $\ell_k = \mathbf{e}_k$ maximizes the ratio subject to $0 = \text{Cov}(\ell_k'\mathbf{X}, \ell_i'\mathbf{X})$, $i < k$, and $\ell_k'\mathbf{X}$ is called the kth discriminant. Also $\text{Var}(\ell_i'\mathbf{X}) = 1$, $i = 1, \ldots, s$.

Proof. We first convert the maximization problem to one already solved. By the spectral decomposition in (2-20), $\mathbf{\Sigma} = \mathbf{P}'\mathbf{\Lambda}\mathbf{P}$ where $\mathbf{\Lambda}$ is a diagonal matrix with positive elements, λ_i. Let $\mathbf{\Lambda}^{1/2}$ denote the diagonal matrix with elements $\sqrt{\lambda_i}$. By (2-22) the symmetric square-root matrix $\mathbf{\Sigma}^{1/2} = \mathbf{P}'\mathbf{\Lambda}^{1/2}\mathbf{P}$ and its inverse $\mathbf{\Sigma}^{-1/2} = \mathbf{P}'\mathbf{\Lambda}^{-1/2}\mathbf{P}$ satisfy $\mathbf{\Sigma}^{1/2}\mathbf{\Sigma}^{1/2} = \mathbf{\Sigma}$, $\mathbf{\Sigma}^{1/2}\mathbf{\Sigma}^{-1/2} = \mathbf{I} = \mathbf{\Sigma}^{-1/2}\mathbf{\Sigma}^{1/2}$ and $\mathbf{\Sigma}^{-1/2}\mathbf{\Sigma}^{-1/2} = \mathbf{\Sigma}^{-1}$. Next set

$$\mathbf{a} = \mathbf{\Sigma}^{1/2}\ell \tag{10-74}$$

so $\mathbf{a}'\mathbf{a} = \ell'\mathbf{\Sigma}^{1/2}\mathbf{\Sigma}^{1/2}\ell = \ell'\mathbf{\Sigma}\ell$ and $\mathbf{a}'\mathbf{\Sigma}^{-1/2}\mathbf{B}_0\mathbf{\Sigma}^{-1/2}\mathbf{a} = \ell'\mathbf{\Sigma}^{1/2}\mathbf{\Sigma}^{-1/2}\mathbf{B}_0\mathbf{\Sigma}^{-1/2}\mathbf{\Sigma}^{1/2}\ell = \ell'\mathbf{B}_0\ell$. Consequently, the problem reduces to maximizing

$$\frac{\mathbf{a}'\mathbf{\Sigma}^{-1/2}\mathbf{B}_0\mathbf{\Sigma}^{-1/2}\mathbf{a}}{\mathbf{a}'\mathbf{a}} \tag{10-75}$$

over $\mathbf{a}$. From (2-51) the maximum of this ratio is λ_1, the largest eigenvalue of $\mathbf{\Sigma}^{-1/2}\mathbf{B}_0\mathbf{\Sigma}^{-1/2}$. This maximum occurs when $\mathbf{a} = \mathbf{e}_1$, the normalized eigenvector associated with λ_1. Because $\mathbf{e}_1 = \mathbf{a} = \mathbf{\Sigma}^{1/2}\ell_1$, or $\ell_1 = \mathbf{\Sigma}^{-1/2}\mathbf{e}_1$, $\text{Var}(\ell_1'\mathbf{X}) = \ell_1'\mathbf{\Sigma}\ell_1 = \mathbf{e}_1'\mathbf{\Sigma}^{-1/2}\mathbf{\Sigma}\mathbf{\Sigma}^{-1/2}\mathbf{e}_1 = \mathbf{e}_1'\mathbf{\Sigma}^{-1/2}\mathbf{\Sigma}^{1/2}\mathbf{\Sigma}^{1/2}\mathbf{\Sigma}^{-1/2}\mathbf{e}_1 = \mathbf{e}_1'\mathbf{e}_1 = 1$. By (2-52), $\mathbf{a} \perp \mathbf{e}_1$ maximizes the ratio (10-75) when $\mathbf{a} = \mathbf{e}_2$, the normalized eigenvector corresponding to λ_2. For this choice, $\ell_2 = \mathbf{\Sigma}^{-1/2}\mathbf{e}_2$ and $\text{Cov}(\ell_2'\mathbf{X}, \ell_1'\mathbf{X}) = \ell_2'\mathbf{\Sigma}\ell_1 = \mathbf{e}_2'\mathbf{\Sigma}^{-1/2}\mathbf{\Sigma}\mathbf{\Sigma}^{-1/2}\mathbf{e}_1 = \mathbf{e}_2'\mathbf{e}_1 = 0$ since $\mathbf{e}_2 \perp \mathbf{e}_1$. Similarly, $\text{Var}(\ell_2'\mathbf{X}) = \ell_2'\mathbf{\Sigma}\ell_2 = \mathbf{e}_2'\mathbf{e}_2 = 1$. Continuing, $\mathbf{a} = \mathbf{e}_k$ maximizes (10-75) subject of $\mathbf{a} \perp \mathbf{e}_1, \ldots, \mathbf{e}_{k-1}$ and $\ell_k = \mathbf{\Sigma}^{-1/2}\mathbf{e}_k$ satisfies $\text{Cov}(\ell_k'\mathbf{X}, \ell_i'\mathbf{X}) = \ell_k'\mathbf{\Sigma}\ell_i = \mathbf{e}_k'\mathbf{\Sigma}^{-1/2}\mathbf{\Sigma}\mathbf{\Sigma}^{-1/2}\mathbf{e}_i = \mathbf{e}_k'\mathbf{e}_i = 0$ if $i < k$ or 1 if $i = k$. Finally, if λ and $\mathbf{e}$ are an eigenvalue and eigenvector pair of $\mathbf{\Sigma}^{-1/2}\mathbf{B}_0\mathbf{\Sigma}^{-1/2}$

$$\mathbf{\Sigma}^{-1/2}\mathbf{B}_0\mathbf{\Sigma}^{-1/2}\mathbf{e} = \lambda\mathbf{e}$$

and multiplication on the left by $\mathbf{\Sigma}^{-1/2}$ gives

$$\mathbf{\Sigma}^{-1/2}\mathbf{\Sigma}^{-1/2}\mathbf{B}_0\mathbf{\Sigma}^{-1/2}\mathbf{e} = \lambda\mathbf{\Sigma}^{-1/2}\mathbf{e} \quad \text{or} \quad \mathbf{\Sigma}^{-1}\mathbf{B}_0(\mathbf{\Sigma}^{-1/2}\mathbf{e}) = \lambda(\mathbf{\Sigma}^{-1/2}\mathbf{e}) \tag{10-76}$$

Thus $\mathbf{\Sigma}^{-1}\mathbf{B}_0$ has the same eigenvalues as $\mathbf{\Sigma}^{-1/2}\mathbf{B}_0\mathbf{\Sigma}^{-1/2}$, but the corresponding eigenvector is proportional to $\mathbf{\Sigma}^{-1/2}\mathbf{e} = \boldsymbol{\ell}$, as asserted. Note $\boldsymbol{\ell}'\mathbf{\Sigma}\boldsymbol{\ell} = \mathbf{e}'\mathbf{\Sigma}^{-1/2}\mathbf{\Sigma}\mathbf{\Sigma}^{-1/2}\mathbf{e} = \mathbf{e}'\mathbf{e} = 1$, and consequently the linear combinations $\boldsymbol{\ell}_1'\mathbf{X}, \boldsymbol{\ell}_2'\mathbf{X}, \ldots, \boldsymbol{\ell}_s'\mathbf{X}$, corresponding to nonzero eigenvalues of $\mathbf{\Sigma}^{-1}\mathbf{B}_0$, form a complete set of discriminants with $\text{Var}(\boldsymbol{\ell}_i'\mathbf{X}) = 1$ and $0 = \text{Cov}(\boldsymbol{\ell}_i'\mathbf{X}, \boldsymbol{\ell}_k'\mathbf{X})$, for $i \neq k$. ■

Ordinarily, $\mathbf{\Sigma}$ and the $\boldsymbol{\mu}_i$ are unavailable, but we have a training set consisting of correctly classified observations. Suppose the training set consists of a random sample of size n_i from population π_i, $i = 1, 2, \ldots, g$. Denote the $p \times n_i$ data set, from population π_i, by $\mathbf{X}_i$ and its jth column by $\mathbf{x}_{ij}$. After first constructing the sample mean vectors

$$\bar{\mathbf{x}}_i = \frac{1}{n_i}\sum_{j=1}^{n_i}\mathbf{x}_{ij}$$

and the covariance matrices $\mathbf{S}_i$, $i = 1, 2, \ldots, g$, we define the "overall average" vector

$$\bar{\mathbf{x}} = \frac{\sum_{i=1}^{g} n_i\bar{\mathbf{x}}_i}{\sum_{i=1}^{g} n_i} = \frac{\sum_{i=1}^{g}\sum_{j=1}^{n_i}\mathbf{x}_{ij}}{\sum_{i=1}^{g} n_i}$$

which is the $p \times 1$ vector average taken over *all* of the sample observations in the training set.

Corresponding to the population between groups matrix $\mathbf{B}_0$ in (10-72), we define the *sample between groups* matrix

$$\hat{\mathbf{B}}_0 = \sum_{i=1}^{g}(\bar{\mathbf{x}}_i - \bar{\mathbf{x}})(\bar{\mathbf{x}}_i - \bar{\mathbf{x}})' \tag{10-77}$$

Also, an estimate of $\mathbf{\Sigma}$ is based on the *sample within groups* matrix.

$$\mathbf{W} = \sum_{i=1}^{g}(n_i - 1)\mathbf{S}_i = \sum_{i=1}^{g}\sum_{j=1}^{n_i}(\mathbf{x}_{ij} - \bar{\mathbf{x}}_i)(\mathbf{x}_{ij} - \bar{\mathbf{x}}_i)' \tag{10-78}$$

Consequently, $\mathbf{W}/(n_1 + n_2 + \cdots + n_g - g) = \mathbf{S}_{\text{pooled}}$ is the estimate of $\mathbf{\Sigma}$. Before presenting the sample discriminants, we note that $\mathbf{W}$ is the constant $(n_1 + n_2 + \cdots + n_g - g)$ times $\mathbf{S}_{\text{pooled}}$, so the same $\hat{\boldsymbol{\ell}}$ that maximizes $\hat{\boldsymbol{\ell}}'\hat{\mathbf{B}}_0\hat{\boldsymbol{\ell}}/\hat{\boldsymbol{\ell}}'\mathbf{S}_{\text{pooled}}\hat{\boldsymbol{\ell}}$ also maximizes $\hat{\boldsymbol{\ell}}'\hat{\mathbf{B}}_0\hat{\boldsymbol{\ell}}/\hat{\boldsymbol{\ell}}'\mathbf{W}\hat{\boldsymbol{\ell}}$. Moreover, we can present the optimizing $\hat{\boldsymbol{\ell}}$ in the more customary form as eigenvectors, $\hat{\mathbf{e}}_i$, of $\mathbf{W}^{-1}\hat{\mathbf{B}}_0$, because if $\mathbf{W}^{-1}\hat{\mathbf{B}}_0\hat{\mathbf{e}} = \hat{\lambda}\hat{\mathbf{e}}$ then $\mathbf{S}_{\text{pooled}}^{-1}\hat{\mathbf{B}}_0\hat{\mathbf{e}} = \hat{\lambda}(n_1 + n_2 + \cdots + n_g - g)\hat{\mathbf{e}}$.

Fisher's Sample Discriminants

Let $\hat{\lambda}_1, \hat{\lambda}_2, \ldots, \hat{\lambda}_s > 0$ denote the $s \le \min(g-1, p)$ nonzero eigenvalues of $\mathbf{W}^{-1}\hat{\mathbf{B}}_0$ and $\hat{\mathbf{e}}_1, \ldots, \hat{\mathbf{e}}_s$ be the corresponding eigenvectors (scaled so that $\hat{\mathbf{e}}'\mathbf{S}_{\text{pooled}}\hat{\mathbf{e}} = 1$). Then the vector of coefficients $\hat{\ell}$ that maximizes the ratio

$$\frac{\hat{\ell}'\hat{\mathbf{B}}_0\hat{\ell}}{\hat{\ell}'\mathbf{W}\hat{\ell}} = \frac{\hat{\ell}'\left(\sum_{i=1}^{g} (\bar{\mathbf{x}}_i - \bar{\mathbf{x}})(\bar{\mathbf{x}}_i - \bar{\mathbf{x}})'\right)\hat{\ell}}{\hat{\ell}'\left[\sum_{i=1}^{g}\sum_{j=1}^{n_i} (\mathbf{x}_{ij} - \bar{\mathbf{x}}_i)(\mathbf{x}_{ij} - \bar{\mathbf{x}}_i)'\right]\hat{\ell}} \tag{10-79}$$

is given by $\hat{\ell}_1 = \hat{\mathbf{e}}_1$. The linear combination $\hat{\ell}_1'\mathbf{x}$ is called the *sample first discriminant*.

The choice $\hat{\ell}_2 = \hat{\mathbf{e}}_2$ produces the *sample second discriminant*, $\hat{\ell}_2'\mathbf{x}$. Continuing, $\hat{\ell}_k'\mathbf{x} = \hat{\mathbf{e}}_k'\mathbf{x}$ is the *sample kth discriminant*, $k \le s$.

Unlike the population result of Result 10.7, the discriminants will not have zero covariance for each random sample $\mathbf{X}_i$. Rather, the condition

$$\hat{\ell}_i'\mathbf{S}_{\text{pooled}}\hat{\ell}_k = \begin{cases} 1 & \text{if } i = k \le s \\ 0 & \text{otherwise} \end{cases} \tag{10-80}$$

will be satisfied. The use of $\mathbf{S}_{\text{pooled}}$ is appropriate because we tentatively assumed that the g population covariance matrices were equal.

Example 10.12

Consider the observations on $p = 2$ variables from $g = 3$ populations given in Example 10.10. Assuming the populations have a common covariance matrix $\mathbf{\Sigma}$, let us obtain the Fisher discriminants.

Data

$$\begin{array}{ccc} \pi_1(n_1 = 3) & \pi_2(n_2 = 3) & \pi_3(n_3 = 3) \\ \mathbf{X}_1 = \begin{bmatrix} -2 & 0 & -1 \\ 5 & 3 & 1 \end{bmatrix}; & \mathbf{X}_2 = \begin{bmatrix} 0 & 2 & 1 \\ 6 & 4 & 2 \end{bmatrix}; & \mathbf{X}_3 = \begin{bmatrix} 1 & 0 & -1 \\ -2 & 0 & -4 \end{bmatrix} \end{array}$$

In Example 10.10 we found

$$\bar{\mathbf{x}}_1 = \begin{bmatrix} -1 \\ 3 \end{bmatrix}; \quad \bar{\mathbf{x}}_2 = \begin{bmatrix} 1 \\ 4 \end{bmatrix}; \quad \bar{\mathbf{x}}_3 = \begin{bmatrix} 0 \\ -2 \end{bmatrix}$$

so

$$\bar{\mathbf{x}} = \begin{bmatrix} 0 \\ 5/3 \end{bmatrix}; \quad \hat{\mathbf{B}}_0 = \sum_{i=1}^{3} (\bar{\mathbf{x}}_i - \bar{\mathbf{x}})(\bar{\mathbf{x}}_i - \bar{\mathbf{x}})' = \begin{bmatrix} 2 & 1 \\ 1 & \frac{62}{3} \end{bmatrix}$$

$$\mathbf{W} = \sum_{i=1}^{3}\sum_{j=1}^{n_i} (\mathbf{x}_{ij} - \bar{\mathbf{x}}_i)(\mathbf{x}_{ij} - \bar{\mathbf{x}}_i)' = (n_1 + n_2 + n_3 - 3)\mathbf{S}_{\text{pooled}} = \begin{bmatrix} 6 & -2 \\ -2 & 24 \end{bmatrix}$$

$$\mathbf{W}^{-1} = \frac{1}{140}\begin{bmatrix} 24 & 2 \\ 2 & 6 \end{bmatrix}; \quad \mathbf{W}^{-1}\hat{\mathbf{B}}_0 = \begin{bmatrix} .3571 & .4667 \\ .0714 & .9000 \end{bmatrix}$$

To solve for the $s \le \min(g-1, p) = \min(2,2) = 2$ nonzero eigenvalues of $\mathbf{W}^{-1}\hat{\mathbf{B}}_0$, we must solve

$$|\mathbf{W}^{-1}\hat{\mathbf{B}}_0 - \lambda \mathbf{I}| = \left| \begin{bmatrix} .3571 - \lambda & .4667 \\ .0714 & .9000 - \lambda \end{bmatrix} \right| = 0$$

or

$$(.3571 - \lambda)(.9000 - \lambda) - (.4667)(.0714) = \lambda^2 - 1.2571\lambda + .2881 = 0$$

Using the quadratic formula, we find $\hat{\lambda}_1 = .9556$ and $\hat{\lambda}_2 = .3015$. The normalized eigenvectors $\hat{\ell}_1$ and $\hat{\ell}_2$ are obtained by solving

$$(\mathbf{W}^{-1}\hat{\mathbf{B}}_0 - \hat{\lambda}_i \mathbf{I})\hat{\ell}_i = \mathbf{0} \qquad i = 1, 2$$

and scaling the results such that $\hat{\ell}_i' \mathbf{S}_{\text{pooled}} \hat{\ell}_i = 1$. For example, the solution of

$$(\mathbf{W}^{-1}\hat{\mathbf{B}}_0 - \hat{\lambda}_1 \mathbf{I})\hat{\ell}_1 = \begin{bmatrix} .3571 - .9556 & .4667 \\ .0714 & .9000 - .9556 \end{bmatrix} \begin{bmatrix} \hat{\ell}_{11} \\ \hat{\ell}_{12} \end{bmatrix} = \begin{bmatrix} 0 \\ 0 \end{bmatrix}$$

is, after the normalization, $\hat{\ell}_1' \mathbf{S}_{\text{pooled}} \hat{\ell}_1 = 1$,

$$\hat{\ell}_1' = [.385 \quad .495]$$

Similarly,

$$\hat{\ell}_2' = [.938 \quad -.112]$$

The two discriminants are

$$\hat{y}_1 = \hat{\ell}_1' \mathbf{x} = [.385 \quad .495] \begin{bmatrix} x_1 \\ x_2 \end{bmatrix} = .385x_1 + .495x_2$$

$$\hat{y}_2 = \hat{\ell}_2' \mathbf{x} = [.938 \quad -.112] \begin{bmatrix} x_1 \\ x_2 \end{bmatrix} = .938x_1 - .112x_2$$ ■

Example 10.13

Gerrild and Lantz [8] collected crude-oil samples from sandstone in the Elk Hills, California, petroleum reserve. These crude oils can be assigned to one of the three stratigraphic units (populations)

π_1: Wilhelm sandstone
π_2: Sub-Mulinia sandstone
π_3: Upper sandstone

on the basis of their chemistry. For illustrative purposes we consider only the five variables:

X_1 = vanadium (in percent ash)

$X_2 = \sqrt{\text{iron (in percent ash)}}$

$X_3 = \sqrt{\text{beryllium (in percent ash)}}$

X_4 = 1/[saturated hydrocarbons (in percent area)]

X_5 = aromatic hydrocarbons (in percent area)

The first three variables are trace elements and the last two are determined from a segment of the curve produced by a gas chromatograph chemical analysis. Table 10.6 (see Exercise 10.23) gives the values of the five original variables (vanadium, iron, berylium, saturated hydrocarbons, and aromatic hydrocarbons) for 56 cases whose population assignment was certain.

A computer calculation yields the summary statistics

$$\bar{\mathbf{x}}_1 = \begin{bmatrix} 3.229 \\ 6.587 \\ .303 \\ .150 \\ 11.540 \end{bmatrix}, \quad \bar{\mathbf{x}}_2 = \begin{bmatrix} 4.445 \\ 5.667 \\ .344 \\ .157 \\ 5.484 \end{bmatrix},$$

$$\bar{\mathbf{x}}_3 = \begin{bmatrix} 7.226 \\ 4.634 \\ .598 \\ .223 \\ 5.768 \end{bmatrix}, \quad \bar{\mathbf{x}} = \begin{bmatrix} 6.180 \\ 5.081 \\ .511 \\ .201 \\ 6.434 \end{bmatrix}$$

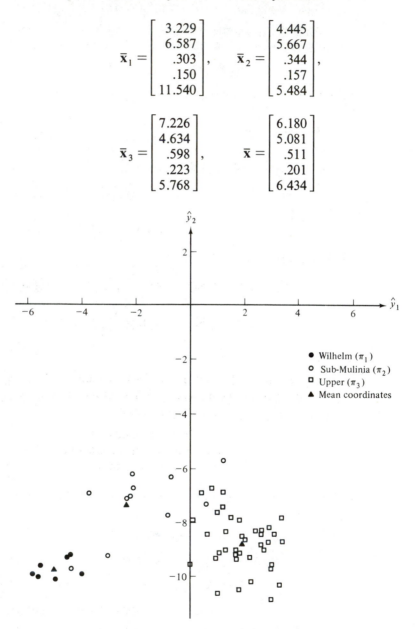

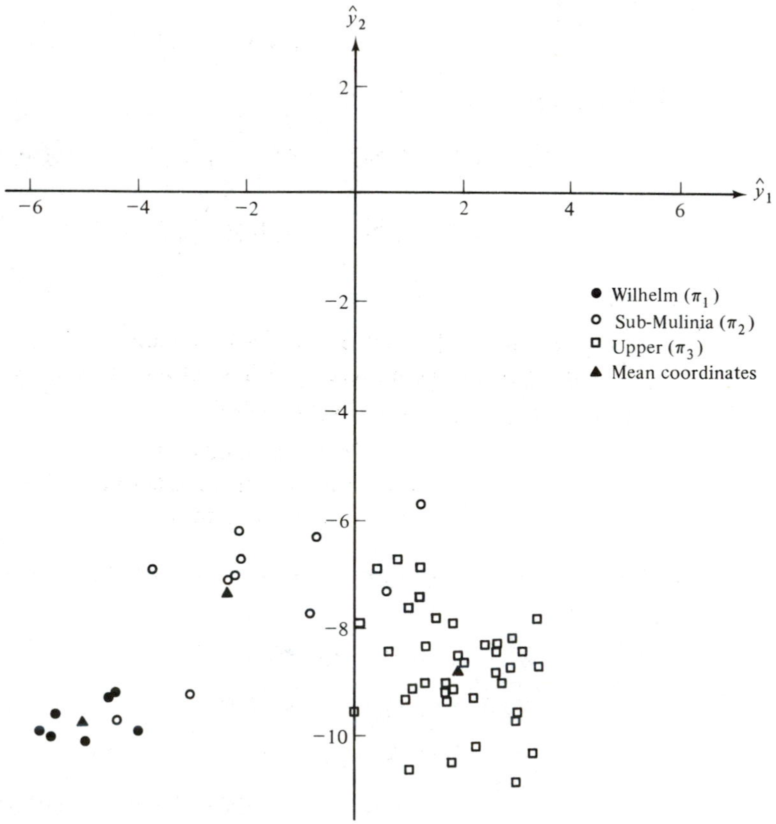

Figure 10.9 Crude-oil samples in discriminant space.

and

$$(n_1 + n_2 + n_3 - 3)\mathbf{S}_{\text{pooled}} = (38 + 11 + 7 - 3)\mathbf{S}_{\text{pooled}}$$

$$= \mathbf{W} = \begin{bmatrix} 187.575 & & & & \\ 1.957 & 41.789 & & & \\ -4.031 & 2.128 & 3.580 & & \\ 1.092 & -.143 & -.284 & .077 & \\ 79.672 & -28.243 & 2.559 & -.996 & 338.023 \end{bmatrix}$$

There are at most $s = \min(g - 1, p) = \min(2, 5) = 2$ positive eigenvalues of $\mathbf{W}^{-1}\hat{\mathbf{B}}_0$, and they are 4.354 and .559. The Fisher linear discriminants are

$$\hat{y}_1 = .312x_1 - .710x_2 + 2.764x_3 + 11.809x_4 - .235x_5$$

$$\hat{y}_2 = .169x_1 - .245x_2 - 2.046x_3 - 24.453x_4 - .378x_5$$

The separation of the three group means is fully explained in the two-dimensional "discriminant space." The group means and the scatter of the individual observations in the discriminant coordinate system are shown in Figure 10.9. The separation is quite good. ■

Example 10.14

Investigators, interested in sports psychology, administered the Minnesota Multiphasic Personality Inventory (MMPI) to 670 letter winners at the University of Wisconsin in Madison. The sports involved and the coefficients in the two discriminant functions are given in Table 10.2.

A plot of the group means using the first two discriminant scores is shown in Figure 10.10. Here the separation on the basis of the MMPI scores is not good, although a test for the equality of means is significant at the 5 percent level. (This is due to the large sample sizes.)

TABLE 10.2

Sport	Sample size	MMPI Scale	1st Discriminant	2nd Discriminant
		QE	.055	−.098
Football	158	L	−.194	.046
Basketball	42	F	−.047	−.099
Baseball	79	K	.053	−.017
Crew	61	Hs	.077	−.076
Fencing	50	D	.049	.183
Golf	28	Hy	−.028	.031
Gymnastics	26	Pd	.001	−.069
Hockey	28	Mf	−.074	−.076
Swimming	51	Pa	.189	.088
Tennis	31	Pt	.025	−.188
Track	52	Sc	−.046	.088
Wrestling	64	Ma	−.103	.053
		Si	.041	.016

SOURCE: W. Morgan and R. W. Johnson.

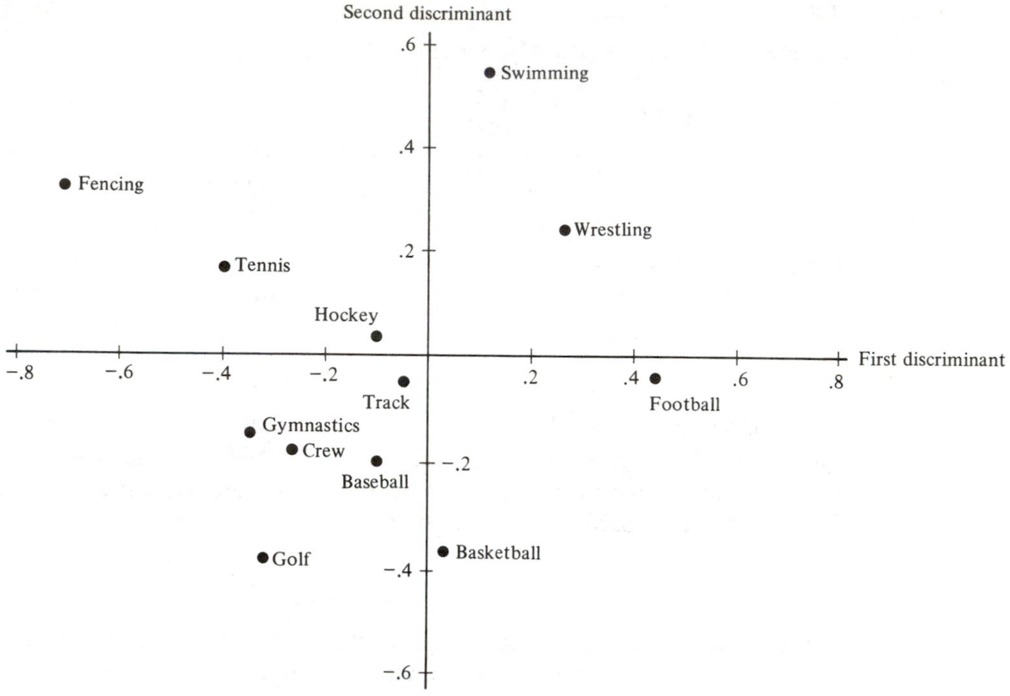

Figure 10.10 The discriminant means $\bar{\mathbf{y}}' = [\bar{y}_1, \bar{y}_2]$ for each sport.

The first discriminant, which accounted for 34.4% of the common variance, was highly correlated with the Mf scale (r = −.78). The second discriminant, which accounted for an additional 18.3% of the variance, was most highly related to scores on the Sc, F, and D scales (r's = .66, .54, and .50, respectively). The investigators suggest that the first discriminant best represents an interest dimension; the second discriminant reflects psychological adjustment.

In general, plots should also be made of other pairs of the first few discriminants. In addition, scatterplots of the discriminant scores for pairs of discriminants can be made for each sport. Under the assumption of multivariate normality, the unit ellipse (circle) centered at the discriminant mean vector $\bar{\mathbf{y}}$ should contain approximately a proportion

$$P[(\mathbf{Y} - \boldsymbol{\mu}_Y)'(\mathbf{Y} - \boldsymbol{\mu}_Y) \leq 1] = P[\chi_2^2 \leq 1] = .39$$

of the points. ■

Using Fisher's Discriminants to Classify

Fisher's discriminants were derived for the purpose of obtaining a low-dimensional representation of the data that separates the populations as much as possible. Although they were derived from separatory considerations, the discriminants also provide the basis for a classification rule. We first explain the connection in terms of the population discriminants $\boldsymbol{\ell}_i'\mathbf{X}$ given in Result 10.7.

Setting

$$Y_k = \boldsymbol{\ell}_k'\mathbf{X} = k\text{th discriminant}, \qquad k \le s \tag{10-81}$$

we conclude from Result 10.7 that

$$\mathbf{Y} = \begin{bmatrix} Y_1 \\ Y_2 \\ \vdots \\ Y_s \end{bmatrix} \quad \text{has mean vector} \quad \boldsymbol{\mu}_{iY} = \begin{bmatrix} \mu_{iY_1} \\ \vdots \\ \mu_{iY_s} \end{bmatrix} = \begin{bmatrix} \boldsymbol{\ell}_1'\boldsymbol{\mu}_i \\ \vdots \\ \boldsymbol{\ell}_s'\boldsymbol{\mu}_i \end{bmatrix}$$

under population π_i and covariance matrix $\mathbf{I}$, for all populations.

Because the components of $\mathbf{Y}$ have unit variances and zero covariances, the appropriate measure of squared distance from $\mathbf{Y} = \mathbf{y}$ to $\boldsymbol{\mu}_{iY}$ is

$$(\mathbf{y} - \boldsymbol{\mu}_{iY})'(\mathbf{y} - \boldsymbol{\mu}_{iY}) = \sum_{j=1}^{s} (y_j - \mu_{iY_j})^2$$

A reasonable classification rule is one that assigns $\mathbf{y}$ to population π_k if the squared distance from $\mathbf{y}$ to $\boldsymbol{\mu}_{kY}$ is smaller than the squared distance from $\mathbf{y}$ to $\boldsymbol{\mu}_{iY}$ for $i \ne k$.

If only r of the discriminants are used for allocation, the rule is:

Allocate $\mathbf{x}$ to π_k if

$$\sum_{j=1}^{r} (y_j - \mu_{kY_j})^2 = \sum_{j=1}^{r} \left[\boldsymbol{\ell}_j'(\mathbf{x} - \boldsymbol{\mu}_k)\right]^2$$
$$\le \sum_{j=1}^{r} \left[\boldsymbol{\ell}_j'(\mathbf{x} - \boldsymbol{\mu}_i)\right]^2 \quad \text{for all } i \ne k \tag{10-82}$$

Before relating this classification procedure to those of Section 10.7, we look more closely at the restriction on the number of discriminants. From Result 10.7 and (10-76)

$$\begin{aligned} s &= \text{number of discriminants} \\ &= \text{number of nonzero eigenvalues of } \boldsymbol{\Sigma}^{-1}\mathbf{B}_0 \text{ or of } \boldsymbol{\Sigma}^{-1/2}\mathbf{B}_0\boldsymbol{\Sigma}^{-1/2} \end{aligned}$$

Now $\boldsymbol{\Sigma}^{-1}\mathbf{B}_0$ is $p \times p$, so $s \le p$. Further, the g vectors

$$\boldsymbol{\mu}_1 - \bar{\boldsymbol{\mu}}, \boldsymbol{\mu}_2 - \bar{\boldsymbol{\mu}}, \ldots, \boldsymbol{\mu}_g - \bar{\boldsymbol{\mu}} \tag{10-83}$$

satisfy $(\boldsymbol{\mu}_1 - \bar{\boldsymbol{\mu}}) + (\boldsymbol{\mu}_2 - \bar{\boldsymbol{\mu}}) + \cdots + (\boldsymbol{\mu}_g - \bar{\boldsymbol{\mu}}) = g\bar{\boldsymbol{\mu}} - g\bar{\boldsymbol{\mu}} = \mathbf{0}$. That is, the first difference $\boldsymbol{\mu}_1 - \bar{\boldsymbol{\mu}}$ can be written as a linear combination of the last $g - 1$ differences. Linear combinations of the g vectors in (10-83) determine a hyperplane of dimension $q \le g - 1$. Taking any vector $\mathbf{e}$ perpendicular to every $\boldsymbol{\mu}_i - \bar{\boldsymbol{\mu}}$, and hence the hyperplane, gives

$$\mathbf{B}_0\mathbf{e} = \sum_{i=1}^{g} (\boldsymbol{\mu}_i - \bar{\boldsymbol{\mu}})(\boldsymbol{\mu}_i - \bar{\boldsymbol{\mu}})'\mathbf{e} = \sum_{i=1}^{g} (\boldsymbol{\mu}_i - \bar{\boldsymbol{\mu}})0 = \mathbf{0}$$

so

$$\boldsymbol{\Sigma}^{-1}\mathbf{B}_0\mathbf{e} = 0\mathbf{e}$$

There are $p - q$ orthogonal eigenvectors corresponding to the zero eigenvalue. This implies that there are q or fewer *nonzero* eigenvalues. Since it is always true that $q \leq g - 1$, the number of nonzero eigenvalues s must satisfy $s \leq \min(p, g - 1)$.

Thus there is no loss of discriminant information by plotting in two dimensions if the following conditions hold.

Number of variables	Number of populations	Maximum number of discriminants
any p	$g = 2$	1
any p	$g = 3$	2
$p = 2$	any g	2

Given the classification rule in (10-82) and the "normal theory" discriminant scores,

$$d_i(\mathbf{x}) = \boldsymbol{\mu}_i'\boldsymbol{\Sigma}^{-1}\mathbf{x} - \tfrac{1}{2}\boldsymbol{\mu}_i'\boldsymbol{\Sigma}^{-1}\boldsymbol{\mu}_i + \ln p_i,$$

or, equivalently,

$$d_i(\mathbf{x}) - \tfrac{1}{2}\mathbf{x}'\boldsymbol{\Sigma}^{-1}\mathbf{x} = -\tfrac{1}{2}(\mathbf{x} - \boldsymbol{\mu}_i)'\boldsymbol{\Sigma}^{-1}(\mathbf{x} - \boldsymbol{\mu}_i) + \ln p_i,$$

obtained by adding the same constant $-\frac{1}{2}\mathbf{x}'\boldsymbol{\Sigma}^{-1}\mathbf{x}$ to each $d_i(\mathbf{x})$, we present the following important result.

Result 10.8. Let $y_j = \boldsymbol{\ell}_j'\mathbf{x}$ where $\boldsymbol{\ell}_j = \boldsymbol{\Sigma}^{-1}\mathbf{e}_j$ and $\mathbf{e}_j$ is an eigenvector of $\boldsymbol{\Sigma}^{-1/2}\mathbf{B}_0\boldsymbol{\Sigma}^{-1/2}$. Then

$$\sum_{j=1}^{p}(y_j - \mu_{iY_j})^2 = \sum_{j=1}^{p}\left[\boldsymbol{\ell}_j'(\mathbf{x} - \boldsymbol{\mu}_i)\right]^2 = (\mathbf{x} - \boldsymbol{\mu}_i)'\boldsymbol{\Sigma}^{-1}(\mathbf{x} - \boldsymbol{\mu}_i)$$
$$= -d_i(\mathbf{x}) + \tfrac{1}{2}\mathbf{x}'\boldsymbol{\Sigma}^{-1}\mathbf{x} + \ln p_i$$

If $\lambda_1 \geq \cdots \geq \lambda_s > 0 = \lambda_{s+1} = \cdots = \lambda_p$,

$\sum_{j=s+1}^{p}(y_j - \mu_{iY_j})^2$ is constant for all populations $i = 1, 2, \ldots, g$ so only the first s y_j, or $\sum_{j=1}^{s}(y_j - \mu_{iY_j})^2$, $i = 1, 2, \ldots, g$ contribute to the classification.

Also, if the prior probabilities are such that $p_1 = p_2 = \cdots = p_g = 1/g$, the rule (10-82) with $r = s$ is equivalent to the minimum TMP rule (10-63).

Proof. The squared distance $(\mathbf{x} - \boldsymbol{\mu}_i)'\boldsymbol{\Sigma}^{-1}(\mathbf{x} - \boldsymbol{\mu}_i) = (\mathbf{x} - \boldsymbol{\mu}_i)'\boldsymbol{\Sigma}^{-1/2}\boldsymbol{\Sigma}^{-1/2}(\mathbf{x} - \boldsymbol{\mu}_i) = (\mathbf{x} - \boldsymbol{\mu}_i)'\boldsymbol{\Sigma}^{-1/2}\mathbf{E}\mathbf{E}'\boldsymbol{\Sigma}^{-1/2}(\mathbf{x} - \boldsymbol{\mu}_i)$, where $\mathbf{E} = [\mathbf{e}_1, \mathbf{e}_2, \ldots, \mathbf{e}_p]$ is the orthogonal matrix whose columns are eigenvectors of $\boldsymbol{\Sigma}^{-1/2}\mathbf{B}_0\boldsymbol{\Sigma}^{-1/2}$ (see proof of Result 10.7).

Since $\boldsymbol{\Sigma}^{-1/2}\mathbf{e}_i = \boldsymbol{\ell}_i$ or $\boldsymbol{\ell}_i' = \mathbf{e}_i'\boldsymbol{\Sigma}^{-1/2}$,

$$\mathbf{E}'\boldsymbol{\Sigma}^{-1/2}(\mathbf{x} - \boldsymbol{\mu}_i) = \begin{bmatrix} \boldsymbol{\ell}_1'(\mathbf{x} - \boldsymbol{\mu}_i) \\ \boldsymbol{\ell}_2'(\mathbf{x} - \boldsymbol{\mu}_i) \\ \vdots \\ \boldsymbol{\ell}_p'(\mathbf{x} - \boldsymbol{\mu}_i) \end{bmatrix}$$

and

$$(\mathbf{x} - \boldsymbol{\mu}_i)'\boldsymbol{\Sigma}^{-1/2}\mathbf{E}\mathbf{E}'\boldsymbol{\Sigma}^{-1/2}(\mathbf{x} - \boldsymbol{\mu}_i) = \sum_{j=1}^{p} \left[\boldsymbol{\ell}_j'(\mathbf{x} - \boldsymbol{\mu}_i)\right]^2$$

Next, each $\boldsymbol{\ell}_j = \boldsymbol{\Sigma}^{-1/2}\mathbf{e}_j$, $j > s$, is an (unscaled) eigenvector of $\boldsymbol{\Sigma}^{-1}\mathbf{B}_0$ with eigenvalue zero. As shown below (10-83), $\boldsymbol{\ell}_j$ is perpendicular to every $\boldsymbol{\mu}_i - \bar{\boldsymbol{\mu}}$ and hence to $(\boldsymbol{\mu}_k - \bar{\boldsymbol{\mu}}) - (\boldsymbol{\mu}_i - \bar{\boldsymbol{\mu}}) = \boldsymbol{\mu}_k - \boldsymbol{\mu}_i$ for $i, k = 1, 2, \ldots, g$. The condition $0 = \boldsymbol{\ell}_j'(\boldsymbol{\mu}_k - \boldsymbol{\mu}_i) = \mu_{kY_j} - \mu_{iY_j}$ implies $y_j - \mu_{kY_j} = y_j - \mu_{iY_j}$ so $\sum_{j=s+1}^{p} (y_j - \mu_{iY_j})^2$ is constant for all $i = 1, 2, \ldots, g$. Therefore, only the first s y_j needs to be used for classification. ■

We now state the classification rule based on the first $r \leq s$ sample discriminants.

> Fisher's Classification Procedure Based on Sample Discriminants
>
> Allocate $\mathbf{x}$ to π_k if
>
> $$\sum_{j=1}^{r} (\hat{y}_j - \bar{y}_{kj})^2 = \sum_{j=1}^{r} \left[\hat{\boldsymbol{\ell}}_j'(\mathbf{x} - \bar{\mathbf{x}}_k)\right]^2 \leq \sum_{j=1}^{r} [\hat{\boldsymbol{\ell}}_j'(\mathbf{x} - \bar{\mathbf{x}}_i)]^2 \quad \text{for all } i \neq k$$
>
> where $\hat{\boldsymbol{\ell}}_j$ is defined in (10-79) and $r \leq s$. (10-84)

When the prior probabilities are such that $p_1 = p_2 = \cdots = p_g = 1/g$ and $r = s$, the rule (10-84) is equivalent to the rule based on the largest linear discriminant score of (10-66). In addition, if $r < s$ discriminants are used for classification, there is a loss of squared distance, or score, of $\sum_{j=r+1}^{p} [\boldsymbol{\ell}_j'(\mathbf{x} - \boldsymbol{\mu}_i)]^2$ for each population π_i where $\sum_{j=r+1}^{s} [\hat{\boldsymbol{\ell}}_j'(\mathbf{x} - \boldsymbol{\mu}_i)]^2$ is the part useful for classification.

Example 10.15

Let us use the Fisher discriminants

$$\hat{y}_1 = \hat{\boldsymbol{\ell}}_1'\mathbf{x} = .385x_1 + .495x_2$$

$$\hat{y}_2 = \hat{\boldsymbol{\ell}}_2'\mathbf{x} = .938x_1 - .112x_2$$

from Example 10.12 to classify the new observation $\mathbf{x}_0' = [1 \quad 3]$ in accordance with (10-84).

Inserting $\mathbf{x}_0' = [x_{01}, x_{02}] = [1 \quad 3]$, we have

$$\hat{y}_1 = .385x_{01} + .495x_{02} = .385(1) + .495(3) = 1.87$$

$$\hat{y}_2 = .938x_{01} - .112x_{02} = .938(1) - .112(3) = .60$$

Moreover, $\bar{y}_{kj} = \hat{\ell}_j'\bar{\mathbf{x}}_k$, so that (see Example 10.12)

$$\bar{y}_{11} = \hat{\ell}_1'\bar{\mathbf{x}}_1 = [.385 \quad .495]\begin{bmatrix} -1 \\ 3 \end{bmatrix} = 1.10$$

$$\bar{y}_{12} = \hat{\ell}_2'\bar{\mathbf{x}}_1 = [.938 \quad -.112]\begin{bmatrix} -1 \\ 3 \end{bmatrix} = -1.27$$

Similarly,

$$\bar{y}_{21} = \hat{\ell}_1'\bar{\mathbf{x}}_2 = 2.37$$
$$\bar{y}_{22} = \hat{\ell}_2'\bar{\mathbf{x}}_2 = .49$$
$$\bar{y}_{31} = \hat{\ell}_1'\bar{\mathbf{x}}_3 = -.99$$
$$\bar{y}_{32} = \hat{\ell}_2'\bar{\mathbf{x}}_3 = .22$$

Finally, the smallest value of

$$\sum_{j=1}^{2} (\hat{y}_j - \bar{y}_{kj})^2 = \sum_{j=1}^{2} \left[\hat{\ell}_j'(\mathbf{x} - \bar{\mathbf{x}}_k)\right]^2$$

for $k = 1, 2, 3$, must be identified. Using the numbers above,

$$\sum_{j=1}^{2} (\hat{y}_j - \bar{y}_{1j})^2 = (1.87 - 1.10)^2 + (.60 + 1.27)^2 = 4.09$$

$$\sum_{j=1}^{2} (\hat{y}_j - \bar{y}_{2j})^2 = (1.87 - 2.37)^2 + (.60 - .49)^2 = .26$$

$$\sum_{j=1}^{2} (\hat{y}_j - \bar{y}_{3j})^2 = (1.87 + .99)^2 + (.60 - .22)^2 = 8.32$$

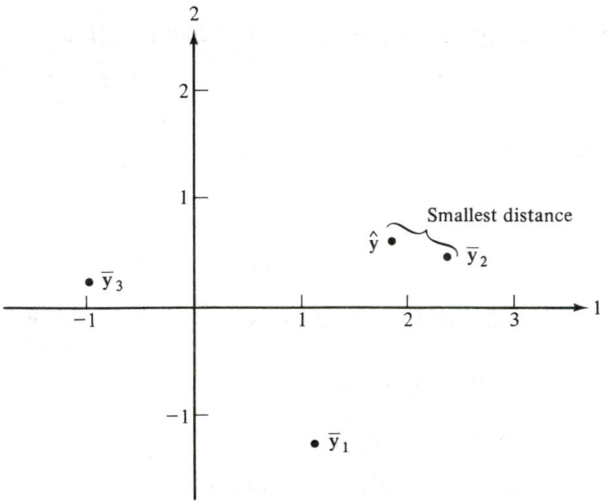

Figure 10.11 The points $\hat{\mathbf{y}}' = [\hat{y}_1, \hat{y}_2]$, $\bar{\mathbf{y}}_1' = [\bar{y}_{11}, \bar{y}_{12}]$, $\bar{\mathbf{y}}_2' = [\bar{y}_{21}, \bar{y}_{22}]$ and $\bar{\mathbf{y}}_3' = [\bar{y}_{31}, \bar{y}_{32}]$ in the classification plane.

Since the minimum of $\sum_{j=1}^{2} (\hat{y}_j - \bar{y}_{kj})^2$ occurs when $k = 2$, we allocate $\mathbf{x}_0$ to population π_2. The situation, in terms of the classifiers $\hat{y}_j$, is illustrated schematically in Figure 10.11. ■

Comment. When two linear discriminant functions are used for classification, observations are assigned to populations based on Euclidean distances in the two-dimensional discriminant space.

Up to this point we have not shown why the first few discriminants are more important than the last few. Their relative importance becomes apparent from their contribution to a numerical measure of spread of the populations. Consider the separatory measure

$$\Delta_S^2 = \sum_{i=1}^{g} (\boldsymbol{\mu}_i - \bar{\boldsymbol{\mu}})' \boldsymbol{\Sigma}^{-1} (\boldsymbol{\mu}_i - \bar{\boldsymbol{\mu}}) \tag{10-85}$$

where

$$\bar{\boldsymbol{\mu}} = \frac{1}{g} \sum_{i=1}^{g} \boldsymbol{\mu}_i$$

and $(\boldsymbol{\mu}_i - \bar{\boldsymbol{\mu}})' \boldsymbol{\Sigma}^{-1} (\boldsymbol{\mu}_i - \bar{\boldsymbol{\mu}})$ is the squared statistical distance from the ith population mean $\boldsymbol{\mu}_i$, to the centroid $\bar{\boldsymbol{\mu}}$. It can be shown (see Exercise 10.18) that $\Delta_S^2 = \lambda_1 + \lambda_2 + \cdots + \lambda_p$ where the $\lambda_1 \geq \lambda_2 \geq \cdots \geq \lambda_s$ are the *nonzero* eigenvalues of $\boldsymbol{\Sigma}^{-1}\mathbf{B}_0$(or $\boldsymbol{\Sigma}^{-1/2}\mathbf{B}_0\boldsymbol{\Sigma}^{-1/2}$) and $\lambda_{s+1}, \ldots, \lambda_p$ are the zero eigenvalues.

The separation given by Δ_S^2 can be reproduced in terms of discriminant means. The first discriminant $Y_1 = \mathbf{e}_1' \boldsymbol{\Sigma}^{-1/2} \mathbf{X}$ has means $\mu_{iY_1} = \mathbf{e}_1' \boldsymbol{\Sigma}^{-1/2} \boldsymbol{\mu}_i$ and the squared distance $\sum_{i=1}^{g} (\mu_{iY_1} - \bar{\mu}_{Y_1})^2$ of the μ_{iY_1}'s from the central value $\bar{\mu}_{Y_1} = \mathbf{e}_1' \boldsymbol{\Sigma}^{-1/2} \bar{\boldsymbol{\mu}}$ is λ_1 (see Exercise 10.18). Since Δ_S^2 can also be written as

$$\Delta_S^2 = \lambda_1 + \lambda_2 + \cdots + \lambda_p = \sum_{i=1}^{g} (\boldsymbol{\mu}_{iY} - \bar{\boldsymbol{\mu}}_Y)'(\boldsymbol{\mu}_{iY} - \bar{\boldsymbol{\mu}}_Y) = \sum_{i=1}^{g} (\mu_{iY_1} - \bar{\mu}_{Y_1})^2$$
$$+ \sum_{i=1}^{g} (\mu_{iY_2} - \bar{\mu}_{Y_2})^2 + \cdots + \sum_{i=1}^{g} (\mu_{iY_p} - \bar{\mu}_{Y_p})^2$$

it follows that the first discriminant makes the largest single contribution, λ_1, to the separatory measure Δ_S^2. In general the rth discriminant, $Y_r = \mathbf{e}_r' \boldsymbol{\Sigma}^{-1/2} \mathbf{X}$, contributes λ_r to Δ_S^2. If the next $s - r$ eigenvalues (recall $\lambda_{s+1} = \lambda_{s+2} = \cdots = \lambda_p = 0$) are such that $\lambda_{r+1} + \lambda_{r+2} + \cdots + \lambda_s$ is small compared to $\lambda_1 + \lambda_2 + \cdots + \lambda_r$, then the last discriminants $Y_{r+1}, Y_{r+2}, \ldots, Y_s$ can be neglected without appreciably decreasing the amount of separation.[14]

Not much is known about the efficacy of the allocation rule in (10-84). Some insight is provided by computer-generated sampling experiments and Lachenbruch

[14] See [11] for further optimal dimension reducing properties.

[14] summarizes its performance in particular cases. The development of the population result in (10-82) required a common covariance matrix $\mathbf{\Sigma}$. If this is essentially true and the samples are reasonably large, the rule of (10-84) should perform fairly well. In any event, performance can be checked by computing estimated error rates. Specifically, Lachenbruch's estimate of the expected actual error rate given by (10-71) should be calculated if the computing resources are available.

10.9 FINAL COMMENTS

Including Qualitative Variables

Our discussion in this chapter assumes that the discriminatory or classifatory variables, $X_1, X_2, \ldots, X_p$, have natural units of measurement. That is, each variable can, in principle, assume any real number and these numbers can be recorded. Often, a *qualitative* or *categorical variable* may be a useful discriminator (classifier). For example, the presence or absence of a characteristic such as the color red may be a worthwhile classifier. This situation is frequently handled by creating a variable, X, whose numerical value is 1 if the object possesses the characteristic and zero if the object does not possess the characteristic. This variable is then treated like the measured variables in the usual discrimination and classification procedures.

There is very little theory available to handle the case in which some variables are continuous and some qualitative. Computer simulation experiments (see Krzanowski [13]) indicate that Fisher's linear discriminant function can perform poorly or satisfactorily depending upon the correlations between the qualitative and continuous variables. As Krzanowski [13] notes: "A low correlation in one population but a high correlation in the other, or a change in the sign of the correlations between the two populations could indicate conditions unfavorable to Fisher's linear discriminant function." This is a troublesome area and one that needs further study.

Selection of Variables

In some applications of discriminant analysis, there is data available on a large number of variables. Mucciardi and Gose [16] discuss a discriminant analysis based on 157 variables.[15] In this case it would obviously be desirable to select a relatively small subset of variables that would contain almost as much information as the original collection. This is the objective of *stepwise discriminant analysis* and several popular commercial computer programs have a capability for stepwise discriminant analyses.

If a stepwise discriminant analysis (or any variable selection method) is employed, the results should be interpreted with caution (see [17]). There is no guarantee that the subset selected is "best," regardless of the criterion used to make the selection. For example, subsets selected on the basis of minimizing the apparent

[15] Imagine the problems of verifying the assumption of 157-variate normality and simultaneously estimating, for example, the 12,403 parameters of the 157×157 presumed common covariance matrix!

error rate or maximizing "discriminatory power" may perform poorly in future samples. Problems associated with variable selection procedures are magnified if there are large correlations among the variables or large correlations between linear combinations of the variables.

Choosing a subset of variables that seems to be optimal for *a given data set* is especially disturbing if classification is the objective. At the very least, the derived classification function should be evaluated with a validation sample. As Murray [17] suggests, a better idea might be to split the sample into a number of groups and determine the "best" subset for each group. The number of times a given variable appears in the best subsets provides a measure of the worth of that variable for future classification.

Testing for Group Differences

We have pointed out in connection with two group classification that effective allocation is probably not possible unless the populations are well separated. The same is true for the many-group situation. Classification is ordinarily not attempted unless the population mean vectors differ significantly from one another. Assuming the data are nearly multivariate normal, with a common covariance matrix, MANOVA can be performed to test for differences in the population mean vectors. Although apparent significant differences do not automatically imply effective classification, testing is a necessary first step. If no significant differences are found, constructing classification rules will probably be a waste of time.

Graphics

Sophisticated computer graphics now allow one to visually examine multivariate data in two and three dimensions. Thus groupings in the variable space for any choice of two or three variables can often be discerned by eye. In this way, potentially important classifying variables are often identified and outlying, or "atypical," observations revealed. Visual displays are important aids in discrimination and classification, and their use is likely to increase as the hardware and associated computer programs become readily available. Frequently, as much can be learned for a visual examination as by a complex numerical analysis.

Practical Considerations Regarding Multivariate Normality

The interplay between the choice of tentative assumptions and the form of the resulting classifier is important. Consider Figure 10.12, which shows the kidney-shaped density contours from two very nonnormal densities. In this case the normal theory linear (or even quadratic) classification rule will be inadequate compared to another choice. That is, linear discrimination here is inappropriate.

Often discrimination is attempted with a large number of variables where some of the variables are qualitative of the presence-absence, or 0–1, type. In these situations and other situations with restricted ranges for the variables, multivariate normality may not be a sensible assumption. As we have seen, classification based

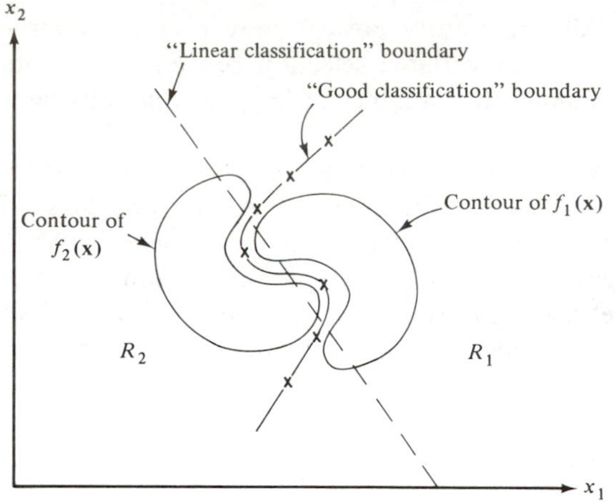

Figure 10.12 Two nonnormal populations for which linear discrimination is inappropriate.

on Fisher's linear discriminants can be optimal from a minimum ECM or minimum TPM point of view only when multivariate normality holds. How are we to interpret these quantities when normality is clearly not viable?

In the absence of multivariate normality, Fisher's linear discriminants can be viewed as providing an approximation to the total sample information. The values of the first few discriminants themselves can be checked for normality, and rule (10-84) employed. Since the discriminants are linear combinations of a large number of variables, they will often be nearly normal. Of course one must keep in mind that the first few discriminants are an *incomplete* summary of the original sample information. Classification rules based on this restricted set may perform poorly, while optimal rules derived from all of the sample information may perform well.

EXERCISES

10.1. Consider the two data sets

$$\mathbf{X}_1 = \begin{bmatrix} 3 & 2 & 4 \\ 7 & 4 & 7 \end{bmatrix} \quad \text{and} \quad \mathbf{X}_2 = \begin{bmatrix} 6 & 5 & 4 \\ 9 & 7 & 8 \end{bmatrix}$$

for which

$$\bar{\mathbf{x}}_1 = \begin{bmatrix} 3 \\ 6 \end{bmatrix}, \quad \bar{\mathbf{x}}_2 = \begin{bmatrix} 5 \\ 8 \end{bmatrix}$$

and

$$\mathbf{S}_{\text{pooled}} = \begin{bmatrix} 1 & 1 \\ 1 & 2 \end{bmatrix}$$

(a) Calculate Fisher's linear discriminant function.

(b) Classify the observation $\mathbf{x}_0' = [2 \quad 7]$ as population π_1 or population π_2 using Rule (10-18).

10.2. (a) Develop a linear classification function for the data in Example 10.3 using Fisher's method.

(b) Using the function in (a), construct the "confusion matrix" by classifying the given observations. Compare your classification results with those of Figure 10.3 where the classification regions were determined "by eye" (see Example 10.7).

(c) Given the results in (b), calculate the apparent error rate (AER).

(d) State any assumptions you make to justify the use of Fisher's method.

10.3. Consider the linear function $Y = \boldsymbol{\ell}'\mathbf{X}$. Let $E(\mathbf{X}) = \boldsymbol{\mu}_1$ and $\text{Cov}(\mathbf{X}) = \boldsymbol{\Sigma}$ if $\mathbf{X}$ belongs to population π_1. Let $E(\mathbf{X}) = \boldsymbol{\mu}_2$ and $\text{Cov}(\mathbf{X}) = \boldsymbol{\Sigma}$ if $\mathbf{X}$ belongs to population π_2. Let $m = \frac{1}{2}(\mu_{1Y} + \mu_{2Y}) = \frac{1}{2}(\boldsymbol{\ell}'\boldsymbol{\mu}_1 + \boldsymbol{\ell}'\boldsymbol{\mu}_2)$. Given $\boldsymbol{\ell}' = (\boldsymbol{\mu}_1 - \boldsymbol{\mu}_2)'\boldsymbol{\Sigma}^{-1}$, show each of the following.

(a) $E(\boldsymbol{\ell}'\mathbf{X} \mid \pi_1) - m = \boldsymbol{\ell}'\boldsymbol{\mu}_1 - m > 0$

(b) $E(\boldsymbol{\ell}'\mathbf{X} \mid \pi_2) - m = \boldsymbol{\ell}'\boldsymbol{\mu}_2 - m < 0$

(*Hint*: Recall $\boldsymbol{\Sigma}$ is of full rank and is positive definite, so $\boldsymbol{\Sigma}^{-1}$ exists and is positive definite.)

10.4. Show that $\hat{\boldsymbol{\ell}}' = (\bar{\mathbf{x}}_1 - \bar{\mathbf{x}}_2)'\mathbf{S}_{\text{pooled}}^{-1}$ maximizes the ratio in (10-19).
[*Hint*: Regard the sample quantities as fixed and use (2-50).]

10.5. Show that $\max_{\hat{\boldsymbol{\ell}}}\left[\dfrac{(\hat{\boldsymbol{\ell}}'\mathbf{d})^2}{\hat{\boldsymbol{\ell}}'\mathbf{S}_{\text{pooled}}\hat{\boldsymbol{\ell}}}\right]$ is D^2, the sample squared Mahalanobis distance, by substituting $\hat{\boldsymbol{\ell}}' = (\bar{\mathbf{x}}_1 - \bar{\mathbf{x}}_2)'\mathbf{S}_{\text{pooled}}^{-1}$ [see Equation (10-21)].

10.6. Show that

$$-\tfrac{1}{2}(\mathbf{x} - \boldsymbol{\mu}_1)'\boldsymbol{\Sigma}^{-1}(\mathbf{x} - \boldsymbol{\mu}_1) + \tfrac{1}{2}(\mathbf{x} - \boldsymbol{\mu}_2)'\boldsymbol{\Sigma}^{-1}(\mathbf{x} - \boldsymbol{\mu}_2)$$
$$= (\boldsymbol{\mu}_1 - \boldsymbol{\mu}_2)'\boldsymbol{\Sigma}^{-1}\mathbf{x} - \tfrac{1}{2}(\boldsymbol{\mu}_1 - \boldsymbol{\mu}_2)'\boldsymbol{\Sigma}^{-1}(\boldsymbol{\mu}_1 + \boldsymbol{\mu}_2)$$

[see Equation (10-36)].

10.7. Derive the expressions in (10-40) from (10-29) when $f_1(\mathbf{x})$ and $f_2(\mathbf{x})$ are multivariate normal densities with means $\boldsymbol{\mu}_1, \boldsymbol{\mu}_2$ and covariances $\boldsymbol{\Sigma}_1, \boldsymbol{\Sigma}_2$, respectively.

10.8. Show that the ratio in (10-73) is proportional to the ratio in (10-6) when $g = 2$. [*Hint*: Note that $(\boldsymbol{\mu}_i - \bar{\boldsymbol{\mu}})(\boldsymbol{\mu}_i - \bar{\boldsymbol{\mu}})' = \frac{1}{4}(\boldsymbol{\mu}_1 - \boldsymbol{\mu}_2)(\boldsymbol{\mu}_1 - \boldsymbol{\mu}_2)'$ for $i = 1, 2$, where $\bar{\boldsymbol{\mu}} = \frac{1}{2}(\boldsymbol{\mu}_1 + \boldsymbol{\mu}_2)$.]

10.9. Suppose that $n_1 = 11$ and $n_2 = 12$ observations are made on two random variables X_1 and X_2, where X_1 and X_2 are assumed to have a bivariate normal distribution with a common covariance matrix, $\boldsymbol{\Sigma}$, but possibly different mean vectors $\boldsymbol{\mu}_1$ and $\boldsymbol{\mu}_2$. The sample mean vectors and pooled covariance matrix are

$$\bar{\mathbf{x}}_1 = \begin{bmatrix} -1 \\ -1 \end{bmatrix}; \quad \bar{\mathbf{x}}_2 = \begin{bmatrix} 2 \\ 1 \end{bmatrix}$$
$$\mathbf{S}_{\text{pooled}} = \begin{bmatrix} 7.3 & -1.1 \\ -1.1 & 4.8 \end{bmatrix}$$

(a) Test for the difference in population mean vectors using Hotelling's two-sample T^2-statistic. Let $\alpha = .10$.

(b) Construct Fisher's (sample) linear discriminant function.

(c) Consider the observation $\mathbf{x}_0' = [0 \quad 1]$. Does this observation belong to population π_1 or π_2? Assume equal costs and equal prior probabilities.

10.10. A researcher wants to determine a procedure for discriminating between two multivariate populations. The researcher has enough data available to estimate the density functions $f_1(\mathbf{x})$ and $f_2(\mathbf{x})$ associated with populations π_1 and π_2, respectively. Let

$c(2 \mid 1) = 50$ (this is cost of assigning items as π_2, given that π_1 is true), and $c(1 \mid 2) = 100$.

In addition, it is known that about 20% of all possible items (for which the measurements **x** can be recorded) belong to π_2.

(a) Give the minimum ECM rule (general form) for assigning a new item to one of the two populations.

(b) Measurements recorded on a new item yield the density values $f_1(\mathbf{x}) = .3$ and $f_2(\mathbf{x}) = .5$. Given the information above, assign this item to population π_1 or population π_2.

10.11. Suppose a univariate random variable X has a normal distribution with variance 4. If X is from population π_1, its mean is 10; if it is from population π_2, its mean is 14. Assume equal prior probabilities for the events $A1 = X$ is from population π_1, and $A2 = X$ is from population π_2, and assume that the misclassification costs $c(2 \mid 1)$ and $c(1 \mid 2)$ are equal (for instance, \$10). We decide that we shall allocate (classify) X to population π_1 if $X \le c$, for some c to be determined, and to population π_2 if $X > c$. Let $B1$ be the event X is classified as from population π_1, and $B2$ be the event X is classified as from population π_2. Make a table showing the following: $P(B1 \mid A2)$, $P(B2 \mid A1)$, $P(A1 \text{ and } B2)$, $P(A2 \text{ and } B1)$, $P(\text{misclassification})$, and expected cost for various values of c. For what choice of c is expected cost minimized? The table should take the following form

c:	$P(B1 \mid A2)$	$P(B2 \mid A1)$	$P(A1 \text{ and } B2)$	$P(A2 \text{ and } B1)$	$P(\text{error})$	Expected cost
10						
⋮						
14						

What is the value of the minimum expected cost?

10.12. Repeat Exercise 10.11 if the prior probabilities of $A1$ and $A2$ are equal, but $c(2 \mid 1) = \$5$ and $c(1 \mid 2) = \$15$.

10.13. Repeat Exercise 10.11 if the prior probabilities of $A1$ and $A2$ are $P(A1) = .25$ and $P(A2) = .75$ and the misclassification costs are as in Exercise 10.12.

10.14. Suppose **x** comes from one of two populations:

π_1: Normal with mean $\boldsymbol{\mu}_1$ and covariance matrix $\boldsymbol{\Sigma}_1$.

π_2: Normal with mean $\boldsymbol{\mu}_2$ and covariance matrix $\boldsymbol{\Sigma}_2$.

If the respective density functions are denoted by $f_1(\mathbf{x})$ and $f_2(\mathbf{x})$, find the expression for the quadratic discriminator Q, where

$$Q = \ln\left[\frac{f_1(\mathbf{x})}{f_2(\mathbf{x})}\right]$$

If $\boldsymbol{\Sigma}_1 = \boldsymbol{\Sigma}_2 = \boldsymbol{\Sigma}$, for instance, verify that Q becomes

$$(\boldsymbol{\mu}_1 - \boldsymbol{\mu}_2)'\boldsymbol{\Sigma}^{-1}\mathbf{x} - \tfrac{1}{2}(\boldsymbol{\mu}_1 - \boldsymbol{\mu}_2)'\boldsymbol{\Sigma}^{-1}(\boldsymbol{\mu}_1 + \boldsymbol{\mu}_2)$$

10.15. Suppose populations π_1 and π_2 are as follows

	Population:	
	π_1	π_2
Distribution:	Normal	Normal
Mean $\boldsymbol{\mu}$:	$[10, 15]'$	$[20, 25]'$
Variance-Covariance $\boldsymbol{\Sigma}$:	$\begin{bmatrix} 18 & 12 \\ 12 & 32 \end{bmatrix}$	$\begin{bmatrix} 20 & -7 \\ -7 & 5 \end{bmatrix}$

Assume equal prior probabilities and misclassifications costs of $c(2 \mid 1) = \$10$ and $c(1 \mid 2) = \$73.89$. Find the posterior probabilities of populations π_1 and π_2, $P(\pi_1 \mid \mathbf{x})$ and $P(\pi_2 \mid \mathbf{x})$, the value of the quadratic discriminator Q in Exercise 10.14, and the classification for each value of $\mathbf{x}$ below.

$\mathbf{x}$	$P(\pi_1 \mid \mathbf{x})$	$P(\pi_2 \mid \mathbf{x})$	Q Classification
$[10, 15]'$			
$[12, 17]'$			
$\vdots$			
$[30, 35]'$			

(*Note*: Use an increment of 2 in each coordinate—11 points in all.)

Show each of the following on a graph of the x_1, x_2 plane.

(a) The mean of each population.

(b) The ellipse of minimal area with probability .95 of containing $\mathbf{x}$ for each population.

(c) The region R_1 (for population π_1) and the region $\Omega - R_1 = R_2$ (for population π_2).

(d) The 11 points classified in the table.

10.16. If $\mathbf{B}$ is defined as $c(\boldsymbol{\mu}_1 - \boldsymbol{\mu}_2)(\boldsymbol{\mu}_1 - \boldsymbol{\mu}_2)'$ for some constant c, verify that $\mathbf{e} = c\boldsymbol{\Sigma}^{-1}(\boldsymbol{\mu}_1 - \boldsymbol{\mu}_2)$ is in fact an (unscaled) eigenvector of $\boldsymbol{\Sigma}^{-1}\mathbf{B}$, where $\boldsymbol{\Sigma}$ is a covariance matrix.

10.17. **(a)** Using the original data sets $\mathbf{X}_1$ and $\mathbf{X}_2$ given in Example 10.8, calculate $\bar{\mathbf{x}}_i, \mathbf{S}_i$, $i = 1, 2$, and $\mathbf{S}_{\text{pooled}}$, verifying the results provided for these quantities in the example.

(b) Using the calculations in Part a, compute Fisher's linear discriminant function and use it to classify the sample observations according to Rule (10-18). Verify that the confusion matrix given in Example 10.8 is correct.

(c) Classify the sample observations on the basis of smallest Mahalanobis distance of the observations from the group means $\bar{\mathbf{x}}_1$ and $\bar{\mathbf{x}}_2$. [See (10-68).] Compare the results with those in Part b. Comment.

10.18. Show that $\Delta_S^2 = \lambda_1 + \lambda_2 + \cdots + \lambda_p = \lambda_1 + \lambda_2 + \cdots + \lambda_s$, where $\lambda_1, \lambda_2, \ldots, \lambda_s$ are the nonzero eigenvalues of $\boldsymbol{\Sigma}^{-1}\mathbf{B}_0$ (or $\boldsymbol{\Sigma}^{-1/2}\mathbf{B}_0\boldsymbol{\Sigma}^{-1/2}$) and Δ_S^2 is given by (10-85). Also show that $\lambda_1 + \lambda_2 + \cdots + \lambda_r$ is the resulting separation when only the first r discriminants, $Y_1, Y_2, \ldots, Y_r$ are used.

[*Hint*: Let $\mathbf{P}$ be the orthogonal matrix whose ith row, $\mathbf{e}_i'$, is the eigenvector of $\mathbf{\Sigma}^{-1/2}\mathbf{B}_0\mathbf{\Sigma}^{-1/2}$ corresponding to the ith largest eigenvalue, $i = 1, 2, \ldots, p$. Consider

$$\underset{(p\times 1)}{\mathbf{Y}} = \begin{bmatrix} Y_1 \\ \vdots \\ Y_s \\ \vdots \\ Y_p \end{bmatrix} = \begin{bmatrix} \mathbf{e}_1'\mathbf{\Sigma}^{-1/2}\mathbf{X} \\ \vdots \\ \mathbf{e}_s'\mathbf{\Sigma}^{-1/2}\mathbf{X} \\ \vdots \\ \mathbf{e}_p'\mathbf{\Sigma}^{-1/2}\mathbf{X} \end{bmatrix} = \mathbf{P}\mathbf{\Sigma}^{-1/2}\mathbf{X}$$

Now $\boldsymbol{\mu}_{iY} = E(\mathbf{Y} \mid \pi_i) = \mathbf{P}\mathbf{\Sigma}^{-1/2}\boldsymbol{\mu}_i$ and $\bar{\boldsymbol{\mu}}_Y = \mathbf{P}\mathbf{\Sigma}^{-1/2}\bar{\boldsymbol{\mu}}$, so

$$\begin{aligned}(\boldsymbol{\mu}_{iY} - \bar{\boldsymbol{\mu}}_Y)'(\boldsymbol{\mu}_{iY} - \bar{\boldsymbol{\mu}}_Y) &= (\boldsymbol{\mu}_i - \bar{\boldsymbol{\mu}})'\mathbf{\Sigma}^{-1/2}\mathbf{P}'\mathbf{P}\mathbf{\Sigma}^{-1/2}(\boldsymbol{\mu}_i - \bar{\boldsymbol{\mu}}) \\ &= (\boldsymbol{\mu}_i - \bar{\boldsymbol{\mu}})'\mathbf{\Sigma}^{-1}(\boldsymbol{\mu}_i - \bar{\boldsymbol{\mu}})\end{aligned}$$

Therefore $\Delta_S^2 = \sum_{i=1}^{g} (\boldsymbol{\mu}_{iY} - \bar{\boldsymbol{\mu}}_Y)'(\boldsymbol{\mu}_{iY} - \bar{\boldsymbol{\mu}}_Y)$. Using Y_1, we have

$$\begin{aligned}\sum_{i=1}^{g} (\mu_{iY_1} - \bar{\mu}_{Y_1})^2 &= \sum_{i=1}^{g} \mathbf{e}_1'\mathbf{\Sigma}^{-1/2}(\boldsymbol{\mu}_i - \bar{\boldsymbol{\mu}})(\boldsymbol{\mu}_i - \bar{\boldsymbol{\mu}})'\mathbf{\Sigma}^{-1/2}\mathbf{e}_1 \\ &= \mathbf{e}_1'\mathbf{\Sigma}^{-1/2}\mathbf{B}_0\mathbf{\Sigma}^{-1/2}\mathbf{e}_1 = \lambda_1\end{aligned}$$

because $\mathbf{e}_1$ has eigenvalue λ_1. Similarly, Y_2 produces

$$\sum_{i=1}^{g} (\mu_{iY_2} - \bar{\mu}_{Y_2})^2 = \mathbf{e}_2'\mathbf{\Sigma}^{-1/2}\mathbf{B}_0\mathbf{\Sigma}^{-1/2}\mathbf{e}_2 = \lambda_2$$

and Y_p produces

$$\sum_{i=1}^{g} (\mu_{iY_p} - \bar{\mu}_{Y_p})^2 = \mathbf{e}_p'\mathbf{\Sigma}^{-1/2}\mathbf{B}_0\mathbf{\Sigma}^{-1/2}\mathbf{e}_p = \lambda_p$$

Thus

$$\begin{aligned}\Delta_S^2 &= \sum_{i=1}^{g} (\boldsymbol{\mu}_{iY} - \bar{\boldsymbol{\mu}}_Y)'(\boldsymbol{\mu}_{iY} - \bar{\boldsymbol{\mu}}_Y) \\ &= \sum_{i=1}^{g} (\mu_{iY_1} - \bar{\mu}_{Y_1})^2 + \sum_{i=1}^{g} (\mu_{iY_2} - \bar{\mu}_{Y_2})^2 + \cdots + \sum_{i=1}^{g} (\mu_{iY_p} - \bar{\mu}_{Y_p})^2 \\ &= \lambda_1 + \lambda_2 + \cdots + \lambda_p = \lambda_1 + \lambda_2 + \cdots + \lambda_s\end{aligned}$$

since $\lambda_{s+1} = \cdots = \lambda_p = 0$. If only the first r discriminants are used, their contribution to Δ_S^2 is $\lambda_1 + \lambda_2 + \cdots + \lambda_r$.]

The following exercises require the use of a computer.

10.19. Consider the data give in the Exercise 1.14.

(a) Check the marginal distributions of the x_i's in both the multiple-sclerosis (MS) group and non–multiple-sclerosis (NMS) group for normality by graphing the corresponding observations as normal probability plots. Suggest appropriate data transformations if the normality assumption is suspect.

(b) Assume $\mathbf{\Sigma}_1 = \mathbf{\Sigma}_2 = \mathbf{\Sigma}$. Construct Fisher's linear discriminant function. Do all the variables in the discriminant function appear to be important? Discuss. Develop a classification rule assuming equal prior probabilities and equal costs of misclassification.

(c) Using the results in (b), calculate the apparent error rate. If computing resources allow, calculate an estimate of the expected actual error rate using Lachenbruch's holdout procedure. Compare the two error rates.

10.20. Annual financial data are collected for firms approximately 2 years prior to bankruptcy and for financially sound firms at about the same point in time. The data on four variables, x_1 = CF/TD = (cash flow)/(total debt), X_2 = NI/TA = (net income)/(total assets), x_3 = CA/CL = (current assets)/(current liabilities), and X_4 = CA/NS = (current assets)/(net sales) are given in Table 10.3.

(a) Plot the data for the pairs of observations (x_1, x_2), (x_1, x_3), and (x_1, x_4). Does it appear as if the data are approximately bivariate normal for any of these pairs of variables?

(b) Using the $n_1 = 21$ pairs of observations (x_1, x_2) for bankrupt firms and the $n_2 = 25$ pairs of observations (x_1, x_2) for nonbankrupt firms, calculate the sample mean vectors $\bar{\mathbf{x}}_1$ and $\bar{\mathbf{x}}_2$ and the sample covariance matrices $\mathbf{S}_1$ and $\mathbf{S}_2$.

(c) Using the results in (b) and assuming that both random samples are from bivariate normal populations, construct the classification rule in (10-42) with $p_1 = p_2$ and $c(1 \mid 2) = c(2 \mid 1)$.

(d) Evaluate the performance of the classification rule developed in (c) by computing the apparent error rate (APER) from (10-47) and the estimated expected actual error rate $\hat{E}$(AER) from (10-49).

(e) Repeat Parts c and d, assuming $p_1 = .05$, $p_2 = .95$, and $c(1 \mid 2) = c(2 \mid 1)$. Is this choice of prior probabilities reasonable? Explain.

(f) Using the results in (b), form the pooled covariance matrix, $\mathbf{S}_{\text{pooled}}$, and construct Fisher's sample linear discriminant function in (10-16). Use this function to classify the sample observations and evaluate the APER. Is Fisher's linear discriminant function a sensible choice for a classifier in this case? Explain.

(g) Repeat Parts b–e using the observation pairs (x_1, x_3) and (x_1, x_4). Do some variables appear to be better classifiers than others? Explain.

(h) Repeat Parts b–e using observations on all four variables (X_1, X_2, X_3, X_4).

10.21. The data in Table 10.4 consists of observations on X_1 = sepal width and X_2 = petal width for samples from three species of iris. There are $n_1 = n_2 = n_3 = 50$ observations in each sample.

(a) Plot the data in the (x_1, x_2) variable space. Do the observations for the three groups appear to be bivariate normal?

(b) Assume the samples are from bivariate normal populations with a common covariance matrix. Test the hypothesis $H_0: \boldsymbol{\mu}_1 = \boldsymbol{\mu}_2 = \boldsymbol{\mu}_3$, versus H_1: at least one $\boldsymbol{\mu}_i$ different from the others at the $\alpha = .05$ significance level. Is the common covariance matrix assumption reasonable in this case? Explain.

(c) Assuming the populations are bivariate normal, construct the quadratic discriminate scores $\hat{d}_i^Q(\mathbf{x})$ given by (10-60) with $p_1 = p_2 = p_3 = \frac{1}{3}$. Using Rule (10-61), classify the new observation $\mathbf{x}_0' = [3.5 \quad 1.75]$ as population π_1, π_2, or π_3.

(d) Assume the covariance matrices $\boldsymbol{\Sigma}_i$ are the same for all three bivariate normal populations. Construct the linear discriminate score $\hat{d}_i(\mathbf{x})$ given by (10-65) and use it to assign $\mathbf{x}_0' = [3.5 \quad 1.75]$ to one of the populations π_i, $i = 1, 2, 3$, according to (10-66). Take $p_1 = p_2 = p_3 = \frac{1}{3}$. Compare the results in Parts c and d. Which approach do you prefer? Explain.

(e) Assuming equal covariance matrices, bivariate normal populations, and $p_1 = p_2 = p_3 = \frac{1}{3}$, allocate $\mathbf{x}_0' = [3.5 \quad 1.75]$ to π_1, π_2, or π_3 using Rule (10-70). Compare the

TABLE 10.3 BANKRUPTCY DATA

Row	$x_1 = \frac{CF}{TD}$	$x_2 = \frac{NI}{TA}$	$x_3 = \frac{CA}{CL}$	$x_4 = \frac{CA}{NS}$	Population π_i
1	−.4485	−.4106	1.0865	.4526	1.
2	−.5633	−.3114	1.5134	.1642	1.
3	.0643	.0156	1.0077	.3978	1.
4	−.0721	−.0930	1.4544	.2589	1.
5	−.1002	−.0917	1.5644	.6683	1.
6	−.1421	−.0651	.7066	.2794	1.
7	.0351	.0147	1.5046	.7080	1.
8	−.0653	−.0566	1.3737	.4032	1.
9	.0724	−.0076	1.3723	.3361	1.
10	−.1353	−.1433	1.4196	.4347	1.
11	−.2298	−.2961	.3310	.1824	1.
12	.0713	.0205	1.3124	.2497	1.
13	.0109	.0011	2.1495	.6969	1.
14	−.2777	−.2316	1.1918	.6601	1.
15	.1454	.0500	1.8762	.2723	1.
16	.3703	.1098	1.9941	.3828	1.
17	−.0757	−.0821	1.5077	.4215	1.
18	.0451	.0263	1.6756	.9494	1.
19	.0115	−.0032	1.2602	.6038	1.
20	.1227	.1055	1.1434	.1655	1.
21	−.2843	−.2703	1.2722	.5128	1.
22	.5135	.1001	2.4871	.5368	2.
23	.0769	.0195	2.0069	.5304	2.
24	.3776	.1075	3.2651	.3548	2.
25	.1933	.0473	2.2506	.3309	2.
26	.3248	.0718	4.2401	.6279	2.
27	.3132	.0511	4.4500	.6852	2.
28	.1184	.0499	2.5210	.6925	2.
29	−.0173	.0233	2.0538	.3484	2.
30	.2169	.0779	2.3489	.3970	2.
31	.1703	.0695	1.7973	.5174	2.
32	.1460	.0518	2.1692	.5500	2.
33	−.0985	−.0123	2.5029	.5778	2.
34	.1398	−.0312	.4611	.2643	2.
35	.1379	.0728	2.6123	.5151	2.
36	.1486	.0564	2.2347	.5563	2.
37	.1633	.0486	2.3080	.1978	2.
38	.2907	.0597	1.8381	.3786	2.
39	.5383	.1064	2.3293	.4835	2.
40	−.3330	−.0854	3.0124	.4730	2.
41	.4785	.0910	1.2444	.1847	2.
42	.5603	.1112	4.2918	.4443	2.
43	.2029	.0792	1.9936	.3018	2.
44	.4746	.1380	2.9166	.4487	2.
45	.1661	.0351	2.4527	.1370	2.
46	.5808	.0371	5.0594	.1268	2.

SOURCE: 1968, 1969, 1970, 1971, 1972 Moody's Industrial Manuals.
LEGEND: π_1: Bankrupt firms π_2: Nonbankrupt firms.

TABLE 10.4 IRIS DATA

π_1: *Iris setosa* x_1 = sepal width	π_1: *Iris setosa* x_2 = petal width	π_2: *Iris versicolor* x_1 = sepal width	π_2: *Iris versicolor* x_2 = petal width	π_3: *Iris virginica* x_1 = sepal width	π_3: *Iris virginica* x_2 = petal width
3.5	0.2	3.2	1.4	3.3	2.5
3.0	0.2	3.2	1.5	2.7	1.9
3.2	0.2	3.1	1.5	3.0	2.1
3.1	0.2	2.3	1.3	2.9	1.8
3.6	0.2	2.8	1.5	3.0	2.2
3.9	0.4	2.8	1.3	3.0	2.1
3.4	0.3	3.3	1.6	2.5	1.7
3.4	0.2	2.4	1.0	2.9	1.8
2.9	0.2	2.9	1.3	2.5	1.8
3.1	0.1	2.7	1.4	3.6	2.5
3.7	0.2	2.0	1.0	3.2	2.0
3.4	0.2	3.0	1.5	2.7	1.9
3.0	0.1	2.2	1.0	3.0	2.1
3.0	0.1	2.9	1.4	2.5	2.0
4.0	0.2	2.9	1.3	2.8	2.4
4.4	0.4	3.1	1.4	3.2	2.3
3.9	0.4	3.0	1.5	3.0	1.8
3.5	0.3	2.7	1.0	3.8	2.2
3.8	0.3	2.2	1.5	2.6	2.3
3.8	0.3	2.5	1.1	2.2	1.5
3.4	0.2	3.2	1.8	3.2	2.3
3.7	0.4	2.8	1.3	2.8	2.0
3.6	0.2	2.5	1.5	2.8	2.0
3.3	0.5	2.8	1.2	2.7	1.8
3.4	0.2	2.9	1.3	3.3	2.1
3.0	0.2	3.0	1.4	3.2	1.8
3.4	0.4	2.8	1.4	2.8	1.8
3.5	0.2	3.0	1.7	3.0	1.8
3.4	0.2	2.9	1.5	2.8	2.1
3.2	0.2	2.6	1.0	3.0	1.6
3.1	0.2	2.4	1.1	2.8	1.9
3.4	0.4	2.4	1.0	3.8	2.0
4.1	0.1	2.7	1.2	2.8	2.2
4.2	0.2	2.7	1.6	2.8	1.5
3.1	0.2	3.0	1.5	2.6	1.4
3.2	0.2	3.4	1.6	3.0	2.3
3.5	0.2	3.1	1.5	3.4	2.4
3.6	0.1	2.3	1.3	3.1	1.8
3.0	0.2	3.0	1.3	3.0	1.8
3.4	0.2	2.5	1.3	3.1	2.1
3.5	0.3	2.6	1.2	3.1	2.4
2.3	0.3	3.0	1.4	3.1	2.3
3.2	0.2	2.6	1.2	2.7	1.9
3.5	0.6	2.3	1.0	3.2	2.3
3.8	0.4	2.7	1.3	3.3	2.5
3.0	0.3	3.0	1.2	3.0	2.3
3.8	0.2	2.9	1.3	2.5	1.9
3.2	0.2	2.9	1.3	3.0	2.0
3.7	0.2	2.5	1.1	3.4	2.3
3.3	0.2	2.8	1.3	3.0	1.8

SOURCE: Fisher [5].

result with that in Part d. Delineate the classification regions $\hat{R}_1$, $\hat{R}_2$, and $\hat{R}_3$ on your graph from Part a determined by the linear functions $\hat{d}_{ki}(\mathbf{x}_0)$ in (10-70).

(f) Using the linear discriminant scores from Part d, classify the sample observations. Calculate the APER and $\hat{E}$(AER). (To calculate the latter, you should use Lachenbruch's holdout procedure [see (10-71)]. This will require a fair amount of computing effort.)

10.22. The GPA and GMAT data alluded to in Example 10.11 are listed in Table 10.5.

(a) Using these data, calculate $\bar{\mathbf{x}}_1$, $\bar{\mathbf{x}}_2$, $\bar{\mathbf{x}}_3$, $\bar{\mathbf{x}}$, and $\mathbf{S}_{\text{pooled}}$ and thus verify the results for these quantities given in Example 10.11.

(b) Calculate $\mathbf{W}^{-1}$ and $\hat{\mathbf{B}}_0$ and the eigenvalues and eigenvectors of $\mathbf{W}^{-1}\hat{\mathbf{B}}_0$ Use the linear discriminants derived from these eigenvectors to classify the new observation $\mathbf{x}_0' = [3.21 \quad 497]$ into one of the populations π_1: admit, π_2: not admit, and π_3: borderline. Does the classification agree with that in Example 10.11? Should it? Explain.

TABLE 10.5 ADMISSION DATA FOR GRADUATE SCHOOL OF BUSINESS

π_1: Admit			π_2: Not admit			π_3: Borderline		
Applicant no.	GPA (x_1)	GMAT (x_2)	Applicant no.	GPA (x_1)	GMAT (x_2)	Applicant no.	GPA (x_1)	GMAT (x_2)
1	2.96	596.	32	2.54	446.	60	2.86	494.
2	3.14	473.	33	2.43	425.	61	2.85	496.
3	3.22	482.	34	2.20	474.	62	3.14	419.
4	3.29	527.	35	2.36	531.	63	3.28	371.
5	3.69	505.	36	2.57	542.	64	2.89	447.
6	3.46	693.	37	2.35	406	65	3.15	313.
7	3.03	626.	38	2.51	412.	66	3.50	402.
8	3.19	663.	39	2.51	458.	67	2.89	485.
9	3.63	447.	40	2.36	399.	68	2.80	444.
10	3.59	588.	41	2.36	482.	69	3.13	416.
11	3.30	563.	42	2.66	420	70	3.01	471.
12	3.40	553.	43	2.68	414.	71	2.79	490.
13	3.50	572.	44	2.48	533.	72	2.89	431.
14	3.78	591.	45	2.46	509.	73	2.91	446.
15	3.44	692.	46	2.63	504.	74	2.75	546.
16	3.48	528.	47	2.44	336.	75	2.73	467.
17	3.47	552.	48	2.13	408.	76	3.12	463.
18	3.35	520.	49	2.41	469.	77	3.08	440.
19	3.39	543.	50	2.55	538.	78	3.03	419.
20	3.28	523.	51	2.31	505.	79	3.00	509.
21	3.21	530.	52	2.41	489.	80	3.03	438.
22	3.58	564.	53	2.19	411.	81	3.05	399.
23	3.33	565.	54	2.35	321.	82	2.85	483.
24	3.40	431.	55	2.60	394.	83	3.01	453.
25	3.38	605.	56	2.55	528	84	3.03	414.
26	3.26	664.	57	2.72	399.	85	3.04	446.
27	3.60	609.	58	2.85	381.			
28	3.37	559.	59	2.90	384.			
29	3.80	521.						
30	3.76	646.						
31	3.24	467.						

10.23. Gerrild and Lantz [8] chemically analyzed crude-oil samples from three zones of sandstone.

π_1: Wilhelm
π_2: Sub-Mulinia
π_3: Upper (Mulinia, second sub-scales, first sub-scales)

The values of the trace elements

$$X_1 = \text{vanadium (in percent ash)}$$

$$X_2 = \text{iron (in percent ash)}$$

$$X_3 = \text{berylium (in percent ash)}$$

and two measures of hydrocarbons

$$X_4 = \text{saturated hydrocarbons (in percent area)}$$

$$X_5 = \text{aromatic hydrocarbons (in percent area)}$$

are presented for 56 cases in Table 10.6. The last two measurements are determined from areas under a gas-liquid chromatography curve.

(a) Obtain the estimated minimum TPM rule assuming normality. Comment on the adequacy of the normal assumption.

(b) Determine the estimate of $E(\text{AER})$ using Lachenbruch's holdout procedure. Also give the confusion matrix.

(c) Consider various transformations of the data to normality (see Example 10.13) and repeat Parts a and b.

TABLE 10.6 CRUDE-OIL DATA

	x_1	x_2	x_3	x_4	x_5
π_1	3.9	51.0	0.20	7.06	12.19
	2.7	49.0	0.07	7.14	12.23
	2.8	36.0	0.30	7.00	11.30
	3.1	45.0	0.08	7.20	13.01
	3.5	46.0	0.10	7.81	12.63
	3.9	43.0	0.07	6.25	10.42
	2.7	35.0	0.00	5.11	9.00
π_2	5.0	47.0	0.07	7.06	6.10
	3.4	32.0	0.20	5.82	4.69
	1.2	12.0	0.00	5.54	3.15
	8.4	17.0	0.07	6.31	4.55
	4.2	36.0	0.50	9.25	4.95
	4.2	35.0	0.50	5.69	2.22
	3.9	41.0	0.10	5.63	2.94
	3.9	36.0	0.07	6.19	2.27
	7.3	32.0	0.30	8.02	12.92
	4.4	46.0	0.07	7.54	5.76
	3.0	30.0	0.00	5.12	10.77

TABLE 10.6 (*continued*)

	x_1	x_2	x_3	x_4	x_5
π_3	6.3	13.0	0.50	4.24	8.27
	1.7	5.6	1.00	5.69	4.64
	7.3	24.0	0.00	4.34	2.99
	7.8	18.0	0.50	3.92	6.09
	7.8	25.0	0.70	5.39	6.20
	7.8	26.0	1.00	5.02	2.50
	9.5	17.0	0.05	3.52	5.71
	7.7	14.0	0.30	4.65	8.63
	11.0	20.0	0.50	4.27	8.40
	8.0	14.0	0.30	4.32	7.87
	8.4	18.0	0.20	4.38	7.98
	10.0	18.0	0.10	3.06	7.67
	7.3	15.0	0.05	3.76	6.84
	9.5	22.0	0.30	3.98	5.02
	8.4	15.0	0.20	5.02	10.12
	8.4	17.0	0.20	4.42	8.25
	9.5	25.0	0.50	4.44	5.95
	7.2	22.0	1.00	4.70	3.49
	4.0	12.0	0.50	5.71	6.32
	6.7	52.0	0.50	4.80	3.20
	9.0	27.0	0.30	3.69	3.30
	7.8	29.0	1.50	6.72	5.75
	4.5	41.0	0.50	3.33	2.27
	6.2	34.0	0.70	7.56	6.93
	5.6	20.0	0.50	5.07	6.70
	9.0	17.0	0.20	4.39	8.33
	8.4	20.0	0.10	3.74	3.77
	9.5	19.0	0.50	3.72	7.37
	9.0	20.0	0.50	5.97	11.17
	6.2	16.0	0.05	4.23	4.18
	7.3	20.0	0.50	4.39	3.50
	3.6	15.0	0.70	7.00	4.82
	6.2	34.0	0.07	4.84	2.37
	7.3	22.0	0.00	4.13	2.70
	4.1	29.0	0.70	5.78	7.76
	5.4	29.0	0.20	4.64	2.65
	5.0	34.0	0.70	4.21	6.50
	6.2	27.0	0.30	3.97	2.97

REFERENCES

[1] Anderson, T. W., *An Introduction to Multivariate Statistical Methods*, New York: John Wiley, 1958.

[2] Bartlett, M. S., "An Inverse Matrix Adjustment Arising in Discriminant Analysis," *Annals of Mathematical Statistics*, **22** (1951), 107–111.

[3] Bouma, B. N., and others, "Evaluation of the Detection Rate of Hemophilia Carriers" in *Statistical Methods for Clinical Decision Making*, **7**, no. 2 (1975), 339–350.

[4] Eisenbeis, R. A., "Pitfalls in the Application of Discriminant Analysis in Business, Finance and Economics," *Journal of Finance*, **32**, no. 3 (1977), 875–900.

[5] Fisher, R. A., "The Use of Multiple Measurements in Taxonomic Problems," *Annals of Eugenics*, **7** (1936), 179–188.

[6] Fisher, R. A., "The Statistical Utilization of Multiple Measurements," *Annals of Eugenics*, **8** (1938), 376–386.

[7] Geisser, S., "Discrimination, Allocatory and Separatory, Linear Aspects" in *Classification and Clustering*, ed. J. Van Ryzin, pp. 301–330. New York: Academic Press, 1977.

[8] Gerrild, P. M., and R. J. Lantz, "Chemical Analysis of 75 Crude Oil Samples from Pliocence Sand Units, Elk Hills Oil Field, California," *U.S. Geological Survey Open-File Report*, 1969.

[9] Gnanadesikan, R., *Methods for Statistical Data Analysis of Multivariate Observations*, New York: John Wiley, 1977.

[10] Hills, M., "Allocation Rules and Their Error Rates," *Journal of the Royal Statistical Society (B)*, **28** (1966), 1–31.

[11] Hudlet, R., and R. A. Johnson, "Linear Discrimination and Some Further Results on Best Lower Dimensional Representations" in *Classification and Clustering*, ed. J. Van Ryzin, pp. 371–394. New York: Academic Press, 1977.

[12] Kendall, M. G., *Multivariate Analysis*, New York: Hafner Press, 1975.

[13] Krzanowski, W. J., "The Performance of Fisher's Linear Discriminant Function Under Non-Optimal Conditions," *Technometrics*, **19**, no. 2 (1977), 191–200.

[14] Lachenbruch, P. A., *Discriminant Analysis*, New York: Hafner Press, 1975.

[15] Lachenbruch, P. A., and M. R. Mickey, "Estimation of Error Rates in Discriminant Analysis," *Technometrics*, **10**, no. 1 (1968), 1–11.

[16] Mucciardi, A. N., and E. E. Gose, "A Comparison of Seven Techniques for Choosing Subsets of Pattern Recognition Properties," *IEEE Trans. Computers*, **C20** (1971), 1023–1031.

[17] Murray, G. D., "A Cautionary Note on Selection of Variables in Discriminant Analysis," *Applied Statistics*, **26**, no. 3 (1977), 246–250.

[18] Wald, A., "On a Statistical Problem Arising in the Classification of an Individual into One of Two Groups," *Annals of Mathematical Statistics*, **15** (1944), 145–162.

[19] Welch, B. L., "Note on Discriminant Functions," *Biometrika*, **31** (1939), 218–220.

11

Clustering

11.1 INTRODUCTION

Rudimentary, exploratory procedures are often quite helpful in understanding the complex nature of multivariate relationships. For example, throughout this book we have emphasized the value of data plots. In this chapter we shall discuss some additional graphical techniques and suggested step-by-step rules (algorithms) for grouping objects (variables or items). Searching the data for a structure of "natural" groupings is an important exploratory technique. Groupings can provide an informal means for assessing dimensionality, identifying outliers and suggesting interesting hypotheses concerning relationships.

Grouping, or clustering, is distinct from the classification methods discussed in the previous chapter. Classification pertains to a *known* number of groups, and the operational objective is to assign new observations to one of these groups. Cluster analysis is a more primitive technique in that no assumptions are made concerning the number of groups or the group structure. Grouping is done on the basis of similarities or distances (dissimilarities). The inputs required are similarity measures or data from which similarities can be computed.

To illustrate the nature of the difficulty in defining a natural grouping, consider sorting the 16 face cards in an ordinary deck of playing cards into clusters of similar objects. Some groupings are illustrated in Figure 11.1. It is immediately clear that meaningful partitions depend on the definition of *similar*.

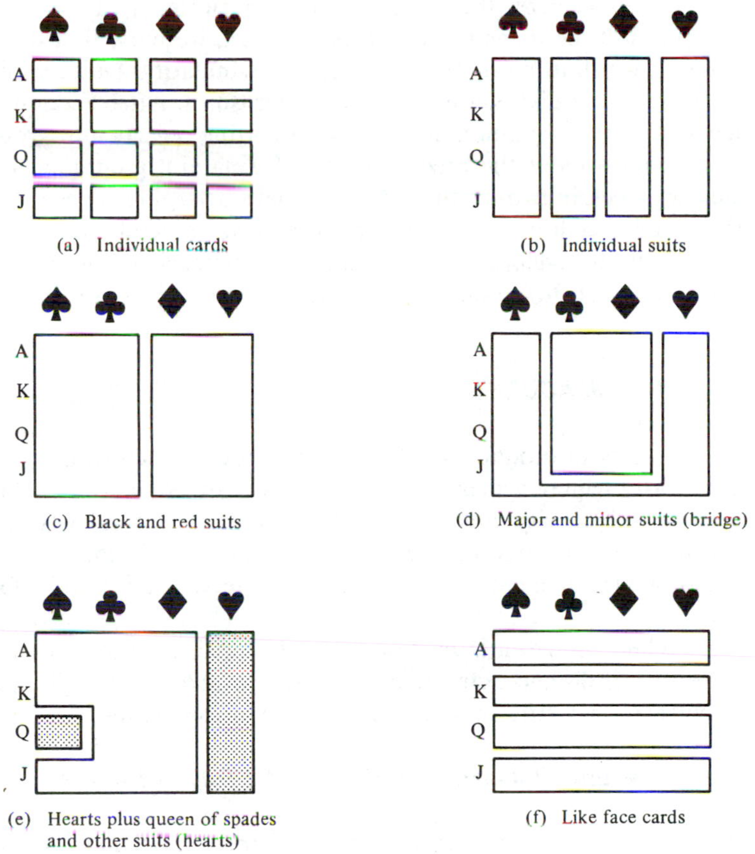

Figure 11.1 Grouping face cards.

In most practical applications of cluster analysis, the investigator knows enough about the problem to distinguish "good" groupings from "bad" groupings. Why not enumerate all possible groupings and select the "best" ones for further study?

For the playing card example, there is one way to form a *single* group of 16 face cards; there are 32,767 ways to partition the face cards into *two* groups (of varying sizes); there are 7,141,686 ways to sort the face cards into *three* groups (of varying sizes), and so on.[1] Obviously time constraints make it impossible to determine the best groupings of similar objects from a list of all possible structures. Even large computers are easily overwhelmed by the typically large number of cases, so one must settle for *algorithms* that search for good, but not necessarily the best, groupings.

[1]The number of ways of sorting n objects into k nonempty groups is a Stirling number of the second kind given by $(1/k!)\sum_{\ell=0}^{k}(-1)^{k-\ell}\binom{k}{\ell}\ell^n$ [see [1]]. Adding these numbers for $k = 1, 2, \ldots, n$ groups, we obtain the total number of possible ways to sort n objects into groups.

To summarize, the basic objective in cluster analysis is to discover natural groupings of the items (or variables). In turn, we must first develop a quantitative scale on which to measure the association (similarity) between objects. Section 11.2 is devoted to a discussion of similarity measures. After Section 11.2 we describe a few of the more common algorithms for sorting objects into groups.

Even without the precise notion of a natural grouping, we are often able to cluster objects in two- or three-dimensional scatterplots by eye. To take advantage of the mind's ability to group similar objects, several graphical procedures have recently been developed for depicting high-dimensional observations in two dimensions. Some of these graphical techniques are considered in Sections 11.5 and 11.6.

11.2 SIMILARITY MEASURES

Most efforts to produce a rather simple group structure from a complex data set necessarily require a measure of "closeness," or "similarity." There is often a great deal of subjectivity involved in the choice of a similarity measure. Important considerations include the nature of the variables (discrete, continuous, binary) or scales of measurement (nominal, ordinal, interval, ratio) and subject matter knowledge.

When *items* (units or cases) are clustered, proximity is usually indicated by some sort of distance. On the other hand, *variables* are usually grouped on the basis of correlation coefficients or like measures of association.

Distances and Similarity Coefficients for Pairs of Items

We discussed the notion of distance in Chapter 1, Section 1.4. Recall the Euclidean (straight-line) distance between two p-dimensional observations (items) $\mathbf{x} = [x_1, x_2, \ldots, x_p]'$ and $\mathbf{y} = [y_1, y_2, \ldots, y_p]'$ is, from (1-12),

$$d(\mathbf{x}, \mathbf{y}) = \sqrt{(x_1 - y_1)^2 + (x_2 - y_2)^2 + \cdots + (x_p - y_p)^2} = \sqrt{(\mathbf{x} - \mathbf{y})'(\mathbf{x} - \mathbf{y})} \tag{11-1}$$

The statistical distance between the same two observations is of the form [see (1-23)]

$$d(\mathbf{x}, \mathbf{y}) = \sqrt{(\mathbf{x} - \mathbf{y})'\mathbf{A}(\mathbf{x} - \mathbf{y})} \tag{11-2}$$

Ordinarily, $\mathbf{A} = \mathbf{S}^{-1}$ where $\mathbf{S}$ contains the sample variances and covariances. However, without prior knowledge of the distinct groups, these quantities cannot be computed. For this reason Euclidean distance is often preferred for clustering.

Another distance measure is the Minkowski metric

$$d(\mathbf{x}, \mathbf{y}) = \left[\sum_{i=1}^{p} |x_i - y_i|^m \right]^{1/m} \tag{11-3}$$

For $m = 1$, $d(\mathbf{x}, \mathbf{y})$ measures the "city-block" distance between two points in p dimensions. For $m = 2$, $d(\mathbf{x}, \mathbf{y})$ becomes the Euclidean distance. In general, varying m changes the weight given to larger and smaller differences.

Whenever possible, it is advisable to use "true" distances, that is, distances satisfying the distance properties of (1-25), for clustering objects. On the other hand, most clustering algorithms will accept subjectively assigned distance numbers that may not satisfy, for example, the triangle inequality.

The following example illustrates how rudimentary groupings can be formed by simply reorganizing the elements of the distance matrix.

Example 11.1

Table 11.1 gives the Euclidean distances between pairs of 22 U.S. public utility companies based on the data listed in Table 11.5 after it has been standardized.

Because the distance matrix is large, it is difficult to visually select firms that are close together (similar). However, the graphical method of *shading* allows us to pick out clusters of similar firms quite easily.

First distances are arranged into several classes (for instance, 15 or fewer) based on their magnitudes. Next all distances within a given class are replaced by a common symbol with a certain shade of gray. Darker symbols correspond to smaller distances. Finally, the distance matrix is reorganized so that items with common symbols appear in contiguous locations along the main diagonal. Groups of similar items correspond to patches of dark shadings.

From Figure 11.2 we see that firms 1, 18, 19, and 14 form a group, firms 22, 10, 13, 20, and 4 form a group, firms 9 and 3 form a group, firms 3 and 6 form a group, and so forth. The groups (9, 3) and (3, 6) overlap, as do other groups in the diagram. Firms 11, 5, and 17 appear to stand alone. ■

When items cannot be represented by meaningful p-dimensional measurements, pairs of items are often compared on the basis of the presence or absence of certain characteristics. Similar items have more characteristics in common than dissimilar items. The presence or absence of a characteristic can be described mathematically by introducing a *binary variable*, which assumes value 1 if the characteristic is present and value 0 if the characteristic is absent. For $p = 5$ binary variables, for instance, the variable "scores" for two items i and k might be arranged as follows.

	Variable				
	1	2	3	4	5
Item i	1	0	0	1	1
Item k	1	1	0	1	0

In this case there are two 1-1 matches, one 0-0 match, and two mismatches.

Let x_{ij} be the score (1 or 0) of the jth binary variable on the ith item and x_{kj} be the score (again, 1 or 0) of the jth variable on the kth item, $j = 1, 2, \ldots, p$. Consequently,

$$(x_{ij} - x_{kj})^2 = \begin{cases} 0 & \text{if } x_{ij} = x_{kj} = 1 \text{ or } x_{ij} = x_{kj} = 0 \\ 1 & \text{if } x_{ij} \neq x_{kj} \end{cases} \tag{11-4}$$

TABLE 11.1 DISTANCES BETWEEN 22 UTILITIES

Firm no.	1	2	3	4	5	6	7	8	9	10	11	12	13	14	15	16	17	18	19	20	21	22
1	.00																					
2	3.10	.00																				
3	3.68	4.92	.00																			
4	2.46	2.16	4.11	.00																		
5	4.12	3.85	4.47	4.13	.00																	
6	3.61	4.22	2.99	3.20	4.60	.00																
7	3.90	3.45	4.22	3.97	4.60	3.35	.00															
8	2.74	3.89	4.99	3.69	5.16	4.91	4.36	.00														
9	3.25	3.96	2.75	3.75	4.49	3.73	2.80	3.59	.00													
10	3.10	2.71	3.93	1.49	4.05	3.83	4.51	3.67	3.57	.00												
11	3.49	4.79	5.90	4.86	6.46	6.00	6.00	3.46	5.18	5.08	.00											
12	3.22	2.43	4.03	3.50	3.60	3.74	1.66	4.06	2.74	3.94	5.21	.00										
13	3.96	3.43	4.39	2.58	4.76	4.55	5.01	4.14	3.66	1.41	5.31	4.50	.00									
14	2.11	4.32	2.74	3.23	4.82	3.47	4.91	4.34	3.82	3.61	4.32	4.34	4.39	.00								
15	2.59	2.50	5.16	3.19	4.26	4.07	2.93	3.85	4.11	4.26	4.74	2.33	5.10	4.24	.00							
16	4.03	4.84	5.26	4.97	5.82	5.84	5.04	2.20	3.63	4.53	3.43	4.62	4.41	5.17	5.18	.00						
17	4.40	3.62	6.36	4.89	5.63	6.10	4.58	5.43	4.90	5.48	4.75	3.50	5.61	5.56	3.40	5.56	.00					
18	1.88	2.90	2.72	2.65	4.34	2.85	2.95	3.24	2.43	3.07	3.95	2.45	3.78	2.30	3.00	3.97	4.43	.00				
19	2.41	4.63	3.18	3.46	5.13	2.58	4.52	4.11	4.11	4.13	4.52	4.41	5.01	1.88	4.03	5.23	6.09	2.47	.00			
20	3.17	3.00	3.73	1.82	4.39	2.91	3.54	4.09	2.95	2.05	5.35	3.43	2.23	3.74	3.78	4.82	4.87	2.92	3.90	.00		
21	3.45	2.32	5.09	3.88	3.64	4.63	2.68	3.98	3.74	4.36	4.88	1.38	4.94	4.93	2.10	4.57	3.10	3.19	4.97	4.15	.00	
22	2.51	2.42	4.11	2.58	3.77	4.03	4.00	3.24	3.21	2.56	3.44	3.00	2.74	3.51	3.35	3.46	3.63	2.55	3.97	2.62	3.01	.00

The distances have been represented below in shaded form according to the following scheme.

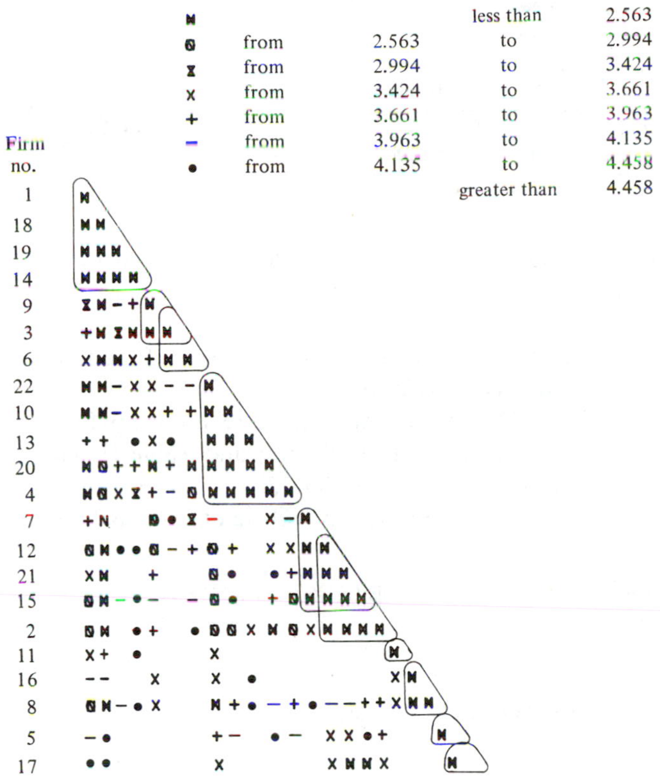

Figure 11.2 Shaded distance matrix for 22 utilities.

and the squared Euclidean distance, $\sum_{j=1}^{p} (x_{ij} - x_{kj})^2$, provides a count of the number of mismatches. A large distance corresponds to many mismatches; that is, dissimilar items. From the display above, the squared distance between items i and k would be

$$\sum_{j=1}^{5} (x_{ij} - x_{kj})^2 = (1-1)^2 + (0-1)^2 + (0-0)^2 + (1-1)^2 + (1-0)^2$$
$$= 2$$

Although a distance based on (11-4) might be used to measure similarity, it suffers from weighting the 1-1 and 0-0 matches equally. In some cases a 1-1 match is a stronger indication of similarity than a 0-0 match. For instance, when grouping people, the evidence that two persons both read ancient Greek is stronger evidence of similarity than the absence of this ability. Thus it might be reasonable to discount the 0-0 matches or even disregard them completely. To allow for differential treatment of the 1-1 matches and the 0-0 matches, several schemes for defining similarity coefficients have been suggested.

To introduce these schemes, let us arrange the frequencies of matches and mismatches for items i and k in the form of a contingency table.

		Item k 1	0	Totals
Item i	1	a	b	$a+b$
	0	c	d	$c+d$
Totals		$a+c$	$b+d$	$p=a+b+c+d$

(11-5)

In these tables, a represents the frequency of 1-1 matches, b is the frequency of 1-0 matches, and so forth. Given the five pairs of binary outcomes above, $a = 2$ and $b = c = d = 1$.

Table 11.2 lists common similarity coefficients defined in terms of the frequencies in (11-5). A short rationale follows each definition.

Coefficients 1, 2, and 3 in Table 11.2 enjoy a monotonic relationship. Suppose coeḟficient 1 is calculated for two contingency tables, Table I and Table II. Then if $(a_{\mathrm{I}} + d_{\mathrm{I}})/p \geq (a_{\mathrm{II}} + d_{\mathrm{II}})/p$, we also have $2(a_{\mathrm{I}} + d_{\mathrm{I}})/[2(a_{\mathrm{I}} + d_{\mathrm{I}}) + b_{\mathrm{I}} + c_{\mathrm{I}}] \geq 2(a_{\mathrm{II}} + d_{\mathrm{II}})/[2(a_{\mathrm{II}} + d_{\mathrm{II}}) + b_{\mathrm{II}} + c_{\mathrm{II}}]$ and coefficient 3 will be at least as large for

TABLE 11.2 SIMILARITY COEFFICIENTS FOR CLUSTERING ITEMS*

Coefficient	Rationale
1. $\dfrac{a+d}{p}$	Equal weights for 1-1 matches and 0-0 matches.
2. $\dfrac{2(a+d)}{2(a+d)+b+c}$	Double weight for 1-1 matches and 0-0 matches.
3. $\dfrac{a+d}{a+d+2(b+c)}$	Double weight for unmatched pairs.
4. $\dfrac{a}{p}$	No 0-0 matches in numerator.
5. $\dfrac{a}{a+b+c}$	No 0-0 matches in numerator or denominator. (The 0-0 matches are treated as irrelevant.)
6. $\dfrac{2a}{2a+b+c}$	No 0-0 matches in numerator or denominator. Double weight for 1-1 matches.
7. $\dfrac{a}{a+2(b+c)}$	No 0-0 matches in numerator or denominator. Double weight for unmatched pairs.
8. $\dfrac{a}{b+c}$	Ratio of matches to mismatches with 0-0 matches excluded.

*[p binary variables; see (11-5).]

Table I as it is for Table II. Coefficients 5, 6, and 7 (Table 11.2) also retain their relative orders (see Exercise 11.4).

Monotonicity is important because some clustering procedures are not affected if the definition of similarity is changed in a manner that leaves the relative orderings of similarities unchanged. The single linkage and complete linkage hierarchical procedures discussed in Section 11.3 are not affected. For these methods any choice of the coefficients 1, 2, and 3 (in Table 11.2) will produce the same groupings. Similarly any choice of the coefficients 5, 6 and 7 will yield identical groupings.

Example 11.2

Suppose five individuals possess the following characteristics:

	Height	Weight	Eye color	Hair color	Handedness	Sex
Individual 1	68 in	140 lb	green	blond	right	female
Individual 2	73 in	185 lb	brown	brown	right	male
Individual 3	67 in	165 lb	blue	blond	right	male
Individual 4	64 in	120 lb	brown	brown	right	female
Individual 5	76 in	210 lb	brown	brown	left	male

Define six binary variables $X_1, X_2, X_3, X_4, X_5, X_6$ as

$$X_1 = \begin{cases} 1 & \text{height} \geq 72 \text{ in.} \\ 0 & \text{height} < 72 \text{ in.} \end{cases} \qquad X_4 = \begin{cases} 1 & \text{blond hair} \\ 0 & \text{brown hair} \end{cases}$$

$$X_2 = \begin{cases} 1 & \text{weight} \geq 150 \text{ lb} \\ 0 & \text{weight} < 150 \text{ lb} \end{cases} \qquad X_5 = \begin{cases} 1 & \text{right handed} \\ 0 & \text{left handed} \end{cases}$$

$$X_3 = \begin{cases} 1 & \text{brown eyes} \\ 0 & \text{otherwise} \end{cases} \qquad X_6 = \begin{cases} 1 & \text{female} \\ 0 & \text{male} \end{cases}$$

The scores for individuals 1 and 2 on the $p = 6$ binary variables are

		X_1	X_2	X_3	X_4	X_5	X_6
Individual	1	0	0	0	1	1	1
	2	1	1	1	0	1	0

and the number of matches and mismatches are indicated in the two-way array

		Individual 2: 1	0	Totals
Individual 1	1	1	2	3
	0	3	0	3
	Totals	4	2	6

Employing similarity coefficient 1, which gives equal weight to matches, we compute

$$\frac{a+d}{p} = \frac{1+0}{6} = \frac{1}{6}$$

Continuing with similarity coefficient 1, we calculate the remaining similarity numbers for pairs of individuals. These are displayed in the 5 × 5 symmetric matrix

$$\text{Individual} \quad \begin{matrix} \\ 1 \\ 2 \\ 3 \\ 4 \\ 5 \end{matrix} \overset{\displaystyle\text{Individual}}{\begin{matrix} 1 & 2 & 3 & 4 & 5 \\ \end{matrix}} \begin{bmatrix} 1 & & & & \\ \frac{1}{6} & 1 & & & \\ \frac{4}{6} & \frac{3}{6} & 1 & & \\ \frac{4}{6} & \frac{3}{6} & \frac{2}{6} & 1 & \\ 0 & \textcircled{\frac{5}{6}} & \frac{2}{6} & \frac{2}{6} & 1 \end{bmatrix}$$

Based on the magnitudes of the similarity coefficient, we should conclude individuals 2 and 5 are most similar and individuals 1 and 5 are least similar. Other pairs fall between these extremes. If we were to divide the individuals into two relatively homogeneous subgroups on the basis of the similarity numbers, we might form the subgroups (1 3 4) and (2 5).

Note that $X_3 = 0$ implies an absence of brown eyes, so the two people with the blue green eyes will yield a 0-0 match. Consequently, it may be inappropriate to use similarity coefficients 1, 2, or 3 because these coefficients give the same weights to 1-1 and 0-0 matches. ■

We have described the construction of distances and similarities. It is always possible to construct similarities from distances. For example, we might set

$$\tilde{s}_{ik} = \frac{1}{1 + d_{ik}} \tag{11-6}$$

where $0 < \tilde{s}_{ik} \leq 1$ is the similarity between items i and k and d_{ik} is the corresponding distance.

However, distances cannot always be constructed from similarities. As Gower [9, 10] has shown, this can be done only if the matrix of similarities is nonnegative definite. With the nonnegative definite condition and with the maximum similarity scaled so that $\tilde{s}_{ii} = 1$,

$$d_{ik} = \sqrt{2(1 - \tilde{s}_{ik})} \tag{11-7}$$

has the properties of a distance.

Similarities and Association Measures for Pairs of Variables

We discussed similarity measures for items above. In some applications, it is the variables rather than the items that must be grouped. Similarity measures for variables often take the form of sample correlation coefficients. Moreover, in some clustering applications, negative correlations are replaced by their absolute values.

When the variables are binary, the data can again be arranged in the form of a contingency table. This time, however, the variables, rather than the items, delineate the categories. For each pair of variables, there are n items categorized in the table. With the usual 0 and 1 coding, the table becomes as follows.

		Variable k 1	0	Totals
Variable i	1	a	b	$a + b$
	0	c	d	$c + d$
	Totals	$a + c$	$b + d$	$n = a + b + c + d$

(11-8)

For instance variable i equals 1 and variable k equals 0 for b of the n items.

The usual product moment correlation formula applied to the binary variables in the contingency table of (11-8) gives (see Exercise 11.3)

$$r = \frac{ad - bc}{[(a + b)(c + d)(a + c)(b + d)]^{1/2}} \tag{11-9}$$

This number can be taken as a measure of the similarity between the two variables.

The correlation coefficient in (11-9) is related to the chi-square statistic ($r^2 = \chi^2/n$) for testing the independence of two categorical variables. For n fixed, a large similarity (or correlation) is consistent with lack of independence.

Given the table in (11-8), measures of association (or similarity) exactly analogous to the ones listed in Table 11.2 can be developed. The only change required is the substitution of n (the number of items) for p (the number of variables).

Concluding Comments on Similarity

To summarize this section, we note that there are many ways to measure the similarity between pairs of objects. It appears that most practitioners use distances [see (11-1), (11-2), and (11-3)] or the coefficients in Table 11.2 to cluster *items* and correlations to cluster *variables*. However, at times, inputs to clustering algorithms may be simple frequencies.

Example 11.3

The meaning of words changes with the course of history. However, the meaning of the numbers 1, 2, 3, ... represents one conspicuous exception. A first comparison of languages might be based on the numerals alone. Table 11.3 gives the first 10 numbers in English, Polish, Hungarian, and 8 other modern European languages. (Only languages that use the Roman alphabet are considered. Certain accent marks, such as cedillas, are omitted.)

TABLE 11.3 NUMERALS IN ELEVEN LANGUAGES

English	Norwegian	Danish	Dutch	German	French	Spanish	Italian	Polish	Hungarian	Finnish
one	en	cn	een	ein	un	uno	uno	jeden	egy	yksi
two	to	to	twee	zwei	deux	dos	due	dwa	ketto	kaksi
three	tre	tre	drie	drei	trois	tres	tre	trzy	harom	kolme
four	fire	fire	vier	vier	quatre	cuatro	quattro	cztery	negy	neua
five	fem	fem	vijf	funf	cinq	cinco	cinque	piec	ot	viisi
six	seks	seks	zes	sechs	six	seix	sei	szesc	hat	kuusi
seven	sju	syv	zeven	sieben	sept	siete	sette	siedem	het	seitseman
eight	atte	otte	acht	acht	huit	ocho	otto	osiem	nyolc	kahdeksan
nine	ni	ni	negen	neun	neuf	nueve	nove	dziewiec	kilenc	yhdeksan
ten	ti	ti	tien	zehn	dix	diez	dieci	dziesiec	tiz	kymmenen

A cursory examination of the spelling of the numerals in Table 11.3 suggests that the first five languages (English, Norwegian, Danish, Dutch, and German) are pretty much alike. French, Spanish, and Italian are in even closer agreement. Hungarian and Finnish seem to stand by themselves, and Polish has some of the characteristics of the languages in each of the larger subgroups.

The words for 1 in French, Spanish, and Italian all begin with *u*. For illustrative purposes, we might compare languages by looking at the *first letters* of the numbers. We call the words for the same number in two different languages *concordant* if they have the same first letter and *discordant* if they do not. Using Table 11.3, the table of concordances (frequencies of matching first initials) for the numbers 1–10 is given in Table 11.4.

TABLE 11.4 CONCORDANT FIRST LETTERS FOR NUMBERS IN ELEVEN LANGUAGES

	E	N	Da	Du	G	Fr	Sp	I	P	H	Fi
E	10										
N	8	10									
Da	8	9	10								
Du	3	5	4	10							
G	4	6	5	5	10						
Fr	4	4	4	1	3	10					
Sp	4	4	5	1	3	8	10				
I	4	4	5	1	3	9	9	10			
P	3	3	4	0	2	5	7	6	10		
H	1	2	2	2	1	0	0	0	0	10	
Fi	1	1	1	1	1	1	1	1	1	2	10

We see that English and Norwegian have the same first letter for 8 of the 10 word pairs. The remaining frequencies were calculated in the same manner.

The results in Table 11.4 confirm our initial visual impressions of Table 11.3. That is, English, Norwegian, Danish, Dutch, and German seem to form a group. French, Spanish, Italian, and Polish might be grouped together, while Hungarian and Finnish appear to stand alone. ■

In our examples so far we have used our visual impression of similarity or distance measures to form groups. We now discuss less subjective schemes for creating clusters.

11.3 HIERARCHICAL CLUSTERING METHODS

We can rarely examine all grouping possibilities, even with the largest and fastest computers. Because of this problem, a wide variety of clustering algorithms have emerged that find "reasonable" clusters without having to look at all configurations.

Hierarchical clustering techniques proceed by either a series of successive mergers or a series of successive divisions. *Agglomerative hierarchical methods* start with the individual objects. Thus there are initially as many clusters as objects. Most similar objects are first grouped, and these initial groups are merged according to their similarities. Eventually, as the similarity decreases, all subgroups are fused into a single cluster.

Divisive hierarchical methods work in the opposite direction. An initial single group of objects is divided into two subgroups such that the objects in one subgroup are "far from" the objects in the other. These subgroups are then further divided into dissimilar subgroups; the process continues until there are as many subgroups as objects; that is, until each object forms a group.

The results of both agglomerative and divisive methods may be displayed in the form of a two-dimensional diagram known as a *dendogram*. As we shall see, the dendogram illustrates the mergers or divisions which have been made at successive levels.

In this section we shall concentrate on agglomerative hierarchical procedures and, in particular, *linkage methods*. Excellent elementary discussions of divisive hierarchical procedures and other agglomerative techniques are available in [2] and [6].

Linkage methods are suitable for clustering items, as well as variables. This is not true for all hierarchical agglomerative procedures. We shall discuss, in turn, *single linkage* (minimum distance or nearest neighbor), *complete linkage* (maximum distance or farthest neighbor), and *average linkage* (average distance). The merging of clusters under the three linkage criteria is illustrated schematically in Figure 11.3.

From Figure 11.3 we see that single linkage results when groups are fused according to the distance between their nearest members. Complete linkage occurs when groups are fused according to the distance between their farthest members. For average linkage, groups are fused according to the average distance between pairs of members in the respective sets.

The following are the steps in the agglomerative hierarchical clustering algorithm for grouping N objects (items or variables).

1. Start with N clusters, each containing a single entity and an $N \times N$ symmetric matrix of distances (or similarities) $\mathbf{D} = \{d_{ik}\}$.
2. Search the distance matrix for the nearest (most similar) pair of clusters. Let the distance between "most similar" clusters U and V be d_{UV}.

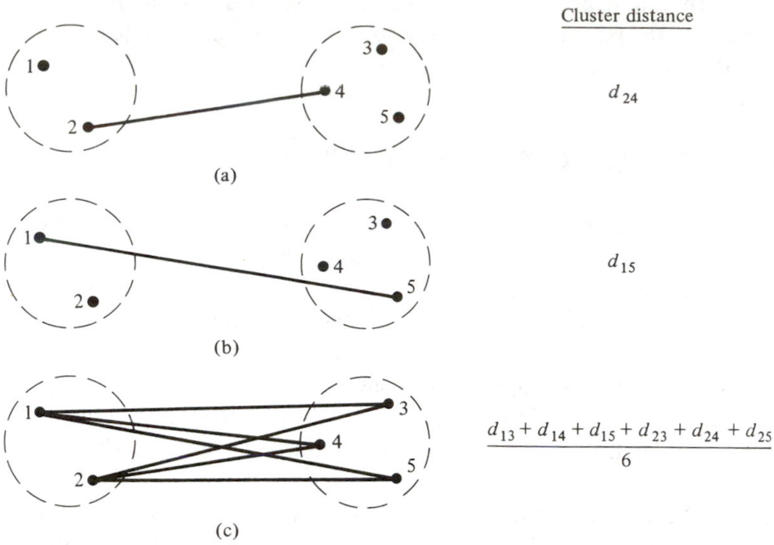

Figure 11.3 Inter-cluster distance (dissimilarity) for (a) single linkage, (b) complete linkage, and (c) average linkage.

3. Merge clusters U and V. Label the newly formed cluster (UV). Update the entries in the distance matrix by (a) deleting the rows and columns corresponding to clusters U and V and (b) adding a row and column giving the distances between cluster (UV) and the remaining clusters. (11-10)
4. Repeat Steps 2 and 3 a total of $N - 1$ times. (All objects will be in a *single* cluster at termination of the algorithm.) Record the identity of clusters that are merged and the levels (distances or similarities) at which the mergers take place.

The ideas behind any clustering procedure are probably best conveyed through examples. After brief discussions of the input and algorithmic components of the linkage methods, several examples will be presented in an attempt to clarify the concepts.

Single Linkage

The inputs to a single linkage algorithm can be distances or similarities between pairs of objects. Groups are formed from the individual entities by merging nearest neighbors, where the term *nearest neighbor* connotes smallest distance or largest similarity.

Initially, we must find the smallest distance in $\mathbf{D} = \{d_{ik}\}$ and merge the corresponding objects, say U and V, to get the cluster (UV). For Step 3 of the general algorithm of (11-10), the distances between (UV) and any other cluster W are computed by

$$d_{(UV)W} = \min\{d_{UW}, d_{VW}\} \tag{11-11}$$

Here quantities d_{UW} and d_{VW} are the distances between the nearest neighbors of clusters U and W and clusters V and W, respectively.

The results of single linkage clustering can be graphically displayed in the form of a dendogram, or tree diagram. The branches in the tree represent clusters. The branches come together (merge) at nodes whose positions along a distance (or similarity) axis indicate the level at which the fusions occur. Dendograms for some specific cases are considered in the following examples.

Example 11.4

To illustrate the single linkage algorithm, we consider the hypothetical distances between pairs of five objects given below.

$$\mathbf{D} = \{d_{ik}\} = \begin{array}{c} \\ 1 \\ 2 \\ 3 \\ 4 \\ 5 \end{array} \begin{array}{c} \begin{array}{ccccc} 1 & 2 & 3 & 4 & 5 \end{array} \\ \left[\begin{array}{ccccc} 0 & & & & \\ 9 & 0 & & & \\ 3 & 7 & 0 & & \\ 6 & 5 & 9 & 0 & \\ 11 & 10 & \textcircled{2} & 8 & 0 \end{array}\right] \end{array}$$

Treating each object as a cluster, the clustering commences by merging the two closest items. Since

$$\min_{i,k}(d_{ik}) = d_{53} = 2$$

objects 5 and 3 are merged to form the cluster (35). To implement the next level of clustering, we need the distances between the cluster (35) and the remaining objects, 1, 2, and 4. The nearest-neighbor distances are

$$d_{(35)1} = \min\{d_{31}, d_{51}\} = \min\{3, 11\} = 3$$
$$d_{(35)2} = \min\{d_{32}, d_{52}\} = \min\{7, 10\} = 7$$
$$d_{(35)4} = \min\{d_{34}, d_{54}\} = \min\{9, 8\} = 8$$

Deleting the rows and columns of **D** corresponding to objects 3 and 5 and adding a row and column for the cluster (35), we obtain the new distance matrix

$$\begin{array}{c} \\ (35) \\ 1 \\ 2 \\ 4 \end{array} \begin{array}{c} \begin{array}{cccc} (35) & 1 & 2 & 4 \end{array} \\ \left[\begin{array}{cccc} 0 & & & \\ \textcircled{3} & 0 & & \\ 7 & 9 & 0 & \\ 8 & 6 & 5 & 0 \end{array}\right] \end{array}$$

The smallest distance between pairs of clusters is now $d_{(35)1} = 3$ and we merge cluster (1) with cluster (35) to get the next cluster, (135). Calculating

$$d_{(135)2} = \min\{d_{(35)2}, d_{12}\} = \min\{7, 9\} = 7$$
$$d_{(135)4} = \min\{d_{(35)4}, d_{14}\} = \min\{8, 6\} = 6$$

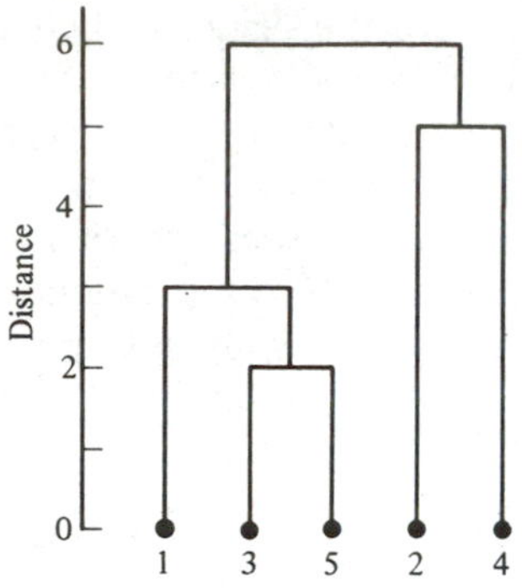

Figure 11.4 Single-linkage dendogram for distances between five objects.

the distance matrix for the next level of clustering is

$$\begin{array}{c} \\ (135) \\ 2 \\ 4 \end{array} \begin{array}{c} \begin{array}{ccc} (135) & 2 & 4 \end{array} \\ \begin{bmatrix} 0 & & \\ 7 & 0 & \\ 6 & \textcircled{5} & 0 \end{bmatrix} \end{array}$$

The minimum nearest neighbor between pairs of clusters is $d_{42} = 5$, and we merge objects 4 and 2 to get the cluster (24).

At this point we have two distinct clusters, (135) and (24). Their nearest neighbor distance is

$$d_{(135)(24)} = \min\{d_{(135)2}, d_{(135)4}\} = \min\{7, 6\} = 6$$

The final distance matrix becomes

$$\begin{array}{c} \\ (135) \\ (24) \end{array} \begin{array}{c} \begin{array}{cc} (135) & (24) \end{array} \\ \begin{bmatrix} 0 & \\ \textcircled{6} & 0 \end{bmatrix} \end{array}$$

Consequently, clusters (135) and (24) are merged to form a single cluster of all five objects, (12345), when the nearest-neighbor distance reaches 6.

The dendogram picturing the hierarchical clustering just concluded is shown in Figure 11.4. The groupings, and the distance levels at which they occur, are clearly illustrated by the dendogram. ■

In typical applications of hierarchical clustering, the intermediate results—where the objects are sorted into a moderate number of clusters—are of chief interest.

Example 11.5

Consider the array of concordances in Table 11.4 representing the closeness between the numbers 1–10 in 11 languages. To develop a matrix of distances, we subtract the concordances from the perfect agreement figure of 10, which each language has with itself. The subsequent assignments of distances are

	E	N	Da	Du	G	Fr	Sp	I	P	H	Fi
E	0										
N	2	0									
Da	2	①	0								
Du	7	5	6	0							
G	6	4	5	5	0						
Fr	6	6	6	9	7	0					
Sp	6	6	5	9	7	2	0				
I	6	6	5	9	7	①	①	0			
P	7	7	6	10	8	5	3	4	0		
H	9	8	8	8	9	10	10	10	10	0	
Fi	9	9	9	9	9	9	9	9	9	8	0

We first search for the minimum distance between pairs of languages (clusters). The minimum distance, 1, occurs between Danish and Norwegian, Italian and French, and Italian and Spanish. Numbering the languages in the order in which they appear across the top of the array, we have

$$d_{32} = 1; \qquad d_{86} = 1; \qquad \text{and} \qquad d_{87} = 1$$

Since $d_{76} = 2$, we can merge only clusters 8 and 6 or clusters 8 and 7. We cannot merge clusters 6, 7, and 8 at level 1. We choose first to merge 8 and 6, and then to update the distance matrix and merge 2 and 3 to obtain the clusters (68) and (23). Subsequent computer calculations produce the dendogram in Figure 11.5.

From the dendogram we see that Norwegian and Danish and also French and Italian cluster at the minimum distance (maximum similarity) level. When the allowable distance is increased, English is added to the Norwegian–Danish group and Spanish merges with the French–Italian group. Notice that Hungarian and Finnish are more similar to each other than to the other clusters of languages. However, these two clusters (languages) do not merge until the distance between nearest neighbors has increased substantially. Finally, all the clusters of languages are merged into a single cluster at the largest nearest-neighbor distance, 9. ■

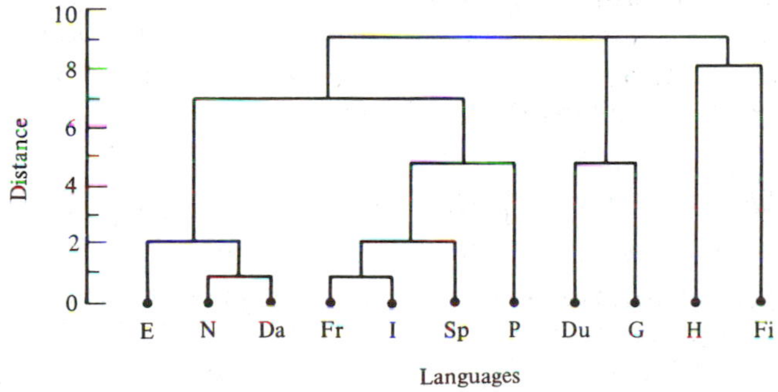

Figure 11.5 Single-linkage dendogram for distances between numbers in 11 languages.

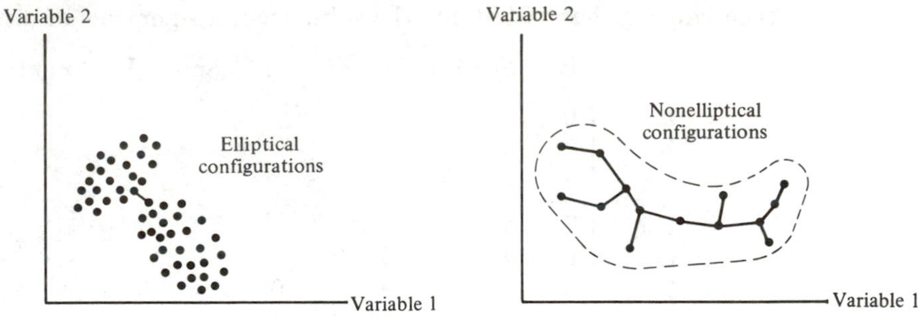

(a) Single linkage confused by near overlap (b) Chaining effect

Figure 11.6 Single-linkage clusters.

Since single linkage joins clusters by the shortest link between them, the technique cannot discern poorly separated clusters [see Figure 11.6(a)]. On the other hand, single linkage is one of the few clustering methods that can delineate nonellipsoidal clusters. The tendency of single linkage to pick out long stringlike clusters is known as *chaining* [see Figure 11.6(b)]. Chaining can be misleading if items at opposite ends of the chain are, in fact, quite dissimilar.

The clusters formed by the single linkage method will be unchanged by any assignment of distance (similarity) that gives the same relative orderings as the initial distances (similarities). In particular, any one of a set of similarity coefficients from Table 11.2 that are monotonic to one another will produce the same clustering.

Complete Linkage

Complete-linkage clustering proceeds in much the same manner as single linkage, with one important exception. At each stage, the distance (similarity) between clusters is determined by the distance (similarity) between the two elements, one from each cluster, that are most distant. Thus complete linkage ensures that all items in a cluster are within some maximum distance (or minimum similarity) of each other.

The general agglomerative algorithm again starts by finding the minimum entry in $\mathbf{D} = \{d_{ik}\}$ and merging the corresponding objects, such as U and V, to get cluster (UV). For Step 3 of the general algorithm in (11-10), the distances between (UV) and any other cluster W are computed by

$$d_{(UV)W} = \max\{d_{UW}, d_{VW}\} \tag{11-12}$$

Here d_{UW} and d_{VW} are the distances between the most distant members of clusters U and W and clusters V and W, respectively.

Example 11.6

Let us return to the distance matrix introduced in Example 11.4. In this case

$$\mathbf{D} = \{d_{ik}\} = \begin{array}{c} \\ 1 \\ 2 \\ 3 \\ 4 \\ 5 \end{array} \begin{array}{c} \begin{array}{ccccc} 1 & 2 & 3 & 4 & 5 \end{array} \\ \begin{bmatrix} 0 & & & & \\ 9 & 0 & & & \\ 3 & 7 & 0 & & \\ 6 & 5 & 9 & 0 & \\ 11 & 10 & \textcircled{2} & 8 & 0 \end{bmatrix} \end{array}$$

At the first stage, objects 3 and 5 are merged since they are most similar. This gives the cluster (35). At stage 2, we compute

$$d_{(35)1} = \max\{d_{31}, d_{51}\} = \max\{3, 11\} = 11$$

$$d_{(35)2} = \max\{d_{32}, d_{52}\} = 10$$

$$d_{(35)4} = \max\{d_{34}, d_{54}\} = 9$$

and the modified distance matrix becomes

$$\begin{array}{c} \\ (35) \\ 1 \\ 2 \\ 4 \end{array} \begin{array}{c} \begin{array}{cccc} (35) & 1 & 2 & 4 \end{array} \\ \begin{bmatrix} 0 & & & \\ 11 & 0 & & \\ 10 & 9 & 0 & \\ 9 & 6 & \textcircled{5} & 0 \end{bmatrix} \end{array}$$

The next merger occurs between the most similar groups, 2 and 4, to give the cluster (24). At stage 3 we have

$$d_{(24)(35)} = \max\{d_{2(35)}, d_{4(35)}\} = \max\{10, 9\} = 10$$

$$d_{(24)1} = \max\{d_{21}, d_{41}\} = 9$$

and the distance matrix

$$\begin{array}{c} \\ (35) \\ (24) \\ 1 \end{array} \begin{array}{c} \begin{array}{ccc} (35) & (24) & 1 \end{array} \\ \begin{bmatrix} 0 & & \\ 10 & 0 & \\ 11 & \textcircled{9} & 0 \end{bmatrix} \end{array}$$

The next merger produces the cluster (124). At the final stage the groups (35) and (124) are merged as the single cluster (12345) at level

$$d_{(124)(35)} = \max\{d_{1(35)}, d_{(24)(35)}\} = \max\{11, 10\} = 11$$

The dendogram is given in Figure 11.7. ■

Comparing Figures 11.4 and 11.7, we see that the dendograms for single linkage and complete linkage differ in the allocation of object 1 to previous groups.

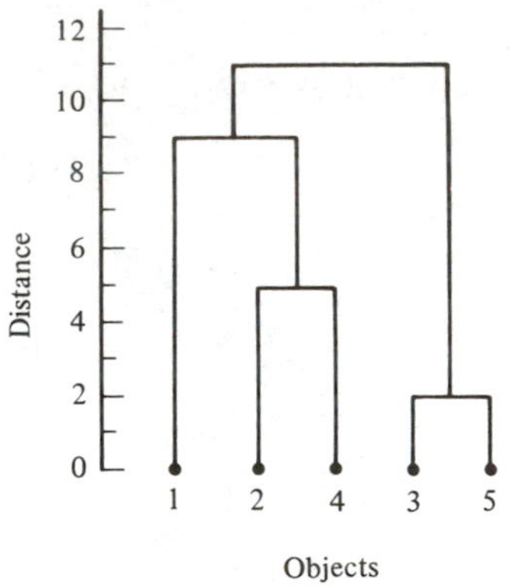

Figure 11.7 Complete-linkage dendogram for distances between five objects.

Example 11.7

In Example 11.5 we presented a distance matrix for numbers in 11 languages. The complete-linkage clustering algorithm applied to this distance matrix produced the dendogram shown in Figure 11.8 .

Comparing Figures 11.8 and 11.5, we see that both hierarchical methods yield the English–Norwegian–Danish and the French–Italian–Spanish language groups. Polish is merged with French-Italian-Spanish at an intermediate level. In addition, both methods merge Hungarian and Finnish only at the pentultimate stage.

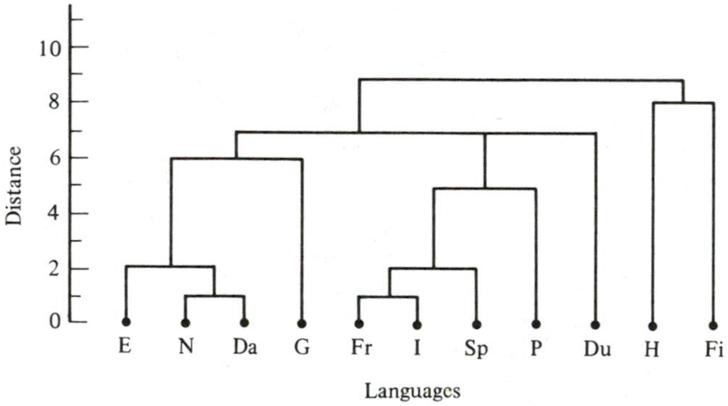

Figure 11.8 Complete-linkage dendogram for distances between numbers in 11 languages.

However, the two methods handle German and Dutch differently. Single linkage merges German and Dutch at an intermediate distance and these two languages remain a cluster until the final merger. Complete linkage merges German with the English–Norwegian–Danish group at an intermediate level. Dutch remains a cluster by itself until it is merged with the English–Norwegian–Danish–German and French–Italian–Spanish–Polish groups at a higher distance level. The final complete linkage merger involves two clusters. The final merger in single linkage involves three clusters. ■

Example 11.8

Data were collected on 22 U.S. public utility companies for the year 1975. These data are listed in Table 11.5.

TABLE 11.5 PUBLIC-UTILITY DATA (1975)

Company	X_1	X_2	X_3	X_4	X_5	X_6	X_7	X_8
	Variables							
1. Arizona Public Service	1.06	9.2	151.	54.4	1.6	9077.	0.	.628
2. Boston Edison Co.	.89	10.3	202.	57.9	2.2	5088.	25.3	1.555
3. Central Louisiana Electric Co.	1.43	15.4	113.	53.0	3.4	9212.	0.	1.058
4. Commonwealth Edison Co.	1.02	11.2	168.	56.0	.3	6423.	34.3	.700
5. Consolidated Edison Co. (N.Y.)	1.49	8.8	192.	51.2	1.0	3300.	15.6	2.044
6. Florida Power & Light Co.	1.32	13.5	111.	60.0	−2.2	11127.	22.5	1.241
7. Hawaiian Electric Co.	1.22	12.2	175.	67.6	2.2	7642.	0.	1.652
8. Idaho Power Co.	1.10	9.2	245.	57.0	3.3	13082.	0.	.309
9. Kentucky Utilities Co.	1.34	13.0	168.	60.4	7.2	8406.	0.	.862
10. Madison Gas & Electric Co.	1.12	12.4	197.	53.0	2.7	6455.	39.2	.623
11. Nevada Power Co.	.75	7.5	173.	51.5	6.5	17441.	0.	.768
12. New England Electric Co.	1.13	10.9	178.	62.0	3.7	6154.	0.	1.897
13. Northern States Power Co.	1.15	12.7	199.	53.7	6.4	7179.	50.2	.527
14. Oklahoma Gas & Electric Co.	1.09	12.0	96.	49.8	1.4	9673.	0.	.588
15. Pacific Gas & Electric Co.	.96	7.6	164.	62.2	−0.1	6468.	.9	1.400
16. Puget Sound Power & Light Co.	1.16	9.9	252.	56.0	9.2	15991.	0.	.620
17. San Diego Gas & Electric Co.	.76	6.4	136.	61.9	9.0	5714.	8.3	1.920
18. The Southern Co.	1.05	12.6	150.	56.7	2.7	10140.	0.	1.108
19. Texas Utilities Co.	1.16	11.7	104.	54.0	−2.1	13507.	0.	.636
20. Wisconsin Electric Power Co.	1.20	11.8	148.	59.9	3.5	7287.	41.1	.702
21. United Illuminating Co.	1.04	8.6	204.	61.0	3.5	6650.	0.	2.116
22. Virginia Electric & Power Co.	1.07	9.3	174.	54.3	5.9	10093.	26.6	1.306

SOURCE: Data courtesy of H. E. Thompson.

KEY: X_1: Fixed charge coverage ratio (income/debt).
X_2: Rate of return on capital.
X_3: Cost per KW capacity in place.
X_4: Annual load factor.
X_5: Peak KWH demand growth from 1974 to 1975.
X_6: Sales (KWH use per year).
X_7: Percent nuclear.
X_8: Total fuel costs (cents per KWH).

Although it is more interesting to group companies, we shall see here how the complete linkage algorithm can be used to cluster variables.

We measure the similarity between pairs of variables by the product moment correlation coefficient. The correlation matrix is given in Table 11.6.

When the sample correlations are used as similarity measures, variables with large negative correlations are regarded as very dissimilar; variables with large positive correlations are regarded as very similar. In this case, the "distance" between clusters is measured as the *smallest* similarity between members of the corresponding clusters. The complete linkage algorithm, applied to the similarity matrix above, yields the dendogram in Figure 11.9.

We see that variables 1 and 2 (fixed-charge coverage ratio and rate of return on capital), variables 4 and 8 (annual load factor and total fuel costs)

TABLE 11.6 CORRELATIONS BETWEEN PAIRS OF VARIABLES (PUBLIC UTILITY DATA)

X_1	X_2	X_3	X_4	X_5	X_6	X_7	X_8
1.000							
.643	1.000						
−.103	−.348	1.000					
−.082	−.086	.100	1.000				
−.259	−.260	.435	.034	1.000			
−.152	−.010	.028	−.288	.176	1.000		
.045	.211	.115	−.164	−.019	−.374	1.000	
−.013	−.328	.005	.486	−.007	−.561	−.185	1.000

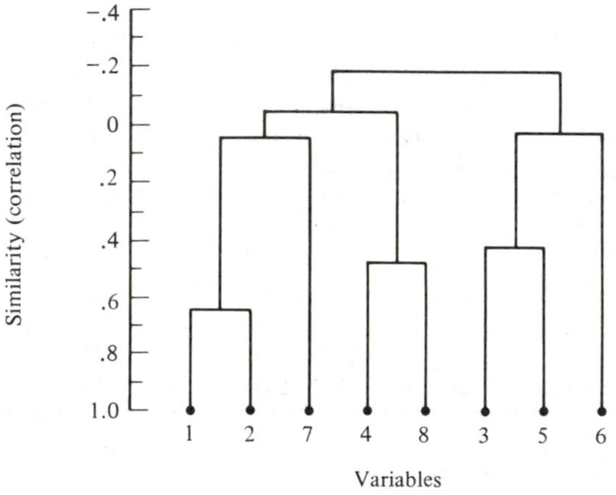

Figure 11.9 Complete-linkage dendogram for similarities between 8 utility company variables.

and variables 3 and 5 (cost per kilowatt capacity in place and peak kilowatt-hour demand growth) cluster at intermediate "similarity" levels. Variables 7 (percent nuclear) and 6 (sales) remain by themselves until the final stages. The final merger brings together the (12478) group and the (356) group. ■

As in single linkage, a "new" assignment of distances (similarities) that have the same relative orderings as the initial distances will not change the configuration of the complete linkage clusters.

Average Linkage

Average linkage treats the distance between two clusters as the average distance between all pairs of items where one member of a pair belongs to each cluster.

Again the input to the average linkage algorithm may be distances or similarities, and the method can be used to group objects or variables. The average linkage algorithm proceeds in the manner of the general algorithm of (11-10). We begin by searching the distance matrix $\mathbf{D} = \{d_{ik}\}$ to find the nearest (most similar) objects, for example, U and V. These objects are merged to form the cluster (UV). For Step 3 of the general agglomerative algorithm, the distances between (UV) and any other cluster W are determined by:

$$d_{(UV)W} = \frac{\sum_i \sum_k d_{ik}}{N_{(UV)} N_W} \tag{11-13}$$

where d_{ik} is the distance between object i in the cluster (UV) and object k in the cluster W, and $N_{(UV)}$ and N_W are the number of items in clusters (UV) and W, respectively.

Example 11.9

The average linkage algorithm was applied to the "distances" between 11 languages given in Example 11.5. The resulting dendogram is displayed in Figure 11.10.

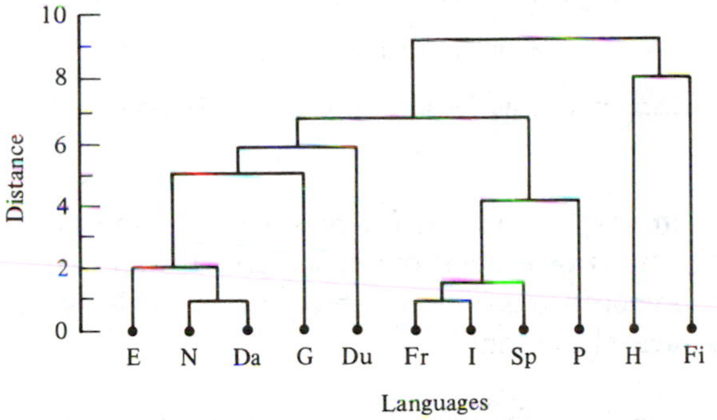

Figure 11.10 Average-linkage dendogram for distances between numbers in 11 languages.

A comparison of the dendogram in Figure 11.10 with the corresponding single-linkage dendogram (Figure 11.5) and complete-linkage dendogram (Figure 11.8) indicates that average linkage yields a configuration very much like the complete-linkage configuration. However, because distance is defined differently for each case, it is not surprising that mergers take place at different levels. ■

Example 11.10

An average-linkage algorithm applied to the Euclidean distances between 22 public utilities (see Table 11.1) produced the dendogram in Figure 11.11.

Concentrating on the intermediate clusters, we see that the utility companies tend to group according to geographical location. For example, one intermediate cluster contains the firms 1 (Arizona Public Service), 18 (The Southern Company—primarily Georgia and Alabama), 19 (Texas Utilities Company), and 14 (Oklahoma Gas and Electric Company). There are some exceptions. The cluster (7, 12, 21, 15, 2) contains firms on the eastern seaboard and in the far west. On the other hand, all of these firms are located near the coasts. Notice that Consolidated Edison Company of New York and San Diego Gas and Electric Company stand by themselves until the final amalgamation stages.

It is, perhaps, not surprising that utility firms with similar locations (or types of locations) cluster. One would expect regulated firms in the same area

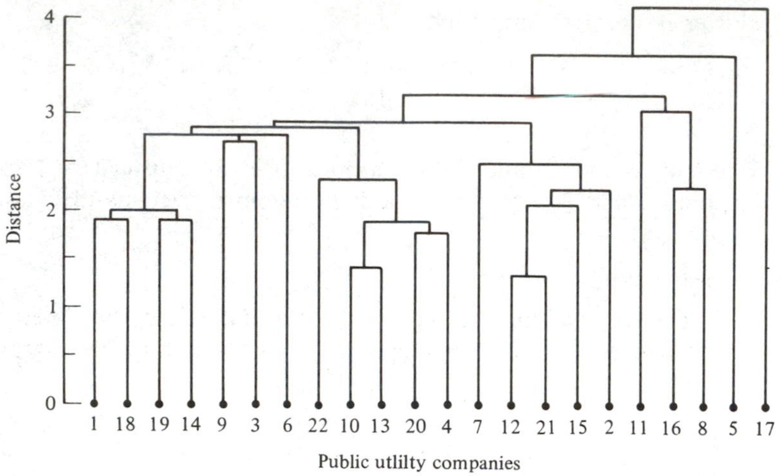

Figure 11.11 Average-linkage dendogram for distances between 22 public utility companies.

to use, basically, the same type of fuel(s) for power plants and face common markets. Consequently, types of generation, costs, growth rates, and so forth should be relatively homogeneous among these firms. This is apparently reflected in the hierarchical clustering. ■

For average linkage clustering, changes in the assignment of distances (similarities) can, even though they preserve relative orderings, affect the final configuration of clusters.

Final Comments

There are many agglomerative hierarchical clustering procedures besides single linkage, complete linkage, and average linkage. However, all of the agglomerative procedures follow the basic algorithm of (11-10).

As with most clustering methods, sources of error and variation are not formally considered in hierarchical procedures. This means that a clustering method will be sensitive to outliers, or "noise points."

There is no provision, in hierarchical clustering, for a reallocation of objects that may have been "incorrectly" grouped at an early stage. Consequently, the final configuration of clusters should always be carefully examined to see if it is sensible.

For a particular problem, it is a good idea to try several clustering methods and, within a given method, a couple different ways of assigning distances (similarities). If the outcomes from the several methods are (roughly) consistent with one another, perhaps a case for "natural" groupings can be advanced.

The *stability* of a hierarchical solution can sometimes be checked by applying the clustering algorithm before and after *small* errors (perturbations) have been added to the data units. If the groups are fairly well distinguished, the clusterings before perturbation and after perturbation should agree.

11.4 NONHIERARCHICAL CLUSTERING METHODS

Nonhierarchical clustering techniques are designed to group *items*, rather than *variables*, into a collection of K clusters. The number of clusters, K, may either be specified in advance or determined as part of the clustering procedure. Because a matrix of distances (similarities) does not have to be determined and the basic data do not have to be stored during the computer run, nonhierarchical methods can be applied to much larger data sets than hierarchical techniques.

Nonhierarchical methods start from either (1) an initial partition of items into groups or (2) an initial set of seed points, which will form the nuclei of clusters. Good choices for starting configurations should be free of overt biases. One way to start is to randomly select seed points from among the items or to randomly partition the items into initial groups.

In this section we discuss one of the more popular nonhierarchical procedures, the K-means method.

K-means Method

MacQueen [15] suggests the term *K-means* for describing his algorithm that assigns each item to the cluster having the nearest centroid (mean). In its simplest version, the process is composed of these three steps.

1. Partition the items into K initial clusters.
2. Proceed through the list of items, assigning an item to the cluster whose centroid (mean) is nearest. (Distance is usually computed using Euclidean distance with either standardized or unstandardized observations.) Recalculate the centroid for the cluster receiving the new item and for the cluster losing the item. (11-14)
3. Repeat Step 2 until no more reassignments take place.

Rather than starting with a partition of all items into K preliminary groups in Step 1, we could specify K initial centroids (seed points) and then proceed to Step 2.

The final assignment of items to clusters will be, to some extent, dependent upon the initial partition or the initial selection of seed points. Experience suggests that most major changes in assignment occur with the first reallocation step.

Example 11.11

Suppose we can measure two variables X_1 and X_2 for each of four items A, B, C, and D. The data are given in the following table.

Item	Observation x_1	x_2
A	5	3
B	-1	1
C	1	-2
D	-3	-2

The objective is to divide these items into $K = 2$ clusters such that the items within a cluster are closer to one another than the items in different clusters. To implement the $K = 2$-means method, we *arbitrarily* partition the items into two clusters, such as (AB) and (CD), and compute the coordinates $(\bar{x}_1, \bar{x}_2)$ of the cluster centroid (mean). Thus, at Step 1, we have:

Cluster	Coordinates of centroid $\bar{x}_1$	$\bar{x}_2$
(AB)	$\frac{5 + (-1)}{2} = 2$	$\frac{3 + 1}{2} = 2$
(CD)	$\frac{1 + (-3)}{2} = -1$	$\frac{-2 + (-2)}{2} = -2$

At Step 2 we compute the Euclidean distance of each item from the group centroids and reassign each item to the nearest group. If an item is moved from the initial configuration, the cluster centroids (means) must be updated before proceeding. We compute the squared distances

$$d^2(A, (AB)) = (5 - 2)^2 + (3 - 2)^2 = 10$$
$$d^2(A, (CD)) = (5 + 1)^2 + (3 + 2)^2 = 61$$

Since A is closer to cluster (AB) than cluster (CD), it is not reassigned. Continuing,

$$d^2(B, (AB)) = (-1 - 2)^2 + (1 - 2)^2 = 10$$
$$d^2(B, (CD)) = (-1 + 1)^2 + (1 + 2)^2 = 9$$

and, consequently, B is reassigned to cluster (CD) giving cluster (BCD) and the following updated centroid coordinates.

Cluster	Coordinates of centroid $\bar{x}_1$	$\bar{x}_2$
A	5	3
(BCD)	−1	−1

Again, each item is checked for reassignment. Computing the squared distances gives the following.

	Squared distances to group centroids			
	Item			
Cluster	A	B	C	D
A	0	40	41	89
(BCD)	52	4	5	5

We see that each item is currently assigned to the cluster with the nearest centroid (mean) and the process stops. The final $K = 2$ clusters are A and (BCD). ■

To check the stability of the clustering, it is desirable to rerun the algorithm with a new initial partition. Once clusters are determined, intuitions concerning their interpretations are aided by rearranging the list of items so that those in the first cluster appear first, those in the second cluster appear next, and so forth. A table of the cluster centroids (means) and within-cluster variances also helps to delineate group differences.

Example 11.12

Let us return to the problem of clustering the public utilities using the data in Table 11.5. The K-means algorithm for several choices of K was run. We present a summary of the results for $K = 4$ and $K = 5$ below. In general, the choice of a particular K is not clear-cut and depends upon subject matter knowledge, as well as data-based appraisals. (Data-based appraisals might include choosing K to maximize the between-cluster variability relative to the within-cluster variability. Relevant measures might include $|\mathbf{W}|/|\mathbf{B}+\mathbf{W}|$ [see (6-38)] and $\text{tr}(\mathbf{W}^{-1}\mathbf{B})$].

$K = 4$

Cluster	Number of firms	Firms
1	5	Idaho Power Co. (8), Nevada Power Co. (11), Puget Sound Power & Light Co. (16), Virginia Electric & Power Co. (22), Kentucky Utilities Co. (9).
2	6	Central Louisiana Electric Co. (3), Oklahoma Gas & Electric Co. (14), The Southern Co. (18), Texas Utilities Co. (19), Arizona Public Service (1), Florida Power and Light Co. (6).
3	5	New England Electric Co. (12), Pacific Gas & Electric Co. (15), San Diego Gas and Electric Co. (17), United Illuminating Co. (21), Hawaiian Electric Co. (7).
4	6	Consolidated Edison Co. (N. Y.) (5), Boston Edison Co. (2), Madison Gas & Electric Co. (10), Northern States Power Co. (13), Wisconsin Electric Power Co. (20), Commonwealth Edison Co. (4).

Distances Between Cluster Centers

$$\begin{array}{c} \\ 1 \\ 2 \\ 3 \\ 4 \end{array}\begin{array}{c} \begin{array}{cccc} 1 & 2 & 3 & 4 \end{array} \\ \left[\begin{array}{cccc} 0 & & & \\ 3.08 & 0 & & \\ 3.29 & 3.56 & 0 & \\ 3.05 & 2.84 & 3.18 & 0 \end{array}\right] \end{array}$$

$K = 5$

Cluster	Number of firms	Firms
1	5	Nevada Power Co. (11), Puget Sound Power & Light Co. (16), Idaho Power Co. (8), Virginia Electric & Power Co. (22), Kentucky Utilities Co. (9).
2	6	Central Louisiana Electric Co. (3), Texas Utilities Co. (19), Oklahoma Gas and Electric Co. (14), The Southern Co. (18), Arizona Public Service (1), Florida Power and Light Co. (6).
3	5	New England Electric Co. (12), Pacific Gas & Electric Co. (15), San Diego Gas & Electric Co. (17), United Illuminating Co. (21), Hawaiian Electric Co. (7).
4	2	Consolidated Edison Co. (N. Y.) (5), Boston Edison Co. (2).
5	4	Commonwealth Edison Co. (4), Madison Gas & Electric Co. (10), Northern States Power Co. (13), Wisconsin Electric Power Co. (20).

Distances Between Cluster Centers

$$\begin{array}{c} \\ 1 \\ 2 \\ 3 \\ 4 \\ 5 \end{array} \begin{array}{c} \begin{array}{ccccc} 1 & 2 & 3 & 4 & 5 \end{array} \\ \begin{bmatrix} 0 & & & & \\ 3.08 & 0 & & & \\ 3.29 & 3.56 & 0 & & \\ 3.63 & 3.46 & 2.63 & 0 & \\ 3.18 & 2.99 & 3.81 & 2.89 & 0 \end{bmatrix} \end{array}$$

The cluster profiles ($K = 5$) shown in Figure 11.12 order the eight variables according to the ratios of their between-cluster variability to their within-cluster variability [for univariate F-ratios, see Section 6.4]. We have

$$F_{\text{nuc}} = \frac{\text{mean square percent nuclear between clusters}}{\text{mean square percent nuclear within clusters}} = \frac{3.335}{.255} = 13.1$$

so firms within different clusters are widely separated with respect to percent nuclear, but firms within the same cluster show little percent nuclear variation. Fuel costs (FUELC) and annual sales (SALES) also seem to be of some importance in distinguishing the clusters.

Reviewing the firms in the five clusters, it is apparent that the K-means method gives results generally consistent with the average linkage hierarchical method (see Example 11.10). Firms with common or compatible geographical locations cluster. Also, the firms in a given cluster seem to be of roughly the same in terms of percent nuclear. ■

We must caution, as we have throughout the book, that the importance of *individual* variables in clustering must be judged from a multivariate perspective. All of the variables (multivariate observations) determine the cluster means and the

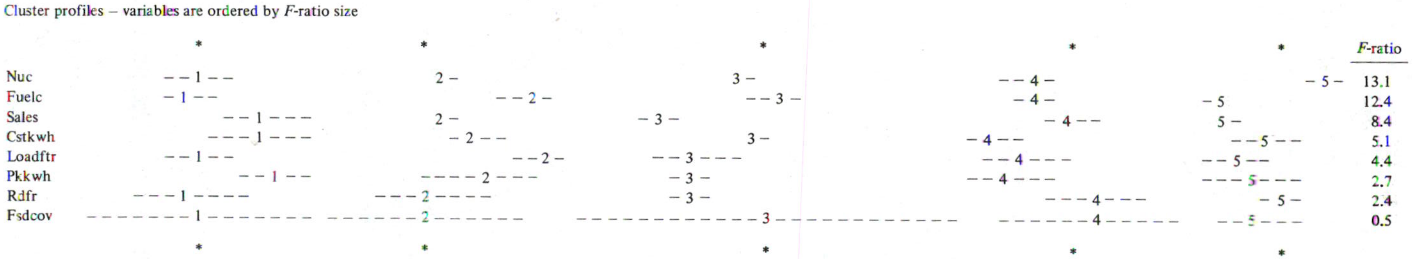
Cluster profiles – variables are ordered by F-ratio size
F-ratio
Nuc 13.1
Fuelc 12.4
Sales 8.4
Cstkwh 5.1
Loadftr 4.4
Pkkwh 2.7
Rdfr 2.4
Fsdcov 0.5

Each column describes a cluster.
The cluster number is printed at the mean of each variable.
Dashes indicate one standard deviation above and below mean.

Figure 11.12 Cluster profiles ($K = 5$) for public-utility data.

reassignment of items. In addition, the values of the descriptive statistics measuring the importance of individual variables are functions of the number of clusters and the final configuration of the clusters. On the other hand, descriptive measures can be helpful, after the fact, in assessing the "success" of the clustering procedure.

Final Comments

There are strong arguments for not fixing the number of clusters, K, in advance. These include the following.

1. If two or more seed points inadvertently lie within a single cluster, their resulting clusters will be poorly differentiated.
2. The existence of an outlier might produce at least one group with very disperse items.
3. Even if the population is known to consist of K groups, the sampling method may be such that data from the rarest group do not appear in the sample. Forcing the data into K groups would lead to nonsensical clusters.

In cases where a single run of the algorithm requires the user to specify K, it is always a good idea to rerun the algorithm for several choices.

Discussions of other nonhierarchical clustering procedures are available in [2], [6], and [12].

11.5 MULTIDIMENSIONAL SCALING

Multidimensional scaling techniques deal with the following problem: For a set of observed similarities (or distances) between every pair of N items, find a representation of the items in few dimensions such that the interitem proximities "nearly match" the original similarities (or distances).

It may not be possible to match exactly the ordering of the original similarities (distances). Consequently, scaling techniques attempt to find configurations in $q \leq N - 1$ dimensions such that the match is as close as possible. The numerical measure of closeness is called the *stress*.

It is possible to arrange the N items in a low-dimensional coordinate system using only the *rank orders* of the $N(N - 1)/2$ original similarities (distances) and not their magnitudes. When only this ordinal information is used to obtain a geometric representation, the process is called *nonmetric multidimensional scaling*. If the actual magnitudes of the original similarities (distances) are used to obtain a geometric representation in q dimensions, the process is called *metric multidimensional scaling*.

Scaling techniques were developed by Shepard (see [16] for a review of early work), Kruskal [13, 14], and others. Multidimensional scaling invariably requires the use of a computer, and several good computer programs are now available for this purpose.

The Basic Algorithm

For N items, there are $M = N(N-1)/2$ similarities (distances) between pairs of different items. These similarities (distances) constitute the basic data. (In cases where the similarities cannot be easily quantified as, for example, the similarity between two colors, the rank orders of the similarities are the basic data.)

Assuming no ties, the similarities can be arranged in a strictly ascending order as

$$s_{i_1k_1} < s_{i_2k_2} < \cdots < s_{i_Mk_M} \tag{11-15}$$

Here $s_{i_1k_1}$ is the smallest of the M similarities. The subscript i_1k_1 indicates the pair of items that are least similar; that is, the items with rank 1 in the similarity ordering. Other subscripts are interpreted in the same manner. We want to find a q-dimensional configuration of the N items such that the distances, $d_{ik}^{(q)}$, between pairs of items match the ordering in (11-15). If the distances are laid out in a manner corresponding to (11-15), a perfect match occurs when

$$d_{i_1k_1}^{(q)} > d_{i_2k_2}^{(q)} > \cdots > d_{i_Mk_M}^{(q)} \tag{11-16}$$

That is, the descending ordering of the distances in q dimensions is exactly analogous to the ascending ordering of the initial similarities. As long as the order in (11-16) is preserved, the magnitudes of the distances are unimportant.

For a given value of q, it may not be possible to find a configuration of points whose pairwise distances are monotonically related to the original similarities. Kruskal [13] proposed a measure of the extent to which a geometrical representation falls short of a perfect match. This measure, the stress, is defined as

$$\text{Stress}(q) = \left\{ \frac{\sum\sum_{i<k} \left(d_{ik}^{(q)} - \hat{d}_{ik}^{(q)}\right)^2}{\sum\sum_{i<k} \left[d_{ik}^{(q)}\right]^2} \right\}^{1/2} \tag{11-17}$$

The $\hat{d}_{ik}^{(q)}$'s in the stress formula are numbers known to satisfy (11-16); that is, they are monotonically related to the similarities. The $\hat{d}_{ik}^{(q)}$'s are *not* distances in the sense that they satisfy the usual distance properties of (1-25). They are merely reference numbers used to judge the nonmonotonicity of the observed $d_{ik}^{(q)}$'s.

The idea is to find a representation of the items as points in q-dimensions such that the stress is as small as possible. Kruskal [13] suggests the stress be informally interpreted according to the following guidelines.

Stress	*Goodness of fit*	
20%	Poor	
10%	Fair	
5%	Good	(11-18)
2.5%	Excellent	
0%	Perfect	

Goodness of fit refers to the monotone relationship between the similarities and final distances.

Once items are located in q dimensions, their $q \times 1$ vectors of coordinates can be treated as multivariate observations. For display purposes it is convenient to represent this q-dimensional scatterplot in terms of its principal component axes (see Chapter 8).

We have written the stress measure as a function of q, the number of dimensions for the geometrical representation. For each q, the configuration leading to the minimum stress can be obtained. As q increases minimum stress will, within rounding error, decrease and will be zero for $q = N - 1$. Beginning with $q = 1$, a plot of these stress(q) numbers versus q can be constructed. The value of q for which this plot begins to level off may be selected as the "best" choice of the dimensionality. That is, we look for an "elbow" in the stress-dimensionality plot.

The entire multidimensional scaling algorithm is summarized by these steps.

1. For N times, obtain the $M = N(N-1)/2$ similarities (distances) between distinct pairs of items. Order the similarities (distances) as in (11-15). (Distances are ordered from largest to smallest. If similarities (distances) cannot be computed, the rank orders must be specified.)
2. Using a trial configuration in q dimensions, determine the inter-item distances $d_{ik}^{(q)}$ and numbers $\hat{d}_{ik}^{(q)}$ where the latter satisfy (11-16) and minimize the stress (11-17). (The $\hat{d}_{ik}^{(q)}$ are frequently determined within scaling computer programs using regression methods designed to produce monotone "fitted" distances.) (11-19)
3. Using the $\hat{d}_{ik}^{(q)}$'s, the points are moved around to obtain an improved configuration. (For q fixed, an improved configuration is determined by a general function minimization procedure applied to the stress. In this context the stress is regarded as a function of the $N \times q$ coordinates of the N items.) A new configuration will have new $d_{ik}^{(q)}$'s, new $\hat{d}_{ik}^{(q)}$'s, and smaller stress. The process is repeated until the best (minimum stress) representation is obtained.
4. Plot minimum stress(q) versus q and choose the best number of dimensions, q^*, from an examination of this plot.

We have assumed the initial similarity values are symmetric ($s_{ik} = s_{ki}$), there are no ties, and there are no missing observations. Kruskal [13, 14] has suggested methods for handling asymmetries, ties and missing observations. In addition, there are now multidimensional scaling computer programs that will handle not only Euclidean distance, but any distance of the Minkowski type [see (11-3)].

The next examples illustrate multidimensional scaling with distances as the initial (dis)similarity measures.

Example 11.13

Table 11.7 displays the airline distances between pairs of selected U.S. cities.

Since the cities naturally lie in a two-dimensional space (a nearly level part of the curved surface of the earth), it is not surprising that multidimensional scaling with $q = 2$ will locate these items about as they occur on a map.

TABLE 11.7 AIRLINE-DISTANCE DATA

	Atlanta (1)	Boston (2)	Cincinnati (3)	Columbus (4)	Dallas (5)	Indianapolis (6)	Little Rock (7)	Los Angeles (8)	Memphis (9)	St. Louis (10)	Spokane (11)	Tampa (12)
(1)	0											
(2)	1068	0										
(3)	461	867	0									
(4)	549	769	107	0								
(5)	805	1819	943	1050	0							
(6)	508	941	108	172	882	0						
(7)	505	1494	618	725	325	562	0					
(8)	2197	3052	2186	2245	1403	2080	1701	0				
(9)	366	1355	502	586	464	436	137	1831	0			
(10)	558	1178	338	409	645	234	353	1848	294	0		
(11)	2467	2747	2067	2131	1891	1959	1988	1227	2042	1820	0	
(12)	467	1379	928	985	1077	975	912	2480	779	1016	2821	0

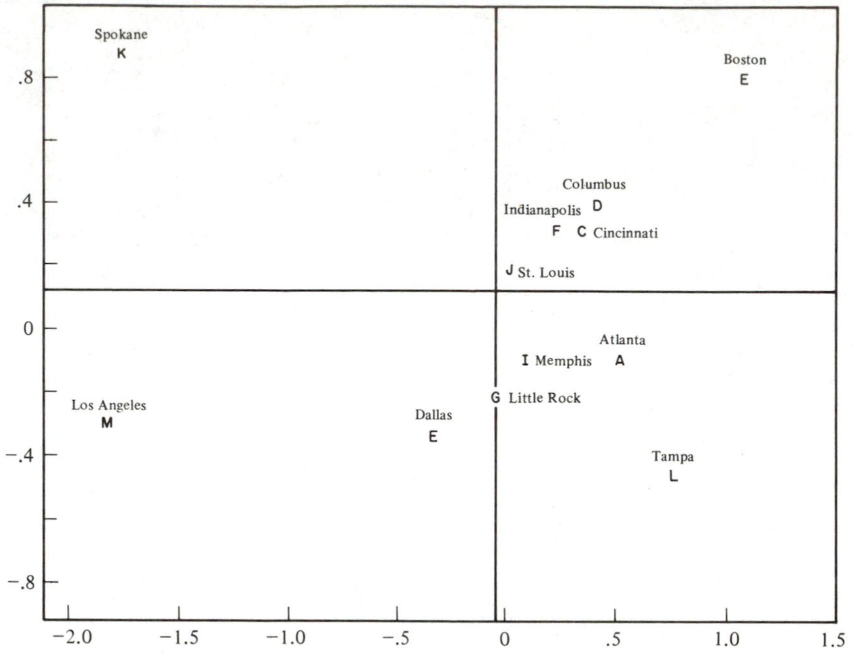

Figure 11.13 A geometrical representation of cities produced by multidimensional scaling.

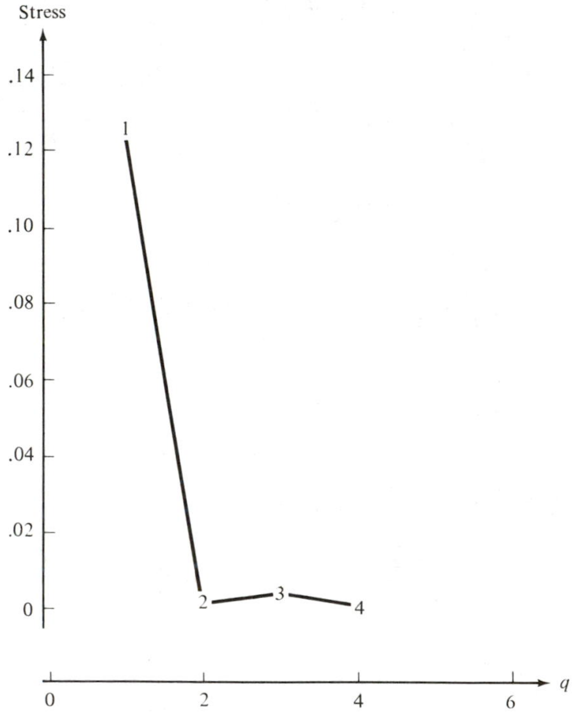

Figure 11.14 Stress function for airline distances between cities.

Note if the distances in Table 11.7 are ordered from largest to smallest—that is, least similar to most similar—the first position is occupied by $d_{\text{Boston, L.A.}} = 3052$.

A multidimensional scaling plot for $q = 2$ dimensions is shown in Figure 11.13. The axes lie along the sample principal components of the scatterplot.

A plot of stress(q) versus q is shown in Figure 11.14. Since stress(1) $\times$ 100% = 12%, a representation of the cities in one dimension (along a single axis) is not unreasonable. The "elbow" of the stress function occurs at $q = 2$. Here stress(2) $\times$ 100% = 0.8% and the "fit" is almost perfect.

The plot in Figure 11.14 indicates $q = 2$ is the best choice for the dimension of the final configuration. Note that the stress actually increases for $q = 3$. This anomaly can occur for extremely small values of stress because of difficulties with the numerical search procedure used to locate the minimum stress. ■

Example 11.14

Let us try to represent the 22 public utility firms discussed in Example 11.8 as points in a low-dimensional space. The measures of (dis)similarities between pairs of firms are the Euclidean distances listed in Table 11.1. Multidimensional scaling in $q = 1, 2, \ldots, 6$ dimensions produced the stress function shown in Figure 11.15.

The stress function in Figure 11.15 has no sharp elbow. The plot appears to level out at "good" values of stress (less than or equal 5%) in the neighborhood of $q = 4$. A good 4-dimensional representation of the utilities is achievable but difficult to display. We show a plot of the utility configuration

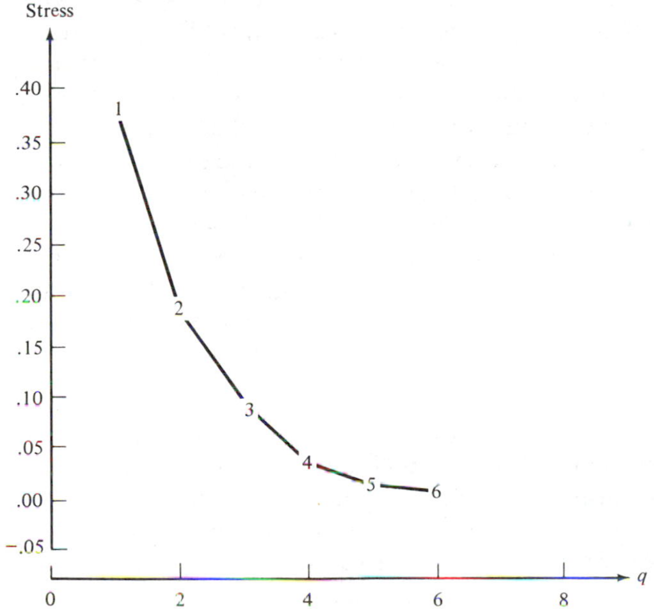

Figure 11.15 Stress function for distances between utilities.

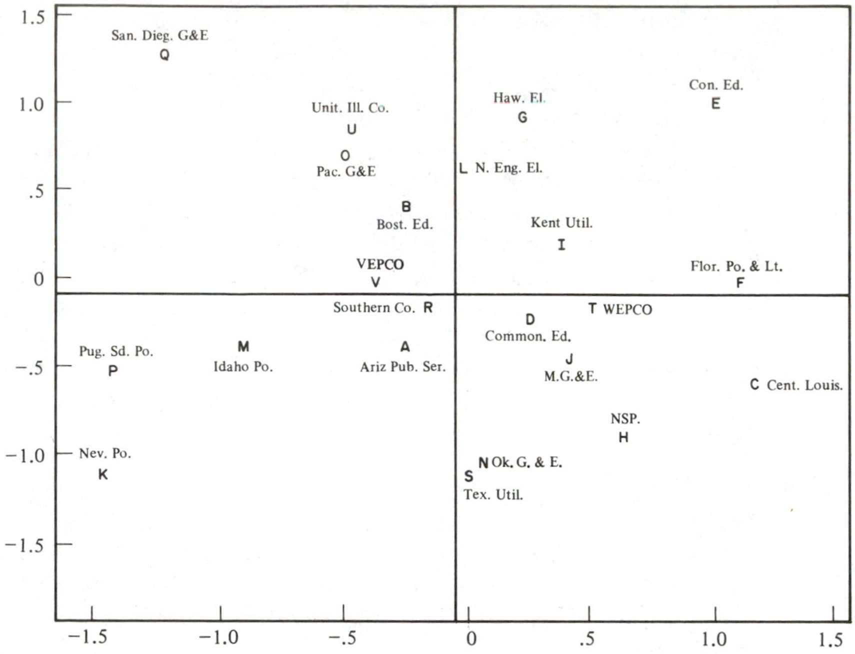

Figure 11.16 A geometrical representation of utilities produced by multidimensional scaling.

obtained in $q = 2$ dimensions in Figure 11.16. The axes lie along the sample principal components of the final scatter.

Although the stress for two dimensions is rather high (stress(2) $\times$ 100% = 19%), the distances between firms in Figure 11.16 are not wildly inconsistent with the clustering results presented earlier in this chapter. For example, the midwest utilities — Commonwealth Edison, Wisconsin Electric Power (WEPCO), Madison Gas and Electric (MG & E) and Northern States Power (NSP)—are close together (similar). Texas Utilities and Oklahoma Gas and Electric (Ok. G & E) are also very close together (similar). Other utilities tend to group according to geographical locations or similar environments.

The utilities cannot be positioned in two dimensions such that the interutility distances $d_{ik}^{(2)}$ are entirely consistent with the original distances in Table 11.1. More flexibility for positioning the points is required and this can only be obtained by introducing additional dimensions. ■

To summarize, the key objective of multidimensional scaling procedures is a low-dimensional picture. Whenever multivariate data can be presented graphically in two or three dimensions, visual inspection can greatly aid interpretations.

When the multivariate observations are naturally numerical and Euclidean distances in p-dimensions, $d_{ik}^{(p)}$, can be computed, we can seek a $q < p$-dimensional representation by minimizing

$$E = \left[\sum\sum_{i<k} \left(d_{ik}^{(p)} - d_{ik}^{(q)}\right)^2 \Big/ d_{ik}^{(p)} \right] \left[\sum\sum_{i<k} d_{ik}^{(p)} \right]^{-1} \tag{11-20}$$

In this alternative approach, the Euclidean distances in p and q dimensions are compared directly. Techniques for obtaining low-dimensional representations by minimizing E are called *nonlinear mappings*.

Final goodness of fit of any low dimensional representation can be depicted graphically by *minimal spanning trees*. You are referred to [7] and [12] for more discussion of these topics.

11.6 PICTORIAL REPRESENTATIONS

If multidimensional observations can be represented in two dimensions, then distinguishable groupings can often be discerned by eye. Several clever methods for representing multivariate data in two dimensions have been proposed. As we have seen in the previous section, multidimensional scaling seeks to represent p-dimensional observations in few dimensions such that the original distances (or similarities) between pairs of observations are (nearly) preserved. Also, plots of pairs of the first few principal components, described in Chapter 8, provide two-dimensional representations of the data.

We shall discuss and illustrate three additional methods for graphing multivariate observations in two dimensions: stars, Andrews plots, and Chernoff faces. One good source for more discussion of graphical methods is [7].

Stars

Suppose each data unit consists of nonnegative observations on $p \geq 2$ variables. In two dimensions we can construct circles of a fixed (reference) radius with p equally spaced rays emanating from the center of the circle. The lengths of the rays represent the values of the variables. The ends of the rays can be connected with straight lines to form a star. Each star represents a multivariate observation and the stars can be grouped according to their (subjective) similarities.

It is often helpful, when constructing the stars, to standardize the observations. In this case some of the observations will be negative. The observations can then be reexpressed so that the center of the circle represents the smallest standardized observation within the entire data set.

Example 11.15

Stars representing the first 5 of the 22 public utility firms in Table 11.5 are shown in Figure 11.17. There are 8 variables; consequently the stars are distorted octogons.

The observations on all variables were standardized. Among the first five utilities, the smallest standardized observation for any variable was -1.6.

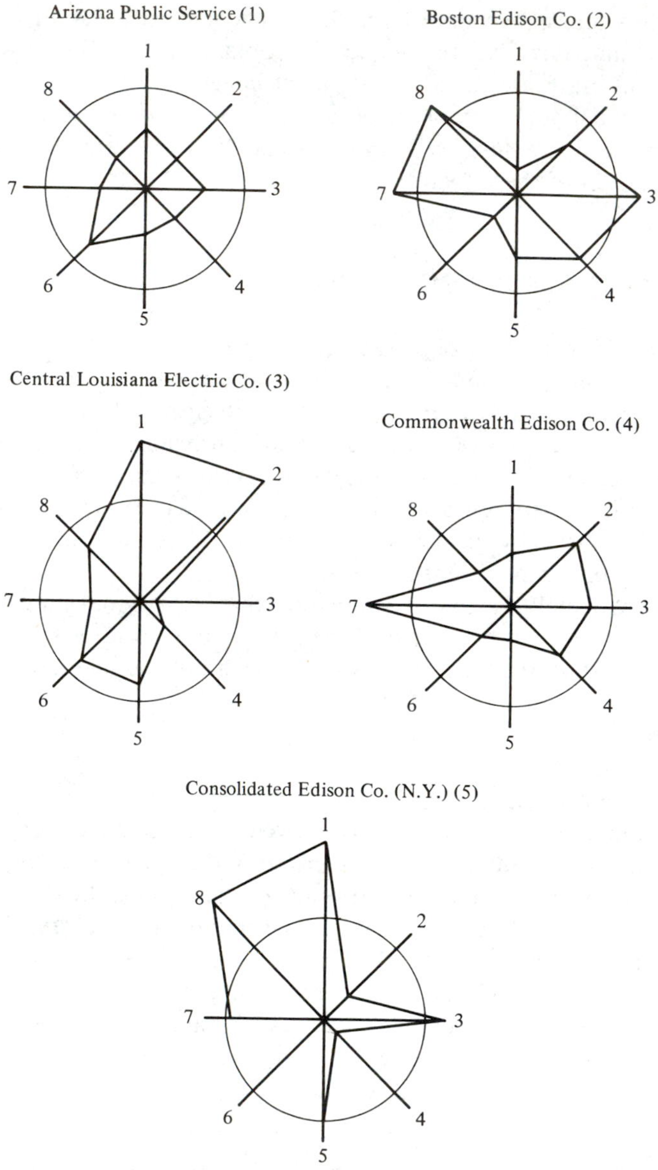

Figure 11.17 Stars for the first five public utilities.

Treating this value as zero, the variables are plotted on identical scales along eight equiangular rays originating from the center of the circle. The variables are ordered in a clockwise direction beginning in the 12 o'clock position.

At first glance none of these utilities appears to be similar. However, because of the way the stars are constructed, each variable gets equal weight in the visual impression. If we concentrate on the variables 6 (sales in KWH use per year) and 8 (total fuel costs in cents per KWH), then Boston Edison and Consolidated Edison are similar (small variable 6, large variable 8) and Arizona Public Service, Central Louisiana Electric, and Commonwealth Edison are similar (moderate variable 6, moderate variable 8). These variables seemed to be important determinants of similarity in the K-means clustering algorithm (see Example 11.12). ■

Andrews Plots

Andrews [3] has suggested that a p-dimensional vector of measurements $[x_1, x_2, \ldots, x_p]'$ be represented by the finite Fourier series

$$f(t) = \frac{x_1}{\sqrt{2}} + x_2 \sin t + x_3 \cos t + x_4 \sin 2t + x_5 \cos 2t + \cdots \qquad -\pi \le t \le \pi \tag{11-21}$$

That is, the measurements become the coefficients in an expression whose graph is a periodic function. For example, a four-dimensional observation $[6, 3, -1, 2]'$ might be converted to the function

$$f(t) = \frac{6}{\sqrt{2}} + 3 \sin t - \cos t + 2 \sin 2t, \qquad -\pi \le t \le \pi$$

and plotted as a function of t.

Plots of the Fourier series representations of the multivariate observations will be curves which can then be visually grouped. Andrews plots are affected by interchanging coordinates (coefficients). Consequently it is desirable to try a variety of displays before deciding on the best one for a given data set.

Experience has indicated that the data should be standardized before forming the Fourier series. Moreover, if the number of items is moderate to large, Andrews plots tend to be tangled. The number of Andrews curves superimposed on one graph should probably be limited to five or six.

Example 11.16

Observations representing the 22 public utilities were plotted according to (11-21) in Figure 11.18. Groups of similar firms are somewhat difficult to see. Motivated by the shaded distance matrix in Figure 11.2 (see Example 11.1), we plotted the group consisting of firms (4, 10, 13, 20, 22). The result is shown in Figure 11.19. Note that firm 22 (Virginia Electric and Power Company) seems to be a bit different from the rest and the Andrews plot is consistent with the average-linkage hierarchical clustering algorithm illustrated in Example 11.10 (see Figure 11.11). ■

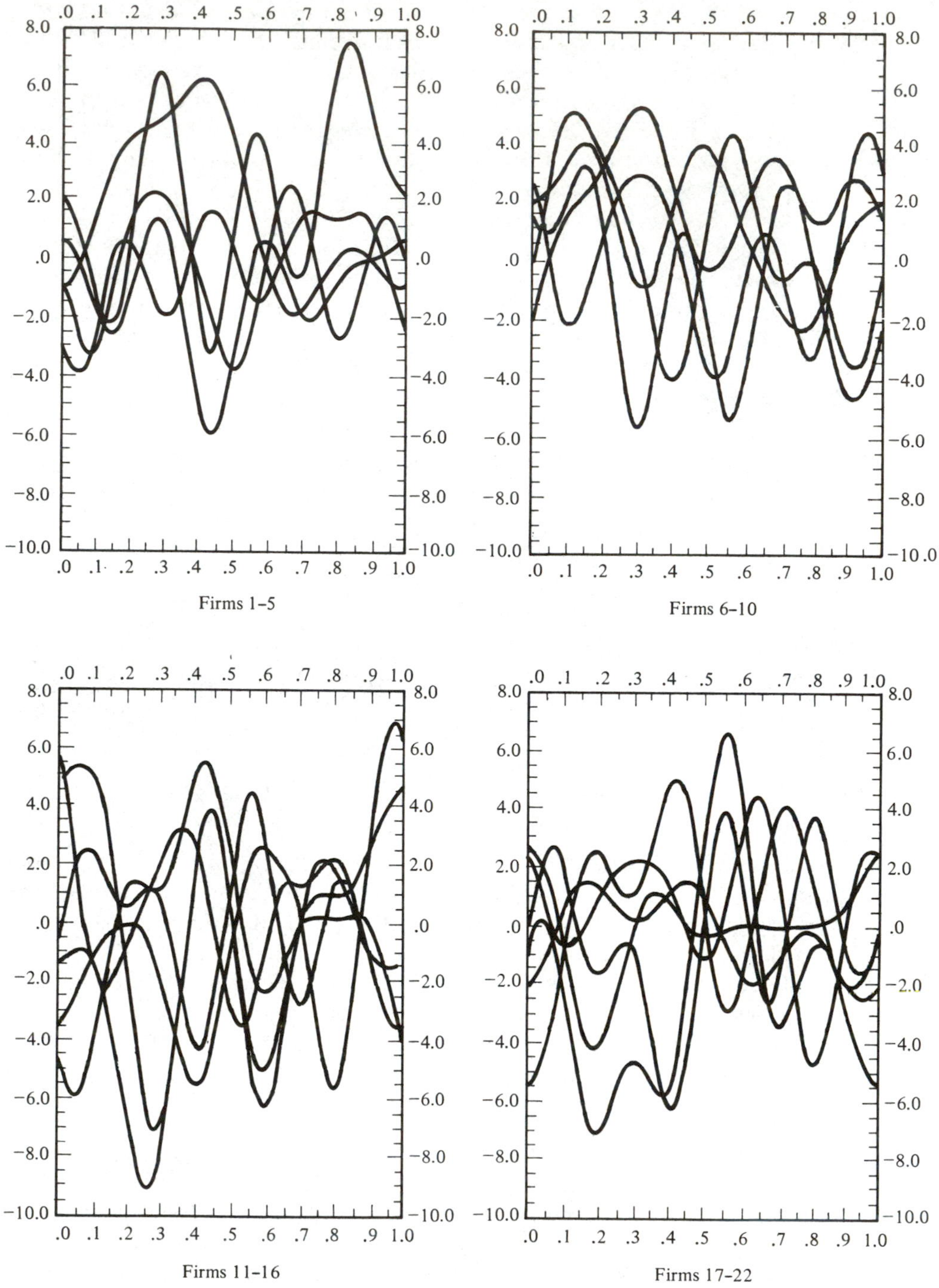

Figure 11.18 Andrews plots for 22 public utilities.

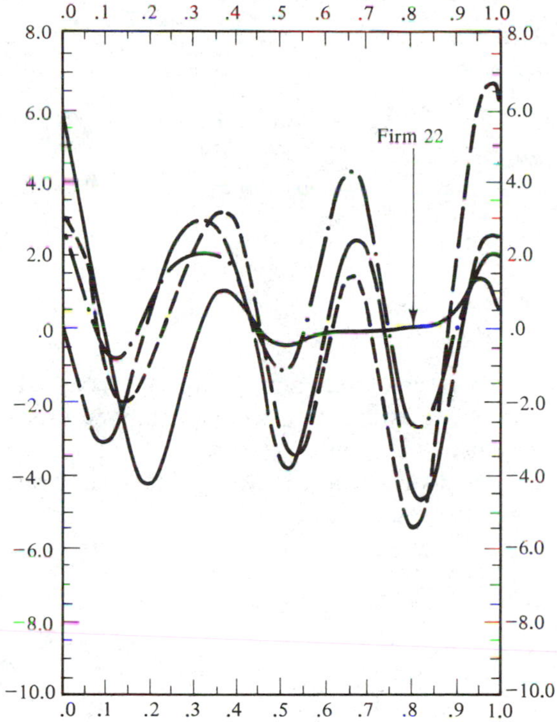

Figure 11.19 Andrews plots for utilities 4, 10, 13, 20, and 22.

Chernoff Faces

People react to faces. Chernoff [4] suggested representing p-dimensional observations as a two-dimensional face whose characteristics (face shape, curvature of mouth, nose length, eye size, pupil position, and so forth) are determined by the measurements on the p variables.

As originally designed, Chernoff faces can handle up to 18 variables. The assignment of variables to facial features is done by the experimenter and different choices produce different results. Some iteration is usually necessary before satisfactory representations are achieved. If the investigator is fairly sure two or three variables are primarily responsible for distinguishing clusters, these variables can be associated with *prominent* facial characteristics. Associating an "important" variable with a characteristic such as nose length, rather than a less-prominent characteristic like pupil position, allows one to select groupings more readily.

Like Andrews plots, Chernoff faces appear to be most useful for verifying (1) an initial grouping suggested by subject matter knowledge and intuition or (2) final groupings produced by clustering algorithms.

Example 11.17

Using the data in Table 11.5, the 22 public utility companies were represented as Chernoff faces. We have the following correspondences.

	Variable		Facial characteristic
X_1:	Fixed charge coverage.	↔	Face half height.
X_2:	Rate of return on capital.	↔	Face width.
X_3:	Cost per KW capacity in place.	↔	Position of mouth center.
X_4:	Annual load factor.	↔	Slant of eyes.
X_5:	Peak KWH demand growth from 1974.	↔	Eccentricity $\left(\frac{\text{height}}{\text{width}}\right)$ of eyes.
X_6:	Sales (KWH use per year).	↔	Half length of eye.
X_7:	Percent nuclear.	↔	Curvature of mouth.
X_8:	Total fuel costs (cents per KWH).	↔	Length of nose.

The Chernoff faces are shown in Figure 11.20. We have subjectively grouped "similar" faces into 7 clusters. If a smaller number of clusters is desired, we might combine clusters 5, 6, and 7 and, perhaps, clusters 2 and 3 to obtain 4 or 5 clusters. For our assignment of variables to facial features, the firms group largely according to geographical location.

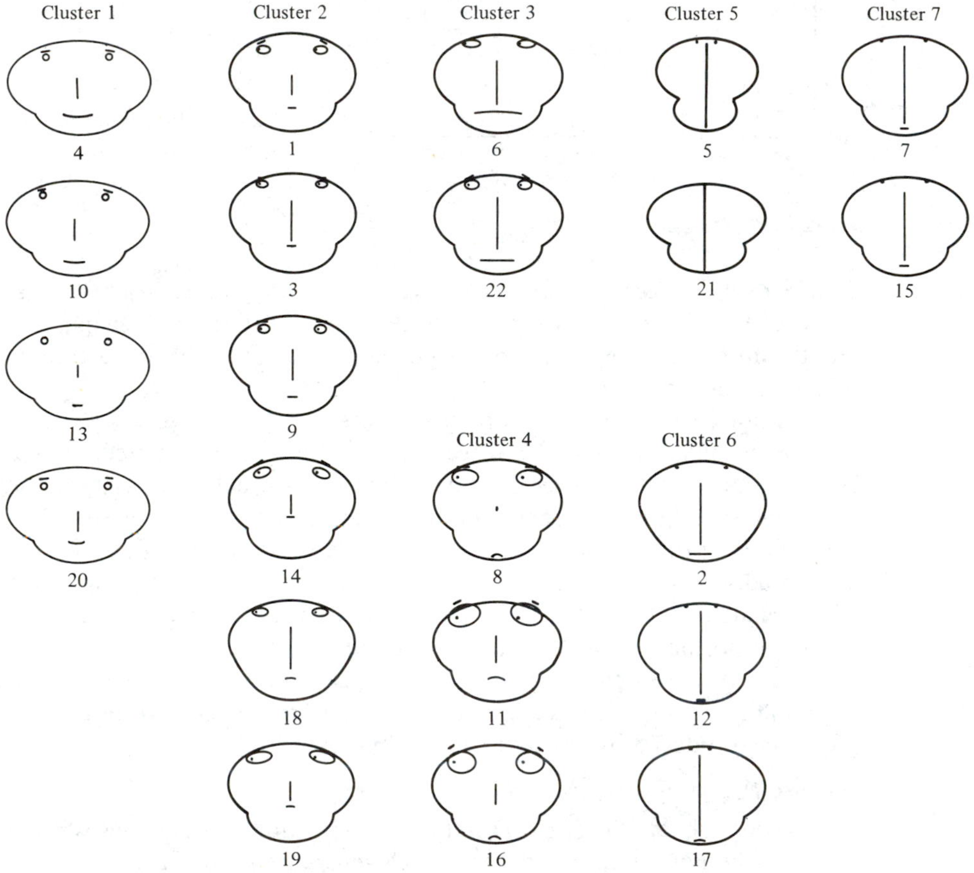

Figure 11.20 Chernoff faces for 22 public utilities.

The Chernoff face groupings are roughly consistent with the results from the K-means clustering discussed in Example 11.12. ■

Constructing Chernoff faces is a task that must be done with the aid of a computer. The data is ordinarily standardized within the computer program as part of the process for determining the locations, sizes and orientations of the facial characteristics. With some training, Chernoff faces can be an effective way to communicate similarities.

Final Comments

There are several ingenious ways to picture multivariate data in two dimensions. We have described some of them. Further advances are possible and will almost certainly take advantage of improved computer graphics.

The effectiveness of the stars, Andrews plots, and Chernoff faces is mixed. Sometimes these pictorial displays can be informative; however, more often than not, they will not clearly distinguish groups. At the present stage of development, a two-dimensional plot of the first two principal components may be more informative than any of the exotic graphical techniques.

EXERCISES

11.1. Certain characteristics associated with a few recent United States presidents are listed in Table 11.8.

TABLE 11.8

President	Birthplace (region of United States)	Elected first term?	Party	Prior U.S. congressional experience?	Served as vice-president?
1. R. Reagan	Midwest	Yes	Republican	No	No
2. J. Carter	South	Yes	Democrat	No	No
3. G. Ford	Midwest	No	Republican	Yes	Yes
4. R. Nixon	West	Yes	Republican	Yes	Yes
5. L. Johnson	South	No	Democrat	Yes	Yes
6. J. Kennedy	East	Yes	Democrat	Yes	No

(a) Introducing appropriate binary variables, calculate similarity coefficient 1 in Table 11.2 for pairs of presidents. (*Hint:* You may use birthplace as South, non-South.)

(b) Proceeding as in Part a, calculate similarity coefficients 2 and 3 in Table 11.2. Verify the monotonicity relation of coefficients 1, 2, and 3 by displaying the order of the 15 similarities for each coefficient.

11.2. Repeat Exercise 11.1 using similarity coefficients 5, 6, and 7 in Table 11.2.

11.3. Show that the sample correlation coefficient, r, can be written as

$$r = \frac{ad - bc}{[(a + b)(a + c)(b + d)(c + d)]^{1/2}} \quad \text{[see (11-9)]}$$

for two 0-1 binary variables with the frequencies

		Variable 2	
		0	1
Variable 1	0	a	b
	1	c	d

11.4. Show that the monotonicity property holds for the similarity coefficients 1, 2, and 3 in Table 11.2. (*Hint:* $(b + c) = p - (a + d)$. So, for instance,

$$\frac{a + d}{a + d + 2(b + c)} = \frac{1}{1 + 2[p/(a + d) - 1]}$$

This equation relates coefficients 3 and 1. Find analogous representations for the other pairs.)

11.5. Consider the matrix of distances

$$
\begin{array}{c c}
 & \begin{array}{cccc} 1 & 2 & 3 & 4 \end{array} \\
\begin{array}{c} 1 \\ 2 \\ 3 \\ 4 \end{array} &
\left[\begin{array}{cccc}
0 & & & \\
1 & 0 & & \\
11 & 2 & 0 & \\
5 & 3 & 4 & 0
\end{array}\right]
\end{array}
$$

Cluster the four items using each procedure.
(a) Single-linkage hierarchical procedure.
(b) Complete-linkage hierarchical procedure.
(c) Average-linkage hierarchical procedure.
Draw the dendograms and compare the results in (a), (b) and (c).

11.6. The distances between pairs of five items are given below.

$$
\begin{array}{c c}
 & \begin{array}{ccccc} 1 & 2 & 3 & 4 & 5 \end{array} \\
\begin{array}{c} 1 \\ 2 \\ 3 \\ 4 \\ 5 \end{array} &
\left[\begin{array}{ccccc}
0 & & & & \\
4 & 0 & & & \\
6 & 9 & 0 & & \\
1 & 7 & 10 & 0 & \\
6 & 3 & 5 & 8 & 0
\end{array}\right]
\end{array}
$$

Cluster the five items using the single-linkage, complete-linkage, and average-linkage hierarchical methods. Draw the dendograms and compare the results.

11.7. The sample correlations for five stocks were given in Example 8.5. These correlations rounded to two decimal places are reproduced below.

	Allied Chemical	DuPont	Union Carbide	Exxon	Texaco
Allied Chemical	1				
DuPont	.58	1			
Union Carbide	.51	.60	1		
Exxon	.39	.39	.44	1	
Texaco	.46	.32	.43	.52	1

Treating the sample correlations as similarity measures, cluster the stocks using the single linkage and complete linkage hierarchical procedures. Draw the dendograms and compare the results.

11.8. Using the distances in Example 11.4, cluster the items using the average linkage hierarchical procedure. Draw the dendogram. Compare the results with those in Examples 11.4 and 11.6.

11.9. The vocabulary "richness" of a text can be quantitatively described by counting the words used once, the words used twice and so forth. Based on these counts a linguist proposed the following distances between chapters of the Old Testament book Lamentations. (Data courtesy of Y. T. Radday and M. A. Pollatschek.)

$$\text{Lamentations chapter}\quad \begin{matrix} 1 \\ 2 \\ 3 \\ 4 \\ 5 \end{matrix} \overset{\begin{matrix}\text{Lamentations chapter} \\ 1 \quad\quad 2 \quad\quad 3 \quad\quad 4 \quad\quad 5\end{matrix}}{\begin{bmatrix} 0 & & & & \\ .76 & 0 & & & \\ 2.97 & .80 & 0 & & \\ 4.88 & 4.17 & .21 & 0 & \\ 3.86 & 1.92 & 1.51 & .51 & 0 \end{bmatrix}}$$

Cluster the chapters of Lamentations using the three agglomerative hierarchical methods we have discussed. Draw the dendograms and compare the results.

11.10. Suppose we can measure two variables X_1 and X_2 for four items A, B, C and D. The data are as follows.

Item	Observations x_1	x_2
A	5	4
B	1	−2
C	−1	1
D	3	1

Use the K-means clustering technique to divide the items into $K = 2$ clusters. Start with the initial groups (AB) and (CD).

11.11. Repeat Example 11.11, starting with the initial groups (AC) and (BD). Compare your solution with the solution in Example 11.11. Are they the same? Graph the items in terms of their (x_1, x_2) coordinates and comment on the solutions.

11.12. Repeat Example 11.11 but start at the bottom of the list of items and proceed up in the order D, C, B, A. Begin with the initial groups (AB) and (CD). [The first potential reassignment will be based on the distances $d^2(D,(AB))$ and $d^2(D,(CD))$.] Compare your solution with the solution in Example 11.11. Are they the same? Should they be the same?

The following exercises require the use of a computer.

11.13. Table 11.9 lists measurements on 5 nutritional variables for 12 breakfast cereals.

(a) Using the data in Table 11.9 calculate the Euclidean distances between pairs of cereal brands.

(b) Treating the distances calculated in (a) as measures of (dis)similarity, cluster the cereals using the single-linkage and complete-linkage hierarchical procedures. Construct dendograms and compare the results.

Table 11.9 BREAKFAST-CEREAL DATA

Cereal	x_1 Protein (gm)	x_2 Carbohydrates (gm)	x_3 Fat (gm)	x_4 Calories (per oz)	x_5 Vitamin A (% daily allowance)[a]
1. Life	6	19	1	110	0
2. Grape Nuts	3	23	0	100	25
3. Super Sugar Crisp	2	26	0	110	25
4. Special K	6	21	0	110	25
5. Rice Krispies	2	25	0	110	25
6. Raisin Bran	3	28	1	120	25
7. Product 19	2	24	0	110	100
8. Wheaties	3	23	1	110	25
9. Total	3	23	1	110	100
10. Puffed Rice	1	13	0	50	0
11. Sugar Corn Pops	1	26	0	110	25
12. Sugar Smacks	2	25	0	110	25

[a]0 indicates less than 2%.

11.14. Input the data in Table 11.9 into a K-means clustering program. Cluster the cereals into $K = 2$, 3, and 4 groups. Compare the results with those in Exercise 11.13.

11.15. Table 11.10 gives the road distances between 12 Wisconsin and neighboring cities. Given the data in Table 11.10, locate the cities in $q = 1$, 2, and 3 dimensions using multidimensional scaling. Plot the minimum stress(q) versus q and interpret the graph. Compare the two-dimensional multidimensional scaling configuration with the locations of the cities on a map from an atlas.

11.16. Using the data in Table 11.9 represent the cereals in each specified way.

(a) Stars.

(b) Chernoff faces (experiment with the assignment of variables to facial characteristics).

(c) Andrews plots.

11.17. Using the utility data in Table 11.5, represent the public utility companies as Chernoff faces with assignments of variables to facial characteristics different from those considered in Example 11.17. Compare your faces with the faces in Figure 11.20. Are different groupings indicated?

11.18. Using the data in Table 11.5, represent the 22 public utility companies as stars. Visually group the companies into 4 or 5 clusters.

TABLE 11.10 DISTANCES BETWEEN CITIES IN WISCONSIN AND CITIES IN NEIGHBORING STATES

	Appleton (1)	Beloit (2)	Fort Atkinson (3)	Madison (4)	Marshfield (5)	Milwaukee (6)	Monroe (7)	Superior (8)	Wausau (9)	Dubuque (10)	St. Paul (11)	Chicago (12)
(1)	0											
(2)	130	0										
(3)	98	33	0									
(4)	102	50	36	0								
(5)	103	185	164	138	0							
(6)	100	73	54	77	184	0						
(7)	149	33	58	47	170	107	0					
(8)	315	377	359	330	219	394	362	0				
(9)	91	186	166	139	45	181	186	223	0			
(10)	196	94	119	95	186	168	61	351	215	0		
(11)	257	304	287	258	161	322	289	162	175	274	0	
(12)	186	97	113	146	276	93	130	467	275	184	395	0

REFERENCES

[1] Abramowitz, M., and I. A. Stegun, eds., *Handbook of Mathematical Functions*, U.S. Department of Commerce, National Bureau of Standards Applied Mathematical Series .55, 1964.

[2] Anderberg, M. R., *Cluster Analysis for Applications*, New York: Academic Press, 1973.

[3] Andrews, D. F., "Plots of High Dimensional Data," *Biometrics*, **28** (1972), 125–136.

[4] Chernoff, H., "Using Faces to Represent Points in *K*-Dimensional Space Graphically," *Journal of the American Statistical Association*, **68**, no. 342 (1973), 361–368.

[5] Cormack, R. M., "A Review of Classification (with discussion)," *Journal of the Royal Statistical Society* (*A*), **134**, no. 3 (1971), 321–367.

[6] Everitt, B., *Cluster Analysis*, London: Heinemann Educational Books, 1974.

[7] Everitt, B., *Graphical Techniques for Multivariate Data*, New York: North-Holland, 1978.

[8] Fienberg, S. E., "Graphical Methods in Statistics," *The American Statistician*, **33**, no. 4 (1979), 165–178.

[9] Gower, J. C., "Some Distance Properties of Latent Root and Vector Methods Used in Multivariate Analysis," *Biometrika*, **53** (1966), 325–338.

[10] Gower, J. C., "Multivariate Analysis and Multidimensional Geometry," *The Statistician*, **17** (1967), 13–25.

[11] Gnanadesikan, R., *Methods for Statistical Data Analysis of Multivariate Observations*, New York: John Wiley, 1977.

[12] Hartigan, J. A., *Clustering Algorithms*, New York: John Wiley, 1975.

[13] Kruskal, J. B., "Multidimensional Scaling by Optimizing Goodness of Fit to a Non-metric Hypothesis," *Psychometrika*, **29**, no. 1 (1964), 1–27.

[14] Kruskal, J. B., "Non-metric Multidimensional Scaling: A Numerical Method," *Psychometrika*, **29**, no. 1 (1964), 115–129.

[15] MacQueen, J. B., "Some Methods for Classification and Analysis of Multivariate Observations," *Proceedings of 5th Berkeley Symposium on Mathematical Statistics and Probability*, **1**, Berkeley, Calif.: University of California Press (1967), 281–297.

[16] Shepard, R. N., "Multidimensional Scaling, Tree-Fitting, and Clustering," *Science*, **210**, no. 4468 (1980), 390–398.

Appendix

TABLE 1 STANDARD NORMAL PROBABILITIES

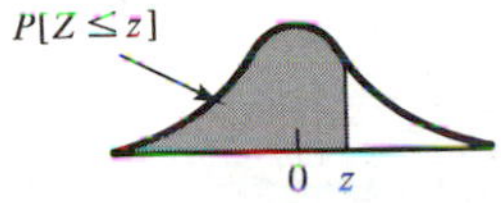

z	.00	.01	.02	.03	.04	.05	.06	.07	.08	.09
.0	.5000	.5040	.5080	.5120	.5160	.5199	.5239	.5279	.5319	.5359
.1	.5398	.5438	.5478	.5517	.5557	.5596	.5636	.5675	.5714	.5753
.2	.5793	.5832	.5871	.5910	.5948	.5987	.6026	.6064	.6103	.6141
.3	.6179	.6217	.6255	.6293	.6331	.6368	.6406	.6443	.6480	.6517
.4	.6554	.6591	.6628	.6664	.6700	.6736	.6772	.6808	.6844	.6879
.5	.6915	.6950	.6985	.7019	.7054	.7088	.7123	.7157	.7190	.7224
.6	.7257	.7291	.7324	.7357	.7389	.7422	.7454	.7486	.7517	.7549
.7	.7580	.7611	.7642	.7673	.7703	.7734	.7764	.7794	.7823	.7852
.8	.7881	.7910	.7939	.7967	.7995	.8023	.8051	.8078	.8106	.8133
.9	.8159	.8186	.8212	.8238	.8264	.8289	.8315	.8340	.8365	.8389
1.0	.8413	.8438	.8461	.8485	.8508	.8531	.8554	.8577	.8599	.8621
1.1	.8643	.8665	.8686	.8708	.8729	.8749	.8770	.8790	.8810	.8830
1.2	.8849	.8869	.8888	.8907	.8925	.8944	.8962	.8980	.8997	.9015
1.3	.9032	.9049	.9066	.9082	.9099	.9115	.9131	.9147	.9162	.9177
1.4	.9192	.9207	.9222	.9236	.9251	.9265	.9279	.9292	.9306	.9319
1.5	.9332	.9345	.9357	.9370	.9382	.9394	.9406	.9418	.9429	.9441
1.6	.9452	.9463	.9474	.9484	.9495	.9505	.9515	.9525	.9535	.9545
1.7	.9554	.9564	.9573	.9582	.9591	.9599	.9608	.9616	.9625	.9633
1.8	.9641	.9649	.9656	.9664	.9671	.9678	.9686	.9693	.9699	.9706
1.9	.9713	.9719	.9726	.9732	.9738	.9744	.9750	.9756	.9761	.9767
2.0	.9772	.9778	.9783	.9788	.9793	.9798	.9803	.9808	.9812	.9817
2.1	.9821	.9826	.9830	.9834	.9838	.9842	.9846	.9850	.9854	.9857
2.2	.9861	.9864	.9868	.9871	.9875	.9878	.9881	.9884	.9887	.9890
2.3	.9893	.9896	.9898	.9901	.9904	.9906	.9909	.9911	.9913	.9916
2.4	.9918	.9920	.9922	.9925	.9927	.9929	.9931	.9932	.9934	.9936
2.5	.9938	.9940	.9941	.9943	.9945	.9946	.9948	.9949	.9951	.9952
2.6	.9953	.9955	.9956	.9957	.9959	.9960	.9961	.9962	.9963	.9964
2.7	.9965	.9966	.9967	.9968	.9969	.9970	.9971	.9972	.9973	.9974
2.8	.9974	.9975	.9976	.9977	.9977	.9978	.9979	.9979	.9980	.9981
2.9	.9981	.9982	.9982	.9983	.9984	.9984	.9985	.9985	.9986	.9986
3.0	.9987	.9987	.9987	.9988	.9988	.9989	.9989	.9989	.9990	.9990
3.1	.9990	.9991	.9991	.9991	.9992	.9992	.9992	.9992	.9993	.9993
3.2	.9993	.9993	.9994	.9994	.9994	.9994	.9994	.9995	.9995	.9995
3.3	.9995	.9995	.9995	.9996	.9996	.9996	.9996	.9996	.9996	.9997
3.4	.9997	.9997	.9997	.9997	.9997	.9997	.9997	.9997	.9997	.9998
3.5	.9998	.9998	.9998	.9998	.9998	.9998	.9998	.9998	.9998	.9998

TABLE 2 STUDENT'S t-DISTRIBUTION CRITICAL POINTS

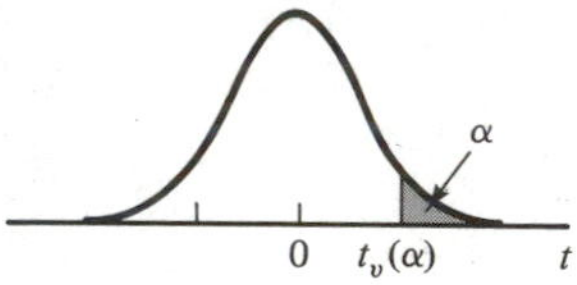

d.f. ν	α .250	.100	.050	.025	.010	.00833	.00625	.005
1	1.000	3.078	6.314	12.706	31.821	38.190	50.923	63.657
2	.816	1.886	2.920	4.303	6.965	7.649	8.860	9.925
3	.765	1.638	2.353	3.182	4.541	4.857	5.392	5.841
4	.741	1.533	2.132	2.776	3.747	3.961	4.315	4.604
5	.727	1.476	2.015	2.571	3.365	3.534	3.810	4.032
6	.718	1.440	1.943	2.447	3.143	3.287	3.521	3.707
7	.711	1.415	1.895	2.365	2.998	3.128	3.335	3.499
8	.706	1.397	1.860	2.306	2.896	3.016	3.206	3.355
9	.703	1.383	1.833	2.262	2.821	2.933	3.111	3.250
10	.700	1.372	1.812	2.228	2.764	2.870	3.038	3.169
11	.697	1.363	1.796	2.201	2.718	2.820	2.981	3.106
12	.695	1.356	1.782	2.179	2.681	2.779	2.934	3.055
13	.694	1.350	1.771	2.160	2.650	2.746	2.896	3.012
14	.692	1.345	1.761	2.145	2.624	2.718	2.864	2.977
15	.691	1.341	1.753	2.131	2.602	2.694	2.837	2.947
16	.690	1.337	1.746	2.120	2.583	2.673	2.813	2.921
17	.689	1.333	1.740	2.110	2.567	2.655	2.793	2.898
18	.688	1.330	1.734	2.101	2.552	2.639	2.775	2.878
19	.688	1.328	1.729	2.093	2.539	2.625	2.759	2.861
20	.687	1.325	1.725	2.086	2.528	2.613	2.744	2.845
21	.686	1.323	1.721	2.080	2.518	2.601	2.732	2.831
22	.686	1.321	1.717	2.074	2.508	2.591	2.720	2.819
23	.685	1.319	1.714	2.069	2.500	2.582	2.710	2.807
24	.685	1.318	1.711	2.064	2.492	2.574	2.700	2.797
25	.684	1.316	1.708	2.060	2.485	2.566	2.692	2.787
26	.684	1.315	1.706	2.056	2.479	2.559	2.684	2.779
27	.684	1.314	1.703	2.052	2.473	2.552	2.676	2.771
28	.683	1.313	1.701	2.048	2.467	2.546	2.669	2.763
29	.683	1.311	1.699	2.045	2.462	2.541	2.663	2.756
30	.683	1.310	1.697	2.042	2.457	2.536	2.657	2.750
40	.681	1.303	1.684	2.021	2.423	2.499	2.616	2.704
60	.679	1.296	1.671	2.000	2.390	2.463	2.575	2.660
120	.677	1.289	1.658	1.980	2.358	2.428	2.536	2.617
∞	.674	1.282	1.645	1.960	2.326	2.394	2.498	2.576

TABLE 3 χ^2 CRITICAL POINTS

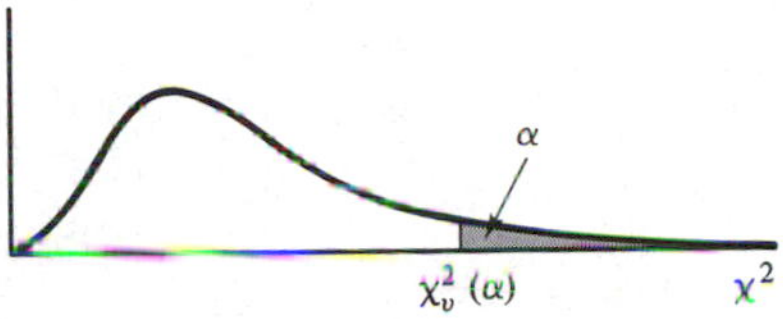

d.f. ν	α .990	.950	.900	.500	.100	.050	.025	.010	.005
1	.0002	.004	.02	.45	2.71	3.84	5.02	6.63	7.88
2	.02	.10	.21	1.39	4.61	5.99	7.38	9.21	10.60
3	.11	.35	.58	2.37	6.25	7.81	9.35	11.34	12.84
4	.30	.71	1.06	3.36	7.78	9.49	11.14	13.28	14.86
5	.55	1.15	1.61	4.35	9.24	11.07	12.83	15.09	16.75
6	.87	1.64	2.20	5.35	10.64	12.59	14.45	16.81	18.55
7	1.24	2.17	2.83	6.35	12.02	14.07	16.01	18.48	20.28
8	1.65	2.73	3.49	7.34	13.36	15.51	17.53	20.09	21.95
9	2.09	3.33	4.17	8.34	14.68	16.92	19.02	21.67	23.59
10	2.56	3.94	4.87	9.34	15.99	18.31	20.48	23.21	25.19
11	3.05	4.57	5.58	10.34	17.28	19.68	21.92	24.72	26.76
12	3.57	5.23	6.30	11.34	18.55	21.03	23.34	26.22	28.30
13	4.11	5.89	7.04	12.34	19.81	22.36	24.74	27.69	29.82
14	4.66	6.57	7.79	13.34	21.06	23.68	26.12	29.14	31.32
15	5.23	7.26	8.55	14.34	22.31	25.00	27.49	30.58	32.80
16	5.81	7.96	9.31	15.34	23.54	26.30	28.85	32.00	34.27
17	6.41	8.67	10.09	16.34	24.77	27.59	30.19	33.41	35.72
18	7.01	9.39	10.86	17.34	25.99	28.87	31.53	34.81	37.16
19	7.63	10.12	11.65	18.34	27.20	30.14	32.85	36.19	38.58
20	8.26	10.85	12.44	19.34	28.41	31.41	34.17	37.57	40.00
21	8.90	11.59	13.24	20.34	29.62	32.67	35.48	38.93	41.40
22	9.54	12.34	14.04	21.34	30.81	33.92	36.78	40.29	42.80
23	10.20	13.09	14.85	22.34	32.01	35.17	38.08	41.64	44.18
24	10.86	13.85	15.66	23.34	33.20	36.42	39.36	42.98	45.56
25	11.52	14.61	16.47	24.34	34.38	37.65	40.65	44.31	46.93
26	12.20	15.38	17.29	25.34	35.56	38.89	41.92	45.64	48.29
27	12.88	16.15	18.11	26.34	36.74	40.11	43.19	46.96	49.64
28	13.56	16.93	18.94	27.34	37.92	41.34	44.46	48.28	50.99
29	14.26	17.71	19.77	28.34	39.09	42.56	45.72	49.59	52.34
30	14.95	18.49	20.60	29.34	40.26	43.77	46.98	50.89	53.67
40	22.16	26.51	29.05	39.34	51.81	55.76	59.34	63.69	66.77
50	29.71	34.76	37.69	49.33	63.17	67.50	71.42	76.15	79.49
60	37.48	43.19	46.46	59.33	74.40	79.08	83.30	88.38	91.95
70	45.44	51.74	55.33	69.33	85.53	90.53	95.02	100.43	104.21
80	53.54	60.39	64.28	79.33	96.58	101.88	106.63	112.33	116.32
90	61.75	69.13	73.29	89.33	107.57	113.15	118.14	124.12	128.30
100	70.06	77.93	82.36	99.33	118.50	124.34	129.56	135.81	140.17

TABLE 4 *F*-DISTRIBUTION CRITICAL POINTS ($\alpha = .10$)

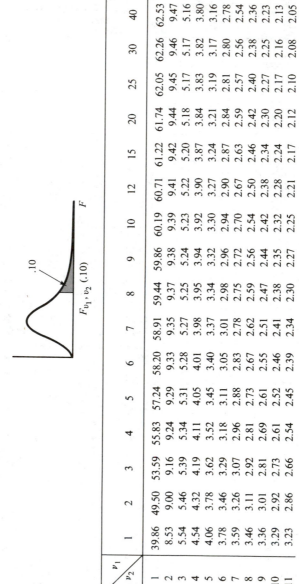

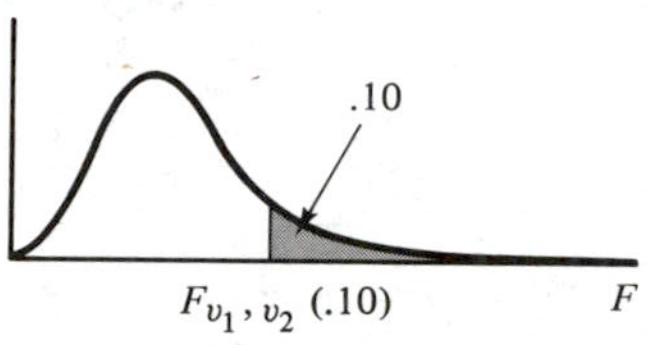

ν_2 \ ν_1	1	2	3	4	5	6	7	8	9	10	12	15	20	25	30	40	60
1	39.86	49.50	53.59	55.83	57.24	58.20	58.91	59.44	59.86	60.19	60.71	61.22	61.74	62.05	62.26	62.53	62.79
2	8.53	9.00	9.16	9.24	9.29	9.33	9.35	9.37	9.38	9.39	9.41	9.42	9.44	9.45	9.46	9.47	9.47
3	5.54	5.46	5.39	5.34	5.31	5.28	5.27	5.25	5.24	5.23	5.22	5.20	5.18	5.17	5.17	5.16	5.15
4	4.54	4.32	4.19	4.11	4.05	4.01	3.98	3.95	3.94	3.92	3.90	3.87	3.84	3.83	3.82	3.80	3.79
5	4.06	3.78	3.62	3.52	3.45	3.40	3.37	3.34	3.32	3.30	3.27	3.24	3.21	3.19	3.17	3.16	3.14
6	3.78	3.46	3.29	3.18	3.11	3.05	3.01	2.98	2.96	2.94	2.90	2.87	2.84	2.81	2.80	2.78	2.76
7	3.59	3.26	3.07	2.96	2.88	2.83	2.78	2.75	2.72	2.70	2.67	2.63	2.59	2.57	2.56	2.54	2.51
8	3.46	3.11	2.92	2.81	2.73	2.67	2.62	2.59	2.56	2.54	2.50	2.46	2.42	2.40	2.38	2.36	2.34
9	3.36	3.01	2.81	2.69	2.61	2.55	2.51	2.47	2.44	2.42	2.38	2.34	2.30	2.27	2.25	2.23	2.21
10	3.29	2.92	2.73	2.61	2.52	2.46	2.41	2.38	2.35	2.32	2.28	2.24	2.20	2.17	2.16	2.13	2.11
11	3.23	2.86	2.66	2.54	2.45	2.39	2.34	2.30	2.27	2.25	2.21	2.17	2.12	2.10	2.08	2.05	2.03

12	3.18	2.81	2.61	2.48	2.39	2.33	2.28	2.24	2.21	2.19	2.15	2.10	2.06	2.03	2.01	1.99	1.96
13	3.14	2.76	2.56	2.43	2.35	2.28	2.23	2.20	2.16	2.14	2.10	2.05	2.01	1.98	1.96	1.93	1.90
14	3.10	2.73	2.52	2.39	2.31	2.24	2.19	2.15	2.12	2.10	2.05	2.01	1.96	1.93	1.91	1.89	1.86
15	3.07	2.70	2.49	2.36	2.27	2.21	2.16	2.12	2.09	2.06	2.02	1.97	1.92	1.89	1.87	1.85	1.82
16	3.05	2.67	2.46	2.33	2.24	2.18	2.13	2.09	2.06	2.03	1.99	1.94	1.89	1.86	1.84	1.81	1.78
17	3.03	2.64	2.44	2.31	2.22	2.15	2.10	2.06	2.03	2.00	1.96	1.91	1.86	1.83	1.81	1.78	1.75
18	3.01	2.62	2.42	2.29	2.20	2.13	2.08	2.04	2.00	1.98	1.93	1.89	1.84	1.80	1.78	1.75	1.72
19	2.99	2.61	2.40	2.27	2.18	2.11	2.06	2.02	1.98	1.96	1.91	1.86	1.81	1.78	1.76	1.73	1.70
20	2.97	2.59	2.38	2.25	2.16	2.09	2.04	2.00	1.96	1.94	1.89	1.84	1.79	1.76	1.74	1.71	1.68
21	2.96	2.57	2.36	2.23	2.14	2.08	2.02	1.98	1.95	1.92	1.87	1.83	1.78	1.74	1.72	1.69	1.66
22	2.95	2.56	2.35	2.22	2.13	2.06	2.01	1.97	1.93	1.90	1.86	1.81	1.76	1.73	1.70	1.67	1.64
23	2.94	2.55	2.34	2.21	2.11	2.05	1.99	1.95	1.92	1.89	1.84	1.80	1.74	1.71	1.69	1.66	1.62
24	2.93	2.54	2.33	2.19	2.10	2.04	1.98	1.94	1.91	1.88	1.83	1.78	1.73	1.70	1.67	1.64	1.61
25	2.92	2.53	2.32	2.18	2.09	2.02	1.97	1.93	1.89	1.87	1.82	1.77	1.72	1.68	1.66	1.63	1.59
26	2.91	2.52	2.31	2.17	2.08	2.01	1.96	1.92	1.88	1.86	1.81	1.76	1.71	1.67	1.65	1.61	1.58
27	2.90	2.51	2.30	2.17	2.07	2.00	1.95	1.91	1.87	1.85	1.80	1.75	1.70	1.66	1.64	1.60	1.57
28	2.89	2.50	2.29	2.16	2.06	2.00	1.94	1.90	1.87	1.84	1.79	1.74	1.69	1.65	1.63	1.59	1.56
29	2.89	2.50	2.28	2.15	2.06	1.99	1.93	1.89	1.86	1.83	1.78	1.73	1.68	1.64	1.62	1.58	1.55
30	2.88	2.49	2.28	2.14	2.05	1.98	1.93	1.88	1.85	1.82	1.77	1.72	1.67	1.63	1.61	1.57	1.54
40	2.84	2.44	2.23	2.09	2.00	1.93	1.87	1.83	1.79	1.76	1.71	1.66	1.61	1.57	1.54	1.51	1.47
60	2.79	2.39	2.18	2.04	1.95	1.87	1.82	1.77	1.74	1.71	1.66	1.60	1.54	1.50	1.48	1.44	1.40
120	2.75	2.35	2.13	1.99	1.90	1.82	1.77	1.72	1.68	1.65	1.60	1.55	1.48	1.45	1.41	1.37	1.32
∞	2.71	2.30	2.08	1.94	1.85	1.77	1.72	1.67	1.63	1.60	1.55	1.49	1.42	1.38	1.34	1.30	1.24

TABLE 5 *F*-DISTRIBUTION CRITICAL POINTS ($\alpha = .05$)

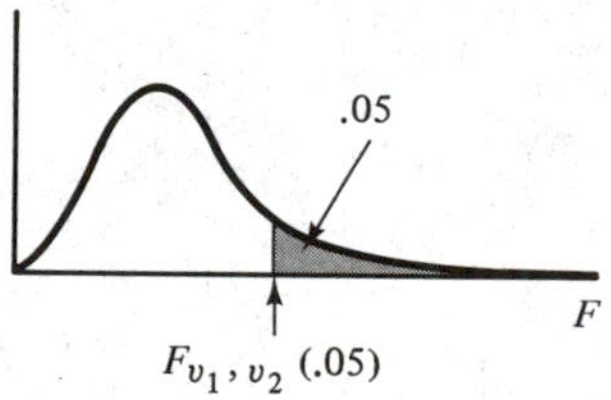

ν_2 \ ν_1	1	2	3	4	5	6	7	8	9	10	12	15	20	25	30	40	60
1	161.5	199.5	215.7	224.6	230.2	234.0	236.8	238.9	240.5	241.9	243.9	246.0	248.0	249.3	250.1	251.1	252.2
2	18.51	19.00	19.16	19.25	19.30	19.33	19.35	19.37	19.38	19.40	19.41	19.43	19.45	19.46	19.46	19.47	19.48
3	10.13	9.55	9.28	9.12	9.01	8.94	8.89	8.85	8.81	8.79	8.74	8.70	8.66	8.63	8.62	8.59	8.57
4	7.71	6.94	6.59	6.39	6.26	6.16	6.09	6.04	6.00	5.96	5.91	5.86	5.80	5.77	5.75	5.72	5.69
5	6.61	5.79	5.41	5.19	5.05	4.95	4.88	4.82	4.77	4.74	4.68	4.62	4.56	4.52	4.50	4.46	4.43
6	5.99	5.14	4.76	4.53	4.39	4.28	4.21	4.15	4.10	4.06	4.00	3.94	3.87	3.83	3.81	3.77	3.74
7	5.59	4.74	4.35	4.12	3.97	3.87	3.79	3.73	3.68	3.64	3.57	3.51	3.44	3.40	3.38	3.34	3.30
8	5.32	4.46	4.07	3.84	3.69	3.58	3.50	3.44	3.39	3.35	3.28	3.22	3.15	3.11	3.08	3.04	3.01
9	5.12	4.26	3.86	3.63	3.48	3.37	3.29	3.23	3.18	3.14	3.07	3.01	2.94	2.89	2.86	2.83	2.79
10	4.96	4.10	3.71	3.48	3.33	3.22	3.14	3.07	3.02	2.98	2.91	2.85	2.77	2.73	2.70	2.66	2.62
11	4.84	3.98	3.59	3.36	3.20	3.09	3.01	2.95	2.90	2.85	2.79	2.72	2.65	2.60	2.57	2.53	2.49

12	4.75	3.89	3.49	3.26	3.11	3.00	2.91	2.85	2.80	2.75	2.69	2.62	2.54	2.50	2.47	2.43	2.38
13	4.67	3.81	3.41	3.18	3.03	2.92	2.83	2.77	2.71	2.67	2.60	2.53	2.46	2.41	2.38	2.34	2.30
14	4.60	3.74	3.34	3.11	2.96	2.85	2.76	2.70	2.65	2.60	2.53	2.46	2.39	2.34	2.31	2.27	2.22
15	4.54	3.68	3.29	3.06	2.90	2.79	2.71	2.64	2.59	2.54	2.48	2.40	2.33	2.28	2.25	2.20	2.16
16	4.49	3.63	3.24	3.01	2.85	2.74	2.66	2.59	2.54	2.49	2.42	2.35	2.28	2.23	2.19	2.15	2.11
17	4.45	3.59	3.20	2.96	2.81	2.70	2.61	2.55	2.49	2.45	2.38	2.31	2.23	2.18	2.15	2.10	2.06
18	4.41	3.55	3.16	2.93	2.77	2.66	2.58	2.51	2.46	2.41	2.34	2.27	2.19	2.14	2.11	2.06	2.02
19	4.38	3.52	3.13	2.90	2.74	2.63	2.54	2.48	2.42	2.38	2.31	2.23	2.16	2.11	2.07	2.03	1.98
20	4.35	3.49	3.10	2.87	2.71	2.60	2.51	2.45	2.39	2.35	2.28	2.20	2.12	2.07	2.04	1.99	1.95
21	4.32	3.47	3.07	2.84	2.68	2.57	2.49	2.42	2.37	2.32	2.25	2.18	2.10	2.05	2.01	1.96	1.92
22	4.30	3.44	3.05	2.82	2.66	2.55	2.46	2.40	2.34	2.30	2.23	2.15	2.07	2.02	1.98	1.94	1.89
23	4.28	3.42	3.03	2.80	2.64	2.53	2.44	2.37	2.32	2.27	2.20	2.13	2.05	2.00	1.96	1.91	1.86
24	4.26	3.40	3.01	2.78	2.62	2.51	2.42	2.36	2.30	2.25	2.18	2.11	2.03	1.97	1.94	1.89	1.84
25	4.24	3.39	2.99	2.76	2.60	2.49	2.40	2.34	2.28	2.24	2.16	2.09	2.01	1.96	1.92	1.87	1.82
26	4.23	3.37	2.98	2.74	2.59	2.47	2.39	2.32	2.27	2.22	2.15	2.07	1.99	1.94	1.90	1.85	1.80
27	4.21	3.35	2.96	2.73	2.57	2.46	2.37	2.31	2.25	2.20	2.13	2.06	1.97	1.92	1.88	1.84	1.79
28	4.20	3.34	2.95	2.71	2.56	2.45	2.36	2.29	2.24	2.19	2.12	2.04	1.96	1.91	1.87	1.82	1.77
29	4.18	3.33	2.93	2.70	2.55	2.43	2.35	2.28	2.22	2.18	2.10	2.03	1.94	1.89	1.85	1.81	1.75
30	4.17	3.32	2.92	2.69	2.53	2.42	2.33	2.27	2.21	2.16	2.09	2.01	1.93	1.88	1.84	1.79	1.74
40	4.08	3.23	2.84	2.61	2.45	2.34	2.25	2.18	2.12	2.08	2.00	1.92	1.84	1.78	1.74	1.69	1.64
60	4.00	3.15	2.76	2.53	2.37	2.25	2.17	2.10	2.04	1.99	1.92	1.84	1.75	1.69	1.65	1.59	1.53
120	3.92	3.07	2.68	2.45	2.29	2.18	2.09	2.02	1.96	1.91	1.83	1.75	1.66	1.60	1.55	1.50	1.43
∞	3.84	3.00	2.61	2.37	2.21	2.10	2.01	1.94	1.88	1.83	1.75	1.67	1.57	1.51	1.46	1.39	1.32

TABLE 6 *F*-DISTRIBUTION CRITICAL POINTS ($\alpha = .01$)

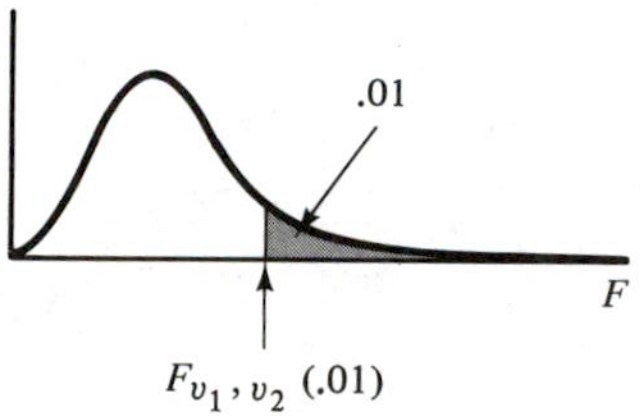

ν_2 \ ν_1	1	2	3	4	5	6	7	8	9	10	12	15	20	25	30	40	60
1	4052.	5000.	5403.	5625.	5764.	5859.	5928.	5981.	6023.	6056.	6106.	6157.	6209.	6240.	6261.	6287.	6313.
2	98.50	99.00	99.17	99.25	99.30	99.33	99.36	99.37	99.39	99.40	99.42	99.43	99.45	99.46	99.47	99.47	99.48
3	34.12	30.82	29.46	28.71	28.24	27.91	27.67	27.49	27.35	27.23	27.05	26.87	26.69	26.58	26.50	26.41	26.32
4	21.20	18.00	16.69	15.98	15.52	15.21	14.98	14.80	14.66	14.55	14.37	14.20	14.02	13.91	13.84	13.75	13.65
5	16.26	13.27	12.06	11.39	10.97	10.67	10.46	10.29	10.16	10.05	9.89	9.72	9.55	9.45	9.38	9.29	9.20
6	13.75	10.92	9.78	9.15	8.75	8.47	8.26	8.10	7.98	7.87	7.72	7.56	7.40	7.30	7.23	7.14	7.06
7	12.25	9.55	8.45	7.85	7.46	7.19	6.99	6.84	6.72	6.62	6.47	6.31	6.16	6.06	5.99	5.91	5.82
8	11.26	8.65	7.59	7.01	6.63	6.37	6.18	6.03	5.91	5.81	5.67	5.52	5.36	5.26	5.20	5.12	5.03
9	10.56	8.02	6.99	6.42	6.06	5.80	5.61	5.47	5.35	5.26	5.11	4.96	4.81	4.71	4.65	4.57	4.48
10	10.04	7.56	6.55	5.99	5.64	5.39	5.20	5.06	4.94	4.85	4.71	4.56	4.41	4.31	4.25	4.17	4.08
11	9.65	7.21	6.22	5.67	5.32	5.07	4.89	4.74	4.63	4.54	4.40	4.25	4.10	4.01	3.94	3.86	3.78

12	9.33	6.93	5.95	5.41	5.06	4.82	4.64	4.50	4.39	4.30	4.16	4.01	3.86	3.76	3.70	3.62	3.54
13	9.07	6.70	5.74	5.21	4.86	4.62	4.44	4.30	4.19	4.10	3.96	3.82	3.66	3.57	3.51	3.43	3.34
14	8.86	6.51	5.56	5.04	4.69	4.46	4.28	4.14	4.03	3.94	3.80	3.66	3.51	3.41	3.35	3.27	3.18
15	8.68	6.36	5.42	4.89	4.56	4.32	4.14	4.00	3.89	3.80	3.67	3.52	3.37	3.28	3.21	3.13	3.05
16	8.53	6.23	5.29	4.77	4.44	4.20	4.03	3.89	3.78	3.69	3.55	3.41	3.26	3.16	3.10	3.02	2.93
17	8.40	6.11	5.19	4.67	4.34	4.10	3.93	3.79	3.68	3.59	3.46	3.31	3.16	3.07	3.00	2.92	2.83
18	8.29	6.01	5.09	4.58	4.25	4.01	3.84	3.71	3.60	3.51	3.37	3.23	3.08	2.98	2.92	2.84	2.75
19	8.18	5.93	5.01	4.50	4.17	3.94	3.77	3.63	3.52	3.43	3.30	3.15	3.00	2.91	2.84	2.76	2.67
20	8.10	5.85	4.94	4.43	4.10	3.87	3.70	3.56	3.46	3.37	3.23	3.09	2.94	2.84	2.78	2.69	2.61
21	8.02	5.78	4.87	4.37	4.04	3.81	3.64	3.51	3.40	3.31	3.17	3.03	2.88	2.79	2.72	2.64	2.55
22	7.95	5.72	4.82	4.31	3.99	3.76	3.59	3.45	3.35	3.26	3.12	2.98	2.83	2.73	2.67	2.58	2.50
23	7.88	5.66	4.76	4.26	3.94	3.71	3.54	3.41	3.30	3.21	3.07	2.93	2.78	2.69	2.62	2.54	2.45
24	7.82	5.61	4.72	4.22	3.90	3.67	3.50	3.36	3.26	3.17	3.03	2.89	2.74	2.64	2.58	2.49	2.40
25	7.77	5.57	4.68	4.18	3.85	3.63	3.46	3.32	3.22	3.13	2.99	2.85	2.70	2.60	2.54	2.45	2.36
26	7.72	5.53	4.64	4.14	3.82	3.59	3.42	3.29	3.18	3.09	2.96	2.81	2.66	2.57	2.50	2.42	2.33
27	7.68	5.49	4.60	4.11	3.78	3.56	3.39	3.26	3.15	3.06	2.93	2.78	2.63	2.54	2.47	2.38	2.29
28	7.64	5.45	4.57	4.07	3.75	3.53	3.36	3.23	3.12	3.03	2.90	2.75	2.60	2.51	2.44	2.35	2.26
29	7.60	5.42	4.54	4.04	3.73	3.50	3.33	3.20	3.09	3.00	2.87	2.73	2.57	2.48	2.41	2.33	2.23
30	7.56	5.39	4.51	4.02	3.70	3.47	3.30	3.17	3.07	2.98	2.84	2.70	2.55	2.45	2.39	2.30	2.21
40	7.31	5.18	4.31	3.83	3.51	3.29	3.12	2.99	2.89	2.80	2.66	2.52	2.37	2.27	2.20	2.11	2.02
60	7.08	4.98	4.13	3.65	3.34	3.12	2.95	2.82	2.72	2.63	2.50	2.35	2.20	2.10	2.03	1.94	1.84
120	6.85	4.79	3.95	3.48	3.17	2.96	2.79	2.66	2.56	2.47	2.34	2.19	2.03	1.93	1.86	1.76	1.66
∞	6.63	4.61	3.78	3.32	3.02	2.80	2.64	2.51	2.41	2.32	2.18	2.04	1.88	1.78	1.70	1.59	1.47

Data Index

Subject Index